VOLUME EIGHTY-SEVEN

VITAMINS AND HORMONES

Stem Cell Regulators

VOLUME EIGHTY-SEVEN

VITAMINS AND HORMONES
Stem Cell Regulators

Editor-in-Chief

GERALD LITWACK

Toluca Lake, North Hollywood, California

ELSEVIER

AMSTERDAM • BOSTON • HEIDELBERG • LONDON
NEW YORK • OXFORD • PARIS • SAN DIEGO
SAN FRANCISCO • SINGAPORE • SYDNEY • TOKYO
Academic Press is an imprint of Elsevier

Cover photo credit:
Collino, F., Bruno, S., Deregibus, M.C., Tetta, C. and Camussi, G.
MicroRNAs and Mesenchymal Stem Cells
Vitamins and Hormones (2011) **87**, pp. 291–320

Academic Press is an imprint of Elsevier
32 Jamestown Road, London, NW1 7BY, UK
Radarweg 29, PO Box 211, 1000 AE Amsterdam, The Netherlands
Linacre House, Jordan Hill, Oxford OX2 8DP, UK
225 Wyman Street, Waltham, MA 02451, USA
525 B Street, Suite 1900, San Diego, CA 92101-4495, USA

First edition 2011

ISBN: 978-0-12-386015-6
ISSN: 0083-6729

Printed and bound by CPI Group (UK) Ltd, Croydon, CR0 4YY

Transferred to digital print 2012

Contents

Contributors *xv*
Preface *xxi*

1. Factors Regulating Pluripotency and Differentiation in Early Mammalian Embryos and Embryo-derived Stem Cells **1**

Aneta Suwinska and Maria A. Ciemerych

I. Introduction 2
II. From Totipotency to Pluripotency 3
III. Inner Cell Mass (ICM): Pluripotent Cells in the Mammalian Embryo 4
IV. Embryo-Derived Stem Cells 15
V. Transcriptional Regulators of Pluripotency in Embryo-Derived Stem Cells 17
VI. Extrinsic Factors and Signaling Pathways Regulating Pluripotency and Differentiation 19
VII. Conclusions 26
Acknowledgments 26
References 27

2. Molecular Mediators of Mesenchymal Stem Cell Biology **39**

Maria P. Alfaro, Sarika Saraswati, and Pampee P. Young

I. Introduction 40
II. Mesenchymal Stem Cells 40
III. Differentiation of MSCs 41
IV. Self-Renewal 41
V. MSC Therapy 42
VI. Immunomodulatory Properties 44
VII. MI Therapy 44
VIII. Molecular Mediators of MSC Biology 45
IX. Enhancing MSC Survival in the Wound 50
X. Secreted Frizzled-Related Proteins 50
XI. Mediating MSC Self-Renewal 51
XII. Conclusions 52
References 53

3. Insulin and Germline Proliferation in *Caenorhabditis elegans* **61**

E. Jane Albert Hubbard

I. Germline Proliferation in *C. elegans*: A Model for Developmental, Physiological, and Environmental Control of Cell Proliferation 62
II. *C. elegans* Germline Development 63
III. The *C. elegans* Germ line "Proliferation Versus Differentiation" Decision Is Mediated from the Soma to the Germ line by a Conserved Notch Signaling Pathway 64
IV. Evidence for Notch-Independent Soma-Germ line Signaling Mechanisms That Modulate Germline Proliferation 65
V. A Counter-Intuitive Assay to Indentify Potential Notch-Independent Mechanisms That Promote the Expansion of the Larval Germline Progenitor Pool 66
VI. Identification of the Insulin/IGF-Like Receptor (IIR) Pathway in Germline Proliferation 68
VII. IIR Signaling in *C. elegans* 69
VIII. Insulin Signaling Promotes the Larval Germline Cell Division Cycle 70
IX. *C. elegans* Insulins 71
X. Many Target Tissues for IIR Signaling 71
XI. Other Germline Roles for the IIR Pathway 73
XII. IIR Role in Larval Germline Proliferation: A Reproductive Timing and Lifespan Connection? 74
XIII. A Current Model and Future Directions 74
Acknowledgments 76
References 76

4. Generating Mature β-Cells from Embryonic Stem Cells: Strategies for Late-Stage Differentiation **79**

Spyridon Champeris Tsaniras

I. Introduction 79
II. Signaling Pathways in β-Cell Differentiation 81
III. Summary and Conclusions 86
Acknowledgments 87
References 87

5. Activation and Regulation of Reserve Liver Progenitor Cells **93**

D. Hunter Best and William B. Coleman

I. Introduction 94
II. Activation and Regulation of Mature Hepatocytes in Normal Liver Regeneration 94
III. Reserve Liver Progenitor Cells 97

IV. Hierarchical Responses in Liver Disease and Regeneration 102
References 103

6. Adult Cardiac-Derived Stem Cells: Differentiation and Survival Regulators **111**

Nanako Kawaguchi

I. Introduction 112
II. c-kit-Positive Cardiac Cells 113
III. Conclusions and Future Prospects 121
Acknowledgments 121
References 121

7. TGF-β1 Regulates Differentiation of Bone Marrow Mesenchymal Stem Cells **127**

Longmei Zhao and Basil M. Hantash

I. Bone Marrow Mesenchymal Stem Cells 128
II. The Role of TGF-β1 in Differentiation of Bone Marrow MSCs 131
III. Summary 137
Acknowledgments 137
References 137

8. Maternal Intake of Folic Acid and Neural Crest Stem Cells **143**

Chandra S. Mayanil, Shunsuke Ichi, Barbara-Mania Farnell, Vanda Boshnjaku, Tadanori Tomita, and David G. McLone

I. Introduction 144
II. Role of FA in Human Health 145
III. Mouse Models of NTD 148
IV. Neural Crest Development and Neural Crest Stem Cells 149
V. Role of FA in Neural Crest Development 151
VI. Folate Nonresponsive Genetic Mouse Models 162
VII. Conclusions and Future Directions 162
Acknowledgments 163
References 163

9. Modulation of the Generation of Dopaminergic Neurons from Human Neural Stem Cells by Bcl-X_L: Mechanisms of Action **175**

Alberto Martínez-Serrano, Claudia G. Castillo, Elise T. Courtois, Elisa García-García, and Isabel Liste

I. Introduction 176
II. Sources of Human DAn for Cell Replacement in PD 178

III. Epigenetic Cues and Genetic Manipulations to Improve hNSCs Differentiation Toward the A9 DA Phenotype 185
IV. Concluding Remarks 197
Acknowledgments 198
References 198

10. Glucocorticoid Hedgehog Agonists in Neurogenesis 207

Jiangbo Wang, Larry S. Barak, Robert A. Mook, and Wei Chen

I. Introduction 208
II. Select Glucocorticoids as Smoothened Agonists: Potential Effects for Neurogenesis 209
III. Mechanism of Action 210
IV. Structure–Activity Relationships (SAR) of Glucocorticoid Smoothened Agonists 211
V. Conclusion 212
References 214

11. Effect of Progesterone on Human Mesenchymal Stem Cells 217

Dobroslav S. Kyurkchiev, Ekaterina Ivanova-Todorova, and Stanimir D. Kyurkchiev

I. Introduction 218
II. Biological Roles of Progesterone 218
III. Mesenchymal Stem Cells 222
IV. Multipotent MSCs in Human Endometrium 224
V. Interaction Between Progesterone and MSCs 226
VI. Conclusions and Future Directions 231
Acknowledgments 231
References 231

12. Regulation of Muscle Stem Cells Activation: The Role of Growth Factors and Extracellular Matrix 239

Edyta Brzoska, Maria A. Ciemerych, Marta Przewozniak, and Małgorzata Zimowska

I. Cells Participating in Muscle Growth and Repair: Mechanisms of Activation 240
II. The Unique Ability of Skeletal Muscles to Regenerate 246
III. Muscle Stem Cells Activation: The Importance of Satellite Cell Niche 249
IV. The Interactions with ECM 252
V. Growth Factors Regulating Activation of Satellite Cells: The Case of HGF 255
VI. Other ECM-bound Growth Factors Regulating Myoblast Proliferation and Differentiation 258

VII. Concluding Remarks 260
Acknowledgments 261
References 261

13. Thymosins and Muscle Regeneration 277

Takahiko Hara

I. Introduction 278
II. Basic Properties of Thymosins 278
III. Physiological Activities of Tβ_4 281
IV. Roles of Tβ_4 in Skin Tissue Regeneration 282
V. Roles of Tβ_4 in Heart Regeneration 282
VI. Roles of Tβ_4 in Skeletal Muscle Regeneration 283
VII. Signaling Mechanism Involved in the Chemotactic Activity of Tβ_4 285
VIII. Concluding Remarks 285
References 286

14. MicroRNAs and Mesenchymal Stem Cells 291

Federica Collino, Stefania Bruno, Maria Chiara Deregibus, Ciro Tetta, and Giovanni Camussi

I. Introduction 292
II. MicroRNAs 293
III. Mesenchymal Stem Cells 296
IV. miRNAs and Stem Cells 296
V. Role of miRNAs in MSC Differentiation 299
VI. The Role of miRNAs in Cell-to-Cell Communication 306
VII. Conclusions 312
Acknowledgments 312
References 312

15. MicroRNA and Vascular Smooth Muscle Cells 321

Changqing Xie, Jifeng Zhang, and Y. Eugene Chen

I. Introduction 322
II. miRNA Biogenesis and Mechanism 323
III. miRNA and VSMC Differentiation 325
IV. miRNA and VSMC Phenotypic Switch 326
V. miRNA and VSMC Neointima Hyperplasia 329
VI. miRNA and VSMC-Related Diseases 330
VII. Prospective Application of miRNAs as Therapeutics for Vascular Disease 333
Acknowledgments 334
References 334

16. Transforming Growth Factor-Beta Superfamily in Mouse Embryonic Stem Cell Self-Renewal **341**

Katherine E. Galvin-Burgess and Jay L. Vivian

I. Introduction 342
II. Overview of TGF-Beta-Related Signaling 343
III. ES Cell Regulation by the BMP Pathway 347
IV. ES Cell Regulation by the Nodal Pathway 350
V. Interactions of TGF-Beta Signaling with Other Pluripotency Pathways 354
VI. TGF-Beta Signaling in Other Pluripotent Cells 354
VII. Conclusions and Future Directions 358
Acknowledgment 359
References 359

17. The Biology of HIFα Proteins in Cell Differentiation and Disease **367**

Inmaculada Royo González, Victoria Moreno-Manzano, Francisco Javier Rodríguez-Jimenez, Pilar Sepúlveda, and José-María Sánchez-Puelles

I. Introduction 368
II. HIF and Cancer 369
III. HIF and Stem 370
IV. HIF and Neurodegenerative Diseases 372
V. HIFs in Cardiac Ischemic Diseases 373
VI. HIFs as Target 374
VII. Conclusion 375
References 375

18. Regulatory Role of Klf5 in Early Mouse Development and in Embryonic Stem Cells **381**

Silvia Parisi and Tommaso Russo

I. Klf5: A Krüppel-Like Transcription Factor 382
II. Multiple Functions of Klf5 382
III. Klfs in Reprogramming 383
IV. Gene Expression Network for the Maintenance of ESC Pluripotent State 384
V. Klf5 Function in ESCs 385
VI. Klf5 Role in Early Embryonic Development 387
VII. Klf5 Targets 390
VIII. Klf5 Connection to the Core Pluripotency Network 391
IX. Conclusions 392
Acknowledgments 392
References 393

19. Bam and Bgcn in *Drosophila* Germline Stem Cell Differentiation 399
Sreejith Perinthottathil and Changsoo Kim

I. Introduction 400
II. Bam Repression of Stem Cell Maintenance Factors 402
III. Conclusion 409
Acknowledgments 412
References 413

20. The Effects of Mechanical Loading on Mesenchymal Stem Cell Differentiation and Matrix Production 417
Robin M. Delaine-Smith and Gwendolen C. Reilly

I. Introduction 418
II. Mesenchymal Stem Cells 420
III. Mechanical Loading 427
IV. Experimental Results 433
V. Conclusions and Future Directions 466
Acknowledgments 469
References 469

Index *481*

Contributors

Maria P. Alfaro
Department of Pathology, Vanderbilt University Medical Center, Nashville, Tennessee, USA

Larry S. Barak
Department of Cell Biology, Duke University Medical Center, Durham, North Carolina, USA

D. Hunter Best
Department of Pathology, University of Utah School of Medicine, and Genetics Division, ARUP Laboratories, Inc., Salt Lake City, Utah, USA

Vanda Boshnjaku
Developmental Biology Program, Children's Memorial Research Center, Department of Pediatric Neurosurgery, Children's Memorial Medical Center and Northwestern University Feinberg School of Medicine, Chicago, Illinois, USA

Stefania Bruno
Department of Internal Medicine and Center for Molecular Biotechnology University of Torino, Torino, and SiS-Ter S.p.A., Palazzo Pignano, Crema, Italy

Edyta Brzoska
Department of Cytology, Institute of Zoology, Faculty of Biology, University of Warsaw, Warsaw, Poland

Giovanni Camussi
Department of Internal Medicine and Center for Molecular Biotechnology University of Torino, Torino, Italy

Claudia G. Castillo
Departamento de Bioquímica, Facultad Medicina, Universidad de San Luis Potosí, San Luis Potosí, México, USA

Wei Chen
Department of Medicine, Duke University Medical Center, Durham, North Carolina, USA

Y. Eugene Chen
Cardiovascular Center, Department of Internal Medicine, University of Michigan Medical Center, Ann Arbor, Michigan, USA

Maria A. Ciemerych
Department of Cytology, Institute of Zoology, Faculty of Biology, University of Warsaw, Warsaw, Poland

William B. Coleman
Department of Pathology and Laboratory Medicine, Curriculum in Toxicology, Program in Translational Medicine, UNC Lineberger Comprehensive Cancer Center, University of North Carolina School of Medicine, Chapel Hill, North Carolina, USA

Federica Collino
Department of Internal Medicine and Center for Molecular Biotechnology University of Torino, Torino, Italy

Elise T. Courtois
Institute of Molecular and Cell Biology, A*STAR, Cancer Genetics group, Proteos, Singapore, Singapore

Robin M. Delaine-Smith
The Kroto Research Institute, Department of Materials Science and Engineering, University of Sheffield, Sheffield, United Kingdom

Maria Chiara Deregibus
Department of Internal Medicine and Center for Molecular Biotechnology University of Torino, Torino, Italy

Barbara-Mania Farnell
Department of Biology, Purdue University at Calumet, Hammond, Indiana, USA

Katherine E. Galvin-Burgess
Department of Pathology and Laboratory Medicine, Institute for Reproductive Health and Regenerative Medicine, University of Kansas Medical Center, Kansas City, USA

Elisa García-García
Departamento de Biología Molecular and Centro de Biología Molecular "Severo Ochoa", Universidad Autónoma de Madrid - C.S.I.C. Cantoblanco, Madrid, Spain

Inmaculada Royo González
Molecular Pharmacology Unit, Molecular Medicine Department, Centro de Investigaciones Biológicas, Consejo Superior de Investigaciones Científicas, Madrid, Spain

Basil M. Hantash
Escape Therapeutics, Inc., San Jose, California, USA

Takahiko Hara
Stem Cell Project, Tokyo Metropolitan Institute of Medical Science, Setagaya-ku, Tokyo, Japan

E. Jane Albert Hubbard
Developmental Genetics Program, Skirball Institute of Biomolecular Medicine, Helen and Martin Kimmel Center for Stem Cell Biology, Department of Pathology, New York University School of Medicine, New York, New York, USA

Shunsuke Ichi
Developmental Biology Program, Children's Memorial Research Center, Department of Pediatric Neurosurgery, Children's Memorial Medical Center and Northwestern University Feinberg School of Medicine, Chicago, Illinois, USA

Ekaterina Ivanova-Todorova
Laboratory of Clinical Immunology, University Hospital "St. I.Rislki", Medical University Sofia, Sofia, Bulgaria

Nanako Kawaguchi
Department of Patriotic Cardiology, Tokyo Women's Medical University, Tokyo, Japan

Changsoo Kim
Hormone Research Center, School of Biological Sciences and Technology, Chonnam National University, Gwangju, South Korea

Dobroslav S. Kyurkchiev
Laboratory of Clinical Immunology, University Hospital "St. I.Rislki", Medical University Sofia, Sofia, Bulgaria

Stanimir D. Kyurkchiev
Institute for Reproductive Health, Ob/Gyn Hospital "Dr. Shterev", Sofia, Bulgaria

Isabel Liste
Departamento de Biología Molecular and Centro de Biología Molecular "Severo Ochoa", Universidad Autónoma de Madrid - C.S.I.C. Cantoblanco, and Área de Biología Celular y Desarrollo, Centro Nacional de Microbiología, Instituto de Salud Carlos III, Majadahonda, Madrid, Spain

Alberto Martínez-Serrano
Departamento de Biología Molecular and Centro de Biología Molecular "Severo Ochoa", Universidad Autónoma de Madrid - C.S.I.C. Cantoblanco, Madrid, Spain

Chandra S. Mayanil
Developmental Biology Program, Children's Memorial Research Center, Department of Pediatric Neurosurgery, Children's Memorial Medical Center and Northwestern University Feinberg School of Medicine, Chicago, Illinois, USA

David G. McLone
Developmental Biology Program, Children's Memorial Research Center, Department of Pediatric Neurosurgery, Children's Memorial Medical Center and Northwestern University Feinberg School of Medicine, Chicago, Illinois, USA

Robert A. Mook
Department of Medicine, Duke University Medical Center, Durham, North Carolina, USA

Victoria Moreno-Manzano
MGM Integrated Solutions, Madrid, Spain

Silvia Parisi
Dipartimento di Biochimica e Biotecnologie Mediche, Università di Napoli "Federico II," CEINGE Biotecnologie Avanzate, European School of Molecular Medicine (SEMM), Naples, Italy

Sreejith Perinthottathil
Hormone Research Center, School of Biological Sciences and Technology, Chonnam National University, Gwangju, South Korea

Marta Przewozniak
Department of Cytology, Institute of Zoology, Faculty of Biology, University of Warsaw, Warsaw, Poland

Gwendolen C. Reilly
The Kroto Research Institute, Department of Materials Science and Engineering, University of Sheffield, Sheffield, United Kingdom

Francisco Javier Rodríguez-Jimenez
MGM Integrated Solutions, Madrid, Spain

Tommaso Russo
Dipartimento di Biochimica e Biotecnologie Mediche, Università di Napoli "Federico II," CEINGE Biotecnologie Avanzate, European School of Molecular Medicine (SEMM), Naples, Italy

José-María Sánchez-Puelles
Molecular Pharmacology Unit, Molecular Medicine Department, Centro de Investigaciones Biológicas, Consejo Superior de Investigaciones Científicas, and Fundación para la Investigación Hospital La Fe, Valencia, Spain

Sarika Saraswati
Department of Pathology, Vanderbilt University Medical Center, Nashville, Tennessee, USA

Pilar Sepúlveda
MGM Integrated Solutions, Madrid, and Fundación Valenciana Centro de Investigación Príncipe Felipe, Valencia, Spain

Aneta Suwinska
Department of Embryology, Faculty of Biology, University of Warsaw, Miecznikowa 1, Warsaw, Poland

Ciro Tetta
SiS-Ter S.p.A., Palazzo Pignano, Crema, Italy, and Fresenius Medical Care, Bad Homburg, Germany

Tadanori Tomita
Developmental Biology Program, Children's Memorial Research Center, Department of Pediatric Neurosurgery, Children's Memorial Medical Center and Northwestern University Feinberg School of Medicine, Chicago, Illinois, USA

Spyridon Champeris Tsaniras
Bionetwork Ltd, Chalandri, Athens, Greece

Jay L. Vivian
Department of Pathology and Laboratory Medicine, Institute for Reproductive Health and Regenerative Medicine, University of Kansas Medical Center, Kansas City, USA

Jiangbo Wang
Department of Medicine, Duke University Medical Center, Durham, North Carolina, USA

Changqing Xie
Cardiovascular Center, Department of Internal Medicine, University of Michigan Medical Center, Ann Arbor, Michigan, USA

Pampee P. Young
Department of Pathology, and Department of Internal Medicine, Vanderbilt University Medical Center; The Department of Veterans Affairs Medical Center, Nashville, Tennessee, USA

Jifeng Zhang
Cardiovascular Center, Department of Internal Medicine, University of Michigan Medical Center, Ann Arbor, Michigan, USA

Longmei Zhao
Escape Therapeutics, Inc., San Jose, California, USA

Małgorzata Zimowska
Department of Cytology, Institute of Zoology, Faculty of Biology, University of Warsaw, Warsaw, Poland

Preface

The distinguishing characteristic of a stem cell is the ability to renew itself and generate new and differentiated cells. It seems possible that virtually every tissue in the human body has a niche of stem cells, some niches more extensive and prominent than others. Thus, current work concentrates on the more prominent clusters of stem cells. When tissue cells are damaged or destroyed, the niche stem cells are called upon to replace the lost tissue cells, a process that is efficient in the growing body. Aging decreases the rate at which stem cells multiply and the destruction of aging tissue cells probably outruns the ability of the stem cells to replace the lost cells (is this the definition of aging?). Obviously there are hosts of factors involved in the life and activities of stem cells. This volume examines many of these factors and concerns embryonic stem cells and tissue stem cells and how the environment affects stem cell differentiation.

Although there are many different factors and topics to consider, the chapters, here, in contrast to previous volumes in this Serial, are not arranged in a specifically characterized order. However, the overall coverage is extensive and should satisfy the interests of researchers and students in this field.

The following 20 chapters range through many topics. The first is entitled "Factors regulating pluripotency and differentiation in early mammalian embryos and embryo-derived stem cells" by A. Suwinska and M.A. Ciemerych. M.P. Alfaro, S. Saraswati, and P.P. Young describe "Molecular mediators of mesenchymal stem cell biology." "Insulin and germline proliferation in *Caenorhabditis elegans*" is the next topic of E.J.A. Hubbard. S.C. Tsaniras offers "Generation of mature β-cells from embryonic stem cells: what are we missing?" D.H. Best and W.B. Coleman discuss "Activation and regulation of reserve liver progenitor cells." "Adult cardiac-derived stem cells: differentiation and survival regulators" is the work of N. Kawaguchi. L. Zhao and B.M. Hantash report on "TGF-beta1 regulates differentiation of bone marrow mesenchymal stem cells." C.S. Mayanil, S. Ichi, B.-M. Farnell, V. Boshnjaku, T. Tomita, and D.G. McLone contribute "Maternal intake of folic acid and neural crest stem cells." "Modulation of the generation of dopaminergic neurons from human neural stem cells by Bcl-X_L. Mechanisms of action" is a chapter by A. Martínez-Serrano, C.G. Castillo, E.T. Courtois, E. García-García, and I. Liste. J. Wang, L.S. Barak, R.A. Mook, Jr., and W. Chen report on "Glucocorticoid hedgehog agonists in neurogenesis." The chapter by D.S. Kyurkchiev, E. Ivanova-Todorova, and S.D. Kyurkchiev is "Effect of progesterone on human mesenchymal stem cells." "Regulation of muscle stem cells activation—the role of growth factors and extracellular matrix" is the chapter by E. Brzoska, M.A. Ciemerych, M. Przewozniak, and M. Zimowska. T. Hara describes Thymosins and muscle regeneration." "MicroRNAs and mesenchymal stem cells" is the concentration of F. Collino, S. Bruno, M.C. Deregibus, C. Tetta, and G. Camussi. C. Xie, J. Zhang, and Y.E. Chen collaborate on "microRNA and vascular smooth muscle cells." "Transforming growth factor-beta superfamily in mouse embryonic stem cell self-renewal" is the subject of K.E. Galvin-Burgess and J.L. Vivian. I.R. González, V. Moreno-Manzano, F.J. Rodríguez-Jimenez, P. Sepúlveda, and J.-M. Sánchez-Puelles review "The biology of HIFα proteins in cell differentiation and disease." "Regulatory role of Klf5 in early mouse development and in embryonic stem cells" is the topic of S. Parisi and T. Russo. "Bam and Bgcn in *Drosophila* germline stem cell differentiation" is the subject of S. Perinthottathil and C. Kim. Finally, in an extensive chapter, R.M. Delaine-Smith and G.C. Reilly review "The effects of mechanical loading on mesenchymal stem cell differentiation and matrix production."

The editor is grateful for the helpful work of Delsy Retcharger and Mohan Murali in the initial stages of recruiting authors for this book and, more recently, Sarah Latham in London. Mary Ann Zimmerman and Sarah Latham continue to smooth out the process of getting manuscripts into published form. Overseeing all of the components is Lisa Tickner, also in London, and my continuing thanks go to them all.

In recent volumes of this Serial, I have resorted to illustrations submitted by the authors and this book is no exception. On the cover is an illustration from Fig. 14.2 of "MicroRNAs and mesenchymal stem cells" by F. Collino, S. Bruno, M.C. Deregibus, C. Tetta, and G. Camussi.

Gerald Litwack
Toluca Lake, CA
August 23, 2011

Apology

In preparing the chapter entitled "Vitamin D and Inflammatory Bowel Disease" (see *Vitamins and Hormones—Vitamins and the Immune System*, Volume 86), I have consulted many papers. In particular, I read with great interest the paper by Verway M. et al. (Expert Rev. Clin. Immunol. 2010; 6:505–508). This paper elegantly summarizes interesting and original results about vitamin D and innate immunity. Given my lack of mastery of the English language, I highlighted the part that I reported in my chapter and saved it as a word file separately. When I wrote the paper, I faithfully used this part of the Verway paper in my chapter, but I neglected to cite the fact that I quoted the information directly from Verway and the other authors.

Taking personal responsibility for what happened, I apologize for this oversight to the editor (Dr. Litwack) and to the authors (Drs. Mark Verway, Marcel A. Behr, and John H. White) whose original work was published in Expert Rev. Clin. Immunol. (2010; 6:505–508).

Dr. Sandro Ardizzone

CHAPTER ONE

Factors Regulating Pluripotency and Differentiation in Early Mammalian Embryos and Embryo-derived Stem Cells

Aneta Suwinska* *and* Maria A. Ciemerych†

Contents

I. Introduction 2
II. From Totipotency to Pluripotency 3
III. Inner Cell Mass (ICM): Pluripotent Cells in the Mammalian Embryo 4
  A. The first differentiation event in the embryo: Specification of ICM and TE 4
  B. The second cell lineage decision: Formation of EPI and PE 11
  C. Differences in cell lineage segregation mechanisms between mammals 13
IV. Embryo-Derived Stem Cells 15
  A. Pluripotent stem cell lines: ES cells 15
  B. Pluripotent stem cell lines: EpiS cells 16
V. Transcriptional Regulators of Pluripotency in Embryo-Derived Stem Cells 17
VI. Extrinsic Factors and Signaling Pathways Regulating Pluripotency and Differentiation 19
  A. LIF/Jak/STAT3 signaling 19
  B. LIF/Ras/MEK/ERK signaling 21
  C. Phosphatidylinositol 3-kinase (PI3K)/Akt and Src signaling 21
  D. BMP/SMAD/Id signaling 21
  E. WNT/β-catenin signaling 22
  F. FGF2 and transforming growth factor β (TGFβ)/ActivinA/Nodal signaling 23
VII. Conclusions 26
Acknowledgments 26
References 27

* Department of Embryology, Faculty of Biology, University of Warsaw, Miecznikowa 1, Warsaw, Poland
† Department of Cytology, Faculty of Biology, University of Warsaw, Miecznikowa 1, Warsaw, Poland

Vitamins and Hormones, Volume 87
ISSN 0083-6729, DOI: 10.1016/B978-0-12-386015-6.00022-6

Abstract

Mammalian development relies on the cellular proliferation and precisely orchestrated differentiation processes. In preimplantation embryos preservation of the pluripotent state and timely onset of differentiation are secured by specific mechanisms involving such factors as OCT4, NANOG, SOX2, or SALL4. The pluripotency-sustaining cellular machinery is operational not only in the cells of preimplantation embryos but also in embryo-derived embryonic stem cells and epiblast stem cells. However, certain variations in the execution of pluripotency exist and result in the differences not only between embryonic cells and stem cells of the same mammalian species, but also between those of different mammalian species, such as mouse, rat, bank vole, or humans. In this review we describe the involvement of exogenous stimuli (e.g., LIF, WNT, BMP, FGF, and Activin) and function of intrinsic factors (e.g., OCT4, NANOG, SOX2, SALL4) in the regulation of pluripotency in mammalian preimplantation embryos and pluripotent stem cells derived from them. We also focus at the existence of species-specific differences at the level of growth factor requirements, signaling pathways, and transcription factors. Thus, we discuss differences in mechanisms which understanding is one of the necessary steps allowing establishment of methods of efficient derivation, defined *in vitro* culture conditions, and possible future therapeutic applications of pluripotent stem cells.

I. Introduction

During mammalian embryogenesis totipotent 1-cell embryo, that is, zygote undergoes cleavage divisions and gives rise to the blastomeres which progressively differentiate into distinct cell lineages. Thus, the unrestricted, pluripotent cells, which are initially able to generate any of the cells of an embryo, gradually lose their developmental potential. The state of pluripotency is transient *in vivo*, but by culturing embryos under strictly defined conditions it is possible to obtain embryo-derived stem cell lines that can sustain pluripotency for the indefinite period of *in vitro* culture. The crucial developmental processes, including cell proliferation, migration, and also maintenance of pluripotency and cell specification, in both the early embryo and embryo-derived stem cells, are precisely controlled by the interplay of extrinsic signaling pathways and intrinsic network of transcription factors. Our current knowledge indicates that the significant differences exist not only between stem cells derived from embryo at different developmental stages, but also between the stem cells of different animal species.

Nowadays, increasing interest is paid to the stem cells, especially of human origin, as a promise for therapy of degenerative diseases and damaged organs. This powers the extensive research aiming at elucidation of mechanisms underlying pluripotency and differentiation. Full understanding of how various factors interact with one another might also provide the explanation of species-specific differences and make possible the standardization of *in vitro* culture of various stem cells. The strict definition of culture conditions is required before the data coming from various sources could be extrapolated to the medicine and has possible future therapeutic applications. It is assumed that the same signals that regulate developmental processes in the pre- and periimplantation embryo will also control the maintenance of stem cells in a pluripotent state *in vitro*. Thus, the knowledge of signaling pathways operating in such embryos might be helpful in efforts to regulate the process of differentiation of embryonic stem cells (ESCs), including human ones. Conversely, an understanding of the regulatory factors responsible for pluripotency, self-renewal, and differentiation of human ESCs may be fundamental to understanding not only animal but also human development and realizing the therapeutic potential of these cells.

II. From Totipotency to Pluripotency

In the mouse, first the zygote and then the blastomeres of 2-cell stage embryo are totipotent, that is, able to give rise to both complete conceptus and extraembryonic structures supporting development. This was manifested by the experiments in that after the destruction of one of the blastomeres of the 2-cell stage embryo the remaining blastomere was able to give rise to a normally developing mouse (Tarkowski, 1959). The totipotency of later stages mouse blastomeres is not clear, as no mouse has been derived from a single blastomere of the 4-cell stage or more advanced embryos (Rossant, 1976). However, single blastomeres derived from 4-, 8-, and even 16-cell stage mouse embryos, when aggregated with "carrier" tetraploid ones, were shown to be able to contribute to all embryonic tissues, suggesting that they are pluripotent (Tarkowski *et al.*, 2001, 2005, 2010; Valer Carstea *et al.*, 2007). Studies devoted to other mammalian species, including domestic animals, involved only the experiments in that embryos were derived from single blastomeres at 4- or 8-cell stage (Johnson *et al.*, 1995; Moore *et al.*, 1968; Willadsen, 1981), and the pluripotency of blastomeres isolated from more advanced embryos was not tested, so far. What is definitively well documented is that during embryogenesis, pluripotency is gradually lost. At the blastocyst stage the embryo consists of cells that are either progenitors of extraembryonic tissues, that is placenta and fetal membranes, or pluripotent progenitors of embryonic tissues.

III. Inner Cell Mass (ICM): Pluripotent Cells in the Mammalian Embryo

The mammalian embryo at the early blastocyst stage is composed of two distinct cell populations differing morphologically and molecularly: the outer layer of cells, that is, the trophectoderm (TE), and a group of inner, pluripotent cells, the ICM. While the TE lineage develops into the embryonic part of the placenta and trophoblast giant cells, the ICM is a source of germline and all somatic cells of the developing embryo. It also contributes to several extraembryonic tissues. Shortly before implantation, the second differentiation process takes place. It results in the formation, within the ICM, of pluripotent epiblast (EPI) and the primitive endoderm (PE; also termed as hypoblast), that is, another, in addition to TE, extraembryonic lineage. PE emerges as a monolayer of cells, localized at that surface of ICM which directly faces the blastocyst cavity. It will give rise to two derivatives: the visceral endoderm contributing to the yolk sac and the parietal endoderm. Cells located deeper within the ICM will form pluripotent EPI, that is, ICM-derived tissue present in postimplantation embryos (E5.5–E7.5 in the mouse) that generates the conceptus and some of the extraembryonic membranes, such as allantois and amnion (Kaufman, 1992).

All these events, comprising the successful embryo development, require regulation by precise signaling networks. Any disruption in these networks may lead to developmental abnormalities, fetal diseases, and as a result, pregnancy failure.

A. The first differentiation event in the embryo: Specification of ICM and TE

Up to the "early" 8-cell stage, that is, soon after third cleavage division, the blastomeres do not differ and retain their unpolarized character. At the so called "late" 8-cell stage embryo undergoes the process of compaction, which results in the cellular adhesion due to expression of E-cadherin, and the formation of adherent as well as the tight junctions between blastomeres (Fleming *et al.*, 2000). This event is considered to play a crucial role for later morphogenetic events and for the proper establishment of the first cell lineages. At the time of compaction all blastomeres polarize along the axis of the cell–cell contact and form outward, apical domains characterized by the localization of such crucial "polarity" factors, such as atypical protein kinase C (aPKC), partitioning defective (PARD3 and PARD6b); (Plusa *et al.*, 2005; Vinot *et al.*, 2005), junctional adhesion molecule 1 (JAM1); (Thomas *et al.*, 2004), and basolateral domains containing EMK1 protein (Vinot *et al.*, 2005). Division of such polarized blastomeres can be either symmetric or asymmetric. Thus, as the embryo cleaves from 8 to 16 cells, blastomeres that divide symmetrically, that is, cleavage plane is parallel to the inside–outside axis, produce two polarized cells that remain localized on the surface of an embryo. Blastomeres that divide asymmetrically, perpendicular to the inside–outside axis, produce one polar daughter cell that remains outside and one nonpolar, inside daughter cell. As a result of these two types of divisions two populations of cells are created: outside, polar cells, and inside, nonpolar cells. The inside cells contribute to ICM, while outside ones are precursors of TE, as proposed in 1967 by Tarkowski in his inside–outside theory (Tarkowski and Wroblewska, 1967). At the 16-cell stage, inside and outside cells are not irreversibly committed to ICM and TE lineages. Since by the manipulation of their position the fate of 16-cell stage blastomeres can be changed it strongly suggests that, at least at this developmental stage, they retain plasticity (Suwinska *et al.*, 2008). Blastomeres are able to "sense" environmental cues and reprogram the route of their differentiation in accordance with their position. Symmetric and asymmetric divisions also occur as the embryo grows from 16 to 32 cells and possibly also from 32 to 64 cells (Morris *et al.*, 2010). At the 32-cell stage, when the outside cells of the embryo become committed to the TE lineage, a fluid-filled cavity of blastocyst begins to form.

ICM is the only pluripotent cell lineage in the blastocyst. The core network that controls maintenance of its pluripotency relies at OCT4 encoded by *Pou-domain class 5, transcription factor 1*, NANOG (*Nanog homeobox*), and SOX2 [*Sex determining region Y (SRY) box containing gene 2*] (Avilion *et al.*, 2003; Chambers *et al.*, 2003; Mitsui *et al.*, 2003; Nichols *et al.*, 1998). These transcription factors are initially expressed ubiquitously in all blastomeres of the cleaving embryo, but in the blastocyst become restricted to inside, ICM cells (Figs. 1.1 and 1.2) (Avilion *et al.*, 2003; Chambers *et al.*, 2003; Mitsui *et al.*, 2003; Nichols *et al.*, 1998; Palmieri *et al.*, 1994). Additionally, existing reports indicate the role of SALL4, a transcription factor belonging to SPALT family, as an activator of *Oct4* and modulator of *Nanog* transcription (Wu *et al.*, 2006). When ICMs of *Sall4*-null blastocysts were cultured *in vitro* in most cells of the ICM expression of TE lineage markers, such as *Cdx2* (caudal type homeobox 2), was detected (Elling *et al.*, 2006). Moreover, introduction of *Sall4* siRNA into mouse zygotes resulted in reduction of the expression of not only of *Sall4* but also of *Oct4* mRNAs and the increase of the expression of TE-specific genes in the ICM at the blastocyst stage (Zhang *et al.*, 2006). OCT4 and SOX2 synergistically activate the transcription of genes important for maintaining pluripotency, including their own expression (Chew *et al.*, 2005), and also *Nanog* (Avilion *et al.*, 2003; Rodda *et al.*, 2005), or *Fgf4* (fibroblast growth factor) (Yuan *et al.*, 1995). Both *Pou5f1*-null and *Nanog*-null embryos develop only up to the blastocyst stage (Nichols *et al.*, 1998; Mitsui *et al.*, 2003). However, *in vitro* cultured ICM cells lacking *Pou5f1* differentiate only into the TE lineage, specifically into trophoblast giant cells (Nichols

Stages of development

Gene expression

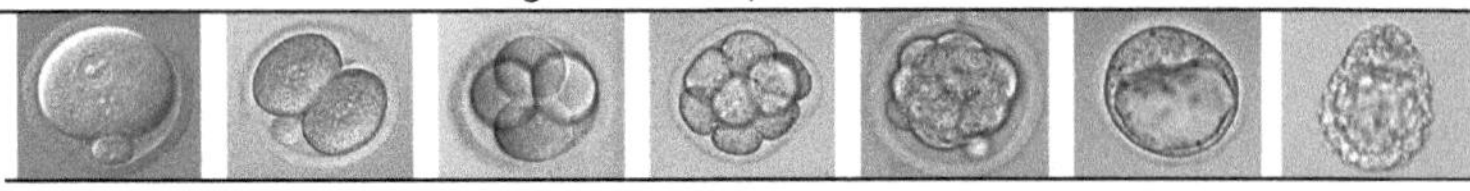

	Zygote	2-cell	4-cell	8-cell	16- to 32-cell		E3.5 blastocyst		E4.5 blastocyst		
					Inner	Outer	ICM	TE	EPI	PE	TE
OCT4	+	+	+	+	+	+	+	+	+	+	–
NANOG	–	–	–	+	+	+	+	–/+	+	–	–
SOX2	+	+	+	+	+	+	+	–	+	+	–
SALL4	–	+	+	+	+	+	+	+	+	+	–
TEAD4	–	+	+	+	+	+	+	+	+	+	+
CDX2	–	–	–	+	+	+	–	+	–	–	+
GATA3	–	–	+	+	+	+	–	+	–	–	+
EOMES	+	+	+	+	+	+	–	+	–	–	+
KLF5	–	+	+	+	+	+	+	+	n.a.	n.a.	n.a.
GATA6	–	–	–	–	+	+	+	–/+	–	+	–
SOX7	–	–	–	–	–	–	–	–	–	+	–
SOX17	–	–	–	–	+	–	+	–	–	+	–

Figure 1.1 Timing of the expression of factors regulating pluripotency and differentiation during preimplantation development of mouse embryo. + indicates expression, –/+ indicates that the expression may vary, – indicates lack of expression of factors involved in the sustaining of pluripotency and embryonic lineage specification, n.a. not analyzed (Artus *et al.*, 2010; Avilion *et al.*, 2003; Chambers *et al.*, 2003; Dietrich and Hiiragi, 2007; Elling *et al.*, 2006; Home *et al.*, 2009; Lin *et al.*, 2010; McConnell *et al.*, 2005; Morris *et al.*, 2009; Nishioka *et al.*, 2008; Palmieri *et al.*, 1994; Ralston and Rossant, 2008; Ralston *et al.*, 2010; Yagi *et al.*, 2007).

et al., 1998). *Nanog*-null ICM cells fate is limited to TE differentiation or apoptosis (Silva *et al.*, 2009). In contrast, *Sox2*-null ICMs are capable of giving rise, at least transiently, to the EPI and PE (Avilion *et al.*, 2003). One of the roles of pluripotency factors is to counteract factors that are involved in the differentiation program (Fig. 1.2). OCT4 and SOX2 proteins form a complex suppressing the expression of *Cdx2* that is responsible for TE differentiation (Niwa *et al.*, 2005). NANOG inhibits the differentiation of PE (Mitsui *et al.*, 2003) and, as it was shown in ESCs, mesoderm (Suzuki *et al.*, 2006), by influencing the expression of genes specific for these lineages, such as *Gata6* and *BrachyuryT*, respectively.

TE is another, in addition to ICM, primary lineage which is specified at the blastocyst stage and is required for embryo implantation within the uterus. One of the factors indispensable for maintenance and proper functioning of the TE is CDX2 (Figs. 1.1 and 1.2). However, its role in the initiation of this lineage specification remains controversial. In the mouse, *Cdx2* is expressed at varying levels in all blastomeres starting from the 8-cell stage. However, prior to blastocyst formation it becomes restricted to the outside, future TE cells (Dietrich and Hiiragi, 2007; Ralston and Rossant, 2008). The heterogeneity of *Cdx2* expression among individual blastomeres, observed at the 8-cell stage, may be a result of differences in the

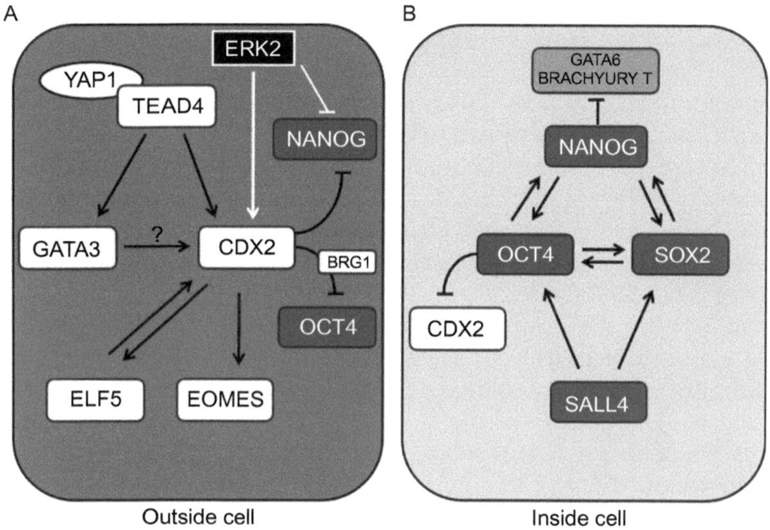

Figure 1.2 Transcriptional networks underlying ICM/TE specification in preimplantation mouse embryo. A. TE specification relies on the activity of TEAD4, CDX2, GATA3, EOMES, and ELF5 transcription factors. Due to inactive HIPPO pathway in outside cells, that is, future TE, TEAD4 cooperates with YAP1 and activates *Cdx2* as well as *Gata3*. CDX2, in turn, upregulates *Eomes* and *Elf5* and suppresses *Nanog* and *Oct4* expression, most likely via BRG1. Moreover, induction of MAP kinase signaling pathway, that is, ERK2 leads to *Cdx2* expression and suppresses *Nanog*, thus promoting TE cell fate. B. OCT4, NANOG, and SOX2 form a core machinery regulating pluripotency in inside, that is, future ICM cells. Additionally, SALL4 activates *Oct4* and modulates *Nanog* expression. OCT4 negatively regulates *Cdx2* involved in TE differentiation, whereas NANOG, as it was shown in ESCs, suppresses *Gata6* and *Brachyury T* expression preventing differentiation into PE and mesoderm, respectively. White box—factors required for TE specification, dark gray box—factors responsible for sustaining of pluripotency, light gray box—markers of endoderm and mesoderm differentiation, black box—factors involved in signal transduction.

order and orientation of the subsequent cleavage divisions preceding this stage. Indeed it has been shown that *Cdx2* regulates cell polarity by enhancing expression and apical localization of aPKC, and thus, promotes symmetric cell divisions starting from 8-cell up the 32-cell stage (i.e., morula stage) embryos, and as a result allocating more blastomeres to the future TE (Jedrusik *et al.*, 2008). However, blastomere polarity also influences CDX2 expression through asymmetric distribution of Cdx2 mRNA to outer blastomeres (Jedrusik *et al.*, 2008). CDX2 becomes spatially restricted earlier than OCT4, NANOG, and SOX2, that is, at the 16-cell stage (Dietrich and Hiiragi, 2007). Since these factors negatively regulate *Cdx2* expression it indicates that this transcription factor may be required to downregulate *Oct4*, *Nanog*, and *Sox2* in outside cells, that is, future TE. In support of this notion,

OCT4 and NANOG fail to be properly restricted in *Cdx2*$^{-/-}$ embryos—these transcription factors are expressed in the inside, as well as in the outside cells at the blastocyst stage, resulting in the death of the embryo at the time of implantation (Strumpf *et al.*, 2005). Chromatin remodeling protein BRG1 (*brahma-related gene 1*) was shown to be required for CDX2-mediated repression of *Oct4* in TE of mouse blastocysts (Fig. 1.2) (Wang *et al.*, 2010). RNA interference (RNAi)-mediated knockdown of *Brg1* in mouse zygotes resulted in the developmental arrest at the blastocyst stage, inability to hatch from *zona pellucida*, and failure to repress *Oct4* and *Nanog* transcription factors in TE cells (Kidder *et al.*, 2009; Wang *et al.*, 2010). Also another findings support the conclusion that during mouse development *Cdx2* indeed plays a crucial role in the maintenance of TE differentiation and function, but neither maternal nor zygotic *Cdx2* expression is essential for initiating the segregation of the ICM and TE lineages (Wu *et al.*, 2010). Elimination of both maternal and zygotic *Cdx2* transcripts by RNAi resulted in the failure of embryo hatching and implantation. However, the developing blastocysts exhibited completely normal morphology, suggesting that TE differentiation had been initiated (Wu *et al.*, 2010). All above results indicate that CDX2 is a factor involved in the first cell lineage decision.

Among the upstream factors that have been proved to regulate *Cdx2* is TEAD4 (TEA domain family member 4) (Nishioka *et al.*, 2009, 2008; Yagi *et al.*, 2007). *Tead4*$^{-/-}$ embryos are characterized by the phenotype more severe than that of *Cdx2*$^{-/-}$ embryos. They fail to express *Cdx2* and to form blastocyst cavity. *Tead4* likely controls TE specifying genes, such as T-box transcription factor *Eomesodermin* (*Eomes*) and *Ets* transcription factor—*Elf5*. ELF5 contributes to the regulation of the TE lineage acting downstream of the initial lineage determination by sustaining *Cdx2* and *Eomes* expression (Ng *et al.*, 2008). Thus, CDX2, EOMES, and ELF5 establish a positive feedback loop which stabilizes the TE lineage. The transcriptional activity of TEAD4 depends on the interaction with YAP1 (yes-associated protein) and is modulated by its cellular localization (Fig. 1.2) (Nishioka *et al.*, 2009). TEAD4 is expressed ubiquitously before the blastocyst stage and then it is localized within the nuclei of TE as well as of ICM (Nishioka *et al.*, 2008). YAP1, starting from the 4-cell stage is present within the nuclei of all blastomeres. From the 16-cell stage it is phosphorylated and as a result translocated from the nuclei to the cytoplasm, but such translocation happens only in the inside blastomeres. At the blastocyst stage, nuclear YAP1 is restricted only to the outside cells of the TE. Thus, in outside cells TEAD4 and YAP1 cooperatively activate and stabilize *Cdx2* expression, but in inside cells TEAD4 is inactive, and in consequence, *Cdx2* expression is silenced (Nishioka *et al.*, 2009). In mammalian cells YAP1 phosphorylation is regulated by the kinases involved in HIPPO signaling cascade. HIPPO pathway members, that is, two closely related kinases LATS1 and LATS2 (large tumor suppressor homolog 1 and 2, LATS1/2), by phosphorylating YAP1 control its subcellular localization (Zhao *et al.*, 2008). In inside cells of an embryo LATS1/2 phosphorylates YAP1, preventing its movement into the nucleus (Nishioka *et al.*, 2009). As a consequence, TEAD4-YAP1 complex fails to form and *Cdx2* is downregulated. HIPPO/LATS1/2 signaling pathway is active in ICM cells only if they are totally enclosed, that is, surrounded by outside cells (Nishioka *et al.*, 2009). However, the molecular mechanism by which such "enclosure" renders HIPPO signaling effective is unclear. Thus, it is possible that activation of this pathway and its influence on cell fate in preimplantation embryos may depend on cell–cell contacts and cell position. TEAD4 can also regulate expression of zinc-finger transcription factor GATA3 which is yet another TE-specific factor. The precise role of GATA3 remains unclear. It can either, similarly to CDX2, promote TE specification (Ralston *et al.*, 2010), or act upstream of *Cdx2* directly regulating its transcription (Home *et al.*, 2009).

All the observations described above indicate a clear role for *Cdx2* in TE development. On the other hand, they also suggest that *Cdx2* is not essential for the initial specification of the TE lineage. This is further supported by studies involving generation of chimeric embryos, during which *Cdx2*-null and wild-type 2-cell stage blastomeres were aggregated. *Cdx2* mutant cells, which show cell-autonomous defects in expression of *Oct4*, *Nanog*, and *Eomes*, did not preferentially contribute to the ICM in chimeric blastocysts (Ralston and Rossant, 2008). In contrast to the data presented above, Jedrusik *et al.* indicated that *Cdx2* participates not only in the maintenance of TE differentiation, but also in its specification (Jedrusik *et al.*, 2010). Using three independent approaches allowing the depletion of both maternal and zygotic *Cdx2*, that is, dsRNA, siRNA, and morpholino oligonucleotides, they documented that the development of embryos deprived of CDX2 was delayed and characterized by defects in cell polarization as well as compaction at the 8- and 16-cell stages. Moreover, expression of TE-specific genes, *Gata3* and *Eomes*, was greatly reduced or even undetectable.

Another protein that was recently shown to play a role not only in development of ICM, but also TE of the preimplantation mouse embryo is SOX2. Knocking down *Sox2* in 2-cell stage mouse embryos using siRNA revealed its role in the establishment of the TE lineage. In the absence of SOX2 embryos arrested at the morula stage, failed to form TE and to cavitate (Keramari *et al.*, 2010). Expression of pluripotency markers, such as OCT4 and NANOG, remained unaffected in such embryos, whereas TE markers, including TEAD4, CDX2, and EOMES were downregulated in outer cells of morulae (Keramari *et al.*, 2010).

Apart from TEAD4, CDX2, GATA3, EOMES and SOX2, another factor, KLF5 (Kruppel-like factor 5), belonging to Kruppel-like transcription factors family of highly related zinc-finger proteins, was shown to be responsible for TE differentiation (Dang *et al.*, 2000; McConnell *et al.*, 2007). In mouse embryos *Klf5* expression is detectable starting from the

2-cell stage, but at morula and blastocyst stage it becomes significantly upregulated in the outer cells. Thus, it was suggested to be crucial for the TE formation (Lin *et al.*, 2010). Lack of *Klf5* gene results in the arrest of the embryonic development at the blastocyst stage, that is, *Klf5*$^{-/-}$ embryos do not form expanded blastocyst cavities and fail to hatch (Ema *et al.*, 2008; Lin *et al.*, 2010). Moreover, expression of TE markers, such as *Cdx2* and *Eomes* was lost and not rescued by exogenous FGF4, which was shown to act as a paracrine signal originated from ICM and required for TE development (Lu *et al.*, 2008). Also in chimeric embryos, generated by aggregation of 2-cell stage embryos wild type and *Klf5*-null, TE marker CDX2 was not detected in *Klf5* mutant cells (Lin *et al.*, 2010). Thus, it suggests that KLF5 functions upstream of *Cdx2*, and downstream of, or in parallel to, *Fgf4* in order to cell-autonomously regulate TE development. *Klf5* is also expressed, at lower levels, in the ICM cells, in that it is required to maintain appropriate transcriptional levels of the pluripotency genes: *Oct4*, *Nanog*, and *Sox2*. Thus, KLF5 plays opposite roles in two primary cell lineages, TE and ICM.

Maintaining the integrity and function of TE also requires the activation of FGF/ERK (fibroblast growth factor/extracellular signal regulated kinases) signaling pathways (Lu *et al.*, 2008). Vertebrate cells express 22 different FGF molecules and five FGF receptors (FGFR 1–5). Interaction of FGF4 with FGFR induces autophosphorylation of tyrosine residues in the receptor's intercellular domain and stimulates activation of ERK1/ERK2 mitogen-activated protein kinase (MAPK) pathway (Dailey *et al.*, 2005; Villegas *et al.*, 2010). A key component of this signaling is the docking protein FRS2a (fibroblast growth factor receptor substrate 2a) that serves as a core of complex consisting of the tyrosine phosphatase SHP2, the adaptor GRB2 (growth factor receptor bound protein 2), and the docking protein GAB1 (lipid anchored Grb2 binding protein). The FRS2 signaling complex recruits the SOS (Son of sevenless guanine nucleotide exchange factor), via binding through GRB2, which activates GTP binding Ras. SOS catalyzes the activation of Ras by inducing exchange of GDP for GTP. Ras, in turn, stimulates the MAPK pathway consisting of Raf, MEK1 (MAP kinase kinase), and ERK1/ERK2 that regulate the activity of downstream kinases or transcription factors (Dailey *et al.*, 2005). Expression of one of the members of FGF family, that is, *Fgf4*, is detectable in all blastomeres of 8- to 16-cell embryos, but later in the development, at expanded blastocyst stage, it becomes restricted to the EPI cells. Its expression is regulated by the pluripotency factors OCT4 and SOX2 (Yuan *et al.*, 1995). ERK2, which is activated via FGF-dependent manner, was shown to be localized asymmetrically, that is, at the apical surface of the 8-cell stage embryo, prior to the establishment of outer and inner blastomeres. Among FGF receptors FGFR2 is predominantly expressed in blastocyst, although it is restricted to the TE and PE lineages (Guo *et al.*, 2010). This pattern of receptor and ligand expression suggests that EPI-produced FGF4 could be important for the establishment or maintenance of extraembryonic tissues, TE and PE. It has also been shown by RT-PCR and immunocytochemistry that other molecules of MAPK pathway, such as FGFR2, FRS2a, GRB2, GAB1, and SOS, are present in the cytoplasmic membrane, cytoplasm, and/or nucleus of ICM and TE cells in E3.5 blastocyst (Wang *et al.*, 2004).

Perturbation of MAPK signaling, by treatment of 8-cell stage mouse embryos with PD98059, which is the inhibitor of MEK1, results not only in inactivation of ERK1/ERK2 MAPKs but also in downregulation of *Cdx2* expression at the morula and blastocyst stage. Moreover, blastocyst development is delayed and TE outgrowth reduced following explanting PD 908059-treated embryos *in vitro*, implicating MAPK signaling in regulating the emergence of TE lineage during mouse development (Lu *et al.*, 2008).

B. The second cell lineage decision: Formation of EPI and PE

The mechanisms responsible for the specification of murine EPI and PE are still poorly understood. It was assumed that positional signals induce the surface cells of ICM, adjacent to the blastocyst cavity, to become PE, and the cells enclosed inside the ICM to become EPI (Yamanaka *et al.*, 2006). However, growing body of evidence indicates that the decision of PE/EPI specification is set much earlier, and that precursors of these two lineages become predetermined before they become separated layers.

Two transcription factors, the closely related GATA family members GATA4 and GATA6, have been shown to be important for the specification of the mouse PE lineage (Fig. 1.1). In mouse embryos lacking the expression of either *Gata4* or *Gata6*, the PE does form, but a later PE derivative, that is, the visceral endoderm, fails to develop (Koutsourakis *et al.*, 1999; Morrisey *et al.*, 1998). When wild-type blastocysts ICM cells were injected with a dominant-negative form of *Gata6*, they contributed in significantly less extent to the PE than those ICM cells in that *Gata6* function was not affected (Meilhac *et al.*, 2009). Importantly, the cells of EPI were also shown to express *Nanog*. It was originally thought that *Nanog*$^{-/-}$ mouse embryos lack EPI, but are able to form PE (Mitsui *et al.*, 2003). More recent studies have shown, however, that these mutant embryos lack EPI as well as PE (Messerschmidt and Kemler, 2010; Silva *et al.*, 2009). Such *Nanog*-null blastocysts contained small number of GATA4-positive cells, and this observation has led to the hypothesis that initially, in the absence of NANOG, PE does form, but its cells cannot survive without support from the neighboring EPI (Silva *et al.*, 2009). Thus, GATA4 and GATA6 appear to be required for promoting the PE fate and suppressing the EPI fate. In addition, *Nanog* expression is required for the formation of the EPI lineage, which in turn, via a non cell-autonomous mechanism, possibly mediated by FGF/ERK signaling (Chazaud *et al.*, 2006; Nichols *et al.*, 2009), impacts at the PE formation (Messerschmidt and Kemler, 2010).

In early mouse blastocysts (up to the 64-cell stage), individual cells of ICM coexpress markers of nascent PE, that is, *Gata4* and *Gata6*, and marker of EPI, that is, *Nanog* (Guo *et al.*, 2010; Plusa *et al.*, 2008). In mid blastocysts (64- to 80-cell stage) ICM is heterogeneous. It means that mutually exclusive mosaic distribution of cells expressing either PE- or EPI-specific genes is established, suggesting that lineage commitment occurs in blastocysts consisting of more than 64 cells (Chazaud *et al.*, 2006; Kurimoto *et al.*, 2006; Plusa *et al.*, 2008). Support for this salt-and-pepper distribution of lineage-committed cells came from the fate mapping of single ICM cells in early blastocysts. Such experiments, involving stable expression of GFP mRNA injected into single ICM cells, revealed that at E5.5 (egg cylinder stage), progeny of GFP-expressing cells were restricted to either PE or EPI, but were rarely present in both lineages (Chazaud *et al.*, 2006). In late blastocyst (80–120 cells) PE precursors translocate to the ICM surface that face blastocyst cavity. Time-lapse imaging of cells expressing *Pdgfrα* (platelet-derived growth factor receptor α), which is considered to be another PE marker, demonstrated that at the mid blastocyst stage precursors of PE that are present inside the ICM translocate, and by the late blastocyst stage end up within the outside monolayer of PE cells (Plusa *et al.*, 2008). In addition, the *Pdgfrα*-expressing cells, that by the late blastocyst stage fail to sort properly into the PE monolayer, undergo apoptosis (Plusa *et al.*, 2008). Except for movement of inside cells to outside positions ICM cells can also migrate from the outside surface of the ICM to take up inside positions (Meilhac *et al.*, 2009). Such cell sorting probably relies on the differences in the adhesive properties of PE and EPI cells, although the molecules and mechanisms responsible for these specific properties have not been identified yet. By the late, expanded blastocyst stage (E4.5), precursor cells of PE and EPI are segregated into these two distinct layers, and their fate is restricted, as in chimeric embryos they were shown to be able to contribute only to their respective lineages (Gardner and Rossant, 1979).

The mechanisms leading to the mosaic "composition" of the ICM and involved in the lineage restriction and positional sorting of EPI and PE cells remain undefined. However, already mentioned FGF/ERK signaling pathway was postulated to be involved in the specification of EPI and PE lineages. Mutations in genes encoding some of the components of this signaling pathway, such as FGF4 (Feldman *et al.*, 1995; Goldin and Papaioannou, 2003), FGFR2 (Arman *et al.*, 1998), GRB2 (Chazaud and Rossant, 2006; Cheng *et al.*, 1998), and FRS2a (Hadari *et al.*, 2001), resulted in complete lack of PE formation *in vitro* in blastocyst outgrowths, and also *in vivo* leading to periimplantation lethality. Although regulation of the FGF/ERK pathway at the blastocyst stage is essential for segregation of EPI and PE lineages, both the source and the timing of FGF expression just begin to be elucidated. Transcriptome studies performed using single ICM cells revealed differences between these cells in the levels of *Fgf4* and *Fgfr2* expression (Kurimoto *et al.*, 2006). This variable expression and position of cells within ICM could determine the cell fate by modulating the intra- and extracellular response to the FGF/ERK signaling pathway (Meilhac *et al.*, 2009). Also the balance between NANOG and GATA6 seems to be regulated by the response of individual ICM cells to the FGF4. Modulation of FGF signaling by either inhibition of the FGF/ERK pathway or addition of exogenous FGF4 shifts the fate of ICM cells—they become either *Nanog*- or *Gata6*-expressing progenitors, respectively (Yamanaka *et al.*, 2010). Inhibition of FGF signaling in embryos cultured *in vitro* from the 8-cell to expanded blastocyst stage by combination of FGFR inhibitor, PD173074, and a MEK1 inhibitor, PD0325901, resulted in complete elimination of the GATA6-positive PE cells. On the other hand, the addition of exogenous FGF4 to the medium in that embryos were cultured from the zygote to blastocyst stage shifted all ICM cells to adopt the PE fate (Yamanaka *et al.*, 2010). The fate of the ICM cells could still be modulated after the mutually exclusive *Nanog* and *Gata6* expression had been established. However, this plasticity is progressively lost by E4.0, that is, at the late blastocyst stage. Exposing embryos during the morula to blastocyst transition to another FGFR inhibitor, SU5402, leads to the downregulation of PE-specific genes, that is, *Gata4*, *Sox17*, and the induction of EPI-specific genes, such as *Nanog* (Guo *et al.*, 2010; Nichols *et al.*, 2009).

Other transcription factors that are involved in PE development belong to SOX family. *Sox7* expression was shown to be limited to PE cells on the surface of expanded blastocyst ICM, while *Sox17* expression was detectable as early as at 32-cell stage (Morris *et al.*, 2010). In blastocyst it is initially expressed within the ICM in the characteristic "salt-and-pepper" pattern and then is gradually restricted to the cells of developing PE, resembling the localization of GATA4 and GATA6 (Artus *et al.*, 2011; Morris *et al.*, 2010; Niakan *et al.*, 2010). However, *Sox17* mutant embryos are able to develop PE layer (Kanai-Azuma *et al.*, 2002), probably due to the compensation by the presence of SOX7 which is specifically expressed in PE cells (Artus *et al.*, 2011). On the other hand, overexpression of *Sox17* in single blastomeres of 8-cell stage embryos biased their commitment towards PE, while its downregulation executed by siRNA biased their commitment towards EPI (Morris *et al.*, 2010).

C. Differences in cell lineage segregation mechanisms between mammals

The function and mechanisms controlling expression of factors regulating lineage specification during preimplantation development of animals other than mouse are definitely less understood. Similarly to mouse, in bank vole (Suwinska *et al.*, unpublished data), cattle (Kuijk *et al.*, 2008), and primates (Harvey *et al.*, 2009), the expression of *Nanog* is restricted only to the ICM of blastocyst. In porcine early blastocyst this protein is undetectable, but becomes apparent in almost all EPI cells in the late blastocyst (du Puy *et al.*, 2011). *Cdx2* expression in bank vole (Suwinska *et al.*, 2010), porcine, bovine (Berg *et al.*, 2011), and primate embryos (Harvey *et al.*, 2009) resembled that of mouse embryos. It is localized specifically to TE cells of the blastocyst, suggesting a conserved role for CDX2 in the TE formation (Kuijk *et al.*, 2008). Although in cattle CDX2 is required for TE maintenance, it does not repress *Oct4* expression, as it was documented in mouse embryos (Berg *et al.*, 2011). In contrary to mouse and bank vole, in which blastocysts *Oct4* expression is restricted to ICM, OCT4 protein is detectable both in ICM and TE of cattle (Degrelle *et al.*, 2005; Kirchhof *et al.*, 2000; Kuijk *et al.*, 2008; van Eijk *et al.*, 1999), pig (du Puy *et al.*, 2011; Kirchhof *et al.*, 2000; Kuijk *et al.*, 2008), sheep (Sanna *et al.*, 2009), rhesus monkey (Harvey *et al.*, 2009), and human blastocysts (Cauffman *et al.*, 2005; Hansis *et al.*, 2000; Kimber *et al.*, 2008). Thus, this factor is probably not involved in the segregation of TE and ICM in these species (Kirchhof *et al.*, 2000; van Eijk *et al.*, 1999). The mechanism of the reciprocal inhibition between OCT4 and CDX2, operating in mouse embryos, might also not be conserved between different mammalian species. GATA6, a transcription factor which specifies PE in mouse blastocysts, is constitutively expressed in cleaving preimplantation human embryos (Kimber *et al.*, 2008). Since *Nanog* mRNA is not expressed until the 8-cell stage, the opposing function of these two transcription factors could operate in the human, similarly to mouse, but would probably be initiated by NANOG rather than GATA6. A random distribution of GATA6 in bovine ICM cells, comparable to that occurring in the mouse embryo, also implies a conserved function of GATA6 in PE formation during mammalian development (Kuijk *et al.*, 2008). In mouse, formation of PE is additionally dependent on a functional EPI, and hence *Nanog* expression (Messerschmidt and Kemler, 2010). The mechanism of EPI and PE formation in pig appears to be slightly different, since in pig embryos PE is already formed before *Nanog* expression can be detected (du Puy *et al.*, 2011).

Above data suggest that mammals differ in early lineage segregation mechanisms. These differences may result from various timing and sequence of pre- and periimplantation development. After implantation, rodent, that is, mouse, rat, and bank vole embryos form a cup-shaped structure termed as egg cylinder, with EPI surrounded by PE. Human, bovine, and porcine embryos display a planar morphology (Behringer *et al.*, 2000), with EPI delaminating as a simple flattened structure called the embryonic disk. Moreover, mouse and human embryos invasively implant at the blastocyst stage, which results in a haemochorial placenta. Porcine and bovine blastocyst strongly elongates before implantation, transforming from sphere of a few millimeters in diameter to a long structure even up to 100 cm in length. This results in a loose diffuse noninvasive epitheliochorial placenta (Enders and Carter, 2004). Presented differences in preimplantation development may potentially influence the formation and characteristics of the pluripotent cell population in the embryo and also explain the difficulty of standardization of derivation and culture methods of ESCs from different mammalian species.

IV. Embryo-Derived Stem Cells

Cells of pre- and postimplantation mammalian embryos may serve as a rich resource of various stem cell lines that preserve the original embryonic lineage identity and developmental potential *in vitro*. Stem cell lines that reflect and retain properties of each of the three lineages: TE, PE, and EPI of the periimplantation mouse embryo can be derived and propagated indefinitely in culture (Rossant, 2008). ESCs represent the preimplantation EPI (Evans and Kaufman, 1981; Martin, 1981), Epiblast stem (EpiS) cells correspond to postimplantation EPI, whereas XEN cells represent the PE (Kunath *et al.*, 2005), and trophoblast stem (TS) cells represent the TE (Roberts and Fisher, 2011; Tanaka *et al.*, 1998). These embryo-derived cell types serve as models for investigating the mechanisms that regulate lineage specification, commitment, and maintenance. Lineage choice is controlled by the expression of key lineage-specific transcription factors that regulate downstream signaling pathways, such that by misexpression of appropriate transcription factors ES cells can be converted into TS-like (Niwa *et al.*, 2005), XEN-like (Shimosato *et al.*, 2007), or EpiS-like cells (Guo *et al.*, 2009). However, it is also possible to convert EpiS cells "back" to ES cells, via either forced expression of *Klf2* and *Klf4* (Guo *et al.*, 2009; Hall *et al.*, 2009) or extended culture in media containing serum and LIF (Bao *et al.*, 2009).

The ability to self-renew is evidenced by the fact that *in vitro*, under special culture conditions, they can be maintained in the form of "immortal," stable cell lines. Developmental potency, in turn, may be assessed by determining either their ability to reintegrate into an embryo proper, that is, to form a chimera, to differentiate *in vivo* after grafting into ectopic sites to form teratomas composed of multiple differentiated cell types, or to differentiate in *in vitro* culture. Whereas TS cells and XEN cells may be passaged for many generations without differentiation, that is, are characterized with the ability to self-renew, they maintain relatively restricted developmental potential, as in chimeras they contribute to TE and PE lineage, respectively.

A. Pluripotent stem cell lines: ES cells

In 1981, two-independent research groups isolated ESCs by culturing the intact blastocysts or immunosurgically isolated ICMs (Evans and Kaufman, 1981; Martin, 1981). Several years later ES cells have been derived from common marmoset (*Callithrix jacchus*) (Sasaki *et al.*, 2005; Thomson *et al.*, 1996), and finally from human embryos (Shamblott *et al.*, 1998; Thomson *et al.*, 1998). The next several years was needed to design the conditions allowing derivation of ES cells from rat blastocysts (Buehr *et al.*, 2008; Li *et al.*, 2008a,b; Ueda *et al.*, 2008).

A number of various studies revealed differences existing between ES cells derived from embryos of different mammalian species, with respect to their morphology, clonogenicity, expression patterns of various markers, and finally developmental potential. In *in vitro* culture mouse ES cells grow as small, compact, multilayered, and domed colonies. In contrast, primate, including human and monkey, as well as rat ES cells form flattened, loosely packed colonies with well-defined edges. The population-doubling time of human ES cells is significantly longer than that of mouse ES cells (30–35 vs. 12–15 h) (Wobus and Boheler, 2005). Moreover, an enzymatic dissociation of human and rat ES cell colonies into single cells can lead to significant decrease of ES cell propagation due to low efficiency of cell attachment, which is not typical for mouse ES cells (Sjogren-Jansson *et al.*, 2005; Ueda *et al.*, 2008). Molecular characteristic of both mouse and human ES cells has been extensively described in multiple studies, which revealed that in general these cells express markers characteristic for the cells present in ICM (see Section III.A). However, they differ in expression pattern of cell surface antigens. Undifferentiated mouse and rat ES cells express SSEA1 (Stage-specific embryonic antigen 1), but not SSEA3 and SSEA4, whereas human ES cells express SSEA3 and SSEA4, and SSEA1 protein appears following their differentiation. It also has to be noted that ES cell lines, regardless of species origin, derived under various conditions, and passaged for different periods of time, can differ, as far as their transcriptome, proteome, or epigenomic characteristics are concerned (e.g., Bock *et al.*, 2011; Humpherys *et al.*, 2001; Newman and Cooper, 2010; Ooi *et al.*, 2010).

As far as the differentiation potential of ES cells is concerned, rodent ES cells are able to form chimeras (Beddington and Robertson, 1989; Nagy *et al.*, 1990) or even whole organism when injected into the blastocysts cavities of tetraploid embryos (Eggan *et al.*, 2001; Nagy *et al.*, 1993). These cells and also human ES cells can form teratomas (Reubinoff *et al.*, 2000; Thomson *et al.*, 1998) and are also able to differentiate *in vitro* in various cell lines (Doetschman *et al.*, 1985). However, mouse and human cells differ in their capability to differentiate into TE-like cells. ES cells of human origin are able to generate trophoblastic cells (Thomson *et al.*, 1998) without any additional genetic manipulation, which are needed for mouse ES cells (Niwa, 2007; Niwa *et al.*, 2000).

B. Pluripotent stem cell lines: EpiS cells

Mouse and rat pluripotent stem cell lines can be obtained not only from ICM of preimplantation embryo, but also from the EPI of postimplantation embryos (E5.5–6.5 at mouse and E7.5–7.75 in case of rat) (Brons *et al.*, 2007; Chenoweth *et al.*, 2010; Tesar *et al.*, 2007). These cells, described as postimplantation EpiS cells (Epiblast stem cells, EpiSCs), resemble human

ES cells in respect to morphology, growing in monolayered, flattened colonies. Like human ES cells they are also sensitive to single-cell dissociation and for this reason they are generally passaged as clusters of cells rather than dispersed. Mouse EpiS cells are characterized by the expression of the genes associated with pluripotent state of ICM and ES cells, including *Oct4*, *Nanog*, *Sox2*, and *Ssea1*. However, in contrast to mouse ES and similarly to human ES cells (Tesar *et al.*, 2007), they also express genes involved in the specification of EPI (*Fgf5* and *Nodal*), and early germ layers—ectoderm, endoderm (*Gata6*, *Sox17*), and mesoderm (*Brachyury T*). Contrary to mouse ES cells, mouse and rat EpiS cells when microinjected into blastocyst cavity hardly incorporate into the ICM of the recipient blastocyst and extremely rarely form chimeras (Brons *et al.*, 2007; Tesar *et al.*, 2007). However, when injected subcutaneously in immunocompromised mice they form teratomas composed of tissues originating from all three germ layers, that is, ecto-, endo-, and mesoderm (Tesar *et al.*, 2007). Recent study revealed that restricted ability of EpiS cells to contribute to chimera development is caused by the existence of two distinct cell subpopulations of EpiS cells. These populations differ in molecular characteristic and developmental potency. A minor one (*circa* 0.5%) consists of cells resembling early-stage EPI, expressing *Oct4*, and capable to form chimeras, while major subpopulation of EpiS cell consists of OCT4-negative cells that correspond to late-stage postimplantation embryonic cells and exhibit limited developmental potential (Han *et al.*, 2010). It has been recently shown that EpiS cells exhibiting molecular features and functional properties consistent with EpiS cells from postimplantation embryos can be derived also from mouse preimplantation embryos at the blastocyst stage (Najm *et al.*, 2011). E3.5 blastocysts were plated onto feeder layer of mouse embryonic fibroblasts (MEF) in medium used routinely for ES cell derivation with no additional growth factor additives. Flat colonies obtained from dissociation of blastocyst outgrowths were propagated by employing conditions developed for the growth of postimplantation-derived EpiS cells, that is, manual dissociation into small clusters and subsequent culture in medium supplemented with FGF2.

V. Transcriptional Regulators of Pluripotency in Embryo-Derived Stem Cells

Various types of embryo-derived stem cells differ in their responses to the extrinsic and intrinsic factors that support their self-renewal and pluripotency. Although both rodent and human ES cells are derived from ICM of blastocysts they exhibit fundamental differences in growth requirements, signaling pathways activated and gene expression profile. Molecular and functional differences between these two stem cell types demonstrate that origin and/or the growth factor milieu may be important determinants of the stem cell identity.

OCT4 transcription factor is a master player in the network of factors regulating mouse ESCs self-renewal and their differentiation into specific lineages. The levels of OCT4 must be very tightly regulated to maintain the ES cell status. Reduction of its expression may result in the induction of trophectodermal differentiation (Hough *et al.*, 2006; Niwa *et al.*, 2000), whereas increase in its level triggers the commitment towards mesoderm and endoderm (Niwa *et al.*, 2000). Similarly, as it was documented for ICM cells, also in ES cells OCT4, together with NANOG and SOX2, constitutes a complex transcriptional circuitry responsible for controlling genes involved in the maintenance of undifferentiated state of mouse ES cells (Catena *et al.*, 2004; Kuroda *et al.*, 2005; Okumura-Nakanishi *et al.*, 2005; Rodda *et al.*, 2005). Among downstream targets of this regulatory circuitry are *Mycn* (V-myc myelocytomatosis viral-related oncogene), *Essrb* (estrogen receptor, β isoform), *Rif1* (Rap1-interacting factor 1) (Loh *et al.*, 2006), *Fgf4*, *Utf1* (undifferentiated embryonic cell transcription factor 1), and *Fbx15* (F-box protein 15) (Rodda *et al.*, 2005). Moreover, KLF4, although dispensable for development of pluripotent embryonic lineages *in vivo*, can cooperate with the OCT4–SOX2 complex to activate ES-specific genes, such as *Lefty1* (Guo *et al.*, 2009; Hall *et al.*, 2009; Jiang *et al.*, 2008; Nakatake *et al.*, 2006). OCT, NANOG, and SOX2 play similar role in preserving the pluripotency and self-renewal of human ES cells, that is, they bind to the promoters of their "own" genes and form interconnected autoregulatory loops (Babaie *et al.*, 2007; Boyer *et al.*, 2005). However, most of the target genes of *Oct4* and *Nanog* do not "overlap" between mouse and human ES cells, suggesting limited conservation of *Oct4*- and *Nanog*-bound sites, and species-specific differences in the networks controlled by these genes (Loh *et al.*, 2006).

Pluripotent cells, both *in vivo* and *in vitro*, that is, ICM cells and ES cells, have long been considered as homogeneous populations of cells. However, recent studies demonstrated that these cells exhibit high degree of heterogeneity in gene expression. Within ICM such variable expression identifies precursors of EPI and PE (Chazaud *et al.*, 2006; Plusa *et al.*, 2008). Morphologically undifferentiated ES cells also constitute heterogeneous cultures comprising cells of different degree of pluripotency corresponding to different developmental stages in the embryo. Expression of surface antigens SSEA1 and PECAM1 (platelet endothelial cell adhesion molecule 1) was shown to positively correlate with ES cell pluripotency (Furusawa *et al.*, 2004). Also NANOG (Singh *et al.*, 2007), REX1 (Toyooka *et al.*, 2008), and markers of the germ cell lineage STELLA or DPPA3 (developmental pluripotency-associated 3) (Hayashi *et al.*, 2008), have been shown to be heterogeneously expressed in phenotypically undifferentiated ES cells. Furthermore, genes associated with differentiation, such as *Brachyury* T (Suzuki *et al.*, 2006), *Gata6* (Singh *et al.*, 2007), *Hex* (Canham *et al.*, 2010), and

Sox17 (Niakan *et al.*, 2010), are also expressed by a subset of undifferentiated ES cells. Thus, ES cells seem to exist in equilibrium between a stable self-renewing ICM-like state, referred to as the "ground state," and a transient intermediate able to either revert to self-renewing state or proceed into differentiation (Chambers *et al.*, 2007; Kalmar *et al.*, 2009; Silva and Smith, 2008). These two states are interconvertible and fluctuate between ICM-like (naïve) "ground state" and "primed for differentiation" intermediate (Nichols *et al.*, 2009; Nichols and Smith, 2009). Accordingly, the pluripotency seems to be determined by the capacity of a mixed population of lineage-biased intermediates to commit to different cell fate, and the most important is to keep the balance between them. This dynamic interchanging is most likely influenced by external stimuli. Therefore, understanding the plasticity of stem cell populations requires the knowledge of their response to extrinsic signaling, as these signals are propagated through intracellular signal-transduction pathways that converge on genetic network controlling pluripotency.

VI. Extrinsic Factors and Signaling Pathways Regulating Pluripotency and Differentiation

A. LIF/Jak/STAT3 signaling

Mouse ESCs have long been derived and maintained in the presence of cytokine LIF and either serum or the growth factor BMP4 (bone morphogenetic protein 4) (Brook and Gardner, 1997; Ying *et al.*, 2003). Although LIF signaling is not required for ICM formation in developing blastocysts, it was thought to be indispensable for promoting self-renewal of mouse ES cells (Fig. 1.3). For this reason, ES cells have been routinely cultured on feeder layer of MEF synthesizing LIF (Rathjen *et al.*, 1990), or in feeder-free conditions in media supplemented with recombinant LIF (Nichols *et al.*, 1990). LIF belongs to interleukin (IL)-6 family of cytokines (Smith *et al.*, 1988; Williams *et al.*, 1988). It signals through LIF receptor (LIFR) consisting of the two subunits: gp130 (glycoprotein 130) and LIFRβ. Binding LIF to LIFRβ results in its heterodimerization with gp130 and activation of Jak-STAT3 (Janus tyrosine kinase—signal transducer and activator of transcription) signaling pathway. Jak kinase phosphorylates the STAT3, transcription factor responsible for activation of genes involved in maintenance of mouse ES cells in undifferentiated state, such as *Klf4*, *Sox2*, and *Oct4* (Bourillot *et al.*, 2009; Burdon *et al.*, 1999; Matsuda *et al.*, 1999; Niwa *et al.*, 1998, 2009). LIF-dependent mechanism supporting the maintenance of undifferentiated state of mouse ES cells is not functional in rat (Buehr *et al.*, 2008; Li *et al.*, 2008b), and also in primate ES cells, including human (Daheron *et al.*, 2004; Humphrey *et al.*, 2004) and

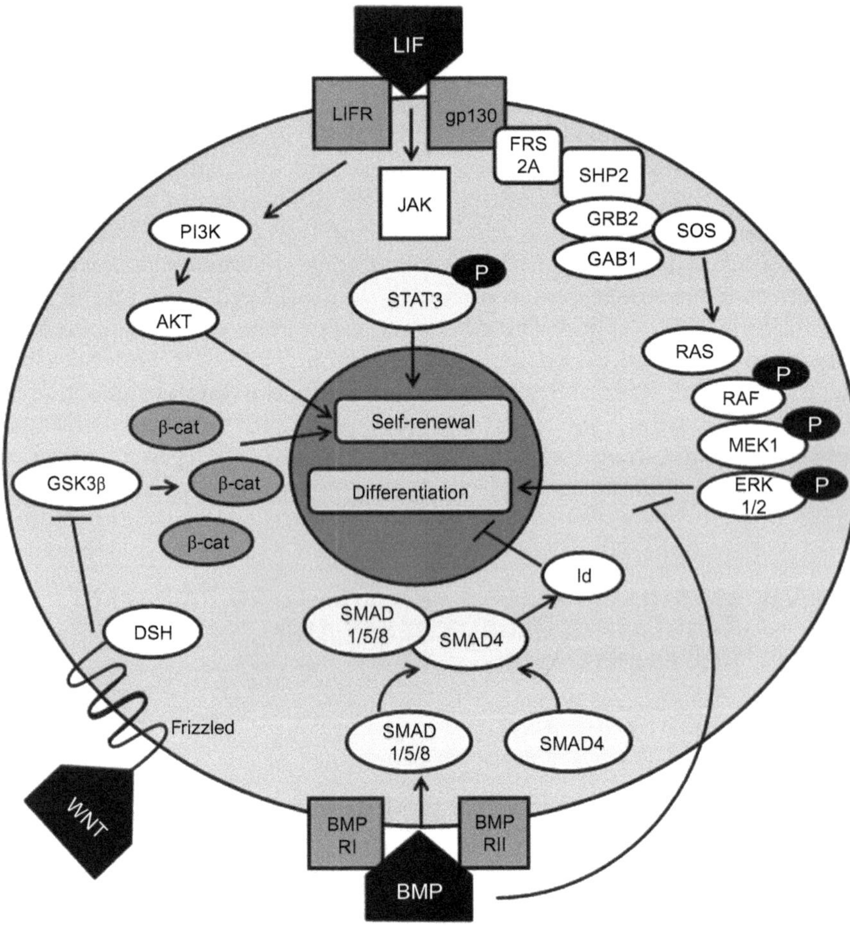

Figure 1.3 Signaling pathways affecting self-renewal and differentiation of mouse ES cells. Self-renewal of mouse ES cells is regulated by: LIF-dependent JAK/STAT3 and PI3K/AKT signaling pathways; BMP-mediated activation of *Id* transcription and suppression of MAP kinases ERK1/ERK2; WNT/GSK3β signaling cascade (detailed description within the text).

monkeys (Sumi *et al.*, 2004). It has been shown that these cells lose pluripotency and rapidly differentiate under feeder layer—free culture conditions, even in the presence of LIF. Although both components of LIFR are expressed in human as well as in cynomolgus monkey ES cells, and LIF can induce STAT3 phosphorylation and nuclear translocation, it is not sufficient to maintain their pluripotent state. LIF/STAT3 pathway is functional also in rat ES cells, but is not sufficient to maintain the pluripotency (Li *et al.*, 2008b). The preservation of undifferentiated state of rodent EpiS cells is also LIF independent, since blocking STAT3 phosphorylation at tyrosine

705 with a Jak inhibitor did not impact on undifferentiated state of these cells (Tesar *et al.*, 2007).

B. LIF/Ras/MEK/ERK signaling

Paradoxically, LIFRβ and gp130 may also activate MAPKs ERK1/2 signaling pathway, directing ES cells to exit from the self-renewal program and to begin differentiation (Fig. 1.3). Activated by MEK1 catalyzed phosphorylation ERK1/ERK2 kinases enter the nucleus and regulate by phosphorylation the transcription factors responsible for differentiation of mouse ES cells (Burdon *et al.*, 1999). Thus, MAPK pathway inhibits self-renewal of mouse ES cells, in contrast to LIF/STAT3 cascade. Inhibition of this signaling pathway using MEK1 inhibitors, such as PD98059 or U0126, increases the efficiency of ES cell derivation (Kunath *et al.*, 2007; Nichols *et al.*, 2009; Ying *et al.*, 2008). Unexpectedly, blocking of MAPK pathway by MEK1 inhibitors, or by RNAi depleting MEK1 and MEK2 encoding mRNA, resulted in rapid loss of self-renewal capacity of human ES cells (Li *et al.*, 2007), indicating the opposite activity of this signaling cascade in ESCs of these two mammalian species. In human ES cells MAPK signaling is implicated in regulating pluripotency, acting downstream of the FGF (Li *et al.*, 2007).

C. Phosphatidylinositol 3-kinase (PI3K)/Akt and Src signaling

In mouse ES cells LIF signaling has been also shown to activate PI3K that plays a role in preserving the ES cells status by negative regulation of ERK activity (Fig. 1.3) (Paling *et al.*, 2004). Whereas in mouse inhibition of PI3K/Akt kinase pathway augments LIF-induced phosphorylation of ERKs and induces differentiation, in human PI3K/Akt and ERK function in a cooperative manner to maintain pluripotency (Li *et al.*, 2007). In mouse, also cYes and Hck of the Src family of nonreceptor tyrosine kinases, which are important for maintaining growth and self-renewal of mouse and human ES cells, are regulated by LIF signaling (Anneren *et al.*, 2004; Ernst *et al.*, 1994). Inhibiting Src proteins, using SU6656, allowed uncovering their role in the maintenance of self-renewal in mouse and also in human ES cells, and showed that Src signaling acts through a JAK/STAT3 and Ras/MEK/ERK-independent pathway. Inhibitor-treated ES cells revealed signs of differentiation, such as a fibroblast-like phenotype and reduced levels of OCT4, NANOG, and FGF4 (Anneren *et al.*, 2004).

D. BMP/SMAD/Id signaling

LIF does not act alone to maintain the undifferentiated state of mouse ES cells. Their proliferation and self-renewal requires supplementation of the culture medium with fetal bovine serum (FBS), which is a rich source of both mitogens and factors sustaining self-renewal and pluripotency. Among them are BMP4 and GDF6 (growth and differentiation factor 6) that can effectively substitute for serum and in conjunction with LIF inhibit neural lineage commitment of mouse ES cells (Fig. 1.3) (Ying *et al.*, 2003). BMP4 protein acts via receptor complex composed of BMPRIa and BMPRII. Binding BMP4 to such complex leads to the phosphorylation of SMAD1/5/8 transcription factors and their interaction with SMAD4. Both BMP4 and GDF6 via SMAD factors influence *Id* (*inhibitors of differentiation*) genes, and thus, suppress differentiation of mouse ES cells (Ying *et al.*, 2003). Alternatively, it has been also reported that the major effect of BMP4 on the self-renewal of mouse ES cells is accomplished by means of the inhibition of ERK signaling cascade (Qi *et al.*, 2004). Again, BMP4 signaling exerts an opposite effect on human ES cells inducing their differentiation into TE lineage (Pera *et al.*, 2004; Xu *et al.*, 2002). In case of EpiS cells BMP stimulation results in their specification into primordial germ cells (Bao *et al.*, 2009; Tesar *et al.*, 2007).

E. WNT/β-catenin signaling

Sato and coworkers defined a role for canonical WNT (wingless typed)-dependent signaling in self-renewal of mouse ES cells, functioning independently of LIF/STAT (Sato *et al.*, 2004). This signal transduction pathway is initiated upon binding WNT to its receptor FRIZZLED (Fig. 1.3). It results in the activation of the protein DSH (dishevelled) counteracting to destruct the complex. Activation of receptor induces the inactivation of GSK3 (glycogen synthase kinase 3) and subsequent nuclear accumulation of β-catenin, which via TCF (T-cell-specific factors) stimulates the expression of *Oct4*, *Nanog*, and *Rex1* (also known as zinc-finger protein 42; ZFP42). Suppression of GSK3, an antagonist of WNT signaling, by specific pharmacological inhibitor BIO (6-bromoindirubin-3′-oxime) is sufficient to maintain self-renewal and pluripotency of mouse ES cells in the absence of LIF and WNT (Sato *et al.*, 2004; Umehara *et al.*, 2007). Human ES cells are also capable of response to WNT/β-catenin activation (Dravid *et al.*, 2005; Sato *et al.*, 2004). However, the precise function of this signaling remains controversial, due to contradictory results of studies either supporting (Sato *et al.*, 2004), or denying the contribution of this pathway to the maintenance of undifferentiated state of human ES cells (Dravid *et al.*, 2005).

The above findings led to the suggestion that mouse ES cell derivation and self-renewal do not necessarily require activating signals from the LIF/STAT3 and BMP/SMAD pathways. ES cells rather need to be secured from the impact of differentiation cues. These findings resulted in identification of exogenous factors driving mouse ES cells into self-renewal versus differentiation pathway. The consequence was designing of chemically defined culture conditions allowing maintenance of highly undifferentiated state and efficient expansion of ES cells for indefinite time. The breakthrough in the development of ES cells derivation methods was caused by the introduction of so called "3i conditions" (Ying *et al.*, 2008). 3i stands for three selective chemical inhibitors: SU5402, FGF receptor inhibitor; PD184352, MEK1 inhibitor; and CHIR99021, GSK3β inhibitor; which allow the manipulation of respective pathways and shielding the ground state of pluripotency from inductive differentiation stimuli. Application of 3i does not require the presence of growth factors, such as LIF or BMP4. By adding GSK3β inhibitor, mimicking WNT/β-catenin signaling (Ying *et al.*, 2008) (Niwa *et al.*, 2009; Sato *et al.*, 2004) and a MEK1 inhibitor to inactivate FGF/ERK signaling (Guo *et al.*, 2009; Ying *et al.*, 2008) mouse ES cells can be derived far more efficiently and from all mouse strains, including those previously considered as "resistant" to ES cell derivation (Hanna *et al.*, 2010; Kiyonari *et al.*, 2010; Nichols *et al.*, 2009; Ying *et al.*, 2008). Moreover, 3i or even 2i cocktail of inhibitors (without SU5402), in combination with LIF, made possible the derivation of rat cell lines that meet all the criteria of true ESCs, that is, are characterized by their ability to long-term self-renewal, expression of pluripotency markers, including OCT4, NANOG, SOX2, and KLF4, and capacity to differentiate into derivatives of all three germ layers. Their pluripotent potential was confirmed by teratoma formation, contribution to the developing chimeric embryo, germline transmission and as result generation of functional gametes (Buehr *et al.*, 2008; Li *et al.*, 2008b). Previously, conventional ES derivation and culture techniques, that is, with feeder layer, LIF, and serum did not ensure derivation of genuine, germline competent rat ES cells (Li *et al.*, 2008a).

F. FGF2 and transforming growth factor β (TGFβ)/ActivinA/ Nodal signaling

Human ES cells require FGF2 and Activin/Nodal signals for self-renewal, and thus, all human ES cells media described to date contain FGF2 (e.g., Dvorak *et al.*, 2005; Furue *et al.*, 2008; Levenstein *et al.*, 2006; Xu *et al.*, 2005). Addition of FGF2 has been associated with pleiotropic-positive effects: increasing human ES cell proliferation, enhancing attachment and survival, moderately stimulating NANOG gene expression, inhibiting spontaneous differentiation, including neural induction (Eiselleova *et al.*, 2009; Greber *et al.*, 2007, 2008; Li *et al.*, 2007; Wang *et al.*, 2009). Exogenous FGF2 activates both MAPK/ERK and PI3K/AKT (Fig. 1.4) (Armstrong *et al.*, 2006; Dvorak *et al.*, 2005; Li *et al.*, 2007). FGF2-stimulated PI3K signaling can activate the GSK3β/β-catenin/WNT signaling pathway resulting in the activation of TCF (Ding *et al.*, 2010). Furthermore, some activity of FGF2 may be indirect via its action on MEFs cocultured with human ESC cells (Bendall *et al.*, 2007; Greber *et al.*,

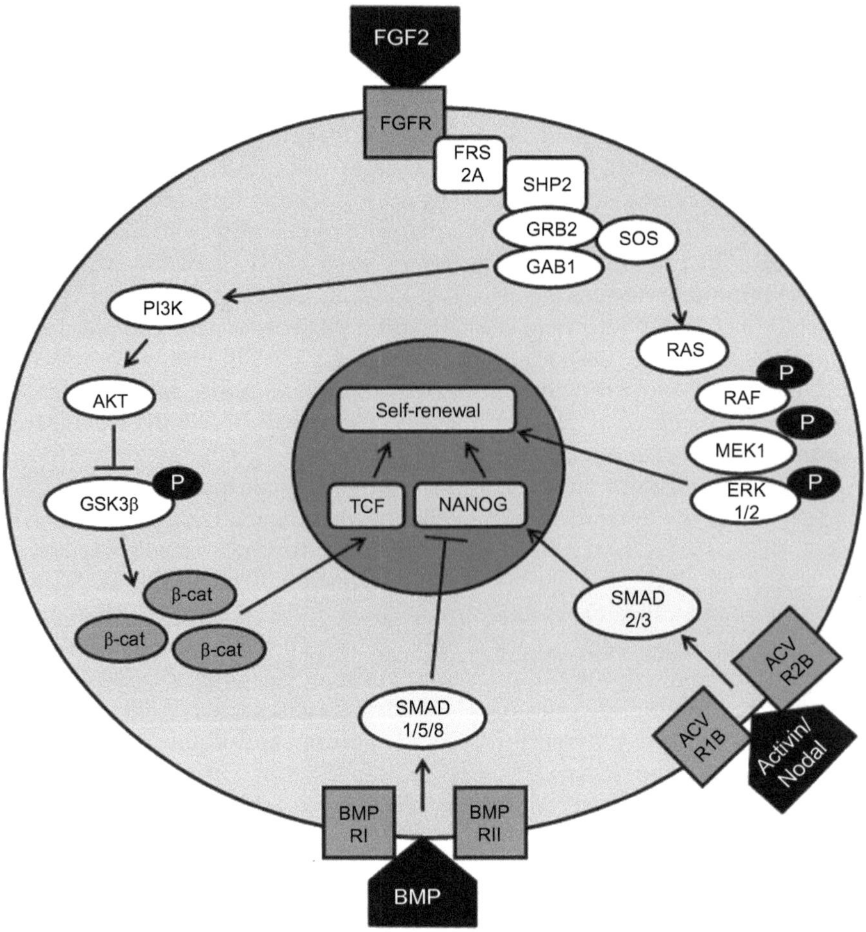

Figure 1.4 Signaling pathways maintaining self-renewal of human ES cells. Human ES cell self-renewal is dependent on FGF2 activating both MAP kinases ERK1/ERK2, PI3K/Akt, and on the activation of Activin/Nodal signaling cascades. On the other hand, BMP signaling induces differentiation. These pathways converge mainly on *Nanog*, which maintains ES cells in undifferentiated state. Inhibition of BMP-mediated signaling balance between pathways involving SMAD2/3 (driven by ActA/Nodal) and that involving SMAD1/5/8 (driven by BMPs) is crucial for self-renewal of human embryonic stem cells.

2007). FGF2 stimulates MEFs to secrete Activin A (ActA) (Greber *et al.*, 2007). Both ActA and Nodal belong to the TGFβ family. They signal through the same receptors (ACVR1B and ACVR2B). They induce the SMAD2/3 transcription factors (James *et al.*, 2005; Vallier *et al.*, 2005) that bind to the NANOG promoter and thereby activate NANOG gene transcription, establishing a direct link between this signaling pathway and the self-renewal network in human ES cells (Fig. 1.4) (Greber *et al.*, 2008). FGF/ERK signaling seems to cooperate with SMAD2/3 in supporting NANOG expression (Greber *et al.*, 2008). *Fgf2* gene appears to be directly controlled by OCT4, suggesting the presence of an autocrine loop that helps to sustain self-renewal of human ES cells. SMAD1/5/8, in turn, which are activated by BMP2, BMP4, or BMP7, bind to the *Nanog* promoter and inhibit its expression. The factor blocking BMP-mediated induction of differentiation is a member of TGFβ family—GDF3 (Levine and Brivanlou, 2006). Thus, inhibition of BMP-mediated signaling and balance between pathways involving SMAD2/3 (driven by ActA and Nodal) and the one involving SMAD1/5/8 (driven by BMPs) seem to be crucial for self-renewal of human stem cells. It has also been shown that ActA alone is sufficient to support long-term feeder-free growth and maintenance of pluripotency of human ES cells, possibly by production of FGF2, and inducing the expression of other regulators, such as *Oct4* and *Nanog*. It also inhibits expression of differentiation markers for all three germ layers, including neurofilament (marker of ectoderm), cardiac muscle actin (mesodermal marker), and α-fetoprotein (marker of endoderm) (Xiao *et al.*, 2006). Multiple signaling pathways, including ActA-dependent one, might also be involved in preserving rat ES-like cells in the undifferentiated state (Li *et al.*, 2008a). However, rat ES cells also require LIF action to sustain the expression of *Oct4* and *Nanog* (Ueda *et al.*, 2008), implicating that these cells share some characteristics with both mouse and human ES cells.

Mouse and rat EpiS cells are characterized by different growth factor requirements than mouse ES cells. Interestingly, they are derived and maintained under the conditions that support human ES cells self-renewal—in FGF2 supplemented culture medium, on inactivated MEFs, or with FGF2 and ActA under feeder-free conditions. Similarly to human ES cells, activation of the ActA pathway promotes self-renewal of EpiS cells, via direct activation of *Nanog* expression (Greber *et al.*, 2010). Inhibition of this pathway induces neuroectodermal differentiation of EpiS cells, manifested by the changes in the morphology of colonies and in gene expression pattern to that characteristic for neural precursors [e.g., expression of *Sox1*, *Pax6* (*Paired box gene 6*), *Nes* (*Nestin*), etc.] (Greber *et al.*, 2010; Tesar *et al.*, 2007). However, in contrast to the role in human ES cells, FGF2 does not cooperate with SMAD2/3 signaling in supporting self-renewal of mouse EpiS cells via *Nanog* expression under feeder-free conditions. It rather supports the EpiS state by inhibiting neural lineage commitment and preventing dedifferentiation (Greber *et al.*, 2010). Direct role of FGF2 signaling in supporting self-renewal with the contribution of NANOG seems to be specific for human ES cells, and not for EpiS cells.

VII. Conclusions

Stem cell fate is governed by the integration of intrinsic and extrinsic positive and negative signals upon inherent transcriptional networks. The ICM and EPI of the mammalian embryo constitute dynamic cell populations whose interactions with extraembryonic tissues (TE and PE) are of great importance for cell fate determination. This dynamism is reflected in the plasticity of pluripotent stem cells *in vitro* in response to changes in cell culture conditions. The above comparisons of rodent ES cells, human ES cells, and rodent EpiS cells illustrate that stem cell pluripotency is not a fixed state but is strongly influenced by both developmental and environmental context. Distinct pluripotent stem cell lines with unique functional characteristics can be derived from different stages of the embryo and under different growth factor conditions. For example, the functional differences in developmental potential between mouse ES cells and EpiS cells may reflect the characteristic of cells/tissue of origin from which the stem cell line is initially derived, ICM versus EPI, or they may be a consequence of their different culture conditions. Pluripotent stem cells are capable to form wide range of differentiated cells. This discovery has opened doors for studies aiming to understand early mammal development and to apply this knowledge in medicine to cure degenerative diseases in a way not previously possible. For this reason, it is extremely important to resolve and define mechanisms underlying pluripotency and differentiation of cells. The knowledge of intrinsic and extrinsic factors that interact with one another may provide explanation to species differences and provide the means to a defined *in vitro* culture of pluripotent cells allowing their possible future medical application. Moreover, the knowledge accumulated during the studies focusing at ES and EpiS cells was of the greatest importance during the difficult scientific "trip" leading to the derivation of induced pluripotent stem cells in 2006 (e.g., Maherali *et al.*, 2007; Meissner *et al.*, 2007; Takahashi *et al.*, 2007; Takahashi and Yamanaka, 2006). It allowed not only designing the experimental conditions necessary to turn somatic cells into pluripotent ones, but also understanding the promises and dangers hidden in the newly created induced pluripotent state.

ACKNOWLEDGMENTS

The authors regret the omission of any relevant literature, which happened either because of space constraint or oversight. The authors thank prof. Marek Maleszewski for his critical comments. During the preparation of this work the MAC was supported by the grant from Polish Ministry of Science and Higher Education N N302 125939, AS by the grant from Polish Ministry of Science and Higher Education N N301 311637 and PARENT-BRIDGE Program from Foundation for Polish Science POMOST/2010-1/9.

REFERENCES

Anneren, C., Cowan, C. A., and Melton, D. A. (2004). The Src family of tyrosine kinases is important for embryonic stem cell self-renewal. *J. Biol. Chem.* **279,** 31590–31598.

Arman, E., Haffner-Krausz, R., Chen, Y., Heath, J. K., and Lonai, P. (1998). Targeted disruption of fibroblast growth factor (FGF) receptor 2 suggests a role for FGF signaling in pregastrulation mammalian development. *Proc. Natl. Acad. Sci. USA* **95,** 5082–5087.

Armstrong, L., Hughes, O., Yung, S., Hyslop, L., Stewart, R., Wappler, I., Peters, H., Walter, T., Stojkovic, P., Evans, J., Stojkovic, M., and Lako, M. (2006). The role of PI3K/AKT, MAPK/ERK and NFkappabeta signalling in the maintenance of human embryonic stem cell pluripotency and viability highlighted by transcriptional profiling and functional analysis. *Hum. Mol. Genet.* **15,** 1894–1913.

Artus, J., Piliszek, A., and Hadjantonakis, A. K. (2011). The primitive endoderm lineage of the mouse blastocyst: Sequential transcription factor activation and regulation of differentiation by Sox17. *Dev. Biol.* **350,** 393–404.

Avilion, A. A., Nicolis, S. K., Pevny, L. H., Perez, L., Vivian, N., and Lovell-Badge, R. (2003). Multipotent cell lineages in early mouse development depend on SOX2 function. *Genes Dev.* **17,** 126–140.

Babaie, Y., Herwig, R., Greber, B., Brink, T. C., Wruck, W., Groth, D., Lehrach, H., Burdon, T., and Adjaye, J. (2007). Analysis of Oct4-dependent transcriptional networks regulating self-renewal and pluripotency in human embryonic stem cells. *Stem Cells* **25,** 500–510.

Bao, S., Tang, F., Li, X., Hayashi, K., Gillich, A., Lao, K., and Surani, M. A. (2009). Epigenetic reversion of post-implantation epiblast to pluripotent embryonic stem cells. *Nature* **461,** 1292–1295.

Beddington, R. S., and Robertson, E. J. (1989). An assessment of the developmental potential of embryonic stem cells in the midgestation mouse embryo. *Development* **105,** 733–737.

Behringer, R. R., Wakamiya, M., Tsang, T. E., and Tam, P. P. (2000). A flattened mouse embryo: Leveling the playing field. *Genesis* **28,** 23–30.

Bendall, S. C., Stewart, M. H., Menendez, P., George, D., Vijayaragavan, K., Werbowetski-Ogilvie, T., Ramos-Mejia, V., Rouleau, A., Yang, J., Bosse, M., Lajoie, G., and Bhatia, M. (2007). IGF and FGF cooperatively establish the regulatory stem cell niche of pluripotent human cells in vitro. *Nature* **448,** 1015–1021.

Berg, D. K., Smith, C. S., Pearton, D. J., Wells, D. N., Broadhurst, R., Donnison, M., and Pfeffer, P. L. (2011). Trophectoderm lineage determination in cattle. *Dev. Cell* **20,** 244–255.

Bock, C., Kiskinis, E., Verstappen, G., Gu, H., Boulting, G., Smith, Z. D., Ziller, M., Croft, G. F., Amoroso, M. W., Oakley, D. H., Gnirke, A., Eggan, K., *et al.* (2011). Reference Maps of human ES and iPS cell variation enable high-throughput characterization of pluripotent cell lines. *Cell* **144,** 439–452.

Bourillot, P. Y., Aksoy, I., Schreiber, V., Wianny, F., Schulz, H., Hummel, O., Hubner, N., and Savatier, P. (2009). Novel STAT3 target genes exert distinct roles in the inhibition of mesoderm and endoderm differentiation in cooperation with Nanog. *Stem Cells* **27,** 1760–1771.

Boyer, L. A., Lee, T. I., Cole, M. F., Johnstone, S. E., Levine, S. S., Zucker, J. P., Guenther, M. G., Kumar, R. M., Murray, H. L., Jenner, R. G., Gifford, D. K., Melton, D. A., *et al.* (2005). Core transcriptional regulatory circuitry in human embryonic stem cells. *Cell* **122,** 947–956.

Brons, I. G., Smithers, L. E., Trotter, M. W., Rugg-Gunn, P., Sun, B., de Sousa, Chuva, Lopes, S. M., Howlett, S. K., Clarkson, A., Ahrlund-Richter, L., Pedersen, R. A., and Vallier, L. (2007). Derivation of pluripotent epiblast stem cells from mammalian embryos. *Nature* **448,** 191–195.

Brook, F. A., and Gardner, R. L. (1997). The origin and efficient derivation of embryonic stem cells in the mouse. *Proc. Natl. Acad. Sci. USA* **94,** 5709–5712.

Buehr, M., Meek, S., Blair, K., Yang, J., Ure, J., Silva, J., McLay, R., Hall, J., Ying, Q. L., and Smith, A. (2008). Capture of authentic embryonic stem cells from rat blastocysts. *Cell* **135,** 1287–1298.

Burdon, T., Stracey, C., Chambers, I., Nichols, J., and Smith, A. (1999). Suppression of SHP-2 and ERK signalling promotes self-renewal of mouse embryonic stem cells. *Dev. Biol.* **210,** 30–43.

Canham, M. A., Sharov, A. A., Ko, M. S., and Brickman, J. M. (2010). Functional heterogeneity of embryonic stem cells revealed through translational amplification of an early endodermal transcript. *PLoS Biol.* **8,** e1000379.

Catena, R., Tiveron, C., Ronchi, A., Porta, S., Ferri, A., Tatangelo, L., Cavallaro, M., Favaro, R., Ottolenghi, S., Reinbold, R., Scholer, H., and Nicolis, S. K. (2004). Conserved POU binding DNA sites in the Sox2 upstream enhancer regulate gene expression in embryonic and neural stem cells. *J. Biol. Chem.* **279,** 41846–41857.

Cauffman, G., Van de Velde, H., Liebaers, I., and Van Steirteghem, A. (2005). Oct-4 mRNA and protein expression during human preimplantation development. *Mol. Hum. Reprod.* **11,** 173–181.

Chambers, I., Colby, D., Robertson, M., Nichols, J., Lee, S., Tweedie, S., and Smith, A. (2003). Functional expression cloning of Nanog, a pluripotency sustaining factor in embryonic stem cells. *Cell* **113,** 643–655.

Chambers, I., Silva, J., Colby, D., Nichols, J., Nijmeijer, B., Robertson, M., Vrana, J., Jones, K., Grotewold, L., and Smith, A. (2007). Nanog safeguards pluripotency and mediates germline development. *Nature* **450,** 1230–1234.

Chazaud, C., and Rossant, J. (2006). Disruption of early proximodistal patterning and AVE formation in Apc mutants. *Development* **133,** 3379–3387.

Chazaud, C., Yamanaka, Y., Pawson, T., and Rossant, J. (2006). Early lineage segregation between epiblast and primitive endoderm in mouse blastocysts through the Grb2-MAPK pathway. *Dev. Cell* **10,** 615–624.

Cheng, A. M., Saxton, T. M., Sakai, R., Kulkarni, S., Mbamalu, G., Vogel, W., Tortorice, C. G., Cardiff, R. D., Cross, J. C., Muller, W. J., and Pawson, T. (1998). Mammalian Grb2 regulates multiple steps in embryonic development and malignant transformation. *Cell* **95,** 793–803.

Chenoweth, J. G., McKay, R. D., and Tesar, P. J. (2010). Epiblast stem cells contribute new insight into pluripotency and gastrulation. *Dev. Growth Differ.* **52,** 293–301.

Chew, J. L., Loh, Y. H., Zhang, W., Chen, X., Tam, W. L., Yeap, L. S., Li, P., Ang, Y. S., Lim, B., Robson, P., and Ng, H. H. (2005). Reciprocal transcriptional regulation of Pou5f1 and Sox2 via the Oct4/Sox2 complex in embryonic stem cells. *Mol. Cell. Biol.* **25,** 6031–6046.

Daheron, L., Opitz, S. L., Zaehres, H., Lensch, M. W., Andrews, P. W., Itskovitz-Eldor, J., and Daley, G. Q. (2004). LIF/STAT3 signaling fails to maintain self-renewal of human embryonic stem cells. *Stem Cells* **22,** 770–778.

Dailey, L., Ambrosetti, D., Mansukhani, A., and Basilico, C. (2005). Mechanisms underlying differential responses to FGF signaling. *Cytokine Growth Factor Rev.* **16,** 233–247.

Dang, D. T., Pevsner, J., and Yang, V. W. (2000). The biology of the mammalian Kruppel-like family of transcription factors. *Int. J. Biochem. Cell Biol.* **32,** 1103–1121.

Degrelle, S. A., Campion, E., Cabau, C., Piumi, F., Reinaud, P., Richard, C., Renard, J. P., and Hue, I. (2005). Molecular evidence for a critical period in mural trophoblast development in bovine blastocysts. *Dev. Biol.* **288,** 448–460.

Dietrich, J. E., and Hiiragi, T. (2007). Stochastic patterning in the mouse pre-implantation embryo. *Development* **134,** 4219–4231.

Ding, V. M., Ling, L., Natarajan, S., Yap, M. G., Cool, S. M., and Choo, A. B. (2010). FGF-2 modulates Wnt signaling in undifferentiated hESC and iPS cells through activated PI3-K/GSK3beta signaling. *J. Cell. Physiol.* **225,** 417–428.

Doetschman, T. C., Eistetter, H., Katz, M., Schmidt, W., and Kemler, R. (1985). The in vitro development of blastocyst-derived embryonic stem cell lines: Formation of visceral yolk sac, blood islands and myocardium. *J. Embryol. Exp. Morphol.* **87,** 27–45.

Dravid, G., Ye, Z., Hammond, H., Chen, G., Pyle, A., Donovan, P., Yu, X., and Cheng, L. (2005). Defining the role of Wnt/beta-catenin signaling in the survival, proliferation, and self-renewal of human embryonic stem cells. *Stem Cells* **23,** 1489–1501.

du Puy, L., Lopes, S. M., Haagsman, H. P., and Roelen, B. A. (2011). Analysis of co-expression of OCT4, NANOG and SOX2 in pluripotent cells of the porcine embryo, in vivo and in vitro. *Theriogenology* **75,** 513–526.

Dvorak, P., Dvorakova, D., Koskova, S., Vodinska, M., Najvirtova, M., Krekac, D., and Hampl, A. (2005). Expression and potential role of fibroblast growth factor 2 and its receptors in human embryonic stem cells. *Stem Cells* **23,** 1200–1211.

Eggan, K., Akutsu, H., Loring, J., Jackson-Grusby, L., Klemm, M., Rideout, W. M., 3rd., Yanagimachi, R., and Jaenisch, R. (2001). Hybrid vigor, fetal overgrowth, and viability of mice derived by nuclear cloning and tetraploid embryo complementation. *Proc. Natl. Acad. Sci. USA* **98,** 6209–6214.

Eiselleova, L., Matulka, K., Kriz, V., Kunova, M., Schmidtova, Z., Neradil, J., Tichy, B., Dvorakova, D., Pospisilova, S., Hampl, A., and Dvorak, P. (2009). A complex role for FGF-2 in self-renewal, survival, and adhesion of human embryonic stem cells. *Stem Cells* **27,** 1847–1857.

Elling, U., Klasen, C., Eisenberger, T., Anlag, K., and Treier, M. (2006). Murine inner cell mass-derived lineages depend on Sall4 function. *Proc. Natl. Acad. Sci. USA* **103,** 16319–16324.

Ema, M., Mori, D., Niwa, H., Hasegawa, Y., Yamanaka, Y., Hitoshi, S., Mimura, J., Kawabe, Y., Hosoya, T., Morita, M., Shimosato, D., Uchida, K., *et al.* (2008). Kruppel-like factor 5 is essential for blastocyst development and the normal self-renewal of mouse ESCs. *Cell Stem Cell* **3,** 555–567.

Enders, A. C., and Carter, A. M. (2004). What can comparative studies of placental structure tell us?—A review. *Placenta* **25**(Suppl. A), S3–S9.

Ernst, M., Gearing, D. P., and Dunn, A. R. (1994). Functional and biochemical association of Hck with the LIF/IL-6 receptor signal transducing subunit gp130 in embryonic stem cells. *EMBO J.* **13,** 1574–1584.

Evans, M. J., and Kaufman, M. H. (1981). Establishment in culture of pluripotential cells from mouse embryos. *Nature* **292,** 154–156.

Feldman, B., Poueymirou, W., Papaioannou, V. E., DeChiara, T. M., and Goldfarb, M. (1995). Requirement of FGF-4 for postimplantation mouse development. *Science* **267,** 246–249.

Fleming, T. P., Papenbrock, T., Fesenko, I., Hausen, P., and Sheth, B. (2000). Assembly of tight junctions during early vertebrate development. *Semin. Cell Dev. Biol.* **11,** 291–299.

Furue, M. K., Na, J., Jackson, J. P., Okamoto, T., Jones, M., Baker, D., Hata, R., Moore, H. D., Sato, J. D., and Andrews, P. W. (2008). Heparin promotes the growth of human embryonic stem cells in a defined serum-free medium. *Proc. Natl. Acad. Sci. USA* **105,** 13409–13414.

Furusawa, T., Ohkoshi, K., Honda, C., Takahashi, S., and Tokunaga, T. (2004). Embryonic stem cells expressing both platelet endothelial cell adhesion molecule-1 and stage-specific embryonic antigen-1 differentiate predominantly into epiblast cells in a chimeric embryo. *Biol. Reprod.* **70,** 1452–1457.

Gardner, R. L., and Rossant, J. (1979). Investigation of the fate of 4–5 day post-coitum mouse inner cell mass cells by blastocyst injection. *J. Embryol. Exp. Morphol.* **52,** 141–152.

Goldin, S. N., and Papaioannou, V. E. (2003). Paracrine action of FGF4 during periimplantation development maintains trophectoderm and primitive endoderm. *Genesis* **36,** 40–47.
Greber, B., Lehrach, H., and Adjaye, J. (2007). Fibroblast growth factor 2 modulates transforming growth factor beta signaling in mouse embryonic fibroblasts and human ESCs (hESCs) to support hESC self-renewal. *Stem Cells* **25,** 455–464.
Greber, B., Lehrach, H., and Adjaye, J. (2008). Control of early fate decisions in human ES cells by distinct states of TGFbeta pathway activity. *Stem Cells Dev.* **17,** 1065–1077.
Greber, B., Wu, G., Bernemann, C., Joo, J. Y., Han, D. W., Ko, K., Tapia, N., Sabour, D., Sterneckert, J., Tesar, P., and Scholer, H. R. (2010). Conserved and divergent roles of FGF signaling in mouse epiblast stem cells and human embryonic stem cells. *Cell Stem Cell* **6,** 215–226.
Guo, G., Yang, J., Nichols, J., Hall, J. S., Eyres, I., Mansfield, W., and Smith, A. (2009). Klf4 reverts developmentally programmed restriction of ground state pluripotency. *Development* **136,** 1063–1069.
Guo, G., Huss, M., Tong, G. Q., Wang, C., Li Sun, L., Clarke, N. D., and Robson, P. (2010). Resolution of cell fate decisions revealed by single-cell gene expression analysis from zygote to blastocyst. *Dev. Cell* **18,** 675–685.
Hadari, Y. R., Gotoh, N., Kouhara, H., Lax, I., and Schlessinger, J. (2001). Critical role for the docking-protein FRS2 alpha in FGF receptor-mediated signal transduction pathways. *Proc. Natl. Acad. Sci. USA* **98,** 8578–8583.
Hall, J., Guo, G., Wray, J., Eyres, I., Nichols, J., Grotewold, L., Morfopoulou, S., Humphreys, P., Mansfield, W., Walker, R., Tomlinson, S., and Smith, A. (2009). Oct4 and LIF/Stat3 additively induce Kruppel factors to sustain embryonic stem cell self-renewal. *Cell Stem Cell* **5,** 597–609.
Han, D. W., Tapia, N., Joo, J. Y., Greber, B., Arauzo-Bravo, M. J., Bernemann, C., Ko, K., Wu, G., Stehling, M., Do, J. T., and Scholer, H. R. (2010). Epiblast stem cell subpopulations represent mouse embryos of distinct pregastrulation stages. *Cell* **143,** 617–627.
Hanna, J. H., Saha, K., and Jaenisch, R. (2010). Pluripotency and cellular reprogramming: Facts, hypotheses, unresolved issues. *Cell* **143,** 508–525.
Hansis, C., Grifo, J. A., and Krey, L. C. (2000). Oct-4 expression in inner cell mass and trophectoderm of human blastocysts. *Mol. Hum. Reprod.* **6,** 999–1004.
Harvey, A. J., Armant, D. R., Bavister, B. D., Nichols, S. M., and Brenner, C. A. (2009). Inner cell mass localization of NANOG precedes OCT3/4 in rhesus monkey blastocysts. *Stem Cells Dev.* **18,** 1451–1458.
Hayashi, K., Lopes, S. M., Tang, F., and Surani, M. A. (2008). Dynamic equilibrium and heterogeneity of mouse pluripotent stem cells with distinct functional and epigenetic states. *Cell Stem Cell* **3,** 391–401.
Home, P., Ray, S., Dutta, D., Bronshteyn, I., Larson, M., and Paul, S. (2009). GATA3 is selectively expressed in the trophectoderm of peri-implantation embryo and directly regulates Cdx2 gene expression. *J. Biol. Chem.* **284,** 28729–28737.
Hough, S. R., Clements, I., Welch, P. J., and Wiederholt, K. A. (2006). Differentiation of mouse embryonic stem cells after RNA interference-mediated silencing of OCT4 and Nanog. *Stem Cells* **24,** 1467–1475.
Humpherys, D., Eggan, K., Akutsu, H., Hochedlinger, K., Rideout, W. M., 3rd., Biniszkiewicz, D., Yanagimachi, R., and Jaenisch, R. (2001). Epigenetic instability in ES cells and cloned mice. *Science* **293,** 95–97.
Humphrey, R. K., Beattie, G. M., Lopez, A. D., Bucay, N., King, C. C., Firpo, M. T., Rose-John, S., and Hayek, A. (2004). Maintenance of pluripotency in human embryonic stem cells is STAT3 independent. *Stem Cells* **22,** 522–530.
James, D., Levine, A. J., Besser, D., and Hemmati-Brivanlou, A. (2005). TGFbeta/activin/nodal signaling is necessary for the maintenance of pluripotency in human embryonic stem cells. *Development* **132,** 1273–1282.

Jedrusik, A., Parfitt, D. E., Guo, G., Skamagki, M., Grabarek, J. B., Johnson, M. H., Robson, P., and Zernicka-Goetz, M. (2008). Role of Cdx2 and cell polarity in cell allocation and specification of trophectoderm and inner cell mass in the mouse embryo. *Genes Dev.* **22,** 2692–2706.

Jedrusik, A., Bruce, A. W., Tan, M. H., Leong, D. E., Skamagki, M., Yao, M., and Zernicka-Goetz, M. (2010). Maternally and zygotically provided Cdx2 have novel and critical roles for early development of the mouse embryo. *Dev. Biol.* **344,** 66–78.

Jiang, J., Chan, Y. S., Loh, Y. H., Cai, J., Tong, G. Q., Lim, C. A., Robson, P., Zhong, S., and Ng, H. H. (2008). A core Klf circuitry regulates self-renewal of embryonic stem cells. *Nat. Cell Biol.* **10,** 353–360.

Johnson, W. H., Loskutoff, N. M., Plante, Y., and Betteridge, K. J. (1995). Production of four identical calves by the separation of blastomeres from an in vitro derived four-cell embryo. *Vet. Rec.* **137,** 15–16.

Kalmar, T., Lim, C., Hayward, P., Munoz-Descalzo, S., Nichols, J., Garcia-Ojalvo, J., and Martinez Arias, A. (2009). Regulated fluctuations in nanog expression mediate cell fate decisions in embryonic stem cells. *PLoS Biol.* **7,** e1000149.

Kanai-Azuma, M., Kanai, Y., Gad, J. M., Tajima, Y., Taya, C., Kurohmaru, M., Sanai, Y., Yonekawa, H., Yazaki, K., Tam, P. P., and Hayashi, Y. (2002). Depletion of definitive gut endoderm in Sox17-null mutant mice. *Development* **129,** 2367–2379.

Kaufman, M. H. (1992). The Atlas of Mouse Development. Academic Press Ltd, London.

Keramari, M., Razavi, J., Ingman, K. A., Patsch, C., Edenhofer, F., Ward, C. M., and Kimber, S. J. (2010). Sox2 is essential for formation of trophectoderm in the preimplantation embryo. *PLoS One* **5,** e13952.

Kidder, B. L., Palmer, S., and Knott, J. G. (2009). SWI/SNF-Brg1 regulates self-renewal and occupies core pluripotency-related genes in embryonic stem cells. *Stem Cells* **27,** 317–328.

Kimber, S. J., Sneddon, S. F., Bloor, D. J., El-Bareg, A. M., Hawkhead, J. A., Metcalfe, A. D., Houghton, F. D., Leese, H. J., Rutherford, A., Lieberman, B. A., and Brison, D. R. (2008). Expression of genes involved in early cell fate decisions in human embryos and their regulation by growth factors. *Reproduction* **135,** 635–647.

Kirchhof, N., Carnwath, J. W., Lemme, E., Anastassiadis, K., Scholer, H., and Niemann, H. (2000). Expression pattern of Oct-4 in preimplantation embryos of different species. *Biol. Reprod.* **63,** 1698–1705.

Kiyonari, H., Kaneko, M., Abe, S., and Aizawa, S. (2010). Three inhibitors of FGF receptor, ERK, and GSK3 establishes germline-competent embryonic stem cells of C57BL/6N mouse strain with high efficiency and stability. *Genesis* **48,** 317–327.

Koutsourakis, M., Langeveld, A., Patient, R., Beddington, R., and Grosveld, F. (1999). The transcription factor GATA6 is essential for early extraembryonic development. *Development* **126,** 723–732.

Kuijk, E. W., Du Puy, L., Van Tol, H. T., Oei, C. H., Haagsman, H. P., Colenbrander, B., and Roelen, B. A. (2008). Differences in early lineage segregation between mammals. *Dev. Dyn.* **237,** 918–927.

Kunath, T., Arnaud, D., Uy, G. D., Okamoto, I., Chureau, C., Yamanaka, Y., Heard, E., Gardner, R. L., Avner, P., and Rossant, J. (2005). Imprinted X-inactivation in extra-embryonic endoderm cell lines from mouse blastocysts. *Development* **132,** 1649–1661.

Kunath, T., Saba-El-Leil, M. K., Almousailleakh, M., Wray, J., Meloche, S., and Smith, A. (2007). FGF stimulation of the Erk1/2 signalling cascade triggers transition of pluripotent embryonic stem cells from self-renewal to lineage commitment. *Development* **134,** 2895–2902.

Kurimoto, K., Yabuta, Y., Ohinata, Y., Ono, Y., Uno, K. D., Yamada, R. G., Ueda, H. R., and Saitou, M. (2006). An improved single-cell cDNA amplification method for efficient high-density oligonucleotide microarray analysis. *Nucleic Acids Res.* **34,** e42.

Kuroda, T., Tada, M., Kubota, H., Kimura, H., Hatano, S. Y., Suemori, H., Nakatsuji, N., and Tada, T. (2005). Octamer and Sox elements are required for transcriptional cis regulation of Nanog gene expression. *Mol. Cell. Biol.* **25,** 2475–2485.

Levenstein, M. E., Ludwig, T. E., Xu, R. H., Llanas, R. A., VanDenHeuvel-Kramer, K., Manning, D., and Thomson, J. A. (2006). Basic fibroblast growth factor support of human embryonic stem cell self-renewal. *Stem Cells* **24,** 568–574.

Levine, A. J., and Brivanlou, A. H. (2006). GDF3, a BMP inhibitor, regulates cell fate in stem cells and early embryos. *Development* **133,** 209–216.

Li, J., Wang, G., Wang, C., Zhao, Y., Zhang, H., Tan, Z., Song, Z., Ding, M., and Deng, H. (2007). MEK/ERK signaling contributes to the maintenance of human embryonic stem cell self-renewal. *Differentiation* **75,** 299–307.

Li, C., Yang, Y., Gu, J., Ma, Y., and Jin, Y. (2008a). Derivation and transcriptional profiling analysis of pluripotent stem cell lines from rat blastocysts. *Cell Res.* **19,** 173–186.

Li, P., Tong, C., Mehrian-Shai, R., Jia, L., Wu, N., Yan, Y., Maxson, R. E., Schulze, E. N., Song, H., Hsieh, C. L., Pera, M. F., and Ying, Q. L. (2008b). Germline competent embryonic stem cells derived from rat blastocysts. *Cell* **135,** 1299–1310.

Lin, S. C., Wani, M. A., Whitsett, J. A., and Wells, J. M. (2010). Klf5 regulates lineage formation in the pre-implantation mouse embryo. *Development* **137,** 3953–3963.

Loh, Y. H., Wu, Q., Chew, J. L., Vega, V. B., Zhang, W., Chen, X., Bourque, G., George, J., Leong, B., Liu, J., Wong, K. Y., Sung, K. W., *et al.* (2006). The Oct4 and Nanog transcription network regulates pluripotency in mouse embryonic stem cells. *Nat. Genet.* **38,** 431–440.

Lu, C. W., Yabuuchi, A., Chen, L., Viswanathan, S., Kim, K., and Daley, G. Q. (2008). Ras-MAPK signaling promotes trophectoderm formation from embryonic stem cells and mouse embryos. *Nat. Genet.* **40,** 921–926.

Maherali, N., Sridharan, R., Xie, W., Utikal, J., Eminli, S., Arnold, K., Stadtfeld, M., Yachechko, R., Tchieu, J., Jaenisch, R., Plath, K., and Hochedlinger, K. (2007). Directly reprogrammed fibroblasts show global epigenetic remodeling and widespread tissue contribution. *Cell Stem Cell* **1,** 55–70.

Martin, G. R. (1981). Isolation of a pluripotent cell line from early mouse embryos cultured in medium conditioned by teratocarcinoma stem cells. *Proc. Natl. Acad. Sci. USA* **78,** 7634–7638.

Matsuda, T., Nakamura, T., Nakao, K., Arai, T., Katsuki, M., Heike, T., and Yokota, T. (1999). STAT3 activation is sufficient to maintain an undifferentiated state of mouse embryonic stem cells. *EMBO J.* **18,** 4261–4269.

McConnell, B. B., Ghaleb, A. M., Nandan, M. O., and Yang, V. W. (2007). The diverse functions of Kruppel-like factors 4 and 5 in epithelial biology and pathobiology. *Bioessays* **29,** 549–557.

McConnell, J., Petrie, L., Stennard, F., Ryan, K., Nichols, J. (2005). Eomesodermin is expressed in mouse oocytes and pre-implantation embryos. *Mol. Reprod. Dev.* **71,** 399–404.

Meilhac, S. M., Adams, R. J., Morris, S. A., Danckaert, A., Le Garrec, J. F., and Zernicka-Goetz, M. (2009). Active cell movements coupled to positional induction are involved in lineage segregation in the mouse blastocyst. *Dev. Biol.* **331,** 210–221.

Meissner, A., Wernig, M., and Jaenisch, R. (2007). Direct reprogramming of genetically unmodified fibroblasts into pluripotent stem cells. *Nat. Biotechnol.* **25,** 1177–1181.

Messerschmidt, D. M., and Kemler, R. (2010). Nanog is required for primitive endoderm formation through a non-cell autonomous mechanism. *Dev. Biol.* **344,** 129–137.

Mitsui, K., Tokuzawa, Y., Itoh, H., Segawa, K., Murakami, M., Takahashi, K., Maruyama, M., Maeda, M., and Yamanaka, S. (2003). The homeoprotein Nanog is required for maintenance of pluripotency in mouse epiblast and ES cells. *Cell* **113,** 631–642.

Moore, N. W., Adams, C. E., and Rowson, L. E. (1968). Developmental potential of single blastomeres of the rabbit egg. *J. Reprod. Fertil.* **17,** 527–531.

Morris, S. A., Teo, R. T., Li, H., Robson, P., Glover, D. M., and Zernicka-Goetz, M. (2010). Origin and formation of the first two distinct cell types of the inner cell mass in the mouse embryo. *Proc. Natl. Acad. Sci. USA* **107,** 6364–6369.

Morrisey, E. E., Tang, Z., Sigrist, K., Lu, M. M., Jiang, F., Ip, H. S., and Parmacek, M. S. (1998). GATA6 regulates HNF4 and is required for differentiation of visceral endoderm in the mouse embryo. *Genes Dev.* **12,** 3579–3590.

Nagy, A., Gocza, E., Diaz, E. M., Prideaux, V. R., Ivanyi, E., Markkula, M., and Rossant, J. (1990). Embryonic stem cells alone are able to support fetal development in the mouse. *Development* **110,** 815–821.

Nagy, A., Rossant, J., Nagy, R., Abramow-Newerly, W., and Roder, J. C. (1993). Derivation of completely cell culture-derived mice from early-passage embryonic stem cells. *Proc. Natl. Acad. Sci. USA* **90,** 8424–8428.

Najm, F. J., Chenoweth, J. G., Anderson, P. D., Nadeau, J. H., Redline, R. W., McKay, R. D., and Tesar, P. J. (2011). Isolation of epiblast stem cells from preimplantation mouse embryos. *Cell Stem Cell* **8,** 318–325.

Nakatake, Y., Fukui, N., Iwamatsu, Y., Masui, S., Takahashi, K., Yagi, R., Yagi, K., Miyazaki, J., Matoba, R., Ko, M. S., and Niwa, H. (2006). Klf4 cooperates with Oct3/4 and Sox2 to activate the Lefty1 core promoter in embryonic stem cells. *Mol. Cell. Biol.* **26,** 7772–7782.

Newman, A. M., and Cooper, J. B. (2010). Lab-specific gene expression signatures in pluripotent stem cells. *Cell Stem Cell* **7,** 258–262.

Ng, R. K., Dean, W., Dawson, C., Lucifero, D., Madeja, Z., Reik, W., and Hemberger, M. (2008). Epigenetic restriction of embryonic cell lineage fate by methylation of Elf5. *Nat. Cell Biol.* **10,** 1280–1290.

Niakan, K. K., Ji, H., Maehr, R., Vokes, S. A., Rodolfa, K. T., Sherwood, R. I., Yamaki, M., Dimos, J. T., Chen, A. E., Melton, D. A., McMahon, A. P., and Eggan, K. (2010). Sox17 promotes differentiation in mouse embryonic stem cells by directly regulating extraembryonic gene expression and indirectly antagonizing self-renewal. *Genes Dev.* **24,** 312–326.

Nichols, J., and Smith, A. (2009). Naive and primed pluripotent states. *Cell Stem Cell* **4,** 487–492.

Nichols, J., Evans, E. P., and Smith, A. G. (1990). Establishment of germ-line-competent embryonic stem (ES) cells using differentiation inhibiting activity. *Development* **110,** 1341–1348.

Nichols, J., Zevnik, B., Anastassiadis, K., Niwa, H., Klewe-Nebenius, D., Chambers, I., Scholer, H., and Smith, A. (1998). Formation of pluripotent stem cells in the mammalian embryo depends on the POU transcription factor Oct4. *Cell* **95,** 379–391.

Nichols, J., Silva, J., Roode, M., and Smith, A. (2009). Suppression of Erk signalling promotes ground state pluripotency in the mouse embryo. *Development* **136,** 3215–3222.

Nishioka, N., Yamamoto, S., Kiyonari, H., Sato, H., Sawada, A., Ota, M., Nakao, K., and Sasaki, H. (2008). Tead4 is required for specification of trophectoderm in pre-implantation mouse embryos. *Mech. Dev.* **125,** 270–283.

Nishioka, N., Inoue, K., Adachi, K., Kiyonari, H., Ota, M., Ralston, A., Yabuta, N., Hirahara, S., Stephenson, R. O., Ogonuki, N., Makita, R., Kurihara, H., *et al.* (2009). The Hippo signaling pathway components Lats and Yap pattern Tead4 activity to distinguish mouse trophectoderm from inner cell mass. *Dev. Cell* **16,** 398–410.

Niwa, H. (2007). How is pluripotency determined and maintained? *Development* **134,** 635–646.

Niwa, H., Burdon, T., Chambers, I., and Smith, A. (1998). Self-renewal of pluripotent embryonic stem cells is mediated via activation of STAT3. *Genes Dev.* **12,** 2048–2060.

Niwa, H., Miyazaki, J., and Smith, A. G. (2000). Quantitative expression of Oct-3/4 defines differentiation, dedifferentiation or self-renewal of ES cells. *Nat. Genet.* **24,** 372–376.

Niwa, H., Toyooka, Y., Shimosato, D., Strumpf, D., Takahashi, K., Yagi, R., and Rossant, J. (2005). Interaction between Oct3/4 and Cdx2 determines trophectoderm differentiation. *Cell* **123,** 917–929.

Niwa, H., Ogawa, K., Shimosato, D., and Adachi, K. (2009). A parallel circuit of LIF signalling pathways maintains pluripotency of mouse ES cells. *Nature* **460,** 118–122.

Okumura-Nakanishi, S., Saito, M., Niwa, H., and Ishikawa, F. (2005). Oct-3/4 and Sox2 regulate Oct-3/4 gene in embryonic stem cells. *J. Biol. Chem.* **280,** 5307–5317.

Ooi, S. K., Wolf, D., Hartung, O., Agarwal, S., Daley, G. Q., Goff, S. P., and Bestor, T. H. (2010). Dynamic instability of genomic methylation patterns in pluripotent stem cells. *Epigenetics Chromatin* **3,** 17.

Paling, N. R., Wheadon, H., Bone, H. K., and Welham, M. J. (2004). Regulation of embryonic stem cell self-renewal by phosphoinositide 3-kinase-dependent signaling. *J. Biol. Chem.* **279,** 48063–48070.

Palmieri, S. L., Peter, W., Hess, H., and Scholer, H. R. (1994). Oct-4 transcription factor is differentially expressed in the mouse embryo during establishment of the first two extraembryonic cell lineages involved in implantation. *Dev. Biol.* **166,** 259–267.

Pera, M. F., Andrade, J., Houssami, S., Reubinoff, B., Trounson, A., Stanley, E. G., Ward-van Oostwaard, D., and Mummery, C. (2004). Regulation of human embryonic stem cell differentiation by BMP-2 and its antagonist noggin. *J. Cell Sci.* **117,** 1269–1280.

Plusa, B., Frankenberg, S., Chalmers, A., Hadjantonakis, A. K., Moore, C. A., Papalopulu, N., Papaioannou, V. E., Glover, D. M., and Zernicka-Goetz, M. (2005). Downregulation of Par3 and aPKC function directs cells towards the ICM in the preimplantation mouse embryo. *J. Cell Sci.* **118,** 505–515.

Plusa, B., Piliszek, A., Frankenberg, S., Artus, J., and Hadjantonakis, A. K. (2008). Distinct sequential cell behaviours direct primitive endoderm formation in the mouse blastocyst. *Development* **135,** 3081–3091.

Qi, X., Li, T. G., Hao, J., Hu, J., Wang, J., Simmons, H., Miura, S., Mishina, Y., and Zhao, G. Q. (2004). BMP4 supports self-renewal of embryonic stem cells by inhibiting mitogen-activated protein kinase pathways. *Proc. Natl. Acad. Sci. USA* **101,** 6027–6032.

Ralston, A., and Rossant, J. (2008). Cdx2 acts downstream of cell polarization to cell-autonomously promote trophectoderm fate in the early mouse embryo. *Dev. Biol.* **313,** 614–629.

Ralston, A., Cox, B. J., Nishioka, N., Sasaki, H., Chea, E., Rugg-Gunn, P., Guo, G., Robson, P., Draper, J. S., and Rossant, J. (2010). Gata3 regulates trophoblast development downstream of Tead4 and in parallel to Cdx2. *Development* **137,** 395–403.

Rathjen, P. D., Toth, S., Willis, A., Heath, J. K., and Smith, A. G. (1990). Differentiation inhibiting activity is produced in matrix-associated and diffusible forms that are generated by alternate promoter usage. *Cell* **62,** 1105–1114.

Reubinoff, B. E., Pera, M. F., Fong, C. Y., Trounson, A., and Bongso, A. (2000). Embryonic stem cell lines from human blastocysts: Somatic differentiation in vitro. *Nat. Biotechnol.* **18,** 399–404.

Roberts, R. M., and Fisher, S. J. (2011). Trophoblast stem cells. *Biol. Reprod.* **84,** 412–421.

Rodda, D. J., Chew, J. L., Lim, L. H., Loh, Y. H., Wang, B., Ng, H. H., and Robson, P. (2005). Transcriptional regulation of nanog by OCT4 and SOX2. *J. Biol. Chem.* **280,** 24731–24737.

Rossant, J. (1976). Postimplantation development of blastomeres isolated from 4- and 8-cell mouse eggs. *J. Embryol. Exp. Morphol.* **36,** 283–290.

Rossant, J. (2008). Stem cells and early lineage development. *Cell* **132,** 527–531.

Sanna, D., Sanna, A., Mara, L., Pilichi, S., Mastinu, A., Chessa, F., Pani, L., and Dattena, M. (2009). Oct4 expression in in-vitro-produced sheep blastocysts and embryonic-stem-like cells. *Cell Biol. Int.* **34,** 53–60.

Sasaki, E., Hanazawa, K., Kurita, R., Akatsuka, A., Yoshizaki, T., Ishii, H., Tanioka, Y., Ohnishi, Y., Suemizu, H., Sugawara, A., Tamaoki, N., Izawa, K., *et al.* (2005). Establishment of novel embryonic stem cell lines derived from the common marmoset (*Callithrix jacchus*). *Stem Cells* **23,** 1304–1313.

Sato, N., Meijer, L., Skaltsounis, L., Greengard, P., and Brivanlou, A. H. (2004). Maintenance of pluripotency in human and mouse embryonic stem cells through activation of Wnt signaling by a pharmacological GSK-3-specific inhibitor. *Nat. Med.* **10,** 55–63.

Shamblott, M. J., Axelman, J., Wang, S., Bugg, E. M., Littlefield, J. W., Donovan, P. J., Blumenthal, P. D., Huggins, G. R., and Gearhart, J. D. (1998). Derivation of pluripotent stem cells from cultured human primordial germ cells. *Proc. Natl. Acad. Sci. USA* **95,** 13726–13731.

Shimosato, D., Shiki, M., and Niwa, H. (2007). Extra-embryonic endoderm cells derived from ES cells induced by GATA factors acquire the character of XEN cells. *BMC Dev. Biol.* **7,** 80.

Silva, J., and Smith, A. (2008). Capturing pluripotency. *Cell* **132,** 532–536.

Silva, J., Nichols, J., Theunissen, T. W., Guo, G., van Oosten, A. L., Barrandon, O., Wray, J., Yamanaka, S., Chambers, I., and Smith, A. (2009). Nanog is the gateway to the pluripotent ground state. *Cell* **138,** 722–737.

Singh, A. M., Hamazaki, T., Hankowski, K. E., and Terada, N. (2007). A heterogeneous expression pattern for Nanog in embryonic stem cells. *Stem Cells* **25,** 2534–2542.

Sjogren-Jansson, E., Zetterstrom, M., Moya, K., Lindqvist, J., Strehl, R., and Eriksson, P. S. (2005). Large-scale propagation of four undifferentiated human embryonic stem cell lines in a feeder-free culture system. *Dev. Dyn.* **233,** 1304–1314.

Smith, A. G., Heath, J. K., Donaldson, D. D., Wong, G. G., Moreau, J., Stahl, M., and Rogers, D. (1988). Inhibition of pluripotential embryonic stem cell differentiation by purified polypeptides. *Nature* **336,** 688–690.

Strumpf, D., Mao, C. A., Yamanaka, Y., Ralston, A., Chawengsaksophak, K., Beck, F., and Rossant, J. (2005). Cdx2 is required for correct cell fate specification and differentiation of trophectoderm in the mouse blastocyst. *Development* **132,** 2093–2102.

Sumi, T., Fujimoto, Y., Nakatsuji, N., and Suemori, H. (2004). STAT3 is dispensable for maintenance of self-renewal in nonhuman primate embryonic stem cells. *Stem Cells* **22,** 861–872.

Suwinska, A., Czolowska, R., Ozdzenski, W., and Tarkowski, A. K. (2008). Blastomeres of the mouse embryo lose totipotency after the fifth cleavage division: Expression of Cdx2 and Oct4 and developmental potential of inner and outer blastomeres of 16- and 32-cell embryos. *Dev. Biol.* **322,** 133–144.

Suwinska, A., Tarkowski, A. K., and Ciemerych, M. A. (2010). Pluripotency of bank vole embryonic cells depends on FGF2 and activin A signaling pathways. *Int. J. Dev. Biol.* **54,** 113–124.

Suzuki, A., Raya, A., Kawakami, Y., Morita, M., Matsui, T., Nakashima, K., Gage, F. H., Rodriguez-Esteban, C., and Izpisua Belmonte, J. C. (2006). Nanog binds to Smad1 and blocks bone morphogenetic protein-induced differentiation of embryonic stem cells. *Proc. Natl. Acad. Sci. USA* **103,** 10294–10299.

Takahashi, K., and Yamanaka, S. (2006). Induction of pluripotent stem cells from mouse embryonic and adult fibroblast cultures by defined factors. *Cell* **126,** 663–676.

Takahashi, K., Tanabe, K., Ohnuki, M., Narita, M., Ichisaka, T., Tomoda, K., and Yamanaka, S. (2007). Induction of pluripotent stem cells from adult human fibroblasts by defined factors. *Cell* **131,** 861–872.

Tanaka, S., Kunath, T., Hadjantonakis, A. K., Nagy, A., and Rossant, J. (1998). Promotion of trophoblast stem cell proliferation by FGF4. *Science* **282,** 2072–2075.

Tarkowski, A. K. (1959). Experiments on the development of isolated blastomers of mouse eggs. *Nature* **184,** 1286–1287.

Tarkowski, A. K., and Wroblewska, J. (1967). Development of blastomeres of mouse eggs isolated at the 4- and 8-cell stage. *J. Embryol. Exp. Morphol.* **18,** 155–180.

Tarkowski, A. K., Ozdzenski, W., and Czolowska, R. (2001). Mouse singletons and twins developed from isolated diploid blastomeres supported with tetraploid blastomeres. *Int. J. Dev. Biol.* **45,** 591–596.

Tarkowski, A. K., Ozdzenski, W., and Czolowska, R. (2005). Identical triplets and twins developed from isolated blastomeres of 8- and 16-cell mouse embryos supported with tetraploid blastomeres. *Int. J. Dev. Biol.* **49,** 825–832.

Tarkowski, A. K., Suwinska, A., Czolowska, R., and Ozdzenski, W. (2010). Individual blastomeres of 16- and 32-cell mouse embryos are able to develop into foetuses and mice. *Dev. Biol.* **348,** 190–198.

Tesar, P. J., Chenoweth, J. G., Brook, F. A., Davies, T. J., Evans, E. P., Mack, D. L., Gardner, R. L., and McKay, R. D. (2007). New cell lines from mouse epiblast share defining features with human embryonic stem cells. *Nature* **448,** 196–199.

Thomas, F. C., Sheth, B., Eckert, J. J., Bazzoni, G., Dejana, E., and Fleming, T. P. (2004). Contribution of JAM-1 to epithelial differentiation and tight-junction biogenesis in the mouse preimplantation embryo. *J. Cell Sci.* **117,** 5599–5608.

Thomson, J. A., Kalishman, J., Golos, T. G., Durning, M., Harris, C. P., and Hearn, J. P. (1996). Pluripotent cell lines derived from common marmoset (*Callithrix jacchus*) blastocysts. *Biol. Reprod.* **55,** 254–259.

Thomson, J. A., Itskovitz-Eldor, J., Shapiro, S. S., Waknitz, M. A., Swiergiel, J. J., Marshall, V. S., and Jones, J. M. (1998). Embryonic stem cell lines derived from human blastocysts. *Science* **282,** 1145–1147.

Toyooka, Y., Shimosato, D., Murakami, K., Takahashi, K., and Niwa, H. (2008). Identification and characterization of subpopulations in undifferentiated ES cell culture. *Development* **135,** 909–918.

Ueda, S., Kawamata, M., Teratani, T., Shimizu, T., Tamai, Y., Ogawa, H., Hayashi, K., Tsuda, H., and Ochiya, T. (2008). Establishment of rat embryonic stem cells and making of chimera rats. *PLoS One* **3,** e2800.

Umehara, H., Kimura, T., Ohtsuka, S., Nakamura, T., Kitajima, K., Ikawa, M., Okabe, M., Niwa, H., and Nakano, T. (2007). Efficient derivation of embryonic stem cells by inhibition of glycogen synthase kinase-3. *Stem Cells* **25,** 2705–2711.

Valer Carstea, B., Catunda Lemos, A. P., Ilie, E. D., Varga, L., Bodo, S., Kovacs, A., Bosze, Z., and Gocza, E. (2007). Production of identical mouse twins and a triplet with predicted gender. *Cloning Stem Cells* **9,** 247–256.

Vallier, L., Alexander, M., and Pedersen, R. A. (2005). Activin/Nodal and FGF pathways cooperate to maintain pluripotency of human embryonic stem cells. *J. Cell Sci.* **118,** 4495–4509.

van Eijk, M. J., van Rooijen, M. A., Modina, S., Scesi, L., Folkers, G., van Tol, H. T., Bevers, M. M., Fisher, S. R., Lewin, H. A., Rakacolli, D., Galli, C., de Vaureix, C., *et al.* (1999). Molecular cloning, genetic mapping, and developmental expression of bovine POU5F1. *Biol. Reprod.* **60,** 1093–1103.

Villegas, S. N., Canham, M., and Brickman, J. M. (2010). FGF signalling as a mediator of lineage transitions—evidence from embryonic stem cell differentiation. *J. Cell. Biochem.* **110,** 10–20.

Vinot, S., Le, T., Ohno, S., Pawson, T., Maro, B., and Louvet-Vallee, S. (2005). Asymmetric distribution of PAR proteins in the mouse embryo begins at the 8-cell stage during compaction. *Dev. Biol.* **282,** 307–319.

Wang, Q. T., Piotrowska, K., Ciemerych, M. A., Milenkovic, L., Scott, M. P., Davis, R. W., and Zernicka-Goetz, M. (2004). A genome-wide study of gene activity reveals developmental signaling pathways in the preimplantation mouse embryo. *Dev. Cell* **6,** 133–144.

Wang, X., Lin, G., Martins-Taylor, K., Zeng, H., and Xu, R. H. (2009). Inhibition of caspase-mediated anoikis is critical for basic fibroblast growth factor-sustained culture of human pluripotent stem cells. *J. Biol. Chem.* **284,** 34054–34064.

Wang, K., Sengupta, S., Magnani, L., Wilson, C. A., Henry, R. W., and Knott, J. G. (2010). Brg1 is required for Cdx2-mediated repression of Oct4 expression in mouse blastocysts. *PLoS One* **5,** e10622.

Willadsen, S. M. (1981). The development capacity of blastomeres from 4- and 8-cell sheep embryos. *J. Embryol. Exp. Morphol.* **65,** 165–172.

Williams, R. L., Hilton, D. J., Pease, S., Willson, T. A., Stewart, C. L., Gearing, D. P., Wagner, E. F., Metcalf, D., Nicola, N. A., and Gough, N. M. (1988). Myeloid leukaemia inhibitory factor maintains the developmental potential of embryonic stem cells. *Nature* **336,** 684–687.

Wobus, A. M., and Boheler, K. R. (2005). Embryonic stem cells: Prospects for developmental biology and cell therapy. *Physiol. Rev.* **85,** 635–678.

Wu, Q., Chen, X., Zhang, J., Loh, Y. H., Low, T. Y., Zhang, W., Zhang, W., Sze, S. K., Lim, B., and Ng, H. H. (2006). Sall4 interacts with Nanog and co-occupies Nanog genomic sites in embryonic stem cells. *J. Biol. Chem.* **281,** 24090–24094.

Wu, G., Gentile, L., Fuchikami, T., Sutter, J., Psathaki, K., Esteves, T. C., Arauzo-Bravo, M. J., Ortmeier, C., Verberk, G., Abe, K., and Scholer, H. R. (2010). Initiation of trophectoderm lineage specification in mouse embryos is independent of Cdx2. *Development* **137,** 4159–4169.

Xiao, L., Yuan, X., and Sharkis, S. J. (2006). Activin A maintains self-renewal and regulates fibroblast growth factor, Wnt, and bone morphogenic protein pathways in human embryonic stem cells. *Stem Cells* **24,** 1476–1486.

Xu, R. H., Chen, X., Li, D. S., Li, R., Addicks, G. C., Glennon, C., Zwaka, T. P., and Thomson, J. A. (2002). BMP4 initiates human embryonic stem cell differentiation to trophoblast. *Nat. Biotechnol.* **20,** 1261–1264.

Xu, R. H., Peck, R. M., Li, D. S., Feng, X., Ludwig, T., and Thomson, J. A. (2005). Basic FGF and suppression of BMP signaling sustain undifferentiated proliferation of human ES cells. *Nat. Methods* **2,** 185–190.

Yagi, R., Kohn, M. J., Karavanova, I., Kaneko, K. J., Vullhorst, D., DePamphilis, M. L., and Buonanno, A. (2007). Transcription factor TEAD4 specifies the trophectoderm lineage at the beginning of mammalian development. *Development* **134,** 3827–3836.

Yamanaka, Y., Ralston, A., Stephenson, R. O., and Rossant, J. (2006). Cell and molecular regulation of the mouse blastocyst. *Dev. Dyn.* **235,** 2301–2314.

Yamanaka, Y., Lanner, F., and Rossant, J. (2010). FGF signal-dependent segregation of primitive endoderm and epiblast in the mouse blastocyst. *Development* **137,** 715–724.

Ying, Q. L., Nichols, J., Chambers, I., and Smith, A. (2003). BMP induction of Id proteins suppresses differentiation and sustains embryonic stem cell self-renewal in collaboration with STAT3. *Cell* **115,** 281–292.

Ying, Q. L., Wray, J., Nichols, J., Batlle-Morera, L., Doble, B., Woodgett, J., Cohen, P., and Smith, A. (2008). The ground state of embryonic stem cell self-renewal. *Nature* **453,** 519–523.

Yuan, H., Corbi, N., Basilico, C., and Dailey, L. (1995). Developmental-specific activity of the FGF-4 enhancer requires the synergistic action of Sox2 and Oct-3. *Genes Dev.* **9,** 2635–2645.

Zhang, J., Tam, W. L., Tong, G. Q., Wu, Q., Chan, H. Y., Soh, B. S., Lou, Y., Yang, J., Ma, Y., Chai, L., Ng, H. H., Lufkin, T., *et al.* (2006). Sall4 modulates embryonic stem cell pluripotency and early embryonic development by the transcriptional regulation of Pou5f1. *Nat. Cell Biol.* **8,** 1114–1123.

Zhao, B., Ye, X., Yu, J., Li, L., Li, W., Li, S., Yu, J., Lin, J. D., Wang, C. Y., Chinnaiyan, A. M., Lai, Z. C., and Guan, K. L. (2008). TEAD mediates YAP-dependent gene induction and growth control. *Genes Dev.* **22,** 1962–1971.

CHAPTER TWO

Molecular Mediators of Mesenchymal Stem Cell Biology

Maria P. Alfaro,*[,1] Sarika Saraswati,*[,1] *and* Pampee P. Young*[,†,‡]

Contents

I.	Introduction	40
II.	Mesenchymal Stem Cells	40
III.	Differentiation of MSCs	41
IV.	Self-Renewal	41
V.	MSC Therapy	42
VI.	Immunomodulatory Properties	43
VII.	MI Therapy	44
VIII.	Molecular Mediators of MSC Biology	44
	A. Wnt pathway	44
	B. Bone morphogenetic pathway	46
	C. Canonical Wnt and BMP signaling and MSC differentiation	46
	D. Involvement of Wnt on survival and proliferation of MSCs	49
	E. Effects of the BMP cascade on MSC survival and proliferation	49
IX.	Enhancing MSC Survival in the Wound	50
X.	Secreted Frizzled-Related Proteins	50
XI.	Mediating MSC Self-Renewal	51
XII.	Conclusions	52
References		53

Abstract

Mesenchymal stem cells (MSCs) have the ability to self-renew and differentiate into multiple lineages making them an appropriate candidate for stem cell therapy. In spite of achieving considerable success in preclinical models, limited success has been achieved in clinical settings with MSCs. A major impediment that is faced is low survival of MSCs in injured tissues following implantation. In order to enhance the reparative properties of MSCs, it is vital to understand the molecular signals that regulate MSC survival and self-renewal. This review assimilates information that characterizes MSCs and mentions their utilization

* Department of Pathology, Vanderbilt University Medical Center, Nashville, Tennessee, USA
† Department of Internal Medicine, Vanderbilt University Medical Center, Nashville, Tennessee, USA
‡ The Department of Veterans Affairs Medical Center, Nashville, Tennessee, USA
1 These authors contributed equally to this work.

Vitamins and Hormones, Volume 87
ISSN 0083-6729, DOI: 10.1016/B978-0-12-386015-6.00023-8

in myocardial infarction therapy. Additionally, our attempt herein is to gather pertinent published information regarding the role of canonical Wnt and BMP signaling in regulating the potential of MSCs to self-renew, proliferate, differentiate, and survive.

I. Introduction

Stem cell therapy is an exciting new field that shows a lot of promise. The efficacy of mesenchymal stem cells (MSCs) as treatment for regeneration and/or repair is currently being clinically tested. However, preclinical models of MSC-directed wound repair show disparate degrees of effectiveness. The greatest barrier faced in these models is the low levels of engraftment of the transplanted cells. Therefore, increasing MSC maintenance in the wound would increase their reparative capabilities. Understanding MSC biology would allow for clinicians to harness the potential these cells have to offer. Several published reports have looked at molecular mediators of MSC biology and this review aims to assemble such data. Particularly, data pertaining to the canonical Wnt and bone morphogenetic protein (BMP) signaling cascades, which have been implicated in regulating MSC differentiation, proliferation, and survival, will be discussed. Signals which amplify the MSC pool, that is, enhance their self-renewal, are largely unknown. This review will also discuss the effects of secreted Frizzled-related protein 2 (sFRP2) on self-renewal and ultimately try to elucidate important molecules involved in MSC biology.

II. Mesenchymal Stem Cells

Stem cells are defined by their ability to self-renew and differentiate into multiple cell types (Fuchs and Segre, 2000). MSCs, traditionally defined as bone marrow-derived fibroblast-like cells, can also be isolated from adipose tissue (Madonna *et al.*, 2009), umbilical cord blood (Harris, 2008; Zhang *et al.*, 2011), and fetal tissues (Pozzobon *et al.*, 2010). MSCs can differentiate into osteogenic, chondrogenic, and adipogenic lineages and give rise to any tissue derived from mesenchyme including bone, muscle, fibroblast, tendon, ligament, and adipose tissue (Pittenger *et al.*, 1999; Tropel *et al.*, 2004). Murine MSCs have been characterized to be negative for CD11b, CD14, CD31, CD34, and CD45 but positive for CD44, CD29, CD73, CD105, CD106, CD166, and Stem Cell Antigen 1 (SCA1) (Baddoo *et al.*, 2003; Rombouts and Ploemacher, 2003; Short *et al.*, 2003). However, the minimal criteria for antigenic identification is shown in Fig. 2.1, as murine MSCs must be lineage negative (Lin^-), $CD45^-$, $CD44^+$, and $SCA\text{-}1^+$.

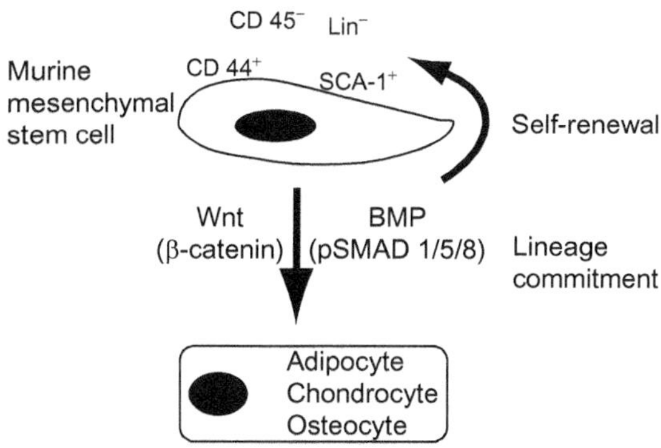

Figure 2.1 Definition and molecular mediators of murine mesenchymal stem cells. Murine MSCs are $CD44^+$ and $SCA\text{-}1^+$, but $CD45^-$ and Lin^-. As stem cells, they must give rise to self (self-renewal) and undergo trilineage differentiation toward the adipogenic, chondrogenic, and osteogenic lineages. The cascades involved in the lineage commitment are labeled with their effector molecules in parentheses.

III. Differentiation of MSCs

The trilineage *in vitro* differentiation of MSCs can be confirmed in different manners. Molecular markers of adipogenic (i.e., peroxisome proliferator-activated receptor gamma, PPAR-γ), chondrogenic (i.e., Collagen XI or Runx2), and osteogenic (i.e., Osteocalcin or Runx2) differentiation may be quantified (Takada *et al.*, 2009). Staining techniques exist for the visual confirmation of lineage commitment. Adipogenesis can be verified with Oil Red-O staining (Delorme and Charbord, 2007). Matrix calcification, which is present following osteogenesis, can be visualized with Alizarin Red or Von Kossa stains (Delorme and Charbord, 2007). Finally, glycosaminoglycans can be stained with Alcian blue (Denker *et al.*, 1999) or dimethylmethylene blue (Farndale *et al.*, 1986) as confirmation of chondrogenic differentiation. Several cytokines, growth factors, adhesion molecules, and extracellular matrix components have been identified as cues that signal MSCs to differentiate (Kratchmarova *et al.*, 2005; Mannello *et al.*, 2006). Included among these are canonical Wnt and BMP signaling which will be discussed in more detail in the following sections.

IV. Self-Renewal

The self-renewal process allows a stem cell to perform symmetric cell division to give rise to two nondifferentiated daughter cells. For self-renewal to occur, stem cells must proliferate in such a way that apoptosis and differentiation are avoided (Satija *et al.*, 2007; Schofield, 1983). Few reports have documented the self-renewal capability of MSCs (Song *et al.*, 2006), and therefore, the signaling involved in this process is largely unknown. On the other hand, self-renewal has been widely documented in the hematopoietic stem cell (HSC) field, and the Wnt, Notch, and BMP signaling cascades are accepted as mediators of the maintenance of a non-differentiated HSC pool (McReynolds *et al.*, 2007; Reya and Clevers, 2005; Suzuki and Chiba, 2005). Canonical Wnt signaling together with Notch and BMP signaling have also been shown to modulate stem cell self-renewal in the intestinal stem cell niche (Haramis *et al.*, 2004; He *et al.*, 2004; van Es *et al.*, 2005). Additionally, Wnt signaling has been shown to direct self-renewal and differentiation of Islet1-expressing precursor cells in neo- and postnatal hearts (Klaus *et al.*, 2007; Qyang *et al.*, 2007; reviewed in reference Klaus and Birchmeier, 2008). Building on this knowledge, our laboratory has recently demonstrated that inhibition of BMP and Wnt by sFRP2 increases MSC self-renewal.

V. MSC Therapy

Although preclinical studies have recognized significant benefits of stem cell therapy, its translation for clinical application is still in its infancy. MSCs have been able to repair infarcted myocardium, bone, and soft tissue (Horwitz *et al.*, 1999; Orlic *et al.*, 2001). MSCs have been reported to induce angiogenesis and secrete paracrine and mitogenic growth factors (Iyer and Rojas, 2008; Kinnaird *et al.*, 2004; Silva *et al.*, 2005; Tang *et al.*, 2004). Two major clinical approaches are utilized for stem cell therapy: endogenous mobilization of progenitor stem cells and exogenous transplantation of culture-expanded stem cells. Presently, there are around 100 clinical trials that involve exogenous human MSCs (http://clinicaltrials.gov/ct2/home; U.S. National Library of Medicine, accessed on 04/28/11). They are at various stages and target a wide variety of pathological conditions such as Crohn's disease and graft versus host disease (GVHD), cardiovascular disease and myocardial infarction (MI), brain and spinal cord injury, ischemic stroke, diabetes, cartilage and bone injury (Phinney and Prockop, 2007). Mixed results have been obtained from these trials, and strategies to enhance MSC engraftment and survival in the regenerating tissues are being developed.

MSC therapy has demonstrated a few promising results. One particular trial documented that intracoronary administration of MSC following MI initially showed significant improvement in one parameter assessing left ventricular function; however, this difference was no longer significant when analyzed after 18 months (Meyer *et al.*, 2006). A phase II clinical trial documented a 2-year reduction in the mortality rate in the GVHD patients treated with MSCs (Le Blanc *et al.*, 2008). Clinically, MSCs homed to the site of injury when injected intravenously into irradiated osteogenesis imperfecta patients, engrafting into bones, skin, and marrow stroma, ultimately stimulating growth (Horwitz *et al.*, 1999, 2002). There have been no adverse events identified in the past and ongoing trials, this being the most advantageous outcome from these trials.

VI. Immunomodulatory Properties

The majority of the clinical trials, approximately 44%, are utilizing MSCs for their immunomodulatory properties ((http://clinicaltrials.gov/ct2/home; U.S. National Library of Medicine, accessed on 04/28/11)). In conditions like GVHD, Crohn's disease, primary Sjogren's syndrome, organ transplantation and rejection, systematic sclerosis, type I diabetes, systemic lupus erythematosus, multiple sclerosis, neuroblastoma, and non-malignant red blood cell disorders, MSCs are being transplanted as treatment by themselves or as adjunct therapy.

The biology behind the effects of MSCs on the immune system is mostly unknown; however, a few experimental models have elucidated some of the key molecular players involved in the anti-inflammatory role of MSCs. One of these models is a rat renal transplantation model where MSC injections increased overall survival of the recipient animals due, in part, to a decrease in interleukin-1α (IL-1), tumor necrosis factor-α (TNF-α), and transforming growth factor (TGF)-β1 (Zhang *et al.*, 2007). The authors of this work remind us that there are still unknowns, as they note that adjunct immunosuppresion (cyclosporine A) treatment that inhibits IL-2 signaling further increases the survival of the rats and further diminishes the levels of the inflammatory cytokines (Laupacis *et al.*, 1982).

Other models have shown a similar effect of MSCs on the downregulation of the immune function. Such is a rodent model of interstitial lung disease, where bleomycin (a cytotoxic glycopeptide antibiotic; Mitchell *et al.*, 1989) treatment induces pulmonary fibrosis and concomitant inflammation (Iyer and Rojas, 2008). Following induction of the disease, Ortiz *et al.*, 2007 treated the mice with MSCs and demonstrated a decrease in the levels of two important proinflammatory cytokines: TNF-α and IL-1α. This group demonstrated the anti-inflammatory capacity of MSCs in this setting due to the expression of interleukin-1 receptor antagonist (IL-1RN). These results were confirmed by *in vitro* assays in which MSC-conditioned media decreased the proliferation of an IL-1 responsive T-cell population (Ortiz *et al.*, 2007).

Studies using an experimental autoimmune encephalomyelitis model demonstrated that MSCs were able to suppress T-cell activation *in vitro* and *in vivo*, and this effect was partially reversible by the addition of IL-2 (Zappia *et al.*, 2005). In this case, the authors suggest that the limited expression of MHC class II molecules as well as lack of costimulatory molecules, such as CD80, CD86, and CD40, on MSCs may be the reason behind the observed suppression. A more detailed study on secreted factors in MSC-conditioned media demonstrated that IL-10, TGF-β1, and prostaglandin E2 (PGE2) were not responsible for the T-cell inhibition (Shi *et al.*, 2000).

In summary, the roles of MSCs in immune suppression have been partially described and the molecular mechanism behind this capacity remains elusive. Further studies on the effects of IL-1RN on T-cells might give insight into how MSCs inhibit inflammation and prevent T-cell activation.

VII. MI Therapy

Treatment for myocardial repair and ventricular dysfunction is the second most common application for MSCs in the clinic (U.S. National Library of Medicine). Recently, Osiris Therapeutics published the safety of utilizing MSCs in the setting of MI. Moreover, they demonstrated that MSC transplantation, compared to placebo control, increased left ventricular ejection fraction and reversed adverse remodeling (Hare *et al.*, 2009). Although the data was only available after a 6-month follow-up, all MSC-treated patients had a statistical improvement in heart function.

In spite of having considerable potential, the clinical application of MSC-based therapy in the context of myocardial repair faces many challenges. Those challenges include poor tissue engraftment, low potency, and low survival of transplanted MSCs. In clinical trials, in spite of MSCs positively affecting myocyte regeneration, poor short-term survival with modest improvement in heart function together with inconsistent outcomes were identified (Lunde *et al.*, 2006; Schachinger *et al.*, 2006; Tendera *et al.*, 2009). In the light of these observations, it is evident that the therapeutic effects of MSCs could only be seen if transplanted MSCs could exert the long-term benefits to the patients. It is therefore very important to understand the biology and the tissue repair mechanism of MSCs in order to overcome the limitations of cell engraftment and survival.

VIII. Molecular Mediators of MSC Biology

A. Wnt pathway

Figure 2.2 depicts a simplified model of the canonical cascade through which secreted Wnt glycoproteins initiate signal transduction upon receptor binding. In the absence of Wnt signaling, β-catenin is in a complex with

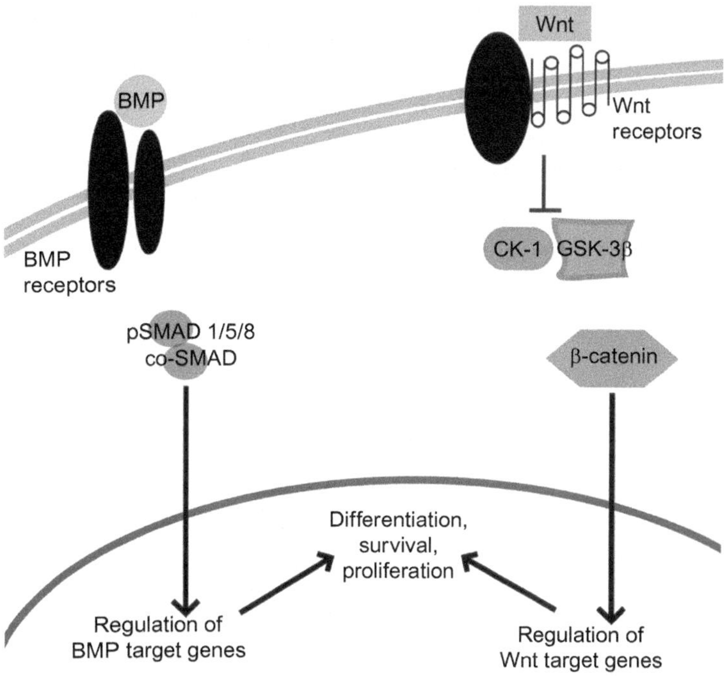

Figure 2.2 Simplified schematic of BMP and Wnt signaling. Activation of BMP and Wnt signaling can lead to the regulation of specific target genes which affect MSC biology. Transcriptional control is exerted on the cell upon the nuclear translocation of signaling molecules: phosphorylated Smads (pSMAD 1/5/8) in the case of BMP signaling and β-catenin in the case of Wnt signaling. (For color version of this figure, the reader is referred to the Web version of this chapter.)

Axin, APC, Dsh, GSK3-β, and CK-1 (Reya and Clevers, 2005). The latter two phosphorylate β-catenin, marking it for ubiquitination and subsequent proteosomal degradation. Occupancy of the transmembrane receptors LRP5/6 and Frizzled by Wnt family members block the kinase activity of the β-catenin destruction complex, allowing the accumulation of β-catenin in the cytoplasm and its translocation to the nucleus. Within the nucleus, this signaling molecule engages TCF/LEF DNA binding proteins to help drive expression of target genes (Moon *et al.*, 2004).

Inhibition of the Wnt pathway can occur at different stages of the cascade. For example, Dikkopf-1 (Dkk-1) antagonizes the signaling cascade when it binds LRP5/6, forms a ternary complex with Kremen, and promotes endocytosis and degradation of the receptor (Mao *et al.*, 2001, 2002). The sFRP family members are capable to act as inhibitors by binding to and sequestering the Wnt ligand from its receptor (Kawano and Kypta, 2003; Uren *et al.*, 2000).

The noncanonical and canonical Wnt signaling pathways are involved during embryonic development (Eisenberg and Eisenberg, 2006; Pandur *et al.*, 2002). Wnt plays a role in cellular proliferation (Kioussi *et al.*, 2002), differentiation, and self-renewal (Reya and Clevers, 2005; Wang and Wynshaw-Boris, 2004). More on the specific role of the Wnt cascade on MSC biology is included in subsequent sections.

B. Bone morphogenetic pathway

Members of the TGF superfamily include TGF-β, activins, and BMPs. These members play important and distinct roles in directing differentiation of stem cells during development and in the adult tissue (Korchynskyi and ten Dijke, 2002; Shi *et al.*, 2000). As seen in Fig. 2.2, this superfamily signals through activation of distinct types of serine/threonine kinase receptors and subsequent activation of signaling molecules called Smads. Upon ligand binding to type II receptors, type I receptors are cross-phosphorylated by the activated type II receptor. The now activated type I receptor phosphorylates its receptor-associated Smad protein (pSMAD). Phosphorylated R-Smads must form a complex with Smad 4 (co-SMAD) to translocate to the nucleus and exert a transcriptional effect on the cell (Miyazono *et al.*, 2005). The specificity of receptor activation and R-Smad activation is such that BMP activity uses only Smads 1, 5, and 8 as signaling molecules and thus directs transcription of target genes responsive to these Smads only. The role of BMP signaling in MSC biology will be expanded upon in the following sections.

C. Canonical Wnt and BMP signaling and MSC differentiation

MSCs possess an extensive potential to self-renew and differentiate into multiple lineages (Beresford, 1989; Caplan, 1991). MSCs can be induced to give rise to osteoblasts, chondrocytes, adipocytes, and myoblasts under suitable culture conditions (Caplan, 1991; Prockop, 1997). The ability to self-renew and differentiate into multiple lineages, easy isolation, and accessibility make MSCs a suitable candidate for therapeutic purposes. To enhance the beneficial effects of MSCs, it is necessary to identify the molecular mechanisms that regulate the self-renewal and differentiation of MSCs (Caplan, 2000). In this section, we will discuss the involvement of Wnt and BMP signaling in MSC differentiation.

1. Osteogenesis

The canonical Wnt signaling is involved in the lineage specification of MSCs (Etheridge *et al.*, 2004). Typically, for osteogenic differentiation in culture conditions, MSCs are treated with ascorbic acid, β-glycerophosphate, and dexamethasone in fetal bovine serum (FBS)-containing medium (Jaiswal *et al.*, 1997; Pittenger *et al.*, 1999) that results in the increase in calcium deposition and alkaline phosphatase activity. The effect of canonical Wnt signaling on osteogenesis is context dependent. It varies from differences in the level of Wnt, type of cell, the species, type and timing of stimulus, and other experimental conditions. Enhanced Wnt signaling either by the addition of high levels of exogenous Wnt3a or through overexpression of LRP5 or stabilization of the mutant form of β-catenin augments osteogenesis in human MSCs (De Boer *et al.*, 2004a; Gong *et al.*, 2001; Qiu *et al.*, 2007). On the other hand, some reports have identified that enhanced activity of Wnt signaling via exogenous addition of Wnt3a or lithium, which is a GSK3β inhibitor, diminishes osteogenesis in dexamethasone-induced human MSCs (de Boer *et al.*, 2004b). In murine pluripotent mesenchymal and osteoprogenitor cells, canonical Wnt signaling promotes the osteoblastogenesis via upregulating RUNX2, Dlx5, or osterix (Bennett *et al.*, 2005; Gaur *et al.*, 2005). It is interesting to observe that the stage of targeted cells also determine the effect of canonical Wnt signaling on osteogenesis in some cases. For example, Wnt signaling enhances the differentiation of MSCs that are committed to osteogenic lineage, on the other hand, inhibiting the terminal differentiation of mature osteoblasts (Eijken *et al.*, 2008; Kahler and Westendorf, 2003; Kahler *et al.*, 2006, 2008).

The BMP cascade, as the name implies, is more clearly implicated in osteogenic commitment of MSCs. For example, adenoviral expression of BMP2, 6, and 9 significantly induced alkaline phosphatase activity in pluripotent C3H10T1/2 cells (Cheng *et al.*, 2003). Similarly the preosteoblastic C2C12 cell line had increased alkaline phosphatase activity in the presence of BMP2, 4, 6, 7, and 9 (Cheng *et al.*, 2003). This data translated to the human MSC field as adenoviral gene expression of BMP2 increased their osteogenic commitment as demonstrated by increased expression of Runx2 and Type I collagen and alkaline phosphatase activity (Koch *et al.*, 2005). Human MSCs derived from osteoporotic patients responded beneficially to BMP2 and 7 as they increased alkaline phosphatase activity and total calcium production (Pountos *et al.*, 2010).

2. Chondrogenesis

MSCs are treated with TGF-β for chondrogenic differentiation induction because it has been shown to be a key regulator of early stages of MSC chondrogenesis (Tuli *et al.*, 2003). The chondrogenic differentiation results in the generation of cartilage-specific highly sulfated proteoglycans and type II collagen (Hwang *et al.*, 2007). The expression of β-catenin is different in different stages of chondrogenesis, indicating a differential role of Wnt signaling during various stages of chondrogenesis (Ling *et al.*, 2009). Canonical Wnt signaling has also been shown to regulate chondrocyte differentiation of MSCs in a Sox9-dependent manner (Yano *et al.*, 2005). A dual role of Wnt signaling has been observed in the process of chondrogenesis, which depends on the particular Wnt ligand and also on the developmental stage. In the chicken limb, during the process of chondrogenesis, Wnt4a promotes the maturation of chondrocytes, whereas Wnt5a negatively affects chondrocyte maturation (Hartmann and Tabin, 2000). The observed opposing effect could be due to the involvement of distinct pathways. The activating affect of Wnt4a on chondrogenesis was possibly due to activation of β-catenin and Wnt receptors FZD1 and 7 (Hartmann and Tabin, 2000). Trabecular bone-derived MSCs induced for TGF-β-mediated chondrogenesis involve a cross talk between mitogen-activated protein kinase and Wnt signaling (Tuli *et al.*, 2003). In the process, the canonical Wnt signaling regulates N-cadherin expression during cellular condensation and chondrogenesis. Recently, Maruyama *et al.* (2010) indicated that Wnt pathway regulates the fate of MSC lineage specification during skeletal development by modulating the balance of the fibroblast growth factor and BMP pathways.

Although BMP2 was implicated in directing osteogenesis of multipotential murine C3H10T1/2 cells, it was also implicated in their chondrogenic lineage commitment (Alcian blue stained cartilage-like matrix and type II collagen; Denker *et al.*, 1999). BMP2-directed chondrogenesis in these cells is presumably due to the induced expression of the cell cycle inhibitory protein/differentiation factor p21/WAF1 (Carlberg *et al.*, 2001). Although involved in noncanonical BMP signaling, BMP13 has been implicated in the chondrogenesis of these cells (Nochi *et al.*, 2004).

3. Adipogenesis

In cell culture conditions, MSCs are induced toward adipogenic differentiation by treatment with dexamethasone, insulin, isobutyl methyl xanthine, and indomethacin in complete media containing serum (Janderova *et al.*, 2003). The differentiation leads to the formation of the lipid vacuoles which are detected by Oil Red-O staining. The commitment of pluripotent MSCs toward adipogenic lineage leads to the formation of preadipocytes before terminally differentiating into mature adipocytes (Bowers and Lane, 2007). Inhibition of Wnt signaling is necessary for the adipogenic differentiation of MSCs (Moldes *et al.*, 2003). The canonical Wnt signaling has been shown to inhibit the expression of adipogenic transcription factor PPARγ in MSCs (Rawadi *et al.*, 2003). Overexpression of Axin2 or dominant-negative TCF4 leads to the inhibition of Wnt signaling which drives MSCs toward adipogenic differentiation (Ross *et al.*, 2000). On the other hand, canonical Wnt1 and Wnt10b and an activated mutant β-catenin averts adipocyte differentiation via reducing the expression of adipogenic transcription factors C/EBPα and PPARγ (Bennett *et al.*, 2005; Ross *et al.*, 2000). It has been demonstrated that in addition to regulating MSC proliferation, cyclin D1 and c-Myc directly inactivate PPARγ and C/EBPα, respectively (Fu *et al.*, 2005; Tetsu and McCormick, 1999). One can speculate that cyclin

D1 and c-Myc might be involved in regulating the inhibitory effects of Wnt signaling on adipogenic transcription factors. Additionally, Wnt10b has also been demonstrated to act as a molecular switch from adipogenesis toward osteogenesis of bipotential mesenchymal precursors (Bennett *et al.*, 2005).

Two separate cell lines (growth-arrested 10T1/2 and preadipocyte cell line, A33 cells, derived from 10T1/2 cells) were utilized to demonstrate the involvement of BMP4 in adipocyte development (Bowers and Lane, 2007). This cascade is deemed necessary for this as disruption of BMP4 signaling by noggin blocks the preadipocyte phenotype (Bowers *et al.*, 2006). BMP2 was also implicated in adipogenesis, and overexpression of constitutively active BMP receptor 1A or 1B induced commitment (Huang *et al.*, 2009).

D. Involvement of Wnt on survival and proliferation of MSCs

Many studies have suggested the role of canonical Wnt signaling in self-renewal and maintenance. Exogenous application of Wnt3a in cell culture enhanced MSC proliferation due to both increased proliferation and inhibition of apoptosis (Boland *et al.*, 2004; Cho *et al.*, 2006). The effect of Wnt3a on MSC proliferation is possibly due to its effect on cell cycle regulators, cyclin D1 and c-Myc (Baek *et al.*, 2003). Additionally, the overexpression of LRP5, a coreceptor involved in Wnt signaling, has been demonstrated to enhance proliferation of MSCs (Baksh *et al.*, 2007). On the other hand, canonical Wnt signaling has also been accounted for the inhibition of human MSC proliferation (Qiu *et al.*, 2007). In addition, inhibition of Wnt signaling through Dkk-1 and LRP inhibitors was necessary for human MSCs to reenter cell cycle and proliferate (Gregory *et al.*, 2003). Our studies also indicated that Wnt inhibition is a key factor required for better proliferation, engraftment, and survival of MSCs (Alfaro *et al.*, 2008). One study has reported inhibition and stimulation of human MSC proliferation by Wnt3a is dose dependent (De Boer *et al.*, 2004a). The differences in the findings about the role of Wnt signaling in MSC survival and proliferation may arise from differences in culture conditions, dose and type of Wnt ligands, and handling of the cells. These studies also indicate that Wnt signaling regulation of MSC biology (with specific regard to self-renewal) is a very complex mechanism and further studies are required to understand this mystery.

E. Effects of the BMP cascade on MSC survival and proliferation

BMPs are widely known to affect the differentiation of MSCs and therefore not much is documented on their roles on survival and proliferation, particularly on murine MSCs. Regardless, a few reports have emerged. Noncanonical BMP3 activated TGF-β signaling in C3H10T1/2 MSCs and 3T3-L1 preadipocytes to cause a threefold increase in their proliferation (Stewart *et al.*, 2010). Human adipose-derived MSCs treated with low-dose BMP4 are less apoptotic and more proliferative (Vicente Lopez *et al.*, 2010). Also, human MSCs derived from osteoporotic bones had increased proliferation after addition of high-dose BMP7 (Pountos *et al.*, 2010). In essence, the effects of BMPs (canonical and noncanonical) on MSC survival and proliferation are largely context dependent and no solid data were found on murine bone marrow-derived MSCs.

IX. Enhancing MSC Survival in the Wound

As was briefly touched upon, the survival of MSCs within the wound microenvironment is a limiting factor on their reparative capabilities (Freyman *et al.*, 2006; Hofmann *et al.*, 2005). Thus, several groups have tried to alter MSCs genetically to increase their survival. Although Wnt and BMP signaling have not been directly tied to this effect, several other molecules which might be involved in such cascades will be addressed briefly. Retroviral expression of the prosurvival gene Akt-1 decreased the apoptotic index of MSCs *in vitro*, an effect also observed *in vivo* following transplantation into infarcted rat myocardium (Jiang *et al.*, 2006; Mangi *et al.*, 2003). A similar effect was observed with retroviral overexpression of erythropoietin by MSCs: increased survival both *in vitro* and *in vivo* in a murine subcutaneous implantation model of matrigel-embedded MSCs (Copland *et al.*, 2008). Our group has shown that overexpression of sFRP2 by MSCs enhances their *in vitro* proliferative index and increases their engraftment in two separate *in vivo* wound models (Alfaro *et al.*, 2008).

X. Secreted Frizzled-Related Proteins

The sFRP family was originally identified by their high homology to the Frizzled Wnt receptor (Rattner *et al.*, 1997; Shirozu *et al.*, 1996). There are eight family members; five of these are mammalian (Bovolenta *et al.*, 2008). The function of the five mammalian members, sFRP1 to sFRP5, has been identified as Wnt inhibition (Chang *et al.*, 1999; Finch *et al.*, 1997; Schumann *et al.*, 2000). sFRPs have a N-terminal conserved cysteine-rich domain (CRD) which shares 30–50% similarity with Frizzled proteins (Melkonyan *et al.*, 1997). The C-terminal domain of sFRPs shares a sequence similarity with the axon guidance protein netrin (NTR). The NTR domain has also been found in a few complement proteins and in the tissue inhibitors of metalloproteases (Banyai and Patthy, 1999). sFRPs inhibit Wnt signaling via interacting through their CRD with

Wnt ligands (Lin *et al.*, 1997), by interaction of NTR-like domain with the Wnt ligands (Uren *et al.*, 2000) or binding to itself and the Frizzled protein (Bafico *et al.*, 1999). The contradiction in the mode of binding of sFRPs to Wnt and other ligands might be due to different affinity of sFRPs towards their binding partners. sFRPs are secreted proteins; however, in cultured cells, they have been found to be associated with the cell and are released into the culture media upon addition of heparin (Finch *et al.*, 1997).

Further documented roles for the sFRP family members include cytoprotection and proliferation. sFRP2 seems to have a positive function in wound repair. This molecule was implicated in skeletal muscle repair in mice and has been identified as a key factor in MSC-based therapy for myocardial repair (Mirotsou *et al.*, 2007; Zhao and Hoffman, 2004). Melkonyan *et al.*, 1997 identified sFRP2 as a prosurvival protein and showed that its overexpression in human breast adenocarcinoma cell line, MCF7, conferred an antiapoptotic effect. Recently, Sfrp2 was identified as one of the key paracrine factors released by Akt-overexpressing MSCs found to play a critical role in the survival of ischemic cardiac myocytes (Mirotsou *et al.*, 2007). Sfrp2 increases the proliferative index of both murine and human MSCs (Alfaro *et al.*, 2008). Identifying a cytoprotective role for sFRP2 in the context of wound repair will provide insight into improving MSC tissue repair. Recently, a new hypothesis for an alternative role for the sFRP family has emerged; the mammalian members may inhibit the BMP signaling cascade much like the nonmammalian Sizzled members (Lee *et al.*, 2006; Muraoka *et al.*, 2006).

XI. Mediating MSC Self-Renewal

The mechanisms by which MSCs modulate Wnt and BMP signaling events during growth and/or lineage commitment are immensely complex. The molecular interplay between these cascades may modulate MSC self-renewal, a hypothesis based on the self-renewal mechanisms of other stem cells. Recent studies performed by our laboratory demonstrated that over-expression of the Wnt inhibitory protein, sFRP2, by MSCs (sFRP2-MSCs) increases *in vitro* proliferation (Alfaro *et al.*, 2008) and survival of MSCs by regulating both the BMP and Wnt signaling pathways (Alfaro *et al.*, 2010). The effects of sFRP2, as depicted in Fig. 2.3, indirectly affect MSC self-renewal.

Regardless of their mode of action, direct regeneration, or paracrine effects, MSCs could not impact the repair process until and unless a substantial amount of cells are generated via MSC self-renewal and proliferation within the wounded area. Although encouraging results have been observed in preclinical models (Hung *et al.*, 2007; Shake *et al.*, 2002; Shi *et al.*, 2007), low survival and poor engraftment significantly restrict the

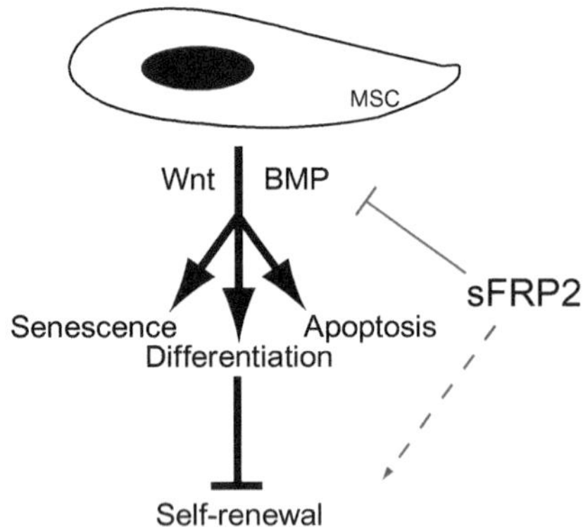

Figure 2.3 Model of the role of sFRP2 in MSC self-renewal. Self-renewal is evaded by senescence, lineage commitment, and apoptosis of MSCs; Wnt and BMP signaling drive these events. sFRP2 can inhibit these cascades to indirectly drive MSC self-renewal (discontinuous line). (For color version of this figure, the reader is referred to the Web version of this chapter.)

efficiency of MSCs for remedial purposes in clinical setting. Some clinical trials address the issue of long-term engraftment of MSCs. For example, the engraftment of donor MSCs in osteogenesis imperfecta patients never exceeded 1% 4–6 weeks following MSC infusion (Horwitz *et al.*, 2002). Enhanced survival of sFRP2-MSCs observed also *in vivo* in two distinct wound repair models still led to a modest degree of engraftment (Alfaro *et al.*, 2008). Achieving a critical mass of MSCs by improving their survival and self-renewal capacity should increase their numbers in the damaged tissue to ultimately increase their reparative abilities. Figure 2.3 demonstrates one molecule that positively affects MSC self-renewal leading to enhanced wound repair.

XII. Conclusions

The focus of this review was BM-derived murine MSCs, with some discussion of their clinical utilization. The data demonstrated gaps in the knowledge of MSC biology, that if better understood would allow clinicians to improve the potential of MSC-directed wound repair. Particularly, we focused on canonical Wnt and BMP signaling to demonstrate how activation and/or inhibition of these two well-understood cascades could affect MSC activity, including cellular differentiation, survival, and proliferation.

The data gathered demonstrate that the effects of both Wnt and BMP signaling on MSCs are variable. The source of the MSCs (tissue of isolation), the passage number, the quantity of ligand, and the timing are all examples of variables that affect the outcome of these cascades on the MSCs. In general, however, BMP and Wnt signaling are thought to drive senescence, differentiation, and apoptosis of MSCs. We and others have recently identified that sFRP2 is a MSC-derived factor that modulates both Wnt and BMP pathways to improve MSC-mediated cardiovascular and wound repair. sFRP2 increased proliferation, prevented apoptosis, and decreased differentiation of MSCs. As a whole, these data suggest sFRP2 is a potent factor involved in MSC self-renewal. Increasing MSC self-renewal is critical to enhancing their therapeutic efficacy; therefore, other molecules that might have similar effects would be of interest to the field. The significance of this finding can be appreciated by looking at the impact self-renewal factors have had on other stem cell fields. Particularly, the *ex vivo* expansion of HSCs ($CD34^+$) through addition of recombinant self-renewal factors is being actively pursued and has yielded important results in several clinical trials of a variety of human diseases (Kelly *et al.*, 2009). Manipulation of the MSC self-renewal factors, including sFRP2 or other modulators of both Wnt and BMP signaling, could profoundly control the extent of their senescence, differentiation, and apoptosis enhancing the regeneration of MSC-mediated therapy.

REFERENCES

Alfaro, M. P., *et al.* (2008). The Wnt modulator sFRP2 enhances mesenchymal stem cell engraftment, granulation tissue formation and myocardial repair. *Proc. Natl. Acad. Sci. USA* **105**(47), 18366–18371.

Alfaro, M. P., *et al.* (2010). sFRP2 suppression of bone morphogenic protein (BMP) and Wnt signaling mediates mesenchymal stem cell (MSC) self-renewal promoting engraftment and myocardial repair. *J. Biol. Chem.* **285**(46), 35645–35653.

Baddoo, M., *et al.* (2003). Characterization of mesenchymal stem cells isolated from murine bone marrow by negative selection. *J. Cell. Biochem.* **89**(6), 1235–1249.

Baek, S. H., *et al.* (2003). Regulated subset of G1 growth-control genes in response to derepression by the Wnt pathway. *Proc. Natl. Acad. Sci. USA* **100**(6), 3245–3250.

Bafico, A., *et al.* (1999). Interaction of frizzled related protein (FRP) with Wnt ligands and the frizzled receptor suggests alternative mechanisms for FRP inhibition of Wnt signaling. *J. Biol. Chem.* **274**(23), 16180–16187.

Baksh, D., Yao, R., and Tuan, R. S. (2007). Comparison of proliferative and multilineage differentiation potential of human mesenchymal stem cells derived from umbilical cord and bone marrow. *Stem Cells* **25**(6), 1384–1392.

Banyai, L., and Patthy, L. (1999). The NTR module: Domains of netrins, secreted frizzled related proteins, and type I procollagen C-proteinase enhancer protein are homologous with tissue inhibitors of metalloproteases. *Protein Sci.* **8**(8), 1636–1642.

Bennett, C. N., *et al.* (2005). Regulation of osteoblastogenesis and bone mass by Wnt10b. *Proc. Natl. Acad. Sci. USA* **102**(9), 3324–3329.

Beresford, J. N. (1989). Osteogenic stem cells and the stromal system of bone and marrow. *Clin. Orthop. Relat. Res.* **240,** 270–280.

Boland, G. M., *et al.* (2004). Wnt 3a promotes proliferation and suppresses osteogenic differentiation of adult human mesenchymal stem cells. *J. Cell. Biochem.* **93**(6), 1210–1230.
Bovolenta, P., *et al.* (2008). Beyond Wnt inhibition: New functions of secreted Frizzled-related proteins in development and disease. *J. Cell Sci.* **121**(Pt 6), 737–746.
Bowers, R. R., and Lane, M. D. (2007). A role for bone morphogenetic protein-4 in adipocyte development. *Cell Cycle* **6**(4), 385–389.
Bowers, R. R., *et al.* (2006). Stable stem cell commitment to the adipocyte lineage by inhibition of DNA methylation: Role of the BMP-4 gene. *Proc. Natl. Acad. Sci. USA* **103**(35), 13022–13027.
Caplan, A. I. (1991). Mesenchymal stem cells. *J. Orthop. Res.* **9**(5), 641–650.
Caplan, A. I. (2000). Mesenchymal stem cells and gene therapy. *Clin. Orthop. Relat. Res.* **379**(Suppl), S67–S70.
Carlberg, A. L., *et al.* (2001). Efficient chondrogenic differentiation of mesenchymal cells in micromass culture by retroviral gene transfer of BMP2. *Differentiation* **67**(4–5), 128–138.
Chang, J. T., *et al.* (1999). Cloning and characterization of a secreted frizzled-related protein that is expressed by the retinal pigment epithelium. *Hum. Mol. Genet.* **8**(4), 575–583.
Cheng, H., *et al.* (2003). Osteogenic activity of the fourteen types of human bone morphogenetic proteins (BMPs). *J. Bone Joint Surg. Am.* **85-A**(8), 1544–1552.
Cho, H. H., *et al.* (2006). Endogenous Wnt signaling promotes proliferation and suppresses osteogenic differentiation in human adipose derived stromal cells. *Tissue Eng.* **12**(1), 111–121.
Copland, I. B., *et al.* (2008). Coupling erythropoietin secretion to mesenchymal stromal cells enhances their regenerative properties. *Cardiovasc. Res.* **79**(3), 405–415.
De Boer, J., Wang, H. J., and Van Blitterswijk, C. (2004a). Effects of Wnt signaling on proliferation and differentiation of human mesenchymal stem cells. *Tissue Eng.* **10**(3–4), 393–401.
de Boer, J., *et al.* (2004b). Wnt signaling inhibits osteogenic differentiation of human mesenchymal stem cells. *Bone* **34**(5), 818–826.
Delorme, B., and Charbord, P. (2007). Culture and characterization of human bone marrow mesenchymal stem cells. *Methods Mol. Med.* **140,** 67–81.
Denker, A. E., *et al.* (1999). Chondrogenic differentiation of murine C3H10T1/2 multipotential mesenchymal cells: I. Stimulation by bone morphogenetic protein-2 in high-density micromass cultures. *Differentiation* **64**(2), 67–76.
Eijken, M., *et al.* (2008). Wnt signaling acts and is regulated in a human osteoblast differentiation dependent manner. *J. Cell. Biochem.* **104**(2), 568–579.
Eisenberg, L. M., and Eisenberg, C. A. (2006). Wnt signal transduction and the formation of the myocardium. *Dev. Biol.* **293**(2), 305–315.
Etheridge, S. L., *et al.* (2004). Expression profiling and functional analysis of wnt signaling mechanisms in mesenchymal stem cells. *Stem Cells* **22,** 849–860.
Farndale, R. W., Buttle, D. J., and Barrett, A. J. (1986). Improved quantitation and discrimination of sulphated glycosaminoglycans by use of dimethylmethylene blue. *Biochim. Biophys. Acta* **883**(2), 173–177.
Finch, P. W., *et al.* (1997). Purification and molecular cloning of a secreted, Frizzled-related antagonist of Wnt action. *Proc. Natl. Acad. Sci. USA* **94**(13), 6770–6775.
Freyman, T., *et al.* (2006). A quantitative, randomized study evaluating three methods of mesenchymal stem cell delivery following myocardial infarction. *Eur. Heart J.* **27**(9), 1114–1122.
Fu, M., *et al.* (2005). Cyclin D1 inhibits peroxisome proliferator-activated receptor gamma-mediated adipogenesis through histone deacetylase recruitment. *J. Biol. Chem.* **280**(17), 16934–16941.
Fuchs, E., and Segre, J. A. (2000). Stem cells: A new lease on life. *Cell* **100**(1), 143–155.
Gaur, T., *et al.* (2005). Canonical WNT signaling promotes osteogenesis by directly stimulating Runx2 gene expression. *J. Biol. Chem.* **280**(39), 33132–33140.

Gong, Y., *et al.* (2001). LDL receptor-related protein 5 (LRP5) affects bone accrual and eye development. *Cell* **107**(4), 513–523.

Gregory, C. A., *et al.* (2003). The Wnt signaling inhibitor dickkopf-1 is required for reentry into the cell cycle of human adult stem cells from bone marrow. *J. Biol. Chem.* **278**(30), 28067–28078.

Haramis, A. P., *et al.* (2004). De novo crypt formation and juvenile polyposis on BMP inhibition in mouse intestine. *Science* **303**(5664), 1684–1686.

Hare, J. M., *et al.* (2009). A randomized, double-blind, placebo-controlled, dose-escalation study of intravenous adult human mesenchymal stem cells (prochymal) after acute myocardial infarction. *J. Am. Coll. Cardiol.* **54**(24), 2277–2286.

Harris, D. T. (2008). Cord blood stem cells: A review of potential neurological applications. *Stem Cell Rev.* **4**(4), 269–274.

Hartmann, C., and Tabin, C. J. (2000). Dual roles of Wnt signaling during chondrogenesis in the chicken limb. *Development* **127**(14), 3141–3159.

He, X. C., *et al.* (2004). BMP signaling inhibits intestinal stem cell self-renewal through suppression of Wnt-beta-catenin signaling. *Nat. Genet.* **36**(10), 1117–1121.

Hofmann, M., *et al.* (2005). Monitoring of bone marrow cell homing into the infarcted human myocardium. *Circulation* **111**(17), 2198–2202.

Horwitz, E. M., *et al.* (1999). Transplantability and therapeutic effects of bone marrow-derived mesenchymal cells in children with osteogenesis imperfecta. *Nat. Med.* **5**(3), 309–313.

Horwitz, E. M., *et al.* (2002). Isolated allogeneic bone marrow-derived mesenchymal cells engraft and stimulate growth in children with osteogenesis imperfecta: Implications for cell therapy of bone. *Proc. Natl. Acad. Sci. USA* **99**(13), 8932–8937.

Huang, H., *et al.* (2009). BMP signaling pathway is required for commitment of C3H10T1/2 pluripotent stem cells to the adipocyte lineage. *Proc. Natl. Acad. Sci. USA* **106**(31), 12670–12675.

Hung, S. C., *et al.* (2007). Short-term exposure of multipotent stromal cells to low oxygen increases their expression of CX3CR1 and CXCR4 and their engraftment in vivo. *PLoS One* **2**(5), e416.

Hwang, N. S., *et al.* (2007). Morphogenetic signals from chondrocytes promote chondrogenic and osteogenic differentiation of mesenchymal stem cells. *J. Cell. Physiol.* **212**(2), 281–284.

Iyer, S. S., and Rojas, M. (2008). Anti-inflammatory effects of mesenchymal stem cells: Novel concept for future therapies. *Expert Opin. Biol. Ther.* **8**(5), 569–581.

Jaiswal, N., *et al.* (1997). Osteogenic differentiation of purified, culture-expanded human mesenchymal stem cells in vitro. *J. Cell. Biochem.* **64**(2), 295–312.

Janderova, L., *et al.* (2003). Human mesenchymal stem cells as an in vitro model for human adipogenesis. *Obes. Res.* **11**(1), 65–74.

Jiang, S., *et al.* (2006). Supportive interaction between cell survival signaling and angiocompetent factors enhances donor cell survival and promotes angiomyogenesis for cardiac repair. *Circ. Res.* **99**(7), 776–784.

Kahler, R. A., and Westendorf, J. J. (2003). Lymphoid enhancer factor-1 and beta-catenin inhibit Runx2-dependent transcriptional activation of the osteocalcin promoter. *J. Biol. Chem.* **278**(14), 11937–11944.

Kahler, R. A., *et al.* (2006). Lymphocyte enhancer-binding factor 1 (Lef1) inhibits terminal differentiation of osteoblasts. *J. Cell. Biochem.* **97**(5), 969–983.

Kahler, R. A., *et al.* (2008). Collagen 11a1 is indirectly activated by lymphocyte enhancer-binding factor 1 (Lef1) and negatively regulates osteoblast maturation. *Matrix Biol.* **27**(4), 330–338.

Kawano, Y., and Kypta, R. (2003). Secreted antagonists of the Wnt signalling pathway. *J. Cell Sci.* **116**(Pt 13), 2627–2634.

Kelly, S. S., *et al.* (2009). Ex vivo expansion of cord blood. *Bone Marrow Transplant.* **44**(10), 673–681.

Kinnaird, T., *et al.* (2004). Marrow-derived stromal cells express genes encoding a broad spectrum of arteriogenic cytokines and promote in vitro and in vivo arteriogenesis through paracrine mechanisms. *Circ. Res.* **94**(5), 678–685.

Kioussi, C., *et al.* (2002). Identification of a Wnt/Dvl/beta-Catenin –> Pitx2 pathway mediating cell-type-specific proliferation during development. *Cell* **111**(5), 673–685.

Klaus, A., and Birchmeier, W. (2008). Wnt signalling and its impact on development and cancer. *Nat. Rev. Cancer* **8**(5), 387–398.

Klaus, A., *et al.* (2007). Distinct roles of Wnt/beta-catenin and Bmp signaling during early cardiogenesis. *Proc. Natl. Acad. Sci. USA* **104**(47), 18531–18536.

Koch, H., *et al.* (2005). Refined adenoviral transduction for controlled gene transfer into human adult mesenchymal stem cells. *Z. Orthop. Ihre Grenzgeb.* **143**(6), 677–683.

Korchynskyi, O., and ten Dijke, P. (2002). Identification and functional characterization of distinct critically important bone morphogenetic protein-specific response elements in the Id1 promoter. *J. Biol. Chem.* **277**(7), 4883–4891.

Kratchmarova, I., *et al.* (2005). Mechanism of divergent growth factor effects in mesenchymal stem cell differentiation. *Science* **308**(5727), 1472–1477.

Laupacis, A., *et al.* (1982). Cyclosporin A: A powerful immunosuppressant. *Can. Med. Assoc. J.* **126**(9), 1041–1046.

Le Blanc, K., *et al.* (2008). Mesenchymal stem cells for treatment of steroid-resistant, severe, acute graft-versus-host disease: A phase II study. *Lancet* **371**(9624), 1579–1586.

Lee, H. X., *et al.* (2006). Embryonic dorsal-ventral signaling: Secreted frizzled-related proteins as inhibitors of tolloid proteinases. *Cell* **124**(1), 147–159.

Lin, K., *et al.* (1997). The cysteine-rich frizzled domain of Frzb-1 is required and sufficient for modulation of Wnt signaling. *Proc. Natl. Acad. Sci. USA* **94**(21), 11196–11200.

Ling, L., Nurcombe, V., and Cool, S. M. (2009). Wnt signaling controls the fate of mesenchymal stem cells. *Gene* **433**(1–2), 1–7.

Lunde, K., *et al.* (2006). Intracoronary injection of mononuclear bone marrow cells in acute myocardial infarction. *N. Engl. J. Med.* **355**(12), 1199–1209.

Madonna, R., Geng, Y. J., and De Caterina, R. (2009). Adipose tissue-derived stem cells: Characterization and potential for cardiovascular repair. *Arterioscler. Thromb. Vasc. Biol.* **29**(11), 1723–1729.

Mangi, A. A., *et al.* (2003). Mesenchymal stem cells modified with Akt prevent remodeling and restore performance of infarcted hearts. *Nat. Med.* **9**(9), 1195–1201.

Mannello, F., *et al.* (2006). Role and function of matrix metalloproteinases in the differentiation and biological characterization of mesenchymal stem cells. *Stem Cells* **24**(3), 475–481.

Mao, B., *et al.* (2001). LDL-receptor-related protein 6 is a receptor for Dickkopf proteins. *Nature* **411**(6835), 321–325.

Mao, B., *et al.* (2002). Kremen proteins are Dickkopf receptors that regulate Wnt/beta-catenin signalling. *Nature* **417**(6889), 664–667.

Maruyama, T., *et al.* (2010). The balance of WNT and FGF signaling influences mesenchymal stem cell fate during skeletal development. *Sci. Signal.* **3**(123), ra40.

McReynolds, L. J., *et al.* (2007). Smad1 and Smad5 differentially regulate embryonic hematopoiesis. *Blood* **110**(12), 3881–3890.

Melkonyan, H. S., *et al.* (1997). SARPs: A family of secreted apoptosis-related proteins. *Proc. Natl. Acad. Sci. USA* **94**(25), 13636–13641.

Meyer, G. P., *et al.* (2006). Intracoronary bone marrow cell transfer after myocardial infarction: Eighteen months' follow-up data from the randomized, controlled BOOST (BOne marrOw transfer to enhance ST-elevation infarct regeneration) trial. *Circulation* **113**(10), 1287–1294.

Mirotsou, M., *et al.* (2007). Secreted frizzled related protein 2 (Sfrp2) is the key Akt-mesenchymal stem cell-released paracrine factor mediating myocardial survival and repair. *Proc. Natl. Acad. Sci. USA* **104**(5), 1643–1648.

Mitchell, J., *et al.* (1989). Alpha-smooth muscle actin in parenchymal cells of bleomycin-injured rat lung. *Lab. Invest.* **60**(5), 643–650.

Miyazono, K., Maeda, S., and Imamura, T. (2005). BMP receptor signaling: Transcriptional targets, regulation of signals, and signaling cross-talk. *Cytokine Growth Factor Rev.* **16**(3), 251–263.

Moldes, M., *et al.* (2003). Peroxisome-proliferator-activated receptor gamma suppresses Wnt/beta-catenin signalling during adipogenesis. *Biochem. J.* **376**(Pt 3), 607–613.

Moon, R. T., *et al.* (2004). WNT and beta-catenin signalling: Diseases and therapies. *Nat. Rev. Genet.* **5**(9), 691–701.

Muraoka, O., *et al.* (2006). Sizzled controls dorso-ventral polarity by repressing cleavage of the Chordin protein. *Nat. Cell Biol.* **8**(4), 329–338.

Nochi, H., *et al.* (2004). Adenovirus mediated BMP-13 gene transfer induces chondrogenic differentiation of murine mesenchymal progenitor cells. *J. Bone Miner. Res.* **19**(1), 111–122.

Orlic, D., *et al.* (2001). Bone marrow cells regenerate infarcted myocardium. *Nature* **410** (6829), 701–705.

Ortiz, L. A., *et al.* (2007). Interleukin 1 receptor antagonist mediates the antiinflammatory and antifibrotic effect of mesenchymal stem cells during lung injury. *Proc. Natl. Acad. Sci. USA* **104**(26), 11002–11007.

Pandur, P., *et al.* (2002). Wnt11 activation of a non-canonical Wnt signalling pathway is required for cardiogenesis. *Nature* **418**(6898), 636–641.

Phinney, D. G., and Prockop, D. J. (2007). Concise review: Mesenchymal stem/multipotent stromal cells: The state of transdifferentiation and modesl of tissue repair-current views. *Stem Cells* **25,** 2896–2902.

Pittenger, M. F., *et al.* (1999). Multilineage potential of adult human mesenchymal stem cells. *Science* **284**(5411), 143–147.

Pountos, I., *et al.* (2010). The effect of bone morphogenetic protein-2, bone morphogenetic protein-7, parathyroid hormone, and platelet-derived growth factor on the proliferation and osteogenic differentiation of mesenchymal stem cells derived from osteoporotic bone. *J. Orthop. Trauma* **24**(9), 552–556.

Pozzobon, M., Ghionzoli, M., and De Coppi, P. (2010). ES, iPS, MSC, and AFS cells. Stem cells exploitation for pediatric surgery: Current research and perspective. *Pediatr. Surg. Int.* **26**(1), 3–10.

Prockop, D. J. (1997). Marrow stromal cells as stem cells for nonhematopoietic tissues. *Science* **276**(5309), 71–74.

Qiu, W., *et al.* (2007). Patients with high bone mass phenotype exhibit enhanced osteoblast differentiation and inhibition of adipogenesis of human mesenchymal stem cells. *J. Bone Miner. Res.* **22**(11), 1720–1731.

Qyang, Y., *et al.* (2007). The renewal and differentiation of Isl1+ cardiovascular progenitors are controlled by a Wnt/beta-catenin pathway. *Cell Stem Cell* **1**(2), 165–179.

Rattner, A., *et al.* (1997). A family of secreted proteins contains homology to the cysteine-rich ligand-binding domain of frizzled receptors. *Proc. Natl. Acad. Sci. USA* **94**(7), 2859–2863.

Rawadi, G., *et al.* (2003). BMP2 controls alkaline phosphatase expression and osteoblast mineralization by a Wnt autocrine loop. *J. Bone Miner. Res.* **18**(10), 1842–1853.

Reya, T., and Clevers, H. (2005). Wnt signalling in stem cells and cancer. *Nature* **434**(7035), 843–850.

Rombouts, W. J., and Ploemacher, R. E. (2003). Primary murine MSC show highly efficient homing to the bone marrow but lose homing ability following culture. *Leukemia* **17**(1), 160–170.

Ross, S. E., *et al.* (2000). Inhibition of adipogenesis by Wnt signaling. *Science* **289**(5481), 950–953.
Satija, N. K., *et al.* (2007). Mesenchymal stem cells: Molecular targets for tissue engineering. *Stem Cells Dev.* **16**(1), 7–23.
Schachinger, V., *et al.* (2006). Intracoronary bone marrow-derived progenitor cells in acute myocardial infarction. *N. Engl. J. Med.* **355**(12), 1210–1221.
Schofield, R. (1983). The stem cell system. *Biomed. Pharmacother.* **37**(8), 375–380.
Schumann, H., *et al.* (2000). Expression of secreted frizzled related proteins 3 and 4 in human ventricular myocardium correlates with apoptosis related gene expression. *Cardiovasc. Res.* **45**(3), 720–728.
Shake, J. G., *et al.* (2002). Mesenchymal stem cell implantation in a swine myocardial infarct model: Engraftment and functional effects. *Ann. Thorac. Surg.* **73**(6), 1919–1925 (Discussion 1926).
Shi, Y., *et al.* (2000). BMP signaling is required for heart formation in vertebrates. *Dev. Biol.* **224**(2), 226–237.
Shi, M., *et al.* (2007). Regulation of CXCR4 expression in human mesenchymal stem cells by cytokine treatment: Role in homing efficiency in NOD/SCID mice. *Haematologica* **92**(7), 897–904.
Shirozu, M., *et al.* (1996). Characterization of novel secreted and membrane proteins isolated by the signal sequence trap method. *Genomics* **37**(3), 273–280.
Short, B., *et al.* (2003). Mesenchymal stem cells. *Arch. Med. Res.* **34**(6), 565–571.
Silva, G. V., *et al.* (2005). Mesenchymal stem cells differentiate into an endothelial phenotype, enhance vascular density, and improve heart function in a canine chronic ischemia model. *Circulation* **111**(2), 150–156.
Song, L., *et al.* (2006). Identification and functional analysis of candidate genes regulating mesenchymal stem cell self-renewal and multipotency. *Stem Cells* **24**(7), 1707–1718.
Stewart, A., Guan, H., and Yang, K. (2010). BMP-3 promotes mesenchymal stem cell proliferation through the TGF-beta/activin signaling pathway. *J. Cell. Physiol.* **223**(3), 658–666.
Suzuki, T., and Chiba, S. (2005). Notch signaling in hematopoietic stem cells. *Int. J. Hematol.* **82**(4), 285–294.
Takada, I., Kouzmenko, A. P., and Kato, S. (2009). Wnt and PPARgamma signaling in osteoblastogenesis and adipogenesis. *Nat. Rev. Rheumatol.* **5**(8), 442–447.
Tang, Y. L., *et al.* (2004). Autologous mesenchymal stem cell transplantation induce VEGF and neovascularization in ischemic myocardium. *Regul. Pept.* **117**(1), 3–10.
Tendera, M., *et al.* (2009). Intracoronary infusion of bone marrow-derived selected CD34+CXCR4+ cells and non-selected mononuclear cells in patients with acute STEMI and reduced left ventricular ejection fraction: Results of randomized, multicentre myocardial regeneration by intracoronary infusion of selected population of stem cells in acute myocardial infarction (REGENT) trial. *Eur. Heart J.* **30**(11), 1313–1321.
Tetsu, O., and McCormick, F. (1999). Beta-catenin regulates expression of cyclin D1 in colon carcinoma cells. *Nature* **398**(6726), 422–426.
Tropel, P., *et al.* (2004). Isolation and characterisation of mesenchymal stem cells from adult mouse bone marrow. *Exp. Cell Res.* **295**(2), 395–406.
Tuli, R., *et al.* (2003). Transforming growth factor-beta-mediated chondrogenesis of human mesenchymal progenitor cells involves N-cadherin and mitogen-activated protein kinase and Wnt signaling cross-talk. *J. Biol. Chem.* **278**(42), 41227–41236.
Uren, A., *et al.* (2000). Secreted frizzled-related protein-1 binds directly to Wingless and is a biphasic modulator of Wnt signaling. *J. Biol. Chem.* **275**(6), 4374–4382.
U.S. National Library of Medicine, U.S.N.I.o.H., U.S. Department of Health & Human Services, *Clinicaltrials.gov*, Lester Hill National Center for Biomedical Communications..
van Es, J. H., *et al.* (2005). Notch/gamma-secretase inhibition turns proliferative cells in intestinal crypts and adenomas into goblet cells. *Nature* **435**(7044), 959–963.

Vicente Lopez, M. A., *et al.* (2010). Low doses of bone morphogenetic protein 4 increase the survival of human adipose-derived stem cells maintaining their stemness and multipotency. *Stem Cells Dev.* **20**(6), 1011–1019.

Wang, J., and Wynshaw-Boris, A. (2004). The canonical Wnt pathway in early mammalian embryogenesis and stem cell maintenance/differentiation. *Curr. Opin. Genet. Dev.* **14**(5), 533–539.

Yano, F., *et al.* (2005). The canonical Wnt signaling pathway promotes chondrocyte differentiation in a Sox9-dependent manner. *Biochem. Biophys. Res. Commun.* **333**(4), 1300–1308.

Zappia, E., *et al.* (2005). Mesenchymal stem cells ameliorate experimental autoimmune encephalomyelitis inducing T-cell anergy. *Blood* **106**(5), 1755–1761.

Zhang, W., Qin, C., and Zhou, Z. M. (2007). Mesenchymal stem cells modulate immune responses combined with cyclosporine in a rat renal transplantation model. *Transplant. Proc.* **39**(10), 3404–3408.

Zhang, X., *et al.* (2011). Isolation and characterization of mesenchymal stem cells from human umbilical cord blood: Reevaluation of critical factors for successful isolation and high ability to proliferate and differentiate to chondrocytes as compared to mesenchymal stem cells from bone marrow and adipose tissue. *J. Cell. Biochem.* **112**(4), 1206–1218.

Zhao, P., and Hoffman, E. P. (2004). Embryonic myogenesis pathways in muscle regeneration. *Dev. Dyn.* **229**(2), 380–392.

CHAPTER THREE

Insulin and Germline Proliferation in *Caenorhabditis elegans*

E. Jane Albert Hubbard

Contents

I. Germline Proliferation in *C. elegans*: A Model for Developmental, Physiological, and Environmental Control of Cell Proliferation 62
II. *C. elegans* Germline Development 63
III. The *C. elegans* Germ line "Proliferation Versus Differentiation" Decision Is Mediated from the Soma to the Germ line by a Conserved Notch Signaling Pathway 64
IV. Evidence for Notch-Independent Soma-Germline Signaling Mechanisms That Modulate Germline Proliferation 65
V. A Counter-Intuitive Assay to Indentify Potential Notch-Independent Mechanisms That Promote the Expansion of the Larval Germline Progenitor Pool 66
VI. Identification of the Insulin/IGF-Like Receptor (IIR) Pathway Role in Germline Proliferation 68
VII. IIR Signaling in *C. elegans* 69
VIII. Insulin Signaling Promotes the Larval Germline Cell Division Cycle 70
IX. *C. elegans* Insulins 71
X. Many Target Tissues for IIR Signaling 71
XI. Other Germline Roles for the IIR Pathway 73
XII. IIR Role in Larval Germline Proliferation: A Reproductive Timing and Lifespan Connection? 74
XIII. A Current Model and Future Directions 74
Acknowledgments 76
References 76

Developmental Genetics Program, Skirball Institute of Biomolecular Medicine, Helen and Martin Kimmel Center for Stem Cell Biology, Department of Pathology, New York University School of Medicine, New York, New York, USA

Vitamins and Hormones, Volume 87
ISSN 0083-6729, DOI: 10.1016/B978-0-12-386015-6.00024-X

Abstract

Germline proliferation in *Caenorhabditis elegans* is emerging as a compelling model system for understanding the molecular basis for the developmental and physiological control of cell proliferation. This review covers the discovery and implications of the role of the insulin/IGF-like signaling pathway in germline proliferation during germline development. This pathway plays a host of important roles in *C. elegans* biology. Its role in germline proliferation is important to generate the proper adult stem/progenitor population and to ensure optimal fecundity. Moreover, in this role, it is restricted to reproductive (as opposed to dauer) larval stages and impinges on the G2 of the cell cycle. Two putative insulin ligands are especially important for the germline role but do not mediate signaling in other tissues. A picture is emerging of a complex web of developmentally and temporally restricted, ligand- and tissue-specific responses to insulin signaling. Avenues for future studies include the regulation of specific insulin-like ligands and the mechanisms for tissue-specific responses to them.

I. Germline Proliferation in *C. elegans*: A Model for Developmental, Physiological, and Environmental Control of Cell Proliferation

Germline proliferation in *Caenorhabditis elegans* is a powerful model for understanding the control of cell proliferation and stem cell biology in general (see reviews by Hansen and Schedl, 2006; Joshi *et al.*, 2010; Kimble and Crittenden, 2005, 2007). Unlike somatic cells in *C. elegans*, the germ line is not under cell lineage constraints and therefore shares features with tissue growth in other organisms. Beyond cell cycle regulation itself, this model is particularly amenable to exploring molecular mechanisms that underlie changes in proliferation rate and in the proliferation/differentiation balance in response to developmental, physiological, and environmental cues (Korta and Hubbard, 2010). Under replete laboratory conditions, *C. elegans* develop to adulthood in ~3 days and produce prodigious broods of ~300 self-progeny over the subsequent 3–5 days. In poor environmental conditions, they can reversibly arrest their developmental program at several stages prior to reproduction and thereby delay reproduction and improve their chances of reproducing in a more favorable environment. Once they have committed to a reproductive mode of development, they can further modulate the proliferation of the germ line. Several well-conserved genetic and cellular mechanisms alter germline proliferation and hence reproductive capacity. Among these, we focus here on the insulin signaling pathway, an important modulator of germline proliferation.

II. *C. elegans* Germline Development

C. elegans hermaphrodite germline development was described by Hirsh *et al.* (1976) and has been extensively reviewed since (see WormBook: http://www.wormbook.org/toc_germline.html for recent reviews). The reader is directed to those reviews as only a brief description is provided here.

The germ line develops over the course of four larval stages (L1–L4) and continues to proliferate in the adult. *C. elegans* exist as either males or hermaphrodites. The latter are modified females that produce and store sperm late in the last larval stage before undergoing a switch to produce exclusively oocytes as adults. Male *C. elegans* produce only sperm and reproduce by mating with hermaphrodites. Unlike many other species, the number of sperm is limiting for brood size. Hence, hermaphrodites in the absence of males will produce broods based on the number of self-made sperm whereas mated hermaphrodites can produce additional progeny.

The hermaphrodite germ line proliferates over the course of development, ultimately establishing a pool of "cells" that remain undifferentiated and divide to give rise to gametes. The germ cell nuclei are surrounded by their own cytoplasm but retain an opening to a central core of cytoplasm. Thus, while they are technically syncytial, they behave as individual cells. For example, adjacent germ nuclei do not divide in synchrony, suggesting that cyclin components are not shared between them. Hence, the field refers to them as germ "cells" for simplicity.

The development of the germ line takes place during the entire larval period of four larval stages and into the adult stage. During early larval stages (L1, L2, and early L3), all germ cells appear undifferentiated (Fig. 3.1). Also during this time, the somatic gonad rearranges such that germ cells are separated into two gonad "arms." Each arm is blind-ended, capped by a distal tip cell (DTC) at the distal end, and open to a common uterus proximally. In the mid-L3 stage, the proximal-most cells reach a distance of 13 cell diameters from the distal tip and germ cells beyond this distance show the first overt signs of meiotic entry (a change from round to crescent-shaped chromatin morphology visible with DAPI or positive staining for anti-HIM-3, an early meiotic fate marker). This zone of early differentiation (leptotene and zygotene of prophase of meiosis I) is called the "transition zone." This first appearance of transition nuclei, an event we call "initial meiosis," sets up the pattern of proliferation and differentiation in the germ line. Once initial meiosis occurs, cells distal to the border are undifferentiated ("proliferative" or "mitotic" zone) and those proximal to it are differentiated (in meiosis and gametogenesis). Although definitive stem cell markers are not yet available in this system, the proliferative zone acts as a stem cell pool and the cells therein can be considered "progenitor" cells in

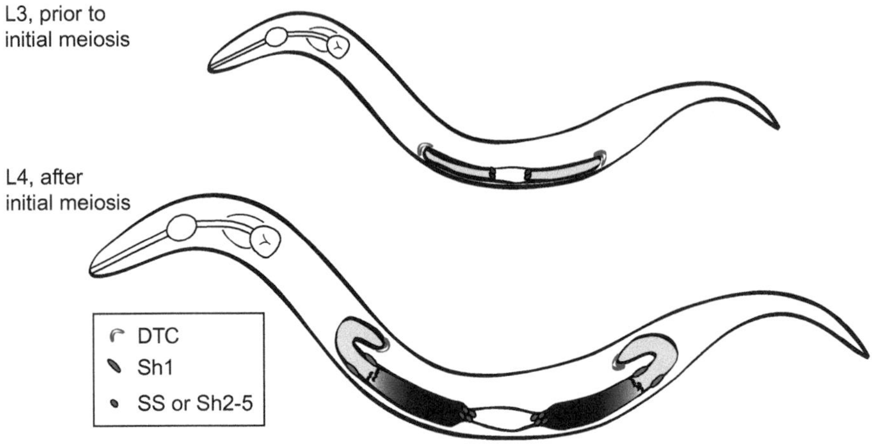

Figure 3.1 Cartoon depictions of L3 and early L4 stage *C. elegans* hermaphrodites. Bilobed gonad is indicated with proliferative germ cells in light gray, meiotic cells dark gray to black. Crescent-shaped nuclei characteristic of the transition zone are depicted. White central portion indicates position of future spermathecae and uterine cells. Distal tip cells are shown in red and distal and proximal sheath cells in lighter and darker blue, respectively. (See Color Insert.)

the sense that they are either stem cells or the "transit amplifying" proliferative progeny of stem cells. The proliferative zone continues to grow from its first establishment in the L3 such that by the early adult there are ~200 nuclei in the zone and the transition zone begins at a distance of ~20–22 cell diameters from the distal tip.

In the late L4, sperm are produced, followed by a switch to oocyte production in adulthood. Mature sperm are pushed into the spermatheca by the first ovulating oocyte where they subsequently reside. Oocytes are fertilized one by one in an assembly like fashion and zygotes initiate development in uterus. Sperm can also be introduced by mating with males that exclusively produce sperm.

III. The *C. elegans* Germ line "Proliferation Versus Differentiation" Decision Is Mediated from the Soma to the Germ line by a Conserved Notch Signaling Pathway

The role of the Notch pathway in *C. elegans* germline development and its downstream effectors in the germ line have been extensively reviewed elsewhere (Byrd and Kimble, 2009; Hansen and Schedl, 2006; Kimble and Crittenden, 2007). Briefly, the somatic DTC is required to maintain the proliferative zone under wild-type conditions in all stages of development. The DTC produces two DSL family ligands: LAG-2 is expressed from the earliest stages, and then APX-1 is expressed in addition starting in the L3 (Nadarajan *et al.*, 2009). These ligands signal to the Notch family receptor GLP-1 in the germ line. Interference with DTC signaling by killing the DTC or by mutations that prevent Notch signaling cause all proliferative germ cells to differentiate. Moreover, Notch signaling is required continuously. That is, manipulations that interfere with this activity any time during the life of the worm will cause all germ cells to differentiate. Interestingly, reducing (but not removing) Notch signaling shortens the distance between the DTC and the transition zone, but does not change the rate of cell proliferation, as measured by mitotic index (Michaelson *et al.*, 2010). In contrast, elevating Notch activity causes germ cells to remain undifferentiated further from the distal end, thereby expanding the proliferative zone. Elevation of Notch activity also elevates the mitotic index in germ cells beyond the first few cell diameters from the DTC (Maciejowski *et al.*, 2006). The need for Notch signaling for the proliferative fate can be bypassed by interfering with the activity of redundant germline-autonomous downstream factors, including two pathways named for their components GLD-1 and GLD-2 that normally promote the meiotic fate and/or interfere with proliferation (see reviews referenced above). In the wild type, however, regardless of environmental conditions, the germ line requires signaling from the DTC to maintain the undifferentiated state and hence the proliferative zone.

IV. Evidence for Notch-Independent Soma-Germline Signaling Mechanisms That Modulate Germline Proliferation

We discovered genes that are required for robust larval germline proliferation but that do not act in the DTC by analysis of mutants from genetic screens that identified "*pro*" genes (Killian and Hubbard, 2004; Voutev *et al.*, 2006). For *pro-1*, the activity was pinpointed to the sheath lineage of the somatic gonad (Fig. 3.1). In addition, time-course analysis of cell–cell interactions and a series of cell-ablation experiments narrowed down the proliferation-promoting activity of the sheath to the distal-most pair of sheath cells (Sh1 cells). Together, our results support a model in which the distal Sh1 cells support or promote robust larval germline proliferation. We therefore became interested in factors that would act in concert with or parallel to Notch signaling to modulate the rate or extent of proliferation, especially in the larval germ line when the proliferative zone is growing to establish the adult progenitor/stem cell pool.

Two immediate questions emerged from these discoveries: (1) what is the molecular basis for the germline proliferation-promoting activity of the distal Sh1 sheath and (2) what mechanisms account for the rapid proliferation of the germ line in larval stages relative to adult stages? The most straightforward genetic approach to address these questions in *C. elegans* would be to identify mutants that reduce but do not eliminate germline proliferation. Such approaches are suboptimal in this circumstance, however, since many housekeeping genes are required for germline proliferation. Instead, we sought a means to identify genes that were not required for germline proliferation *per se* but were required for robust proliferation of the larval germ line.

V. A Counter-Intuitive Assay to Indentify Potential Notch-Independent Mechanisms That Promote the Expansion of the Larval Germline Progenitor Pool

We used a somewhat counterintuitive strategy to identify genes that would promote larval germline proliferation but the loss of which not interfere with proliferation in general. In short, we knew that reducing larval proliferation (either by *pro* gene mutation or by ablation of distal sheath cells) had the secondary effect of delaying initial meiosis. This is due to the expandable nature of the gonad as it grows during larval stages. The distance between the DTC and the gonad center is a function of two processes: a DTC-intrinsic migration program that is at least partially dependent on *gon-1* and *hlh-12/mig-24*, and on the number of germ cells. Importantly, if too-few germ cells are generated in larval stages, the DTC does not move centripetally at the correct rate. As a result, initial meiosis is delayed since Notch signaling exerts its influence over the entire population of germ cells that are now "stuck" within its range of influence. This scenario eventually results in a heterochronic disjunct between the germ line and the somatic gonad in which germ line development is delayed while somatic gonad development continues normally (Fig. 3.2; Killian and Hubbard, 2005; McGovern *et al.*, 2009).

We further discovered a bizarre but opportune twist of development that provided the necessary assay. Certain somatic gonadal cells that are born adjacent to the *proximal* germ line express DSL family ligands that activate Notch receptor (McGovern *et al.*, 2009). Under normal conditions, these cells are born and express these ligands after initial meiosis has occurred in the germ line such that differentiated germ cells act as a barrier preventing inappropriate signaling to proliferative germ cells that express GLP-1/Notch on their surface. If, however, initial meiosis is delayed for any reason, including failure of robust larval germline proliferation, undifferentiated germ cells are inappropriately juxtaposed to ligand-producing proximal

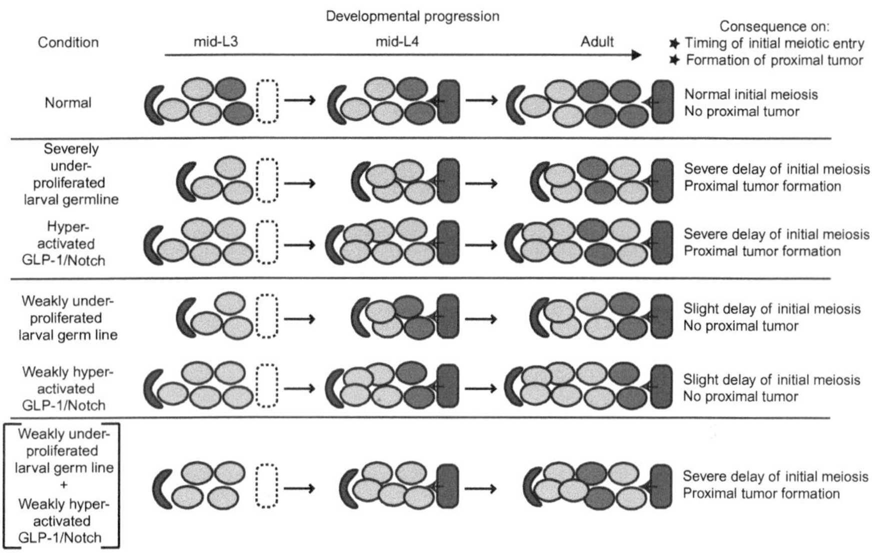

Figure 3.2 Latent niche concept and related screening strategy. See text for details. (For color version of this figure, the reader is referred to the Web version of this chapter.)

somatic gonad cells. Undifferentiated germ cells express the GLP-1/Notch receptor and therefore respond to the new source of ligands and form a proximally located ectopic "tumor" of undifferentiated germ cells. Thus, counter-intuitively, interfering with genes that are required for timely and robust germline proliferation in larval stages delays initial meiosis and can cause the formation of a proximal germline tumor later in development.

This scenario provided a convenient screening strategy. The etiology of the proximal tumor phenotype suggested that we could identify key control genes while ensuring that germ cells were still capable of proliferation, allowing us to identify genes required for robust larval germline proliferation—that is, key developmental or physiological regulators—while avoiding genes required for all aspects of cell proliferation. We reasoned that if we combined two independent mechanisms to slightly delay meiotic entry without causing tumor formation, one germline-autonomous and one potentially sheath-autonomous, they might together cause a severe delay in meiotic entry that would cause tumor formation. To test this hypothesis, we started with a mutant background that causes a temperature-dependent delay in initial meiosis, *glp-1(ar202)*. In this mutant, initial meiosis is severely delayed at the restrictive temperature of 25 °C and proximal tumors form. At the lower permissive temperature of 15 °C, however, initial meiosis is only slightly delayed and tumors do not form since meiotic entry occurs prior to the birth of the proximal sheath cells and the inappropriate cell–cell interaction does not occur. When we combined this mutant background with laser-surgical removal of the distal sheath cells, however, proximal tumor formation occurs at the permissive of 15 °C (Fig. 3.2; Killian and Hubbard, 2005). Thus, if we interfere with the activity of genes that, like the distal sheath cells, are required for robust larval germline proliferation, they too should cause proximal tumor formation in this mutant background at the permissive temperature. In genetic parlance, such mutations would be termed enhancers of the proximal tumor phenotype (Pro) of *glp-1(ar202)*, or enhancers of Pro. Because this screening strategy requires germ cell proliferation in the proximal tumor as a read-out, it was unlikely to identify genes required for cell proliferation *per se*.

VI. Identification of the Insulin/IGF-Like Receptor (IIR) Pathway Role in Germline Proliferation

Using this strategy, we tested mutations in genes representing several candidate signaling pathways looking for mutations that would cause a tumor to form in the sensitized *glp-1(ar202)* genetic background at the permissive temperature. Reducing the function of *daf-2,* the gene encoding the sole insulin/IGF-like receptor (IIR) ortholog in *C. elegans* (Fig. 3.3) markedly elevated the percentage of worms that formed a proximal tumor at the permissive temperature. As summarized below, subsequent

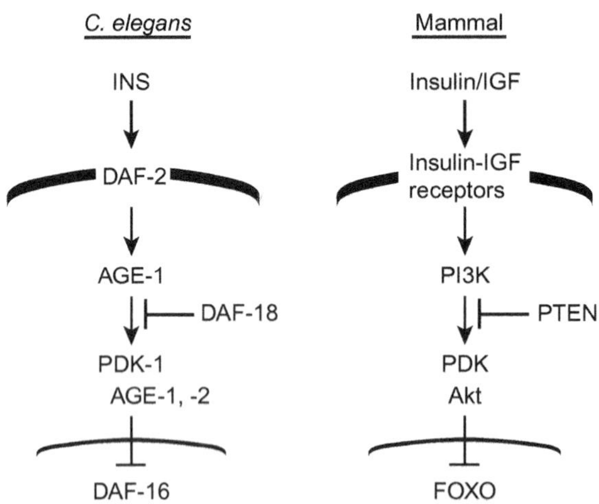

Figure 3.3 Insulin/IGF signaling pathways in *C. elegans* and mammals. Corresponding proteins or classes of proteins are indicated. Broad black line indicates the cell membrane and thin line indicates the nuclear membrane.

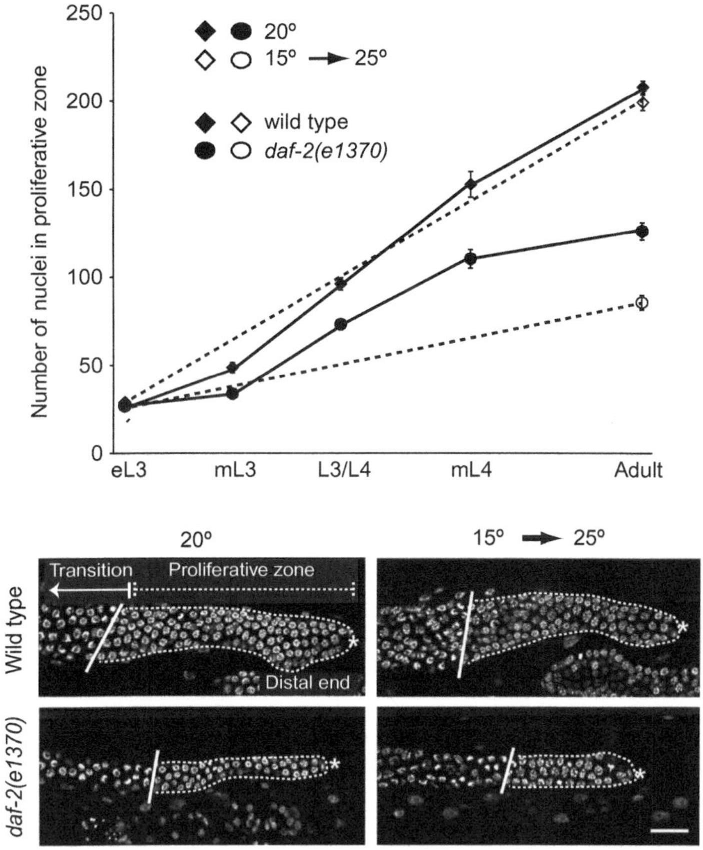

Figure 3.4 Influence of DAF-2 activity on the proliferative zone. Top: number of nuclei in the proliferative zone in gonad arms of animals of the indicated ages in wild-type and in a *daf-2* mutant under two temperature conditions, continuous growth at 20 °C or a shift from 15 to 25 °C. Bottom: images of distal germline nuclei of young adult wild-type and *daf-2* mutants. Positions of the proliferative and transition zones are indicated. Asterisk marks the distal end in all panels. Scale bar indicates 20 μm. Figure modified from (Michaelson *et al.*, 2010). (For color version of this figure, the reader is referred to the Web version of this chapter.)

experiments confirmed that the number of larval cells in the proliferative zone was decreased in the mutants alone (Fig. 3.4).

VII. IIR Signaling in *C. elegans*

The roles of the conserved IIR pathway in *C. elegans* have been extensively reviewed elsewhere (Fielenbach and Antebi, 2008; Kenyon, 2011, 2010; Kleemann and Murphy, 2009; Landis and Murphy, 2010).

Briefly, the *daf-2* gene was initially discovered and named for its role in the dauer decision. Dauer is a special pre-reproductive larval stage that worms enter if they encounter poor environmental conditions. The decision to enter dauer occurs in the late L1 and is largely dependent on an assessment of population density and food availability. Certain mutations in *daf-2* cause worms to enter dauer constitutively, regardless of the environmental conditions. DAF-2 has also been well studied for its role in modulation of lifespan. While it is best characterized for its roles in dauer formation and lifespan, insulin signaling in worms also affects reproductive timing, metabolism, response to toxins, hypoxia, immunity, and L1 diapause. In each case, IIR signaling acts through a highly conserved PI3K-dependent pathway and negatively regulates the activity of a FOXO transcription factor DAF-16. The relationship between IIR signaling and FOXO was discovered in *C. elegans*. Though *daf-2*/IR and *daf-16*/FOXO exhibit both cell autonomous and nonautonomous activity, both are required primarily in neurons for dauer control while intestinal activity of *daf-16*/ FOXO accounts for most of the lifespan phenotype (Apfeld and Kenyon, 1998; Libina *et al.*, 2003).

VIII. Insulin Signaling Promotes the Larval Germline Cell Division Cycle

Our analysis of *daf-2*/IIR mutants required that we avoid early larval arrest and dauer arrest phenotypes (Michaelson *et al.*, 2010). Therefore, we carried out analyses using temperature-sensitive mutations. We examined the germ line under both semipermissive conditions (20 °C, when few animals enter dauer) and after a shift to the restrictive temperature after the dauer decision (L3 shift to 25 °C). We found that the number of adult germ cell progenitors was significantly reduced in the *daf-2*/IIR mutants and that the reduction occurred during the time that the population of these cells normally expands (Fig. 3.4). Moreover, we found that this phenotype could be completely suppressed by simultaneously reducing the activity of *daf-18*/ PTEN or *daf-16*/FOXO, indicating that the canonical pathway is used in this context. We examined cell cycle parameters in the larval germ line and established that reducing *daf-2* activity reduced the proportion of cells in M-phase and in S-phase. Moreover, measurements of the DNA content of larval germ nuclei indicated a shift to the 4N content, suggesting a delay in the G2 of the cell cycle. Finally, by tissue-specific restriction of gene activity, transgene expression and mosaic analysis, we determined that the receptor pathway and its downstream components are largely required in the germ line to promote robust larval proliferation. Taken together, our studies support the model that the DAF-2/IIR signaling cascade is reused after the dauer decision within the germ line to promote progression through the G2.

IX. *C. elegans* Insulins

Despite the high degree of conservation between IIR signaling in many organisms, some features of the *C. elegans* pathway are noteworthy. First, although there appears to be only one gene encoding an IIR (*daf-2*), at least 40 genes encode putative ligands, some of which behave as antagonists, and some as agonists. Interestingly, while few of the putative insulin peptides have been characterized, they appear to largely mediate different activities of the insulin receptor. For example, the ligands *ins-1*, *ins-6*, *ins-7*, and *daf-28* are important and dynamically regulated for the *daf-2* role in the dauer decision (Cornils *et al.*, 2011). Recent studies suggest a widely dynamic use of these ligands during many stages of development (Baugh *et al.*, 2011)

To identify insulin-like ligands that might be important for promoting larval germline proliferation, we used the same general screening strategy described above, targeting each ligand gene individually by RNAi (Michaelson *et al.*, 2010). Given the distinct possibility of redundancy among the ligand-encoding genes, we relied on the extreme sensitivity of the genetic assay to help identify weak effects. We identified two ligand genes, *ins-3* and *ins-33* that, when knocked down individually, caused a phenotype similar to that of *daf-2* at the semipermissive temperature. Reporters for these two genes are expressed in head neurons, and head and uterine cells, respectively, and their RNAi phenotypes are dependent on the activity of *rrf-1*, a gene required for somatic, but not germline RNAi. Therefore, they are required in the soma to influence germline proliferation. Interestingly, reduction of activity of each gene alone yields the same phenotype as reducing both together. This intriguing result is hard to reconcile with a simple ligand–receptor interaction model and future studies will be required to determine precisely how these ligands are working. One hypothesis is that a physical interaction is required between them; an alternate hypothesis is that a ligand–receptor relay exists, possibly including an alternate receptor. Indeed, biochemical and functional assays for tissue-specific aspects of ligand–receptor interaction would advance the field.

X. Many Target Tissues for IIR Signaling

In a curious parallel to the distribution of ligands associated with distinct phenotypes, the activity of the FOXO transcription factor *daf-16* also appears to be required more stringently in some tissue types versus others, depending on the phenotype. For example, for dauer it is required primarily in neurons and for lifespan primarily in the intestine (Kwon *et al.*, 2010; Libina *et al.*, 2003). For the role of the IIR signaling pathway in promoting larval germline proliferation, the target appears to be the germ line itself. We found that replacing *daf-16*/FOXO in neurons or intestine did not rescue the *daf-16* mutant phenotype (in this case, suppression of *daf-2* mutant defects) though replacing it in the muscle had a mild effect. However, replacing it in the germ line had the strongest effect. Despite being tagged with GFP, the transgene used for these assays was not visible in germ cell nuclei, so we could not assess the change in localization of *daf-16* upon removal of *daf-2* in the germ line. Moreover, the transgene was very difficult to generate and has, indeed, been lost to germline silencing. Newer methods that allow for single-copy insertions should facilitate the generation of a more stable transgene for future studies.

This result that *daf-16* activity is required in the germ line for the larval germline proliferation phenotype was important for several reasons. First, it gave us the site of action of the receptor pathway and thereby provides a target tissue for approaches to identify downstream targets. Second, it helped distinguish between the activity of insulin signaling in the dauer decision and in reproductive larval stages. During the dauer stage, the germ line does not proliferate well (Narbonne and Roy, 2006). Therefore, in addition to establishing that the effect of *daf-2* on the germ line could occur after the dauer decision by temperature shifts, we established that the receptor pathway is required in a different tissue. Second, it gave us a way to determine whether the distal sheath cells exert their proliferation-promoting function by way of the insulin pathway. For this experiment, we performed the sheath ablation under conditions that fully reversed the *daf-2* dauer and germline proliferation defects using potent, independently verified *daf-16(RNAi)* and asked whether the sheath ablation still had any effect. This experiment is reported in Michaelson *et al.* (2010) and is summarized in Fig. 3.5. The results clearly indicated that the IIR signaling pathway cannot be the sole mediator of the sheath effect since depleting *daf-16* did not prevent the deleterious effects of the sheath ablation on the proliferative zone.

Given that we had identified ligands that appeared to mediate the effects of *daf-2* on the germ line but not on other phenotypes, we investigated the effects of the "larval germline phenotype" ligands, *ins-3* and *ins-33* on *daf-16* activity (Michaelson *et al.*, 2010). We knew that the germline proliferation defect resulting from reducing *ins-3* or *ins-33* activity could be completely suppressed by loss of *daf-16*/FOXO. We therefore wondered whether reducing the activity of these ligands would modulate the localization of DAF-16 in all tissues. For these assays, we used the intestine, where nuclear localization of a *daf-16::GFP* in response to loss of *daf-2* activity is well-documented. Our results were striking: depleting *ins-3* and *ins-33* is not equivalent to reducing *daf-2* activity as measured by nuclear localization of *daf-16* or target gene activation in the intestine. That is, while *daf-2* depletion caused nuclear localization of DAF-16 and target gene activation in the intestine, depletion of *ins-3* or *ins-33* did not. This result provides further

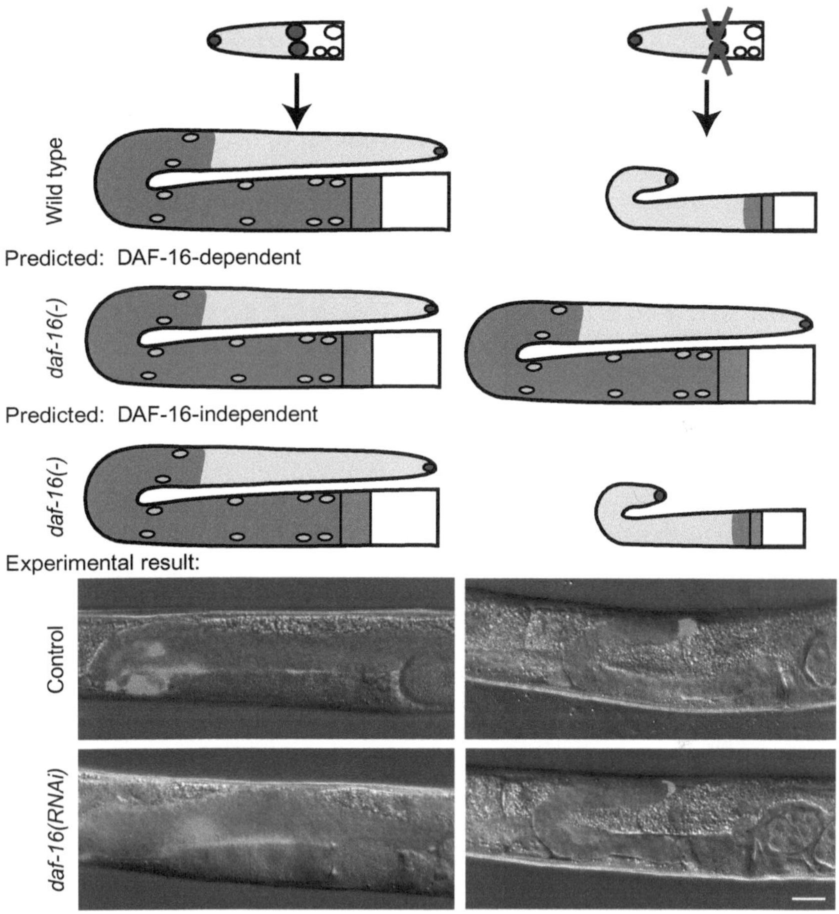

Figure 3.5 Experimental strategy to test dependence of sheath cell-ablation phenotype on *daf-16* activity. Left: unablated conditions. Right: ablation conditions. See text for details. Bottom: Sheath cells and distal tip cell are marked with GFP. Scale bar indicates 20 μm. Modified from Michaelson *et al.* (2010). (For color version of this figure, the reader is referred to the Web version of this chapter.)

evidence for complexity between ligand-receptor interaction in the context of specific target tissues.

XI. Other Germline Roles for the IIR Pathway

Insulin signaling also influences the germ line in earlier stages of development, most notably in the L1 and during dauer. In L1 larvae that hatch in the absence of food both somatic and germline development are arrested in a *daf-18*/PTEN dependent manner (Baugh and Sternberg, 2006; Fukuyama *et al.*, 2006). In contrast to somatic cell arrest, however, germline cell cycle arrest occurs in the G2, and is not dependent on *daf-16*. In addition, germline proliferation is kept in check during dauer by pathways that regulate the dauer decision (IIR and TGFβ signaling) and this repression of proliferation requires *daf-18*/PTEN and the AMPK orthologs *aak-1* and *aak-2* (Narbonne and Roy, 2006).

XII. IIR Role in Larval Germline Proliferation: A Reproductive Timing and Lifespan Connection?

Insulin signaling influences other aspects of worm life history that potentially intersect with germline proliferation: reproductive timing and lifespan. Our studies to date suggest that these roles for the insulin pathway are separable from its role in promoting larval germline proliferation. Reproductive timing refers to the age and span of time over which *C. elegans* produce progeny. Dillin *et al.* (2002) described a *daf-2* mutant phenotype in which progeny production spreads over approximately twice the time-frame of adulthood compared to wild-type animals. With respect to aging, loss of IIR pathway activity extends lifespan. Another way to extend lifespan is to eliminate germ cells or curtail germ cell proliferation. However, this manipulation leads to an additive effect on lifespan extension, suggesting separate mechanisms (Mukhopadhyay and Tissenbaum, 2007). Interestingly, Dillin *et al.* (2002) reported that the reproductive timing phenotype is dependent on activity of the IIR pathway prior to adulthood even though progeny are produced exclusively in adulthood, while the lifespan phenotype is sensitive only in the adult period. While the degree of larval germ cell proliferation could conceivably delay and extend reproductive timing, we found that *ins-3* and *ins-33* do not influence reproductive timing. We also found that the role of IIR signaling in promoting a more rapid cell cycle is limited to the larval stages, not the adult stage. Together these data do not support a causal link between the role of IIR in larval germline proliferation and its role in reproductive timing or lifespan regulation.

XIII. A Current Model and Future Directions

Figure 3.6 depicts a model for the control of larval germline proliferation in which Notch signaling is required primarily for the mitosis/meiosis cell fate decision and independent inputs from the distal sheath and insulin ligands help amplify the size of the germline progenitor pool during larval stages.

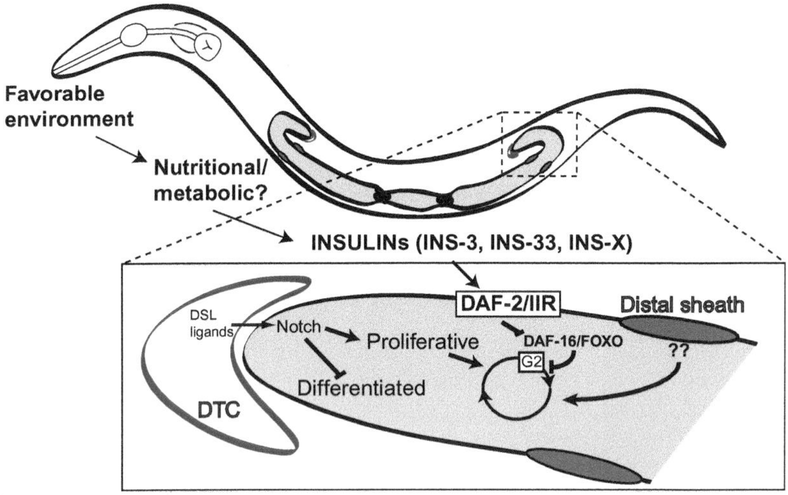

Figure 3.6 Model for the control of larval germline proliferation in response to GLP-1/Notch, DAF-2/IIR, and distal sheath signals. (For color version of this figure, the reader is referred to the Web version of this chapter.)

Nutrition impacts many facets of reproduction and cell proliferation in animals (Pinelli and Tagliabue, 2007). Moreover, the insulin pathway mediates nutrition-sensitive signaling in a variety of contexts in animal development and disease. Our findings suggest that *C. elegans* germline proliferation is a useful model to investigate the mechanistic connection between nutrition and the control of cell proliferation. Further, despite the evolutionary distance between *C. elegans* and mammals, many of the signaling pathways that control cell proliferation are conserved. Therefore, worm studies offer the possibility of broadly applicable results. Future studies may also provide insights into the nutritional control of cell proliferation in humans with implications for fertility, cancer, and stem cell biology.

There are many interesting questions remaining to be addressed in this system. At the organismal level, how is diet or metabolic signaling linked to the insulin pathway? Our model for insulin signaling in the germ line parallels that seen in Drosophila, suggesting that this role may be highly conserved across species. In particular, there is conservation of the features that multiple inputs that impinge on cell cycle, and that IIR signaling acts on the G2. In Drosophila, the work of the Drummond-Barbosa lab (Drummond-Barbosa, 2008) has connected this role to diet, particularly to protein levels. In certain mouse cancer models, short-term caloric restriction reduces tumor growth in an insulin/IGF-PI3K signaling-dependent manner (Kalaany and Sabatini, 2009). The molecular mechanisms by which nutritionally sensitive molecular pathways control germ cell proliferation are best studied in a whole-organism context. *C. elegans* is a compelling and facile model for these studies.

Additional mechanistic questions include: How is ligand activity regulated such that different responses occur in different tissues? Do INS-3 and INS-33 bind DAF-2 at the germline membrane? How is it that depleting *ins-3* and *ins-33* activity causes identical phenotypes that are not further enhanced by loss of both genes? Since loss of these ligands does not cause as strong a phenotype as loss of receptor activity, what other ligands are involved? What are downstream targets of *daf-16*/FOXO in the germ line?

These and many other related questions will provide much work for the future and will likely contribute to our understanding of conserved connections between insulin/IGF signaling and cell proliferation.

ACKNOWLEDGMENTS

Support for this work is from NIH R01 GM061706. Thanks to members of my laboratory, especially David Michaelson and Dorota Korta, for discussion and comments on the manuscript.

REFERENCES

Apfeld, J., and Kenyon, C. (1998). Cell nonautonomy of C. elegans daf-2 function in the regulation of diapause and life span. *Cell* **95,** 199–210.

Baugh, L. R., and Sternberg, P. W. (2006). DAF-16/FOXO regulates transcription of cki-1/Cip/Kip and repression of lin-4 during C. elegans L1 arrest. *Curr. Biol.* **16,** 780–785.

Baugh, L. R., Kurhanewicz, N., and Sternberg, P. W. (2011). Sensitive and precise quantification of insulin-like mRNA expression in Caenorhabditis elegans. *PLoS One* **6,** e18086.

Byrd, D. T., and Kimble, J. (2009). Scratching the niche that controls Caenorhabditis elegans germline stem cells. *Semin. Cell Dev. Biol.* **20**(9), 1–7.

Cornils, A., Gloeck, M., Chen, Z., Zhang, Y., and Alcedo, J. (2011). Specific insulin-like peptides encode sensory information to regulate distinct developmental processes. *Development* **138,** 1183–1193.

Dillin, A., Crawford, D. K., and Kenyon, C. (2002). Timing requirements for insulin/IGF-1 signaling in C. elegans. *Science* **298,** 830–834.

Drummond-Barbosa, D. (2008). Stem cells, their niches and the systemic environment: An aging network. *Genetics* **180,** 1787–1797.

Fielenbach, N., and Antebi, A. (2008). C. elegans dauer formation and the molecular basis of plasticity. *Genes Dev.* **22,** 2149–2165.

Fukuyama, M., Rougvie, A. E., and Rothman, J. H. (2006). C. elegans DAF-18/PTEN mediates nutrient-dependent arrest of cell cycle and growth in the germline. *Curr. Biol.* **16,** 773–779.

Hansen, D., and Schedl, T. (2006). The regulatory network controlling the proliferation-meiotic entry decision in the Caenorhabditis elegans germ line. *Curr. Top. Dev. Biol.* **76,** 185–215.

Hirsh, D., Oppenheim, D., and Klass, M. (1976). Development of the reproductive system of *Caenorhabditis elegans*. *Dev. Biol.* **49,** 200–219.

Joshi, P. M., Riddle, M. R., Djabrayan, N. J. V., and Rothman, J. H. (2010). *Caenorhabditis elegans* as a model for stem cell biology. *Dev. Dyn.* **239,** 1539–1554.

Kalaany, N. Y., and Sabatini, D. M. (2009). Tumours with PI3K activation are resistant to dietary restriction. *Nature* **458,** 725–731.

Kenyon, C. J. (2010). The genetics of ageing. *Nature* **464,** 504–512.

Kenyon, C. (2011). The first long-lived mutants: Discovery of the insulin/IGF-1 pathway for ageing. *Philos. Trans. R. Soc. Lond. B Biol. Sci.* **366,** 9–16.

Killian, D. J., and Hubbard, E. J. A. (2004). C. elegans pro-1 activity is required for soma/germline interactions that influence proliferation and differentiation in the germ line. *Development* **131,** 1267–1278.

Killian, D. J., and Hubbard, E. J. A. (2005). *Caenorhabditis elegans* germline patterning requires coordinated development of the somatic gonadal sheath and the germ line. *Dev. Biol.* **279,** 322–335.

Kimble, J., and Crittenden, S. L. (2005). Germline proliferation and its control. *WormBook* 1–14.

Kimble, J., and Crittenden, S. L. (2007). Controls of germline stem cells, entry into meiosis, and the sperm/oocyte decision in Caenorhabditis elegans. *Annu. Rev. Cell Dev. Biol.* **23,** 405–433.

Kleemann, G. A., and Murphy, C. T. (2009). The endocrine regulation of aging in Caenorhabditis elegans. *Mol. Cell. Endocrinol.* **299,** 51–57.

Korta, D. Z., and Hubbard, E. J. A. (2010). Soma-germline interactions that influence germline proliferation in Caenorhabditis elegans. *Dev. Dyn.* **239,** 1449–1459.

Kwon, E. S., Narasimhan, S. D., Yen, K., and Tissenbaum, H. A. (2010). A new DAF-16 isoform regulates longevity. *Nature* **466,** 498–502.

Landis, J. N., and Murphy, C. T. (2010). Integration of diverse inputs in the regulation of Caenorhabditis elegans DAF-16/FOXO. *Dev. Dyn.* **239,** 1405–1412.

Libina, N., Berman, J. R., and Kenyon, C. (2003). Tissue-specific activities of C. elegans DAF-16 in the regulation of lifespan. *Cell* **115,** 489–502.

Maciejowski, J., Ugel, N., Mishra, B., Isopi, M., and Hubbard, E. J. A. (2006). Quantitative analysis of germline mitosis in adult C. elegans. *Dev. Biol.* **292,** 142–151.

McGovern, M., Voutev, R., Maciejowski, J., Corsi, A. K., and Hubbard, E. J. A. (2009). A "latent niche" mechanism for tumor initiation. *Proc. Natl. Acad. Sci. USA* **106,** 11617–11622.

Michaelson, D., Korta, D. Z., Capua, Y., and Hubbard, E. J. A. (2010). Insulin signaling promotes germline proliferation in C. elegans. *Development* **137,** 671–680.

Mukhopadhyay, A., and Tissenbaum, H. A. (2007). Reproduction and longevity: Secrets revealed by C. elegans. *Trends Cell Biol.* **17,** 65–71.

Nadarajan, S., Govindan, J. A., McGovern, M., Hubbard, E. J. A., and Greenstein, D. (2009). MSP and GLP-1/Notch signaling coordinately regulate actomyosin-dependent cytoplasmic streaming and oocyte growth in C. elegans. *Development* **136,** 2223–2234.

Narbonne, P., and Roy, R. (2006). Inhibition of germline proliferation during C. elegans dauer development requires PTEN, LKB1 and AMPK signalling. *Development* **133,** 611–619.

Pinelli, G., and Tagliabue, A. (2007). Nutrition and fertility. *Minerva Gastroenterol. Dietol.* **53,** 375–382.

Voutev, R., Killian, D. J., Ahn, J. H., and Hubbard, E. J. A. (2006). Alterations in ribosome biogenesis cause specific defects in C. elegans hermaphrodite gonadogenesis. *Dev. Biol.* **298,** 45–58.

CHAPTER FOUR

Generating Mature β-Cells From Embryonic Stem Cells: Strategies for Late-Stage Differentiation

Spyridon Champeris Tsaniras

Contents

I. Introduction 79
II. Signaling Pathways in β-Cell Differentiation 81
  A. PI3K signaling and the generation of β-cells 81
  B. Hedgehog signaling and the generation of β-cells 82
  C. Calcineurin/NFAT signaling and the generation of β-cells 83
  D. Epac signaling and the generation of β-cells 83
  E. BMP signaling and the generation of β-cells 84
III. Summary and Conclusions 86
Acknowledgments 87
References 87

Abstract

Generation of mature pancreatic β-cells from embryonic stem (ES) cells *in vitro* could provide a therapy for insulin-dependent diabetes mellitus. Recent ES cell differentiation protocols have improved the differentiation efficiency toward β-cells by recapitulating *in vivo* pancreatic development. Toward this end, there is a large number of developmental and β-cell functional studies that could guide the design of more efficient differentiation protocols, predominantly during the late stages. In this review, I have classified these studies according to the intracellular signaling pathways they relate to—phosphatidylinositol 3-kinase, Hedgehog, Calcineurin/NFAT, Epac, and bone morphogenetic protein.

I. Introduction

Type 1 (insulin-dependent) diabetes mellitus is caused by the autoimmune destruction of the insulin-producing β-cells which are present exclusively in the pancreatic islets of Langerhans (Eiselein *et al.*, 2004). The clinical

Bionetwork Ltd, Chalandri, Athens, Greece

Vitamins and Hormones, Volume 87
ISSN 0083-6729, DOI: 10.1016/B978-0-12-386015-6.00025-1

symptoms of diabetes appear when insulin production is no longer sufficient to maintain whole-body glucose homeostasis resulting in hyperglycemia. In the United States, type 1 diabetes affects ~20 per 100,000 individuals each year with a steady increase in incidence rates (Centers for Disease Control and Prevention, 2011). Worldwide, the increase is around 3% yearly (Onkamo *et al.*, 1999) and prevalence is increasing both in developed and developing countries (Bruno and Landi, 2011).

Since the successful isolation of biologically active insulin in the early 1920s, insulin therapy has remained the main treatment for this disorder. However, it is now known that the long-term dysregulation of blood glucose levels associated with insulin-treated diabetes contributes toward many of the secondary complications of diabetes, including cardiovascular disease, cerebral and peripheral vascular disease, renal disease, autonomic and peripheral neuropathy, as well as visual impairment which can progress to blindness (Bluestone *et al.*, 2010; Eiselein *et al.*, 2004).

As a result, research has focused on providing a permanent cure for diabetes by replenishing the functional β-cell mass. Islet transplantation has been successfully applied to a handful of patients and allowed them to become insulin-independent (Ryan *et al.*, 2005; Shapiro *et al.*, 2000), but its use has been severely limited by the scarcity of donor tissue. Other sources of β-cells have therefore been proposed and embryonic stem (ES) cells have so far provided the most promising results.

Research in this area was initiated a decade ago, when Soria *et al.* (2000) first reported the generation of insulin-producing cells from mouse ES cells using genetic manipulation, followed by Lumelsky *et al.* (2001) who published the first differentiation protocol for obtaining insulin-producing cells *in vitro*. Since then, several studies built on these early approaches and improved the differentiation efficiency (e.g., see Hori *et al.*, 2002; Ku *et al.*, 2004). However, it was not until 2006 when D'Amour and colleagues designed a differentiation protocol that mimicked the developmental stages of pancreatic organogenesis and derived insulin-expressing cells from human ES cells with a very high insulin content. This strategy was based on employing growth factors together with small molecules and was subsequently adopted by several other groups who have since followed developmental cues in order to drive differentiation toward β-cells (Blyszczuk and Wobus, 2007; Jiang *et al.*, 2007*a,b*; Nostro *et al.*, 2011; Schroeder *et al.*, 2006; Zhang *et al.*, 2009).

Nevertheless, no study has so far managed to generate functionally mature β-cells. D'Amour *et al.* (2006) generated endocrine progenitors that responded only minimally to glucose. In a more recent study, Zhang *et al.* (2009) reported the generation of mature insulin-producing cells from human ES cells that, upon glucose stimulation, released c-peptide (a byproduct of insulin) at relatively high levels (~50% lower than adult β-cells) but did not perform any further insulin-response assays and/or electrophysiological tests. It is also worth noting that no study has yet yielded cells with insulin gene expression levels close to those of mature β-cells.

It is clear that, even though differentiation protocols have dramatically improved over the past few years, recapitulating the final stages of β-cell development has yet to be achieved *in vitro*. Nevertheless, there are late-stage developmental evidence and β-cell functional studies that could aid in the design of more sophisticated and efficient differentiation protocols. This review will attempt to classify these studies by focusing on the intracellular signaling pathways that underpin them; namely the phosphatidylinositol 3-kinase (PI3K), Hedgehog, Calcineurin/NFAT, Epac, and bone morphogenetic protein (BMP).

II. Signaling Pathways in β-Cell Differentiation

A. PI3K signaling and the generation of β-cells

The family of phosphatidylinositol 3-kinases (PI3Ks) is implicated in a plethora of cellular responses, including cell growth and differentiation, proliferation, migration, and apoptosis. Upon activation, the enzymes produce phosphoinositides which act as second messenger molecules and transmit the signals downstream to various effector molecules such as AKT and BTK (Qiu and Kung, 2000; Vanhaesebroeck *et al.*, 2001).

Even though PI3K activators have been employed in the final stages of several ES pancreatic differentiation protocols (Cho *et al.*, 2008; D'Amour *et al.*, 2006; Jiang *et al.*, 2007a; Ku *et al.*, 2004; Zhang *et al.*, 2009), the pathway has not been specifically targeted in these studies. In other words, compounds such as IGF, EGF, and exendin-4, which are known to act through multiple signaling pathways (Butler *et al.*, 1998; Hui *et al.*, 2009; Montrose-Rafizadeh *et al.*, 1999; Sabri *et al.*, 2011), have been widely used in ES cell studies but it was not until recently that these compounds were reported to exert most of their prodifferentiating effects through PI3K activation (Koblas *et al.*, 2010; Watanabe *et al.*, 2008). PI3Ks are believed to promote β-cell differentiation by phosphorylating the FoxO1 transcription factor, resulting in its translocation from the nucleus into the cytoplasm (Biggs *et al.*, 1999; Kops *et al.*, 1999). This enhances the expression of transcription factors critical for β-cell development and mature β-cell function such as *Pdx1* (Kitamura *et al.*, 2002), *Ngn3*, *Nkx6.1*, and insulin (Al-Masri *et al.*, 2010).

However, the role of PI3Ks in β-cell maturation is far from clear. In a study using human fetal pancreatic cells, Ptasznik *et al.* (1997) reported that suppressing this pathway enhanced the endocrine differentiation of islet precursors. This result appears conflicting but it might indicate that, PI3K repression also has a role to play during late-stage β-cell differentiation.

Nevertheless, the requisite for active PI3K signaling in adult β-cell function is unambiguous. Several studies have shown that PI3Ks are regulated by insulin and are implicated in transmission of insulin signaling, including glucose transport and insulin-specific gene activation (Backer *et al.*, 1993; Kaburagi *et al.*, 1999; Ohsugi *et al.*, 2004; Soos *et al.*, 2001). It has been suggested that PI3Ks regulate several aspects of the β-cell stimulus–response coupling, such as K_{ATP} channel and L-type Ca^{2+} channel activity, and endocytosis and exocytosis (reviewed by Barker *et al.*, 2002). In this regard, a sustained activation of the pathway during terminal differentiation of ES cells to β-cells could possibly improve the insulin secretory response.

B. Hedgehog signaling and the generation of β-cells

The Hedgehog signaling pathway is a key regulator of cell growth and differentiation during embryonic development. Hedgehog ligands, namely Sonic, Indian, and Desert Hedgehog (Echelard *et al.*, 1993), are secreted proteins known to control patterning and morphogenesis of developing tissues in a concentration- and time-dependant manner. They are understood to bind to Patched1 receptors in target cells leading to activation of Smoothened, a G-protein coupled receptor. This results in nuclear translocation of Gli transcription factors, which regulate the expression of Hedgehog-responsive genes (Ingham and McMahon, 2001; Varjosalo and Taipale, 2008).

Downregulation of Sonic Hedgehog (Shh) signaling during early pancreatic development has long been known to permit *Pdx1* and insulin expression, thereby allowing specification of the pancreatic domain (Hebrok *et al.*, 1998). In this regard, Shh inhibitors, such as bFGF and cyclopamine, have been used extensively in ES cell studies to promote a pancreatic fate (reviewed by Champeris Tsaniras and Jones, 2010).

However, it is now known that Shh is also implicated in the formation of the pancreatic epithelium, endocrine cell development, and adult β-cell function. In a recent study, Lau and Hebrok (2010) reported that downregulation of Hedgehog signaling in the mouse pancreatic epithelium resulted in delayed epithelial expansion and β-cell differentiation. In line with transcriptional profiling studies (Hebrok *et al.*, 2000; Kawahira *et al.*, 2003), these findings suggest that low-level Hedgehog signaling is required for proper pancreas morphogenesis. Moreover, this pathway is highly active in mature β-cells (Lau and Hebrok, 2010; Thomas *et al.*, 2000) where its disruption has been shown to impact on their function. In a study by Thomas *et al.* (2000), inhibition of Hedgehog signaling by cyclopamine in the INS-1 and MIN6 insulinoma cell lines was reported to diminish transcription of the insulin gene in a dose-dependent manner, with the highest dose dramatically reducing expression by 80%. Conversely, ectopic Shh activation increased the transcriptional activation of the insulin promoter, thereby increasing insulin gene expression. In another study, downregulation of Hedgehog signaling in murine primary β-cells *in vivo* was reported to cause a 40% reduction in insulin gene expression and protein levels whereas cells exhibited increased insulin sensitivity (Lau and Hebrok, 2010). The mechanism through which

Hedgehog regulates these processes is not yet clear, with some evidence suggesting that the pathway directly activates the insulin promoter (Lau and Hebrok, 2010) whereas others have reported that it acts indirectly through *Pdx1* activation (Thomas *et al.*, 2001).

In line with the above, a low-level activation of the Hedgehog pathway after *Pdx1* activation and formation of the pancreatic domain would be expected to enhance pancreatic differentiation. During the final stage of differentiation, the level of Hedgehog signaling could be significantly increased, in order to increase insulin promoter activation and gene expression.

C. Calcineurin/NFAT signaling and the generation of β-cells

The Calcineurin/NFAT signaling pathway was first identified almost two decades ago (Clipstone and Crabtree, 1992), but its study has since been hampered by the high level of genetic redundancy among its family members. Calcineurin is a serine/threonine phosphatase that is activated upon intracellular Ca^{2+} entry (Stemmer and Klee, 1994). It facilitates dephosphorylation of the cytoplasmic subunits of the NFAT transcription factors (NFATc) resulting in their nuclear translocation (Beals *et al.*, 1997) and subsequent regulation of target genes. This pathway has been implicated in patterning and differentiation mainly by regulating cell-cell interactions (Crabtree and Olson, 2002).

Calcineurin/NFAT signaling has recently emerged as an important regulator of β-cell function. NFATc proteins have been shown to associate with cis-regulatory elements of insulin, *Glut2* and *Pdx1*, directly influencing their expression. According to Heit *et al.* (2006), disruption of this pathway in primary β-cells *in vivo* downregulated the *Glut2* glucose sensor and insulin as well as *Pdx1*, *MafA*, and *NeuroD1*, which are involved in regulating the forementioned genes. This resulted in reduced insulin synthesis and impaired secretion. Conversely, active NFATc was reported to have the opposite effects (Heit *et al.*, 2006). In line with these, other studies have reported that the side-effects of FK506, a drug that is known to impair β-cell insulin production, are exerted via the Calcineurin/NFAT signaling pathway (Lawrence *et al.*, 2001; Redmon *et al.*, 1996). In this regard, activation of this pathway during the final stages of an ES cell pancreatic differentiation protocol might increase insulin expression and improve glucose sensitivity.

D. Epac signaling and the generation of β-cells

Epac, an abbreviation for exchange protein directly activated by cAMP, belongs to a family of guanine nucleotide exchange factors and has only recently emerged as an important cAMP effector. cAMP is a second messenger produced by adenylate cyclase upon receptor activation. Its effects are exerted via protein kinase A (PKA), cAMP-gated ion channels, or Epac. Two variants of Epac have been identified (Epac1 and Epac2; de Rooij *et al.*, 1998; Kawasaki *et al.*, 1998) whose downstream signaling cascades are not completely understood but can be broadly classified into Rap GTPase-dependant and -independent pathways (Holz *et al.*, 2006). Recent studies have linked Epac signaling to diverse cellular functions, ranging from differentiation and cellular outgrowth (Kiermayer *et al.*, 2005) to ion transport (Honegger *et al.*, 2006) and exocytosis (Eliasson *et al.*, 2003).

Several studies have linked this pathway to the regulation of insulin secretion in pancreatic β-cells. Epac is known to induce insulin exocytosis in a glucose-dependent manner (Hashiguchi *et al.*, 2006; Kang *et al.*, 2003) but the mechanisms behind this are not clearly understood. It is believed to act through multiple mechanisms; by exerting direct (Kang *et al.*, 2006) and indirect (Tsuboi *et al.*, 2003) inhibitory effects on K_{ATP} channel activity, by mediating Ca^{2+}-induced Ca^{2+} release from the endoplasmic reticulum (Kang *et al.*, 2001, 2003), by regulating components of the exocytotic apparatus (Fujimoto *et al.*, 2002; Ozaki *et al.*, 2000) and by establishing pulsatility in insulin secretion (Idevall-Hagren *et al.*, 2010).

In this regard, Epac has been suggested to enhance β-cell differentiation from ES cells. Li *et al.* (2010) recently reported that high dosage of exendin-4, a GLP-1 analog known to activate Epac in mature β-cells (Kang *et al.*, 2001; Kashima *et al.*, 2001), induced maturation of insulin-producing cells from mouse ES cells by increasing early insulin release (EIR). In this study, exendin-4 was shown to increase *Epac1* and *Epac2* expression, among others, suggesting that differentiation was at least partly driven by activation of the Epac signaling pathway. *SUR1*, a K_{ATP} channel subunit that is essential for Epac-mediated K_{ATP} channel inhibition (Kang *et al.*, 2006) as well as for priming of the insulin secretory granules (Eliasson *et al.*, 2003), was also upregulated, further supporting a prodifferentiating role for Epac. Therefore, if these results are reproducible, Epac activation during the final differentiation steps will become an indispensable tool for improving glucose-responsive insulin secretion.

E. BMP signaling and the generation of β-cells

BMPs belong to the TGF-β superfamily and were initially named after their ability to induce *de novo* formation of bone (Urist, 1965; Wozney *et al.*, 1998). Later studies unveiled a diverse role for BMPs in multiple aspects of embryonic development and adult tissue homeostasis (Goumans and Mummery, 2000). BMP signaling is mediated mainly through the canonical pathway, which involves serine/threonine kinase receptors inducing prosphorylation of receptor-regulated SMAD proteins (SMADS 1, 5, or 8). The latter, together with a common mediator SMAD (SMAD4), form a complex which is transported into the nucleus to regulate transcription of target genes (reviewed by Nohe *et al.*, 2004).

During development, restriction of BMP signaling in the dorsal endoderm is required for establishing a pancreatic fate (Chung *et al.*, 2008; Rossi *et al.*, 2001; Spagnoli and Brivanlou, 2008) and BMP inhibitors such as Noggin have been used in several ES cell pancreatic differentiation protocols toward this end (Kroon *et al.*, 2008; Mfopou *et al.*, 2010; Zhang *et al.*, 2009). Thereafter, developmental evidence suggests that further pancreatic differentiation is associated with active BMP signaling. More specifically, disruption of the BMP pathway has been reported to abrogate *Pdx1* induction and impair expression of *Hnf6* (a marker of pancreatic epithelial cells; Jacquemin *et al.*, 2000), in mouse embryos and quail-chick chimeras (Kumar *et al.*, 2003; Wandzioch and Zaret, 2009). In line with these, a recent study reported that BMP2 administration enhanced pancreatic differentiation of mouse ES cells by upregulating *Pdx1* and insulin expression (Talavera-Adame *et al.*, 2011). Similarly, Jiang *et al.* (2002) have shown that BMP4, 5 and 6 promote differentiation of murine fetal pancreatic epithelial cells into insulin-positive colonies whereas Yew *et al.* (2005) reported that insulin-positive differentiation of a pancreatic exocrine cell line (AR42J) with exendin-4 entails upregulation of *Smad1*, *Smad8*, *Alk1* (SMAD1 and SMAD8 receptor/activator), and endogenous *Bmp2*. In addition, BMP signaling has also been implicated in adult β cell function as reduced BMP4/BMP receptor 1A signaling in murine β cells results in decreased expression of several essential factors including insulin and the proinsulin processing enzymes *PC1/3* and *PC2* as well as *Pdx1*, *Glut2*, and the glucokinase regulatory enzyme, negatively impacting on glucose-stimulated insulin secretion (GSIS). Conversely, exogenous BMP4 administration significantly increases GSIS and improves glucose tolerance (Goulley *et al.*, 2007).

However, BMP signaling has also been reported to inhibit pancreatic endocrine differentiation. Chung *et al.* (2010) reported that in zebrafish the BMP signaling pathway has to be downregulated in endodermal cells and islet precursors to allow for β-cell differentiation. Moreover, Hua *et al.* (2006) reported that BMP4 induces binding of Id2 (inhibitor of DNA binding protein 2) to NeuroD, a transcription factor required for pancreatic endocrine differentiation, and therefore inhibits differentiation in the AR42J cell line. In line with these, a recent human ES cell differentiation study reported that BMP inhibition after *Pdx1* induction resulted in an impressive 19-fold increase in insulin gene expression (Nostro *et al.*, 2011). All these results seem to be conflicting with the developmental and *in vitro* studies described in the previous paragraph. A possible explanation might relate to the distinct responses of different cell types employed in these studies (primary cell cultures vs. cancer cell lines) and/or interspecies variations. However, these discrepancies might also indicate the existence of a temporal window in endocrine cell development during which BMP signaling exerts a suppressive role.

Based on the above, manipulation of the BMP signaling pathway might enhance differentiation toward mature β-cells. After an initial downregulation and establishment of a pancreatic fate, BMP signaling can be upregulated in order to increase *Pdx1* and *Hnf6* expression. It is also worth investigating whether a subsequent downregulation, in the endocrine progenitor stage, can enhance differentiation according to the "temporal window" effects described previously. Thereafter, evidence is in favor of activating BMP signaling in order to increase expression of critical β-cell transcription factors and enhance GSIS.

III. Summary and Conclusions

Recent pancreatic ES cell differentiation protocols demonstrate that mimicking pancreatic organogenesis offers a promising strategy for β-cell generation. However, recapitulating the final stages of β-cell development has yet to be achieved *in vitro*. In this review, I suggest that this might be accomplished by focusing on the intracellular signaling pathways that guide late-stage developmental transitions and govern adult β-cell function; namely the PI3K, Hedgehog, Calcineurin/NFAT, Epac, and BMP pathways. Recent studies suggest that these pathways could promote β cell maturation at the terminal stage of differentiation. Evidence also suggests that Shh must be suppressed to allow for pancreatic endoderm formation whereas its subsequent activation could promote the generation of islet precursors. Conversely, active BMP signaling would be expected to drive differentiation toward the pancreatic endoderm stage. In addition, a temporal window during which PI3K and BMP have the opposite effects might exist after the islet precursor stage but before terminal differentiation, possibly warranting their downregulation. In the case of BMP, this might extend before the specification of islet precursors (Fig. 4.1). Taking all these

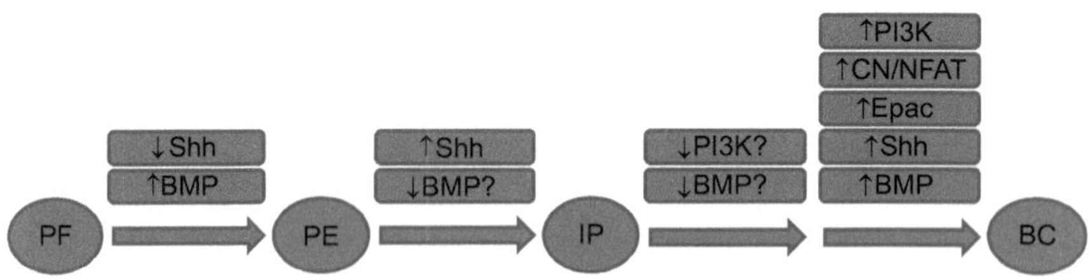

Figure 4.1 Proposed manipulations of the signaling pathways for generating mature β-cells from ES cells during the late stages of differentiation, based on developmental evidence and β-cell functional studies. PF, posterior foregut; PE, pancreatic endoderm; IP, islet precursors; BC, β-cells; CN/NFAT, Calcineurin/NFAT. The role of PI3K and BMP pathway during specific time-points is ambiguous (see question marks). (For color version of this figure, the reader is referred to the Web version of this chapter.)

into account and better understanding β-cell development and physiology will be of paramount importance in the quest for functionally mature *bona fide* β-cells.

ACKNOWLEDGMENTS

I would like to thank Prof. Peter Jones and Dr. Stephen Minger for allowing me to work in their lab and teaching me the skills necessary to become a molecular biologist.

Declaration of interest. The author discloses no potential conflict of interest related to this review chapter.

Funding. The generation of this chapter did not receive any funding from any private or public source.

REFERENCES

Al-Masri, M., Krishnamurthy, M., Li, J., Fellows, G. F., Dong, H. H., Goodyer, C. G., and Wang, R. (2010). Effect of forkhead box O1 (FOXO1) on beta cell development in the human fetal pancreas. *Diabetologia* **53,** 699–711.

Backer, J. M., Myers, M. G., Jr., Sun, X. J., Chin, D. J., Shoelson, S. E., Miralpeix, M., and White, M. F. (1993). Association of IRS-1 with the insulin receptor and the phosphatidylinositol 3'-kinase. Formation of binary and ternary signaling complexes in intact cells. *J. Biol. Chem.* **268,** 8204–8212.

Barker, C. J., Leibiger, I. B., Leibiger, B., and Berggren, P. O. (2002). Phosphorylated inositol compounds in beta-cell stimulus–response coupling. *Am. J. Physiol. Endocrinol. Metab.* **283,** E1113–E1122.

Beals, C. R., Clipstone, N. A., Ho, S. N., and Crabtree, G. R. (1997). Nuclear localization of NF-ATc by a calcineurin-dependent, cyclosporin-sensitive intramolecular interaction. *Genes Dev.* **11,** 824–834.

Biggs, W. H., 3rd, Meisenhelder, J., Hunter, T., Cavenee, W. K., and Arden, K. C. (1999). Protein kinase B/Akt-mediated phosphorylation promotes nuclear exclusion of the winged helix transcription factor FKHR1. *Proc. Natl. Acad. Sci. USA* **96,** 7421–7426.

Bluestone, J. A., Herold, K., and Eisenbarth, G. (2010). Genetics, pathogenesis and clinical interventions in type 1 diabetes. *Nature* **464,** 1293–1300.

Blyszczuk, P., and Wobus, A. M. (2007). *In vitro* differentiation of embryonic stem cells into the pancreatic lineage. *Methods Mol. Biol.* **330,** 373–385.

Bruno, G., and Landi, A. (2011). Epidemiology and costs of diabetes. *Transplant. Proc.* **43,** 327–329.

Butler, A. A., Yakar, S., Gewolb, I. H., Karas, M., Okubo, Y., and LeRoith, D. (1998). Insulin-like growth factor-I receptor signal transduction: At the interface between physiology and cell biology. *Comp. Biochem. Physiol. B Biochem. Mol. Biol.* **121,** 19–26.

Centers for Disease Control and Prevention (2011). National Diabetes Fact Sheet: National estimates and general information on diabetes and prediabetes in the United States, 2011. [Fact sheet] U.S. Department of Health and Human Services, Centers for Disease Control and Prevention, Atlanta, GA.

Champeris Tsaniras, S., and Jones, P. M. (2010). Generating pancreatic beta-cells from embryonic stem cells by manipulating signaling pathways. *J. Endocrinol.* **206,** 13–26.

Cho, Y. M., Lim, J. M., Yoo, D. H., Kim, J. H., Chung, S. S., Park, S. G., Kim, T. H., Oh, S. K., Choi, Y. M., Moon, S. Y., Park, K. S., and Lee, H. K. (2008). Betacellulin and

nicotinamide sustain PDX1 expression and induce pancreatic beta-cell differentiation in human embryonic stem cells. *Biochem. Biophys. Res. Commun.* **366,** 129–134.
Chung, W. S., Shin, C. H., and Stainier, D. Y. (2008). Bmp2 signaling regulates the hepatic versus pancreatic fate decision. *Dev. Cell* **15,** 738–748.
Chung, W. S., Andersson, O., Row, R., Kimelman, D., and Stainier, D. Y. (2010). Suppression of Alk8-mediated Bmp signaling cell-autonomously induces pancreatic beta-cells in zebrafish. *Proc. Natl. Acad. Sci. USA* **107,** 1142–1147.
Clipstone, N. A., and Crabtree, G. R. (1992). Identification of calcineurin as a key signalling enzyme in T-lymphocyte activation. *Nature* **357,** 695–697.
Crabtree, G. R., and Olson, E. N. (2002). NFAT signaling: Choreographing the social lives of cells. *Cell* **109,** S67–S79.
D'Amour, K. A., Bang, A. G., Eliazer, S., Kelly, O. G., Agulnick, A. D., Smart, N. G., Moorman, M. A., Kroon, E., Carpenter, M. K., and Baetge, E. E. (2006). Production of pancreatic hormone-expressing endocrine cells from human embryonic stem cells. *Nat. Biotechnol.* **24,** 1392–1401.
de Rooij, J., Zwartkruis, F. J., Verheijen, M. H., Cool, R. H., Nijman, S. M., Wittinghofer, A., and Bos, J. L. (1998). Epac is a Rap1 guanine-nucleotide-exchange factor directly activated by cyclic AMP. *Nature* **396,** 474–477.
Echelard, Y., Epstein, D. J., St-Jacques, B., Shen, L., Mohler, J., McMahon, J. A., and McMahon, A. P. (1993). Sonic Hedgehog, a member of a family of putative signaling molecules, is implicated in the regulation of CNS polarity. *Cell* **75,** 1417–1430.
Eiselein, L., Schwartz, H. J., and Rutledge, J. C. (2004). The challenge of type 1 diabetes mellitus. *ILAR J.* **45,** 231–236.
Eliasson, L., Ma, X., Renström, E., Barg, S., Berggren, P. O., Galvanovskis, J., Gromada, J., Jing, X., Lundquist, I., Salehi, A., Sewing, S., and Rorsman, P. (2003). SUR1 regulates PKA-independent cAMP-induced granule priming in mouse pancreatic B-cells. *J. Gen. Physiol.* **121,** 181–197.
Fujimoto, K., Shibasaki, T., Yokoi, N., Kashima, Y., Matsumoto, M., Sasaki, T., Tajima, N., Iwanaga, T., and Seino, S. (2002). Piccolo, a Ca2+ sensor in pancreatic beta-cells. Involvement of cAMP-GEFII.Rim2. Piccolo complex in cAMP-dependent exocytosis. *J. Biol. Chem.* **277,** 50497–50502.
Goulley, J., Dahl, U., Baeza, N., Mishina, Y., and Edlund, H. (2007). BMP4-BMPR1A signaling in beta cells is required for and augments glucose-stimulated insulin secretion. *Cell Metab.* **5,** 207–219.
Goumans, M. J., and Mummery, C. (2000). Functional analysis of the TGFbeta receptor/Smad pathway through gene ablation in mice. *Int. J. Dev. Biol.* **44,** 253–265.
Hashiguchi, H., Nakazaki, M., Koriyama, N., Fukudome, M., Aso, K., and Tei, C. (2006). Cyclic AMP/cAMP-GEF pathway amplifies insulin exocytosis induced by Ca2+ and ATP in rat islet beta-cells. *Diabetes Metab. Res. Rev.* **22,** 64–71.
Hebrok, M., Kim, S. K., and Melton, D. A. (1998). Notochord repression of endodermal Sonic Hedgehog permits pancreas development. *Genes Dev.* **12,** 1705–1713.
Hebrok, M., Kim, S. K., St Jacques, B., McMahon, A. P., and Melton, D. A. (2000). Regulation of pancreas development by hedgehog signaling. *Development* **127,** 4905–4913.
Heit, J. J., Apelqvist, A. A., Gu, X., Winslow, M. M., Neilson, J. R., Crabtree, G. R., and Kim, S. K. (2006). Calcineurin/NFAT signalling regulates pancreatic beta-cell growth and function. *Nature* **443,** 345–349.
Holz, G. G., Kang, G., Harbeck, M., Roe, M. W., and Chepurny, O. G. (2006). Cell physiology of cAMP sensor Epac. *J. Physiol.* **577,** 5–15.
Honegger, K. J., Capuano, P., Winter, C., Bacic, D., Stange, G., Wagner, C. A., Biber, J., Murer, H., and Hernando, N. (2006). Regulation of sodium-proton exchanger isoform 3 (NHE3) by PKA and exchange protein directly activated by cAMP (EPAC). *Proc. Natl. Acad. Sci. USA* **103,** 803–808.

Hori, Y., Rulifson, I. C., Tsai, B. C., Heit, J. J., Cahoy, J. D., and Kim, S. K. (2002). Growth inhibitors promote differentiation of insulin-producing tissue from embryonic stem cells. *Proc. Natl. Acad. Sci. USA* **99,** 16105–16110.

Hua, H., Zhang, Y. Q., Dabernat, S., Kritzik, M., Dietz, D., Sterling, L., and Sarvetnick, N. (2006). BMP4 regulates pancreatic progenitor cell expansion through Id2. *J. Biol. Chem.* **281,** 13574–13580.

Hui, H., Tang, Y., Zhu, L., Khoury, N., Hui, Z., Wang, K., Perfetti, R., and Go, V. (2009). Glucagon like peptide-1-directed human embryonic stem cells differentiation into insulin-producing cells via Hedgehog, cAMP, and PI3K pathways. *Pancreas* **39,** 315–322.

Idevall-Hagren, O., Barg, S., Gylfe, E., and Tengholm, A. (2010). cAMP mediators of pulsatile insulin secretion from glucose-stimulated single beta-cells. *J. Biol. Chem.* **285,** 23007–23018.

Ingham, P. W., and McMahon, A. P. (2001). Hedgehog signaling in animal development: Paradigms and principles. *Genes Dev.* **15,** 3059–3087.

Jacquemin, P., Durviaux, S. M., Jensen, J., Godfraind, C., Gradwohl, G., Guillemot, F., Madsen, O. D., Carmeliet, P., Dewerchin, M., Collen, D., Rousseau, G. G., and Lemaigre, F. P. (2000). Transcription factor hepatocyte nuclear factor 6 regulates pancreatic endocrine cell differentiation and controls expression of the proendocrine gene ngn3. *Mol. Cell. Biol.* **20,** 4445–4454.

Jiang, F. X., Stanley, E. G., Gonez, L. J., and Harrison, L. C. (2002). Bone morphogenetic proteins promote development of fetal pancreas epithelial colonies containing insulin-positive cells. *J. Cell Sci.* **115,** 753–760.

Jiang, J., Au, M., Lu, K., Eshpeter, A., Korbutt, G., Fisk, G., and Majumdar, A. S. (2007a). Generation of insulin-producing islet-like clusters from human embryonic stem cells. *Stem Cells* **25,** 1940–1953.

Jiang, W., Shi, Y., Zhao, D., Chen, S., Yong, J., Zhang, J., Qing, T., Sun, X., Zhang, P., Ding, M., Li, D., and Deng, H. (2007b). *In vitro* derivation of functional insulin-producing cells from human embryonic stem cells. *Cell Res.* **17,** 333–344.

Kaburagi, Y., Yamauchi, T., Yamamoto-Honda, R., Ueki, K., Tobe, K., Akanuma, Y., Yazaki, Y., and Kadowaki, T. (1999). The mechanism of insulin-induced signal transduction mediated by the insulin receptor substrate family. *Endocr. J.* **46,** S25–S34.

Kang, G., Chepurny, O. G., and Holz, G. G. (2001). cAMP-regulated guanine nucleotide exchange factor II (Epac2) mediates Ca^{2+}-induced Ca^{2+} release in INS-1 pancreatic beta-cells. *J. Physiol.* **536,** 375–385.

Kang, G., Joseph, J. W., Chepurny, O. G., Monaco, M., Wheeler, M. B., Bos, J. L., Schwede, F., Genieser, H. G., and Holz, G. G. (2003). Epac-selective cAMP analog 8-pCPT-2'-O-Me-cAMP as a stimulus for Ca2+—induced Ca2+ release and exocytosis in pancreatic beta-cells. *J. Biol. Chem.* **278**(10), 8279–8285.

Kang, G., Chepurny, O. G., Malester, B., Rindler, M. J., Rehmann, H., Bos, J. L., Schwede, F., Coetzee, W. A., and Holz, G. G. (2006). cAMP sensor Epac as a determinant of ATP-sensitive potassium channel activity in human pancreatic beta cells and rat INS-1 cells. *J. Physiol.* **573,** 595–609.

Kashima, Y., Miki, T., Shibasaki, T., Ozaki, N., Miyazaki, M., Yano, H., and Seino, S. (2001). Critical role of cAMP-GEFII–Rim2 complex in incretin-potentiated insulin secretion. *J. Biol. Chem.* **276,** 46046–46053.

Kawahira, H., Ma, N. H., Tzanakakis, E. S., McMahon, A. P., Chuang, P. T., and Hebrok, M. (2003). Combined activities of hedgehog signaling inhibitors regulate pancreas development. *Development* **130,** 4871–4879.

Kawasaki, H., Springett, G. M., Mochizuki, N., Toki, S., Nakaya, M., Matsuda, M., Housman, D. E., and Graybiel, A. M. (1998). A family of cAMP-binding proteins that directly activate Rap1. *Science* **282,** 2275–2279.

Kiermayer, S., Biondi, R. M., Imig, J., Plotz, G., Haupenthal, J., Zeuzem, S., and Piiper, A. (2005). Epac activation converts cAMP from a proliferative into a differentiation signal in PC12 cells. *Mol. Biol. Cell* **16,** 5639–5648.

Kitamura, T., Nakae, J., Kitamura, Y., Kido, Y., Biggs, W. H., 3rd, Wright, C. V., White, M. F., Arden, K. C., and Accili, D. (2002). The forkhead transcription factor Foxo1 links insulin signaling to Pdx1 regulation of pancreatic beta cell growth. *J. Clin. Invest.* **110,** 1839–1847.

Koblas, T., Zacharovova, K., Berkova, Z., Girman, P., and Saudek, F. (2010). An acidic pH and activation of phosphoinositide 3-kinase stimulate differentiation of pancreatic progenitors into insulin-producing cells. *Transplant. Proc.* **42,** 2075–2080.

Kops, G. J., de Ruiter, N. D., De Vries-Smits, A. M., Powell, D. R., Bos, J. L., and Burgering, B. M. (1999). Direct control of the Forkhead transcription factor AFX by protein kinase B. *Nature* **398,** 630–634.

Kroon, E., Martinson, L. A., Kadoya, K., Bang, A. G., Kelly, O. G., Eliazer, S., Young, H., Richardson, M., Smart, N. G., Cunningham, J., Agulnick, A. D., D'Amour, K. A., *et al.* (2008). Pancreatic endoderm derived from human embryonic stem cells generates glucose-responsive insulin-secreting cells in vivo. *Nat. Biotechnol.* **26,** 443–452.

Ku, H. T., Zhang, N., Kubo, A., O'Connor, R., Mao, M., Keller, G., and Bromberg, J. S. (2004). Committing embryonic stem cells to early endocrine pancreas *in vitro*. *Stem Cells* **22,** 1205–1217.

Kumar, M., Jordan, N., Melton, D., and Grapin-Botton, A. (2003). Signals from lateral plate mesoderm instruct endoderm toward a pancreatic fate. *Dev. Biol.* **259,** 109–122.

Lau, J., and Hebrok, M. (2010). Hedgehog signaling in pancreas epithelium regulates embryonic organ formation and adult beta-cell function. *Diabetes* **59,** 1211–1221.

Lawrence, M. C., Bhatt, H. S., Watterson, J. M., and Easom, R. A. (2001). Regulation of insulin gene transcription by a Ca(2+)-responsive pathway involving calcineurin and nuclear factor of activated T cells. *Mol. Endocrinol.* **15,** 1758–1767.

Li, H., Lam, A., Xu, A. M., Lam, K. S., and Chung, S. K. (2010). High dosage of Exendin-4 increased early insulin secretion in differentiated beta cells from mouse embryonic stem cells. *Acta Pharmacol. Sin.* **31,** 570–577.

Lumelsky, N., Blondel, O., Laeng, P., Velasco, I., Ravin, R., and McKay, R. (2001). Differentiation of embryonic stem cells to insulin-secreting structures similar to pancreatic islets. *Science* **292,** 1389–1394.

Mfopou, J. K., Chen, B., Mateizel, I., Sermon, K., and Bouwens, L. (2010). Noggin, retinoids, and fibroblast growth factor regulate hepatic or pancreatic fate of human embryonic stem cells. *Gastroenterology* **138,** 2233–2245.

Montrose-Rafizadeh, C., Avdonin, P., Garant, M. J., Rodgers, B. D., Kole, S., Yang, H., Levine, M. A., Schwindinger, W., and Bernier, M. (1999). Pancreatic glucagon-like peptide-1 receptor couples to multiple G proteins and activates mitogen-activated protein kinase pathways in Chinese hamster ovary cells. *Endocrinology* **140,** 1132–1140.

Nohe, A., Keating, E., Knaus, P., and Petersen, N. O. (2004). Signal transduction of bone morphogenetic protein receptors. *Cell. Signal.* **16,** 291–299.

Nostro, M. C., Sarangi, F., Ogawa, S., Holtzinger, A., Corneo, B., Li, X., Micallef, S. J., Park, I. H., Basford, C., Wheeler, M. B., Daley, G. Q., Elefanty, A. G., *et al.* (2011). Stage-specific signaling through TGFβ family members and WNT regulates patterning and pancreatic specification of human pluripotent stem cells. *Development* **138,** 861–871.

Ohsugi, M., Cras-Méneur, C., Zhou, Y., Warren, W., Bernal-Mizrachi, E., and Permutt, M. A. (2004). Glucose and insulin treatment of insulinoma cells results in transcriptional regulation of a common set of genes. *Diabetes* **53,** 1496–1508.

Onkamo, P., Väänänen, S., Karvonen, M., and Tuomilehto, J. (1999). Worldwide increase in incidence of type I diabetes—The analysis of the data on published incidence trends. *Diabetologia* **42,** 1395–1403.

Ozaki, N., Shibasaki, T., Kashima, Y., Miki, T., Takahashi, K., Ueno, H., Sunaga, Y., Yano, H., Matsuura, Y., Iwanaga, T., Takai, Y., and Seino, S. (2000). cAMP-GEFII is a direct target of cAMP in regulated exocytosis. *Nat. Cell Biol.* **2,** 805–811.

Ptasznik, A., Beattie, G. M., Mally, M. I., Cirulli, V., Lopez, A., and Hayek, A. (1997). Phosphatidylinositol 3-kinase is a negative regulator of cellular differentiation. *J. Cell Biol.* **137,** 1127–1136.

Qiu, Y., and Kung, H. J. (2000). Signaling network of the Btk family kinases. *Oncogene* **19,** 5651–5661.

Redmon, J. B., Olson, L. K., Armstrong, M. B., Greene, M. J., and Robertson, R. P. (1996). Effects of tacrolimus (FK506) on human insulin gene expression, insulin mRNA levels, and insulin secretion in HIT-T15 cells. *J. Clin. Invest.* **98,** 2786–2793.

Rossi, J. M., Dunn, N. R., Hogan, B. L., and Zaret, K. S. (2001). Distinct mesodermal signals, including BMPs from the septum transversum mesenchyme, are required in combination for hepatogenesis from the endoderm. *Genes Dev.* **15,** 1998–2009.

Ryan, E. A., Paty, B. W., Senior, P. A., Bigam, D., Alfadhli, E., Kneteman, N. M., Lakey, J. R., and Shapiro, A. M. (2005). Five-year follow-up after clinical islet transplantation. *Diabetes* **54,** 2060–2069.

Sabri, A., Ziaee, A. A., Ostad, S. N., Alimoghadam, K., and Ghahremani, M. H. (2011). Crosstalk of EGF-directed MAPK signalling pathways and its potential role on EGF-induced cell proliferation and COX-2 expression in human mesenchymal stem cells. *Cell Biochem. Funct.* **29,** 64–70.

Schroeder, I. S., Rolletschek, A., Blyszczuk, P., Kania, G., and Wobus, A. M. (2006). Differentiation of mouse embryonic stem cells to insulin-producing cells. *Nat. Protoc.* **1,** 495–507.

Shapiro, A. M., Lakey, J. R., Ryan, E. A., Korbutt, G. S., Toth, E., Warnock, G. L., Kneteman, N. M., and Rajotte, R. V. (2000). Islet transplantation in seven patients with type 1 diabetes mellitus using a glucocorticoid-free immunosuppressive regimen. *N. Engl. J. Med.* **343,** 230–238.

Soos, M. A., Jensen, J., Brown, R. A., O'Rahilly, S., Shepherd, P. R., and Whitehead, J. P. (2001). Class II phosphoinositide 3-kinase is activated by insulin but not by contraction in skeletal muscle. *Arch. Biochem. Biophys.* **396,** 244–248.

Soria, B., Roche, E., Berná, G., León-Quinto, T., Reig, J. A., and Martín, F. (2000). Insulin-secreting cells derived from embryonic stem cells normalize glycemia in streptozotocin-induced diabetic mice. *Diabetes* **49,** 157–162.

Spagnoli, F. M., and Brivanlou, A. H. (2008). The Gata5 target, TGIF2, defines the pancreatic region by modulating BMP signals within the endoderm. *Development* **135,** 451–461.

Stemmer, P. M., and Klee, C. B. (1994). Dual calcium ion regulation of calcineurin by calmodulin and calcineurin B. *Biochemistry* **33,** 6859–6866.

Talavera-Adame, D., Wu, G., He, Y., Ng, T. T., Gupta, A., Kurtovic, S., Hwang, J. Y., Farkas, D. L., and Dafoe, D. C. (2011). Endothelial cells in co-culture enhance embryonic stem cell differentiation to pancreatic progenitors and insulin-producing cells through BMP signaling. *Stem Cell Rev.* **7,** 532–543.

Thomas, M. K., Rastalsky, N., Lee, J. H., and Habener, J. F. (2000). Hedgehog signaling regulation of insulin production by pancreatic beta-cells. *Diabetes* **49,** 2039–2047.

Thomas, M. K., Lee, J. H., Rastalsky, N., and Habener, J. F. (2001). Hedgehog signaling regulation of homeodomain protein islet duodenum homeobox-1 expression in pancreatic beta-cells. *Endocrinology* **142,** 1033–1040.

Tsuboi, T., da Silva Xavier, G., Holz, G. G., Jouaville, L. S., Thomas, A. P., and Rutter, G. A. (2003). Glucagon-like peptide-1 mobilizes intracellular Ca2+ and stimulates mitochondrial ATP synthesis in pancreatic MIN6 beta-cells. *Biochem. J.* **369,** 287–299.

Urist, M. R. (1965). Bone: Formation by autoinduction. *Science* **150,** 893–899.

Vanhaesebroeck, B., Leevers, S. J., Ahmadi, K., Timms, J., Katso, R., Driscoll, P. C., Woscholski, R., Parker, P. J., and Waterfield, M. D. (2001). Synthesis and function of 3-phosphorylated inositol lipids. *Annu. Rev. Biochem.* **70,** 535–602.

Varjosalo, M., and Taipale, J. (2008). Hedgehog: Functions and mechanisms. *Genes Dev.* **22,** 2454–2472.

Wandzioch, E., and Zaret, K. S. (2009). Dynamic signaling network for the specification of embryonic pancreas and liver progenitors. *Science* **324,** 1707–1710.

Watanabe, H., Saito, H., Ueda, J., and Evers, B. M. (2008). Regulation of pancreatic duct cell differentiation by phosphatidylinositol-3 kinase. *Biochem. Biophys. Res. Commun.* **370,** 33–37.

Wozney, J. M., Rosen, V., Celeste, A. J., Mitsock, L. M., Whitters, M. J., Kriz, R. W., Hewick, R. M., and Wang, E. A. (1998). Novel regulators of bone formation: Molecular clones and activities. *Science* **242,** 1528–1534.

Yew, K. H., Hembree, M., Prasadan, K., Preuett, B., McFall, C., Benjes, C., Crowley, A., Sharp, S., Tulachan, S., Mehta, S., Tei, E., and Gittes, G. (2005). Cross-talk between bone morphogenetic protein and transforming growth factor-beta signaling is essential for exendin-4-induced insulin-positive differentiation of AR42J cells. *J. Biol. Chem.* **280,** 32209–32217.

Zhang, D., Jiang, W., Liu, M., Sui, X., Yin, X., Chen, S., Shi, Y., and Deng, H. (2009). Highly efficient differentiation of human ES cells and iPS cells into mature pancreatic insulin-producing cells. *Cell Res.* **19,** 429–438.

CHAPTER FIVE

Activation and Regulation of Reserve Liver Progenitor Cells

D. Hunter Best*,† *and* William B. Coleman‡

Contents

I. Introduction 94
II. Activation and Regulation of Mature Hepatocytes in Normal Liver Regeneration 94
III. Reserve Liver Progenitor Cells 97
A. Activation, regulation and proliferation of oval cells 98
B. Activation, regulation, and proliferation of small hepatocyte-like progenitor cells 100
IV. Hierarchical Responses in Liver Disease and Regeneration 102
References 103

Abstract

The mammalian liver possesses an extraordinary capacity for regeneration of tissue mass and cell numbers following loss of hepatocytes due to partial tissue loss (surgical resection) or hepatotoxic injury (necrosis). Restoration of liver mass can be obtained through the outgrowth and expansion of a number of different cell types depending upon the nature and extent of injury and/or tissue deficit. In an otherwise healthy liver, the replacement of hepatocytes (and tissue mass) is achieved through the proliferation of differentiated, normally quiescent hepatocytes contained in the residual (viable) tissue. However, in certain forms of liver injury, the capacity of differentiated hepatocytes to proliferate in response to liver tissue deficit is significantly impaired. When this occurs, reserve stem-like progenitor cells are activated to proliferate and replace lost hepatocytes. In this review, we will discuss the currently available information regarding the activation and outgrowth of each of these liver progenitor cell populations.

* Department of Pathology, University of Utah School of Medicine, Salt Lake City, Utah, USA
† Genetics Division, ARUP Laboratories, Inc., Salt Lake City, Utah, USA
‡ Department of Pathology and Laboratory Medicine, Curriculum in Toxicology, Program in Translational Medicine, UNC Lineberger Comprehensive Cancer Center, University of North Carolina School of Medicine, Chapel Hill, North Carolina, USA

Vitamins and Hormones, Volume 87
ISSN 0083-6729, DOI: 10.1016/B978-0-12-386015-6.00026-3

I. Introduction

Unlike rapidly renewing epithelial tissues (such as the intestinal mucosa or skin), in which an active stem cell lineage system continually initiates replacement of differentiated cells, the liver is a quiescent organ with minimal rates of cell turnover in the adult. Nonetheless, the liver possesses an extraordinary capacity for regeneration of tissue mass and cell numbers following loss of hepatocytes due to partial tissue loss (surgical resection) or hepatotoxic injury (necrosis). A number of different cell types can be activated to repair or regenerate the liver depending upon the nature and extent of injury and/or tissue deficit. In an otherwise healthy liver, the replacement of hepatocytes (and tissue mass) lost to surgical resection or toxic injury is achieved through the proliferation of differentiated, normally quiescent hepatocytes contained in the residual (viable) tissue. However, in certain forms of liver injury, the capacity of differentiated hepatocytes to proliferate in response to liver tissue deficit is significantly impaired. When this occurs, reserve stem-like progenitor cells are activated to proliferate and replace lost hepatocytes. In this review, we will discuss the currently available information regarding the activation and outgrowth of liver progenitor cell populations.

II. Activation and Regulation of Mature Hepatocytes in Normal Liver Regeneration

The liver retains the capacity to renew itself following cell loss caused by surgical resection or necrosis (Fausto, 1994, 2000, 2004; Fausto *et al.*, 2006; Michalopoulos, 1990, 2007, 2010; Michalopoulos and DeFrances, 1997). Liver regeneration in the rat (and other mammalian species) is triggered by a deficit of liver functional capacity, which typically reflects the extent of hepatocyte injury (surgical removal or necrosis). Surgical partial hepatectomy (PH) in the rat (Higgins and Anderson, 1931) is a widely studied model of liver regeneration. This simple and well-tolerated procedure produces a reproducible liver deficit reflecting 68% tissue loss (Bucher and Farmer, 1998). Rats are capable of regenerating their normal liver mass after greater resection (such as 90% PH), but regenerative responses are dampened in rats subjected to more modest liver injury (such as 30% PH) (Bucher and Farmer, 1998). In rats subjected to surgical PH, the residual (viable) hepatocytes undergo a rapid burst of proliferation that ultimately restores the normal hepatocyte number (Fabrikant, 1968a,b; Grisham, 1962, 1969; Wright and Alison, 1984). Likewise, biliary epithelial cells proliferate after PH to form expansions of the intrahepatic duct system (Grisham, 1962;

Marucci *et al.*, 1993; Polimeno *et al.*, 1995; Wright and Alison, 1984). Activation of stem-like progenitor cells does not occur in liver regeneration where the mature hepatocytes and biliary epithelial cells are capable of proliferating to restore the normal liver mass and structure (Dabeva and Shafritz, 1993; Klinman and Erslev, 1963). Hepatocytes in the periportal one-third of the parenchyma are the first cells to enter S-phase, 12–14 h after surgery (Fabrikant, 1968a; Grisham, 1962; Rabes and Tuczek, 1970). However, the periportal location of early hepatocyte proliferation after PH does not indicate that hepatocyte formation is limited to a periportal "proliferation compartment." Rather, periportal hepatocytes enter S-phase earlier than hepatocytes in other lobular locations due to their relatively shorter G_1-phase (Rabes *et al.*, 1976). Irrespective of their parenchymal location, virtually all hepatocytes proliferate and divide at least once during liver restoration (Fabrikant, 1969; Stocker and Heine, 1971). The "average" hepatocyte divides about 1.7 times during the accelerated growth period following hepatic resection (Fabrikant, 1969; Stocker and Heine, 1971).

The expression of a large number of genes is induced after PH in the rat (Haber *et al.*, 1993). The initial phase of gene expression occurs in the first several hours after PH and involves immediate-early genes (Fausto, 2000), some of which are expressed as early as several minutes after surgery (Taub, 1995, 1996). These genes encode for proteins with many diverse functions, including some cellular proto-oncogenes, transcription factors, tyrosine phosphatases, and metabolic proteins (Alcorn *et al.*, 1990; Diamond *et al.*, 1994; Hsu *et al.*, 1992; Peng *et al.*, 1999; Taub, 1996; Thompson *et al.*, 1986). The second phase of gene expression occurs from about 4 h post-PH until the onset of S-phase at about 20 h post-PH. Delayed-early genes and genes encoding cell cycle regulatory proteins are expressed during this period, including genes for anti-apoptotic proteins, cyclins, and cyclin-dependent kinases (Albrecht and Hansen, 1999; Albrecht *et al.*, 1998; Loyer *et al.*, 1994). The last phase of gene expression occurs coordinate with DNA replication and mitosis, involving additional cell cycle control genes (Fausto, 2000).

Hepatocyte proliferation is tightly regulated during liver regeneration after PH. Quiescent hepatocytes must be primed for proliferation prior to becoming growth factor responsive (Webber *et al.*, 1994, 1998). This is achieved through a cascade of cytokine release in the liver following PH (Diehl, 1999). TNFα is released within minutes of PH (Rai *et al.*, 1997), resulting in the subsequent release of IL6 and IL1 (Akerman *et al.*, 1992; Rai *et al.*, 1997). The primary sources of TNFα are the biliary epithelial cells and portal and venous endothelial cells (Loffreda *et al.*, 1997), while Kupffer cells, endothelial cells, and hepatic stellate cells are sources for release of IL6 and IL1 after TNFα stimulation (Taub, 1999). When anti-TNFα antibodies are employed to neutralize the effects of this cytokine, liver regeneration after PH is impaired (Akerman *et al.*, 1992), suggesting that TNFα is necessary for initiation of liver regeneration. Coordinate with the release of cytokines after PH is the rapid activation of several transcription factors in the remnant liver, including NFκB, AP-1, C/EBPβ, and STAT3 (Cressman *et al.*, 1995; Diehl and Yang, 1994; FitzGerald *et al.*, 1995; Greenbaum *et al.*, 1998; Rana *et al.*, 1995). A number of elegant studies using genetically modified mice have dissected the cytokine-mediated cell signaling pathways that involve these transcription factors (Fausto, 1999). Mice lacking TNFR-1 demonstrate deficient liver regeneration after PH, activation of NFκB and STAT3 does not occur, but activation of C/EBPβ is not altered (Yamada *et al.*, 1997). Injection of IL6 into these mice corrected the defect in hepatocyte replication and restored STAT3 activation, but had no effect on NFκB (Yamada *et al.*, 1997). In IL6 knockout mice, liver regeneration is impaired and activation of STAT3 does not occur, but activation of C/EBPβ is not affected (Cressman *et al.*, 1996). Injection of IL6 prior to PH in these mice corrects STAT3 activation and restores hepatocyte proliferation to near normal levels following PH (Cressman *et al.*, 1996). C/EBPβ is critical for liver regeneration after PH as C/EBPβ-deficient mice exhibit defective liver regeneration (Greenbaum *et al.*, 1998). However, the activation of C/EBPβ in liver regeneration after PH is IL6-independent (Taub, 1999) since its activation is not affected in mice lacking cytokine signaling though TNFα and IL6 (Cressman *et al.*, 1996; Yamada *et al.*, 1997). These studies combine to suggest that TNFα signaling through the TNFR-1 activates NFκB, leading to the release of IL6 which activates STAT3, priming the quiescent hepatocytes to enter the cell cycle and become responsive to growth factors.

Once hepatocytes enter the cell cycle in response to cytokine stimulation, cell cycle progression is driven by mitogenic growth factors, including EGF, HGF, and TGFα (Evarts *et al.*, 1992; Fausto *et al.*, 1995). EGF is present in the blood of normal rodents but is mitogenic for hepatocytes only in liver regeneration, possibly related to the necessity for hepatocyte priming by cytokines. TNFα and IL6 induce the production of HGF by hepatic stellate cells (Diehl and Rai, 1996), which stimulates hepatocyte proliferation in an paracrine manner. In contrast, TGFα is produced by regenerating (activated) hepatocytes and acts as an autocrine stimulus for cell division (Webber *et al.*, 1993). Some studies suggest that TGFα is produced by hepatocytes in response to TNFα stimulation (Gallucci *et al.*, 2000). Under certain conditions *in vivo*, hepatocytes respond to other growth factors, including IGFII, HSS, and insulin (Francavilla *et al.*, 1991).

TGFβ and activin are potent inhibitors of hepatocyte proliferation (Fausto *et al.*, 1995). The mRNAs for TGFβ and activin are expressed at negligible levels in the quiescent rat liver, but their expression is significantly induced several hours after PH (Braun *et al.*, 1988; Yasuda *et al.*, 1993). The expression of TGFβ1 mRNA persist during hepatocyte replication, reaching a maximum 36–72 h after PH (Jakowlew *et al.*, 1991). Nonparenchymal cells of the liver synthesize TGFβ, which acts on hepatocytes in a paracrine manner, while activin is synthesized by hepatocytes (Fausto *et al.*, 1995). Treatment of rats with TGFβ at the time of PH significantly inhibits hepatocyte entry into S-phase in a reversible manner (Russell *et al.*, 1988). This negative growth factor acts through one of several receptors (TGFβRI, TGFβRII, TGFβRIII) and ultimately inhibits the phosphorylation of pRb (late in G1), which prevents progression through the cell cycle into S-phase (DuBois *et al.*, 1998). It has been suggested that the balance between stimulators of hepatocyte proliferation (such as TGFα) and inhibitors of hepatocyte proliferation (TGFβ) may govern the hepatocyte responses during liver regeneration (Fausto *et al.*, 1995).

III. Reserve Liver Progenitor Cells

The accumulating evidence from the literature suggests that the liver possesses multiple progenitor cell populations that can be activated for regeneration of lost cell numbers and tissue mass related to surgical resection or necrosis. Whereas tissue regeneration results from the proliferation of mature (differentiated) hepatocytes and biliary epithelial cells in the normal liver, in the pathological liver the replication capacity of mature hepatocytes is often inhibited or impaired. In this circumstance, undifferentiated progenitor cell populations are activated to replace hepatocyte numbers and correct the deficit of liver function. At least two such undifferentiated cell populations have been identified: (i) oval cells (Sell *et al.*, 1981b) and (ii) small hepatocyte-like progenitor cells (SHPCs) (Gordon *et al.*, 2000c) (Fig. 5.1). Oval cells proliferate abundantly in the 2-acetylaminofluorene/PH model of liver regeneration (Evarts *et al.*, 1989; Tatematsu *et al.*, 1984). In this model, undifferentiated oval cells are refractory to the mitoinhibitory effects of 2-acetylaminofluorene due to lack of the proper enzymes for its metabolism (Grisham and Thorgeirsson, 1997). SHPCs proliferate after PH in retrorsine-exposed rats to give rise to expanding clusters of cells that eventually merge and repopulate the entire liver (Gordon *et al.*, 2000c). SHPCs are refractory to the mitoinhibitory effects of retrorsine due to the lack of the proper CYP enzymes required for its metabolism (Gordon *et al.*, 2000a).

It is of great interest to note that significant proliferation of oval cells does not occur in the retrorsine model of liver regeneration (Gordon *et al.*, 2000c), and that SHPCs have not been recognized in the 2AAF/PH model (Alison *et al.*, 1996; Evarts *et al.*, 1989; Tatematsu *et al.*, 1984). This observation suggests that the specific cell types involved in liver

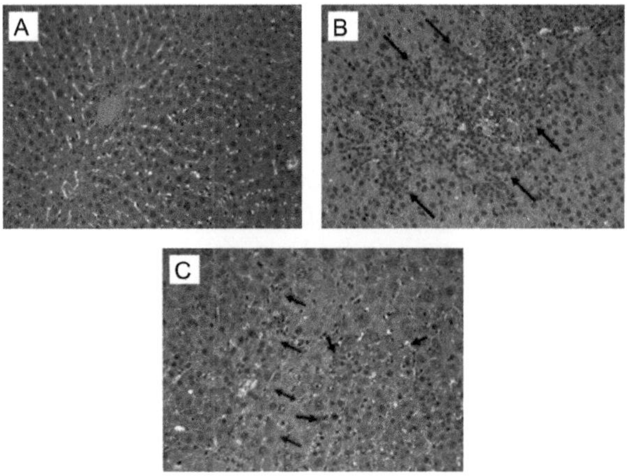

Figure 5.1 Progenitor cells of the adult rodent liver. Liver regeneration after surgical partial hepatectomy primarily involves the proliferation of (A) mature hepatocytes. However, under certain pathological conditions mature hepatocytes are incapable of proliferation and liver mass is restored through the proliferation of either (B) oval cells or (C) small hepatocyte-like progenitor cells. Arrows denote (B) oval cells and (C) small hepatocyte-like progenitor cells. Original magnification: 10×.

regeneration may be selectively activated through a hierarchical control mechanism to proliferate, depending upon the pathophysiology of the tissue at the time that a deficit in liver function occurs and the capacity of the individual progenitor cell populations to participate. Consistent with this suggestion, differentiated hepatocytes represent the primary source of cells for hepatocyte replacement, allowing for tissue renewal and concurrent (continued) hepatocyte function. When mature hepatocytes are impaired, SHPCs are activated to proliferate (perhaps reflecting a primary reserve compartment) and produce new hepatocytes. If SHPCs cannot respond to the liver deficit, undifferentiated oval cells are activated to proliferate, differentiate, and remodel the tissue (perhaps as a second line reserve compartment).

A. Activation, regulation and proliferation of oval cells

Oval cells, which proliferate in several hepatocarcinogenesis models (Farber and Cameron, 1980; Rogers, 1978; Sell and Leffert, 1982) and in some forms of noncarcinogenic liver damage (Dabeva and Shafritz, 1993; Lesch *et al.*, 1970; Smith *et al.*, 1996; Tournier *et al.*, 1988; Yavorkovsky *et al.*, 1995), have been suggested to be related to liver stem-like cells. The similarities between the various pathophysiological circumstances necessary for the induction of oval cell proliferation suggest two necessary conditions: (i) a stimulus for liver growth and (ii) inhibition of normal regenerative mechanisms for tissue restoration (i.e., blockade of the hepatocyte proliferation). The first condition can be satisfied through several different methods, including surgical resection, nutritional stress, or chemically induced necrosis. The second condition is frequently achieved through the use of chemicals (such as 2-acetylaminofluorene) that impede or prevent the proliferation of mature hepatocytes (Tatematsu *et al.*, 1984). The most frequently used experimental models for oval cell proliferation in rats include (i) treatment with azo dyes (Inaoka, 1967), (ii) feeding a choline-deficient diet, with or without supplements of ethionine (Shinozuka *et al.*, 1978) or 2-acetylaminofluorene (Sell *et al.*, 1981a,b), (iii) treatment with 2-acetylaminofluorene combined with PH (Tatematsu *et al.*, 1984), and (iv) treatment with D-galactosamine (Lesch *et al.*, 1970). In addition to these experimental models for induction of oval cell proliferation, similar oval cell responses occur in pathological rat livers related to chronic ethanol exposure (Smith *et al.*, 1996), and chronic hepatitis associated with abnormal copper accumulation in hepatocytes (Betto *et al.*, 1996; Yasui *et al.*, 1997). The cellular response common to each of these models involves the proliferation of small cells with scant cytoplasm and ovoid nuclei that are morphologically described as oval cells (Farber, 1956).

The timing of cellular events differs, sometimes dramatically, among the various models of oval cell proliferation (Coleman and Grisham, 1998). However, the majority of oval cell proliferation models share a common sequence of events: (i) proliferation of oval cells in or around the portal spaces, (ii) invasion of the lobular parenchyma by the proliferating oval cells, (iii) appearance of transitional cell types and immature hepatocytes, and (iv) maturation of hepatocytes and restoration of normal liver structure. Oval cells are initially seen in the portal zones of the liver lobule in the regions of terminal bile ductules or cholangioles (Fujio *et al.*, 1994; Sell, 1994). Proliferating oval cells are recognized to represent a collection of phenotypically distinct cells that compose a heterogeneous cell population or "compartment" (Fausto, 1994; Fausto *et al.*, 1992; Yaswen *et al.*, 1984). Morphologically, the typical oval cell possesses characteristics similar to those of cells of terminal bile ductules (Grisham and Hartroft, 1961; Lenzi *et al.*, 1992; Sarraf *et al.*, 1994). However, the oval cell compartment also contains transitional cells that display morphologic features intermediate between oval cells and hepatocytes (Inaoka, 1967; Sarraf *et al.*, 1994). Proliferating oval cells form irregular duct-like structures that are connected to pre-existing bile ducts (Alpini *et al.*, 1992; Betto *et al.*, 1996; Dunsford *et al.*, 1985; Makino *et al.*, 1988; Sarraf *et al.*, 1994). As they proliferate, oval cells migrate from the portal regions into the lobular parenchyma, sometimes occupying a large percentage of the liver mass. Groups of small basophilic hepatocytes appear among oval cells, followed by gradual disappearance of the oval cells and restoration of normal liver structure (Dabeva and Shafritz, 1993; Evarts *et al.*, 1987, 1989; Factor *et al.*, 1994; Lemire *et al.*, 1991; Sarraf *et al.*, 1994).

The possibility that oval cells might possess stem-like properties and give rise to hepatocytes and/or biliary epithelial cells has been recognized for some time (Farber, 1956; Kinosita, 1937; Price *et al.*, 1952; Wilson and Leduc, 1958). Using the modified Solt-Farber model of oval cell proliferation, Evarts *et al.* (1989) demonstrated that oval cells radiolabeled with ^{3}H-thymidine could give rise to tagged basophilic hepatocytes. In these studies, morphologically identifiable oval cells that expressed oval cell-associated antigens were specifically radiolabeled under conditions that limited the radiolabeling of residual hepatocytes. Subsequent examination of the regenerating liver revealed basophilic foci containing small radiolabeled hepatocytes that expressed markers of hepatocytic differentiation (Evarts *et al.*, 1989). Employing the D-glactosamine model of oval cell proliferation in rats, Lemire *et al.* (1991) and Dabeva and Shafritz (1993) demonstrated the transfer of radiolabel from oval cells to small hepatocytes. In both of these studies transition from oval cells to hepatocytes was accompanied by a shift from the biliary epithelial/oval cell phenotype (expression of α-fetoprotein, γ-glutamyltranspeptidase, and biliary epithelial type cytokeratins) to a cellular phenotype characteristic of hepatocyte differentiation (expression of albumin, glucose-6-phosphatase, and other hepatocyte markers; reduction of α-fetoprotein expression) (Dabeva and Shafritz, 1993; Lemire *et al.*, 1991).

Analogous to mature hepatocytes, the oval cell reaction requires a cytokine-mediated priming stimulus (Nagy *et al.*, 1998). Blockade of cytokine release after PH in rats treated with 2-acetylaminofluorene leads to inhibition of oval cell activation, but this inhibition can be overcome through the administration of exogenous IL6 (Knight *et al.*, 2000; Nagy *et al.*, 1998). Thus, liver regeneration by hepatocytes and oval cells represent different cellular reactions to liver injury and associated functional deficit, but require similar cytokines and growth factors.

B. Activation, regulation, and proliferation of small hepatocyte-like progenitor cells

Retrorsine is a member of the pyrrolizidine alkaloid (PA) family of naturally occurring compounds that are toxic to various mammalian tissues, including liver, lung, kidney, brain, muscle, heart, thymus, lymph nodes, and blood vessels (Bull and Dick, 1959; McLean, 1970; Peterson, 1965; Schoental, 1968; Schoental and Magee, 1957, 1959). The hepatotoxic effects of PAs are long-lasting (Hayes *et al.*, 1985; Laconi *et al.*, 1995, 1998) and include inhibition of hepatocyte cell division coupled with induction of polyploidy and megalocytosis (Laconi *et al.*, 1995, 1998). The acute development of megalocytosis in the livers of retrorsine-exposed rats results from the anti-mitotic actions of PAs and their metabolites on hepatocytes that are stimulated to divide, such as in response to PH or hepatocellular necrosis (Jago, 1969).

The cellular responses and time course for liver regeneration after surgical PH in rats with retrorsine-induced hepatocellular injury have been well described (Gordon *et al.*, 2000c). Similar to other models of chemical liver injury (Coleman and Grisham, 1998; Grisham and Thorgeirsson, 1997), systemic exposure to retrorsine results in a severe inhibition of the replicative capacity of differentiated hepatocytes (Dabeva *et al.*, 1998; Gordon *et al.*, 2000c; Laconi *et al.*, 1998; Oren *et al.*, 1999). When confronted with a strong proliferative stimulus such as PH (Dabeva *et al.*, 1998; Gordon *et al.*, 2000c; Laconi *et al.*, 1997, 1998) or hepatocellular necrosis (Best and Coleman, 2010; Jago, 1969), retrorsine-injured hepatocytes synthesize DNA but are unable to complete mitosis, and arrest as nonproliferative giant cells (megalocytes). In this model, neither retrorsine-injured differentiated hepatocytes nor oval cells proliferate sufficiently to contribute significantly to the restoration of liver mass after PH. Instead, the entire liver mass is reconstituted after PH through a novel cellular response that is mediated by the emergence and rapid expansion of a population of SHPCs, which share some phenotypic traits with fetal hepatoblasts, oval cells, and fully differentiated hepatocytes, but are morphologically and phenotypically distinct from each (Gordon *et al.*, 2000c). SHPCs emerge from all regions of the liver lobule after PH and are not associated with modest oval cell outgrowths, suggesting that SHPCs represent a novel cell population (Best and Coleman, 2007a, 2010; Gordon *et al.*, 2000c). SHPCs morphologically most closely resemble differentiated (but small) hepatocytes at early time-points after PH, perhaps indicating that SHPCs are a subset of retrorsine-resistant hepatocytes and not a novel progenitor cell population. However, the phenotype of SHPCs suggest that they are in fact distinct from fully differentiated hepatocytes, since a subset of SHPCs express the oval cell/bile duct/fetal liver markers OC.2 and OC.5 through 5 days post-PH (Gordon *et al.*, 2000b,c). Co-expression of hepatocyte markers and oval cell markers by early appearing SHPCs suggest that these cells are either not fully differentiated or display a transitional phenotype similar to that expected for a cell type between the bipotential hepatoblast (E14) and the fetal hepatocyte (E18–E20). While some have suggested that SHPCs are a transitional cell type between the oval cell and "mature" hepatocytes, recent data have shown that when retrorsine-exposed rats are also treated with 2-acetylaminofluorene the regenerative response is entirely mediated by oval cells (Best and Coleman, 2007b). In addition, other studies using the biliary toxin diaminodiphenylmethane (DAPM) demonstrated that SHPCs proliferate in the absence of oval cell outgrowth, providing definitive evidence that SHPCs are an independent progenitor cell population (Best and Coleman, 2007a).

Until recently, the molecular mechanisms governing the activation of SHPCs after liver injury remained unknown. In a recent study, retrorsine-exposed rats were treated with the cytokine inhibitor dexamethasone over the course of a 5 day period resulting in a marked decrease in the expansion and outgrowth of the SHPCs at all time points after PH (Best *et al.*, 2010). In this study, the SHPC-mediated regenerative response was rescued through the intravenous administration of recombinant IL6 protein strongly suggesting that this response is IL6-mediated (Best *et al.*, 2010). However, additional studies are necessary to further elucidate all of the regulatory pathways and factors involved in the SHPC response to liver injury.

IV. Hierarchical Responses in Liver Disease and Regeneration

The mammalian liver possesses tremendous flexibility in its capacity to respond to injury and loss of cell numbers (function). At least three different cell populations have been implicated in liver regeneration: (i) differentiated hepatocytes (in otherwise normal liver), (ii) SHPC (observed in retrorsine-exposed rats), and (iii) oval cells (observed in numerous models of liver injury). Given multiple sources of regenerative cells and the observed differences in timing of activation of these cell populations in response to liver deficit, it is intriguing to speculate that there is a hierarchy of potential cellular responses and that the nature of the regenerative cell population activated is determined by (a) the presence or absence of liver injury, (b) the type and extent of injury, and (c) the capacity of each cell population to respond. In this model, activation and proliferation of mature hepatocytes represents the primary (preferential) cellular response, and the activation, emergence, and proliferation of reserve stem-like progenitor cell populations (SHPC or oval cells) represent secondary cellular responses, occurring only when the primary response is blocked or impaired. Several studies from the literature suggest that IL6 may be a common signaling molecule that functions to regulate the liver's response to injury (Best *et al.*, 2010; Cressman *et al.*, 1996; Nagy *et al.*, 1998; Sakamoto *et al.*, 1999). There may be redundancy in the cellular reactions to liver injury, but a common regulatory pathway that directs the process of regeneration. Thus, following a signal for liver regeneration, all possible progenitor cell populations may be concurrently primed for activation (IL6-mediated), but that the secondary reserve progenitor cell responses do not manifest unless the primary response (hepatocyte-mediated) fails (Fig. 5.2). Further studies are required to address these possibilities.

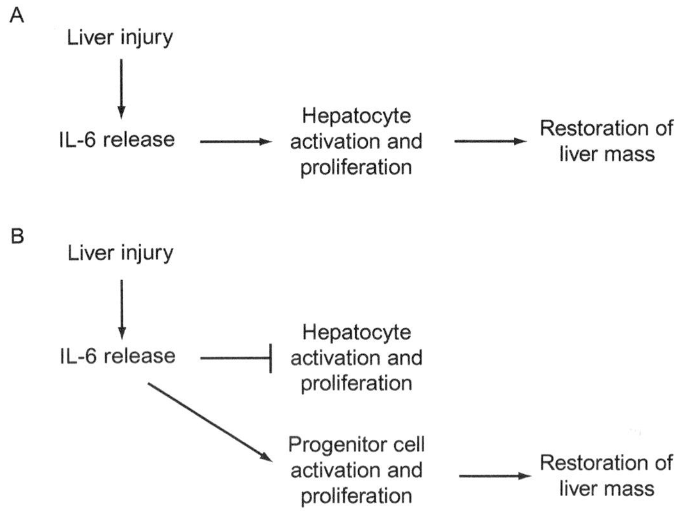

Figure 5.2 Hierarchical responses in liver disease and regeneration. Evidence suggests that there is likely to be redundancy in the mechanisms activating liver progenitor cells. (A) Under otherwise normal conditions, liver deficit results in the release of cytokines that activate mature hepatocytes, the liver's primary progenitor cell. However, (B) under certain pathological circumstances liver injury still results in the release of cytokines but regeneration by mature hepatocytes is blocked. In these conditions, reserve progenitor cell compartments are activated and liver mass is restored.

REFERENCES

Akerman, P., Cote, P., Yang, S. Q., McClain, C., Nelson, S., Bagby, G. J., and Diehl, A. M. (1992). Antibodies to tumor necrosis factor-alpha inhibit liver regeneration after partial hepatectomy. *Am. J. Physiol.* **263,** G579–G585.

Albrecht, J. H., and Hansen, L. K. (1999). Cyclin D1 promotes mitogen-independent cell cycle progression in hepatocytes. *Cell Growth Differ.* **10,** 397–404.

Albrecht, J. H., Poon, R. Y., Ahonen, C. L., Rieland, B. M., Deng, C., and Crary, G. S. (1998). Involvement of p21 and p27 in the regulation of CDK activity and cell cycle progression in the regenerating liver. *Oncogene* **16,** 2141–2150.

Alcorn, J. A., Feitelberg, S. P., and Brenner, D. A. (1990). Transient induction of c-jun during hepatic regeneration. *Hepatology* **11,** 909–915.

Alison, M. R., Golding, M., Sarraf, C. E., Edwards, R. J., and Lalani, E. N. (1996). Liver damage in the rat induces hepatocyte stem cells from biliary epithelial cells. *Gastroenterology* **110,** 1182–1190.

Alpini, G., Aragona, E., Dabeva, M., Salvi, R., Shafritz, D. A., and Tavoloni, N. (1992). Distribution of albumin and alpha-fetoprotein mRNAs in normal, hyperplastic, and preneoplastic rat liver. *Am. J. Pathol.* **141,** 623–632.

Best, D. H., and Coleman, W. B. (2007a). Bile duct destruction by 4,4'-diaminodiphenylmethane does not block the small hepatocyte-like progenitor cell response in retrorsine-exposed rats. *Hepatology* **46,** 1611–1619.

Best, D. H., and Coleman, W. B. (2007b). Treatment with 2-AAF blocks the small hepatocyte-like progenitor cell response in retrorsine-exposed rats. *J. Hepatol.* **46,** 1055–1063.

Best, D. H., and Coleman, W. B. (2010). Liver regeneration by small hepatocyte-like progenitor cells after necrotic injury by carbon tetrachloride in retrorsine-exposed rats. *Exp. Mol. Pathol.* **89,** 92–98.

Best, D. H., Butz, G. M., and Coleman, W. B. (2010). Cytokine-dependent activation of small hepatocyte-like progenitor cells in retrorsine-induced rat liver injury. *Exp. Mol. Pathol.* **88,** 7–14.

Betto, H., Kaneda, K., Yamamoto, T., Kojima, A., and Sakurai, M. (1996). Development of intralobular bile ductules after spontaneous hepatitis in Long-Evans mutant rats. *Lab. Invest.* **75,** 43–53.

Braun, L., Mead, J. E., Panzica, M., Mikumo, R., Bell, G. I., and Fausto, N. (1988). Transforming growth factor beta mRNA increases during liver regeneration: A possible paracrine mechanism of growth regulation. *Proc. Natl. Acad. Sci. USA* **85,** 1539–1543.

Bucher, N. L. R., and Farmer, S. R. (1998). Liver regeneration following partial hepatectomy: Genes and metabolism. *In* "Liver Growth and Repair," (A. J. Strain and A. M. Diehl, Eds.), pp. 3–27. Chapman and Hall, New York.

Bull, L. B., and Dick, A. T. (1959). The chronic pathological effects on the liver of the rat of the pyrrolizidine alkaloids heliotrine, lasiocarpine, and their N-oxides. *J. Pathol. Bacteriol.* **78,** 483–502.

Coleman, W. B., and Grisham, J. W. (1998). Epithelial stem-like cells of the rodent liver. *In* "Liver Growth and Repair," (A. J. Strain and A. M. Diehl, Eds.), pp. 50–99. Chapman and Hall, London.

Cressman, D. E., Diamond, R. H., and Taub, R. (1995). Rapid activation of the Stat3 transcription complex in liver regeneration. *Hepatology* **21,** 1443–1449.

Cressman, D. E., Greenbaum, L. E., DeAngelis, R. A., Ciliberto, G., Furth, E. E., Poli, V., and Taub, R. (1996). Liver failure and defective hepatocyte regeneration in interleukin-6-deficient mice. *Science* **274,** 1379–1383.

Dabeva, M. D., and Shafritz, D. A. (1993). Activation, proliferation, and differentiation of progenitor cells into hepatocytes in the D-galactosamine model of liver regeneration. *Am. J. Pathol.* **143,** 1606–1620.

Dabeva, M. D., Laconi, E., Oren, R., Petkov, P. M., Hurston, E., and Shafritz, D. A. (1998). Liver regeneration and alpha-fetoprotein messenger RNA expression in the retrorsine model for hepatocyte transplantation. *Cancer Res.* **58,** 5825–5834.

Diamond, R. H., Cressman, D. E., Laz, T. M., Abrams, C. S., and Taub, R. (1994). PRL-1, a unique nuclear protein tyrosine phosphatase, affects cell growth. *Mol. Cell. Biol.* **14,** 3752–3762.

Diehl, A. (1999). Cytokine regulation of liver regeneration. *In* "Normal and Malignant Liver Cell Growth," (W. E. Fleig, Ed.), pp. 47–55. Kluwer Academic Publishers, Dordrecht.

Diehl, A. M., and Rai, R. (1996). Review: Regulation of liver regeneration by pro-inflammatory cytokines. *J. Gastroenterol. Hepatol.* **11,** 466–470.

Diehl, A. M., and Yang, S. Q. (1994). Regenerative changes in C/EBP alpha and C/EBP beta expression modulate binding to the C/EBP site in the c-fos promoter. *Hepatology* **19,** 447–456.

DuBois, R. N., Myers, D., and Beauchamp, R. D. (1998). Growth inhibitory signals and liver regeneration: The TGF-beta superfamily. *In* "Liver Growth and Repair," (A. J. Strain and A. M. Diehl, Eds.), pp. 261–282. Chapman and Hall, New York.

Dunsford, H. A., Maset, R., Salman, J., and Sell, S. (1985). Connection of ductlike structures induced by a chemical hepatocarcinogen to portal bile ducts in the rat liver detected by injection of bile ducts with a pigmented barium gelatin medium. *Am. J. Pathol.* **118,** 218–224.

Evarts, R. P., Nagy, P., Marsden, E., and Thorgeirsson, S. S. (1987). A precursor-product relationship exists between oval cells and hepatocytes in rat liver. *Carcinogenesis* **8,** 1737–1740.
Evarts, R. P., Nagy, P., Nakatsukasa, H., Marsden, E., and Thorgeirsson, S. S. (1989). In vivo differentiation of rat liver oval cells into hepatocytes. *Cancer Res.* **49,** 1541–1547.
Evarts, R. P., Nakatsukasa, H., Marsden, E. R., Hu, Z., and Thorgeirsson, S. S. (1992). Expression of transforming growth factor-alpha in regenerating liver and during hepatic differentiation. *Mol. Carcinog.* **5,** 25–31.
Fabrikant, J. I. (1968a). The kinetics of cellular proliferation in regenerating liver. *J. Cell Biol.* **36,** 551–565.
Fabrikant, J. I. (1968b). Rate of cell proliferation in the regenerating liver. *Br. J. Radiol.* **41,** 71.
Fabrikant, J. I. (1969). Size of proliferating pools in regenerating liver. *Exp. Cell Res.* **55,** 277–279.
Factor, V. M., Radaeva, S. A., and Thorgeirsson, S. S. (1994). Origin and fate of oval cells in dipin-induced hepatocarcinogenesis in the mouse. *Am. J. Pathol.* **145,** 409–422.
Farber, E. (1956). Similarities in the sequence of early histological changes induced in the liver of the rat by ethionine, 2-acetylamino-fluorene, and 3'-methyl-4-dimethylaminoazobenzene. *Cancer Res.* **16,** 142–148.
Farber, E., and Cameron, R. (1980). The sequential analysis of cancer development. *Adv. Cancer Res.* **31,** 125–226.
Fausto, N. (1994). Liver stem cells. *In* "The Liver: Biology and Pathobiology," (I. M. Arias, J. L. Boyer, N. Fausto, W. B. Jakoby, D. Shachter, and D. A. Shafritz, Eds.), pp. 1501–1518. Raven Press, New York.
Fausto, N. (1999). Lessons from genetically engineered animal models. V. Knocking out genes to study liver regeneration: Present and future. *Am. J. Physiol.* **277,** G917–G921.
Fausto, N. (2000). Liver regeneration. *J. Hepatol.* **32,** 19–31.
Fausto, N. (2004). Liver regeneration and repair: Hepatocytes, progenitor cells, and stem cells. *Hepatology* **39,** 1477–1487.
Fausto, N., Lemire, J. M., and Shiojiri, N. (1992). Oval cells in liver carcinogenesis: Cell linearges in hepatic development and the identification of facultative stem cells in normal liver. *In* "The Role of Cell Types in Carcinogenesis," (A. E. Sirica, Ed.), pp. 89–108. CRC Press, Boca Raton.
Fausto, N., Laird, A. D., and Webber, E. M. (1995). Liver regeneration. 2. Role of growth factors and cytokines in hepatic regeneration. *FASEB J.* **9,** 1527–1536.
Fausto, N., Campbell, J. S., and Riehle, K. J. (2006). Liver regeneration. *Hepatology* **43,** S45–S53.
FitzGerald, M. J., Webber, E. M., Donovan, J. R., and Fausto, N. (1995). Rapid DNA binding by nuclear factor kappa B in hepatocytes at the start of liver regeneration. *Cell Growth Differ.* **6,** 417–427.
Francavilla, A., Starzl, T. E., Porter, K., Foglieni, C. S., Michalopoulos, G. K., Carrieri, G., Trejo, J., Azzarone, A., Barone, M., and Zeng, Q. H. (1991). Screening for candidate hepatic growth factors by selective portal infusion after canine Eck's fistula. *Hepatology* **14,** 665–670.
Fujio, K., Evarts, R. P., Hu, Z., Marsden, E. R., and Thorgeirsson, S. S. (1994). Expression of stem cell factor and its receptor, c-kit, during liver regeneration from putative stem cells in adult rat. *Lab. Invest.* **70,** 511–516.
Gallucci, R. M., Simeonova, P. P., Toriumi, W., and Luster, M. I. (2000). TNF-alpha regulates transforming growth factor-alpha expression in regenerating murine liver and isolated hepatocytes. *J. Immunol.* **164,** 872–878.
Gordon, G. J., Coleman, W. B., and Grisham, J. W. (2000a). Induction of cytochrome P450 enzymes in the livers of rats treated with the pyrrolizidine alkaloid retrorsine. *Exp. Mol. Pathol.* **69,** 17–26.

Gordon, G. J., Coleman, W. B., and Grisham, J. W. (2000b). Temporal analysis of hepatocyte differentiation by small hepatocyte-like progenitor cells during liver regeneration in retrorsine-exposed rats. *Am. J. Pathol.* **157,** 771–786.

Gordon, G. J., Coleman, W. B., Hixson, D. C., and Grisham, J. W. (2000c). Liver regeneration in rats with retrorsine-induced hepatocellular injury proceeds through a novel cellular response. *Am. J. Pathol.* **156,** 607–619.

Greenbaum, L. E., Li, W., Cressman, D. E., Peng, Y., Ciliberto, G., Poli, V., and Taub, R. (1998). CCAAT enhancer-binding protein beta is required for normal hepatocyte proliferation in mice after partial hepatectomy. *J. Clin. Invest.* **102,** 996–1007.

Grisham, J. W. (1962). A morphologic study of deoxyribonucleic acid synthesis and cell proliferation in regenerating rat liver; autoradiography with thymidine-H3. *Cancer Res.* **22,** 842–849.

Grisham, J. W. (1969). Cellular proliferation in the liver. *Recent Results Cancer Res.* **17,** 28–43.

Grisham, J. W., and Hartroft, W. S. (1961). Morphologic identification by electron microscopy of "oval" cells in experimental hepatic degeneration. *Lab. Invest.* **10,** 317–332.

Grisham, J. W., and Thorgeirsson, S. S. (1997). Liver stem cells. *In* "Stem Cells," (C. S. Potten, Ed.), pp. 233–282. Academic Press, London.

Haber, B. A., Mohn, K. L., Diamond, R. H., and Taub, R. (1993). Induction patterns of 70 genes during nine days after hepatectomy define the temporal course of liver regeneration. *J. Clin. Invest.* **91,** 1319–1326.

Hayes, M. A., Roberts, E., and Farber, E. (1985). Initiation and selection of resistant hepatocyte nodules in rats given the pyrrolizidine alkaloids lasiocarpine and senecionine. *Cancer Res.* **45,** 3726–3734.

Higgins, G. M., and Anderson, R. M. (1931). Experimental pathology of the liver. I. Restoration of the liver of the white rat following partial surgical removal. *Arch. Pathol.* **12,** 186–202.

Hsu, J. C., Bravo, R., and Taub, R. (1992). Interactions among LRF-1, JunB, c-Jun, and c-Fos define a regulatory program in the G1 phase of liver regeneration. *Mol. Cell. Biol.* **12,** 4654–4665.

Inaoka, Y. (1967). Significance of the so-called oval cell proliferation during azo-dye hepatocarcinogenesis. *Gann* **58,** 355–366.

Jago, M. V. (1969). The development of the hepatic megalocytosis of chronic pyrrolizidine alkaloid poisoning. *Am. J. Pathol.* **56,** 405–421.

Jakowlew, S. B., Mead, J. E., Danielpour, D., Wu, J., Roberts, A. B., and Fausto, N. (1991). Transforming growth factor-beta (TGF-beta) isoforms in rat liver regeneration: Messenger RNA expression and activation of latent TGF-beta. *Cell Regul.* **2,** 535–548.

Kinosita, R. (1937). Sudies on the cancergenic chemical substances. *Trans. Soc. Pathol. Jpn.* **27,** 665–727.

Klinman, N. R., and Erslev, A. J. (1963). Cellular response to partial hepatectomy. *Proc. Soc. Exp. Biol. Med.* **112,** 338–340.

Knight, B., Yeoh, G. C., Husk, K. L., Ly, T., Abraham, L. J., Yu, C., Rhim, J. A., and Fausto, N. (2000). Impaired preneoplastic changes and liver tumor formation in tumor necrosis factor receptor type 1 knockout mice. *J. Exp. Med.* **192,** 1809–1818.

Laconi, E., Sarma, D. S., and Pani, P. (1995). Transplantation of normal hepatocytes modulates the development of chronic liver lesions induced by a pyrrolizidine alkaloid, lasiocarpine. *Carcinogenesis* **16,** 139–142.

Laconi, E., Dabeva, M., Laconi, S., Pani, P., Mukhopadhyay, D., Oren, R., and Shafritz, D. A. (1997). Rapid proliferation of transplanted syngeneic DPPIV+ hepatocytes in the liver of pyrrolizidine alkaloid treated DPPIV-Fischer rats. *FASEB J.* **11,** A304.

Laconi, E., Oren, R., Mukhopadhyay, D. K., Hurston, E., Laconi, S., Pani, P., Dabeva, M. D., and Shafritz, D. A. (1998). Long-term, near-total liver replacement by transplantation of isolated hepatocytes in rats treated with retrorsine. *Am. J. Pathol.* **153,** 319–329.

Lemire, J. M., Shiojiri, N., and Fausto, N. (1991). Oval cell proliferation and the origin of small hepatocytes in liver injury induced by D-galactosamine. *Am. J. Pathol.* **139,** 535–552.

Lenzi, R., Liu, M. H., Tarsetti, F., Slott, P. A., Alpini, G., Zhai, W. R., Paronetto, F., Lenzen, R., and Tavoloni, N. (1992). Histogenesis of bile duct-like cells proliferating during ethionine hepatocarcinogenesis. Evidence for a biliary epithelial nature of oval cells. *Lab. Invest.* **66,** 390–402.

Lesch, R., Reutter, W., Keppler, D., and Decker, K. (1970). Liver restitution after acute galactosamine hepatitis: Autoradiographic and biochemical studies in rats. *Exp. Mol. Pathol.* **12,** 58–69.

Loffreda, S., Rai, R., Yang, S. Q., Lin, H. Z., and Diehl, A. M. (1997). Bile ducts and portal and central veins are major producers of tumor necrosis factor alpha in regenerating rat liver. *Gastroenterology* **112,** 2089–2098.

Loyer, P., Glaise, D., Cariou, S., Baffet, G., Meijer, L., and Guguen-Guillouzo, C. (1994). Expression and activation of cdks (1 and 2) and cyclins in the cell cycle progression during liver regeneration. *J. Biol. Chem.* **269,** 2491–2500.

Makino, Y., Yamamoto, K., and Tsuji, T. (1988). Three-dimensional arrangement of ductular structures formed by oval cells during hepatocarcinogenesis. *Acta Med. Okayama* **42,** 143–150.

Marucci, L., Baroni, G. S., Mancini, R., Benedetti, A., Jezequel, A. M., and Orlandi, F. (1993). Cell proliferation following extrahepatic biliary obstruction. Evaluation by immunohistochemical methods. *J. Hepatol.* **17,** 163–169.

McLean, E. K. (1970). The toxic actions of pyrrolizidine (senecio) alkaloids. *Pharmacol. Rev.* **22,** 429–483.

Michalopoulos, G. K. (1990). Liver regeneration: Molecular mechanisms of growth control. *FASEB J.* **4,** 176–187.

Michalopoulos, G. K. (2007). Liver regeneration. *J. Cell. Physiol.* **213,** 286–300.

Michalopoulos, G. K. (2010). Liver regeneration after partial hepatectomy: Critical analysis of mechanistic dilemmas. *Am. J. Pathol.* **176,** 2–13.

Michalopoulos, G. K., and DeFrances, M. C. (1997). Liver regeneration. *Science* **276,** 60–66.

Nagy, P., Kiss, A., Schnur, J., and Thorgeirsson, S. S. (1998). Dexamethasone inhibits the proliferation of hepatocytes and oval cells but not bile duct cells in rat liver. *Hepatology* **28,** 423–429.

Oren, R., Dabeva, M. D., Karnezis, A. N., Petkov, P. M., Rosencrantz, R., Sandhu, J. P., Moss, S. F., Wang, S., Hurston, E., Laconi, E., Holt, P. R., Thung, S. N., *et al.* (1999). Role of thyroid hormone in stimulating liver repopulation in the rat by transplanted hepatocytes. *Hepatology* **30,** 903–913.

Peng, Y., Du, K., Ramirez, S., Diamond, R. H., and Taub, R. (1999). Mitogenic up-regulation of the PRL-1 protein-tyrosine phosphatase gene by Egr-1. Egr-1 activation is an early event in liver regeneration. *J. Biol. Chem.* **274,** 4513–4520.

Peterson, J. E. (1965). Effects of the pyrrolizidine alkaloid, lasiocarpine N-oxide, on nuclear and cell division in the liver rats. *J. Pathol. Bacteriol.* **89,** 153–171.

Polimeno, L., Azzarone, A., Zeng, Q. H., Panella, C., Subbotin, V., Carr, B., Bouzahzah, B., Francavilla, A., and Starzl, T. E. (1995). Cell proliferation and oncogene expression after bile duct ligation in the rat: Evidence of a specific growth effect on bile duct cells. *Hepatology* **21,** 1070–1078.

Price, J. M., Harman, J. W., Miller, E. C., and Miller, J. A. (1952). Progressive microscopic alterations in the livers of rats fed the hepatic carcinogens 3'-methyl-4-dimethylaminoazobenzene and 4'-fluoro-4-dimethylaminoazobenzene. *Cancer Res.* **12,** 192–200.

Rabes, H., and Tuczek, H. V. (1970). Quantitative autoradiographic study of heterogeneity of liver cell proliferation after partial hepatectomy. *Virchows Arch. B Cell Pathol.* **6,** 302–312.

Rabes, H. M., Wirsching, R., Tuczek, H. V., and Iseler, G. (1976). Analysis of cell cycle compartments of hepatocytes after partial hepatecomy. *Cell Tissue Kinet.* **9,** 517–532.

Rai, R. M., Loffreda, S., Karp, C. L., Yang, S. Q., Lin, H. Z., and Diehl, A. M. (1997). Kupffer cell depletion abolishes induction of interleukin-10 and permits sustained over-expression of tumor necrosis factor alpha messenger RNA in the regenerating rat liver. *Hepatology* **25,** 889–895.

Rana, B., Xie, Y., Mischoulon, D., Bucher, N. L., and Farmer, S. R. (1995). The DNA binding activity of C/EBP transcription factor is regulated in the G1 phase of the hepatocyte cell cycle. *J. Biol. Chem.* **270,** 18123–18132.

Rogers, A. E. (1978). Dietary effects on chemical carcinogenesis in the livers of rats. *In* "Rat Hepatic Neoplasia," (P.M Butler and W.H. Newberne, Eds.), pp. 243–262. MIT Press, Cambridge.

Russell, W. E., Coffey, R. J., Jr., Ouellette, A. J., and Moses, H. L. (1988). Type beta transforming growth factor reversibly inhibits the early proliferative response to partial hepatectomy in the rat. *Proc. Natl. Acad. Sci. USA* **85,** 5126–5130.

Sakamoto, T., Liu, Z., Murase, N., Ezure, T., Yokomuro, S., Poli, V., and Demetris, A. J. (1999). Mitosis and apoptosis in the liver of interleukin-6-deficient mice after partial hepatectomy. *Hepatology* **29,** 403–411.

Sarraf, C., Lalani, E. N., Golding, M., Anilkumar, T. V., Poulsom, R., and Alison, M. (1994). Cell behavior in the acetylaminofluorene-treated regenerating rat liver. Light and electron microscopic observations. *Am. J. Pathol.* **145,** 1114–1126.

Schoental, R. (1968). Toxicology and carcinogenic action of pyrrolizidine alkaloids. *Cancer Res.* **28,** 2237–2246.

Schoental, R., and Magee, P. N. (1957). Chronic liver changes in rats after a single dose of lasiocarpine, a pyrrolizidine (senecio) alkaloid. *J. Pathol. Bacteriol.* **74,** 305–319.

Schoental, R., and Magee, P. N. (1959). Further observations on the subacute and chronic liver changes in rats after a single dose of various pyrrolizidine (Senecio) alkaloids. *J. Pathol. Bacteriol.* **78,** 471–482.

Sell, S. (1994). Liver stem cells. *Mod. Pathol.* **7,** 105–112.

Sell, S., and Leffert, H. L. (1982). An evaluation of cellular lineages in the pathogenesis of experimental hepatocellular carcinoma. *Hepatology* **2,** 77–86.

Sell, S., Leffert, H. L., Shinozuka, H., Lombardi, B., and Gochman, N. (1981a). Rapid development of large numbers of alpha-fetoprotein-containing "oval" cells in the liver of rats fed N-2-fluorenylacetamide in a choline-devoid diet. *Gann* **72,** 479–487.

Sell, S., Osborn, K., and Leffert, H. L. (1981b). Autoradiography of "oval cells" appearing rapidly in the livers of rats fed N-2-fluorenylacetamide in a choline devoid diet. *Carcinogenesis* **2,** 7–14.

Shinozuka, H., Lombardi, B., Sell, S., and Iammarino, R. M. (1978). Early histological and functional alterations of ethionine liver carcinogenesis in rats fed a choline-deficient diet. *Cancer Res.* **38,** 1092–1098.

Smith, P. G., Tee, L. B., and Yeoh, G. C. (1996). Appearance of oval cells in the liver of rats after long-term exposure to ethanol. *Hepatology* **23,** 145–154.

Stocker, E., and Heine, W. D. (1971). Regeneration of liver parenchyma under normal and pathological conditions. *Beitr. Pathol.* **144,** 400–408.

Tatematsu, M., Ho, R. H., Kaku, T., Ekem, J. K., and Farber, E. (1984). Studies on the proliferation and fate of oval cells in the liver of rats treated with 2-acetylaminofluorene and partial hepatectomy. *Am. J. Pathol.* **114,** 418–430.

Taub, R. (1995). Expression and function of growth-induced genes during liver regeneration. *In* "Liver Regneration and Carcinogenesis: Molecular and Cellular Interations," (R. L. Jirtle, Ed.), pp. 71–97. Academic Press, San Diego.

Taub, R. (1996). Liver regeneration 4: Transcriptional control of liver regeneration. *FASEB J.* **10,** 413–427.

Taub, R. (1999). IL-6 and liver growth regulation. *In* "Normal and Malignant Liver Cell Growth," (W. E. Fleig, Ed.), pp. 68–74. Kluwar Academic Publishers, Dordrecht.

Thompson, N. L., Mead, J. E., Braun, L., Goyette, M., Shank, P. R., and Fausto, N. (1986). Sequential protooncogene expression during rat liver regeneration. *Cancer Res.* **46,** 3111–3117.

Tournier, I., Legres, L., Schoevaert, D., Feldmann, G., and Bernuau, D. (1988). Cellular analysis of alpha-fetoprotein gene activation during carbon tetrachloride and D-galactosamine-induced acute liver injury in rats. *Lab. Invest.* **59,** 657–665.

Webber, E. M., FitzGerald, M. J., Brown, P. I., Bartlett, M. H., and Fausto, N. (1993). Transforming growth factor-alpha expression during liver regeneration after partial hepatectomy and toxic injury, and potential interactions between transforming growth factor-alpha and hepatocyte growth factor. *Hepatology* **18,** 1422–1431.

Webber, E. M., Godowski, P. J., and Fausto, N. (1994). In vivo response of hepatocytes to growth factors requires an initial priming stimulus. *Hepatology* **19,** 489–497.

Webber, E. M., Bruix, J., Pierce, R. H., and Fausto, N. (1998). Tumor necrosis factor primes hepatocytes for DNA replication in the rat. *Hepatology* **28,** 1226–1234.

Wilson, J. W., and Leduc, E. H. (1958). Role of cholangioles in restoration of the liver of the mouse after dietary injury. *J. Pathol. Bacteriol.* **76,** 441–449.

Wright, N., and Alison, M. (1984). The Biology of Epithelial Cell Populations. Clarendon Press, Oxford.

Yamada, Y., Kirillova, I., Peschon, J. J., and Fausto, N. (1997). Initiation of liver growth by tumor necrosis factor: Deficient liver regeneration in mice lacking type I tumor necrosis factor receptor. *Proc. Natl. Acad. Sci. USA* **94,** 1441–1446.

Yasuda, H., Mine, T., Shibata, H., Eto, Y., Hasegawa, Y., Takeuchi, T., Asano, S., and Kojima, I. (1993). Activin A: An autocrine inhibitor of initiation of DNA synthesis in rat hepatocytes. *J. Clin. Invest.* **92,** 1491–1496.

Yasui, O., Miura, N., Terada, K., Kawarada, Y., Koyama, K., and Sugiyama, T. (1997). Isolation of oval cells from Long-Evans Cinnamon rats and their transformation into hepatocytes in vivo in the rat liver. *Hepatology* **25,** 329–334.

Yaswen, P., Hayner, N. T., and Fausto, N. (1984). Isolation of oval cells by centrifugal elutriation and comparison with other cell types purified from normal and preneoplastic livers. *Cancer Res.* **44,** 324–331.

Yavorkovsky, L., Lai, E., Ilic, Z., and Sell, S. (1995). Participation of small intraportal stem cells in the restitutive response of the liver to periportal necrosis induced by allyl alcohol. *Hepatology* **21,** 1702–1712.

CHAPTER SIX

Adult Cardiac-Derived Stem Cells: Differentiation and Survival Regulators

Nanako Kawaguchi

Contents

I. Introduction 112
II. c-kit-Positive Cardiac Cells 113
A. Characterization of long-term cultures of c-kit-positive cardiac cells 113
B. Left atrium-derived c-kit-positive cells 115
C. Comparison between LA-PCs and other stem cells 115
D. Transcription factors 116
E. Microarray and pathway analysis 116
F. Cross talk between Wnt and TGF-β 118
G. Effect of TGF-β on myogenesis and adipogenesis 118
H. Noggin and cardiac myocyte development 119
I. GATA4 120
J. IGF-1 and the Akt pathway 120
III. Conclusions and Future Prospects 121
Acknowledgments 121
References 121

Abstract

At present, heart failure is one of the most concerning diseases worldwide. To develop efficient treatments, it is necessary to gain a better understanding of the biological characteristics of stem cells in the heart. We recently established and characterized c-kit-positive cardiac stem cells obtained from adult rats. Moreover, we established left atrium-derived pluripotent cells that can differentiate either into skeletal/cardiac myocytes or adipocytes in a methylcellulose-based Methocult medium with almost 100% purity. Microarray and signaling pathway analyses showed that transforming growth factor (TGF)-β is a key molecule in the regulation of the differentiation switch. Indeed, TGF-β1

Department of Patriotic Cardiology, Tokyo Women's Medical University, Tokyo, Japan

Vitamins and Hormones, Volume 87
ISSN 0083-6729, DOI: 10.1016/B978-0-12-386015-6.00041-X

simultaneously inhibits adipogenesis and activates myogenesis in a dose-dependent manner. However, the effect of TGF-β varies with the developmental stage, dosage, and timing of the treatment.

I. Introduction

Myocardial infarction is one of the leading causes of death in the Western world (Braunwald, 2001; Braunwald and Bristow, 2000). Heart failure is induced by a reduction in the number of viable cardiac myocytes (Braunwald and Pfeffer, 1991). Therefore, stem cell therapy was designed to compensate for the damaged cardiac myocytes by using stem cells that are capable of differentiating into cardiac myocytes or to compensate for the function of the surrounding cardiac myocytes by enhancing their viability through the release of soluble factors. The recently established inducible pluripotent stem (iPS) cells are considered the most useful cell source for cardiac myocyte differentiation, because they are autograft cells with a high ability of differentiation, similar to that of embryonic stem (ES) cells (Takahashi and Yamanaka, 2006; Takahashi *et al.*, 2007). Adult somatic stem cells have also gained increasing attention, and bone marrow stem cells are being clinically studied (Rubart and Field, 2008). Positive effects of bone marrow stem cell therapy have been reported (Tousoulis *et al.*, 2008). Remarkably, the paracrine factors secreted by the stem cells act on the surrounding myocytes and contribute to the therapeutic effects (Uemura *et al.*, 2006), suggesting that those stem cells enhance cardiac myocyte viability through the release of soluble factors. Adipose tissue-derived stem cells have been recently characterized, and their use as a source for cell therapy has been enthusiastically discussed (Gimble *et al.*, 2007; Madonna *et al.*, 2009; Planat-Bénard *et al.*, 2004; Yamada *et al.*, 2006).

Several groups have isolated resident stem cells from adult hearts (Barile *et al.*, 2007; Beltrami *et al.*, 2003; Bergmann *et al.*, 2009). These cells differentiate into cardiac myocytes *in vitro* as well as in an *in vivo* myocardial infarct model. Cardiac stem-like cells have been isolated from mouse, rat, dog, and human hearts based on the presence of cell surface antigens, namely stem cell antigen-1 (Sca-1; Matsuura *et al.*, 2004; Oh *et al.*, 2003), Abcg2 (Martin *et al.*, 2004; Pfister *et al.*, 2008), and c-kit. Spheroid-forming cells derived from endomyocardial biopsy specimens have also been used as a cell source in regenerative medicine (Smith *et al.*, 2007). Among the various previously reported cardiac stem cells (CSCs), c-kit-positive CSCs closely resemble *bona fide* stem cells with regard to their clonogenic, self-renewal, and multipotent characteristics. Based on the hypothesis that CSCs sustain stemness by overcoming senescence, we hypothesized that long-term cultures of more than 40 passages would favor the survival of

CSCs while eliminating contaminating fibroblasts. We have isolated and characterized long-term cultures of c-kit-positive cardiac cells (Miyamoto *et al.*, 2010; Machida *et al.*, in press).

Herein, we review the most recent studies conducted at other institutes and ours on stem cell biology, particularly those on c-kit-positive CSCs. Further, the role of TGF-β and other signaling molecules on differentiation, especially on myogenesis and adipogenesis, and their importance in regenerative medicine are discussed. The paracrine and autocrine regulators of stem cell function are also discussed.

II. c-kit-Positive Cardiac Cells

A. Characterization of long-term cultures of c-kit-positive cardiac cells

c-kit-positive cardiac cells are proliferative and sustain proliferation for over 50 passages (Beltrami *et al.*, 2003). The division of somatic cells such as fibroblasts or other differentiated cells is considered to terminate after a certain number of critical passages. We performed reverse transcriptase-polymerase chain reaction (RT-PCR) to study the stem cell capacities of c-kit-positive cardiac cells. In particular, we determined their multipotency at passage 40. We isolated and cultured c-kit-positive stem cells in a bulk 11 times and characterized them. These bulk-cultured cardiac stem cells (CSC-BC) express various cell-specific genes, suggesting heterogeneity. Not only do they express markers of three different cardiac lineages (cardiac myocytes, smooth muscle cells, and endothelial cells), as previously published, but they also express other cell type markers such as preadipocyte and skeletal/cardiac muscle lineages at various levels after long-term culture (Miyamoto *et al.*, 2010).

We also showed that 2 of 11 CSC-BCs had a preadipocyte and skeletal/cardiac myocyte phenotype (Table 6.1). Six of 11 CSC-BCs exhibited adipocyte characteristics. Based on these results, cultured c-kit-positive cells were classified as mesenchymal stem cells (MSCs) (Table 6.1). In addition to our group, another group has reported that c-kit-positive cells exhibit mesenchymal markers and prefer to commit to cardiovascular cells (Gambini *et al.*, 2011). Using the c-EGFP knock-in (at the *c-kit* locus) mice, c-kit-expressing cells were cardiac myocyte progenitors throughout the embryonic and neonatal period but were quite rarely detected in the adulthood (Tallini *et al.*, 2009). The reasons for the different results are described elsewhere (Kawaguchi, Submitted). Differences in methodology or culture conditions (e.g., medium and serum) may contribute to the different results, especially those obtained from *in vitro* studies (Anversa *et al.*, 2007; Hosoda *et al.*, 2010).

Table 6.1 Gene expression characterization of CSC-BCs after long-term culture by RT-PCR

Gene/CSC-BC	1	2	3	4	5	6	7	8	9	10	11
Oct4	−	−	−	+	+	±	−	−	++	−	−
GATA4	−	+	−	+++	+	−	−	−	−	++	++
Desmin	+	++	++	±	++	−	−	−	−	+++	−
MLC2v	−	−	−	−	+++	−	−	±	−	−	−
connexin43	−	+	+	+	+	+	+	+	++	++	−
cActin	−	−	−	−	++	−	−	+	−	−	−
SMA	−	−	+	++	++	++	++	++	++	++	−
PPARα	++	++	+	−	+	−	−	+	++	++	−
Myogenin	−	−	−	−	+++	−	−	±	−	−	−
Pecam	−	−	−	−	−	−	−	−	−	−	+++
GAPDH	+++	+++	+++	+++	+++	+++	+++	+++	+++	+++	+++
Cell type	Adi	Adi/Car	Adi	Car	Mes/Car	Car	Car	Mes	Adi	Car	Car

CSC-BC indicates cardiac stem cell-bulk cultured cells; Adi, preadipocyte type; Sph, sphere-forming; Car, cardiac. "–" indicates no band observed; ±, faint; +, visible; ++, clearly visible; +++, very visible.

B. Left atrium-derived c-kit-positive cells

We isolated c-kit-positive cells from various parts of the heart such as the left atrium, right atrium, left ventricle, right ventricle, septum, and apex. Cells isolated from all of the parts proliferated. However, c-kit-positive cells isolated from the left atrium grew better and could be cultured for a long period. A very small fraction of these cells was myoD and GATA4 positive, and differentiated into skeletal/cardiac myocytes when grown to confluence or in myocyte differentiation medium (Hasan *et al.*, 2010). Because these cells can differentiate into many types of cells and express nestin and other neural markers (as determined by microarray), we named these cells left atrium-derived pluripotent cells (LA-PCs). Further, unique skeletal muscle-like spindle-shaped cells appeared in cultures when the LA-PCs became confluent or were grown in the myocyte differentiation medium (Hasan *et al.*, 2010; Miyamoto *et al.*, 2010). Those cells could be differentiated into both cardiac and skeletal muscle cells in a methylcellulose-based Methocult culture medium (Stem Cell Technologies) supplemented with interleukin (IL)-3, IL-6, and stem cell factor (SCF). IL-3 and SCF or a combination of the 2 factors has been reported to contribute to myogenesis (Hasan *et al.*, 2010).

C. Comparison between LA-PCs and other stem cells

As discussed above, c-kit-positive stem cells obtained from the heart of adult rats resembled MSCs after long-term culturing. Further, carcinogenic P19 cells show similar differentiation abilities (Danalache *et al.*, 2007) and differentiate into cardiac/skeletal myocytes and adipocytes. Mesoangioblasts, which can differentiate into skeletal myocytes, vascular cells, and other mesodermal cells, also exhibit similar characteristics. Because skeletal and cardiac muscles have the same origin, namely the mesoderm, stem cells may also have the same origin. Interestingly, these mesoangioblasts are derived from the dorsal aorta or dorsal vessel (Minasi *et al.*, 2002), and express KIT, CD34, and Flk-1. LA-PCs and other CSC-BCs also express these genes, as determined by microarray (our unpublished results). Moreover, cardiac mesoangioblasts have been isolated from human cardiac myocyte biopsies (Galvez *et al.*, 2009). These cardiac mesoangioblasts are similar to the CSCs, which differentiate into three different cardiac lineages, as previously described by Beltrami *et al.* (2003). Mutations in c-kit—the receptor for SCF—induce anemia. Because the bone marrow is rich in c-kit-positive cells, which enter the circulation, it is possible that bone marrow-originating c-kit-positive cells participate in the recovery of the damaged heart (Fazel *et al.*, 2006).

Similar to LA-PCs, peripheral blood mononuclear cells described by Wojakowski *et al.* (2004) express CD34/CXCR4, CD34/CD117, c-met, early cardiac myocyte markers (GATA4 and MEF2C), skeletal muscle progenitor markers (Myf5, MyoD, and myogenin), and the endothelial specific marker VE-cadherin. They also exhibit the capacity to differentiate into cardiac and skeletal myocytes. They showed that in acute myocardial infarct patients, bone marrow CXCR4-expressing cells receive chemoattractant signals from stromal cell-derived factor (SDF)-1 and enter circulation via the bloodstream. Those stem-like cells function to repair damaged tissues such as the infarcted heart. Cytokines are important for induction of those signals (Galvez *et al.*, 2009). From our microarray data, LA-PCs express CXCR4 (both in undifferentiated cells and differentiated cells) and CD34 (only in differentiated cells). We also found SDF-1 expression by RT-PCR (data not shown).

D. Transcription factors

Avian MyoD-positive cells have been detected in the embryonic heart (Takebayashi-Suzuki *et al.*, 2001). These cells were identified as Purkinje cells that contribute to the heart conduction system. LA-PCs do not presumably express myoD or myogenin *in vivo*. We presume that these cells exist in the heart as highly immature cells. In fact, no study has indicated the presence of myoD/GATA4 double-positive cells in an adult rat heart. We believe that, after isolation or *in vitro* culture, some of these cells become myogenin/GATA4 positive. Therefore, cells may not exhibit the same phenotype as they do *in vivo*. However, not only LA-PCs but also other CSC-BCs show a phenotype similar to the phenotype of cells expressing myogenin after long-term culture (CSC-BC8 in Table 6.1; Miyamoto *et al.*, 2010), suggesting that the phenomenon reflects the nature of the CSCs. An explanation for the differentiation of cardiac/skeletal myocytes from LA-PCs and other cells is that the commonly used transcription factors (TFs) such as Mef2 can not only undergo regulation during early development (Potthoff and Olson, 2007) but also function when the somatic stem cells differentiate into cardiac and skeletal myocytes. Cells such as the P19 or the C2C12 cells might prove to be useful as a myocyte differentiation model. We found that these cells could differentiate into adipocytes or myocytes under certain conditions, as reported previously (Kawaguchi *et al.*, 2010a). Therefore, we tried to identify the signaling mediators that regulate the differentiation into adipocytes or skeletal/cardiac myocytes.

E. Microarray and pathway analysis

LA-PCs cultured in growth medium, (DMEM containing 10% ES cell-grade fetal bovine, 5% horse serum, 10 ng/ml of leukemia inhibitory factor (LIF), 5U/L erythropoietin (EPO)) are an undifferentiated (UND) and heterogeneous population in terms of stemness (Fig. 6.1). As previously reported by our group (Kawaguchi *et al.*, 2010a), a fraction of LA-PCs

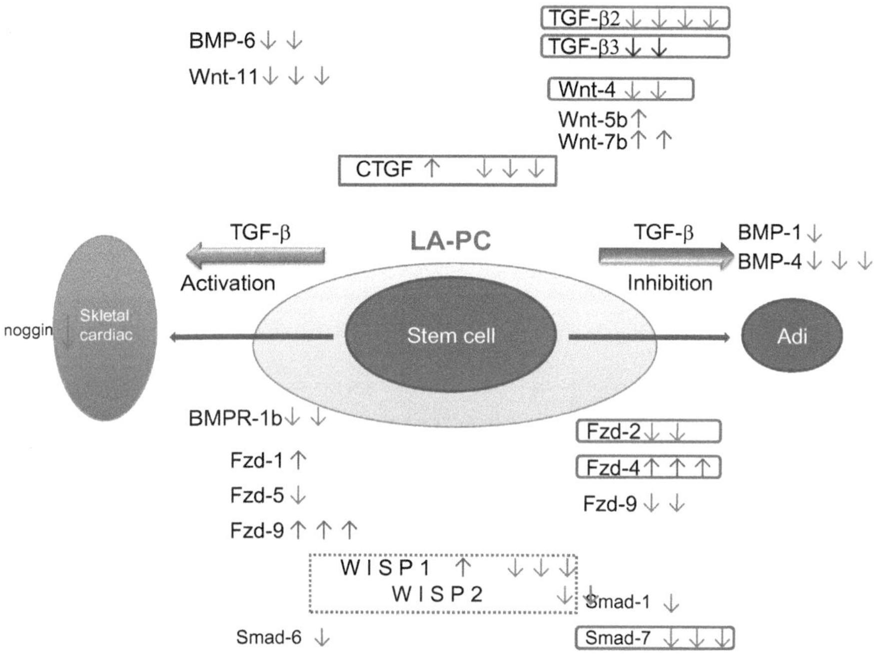

Figure 6.1 Transforming growth factor (TGF)-β and Wnt signal transduction in left atrium-derived pluripotent cell (LA-PC) differentiation. Genes marked with a red line were differentially expressed, as determined using both Agilent and Torey 3D-GENE methods. Blue arrows indicate expression levels lower than the levels in the control undifferentiated cells (↓ twofold, ↓↓ fourfold, and ↓↓↓ eightfold lower). Red arrows indicate expression levels higher than the levels in the control (↑ twofold, ↑↑ fourfold, and ↑↑↑ eightfold higher). (See Color Insert.)

preferably differentiates into mature adipocytes (Adi). In contrast, others differentiate into skeletal/cardiac myocytes (Myo) in the same Methocult medium (Stem Cell Technologies) supplemented with IL-3, IL-6, and SCF. Thus, LA-PCs are a mixed population consisting of UND cells and progenitors of Adi and/or Myo. To identify the signal transduction pathway or the signaling molecule responsible for the differentiation switch between these 2 cell groups (Adi and Myo), we performed microarray analyses for the three cell groups (UND, Adi, and Myo). Microarrays were performed using both Agilent and 3D-GENE (Torey) chips. Differentially expressed genes, as determined by both methods, were selected. We found that striated muscle contraction signaling is upregulated in Myo, whereas adipogenesis and the fatty acid metabolic pathway are upregulated in Adi (Kawaguchi *et al.*, 2010a). Hence, our *in vitro* model reflects the differentiation process. The TGF-β pathway was found to be a key pathway. Indeed,

TGF-β1 induces myogenesis and simultaneously inhibits adipogenesis in a dose-dependent manner. Our other data also support this; for example, TGF-R1 expression, as measured by RT-PCR, was inhibited in Adi, which is in accordance with the positive effect of TGF-β1 on myogenesis and its negative effect on adipogenesis. Further, follistatin (Fst) expression was shown to be upregulated in Myo but downregulated in Adi by RT-PCR, consistent with the microarray data.

F. Cross talk between Wnt and TGF-β

The results from the microarray analysis are shown in Fig. 6.1. Genes differently expressed in both analyses (Agilent and Torey 3D-GENE chips) are marked with a red bold line. Our data showed that TGF-β 2 and TGF-β3 were strongly downregulated during differentiation into Adi. Further, the expression of Wnt and its receptor, frizzled (fzd), was altered in Adi, suggesting the involvement of Wnt signaling. Indeed, Wnt-4 was also downregulated in Adi. However, although Fzd-2 was downregulated, Fzd-4 was upregulated. Smad-7 was strongly downregulated in Adi. Cross talk between Wnt and TGF-β has been previously suggested (Gilbert, 2003; Singh *et al.*, 2009). Human MSCs were reported to stimulate chondrocyte differentiation and inhibit adipocyte differentiation, based on microarray data, and these results were confirmed by RT-PCR and western blotting (Zhou *et al.*, 2004). Connective tissue growth factor (CTGF) and Wnt 1-inducible signaling pathway protein 1 (WISP1) might participate as the molecular switches for differentiation into either myocytes or adipocytes, because their expression was upregulated in Myo and downregulated in Adi. However, in myogenesis, it has been suggested that TGF-β induces CTGF, which acts negatively (Vial *et al.*, 2008). In adipocyte differentiation, it was reported that, similar to TGF-β, CTGF also inhibits adipogenesis (Kirkland *et al.*, 2002). It is necessary to deeply investigate the molecules involved in these differentiation pathways.

G. Effect of TGF-β on myogenesis and adipogenesis

We found that TGF-β1 could be a possible positive regulator for myogenesis. However, it has been shown that it works—at least in the myoblastic C2C12 cell line—as a negative regulator for myogenesis during myogenic maturation via a Smad3-mediated mechanism (Krueger and Hoffmann, 2010). In contrast, Smad7 activation affects myogenesis (Miyake *et al.*, 2010) because of the inhibitory function of Smad7 on TGF-β.

The different functions can result from the different doses of TGF-β, because the function is dose-dependent. Further, we assume that differences in the initial stemness stage affect the outcome. TGF-β maintains the stemness capacities of not only the adult stem cells but also the primitive stem cells such as ES cells (Kitisin *et al.*, 2007; Pucceat, 2006; Wu and Hill, 2009). The dose and timing of the treatment can yield completely different results (Pucceat, 2006). The effect of TGF-β was different with a different cell source. In embryonic myoblast differentiation, TGF-β acts positively (Slager *et al.*, 1993). However, on more mature types of myoblasts such as the C2C12 cells, TGF-β acts negatively (Massague *et al.*, 1986; Olson *et al.*, 1986). Interestingly, TGF-β1 enhances cardiac myocyte differentiation of skeletal muscle-derived adult primitive cells, especially in the fraction of Sca-1-positive cells (Abdel-Latif *et al.*, 2008). Thus, TGF-β can act as a differentiation switch in many ways. TGF-β signaling in adipogenesis can negatively affect differentiation through binding of Smad3/4 to C/EBP TFs (Choy and Derynck, 2003). Smad-mediated signaling mechanisms have been described in detail by Diike and Heiden (2003).

LA-PCs are considered more primitive with respect to stemness. Therefore, TGF-β1 acts as a positive regulator of myogenesis. Alternatively, epithelial–mesenchymal transition (EMT) can contribute to the switch. TGF-β was found to play an important role in EMT, more specifically in myofibrosis of epithelial cells. EMT is very important for the induction of multiple organogenesis in the embryo as well. This transition accompanies the differentiation of pluripotent cells and may be related to cancer (Mani *et al.*, 2008; Yang and Weinberg, 2008). Myogenesis induced by TGF-β1 may concomitantly occur with myofibrosis. Further, CTGF was upregulated during Myo differentiation, and CTGF is involved in fibrosis. EMT can be included as one of the differentiation switches. TGF-β can regulate cell cycle proteins such as CDK2 and Id (Derynck and Miyazono, 2008). Taken together, TGF-β can act on certain cells, including endothelial cells, or immature cells, c-kit-positive cells, for induction of pluripotency and induction of mesenchymal cells.

H. Noggin and cardiac myocyte development

Although we have failed to obtain a large amount of cardiac myocytes from LA-PCs at the first report (Hasan *et al.*, 2010), we have found that noggin was downregulated in our microarray analysis. To confirm the function of noggin, we added noggin to the Methocult culture. To our surprise, noggin had a significant effect on the induction of cardiac myocytes rather than skeletal myocytes. Another group also reported that noggin was an activator of cardiac myogenesis (Yuasa *et al.*, 2005). Considering that TGF-β is expressed in early cardiac development (Gilbert, 2003), and that BMP-2 and BMP-4 play key roles in mesoderm morphogenesis, noggin and BMP proteins as well as TGF-β potentially play roles in regulating mesoderm-derived cells. Thus, the TGF-β superfamily can work as a differentiation switch in various ways.

I. GATA4

GATA4 is important not only in cardiac development (Molkentin *et al.*, 2000) but also in various cardiac functions. GATA4 has been related to cardiac myocyte hypertrophy (Kobayashi *et al.*, 2007; Liang *et al.*, 2001; Oka *et al.*, 2006). Recently, the *GATA4* gene was reported to be required for the direct reprogramming of cardiac myocytes from cardiac fibroblasts (Ieda *et al.*, 2010). On the other hand, GATA4 was reported to regulate cell survival and the expression of survival-promoting cytokines such as fibroblast growth factor (FGF)-4 (Iwahori *et al.*, 2004). GATA4 is expressed in cardiac myocytes but not in adult skeletal muscles, and at high levels in neonatal myocytes but at relatively low levels in mature adult cardiac myocytes (data not shown). One of the cultures of cardiac c-kit-positive-derived cells (CSC-BCs) showed enhanced expression of GATA4. We found that these cells had higher ability to transform into troponin I-positive cells than did CSC-BCs with low GATA4 expression, suggesting that GATA4 can be a TF that induces cardiac myocytes. Interestingly, the clones with high expression of GATA4 showed enhanced sustaining of the beating of the cocultured adult cardiac myocytes despite the use of inserts. Therefore, the cytokines released from $GATA4^{high}$ CSCs enhance the survival of cardiac myocytes. This effect has been suggested to be another role of CSCs.

J. IGF-1 and the Akt pathway

The enhancement of cardiac myocytes survival has been suggested to be associated with IGF-1 secretion level (Kawaguchi *et al.*, 2010b). Moreover, $GATA4^{high}$ CSCs were able to enhance the IGF-1 protein level in the medium cocultured with adult cardiac myocytes by 4–10-fold the level induced by the cardiac myocytes alone. There was almost no IGF-1 detected, neither at the RNA nor at the protein level, in $GATA4^{high}$ CSCs alone, suggesting that $GATA4^{high}$ CSCs can release some cytokines that promote IGF-1 release from cardiac myocytes (Kawaguchi *et al.*, 2010a,b; Miyamoto *et al.*, 2010). Our preliminary data, based on microarray data only, suggested that the matrix gla protein is a potential candidate mediator.

The role of IGF-1 was not clear, but IGF-1 was reported to enhance proliferation of CSCs (Torella *et al.*, 2004). Other researchers and we showed that IGF-1 works through the Akt pathway (Gude *et al.*, 2006; Kawaguchi *et al.*, 2010a,b; Mangi *et al.*, 2003; Mcdevitt *et al.*, 2005).

III. Conclusions and Future Prospects

Atrium tissue-derived LA-PCs are considered more immature than other previously reported CSCs, which mainly differentiate into three different cardiac lineages. In avian species, MyoD/GATA4-positive cells appear only at early stages of heart development. Therefore, TGF-β functions as a positive regulator of myogenesis, in contrast to the adult myoblasts such as C2C12 cells, wherein TGF-β1 can work as a negative regulator of myogenesis. In LA-PCs, noggin induces differentiation into cardiac myocytes rather than into skeletal myocytes. This may help obtain a high number of cardiac myocytes that will be adequate to function in regenerative medicine in the event of heart failure.

Rodent and murine stem cells serve as sources for cell therapy research. Stem cells not only supply the differentiated cells but also have the potential to supply paracrine factors for damaged tissue repair. The molecular biological approach using these cells for differentiation or for maintenance of stemness also provides knowledge for further applications such as drug discovery. Further, this knowledge may help the development of novel therapies. Discussion and exchange of information will contribute to rapid progress in this field.

Acknowledgments

This project was supported by the Encouraging Development of Strategic Research Centers, Special Coordination Funds for Promoting Science and Technology, and Grant in Aid for Scientific Research (A), Ministry of Education, Culture, Sports, Science, and Technology, Japan.

References

Abdel-Latif, A., Zuna-Surma, E. K., Case, J., Tiwari, S., Hunt, G., Ranjan, S., Vincent, R. J., Srour, E. F., Bolli, R., and Dawn, B. (2008). TGF-β1 enhances cardiomyogenic differentiation of skeletal muscle-derived adult primitive cells. *Basic Res. Cardiol.* **103,** 514–524.

Anversa, P., Leri, A., Rota, M., Hosoda, T., Bearzi, C., Urbanek, K., Kajstura, J., and Bolli, R. (2007). Concise review: Stem cells myocardial regeneration and methodological artifacts. *Stem Cells* **25,** 589–601.

Barile, L., Messina, E., Giacomello, A., and Marbán, E. (2007). Endogenous cardiac stem cells. *Prog. Cardiovasc. Dis.* **50,** 31–48.

Beltrami, A., Barlucchi, L., Torella, D., Baker, M., Limana, F., Chimenti, S., Kasahara, H., Rota, M., Musso, E., Urbanek, E., Leri, A., Kajstura, J., *et al.* (2003). Adult cardiac stem cells are multipotent and support myocardial regeneration. *Cell* **114,** 763–776.

Bergmann, O., Bhardwaj, R. D., Bernard, S., Zdunek, S., Barnabé-Heider, F., Walsh, S., Zupicich, J., Alkass, K., Buchholz, B. A., Druid, H., Jovinge, S., and Frisén, J. (2009). Evidence for cardiomyocyte renewal in humans. *Science* **3**(24), 98–102.

Braunwald, E. (2001). The Denolin lecture. Congestive heart failure: A half century perspective. *Eur. Heart J.* **22,** 825–836.
Braunwald, E., and Bristow, M. (2000). Congestive heart failure: Fifty years of progress. *Circulation* **102,** IV14–IV23.
Braunwald, E., and Pfeffer, M. (1991). Ventricular enlargement and remodeling following acute myocardial infarction: Mechanisms and management. *Am. J. Cardiol.* **68,** 1D–6D.
Choy, L., and Derynck, R. (2003). Transforming growth factor-beta inhibits adipocyte differentiation by Smad3 interacting with CCAAT/enhancer-binding protein (C/EBP) and repressing C/EBP transactivation function. *J. Biol. Chem.* **278,** 9609–9619.
Danalache, B., Paquin, J., Donghao, W., Gyrgorczyk, R., Moore, J. C., Mummery, C. L., Gutowska, J., and Jankowski, M. (2007). Nitric oxide signaling in oxytocin-mediated cardiomyogenesis. *Stem Cells* **25,** 679–688.
Derynk, R. and Miyazono, K. (Eds.) (2008). The TGF-β Family. Cold Spring Harbor Laboratory, New York, USA.
Diike, P. and Heiden, C. H. (Eds.) (2003). Smad Signal Transduction. Springer, Netherlands.
Fazel, S., Cimini, M., Chen, L., Li, S., Angoulvant, D., Fedak, P., Verma, S., Weisel, R. D., Keating, A., and Li, R. K. (2006). Cardioprotective c-kit+ cells are from the bone marrow and regulate the myocardial balance of angiogenic cytokines. *J. Clin. Invest.* **116,** 1865–1877.
Galvez, B. R., Covarello, D., Tolorenzi, R., Brunelli, S., Dellavalle, A., Crippa, S., Mohammed, S. A., Scialla, L., Cuccovillo, I., Molla, F., Staszewsky, L., Maisano, F., *et al.* (2009). Human cardiac mesoangioblasts isolated from hypertrophic cardiomyopathies are greatly reduced in proliferation and differentiation potency. *Cardiovasc. Res.* **83,** 707–716.
Gambini, E., Pompilio, G., Biondi, A., Alamanni, F., Capogrossi, M. C., Agrifoglio, M., and Pesce, M. (2011). c-kit + cardiac progenitors exhibit mesenchymal markers and preferential cardiovascular commitment. *Cardiovasc. Res.* **89,** 362–373.
Gilbert, S. F. (2003). Developmental Biology. 7th edn. Sinauer Associates, Inc. Publishers.
Gimble, J., Katz, A., and Bunnell, B. (2007). Adipose-derived stem cells for regenerative medicine. *Circ. Res.* **100,** 1249–1260.
Gude, N., Muraski, J., Rubio, M., Kajstura, J., Schaefer, E., Anversa, P., and Sussman, M. A. (2006). Akt promotes increased cardiomyocyte cycling and expansion of the cardiac progenitor cell population. *Circ. Res.* **99,** 381–388.
Hasan, M. K., Komoike, Y., Tsunesumi, S., Nakao, R., Nagao, H., Matsuoka, R., and Kawaguchi, N. (2010). Myogenic differentiation in atrium-derived adult cardiac pluripotent cells and the transcriptional regulation of GATA4 and myogenin on ANP promoter. *Genes Cells* **15,** 439–454.
Hosoda, T., Kajstura, J., Leri, A., and Anversa, P. (2010). Mechanism of myocardial regeneration. *Circ. J.* **74,** 13–17.
Ieda, M., Fu, J., Delgado-Olguin, P., Vadantham, V., Hayashi, Y., Bruneau, B. G., and Srivastava, D. (2010). Direct reprogramming of fibroblasts into functional cardiomyocytes by defined factors. *Cell* **142,** 375–386.
Iwahori, A., Fraidenraich, D., and Basilico, C. (2004). A conserved enhancer element that drives FGF4 gene expression in the embryonic myotomes is synergistically activated by GATA and bHLH proteins. *Dev. Biol.* **270,** 525–537.
Kawaguchi, N. Stem cells for cardiac regeneration and possible roles of the transforming growth factor-β family. (Submitted).
Kawaguchi, N., Nakao, R., Yamaguchi, M., Ogawa, D., and Matsuoka, R. (2010a). TGF-beta superfamily regulates a switch that mediates differentiation either into adipocytes or myocytes in left atrium derived pluripotent cells (LA-PCS). *Biochem. Biophys. Res. Commun.* **396,** 619–625.

Kawaguchi, N., Smith, A. J., Waring, C. D., Hasan, M. K., Miyamoto, S., Matsuoka, R., and Ellison, G. M. (2010b). c-kitpos GATA-4 high rat cardiac stem cells foster adult cardiomyocyte survival through IGF-1 paracrine signalling. *PLoS One* **5,** e14297.

Kirkland, J. L., Tchkonia, T., Pirtskhalava, T., Han, J., and Karagiannides, I. (2002). Adipogenesis and aging: Does aging make fat go MAD? *Exp. Gerontol.* **37,** 757–767.

Kitisin, K., Saha, T., Blake, T., Golestaneh, N., Deng, M., Kim, C., Tang, Y., Shetty, K., Mishra, B., and Mishra, L. (2007). TGF-beta signaling in development. *Sci. STKE* **399,** cm1.

Kobayashi, S., Mao, K., Zheng, H., Wang, X., Patterson, C., O'Connell, T. D., and Liang, Q. (2007). Diminished GATA4 protein levels contribute to hyperglycemia-induced cardiomyocyte injury. *J. Biol. Chem.* **282,** 21945–21952.

Krueger, C., and Hoffmann, F. M. (2010). Identification of retinoic acid in a high content screen for agents that overcome the anti-myogenic effect of TGF-β1. *PLoS One* **5,** e15511.

Liang, Q., De Windt, L., Witt, S., Kimball, T., Markham, B., and Molkentin, J. (2001). The transcription factors GATA4 and GATA6 regulate cardiomyocyte hypertrophy in vitro and in vivo. *J. Biol. Chem.* **276,** 30245–30253.

Machida, M., Takagaki, Y., Matsuoka R., and Kawaguchi N. Proteomic comparison of spherical aggregates and adherent cells of cardiac stem cells. *Int. J. Cardiol.*, in press.

Madonna, R., Geng, Y., and De Caterina, R. (2009). Adipose tissue-derived stem cells characterization and potential for cardiovascular repair. *Arterioscler. Thromb. Vasc. Biol.* **29,** 1723–1729.

Mangi, A., Noiseux, N., Kong, D., He, H., Rezvani, M., Ingwall, J. S., and Dzau, V. J. (2003). Mesenchymal stem cells modified with Akt prevent remodeling and restore performance of infarcted hearts. *Nat. Med.* **9,** 1195–1201.

Mani, S. A., Guo, W., Liao, M. J., Eaton, E. N., Ayyanan, A., Zhou, A. Y., Brooks, M., Reinhard, F., Zhang, C. C., Shiptisin, M., Campbell, L. L., Polyak, K., *et al.* (2008). The epithelial-mesenchymal transition generates cells with properties of stem cells. *Cell* **133,** 704–715.

Martin, C., Meeson, A., Robertson, S. M., Hawke, T. J., Richardson, J. A., Bates, S., Goetsch, S. C., Gallardo, T. D., and Garry, D. J. (2004). Persistent expression of the ATP-binding cassette transporter Abcg2 identifies cardiac SP cells in the developing and adult heart. *Dev. Biol.* **265,** 262–275.

Massague, J., Checifetz, S., Endo, T., and Nadal-Ginard, B. (1986). Type βtransforming growth factor is an inhibitor of myogenic differentiation. *Proc. Natl. Acad. Sci. USA* **83,** 8206–8210.

Matsuura, K., Nagai, T., Nishigaki, N., Oyama, T., Nishi, J., Wada, H., Sano, M., Toko, H., Akazawa, H., Sato, T., Nakaya, H., Kasanuki, H., *et al.* (2004). Adult cardiac Sca-1-positive cells differentiate into beating cardiomyocytes. *J. Biol. Chem.* **279,** 11384–11391.

Mcdevitt, T., Laflamme, M., and Murry, C. (2005). Proliferation of cardiomyocytes derived from human embryonic stem cells is mediated via the IGF/PI 3-kinase/Akt signaling pathway. *J. Mol. Cell. Cardiol.* **39,** 865–873.

Minasi, M. G., Riminucci, M., De Angelis, L., Borello, U., Berarducci, B., Innocenzi, A., Caprioli, A., Sirabella, D., Baiocchi, M., Maria, R. D., Boratto, R., Jaffredo, T., *et al.* (2002). The meso-angioblast: A multipotent, self-renewing cell that originates from the dorsal aorta and differentiates into most mesodermal tissues. *Development* **129,** 2773–2783.

Miyake, T., Alli, N. S., and McDermott, C. M. (2010). Nuclear function of Smad7 promotes myogenesis. *Mol. Cell. Biol.* **30,** 722–735.

Miyamoto, S., Kawaguchi, N., Ellison, G., Matsuoka, R., Shin'oka, T., and Kurosawa, H. (2010). Characterization of long-term cultured c-kitpos cardiac stem cells (CSCs) derived from adult rat hearts. *Stem Cells Dev.* **19,** 105–116.

Molkentin, J., Antos, C., Mercer, B., Taigen, T., Miano, J., and Olson, E. (2000). Direct activation of a GATA6 cardiac enhancer by Nkx25: Evidence for a reinforcing regulatory network of Nkx25 and GATA transcription factors in the developing heart. *Dev. Biol.* **217,** 301–309.

Oh, H., Bradfute, S., Gallardo, T. D., Nakamura, T., Gaussin, V., Mishina, Y., Pocius, J., Micheal, L. H., Behringer, R. R., Garry, D. J., Entman, M. L., and Schneider, M. D. (2003). Cardiac progenitor cells from adult myocardium: Homing differentiation and fusion after infarction. *Proc. Natl. Acad. Sci. USA* **100,** 12313–12318.

Oka, T., Maillet, M., Watt, A. J., Schwartz, R. J., Aronow, B. J., Duncan, S. A., and Molkentin, J. D. (2006). Cardiac-specific deletion of Gata4 reveals its requirement for hypertrophy compensation and myocyte viability. *Circ. Res.* **98,** 837–845.

Olson, E. N., Sternberg, E., Hu, J. S., Spizz, G., and Wilcox, C. (1986). Regulation of myogenic differentiation by type β transforming growth factor. *J. Cell Biol.* **103,** 1799–1805.

Pfister, O., Oikonomopoulos, A., Sereti, K. I., Sohn, R. L., Cullen, D., Fine, G. C., Monquet, F., Westerman, K., and Liao, R. (2008). Role of the ATP-binding cassette transporter Abcg2 in the phenotype and function of cardiac side population cells. *Circ. Res.* **103,** 825–835.

Planat-Bénard, V., Menard, C., André, M., Puceat, M., Perez, A., Garcia-Verdugo, J. M., Pénicaud, L., and Casteilla, L. (2004). Spontaneous cardiomyocyte differentiation from adipose tissue stroma cells. *Circ. Res.* **94,** 223–229.

Potthoff, M., and Olson, E. (2007). MEF2: A central regulator of diverse developmental programs. *Development* **134,** 4131–4140.

Pucceat, M. (2006). TGF-β in the differentiation of embryonic stem cells. *Cardiovasc. Res.* **74,** 256–261.

Rubart, M., and Field, J. L. (2008). Stem cell differentiation: Cardiac repair. *Cells Tissue Organs* **188,** 202–211.

Singh, R., Bhasin, S., Braga, M., Artaza, J. N., Pervin, S., Taylor, W. E., Krishnan, V., Sinha, S. K., Rajavashisth, T. B., and Jasuja, R. (2009). Regulation of myogenic differentiation by androgens: Cross talk between androgen receptor/beta-catenin and follistatin/transforming growth factor-beta signaling pathways. *Endocrinology* **150,** 1259–1268.

Slager, H. G., Van Inzen, W., Freund, E., Van den Eijnden-Van Raaij, A. J., and Mummery, C. L. (1993). Transforming growth factor-β in the early mouse embryo: Implications for the regulation of muscle formation and implantation. *Dev. Genet.* **14,** 212–224.

Smith, R., Barile, L., Cho, H., Leppo, M. K., Hare, J. M., Messina, E., Giacommolo, A., Abraham, M. R., and Marban, E. (2007). Regenerative potential of cardiosphere-derived cells expanded from percutaneous endomyocardial biopsy specimens. *Circulation* **115,** 896–908.

Takahashi, K., and Yamanaka, S. (2006). Induction of pluripotent stem cells from mouse embryonic and adult fibroblast cultures by defined factors. *Cell* **126,** 663–676.

Torella, D., Rota, M., Nurzynska, D., Musso, E., Monsen, A., Shiraishi, I., Zias, E., Walsh, K., Rosenzweig, A., Sussman, M. A., Urbanek, K., Nadal-Ginard, B., *et al.* (2004). Cardiac stem cell and myocyte aging, heart failure, and insulin-like growth factor-1 overexpression. *Circ. Res.* **94,** 514–524.

Takahashi, K., Tanabe, K., Ohnuki, M., Narita, M., Ichisaka, T., Tomoda, K., and Yamanaka, S. (2007). Induction of pluripotent stem cells from adult human fibroblasts by defined factors. *Cell* **131,** 861–872.

Takebayashi-Suzuki, K., Pauliks, L., Eltsefon, Y., and Mikawa, T. (2001). Purkinje fibers of the avian heart express a myogenic transcription factor program distinct from cardiac and skeletal muscle. *Dev. Biol.* **234,** 390–401.

Tallini, Y., Greene, K., Craven, M., Spealman, A., Breitbach, M., Smith, J., Fisher, P. J., Steffey, M., Hesse, M., Doran, R. M., Woods, A., Singh, B., *et al.* (2009). c-Kit expression identifies cardiovascular precursors in the neonatal heart. *Proc. Natl. Acad. Sci. USA* **106,** 1808–1813.

Tousoulis, D., Briasoulis, A., Antoniades, C., Stefanadi, E., and Stefanadis, C. (2008). Heart regeneration: What cells to use and how? *Curr. Opin. Pharmacol.* **8,** 211–218.

Uemura, R., Xu, M., Ahmad, N., and Ashraf, M. (2006). Bone marrow stem cells prevent left ventricular remodeling of ischemic heart through paracrine signaling. *Circ. Res.* **98,** 1414–1421.

Vial, C., Zúñiga, L. M., Cabello-Verrugio, C., Cañón, P., Fadic, R., and Brandan, E. (2008). Skeletal muscle cells express the profibrotic cytokine connective tissue growth factor (CTGF/CCN2), which induces their dedifferentiation. *J. Cell. Physiol.* **215,** 410–421.

Wojakowski, W., Tendera, M., Michalowska, A., Majka, M., Kucia, M., Malanskiewicz, K., Wyderka, R., Ochala, A., and Ratajczak, M. Z. (2004). Mobilization of CD34/CXCR4+, CD34/CD117+, c-met+ stem cells, and mononuclear cells expressing early cardiac, muscle, and endothelin markers into peripheral blood in patients with acute myocardial infarction. *Circulation* **110,** 3213–3220.

Wu, M. Y., and Hill, C. S. (2009). TGF-β superfamily signaling in embryonic development and homeostasis. *Dev. Cell* **16,** 329–343.

Yamada, Y., Wang, X., Yokoyama, S., Fukuda, N., and Takakura, N. (2006). Cardiac progenitor cells in brown adipose tissue repaired damaged myocardium. *Biochem. Biophys. Res. Commun.* **342,** 662–670.

Yang, J., and Weinberg, R. A. (2008). Epithelial-mesenchymal transition: At the crossroads of development and tumor metastasis. *Dev. Cell* **14,** 818–829.

Yuasa, S., Itabashi, Y., Koshimizu, U., Tanaka, T., Sugimura, K., Kinoshita, M., Hattori, F., Fukami, S., Shimazaki, T., Ogawa, S., Okano, H., and Fukuda, K. (2005). Transient inhibition of BMP signaling bu noggin induces cardiomyocyte differentiation of mouse embryonic stem cells. *Nat. Biotechnol.* **23,** 607–611.

Zhou, S., Eid, K., and Glowacki, J. (2004). Cooporation between TGF-beta and Wnt pathways during chondrocyte and adipocyte differentiation of human marrow stromal cells. *J. Bone Miner. Res.* **19,** 463–470.

CHAPTER SEVEN

TGF-β1 Regulates Differentiation of Bone Marrow Mesenchymal Stem Cells

Longmei Zhao *and* Basil M. Hantash

Contents

I. Bone Marrow Mesenchymal Stem Cells 128
A. Overview 128
B. Heterogeneous nature 128
C. Differentiation potential 129
II. The Role of TGF-β1 in Differentiation of Bone Marrow MSCs 131
A. TGF-β1 and TGF-β signaling 131
B. TGF-β1 induces chondrogenic differentiation of bone marrow MSCs 133
C. TGF-β1 regulates osteogenic differentiation of bone marrow MSCs 134
D. TGF-β1 inhibits adipogenic differentiation of bone marrow MSCs 135
E. TGF-β1 mediates bone marrow MSC differentiation into other lineages 135
F. The molecular mechanism underlying TGF-β1-mediated MSC differentiation 136
III. Summary 137
Acknowledgments 137
References 137

Abstract

Mesenchymal stromal/stem cells (MSCs) are a small population of stromal cells present in most adult connective tissues, such as bone marrow, fat tissue, and umbilical cord blood. MSCs are maintained in a relative state of quiescence *in vivo* but, in response to a variety of physiological and pathological stimuli, are capable of proliferating then differentiating into osteoblasts, chondrocytes, adipocytes, or other mesoderm-type lineages like smooth muscle cells (SMCs) and cardiomyocytes. Multiple signaling networks orchestrate MSCs differentiating into functional

Escape Therapeutics, Inc., San Jose, California, USA

Vitamins and Hormones, Volume 87
ISSN 0083-6729, DOI: 10.1016/B978-0-12-386015-6.00042-1

mesenchymal lineages. Among these, transforming growth factor-β1 (TGF-β1) has emerged as a key player. Hence, we summarize the effects of TGF-β1 on differentiation of MSCs toward different lineages. TGF-β1 can induce either chondrogenic or SMC differentiation of MSCs *in vitro*. However, it requires cell–cell and cell–matrix interactions, similar to development of these tissues *in vivo*. The effect of TGF-β1-regulated osteogenic differentiation of MSCs *in vitro* depends on the specific culture conditions involved. TGF-β1 inhibits adipogenic differentiation of MSCs in monolayer culture. Using this information, we may optimize the culture conditions to differentiate MSCs into desired lineages.

I. Bone Marrow Mesenchymal Stem Cells

A. Overview

Stem cells are undifferentiated multipotent precursor cells that share two characteristic properties: unlimited or prolonged self-renewal and potential for differentiation. Multipotent stem cell populations found in adult tissues have been of great interest because they serve as reservoirs for tissue repair and regeneration after trauma, disease, and aging. One important type of adult stem cell is mesenchymal stromal/stem cells (MSCs), a small population of stromal cells present in most adult connective tissues. MSCs were first indentified by Friedenstein *et al.* who demonstrated that a rare population of plastic-adherent cells (~1 in 10,000 nucleated cells) in bone marrow were able to form single cell-derived colonies at low cell density (Friedenstein *et al.*, 1970). The colonies consist of spindle-shaped cells known as colony-forming unit-fibroblasts (Fig. 7.1).

Although MSCs were originally isolated from bone marrow, similar populations have been reported in other tissues. Besides bone marrow, human MSCs have been isolated from adipose tissue (Fig. 7.1) (Zuk *et al.*, 2001), umbilical cord blood (Erices *et al.*, 2000), peripheral blood (Marinova-Mutafchieva *et al.*, 2000), amniotic fluid (In't Anker *et al.*, 2003b), placenta (In't Anker *et al.*, 2004), liver (Campagnoli *et al.*, 2001), lung (In't Anker *et al.*, 2003a), dermis (Toma *et al.*, 2001), skeletal muscle (Jiang *et al.*, 2002), and others. However, the most well-characterized MSCs remain adult bone marrow MSCs.

B. Heterogeneous nature

Bone marrow is a complex tissue comprising hematopoietic precursors and stromal cells, the latter of which are a heterogeneous mixture of cells including adipocytes, reticulocytes, endothelial cells, and fibroblastic cells (Bruder *et al.*, 1997). Direct plating from bone marrow aspirates is the generally accepted method of bone marrow MSC isolation and expansion. The direct plating

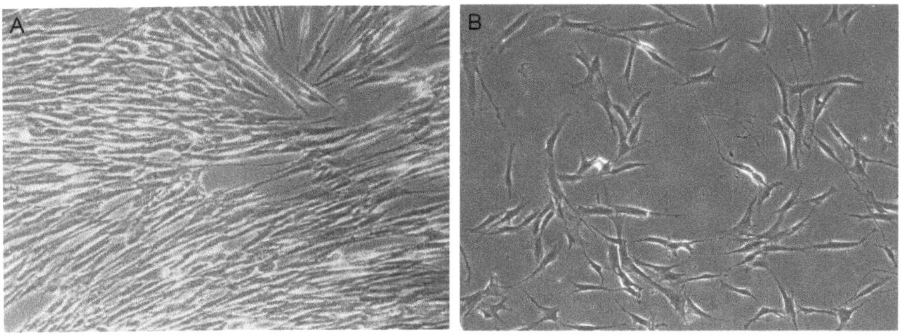

Figure 7.1 Morphology of human bone marrow MSCs (A) and adipose-derived MSCs (B). Cells were cultured in α-MEM (A) or in DMEM (B) containing 1% penicillin/streptomycin and 10% FBS.

method and heterogeneous nature of bone marrow lead to heterogeneity of bone marrow MSCs with respect to cell phenotype, colony size, and differentiation potential. Four types of cells are observed in primary culture of bone marrow MSCs: spindle-shaped cells, star-shaped cells, large flat cells (Xiao *et al.*, 2010), and small round cells (Colter *et al.*, 2001). Even single-cell-derived clonal MSC populations are also highly heterogeneous in their proliferative and differentiation potentials (De Bari *et al.*, 2008; Phinney and Prockop, 2007). In addition, the lack of unified practice for the culture and propagation of bone marrow MSCs results in a wide diversity in MSC isolates across the many research laboratories and clinics that handle them. At present, there remains no single unique specific cell surface marker to identify these cell populations. Thus, The International Society for Cellular Therapy provided the following minimum criteria for defining multipotent human MSCs: (1) plastic-adherent under standard culture conditions; (2) positive for expression of CD105, CD73, and CD90, and negative for expression of hematopoietic cell surface markers CD34, CD45, CD11a, CD19, and HLA-DR; (3) under specific stimuli, cells should differentiate into adipocytes, osteoblasts, and chondrocytes *in vitro* (Horwitz *et al.*, 2005).

C. Differentiation potential

Bone marrow MSCs are maintained in a relative state of quiescence *in vivo* but, in response to a variety of physiological and pathological stimuli, are capable of proliferation then differentiation. It is well established that MSCs are able to differentiate into the chondrogenic, osteogenic, and adipogenic lineages (Fig. 7.2). During chondrogenic differentiation, bone marrow MSCs change from a characteristic fibroblast-like morphology to a large round shape and produce extracellular matrix (ECM), containing a highly

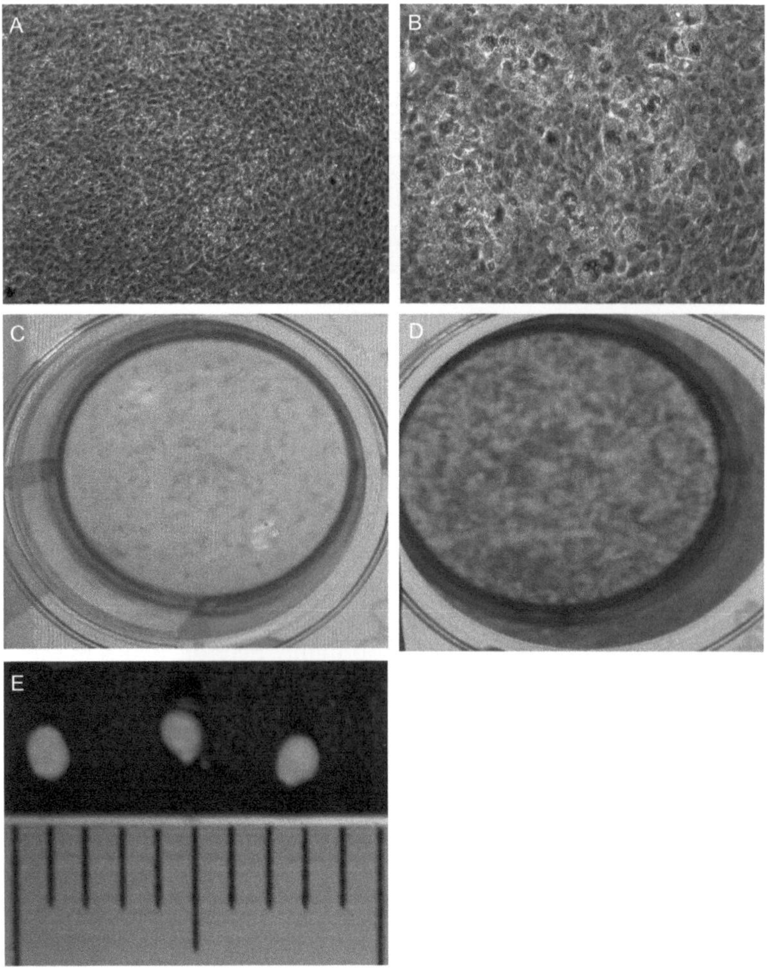

Figure 7.2 Differentiation potential of MSCs. Human bone marrow MSCs were seeded on tissue culture plate and induced to differentiate into adipocytes by culturing the cells in adipogenic media (Cell Applications, San Diego, CA) (B), or regular α-MEM media (A, control) for 14 days and then stained with Oil red O. For osteogenic differentiation, human adipose-derived MSCs were seeded on tissue culture plates and cultured in osteogenic media (Cell Applications) (D), or regular DMEM media (C, control) for 21 days and then stained with alizarin red. Cellular nuclei were count-stained with hematoxylin (C and D). To induce chondrogenic differentiation, human adipose-derived MSCs were cultured as pellets in chondrogenic media (Cell Applications) (E). After 21 days, differentiation into the chondrogenic lineage was visualized. (See Color Insert.)

organized type II collagen and proteoglycans and glycosaminoglycans (Vater *et al.*, 2011). When being differentiated into osteoblasts, bone marrow MSCs transform to a cubic shape and produce ECM, mainly composed of type I collagen (Vater *et al.*, 2011). Increased expression of alkaline phosphatase (ALP) and calcium accumulation is observed in MSCs during osteogenic differentiation. When being differentiated into adipocytes, fibroblastic MSCs are converted to a spherical shape expressing several types of ECM proteins, including fibronectin, laminin, and types I, III, IV, V, and VI collagen (Vater *et al.*, 2011). Accumulation of intracellular lipid-rich vacuoles inside the cells can be stained positively by oil red O.

Under appropriate conditions, MSCs can also differentiate into other mesenchymal lineages such as smooth muscle cells (SMCs) (Gong and Niklason, 2008; Kanematsu *et al.*, 2005; Kinner *et al.*, 2002; Seruya *et al.*, 2004), skeletal myocytes (Wakitani *et al.*, 1995), cardiomyocytes (Gwak *et al.*, 2009), and tenocytes (De Bari *et al.*, 2003; Hoffmann *et al.*, 2006). In addition, researchers have recently transdifferentiated MSCs into nonmesodermal cell types such as neuronal-like cells (Black and Woodbury, 2001) and pancreatic cell progenitors (Moriscot *et al.*, 2005; Timper *et al.*, 2006). The clinical relevance of the presumptive nonmesenchymal potency of MSCs is, however, questioned because, for example, MSC-derived neuron-like cells were unable to generate action potentials and, therefore, to function as neurons (Hofstetter *et al.*, 2002).

Multiple signaling networks orchestrate the development and differentiation of MSCs into functional mesenchymal lineages. Among these, transforming growth factor-β (TGF-β) proteins have emerged as key players in the self-renewal, maintenance of stem cells in their undifferentiated state, and the progression of differentiation along an individual lineage. Here, we illustrate the role of TGF-β1 in differentiation of bone marrow MSCs (see Fig. 7.3 for overview).

II. The Role of TGF-β1 in Differentiation of Bone Marrow MSCs

A. TGF-β1 and TGF-β signaling

TGF-β1 is a 25-kDa disulfide-linked homodimeric peptide, belonging to the TGF-β family. TGF-β1 has two closely related mammalian isoforms (TGF-β2 and -β3) and shares a 64–85% amino acid sequence homology with them. Despite this high-sequence homology, they are functionally nonredundant (Dickson *et al.*, 1995; Geiser *et al.*, 1998; Kulkarni *et al.*, 1993; Proetzel *et al.*, 1995; Shull *et al.*, 1992). The gene encoding TGF-β1 is located in 19q13. Unlike the other two isoforms, TGF-β1 is extensively expressed in almost all tissues (Massague, 1990; Moses and Serra, 1996).

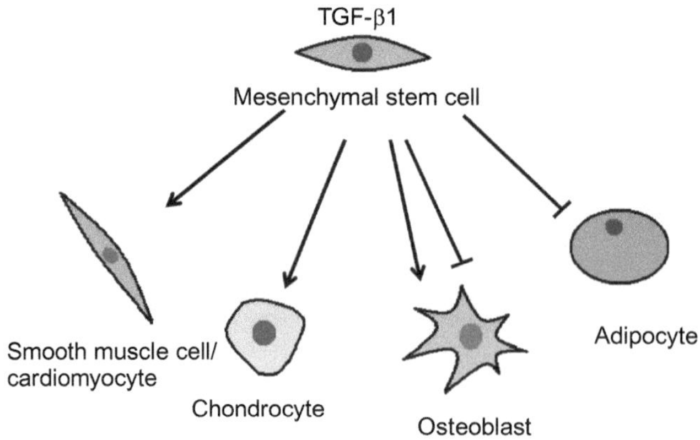

Figure 7.3 Schematic overview of the effects (stimulation or inhibition) of TGF-β1 on differentiation of MSCs toward different lineage. ↓ indicates stimulation, ⊥ indicates inhibition. (For color version of this figure, the reader is referred to the Web version of this chapter.)

TGF-β1 is a multifunctional growth factor, which regulates a broad range of biological processes, including cell proliferation, cell survival, cell differentiation, cell migration, and production of ECM (Massague *et al.*, 2000; Siegel and Massague, 2003). These combined actions mediate the global effect of TGF-β1 on many developmental processes and maintenance of normal tissue homeostasis (Massague, 1998). Combined with its stimulatory effect on MSC proliferation, TGF-β1 signaling thus allows for expansion of MSCs and their progenitors (Chen *et al.*, 2004).

TGF-β1 initiates signaling by its extracellular domain binding two types (type I and type II) of transmembrane receptor serine–threonine kinases, which form a complex at the cell surface. Ligand binding to this complex induces a conformational change that induces phosphorylation and activation of type I receptors by type II receptors. The activated receptors subsequently phosphorylate the effectors Smad2/Smad3. Phosphorylated Smad2/Smad3 then form complexes with the common Smad (Smad4) and translocate into the nucleus, where they interact at the promoter with other transcription factors at DNA sequence-specific binding sites ATF2 (activating transcription factor-2) and SBE (Smad binding element) to regulate gene expression. The heteromeric Smad complex in the nucleus also interacts with various transcriptional coactivators or corepressors resulting in the activation or the repression of downstream target genes (Derynck, 1998; Massague, 1998). TGF-β1 also induces non-SMAD signaling pathways by activation of the mitogen-activated protein kinase (MAPK) pathway (extracellular signal-regulated kinase-1 (ERK-1), c-Jun N-terminal kinase, and p38) through upstream mediators such as TGF-β-activated kinase (TAK1) (Hocevar *et al.*, 1999; Yu *et al.*, 2002).

B. TGF-β1 induces chondrogenic differentiation of bone marrow MSCs

The "natural" mesenchymal propensity of MSCs has prompted researchers to devote attention to their chondrogenic and osteogenic differentiation potential. TGF-β1 plays an important role in cartilage development and is a well-documented potent chondrogenic factor. One of the earliest identified activities of TGF-β1 was the induction of chondrogenesis in primitive rat mesenchymal cells *in vitro* (Rosen *et al.*, 1986; Seyedin *et al.*, 1986). This chondrogenic effect induced by TGF-β1 has since been observed in rabbit chondrocyte cultures (Kato *et al.*, 1988), chicken mesenchymal cells (Leonard *et al.*, 1991), and bovine nasal and articular chondrocytes culture (Xu *et al.*, 1996).

In 1998, Johnstone and colleagues first demonstrated that TGF-β1 induces chondrogenic differentiation of bone marrow MSCs *in vitro* (Johnstone *et al.*, 1998). They developed a "pellet" culture system that allows cell–cell interactions analogous to those that occur in precartilage condensation during embryonic development. Rabbit bone marrow cell pellet preparations was cultured in the defined medium supplied with TGF-β1 and 10% fetal bovine serum. The induction of chondrogenesis was accompanied by enhanced mRNA levels of both type IIA and IIB collagen, two of the most important ECM components in cartilage, and increased ALP activity of the aggregated cells (Johnstone *et al.*, 1998). Since then, the role of TGF-β1 in cartilage tissue engineering has been investigated extensively by cultivating bone marrow MSCs with various biomaterials in three-dimensional (3D) systems. Bosnakovski *et al.* demonstrated that chondrogenic capacity of bovine bone marrow MSCs was greatly enhanced when cultured in type II collagen hydrogel with media containing TGF-β1, even though the hydrogel alone had the potential to induce and maintain MSC chondrogenesis (Bosnakovski *et al.*, 2006). The study of Park's group showed that chondrogenesis was only evident in rabbit bone marrow MSCs encapsulated in the hydrogel containing TGF-β1-loaded gelatin microparticles, and chondrocyte-specific gene expression was varied with TGF-β1 concentration in a dose-dependent manner (Park *et al.*, 2007). The hydrogel may function as both the scaffold of MSCs and the matrix of TGF-β1 release, resulting in enhanced MSC aggregation and the consequent promotion of cell proliferation and chondrogenic differentiation (Ogawa *et al.*, 2010). In the study by Xia *et al.*, bone marrow MSCs adenovirally transduced with the human TGF-β1 gene were encapsulated into a biodegraded scaffold and implanted into the mouse dorsa subcutaneous tissue (Xia *et al.*, 2009). At 3-weeks post-implantation, TGF-β1 significantly increased the volume of neocartilage tissue and the amount of type II collagen and sulfated proteoglycans (a late chondrogenic marker) in neocartilage tissue. *In vivo*, sustained production of TGF-β1, albeit at lower levels, was sufficient for the induction of chondrogenic differentiation of bone marrow MSCs.

C. TGF-β1 regulates osteogenic differentiation of bone marrow MSCs

TGF-β1 plays a pivotal role in bone regeneration because it was proven to affect both bone formation and bone resorption (Janssens *et al.*, 2005). TGF-β1 is secreted by osteoblasts as well as by bone marrow MSCs and is stored in bone matrix (Liu *et al.*, 1999; Robey *et al.*, 1987). TGF-β1 effectively stimulates the formation of collagen I (Arnold *et al.*, 2002; Hock *et al.*, 1990; Liu *et al.*, 1999), the main matrix protein of bone (Anselme, 2000), which already was proved by various studies. Bone marrow MSCs are a major source of osteoprogenitor cells (Owen and Friedenstein, 1988). Thus, TGF-β1 was believed to promote bone formation through stimulation of proliferation and differentiation of bone marrow MSCs, the osteoblast precursors. Although this seems straightforward, the story is much more complicated because many seemingly contradictory reports have been published.

The direct positive effect of TGF-β1 on osteogenic differentiation of bone marrow MSCs *in vitro* was reported by Zhao *et al.* (2010). After 14 days of treatment with TGF-β1 (10 ng/mL), murine bone marrow MSCs underwent osteogenic differentiation by increasing Runx-2 (a global regulator of osteogenesis), type I collagen and osteopontin (two osteoblast differentiation markers) as well as ALP activity in monolayer culture (Zhao *et al.*, 2010). Several investigators reported a biphasic effect of TGF-β1 on osteogenic capacity of bone marrow MSCs synergistically with other osteogenic inducers. Low concentrations of TGF-β1 (0.1–1 ng/mL) stimulated osteogenic differentiation, while high concentrations of TGF-β1 (10 ng/mL) were inhibitory (Lieb *et al.*, 2004; Liu *et al.*, 1999). Liu *et al.* reported that ALP activity was enhanced in human bone marrow MSCs by cotreatment with TGF-β1 (0.1–10 ng/mL) plus Vitamin D_3 [1,25$(OH)_2D_3$] (Liu *et al.*, 1999). Both ALP activity and osteocalcin expression were suppressed by high doses of TGF-β1 (single treatment at 10 ng/mL) in rat bone marrow MSCs cultivated with dexamethasone, ascorbic acid, β-glycerol phosphate, common chemical osteogenic inducers (Lieb *et al.*, 2004). From the above, it is clear that the effect of TGF-β1 on *in vitro* osteogenic differentiation of bone marrow MSCs is highly dependent on a broad range of experimental conditions such as cell density, the dosage of TGF-β1, the presence of serum, amongst others.

D. TGF-β1 inhibits adipogenic differentiation of bone marrow MSCs

TGF-β1 has been shown to be a strong inhibitor of adipogenesis in 3T3 fibroblasts (Ignotz and Massague, 1985). TGF-β1 pathway was reported to mediate the suppressive effects of genistein (and 17β-estradiol) on adipogenic differentiation of human bone marrow MSCs. Blocking the TGF-β1 pathway abolished the genistein-induced decrease in protein levels of adipocyte-specific peroxisome proliferation-activated receptor γ2 (PPARγ2), one of the transcription factors that regulates expression of genes responsible for induction and progression of adipogenesis (Devine *et al.*, 1999; Rosen *et al.*, 1999). This led to a reduction in the proliferation rate of precursor cells (Heim *et al.*, 2004). Recently, Zhao *et al.* demonstrated a direct inhibition effect of TGF-β1 on adipogenic differentiation of murine bone marrow MSCs (Zhao *et al.*, 2010). PPARγ2 and adipsin (a late adipogenic marker) were decreased by TGF-β1 in murine bone marrow MSC monolayer cultures (Zhao *et al.*, 2010).

E. TGF-β1 mediates bone marrow MSC differentiation into other lineages

TGF-β1 modulation of MSC differentiation involves not only the three lineages mentioned above but also other lineages such as SMCs and cardiomyocytes. TGF-β1 signaling contributes to development of SMCs from embryonic stem cells (Becker *et al.*, 2008) and upregulates a variety of SMC differentiation markers in cultured SMCs derived from mature blood vessels (Kennard *et al.*, 2008). The first evidence of TGF-β1's role in SMC differentiation of bone marrow MSCs was from Kinner's study. They found that TGF-β1 significantly increased alpha-smooth muscle actin (α-SMA, an early marker of SMC differentiation) expression and the contractility of human bone marrow MSCs (Kinner *et al.*, 2002). Gong *et al.* showed that a concentration of 0.1–10 ng/mL TGF-β1 inhibited human MSC proliferation but increased calponin (a late-stage SMC differentiation marker) expression in a dose-dependent manner (Gong and Niklason, 2008), indicating that TGF-β1 not only initiates SMC differentiation but also promotes further differentiation. Similar to chondrogenesis, cell–cell contact plays an important role in TGF-β 1-induced SMC differentiation. Rat bone marrow MSCs plated on type IV collagen-coated surfaces and exposed to TGF-β1 differentiated into a homogeneous population expressing α-SMA and calponin (Seruya *et al.*, 2004).

TGF-β1 is also known to play a key role in embryonic heart development (Akhurst *et al.*, 1990) and to induce cardiomyocyte differentiation of mouse embryonic stem cells (Behfar *et al.*, 2002). When treated with TGF-β1, both murine bone marrow MSCs and rat adipose-derived MSCs increased expression of cardiac-specific markers, such as troponin I, troponin T, cardiac myosin heavy chain, and α-sarcomeric actin, suggesting that TGF-β1 may also promote MSC cardiomyogenic differentiation *in vitro* (Gwak *et al.*, 2009; Li *et al.*, 2005).

F. The molecular mechanism underlying TGF-β1-mediated MSC differentiation

TGF-β1-mediated MSC chondrogenic differentiation was revealed through non-SMAD signaling pathways, MAPK signaling and Wnt signaling cascades (Tuli *et al.*, 2003). During chondrogenic differentiation, TGF-β1 treatment activates p38 and ERK-1 to promote sox9 expression, which in turn form transactivating complexes with other proteins, for example, Sox5/Sox6 (Vater *et al.*, 2011), to control expression of the chondrocyte-specific genes (collagen, aggrecan, and cartilage link proteins). TGF-β1-mediated MAPK activation also controls *WNT-7A* gene expression and Wnt-mediated signaling through the intracellular β-catenin–TCF pathway, which regulates N-cadherin expression and subsequent N-cadherin-mediated cell–adhesion complexes during the early steps of MSC chondrogenesis (Tuli *et al.*, 2003).

At the molecular level, the central regulation of bone differentiation and formation is controlled by the transcriptional activity of Runx2 and TAZ (transcriptional coactivator with PDZ-binding motif). Runx2 is a transcription factor required for bone formation (Komori *et al.*, 1997; Otto *et al.*, 1997) and is a common target of bone morphogenetic protein- and TGF-β-induced osteoblast-specific genes expression in pluripotent mesenchymal precursor cells (Lee *et al.*, 2000; Zhao *et al.*, 2010). TAZ functions as a transcriptional modulator to control R-Smad–Smad4 complex nucleocytoplasmic shuttling as well as stem cell self-renewal and differentiation (Hong *et al.*, 2005; Varelas *et al.*, 2008). The net effect of TGF-β1 stimulating osteogenic differentiation while simultaneously blocking the differentiation of MSCs into fat occurs through activating TAZ by the Smad-dependent pathway. The direct interaction between TAZ and the transcription factors Runx2 and PPARγ results in transcriptional enhancement and repression, respectively, of selective programs of gene expression (Hong and Yaffe, 2006; Hong *et al.*, 2005; Zhao *et al.*, 2010).

SMC-specific transcription is regulated by transcription factors, GATA-binding protein 6 (GATA-6), and serum response factor (SRF), the latter binds to CArG (CC(A=T)6GG) *cis* elements that are found in the promoters of almost all SMC marker genes. Deaton *et al.* reported that TGF-β1 induced SMC differentiation of MSCs through the activation of a small GTPase RhoA-driven signaling cascade (Deaton *et al.*, 2005). The signaling pathway involves RhoA/threonine protein kinase N-mediated activation of p38 MAPK, which in turn activates GATA and SRF leading to upregulation of SMC marker gene expression. Other downstream targets of TGF-β1 and RhoA, such as Rho kinase and Smads, may also play a role in mediating these effects on MSCs (Deaton *et al.*, 2005).

III. Summary

MSCs are a promising source of precursor cells which may be applied in various tissue engineering strategies. By using differentiation-specific protocols, MSCs can be induced to differentiate towards a variety of mature target cells. TGF-β1 signaling plays an important role in the regulation of MSCs at both transcriptional and posttranscriptional levels (Kurpinski *et al.*, 2009), and a precise combination of microenvironmental cues may promote or inhibit MSC differentiation. In general, TGF-β1 can induce either chondrogenic or SMC differentiation of MSCs *in vitro*. However, it requires cell–cell and cell–matrix interactions, similar to development of these tissues *in vivo*. The effect of TGF-β1-regulated osteogenic differentiation of MSCs *in vitro* depends on the specific culture conditions involved. TGF-β1 inhibits adipogenic differentiation of MSCs in monolayer culture. The presence of other growth factors in the environment influences the exact outcome of TGF-β1 functioning because TGF-β1 signaling through Smads cross talk with many other signaling pathways such as Wnt, and Smads interact with a multitude of DNA-binding transcription factors, which themselves are targeted by signaling pathways. In order to develop more effective regenerative therapies with MSCs the combination of serum and supplementary agents such as TGF-β1 or other growth factors and cytokines and their concentration should be optimized to meet each individual need.

ACKNOWLEDGMENTS

We apologize to those researchers whose work was not included in the review because of space constraints. We thank Chris Nye for providing the pictures of MSC differentiation. We thank Dr. Yiou Li for assistance with preparing Fig. 7.3.

REFERENCES

Akhurst, R. J., *et al.* (1990). TGF beta in murine morphogenetic processes: The early embryo and cardiogenesis. *Development* **108,** 645–656.

Anselme, K. (2000). Osteoblast adhesion on biomaterials. *Biomaterials* **21,** 667–681.

Arnold, U., *et al.* (2002). In vitro-cultivation of human periosteum derived cells in bioresorbable polymer-TCP-composites. *Biomaterials* **23,** 2303–2310.

Becker, C., *et al.* (2008). TGFbeta-1 and epithelial-mesenchymal interactions promote smooth muscle gene expression in bone marrow stromal cells: Possible application in therapies for urological defects. *Int. J. Artif. Organs* **31,** 951–959.

Behfar, A., *et al.* (2002). Stem cell differentiation requires a paracrine pathway in the heart. *FASEB J.* **16,** 1558–1566.

Black, I. B., and Woodbury, D. (2001). Adult rat and human bone marrow stromal stem cells differentiate into neurons. *Blood Cells Mol. Dis.* **27,** 632–636.

Bosnakovski, D., *et al.* (2006). Chondrogenic differentiation of bovine bone marrow mesenchymal stem cells (MSCs) in different hydrogels: Influence of collagen type II extracellular matrix on MSC chondrogenesis. *Biotechnol. Bioeng.* **93,** 1152–1163.

Bruder, S. P., *et al.* (1997). Monoclonal antibodies reactive with human osteogenic cell surface antigens. *Bone* **21,** 225–235.

Campagnoli, C., *et al.* (2001). Identification of mesenchymal stem/progenitor cells in human first-trimester fetal blood, liver, and bone marrow. *Blood* **98,** 2396–2402.

Chen, D., *et al.* (2004). Bone morphogenetic proteins. *Growth Factors* **22,** 233–241.

Colter, D. C., *et al.* (2001). Identification of a subpopulation of rapidly self-renewing and multipotential adult stem cells in colonies of human marrow stromal cells. *Proc. Natl. Acad. Sci. USA* **98,** 7841–7845.

De Bari, C., *et al.* (2003). Skeletal muscle repair by adult human mesenchymal stem cells from synovial membrane. *J. Cell Biol.* **160,** 909–918.

De Bari, C., *et al.* (2008). A biomarker-based mathematical model to predict bone-forming potency of human synovial and periosteal mesenchymal stem cells. *Arthritis Rheum.* **58,** 240–250.

Deaton, R. A., *et al.* (2005). Transforming growth factor-beta1-induced expression of smooth muscle marker genes involves activation of PKN and p38 MAPK. *J. Biol. Chem.* **280,** 31172–31181.

Derynck, R. (1998). SMAD proteins and mammalian anatomy. *Nature* **393,** 737–739.

Devine, J. H., *et al.* (1999). Adipose expression of the phosphoenolpyruvate carboxykinase promoter requires peroxisome proliferator-activated receptor gamma and 9-cis-retinoic acid receptor binding to an adipocyte-specific enhancer in vivo. *J. Biol. Chem.* **274,** 13604–13612.

Dickson, M. C., *et al.* (1995). Defective haematopoiesis and vasculogenesis in transforming growth factor-beta 1 knock out mice. *Development* **121,** 1845–1854.

Erices, A., *et al.* (2000). Mesenchymal progenitor cells in human umbilical cord blood. *Br. J. Haematol.* **109,** 235–242.

Friedenstein, A. J., *et al.* (1970). The development of fibroblast colonies in monolayer cultures of guinea-pig bone marrow and spleen cells. *Cell Tissue Kinet.* **3,** 393–403.

Geiser, A. G., *et al.* (1998). Decreased bone mass and bone elasticity in mice lacking the transforming growth factor-beta1 gene. *Bone* **23,** 87–93.

Gong, Z., and Niklason, L. E. (2008). Small-diameter human vessel wall engineered from bone marrow-derived mesenchymal stem cells (hMSCs). *FASEB J.* **22,** 1635–1648.

Gwak, S. J., *et al.* (2009). In vitro cardiomyogenic differentiation of adipose-derived stromal cells using transforming growth factor-beta1. *Cell Biochem. Funct.* **27,** 148–154.

Heim, M., *et al.* (2004). The phytoestrogen genistein enhances osteogenesis and represses adipogenic differentiation of human primary bone marrow stromal cells. *Endocrinology* **145,** 848–859.

Hocevar, B. A., *et al.* (1999). TGF-beta induces fibronectin synthesis through a c-Jun N-terminal kinase-dependent, Smad4-independent pathway. *EMBO J.* **18,** 1345–1356.

Hock, J. M., *et al.* (1990). Transforming growth factor-beta stimulates bone matrix apposition and bone cell replication in cultured fetal rat calvariae. *Endocrinology* **126,** 421–426.

Hoffmann, A., *et al.* (2006). Neotendon formation induced by manipulation of the Smad8 signalling pathway in mesenchymal stem cells. *J. Clin. Invest.* **116,** 940–952.

Hofstetter, C. P., *et al.* (2002). Marrow stromal cells form guiding strands in the injured spinal cord and promote recovery. *Proc. Natl. Acad. Sci. USA* **99,** 2199–2204.

Hong, J. H., and Yaffe, M. B. (2006). TAZ: A beta-catenin-like molecule that regulates mesenchymal stem cell differentiation. *Cell Cycle* **5,** 176–179.

Hong, J. H., *et al.* (2005). TAZ, a transcriptional modulator of mesenchymal stem cell differentiation. *Science* **309,** 1074–1078.

Horwitz, E. M., *et al.* (2005). Clarification of the nomenclature for MSC: The International Society for Cellular Therapy position statement. *Cytotherapy* **7,** 393–395.

Ignotz, R. A., and Massague, J. (1985). Type beta transforming growth factor controls the adipogenic differentiation of 3T3 fibroblasts. *Proc. Natl. Acad. Sci. USA* **82,** 8530–8534.

In't Anker, P. S., *et al.* (2003a). Amniotic fluid as a novel source of mesenchymal stem cells for therapeutic transplantation. *Blood* **102,** 1548–1549.

In't Anker, P. S., *et al.* (2003b). Nonexpanded primary lung and bone marrow-derived mesenchymal cells promote the engraftment of umbilical cord blood-derived CD34(+) cells in NOD/SCID mice. *Exp. Hematol.* **31,** 881–889.

In't Anker, P. S., *et al.* (2004). Isolation of mesenchymal stem cells of fetal or maternal origin from human placenta. *Stem Cells* **22,** 1338–1345.

Janssens, K., *et al.* (2005). Transforming growth factor-beta1 to the bone. *Endocr. Rev.* **26,** 743–774.

Jiang, Y., *et al.* (2002). Multipotent progenitor cells can be isolated from postnatal murine bone marrow, muscle, and brain. *Exp. Hematol.* **30,** 896–904.

Johnstone, B., *et al.* (1998). In vitro chondrogenesis of bone marrow-derived mesenchymal progenitor cells. *Exp. Cell Res.* **238,** 265–272.

Kanematsu, A., *et al.* (2005). Induction of smooth muscle cell-like phenotype in marrow-derived cells among regenerating urinary bladder smooth muscle cells. *Am. J. Pathol.* **166,** 565–573.

Kato, Y., *et al.* (1988). Terminal differentiation and calcification in rabbit chondrocyte cultures grown in centrifuge tubes: Regulation by transforming growth factor beta and serum factors. *Proc. Natl. Acad. Sci. USA* **85,** 9552–9556.

Kennard, S., *et al.* (2008). Transforming growth factor-beta (TGF-1) down-regulates Notch3 in fibroblasts to promote smooth muscle gene expression. *J. Biol. Chem.* **283,** 1324–1333.

Kinner, B., *et al.* (2002). Regulation of smooth muscle actin expression and contraction in adult human mesenchymal stem cells. *Exp. Cell Res.* **278,** 72–83.

Komori, T., *et al.* (1997). Targeted disruption of Cbfa1 results in a complete lack of bone formation owing to maturational arrest of osteoblasts. *Cell* **89,** 755–764.

Kulkarni, A. B., *et al.* (1993). Transforming growth factor beta 1 null mutation in mice causes excessive inflammatory response and early death. *Proc. Natl. Acad. Sci. USA* **90,** 770–774.

Kurpinski, K., *et al.* (2009). Proteomic profiling of mesenchymal stem cell responses to mechanical strain and TGF-beta1. *Cell Mol. Bioeng.* **2,** 606–614.

Lee, K. S., *et al.* (2000). Runx2 is a common target of transforming growth factor beta1 and bone morphogenetic protein 2, and cooperation between Runx2 and Smad5 induces osteoblast-specific gene expression in the pluripotent mesenchymal precursor cell line C2C12. *Mol. Cell. Biol.* **20,** 8783–8792.

Leonard, C. M., *et al.* (1991). Role of transforming growth factor-beta in chondrogenic pattern formation in the embryonic limb: Stimulation of mesenchymal condensation and fibronectin gene expression by exogenous TGF-beta and evidence for endogenous TGF-beta-like activity. *Dev. Biol.* **145,** 99–109.

Li, T. S., *et al.* (2005). Regeneration of infarcted myocardium by intramyocardial implantation of ex vivo transforming growth factor-beta-preprogrammed bone marrow stem cells. *Circulation* **111,** 2438–2445.

Lieb, E., *et al.* (2004). Effects of transforming growth factor beta1 on bonelike tissue formation in three-dimensional cell culture. II: Osteoblastic differentiation. *Tissue Eng.* **10,** 1414–1425.

Liu, P., *et al.* (1999). Regulation of osteogenic differentiation of human bone marrow stromal cells: Interaction between transforming growth factor-beta and 1,25(OH)(2) vitamin D(3) in vitro. *Calcif. Tissue Int.* **65,** 173–180.

Marinova-Mutafchieva, L., *et al.* (2000). Mesenchymal cells expressing bone morphogenetic protein receptors are present in the rheumatoid arthritis joint. *Arthritis Rheum.* **43,** 2046–2055.

Massague, J. (1990). The transforming growth factor-beta family. *Annu. Rev. Cell Biol.* **6,** 597–641.

Massague, J. (1998). TGF-beta signal transduction. *Annu. Rev. Biochem.* **67,** 753–791.

Massague, J., *et al.* (2000). TGFbeta signaling in growth control, cancer, and heritable disorders. *Cell* **103,** 295–309.

Moriscot, C., *et al.* (2005). Human bone marrow mesenchymal stem cells can express insulin and key transcription factors of the endocrine pancreas developmental pathway upon genetic and/or microenvironmental manipulation in vitro. *Stem Cells* **23,** 594–603.

Moses, H. L., and Serra, R. (1996). Regulation of differentiation by TGF-beta. *Curr. Opin. Genet. Dev.* **6,** 581–586.

Ogawa, T., *et al.* (2010). In vitro proliferation and chondrogenic differentiation of rat bone marrow stem cells cultured with gelatin hydrogel microspheres for TGF-beta1 release. *J. Biomater. Sci. Polym. Ed.* **21,** 609–621.

Otto, F., *et al.* (1997). Cbfa1, a candidate gene for cleidocranial dysplasia syndrome, is essential for osteoblast differentiation and bone development. *Cell* **89,** 765–771.

Owen, M., and Friedenstein, A. J. (1988). Stromal stem cells: Marrow-derived osteogenic precursors. *Ciba Found. Symp.* **136,** 42–60.

Park, H., *et al.* (2007). Injectable biodegradable hydrogel composites for rabbit marrow mesenchymal stem cell and growth factor delivery for cartilage tissue engineering. *Biomaterials* **28,** 3217–3227.

Phinney, D. G., and Prockop, D. J. (2007). Concise review: Mesenchymal stem/multipotent stromal cells: The state of transdifferentiation and modes of tissue repair—Current views. *Stem Cells* **25,** 2896–2902.

Proetzel, G., *et al.* (1995). Transforming growth factor-beta 3 is required for secondary palate fusion. *Nat. Genet.* **11,** 409–414.

Robey, P. G., *et al.* (1987). Osteoblasts synthesize and respond to transforming growth factor-type beta (TGF-beta) in vitro. *J. Cell Biol.* **105,** 457–463.

Rosen, D. M., *et al.* (1986). Differentiation of rat mesenchymal cells by cartilage-inducing factor. Enhanced phenotypic expression by dihydrocytochalasin B. *Exp. Cell Res.* **165,** 127–138.

Rosen, E. D., *et al.* (1999). PPAR gamma is required for the differentiation of adipose tissue in vivo and in vitro. *Mol. Cell* **4,** 611–617.

Seruya, M., *et al.* (2004). Clonal population of adult stem cells: Life span and differentiation potential. *Cell Transplant.* **13,** 93–101.

Seyedin, S. M., *et al.* (1986). Cartilage-inducing factor-A. Apparent identity to transforming growth factor-beta. *J. Biol. Chem.* **261,** 5693–5695.

Shull, M. M., *et al.* (1992). Targeted disruption of the mouse transforming growth factor-beta 1 gene results in multifocal inflammatory disease. *Nature* **359,** 693–699.

Siegel, P. M., and Massague, J. (2003). Cytostatic and apoptotic actions of TGF-beta in homeostasis and cancer. *Nat. Rev. Cancer* **3,** 807–821.

Timper, K., *et al.* (2006). Human adipose tissue-derived mesenchymal stem cells differentiate into insulin, somatostatin, and glucagon expressing cells. *Biochem. Biophys. Res. Commun.* **341,** 1135–1140.

Toma, J. G., *et al.* (2001). Isolation of multipotent adult stem cells from the dermis of mammalian skin. *Nat. Cell Biol.* **3,** 778–784.

Tuli, R., *et al.* (2003). Transforming growth factor-beta-mediated chondrogenesis of human mesenchymal progenitor cells involves N-cadherin and mitogen-activated protein kinase and Wnt signaling cross-talk. *J. Biol. Chem.* **278,** 41227–41236.

Varelas, X., *et al.* (2008). TAZ controls Smad nucleocytoplasmic shuttling and regulates human embryonic stem-cell self-renewal. *Nat. Cell Biol.* **10,** 837–848.

Vater, C., *et al.* (2011). Culture media for the differentiation of mesenchymal stromal cells. *Acta Biomater.* **7,** 463–477.

Wakitani, S., *et al.* (1995). Myogenic cells derived from rat bone marrow mesenchymal stem cells exposed to 5-azacytidine. *Muscle Nerve* **18,** 1417–1426.

Xia, W., *et al.* (2009). Adenoviral transduction of hTGF-beta1 enhances the chondrogenesis of bone marrow derived stromal cells. *Biotechnol. Lett.* **31,** 639–646.

Xiao, Y., *et al.* (2010). Clonal characterization of bone marrow derived stem cells and their application for bone regeneration. *Int. J. Oral Sci.* **2,** 127–135.

Xu, C., *et al.* (1996). Effects of growth factors and interleukin-1 alpha on proteoglycan and type II collagen turnover in bovine nasal and articular chondrocyte pellet cultures. *Endocrinology* **137,** 3557–3565.

Yu, L., *et al.* (2002). TGF-beta receptor-activated p38 MAP kinase mediates Smad-independent TGF-beta responses. *EMBO J.* **21,** 3749–3759.

Zhao, L., *et al.* (2010). Transforming growth factor beta1 induces osteogenic differentiation of murine bone marrow stromal cells. *Tissue Eng. Part A* **16,** 725–733.

Zuk, P. A., *et al.* (2001). Multilineage cells from human adipose tissue: Implications for cell-based therapies. *Tissue Eng.* **7,** 211–228.

CHAPTER EIGHT

Maternal Intake of Folic Acid and Neural Crest Stem Cells

Chandra S. Mayanil,* Shunsuke Ichi,* Barbara-Mania Farnell,† Vanda Boshnjaku,* Tadanori Tomita,* *and* David G. McLone*

Contents

I. Introduction 144
II. Role of FA in Human Health 145
  A. Neural tube defects 145
  B. Autism 146
  C. Alzheimer's disease 146
  D. Parkinson's disease 147
  E. Depression 147
  F. Cerebral ischemia 147
  G. Cardiac diseases 147
  H. Asthma 148
III. Mouse Models of NTD 148
  A. Folate responsive NTD mutant mice 148
  B. FA nonresponsive NTD mutant mice 148
IV. Neural Crest Development and Neural Crest Stem Cells 149
V. Role of FA in Neural Crest Development 151
  A. Cranial neural crest development 151
  B. Cardiac neural crest development 155
  C. Caudal neural crest development 156
VI. Folate Nonresponsive Genetic Mouse Models 162
VII. Conclusions and Future Directions 162
Acknowledgments 163
References 163

* Developmental Biology Program, Children's Memorial Research Center, Department of Pediatric Neurosurgery, Children's Memorial Medical Center and Northwestern University Feinberg School of Medicine, Chicago, Illinois, USA
† Department of Biology, Purdue University at Calumet, Hammond, Indiana, USA

Vitamins and Hormones, Volume 87
ISSN 0083-6729, DOI: 10.1016/B978-0-12-386015-6.00028-7

Abstract

Maternal folic acid (FA) intake has beneficial effects in preventing neural tube defects and may also play a role in the prevention of adult onset diseases such as Alzheimer's disease, dementia, neuropsychiatric disorders, cardiovascular diseases, and cerebral ischemia. This review will focus on the effects of maternal FA intake on neural crest stem cell proliferation and differentiation. Although FA is generally considered beneficial, it has the potential of promoting cell proliferation at the expense of differentiation. In some situations, this may lead to miscarriage or postnatal developmental abnormalities. Therefore, a blind approach such as "FA for everyone" is not necessarily the best course of action. Ultimately, the best approach for FA supplementation, and potentially other nutritional supplements, will include customized patient genomic profiles for determining dose and duration.

I. Introduction

Neural crest cells (NCCs) develop from a pool of self-renewing multipotent progenitors that differentiate into various cell types to form mature tissue. These progenitors or neural crest stem cells (stem cells derived from dorsal neural folds at the border of neural plate and epidermal ectoderm/NCSCs) are spatially and functionally segregated from surrounding tissues in early development, waiting for extrinsic cues for initiation of specific transcriptional programs. Following induction, the cells delaminate from the neural tube and migrate into the periphery where they differentiate into neurons and glia of the peripheral nervous system, melanocytes of the skin (Knecht and Bronner-Fraser, 2002), and various other structures. Basically, the life history of NCSCs includes (i) stem cell specification, (ii) proliferation, (iii) migration, (iv) differentiation, and (v) integration. These stages involve distinct transcriptional programs, initiated by extrinsic and intrinsic factors. Extrinsic factors include: members of the BMP and Wnt families (García-Castro *et al.*, 2002; Selleck and Bronner-Fraser, 1995), fibroblast growth factors (Fgfs) (Mayor *et al.*, 1997), and sonic hedgehog (Shh) (Calloni *et al.*, 2007; Washington *et al.*, 2005).

Although NCC development has been described extensively, little is known regarding the role of maternal folate on NCSC specification, proliferation, migration, differentiation, and integration. Abnormally low or high levels of folate can severely affect any of these stages. Folate or 5′-methyltetrahydrofolate (5-MTHF) provides the methyl group for the conversion of methionine to *S*-adenosylmethionine (SAM), a major methyl group donor for many methyltransferase reactions. When folate levels are depleted, SAM levels decrease and as a consequence DNA cytosine methylation also decreases. This result in DNA strand breaks and impairs DNA repair, followed by genetic mutations and apoptosis (Pogribny *et al.*, 1995; Scott and Weir, 1998; Wainfan and Poirier, 1992). Low folate intake has been linked with failure of cytoskeletal protein posttranslational methylation, resulting in neural tube defects (NTDs) (Bjorklund and Gordon, 2006). Low extracellular folic acid (FA) induces human α folate receptor (Antony, 1996) and is associated with gene-specific methylation/demethylation, rather than global DNA methylation and transcriptional activation. FA supplementation affects apoptosis and differentiation of embryonic neural stem cells exposed to high glucose (Jia *et al.*, 2008). Global DNA and p53 region-specific hypomethylation in human colonic cells is induced by folate depletion and reversed by folate supplementation (Wasson *et al.*, 2006).

Low folate also increases levels of homocysteine (Hcy), a potentially toxic substance produced by methionine demethylation (Latacha and Rosenquist, 2005; van der Put *et al.*, 2001). Hyperhomocysteinaemia impacts a variety of cellular functions in early development, including apoptosis (Boot *et al.*, 2003), cell division, differentiation (Boot *et al.*, 2006; Rosenquist *et al.*, 1996), and migration (Boot *et al.*, 2006). Elevated Hcy can disrupt derivatives of the neural ectoderm. Although embryotoxic, elevated Hcy does not cause NTDs in mouse embryos (Greene *et al.*, 2003), it perturbs neural crest migration *in vitro* and *in vivo* (Rosenquist *et al.*, 2007). Overall, folate is important in NCSC behavior and function and hence a detailed understanding of the mechanisms of FA action and the stages at which it impacts NCSC development is critical.

In addition to impacts on NCSC development, the main focus of this review, FA acid may have developmental roles in other regions of the central nervous system, and it appears to be also involved in adult onset diseases. The incidence of some of these diseases may be affected by FA levels early in life. These aspects of FA function will be briefly discussed in the following section.

II. Role of FA in Human Health

A. Neural tube defects

The incidence of spina bifida doubled in children conceived during the Dutch famine of 1944 (Hibbard *et al.*, 1965; Smithells, 1985), suggesting that diet may impact neural tube development. Currently, NTDs are seen in 1–5/1000 births worldwide (Frey and Hauser, 2003). In NTD, neural folds do not rise and fuse properly during embryonic development, 21 days after conception in humans (Northrup and Volcik, 2000). FA is a biologically inactive water-soluble B vitamin that is transformed to bioactive forms *in vivo* (Shane, 1995). Dietary supplementation with 400 μg of FA per day prevents 70% of NTD in women with a past history of carrying a fetus with

NTDs and 50% of NTDs in the general population. The remaining 30–50% of the population is nonresponsive to maternal folate, indicating folate responsive and nonresponsive individuals in the human population (Berry and Li, 2002; Berry *et al.*, 1999; Czeizel and Dudas, 1992).

B. Autism

Can FA be detrimental? If low levels of FA result in NTDs, then can too much FA result in neural tissue damage, potentially damage associated with autism (Beard *et al.*, 2011). In the United States, reported cases of autism have increased significantly, coincidentally with the fortification of food with FA (James *et al.*, 2006). Interestingly, FA supplementation enables neural tube development at approximately the time point considered critical in the development of autism (Rodier *et al.*, 1997). Is it plausible that children diagnosed with autism spectrum disorders receive high doses of FA *in utero*, and after birth? Rogers (2008) has hypothesized that enhanced maternal folate status may alter natural selection in favor of an adverse gene polymorphism (MTHFR, C677T), found in high frequency in Autistic children. For normal neurodevelopment to continue enhanced levels of FA may have to be maintained during embryonic development and postnatally. Fetuses possessing MTHFR C677T mutations with decreased maternal FA intake may be at increased risk of miscarriage. Infants possessing the polymorphism, who do not maintain enhanced folate status after birth (as experienced *in utero*) may be at increased risk for diminished methylation and abnormal neurodevelopment, including the development of Autism (Rogers, 2008). The link between maternal folate intake and Autism has not been confirmed.

C. Alzheimer's disease

Increasing evidence supports a role for epigenetic modifications in the pathogenesis of Alzheimer's disease (AD). Individuals with AD have decreased plasma folate levels, increased plasma Hcy levels, and SAM levels may be impaired (Mattson and Shea, 2003). Polymorphisms of genes participating in one-carbon metabolism have been associated with AD risk and/or with increased Hcy levels in AD individuals (Coppedè, 2010). Epidemiological studies show a dose-dependent relationship between mild-to-moderate increases in total plasma Hcy concentrations and the risk of neurodegenerative diseases, such as Alzheimer's, vascular dementia, cognitive impairment, or stroke (Herrmann and Obeid, 2011). Studies in rodents suggest that early life exposure to neurotoxicants or dietary restriction of folate and other B vitamins produces epigenetic modifications in AD related genes (Fuso *et al.*, 2008). Human neuronal cells cultured in media deprived of folate and other B vitamins showed epigenetic modifications on the presenilin (PSEN1) gene (Chan and Shea, 2007; Fuso *et al.*, 2005, 2007, 2011), which is associated with early onset familial AD.

D. Parkinson's disease

Hyperhomocysteinemia has been associated with Parkinson's disease (PD). The increase is primarily due to methylated catabolism of L-DOPA, the main pharmacological treatment for PD (Martignoni *et al.*, 2007), and may support progression of the disease due to concomitant onset of neuropsychiatric symptoms and comorbidities (i.e., vascular disease). Since FA reduces Hcy, it has been used in clinical trials as an addition to dopaminergic medication (Klivenyi and Vecsei, 2010). In the periphery, therapeutic approaches for this L-DOPA-mediated Hcy increase are vitamin supplementation (i.e., FA or application of L-DOPA with an inhibitor of catechol-*O*-methyltransferase) (Müller, 2008).

E. Depression

Folate is used in the brain for the synthesis of norepinephrine, serotonin, and dopamine. FA augmentation may benefit some depressed patients, reducing residual symptoms. Of the three forms of folate commonly used: FA, 5-MTHF, and folinic acid, some may be more bioavailable than others to patients with a genetic polymorphism and in those who take particular medications or use alcohol (Farah, 2009). 5-MTHF in patients with normal or low folate levels improves cognitive function and reduces depressive symptoms in elderly patients with dementia and folate deficiency (Buhr and Bales, 2010); it also reduces depressive and somatic symptoms in patients suffering from depression and alcoholism (Fava and Mischoulon, 2009).

F. Cerebral ischemia

Plasma from stroke patients is characterized by increased excretion of lactate, pyruvate, glycolate, and formate, and decreased excretion of glutamine and methanol; urine is characterized by decreased levels of citrate, hippurate, and glycine. These metabolites are associated with anaerobic glycolysis, FA deficiency, and hyperhomocysteinemia (Jung *et al.*, 2011). Low folate status and elevated Hcy increase the generation of reactive oxygen species (ROS) and contribute to excitotoxicity and mitochondrial dysfunction which may lead to apoptosis (Kronenberg *et al.*, 2009; Pezzini *et al.*, 2007).

G. Cardiac diseases

Cardiac NCCs emerge from the axial level between the midotic placode to the third somite. A subset of these cells contributes to the formation of the developing heart (Kirby and Waldo, 1995). This population of NCCs is particularly susceptible to the teratogenic effects of Hcy (Kapusta *et al.*, 1999; Rosenquist *et al.*, 1996; Wenstrom *et al.*, 2001). Elevated Hcy inhibits cardiac NCC formation and morphogenesis *in vivo* (Tierney *et al.*, 2004). FA has a protective effect on these cells due to its ability to reduce serum Hcy levels by shunting excess Hcy into the methionine metabolic pathway (Epeldegui *et al.*, 2002; Maestro de las Casas *et al.*, 2003; Rosenquist *et al.*, 1996; Vanaerts *et al.*, 1994).

H. Asthma

A methyl-rich maternal diet enhances the severity of heritable asthma in mice, and studies have reported increased asthma risk in children exposed *in utero* to FA supplementation (Håberg *et al.*, 2009; Hollingsworth *et al.*, 2008; Whitrow *et al.*, 2009). Prenatal FA may alter DNA methylation and thus aberrant gene expression, changing the risk of asthma through epigenetic mechanisms operating during gestation (Hollingsworth *et al.*, 2008). Prenatal FA may also be involved in increasing prevalence of food allergies (Branum and Lukacs, 2008; CDC report, 2004).

III. Mouse Models of NTD

Folate responsive and nonresponsive NTD mouse models have been used to help understand folate action in the prevention of NTDs (Gray and Ross, 2009; Harris and Juriloff, 2010).

A. Folate responsive NTD mutant mice

Cart1 null (Zhao *et al.*, 1996), *Cited2* null (Barbera *et al.*, 2002), *Cd*: Lrp6 gain of function (Carter *et al.*, 2005), *Folr1* null (Piedrahita *et al.*, 1999), *RFC1:* reduced folate carrier null (Gelineau-van Waes *et al.*, 2008), Sp^{2H}: Pax3 loss of function (Fleming and Copp, 1998), *Splotch Sp* (also known as *Sp/Sp* or *Sp−/−*): Pax3 loss of function (Wlodarczyk *et al.*, 2006), *kat2a:* Gcn5 loss of function (Lin *et al.*, 2008).

B. FA nonresponsive NTD mutant mice

Axd mutant (Essien and Wannberg, 1993), *Fkbp8* null (Wong *et al.*, 2008), *ct*: Grhl3 loss of function (De Castro *et al.*, 2010), *Grhl3* null (Ting *et al.*, 2003), *SELH/Bc* strain (Hofmann *et al.*, 1998), *Map3k4* null (Abell *et al.*, 2005), *Nog* (Anderson *et al.*, 2006), *Ski* null (Ernest *et al.*, 2006), Alx-3 null (Lakhwani *et al.*, 2010).

The best characterized mouse models are *Pax3* or *Splotch* (Fleming and Copp, 1998; Wlodarczyk *et al.*, 2006) and *Cd* loss-of-function mice, *Lrp6* (Wnt coreceptor) gain of function (Carter *et al.*, 2005), and *Folr1* null

(Piedrahita *et al.*, 1999). Pax3 and the Wnt coreceptor Lrp6 are important for NCSC development and function (Chalepakis *et al.*, 1993; Murphy and Bartlett, 1993; Tamai *et al.*, 2000) and stem cell proliferation (Davidson and Niehrs, 2010). Pax3 is also important for migration of NC precursors (Maschhoff and Baldwin, 2000; Wang *et al.*, 2008).

Studies using mouse models have shown that the etiological basis of NTDs is based on a multifactorial model in which both genetic and environmental factors interact in a complex manner (Harris and Juriloff, 2010; Zohn and Sarkar, 2008). Genes whose deficiencies cause NTD in mice are potential candidates for human NTD. Despite this relatively large and growing list, most human NTDs have not been linked with specific genetic alterations.

IV. Neural Crest Development and Neural Crest Stem Cells

NCCs are found at the boundary between the neural plate and the prospective epidermis; these cells develop as a result of the regulated action of several morphogen proteins and transcription factors (Basch and Bronner-Fraser, 2006). Following epithelial to mesenchymal transformation (EMT), NCCs migrate throughout the embryo and contribute to the formation of various tissues (Morales *et al.*, 2005): (i) cranial NC gives rise to several components in the head including most of the facial bones and cranial vault; (ii) cardiac NC, a subregion of cranial NC, contributes to a number of cell types and cardiac outflow tracts; (iii) trunk NC gives rise to melanocytes and dorsal root ganglion; and (iv) sacral and vagal NC generates parasympathetic (enteric) ganglia of the gut (Graham *et al.*, 2004; Le Douarin and Teillet, 1973; Verberne *et al.*, 2000).

NC development, evolution, and determination are directly regulated by gene regulatory networks (GRNs) (Betancur *et al.*, 2010; Meulemans and Bronner-Fraser, 2004; Sauka-Spengler and Bronner-Fraser, 2008). These genes code for morphogens, transcription factors, morphogen membrane receptors, cell adhesion molecules, and transcriptional modifiers. Several reviews have described the role of morphogens in NCC determination (Gammill and Bronner-Fraser, 2003). BMP, Wnt, and FGFs segregate neural from non-neural ectoderm during neural tube induction (de Crozé *et al.*, 2011; Sasai and De Robertis, 1997; Weinstein and Hemmati-Brivanlou, 1999) and induce epidermal, neural plate, and NC gene expression (Meulemans and Bronner-Fraser, 2004). Wnt, FGF, and BMP signals emanating from the paraxial mesoderm, neural plate, and non-neural ectoderm activate NC specifiers (e.g., *snail2*, *foxd3*, *sox10*), expressed in premigratory NC and essential for further NC devel*opment* (LaBonne and Bronner-Fraser,

Table 8.1 Neural crest specific transcription factors with different functions

Transcription Factors	References
Stimulate proliferation and prevent differentiation	
Zic1	Brewster *et al.* (1998)
	Aruga *et al.* (2002)
	Ebert *et al.* (2003)
Pax3	Reeves *et al.* (1999)
c-myc	Cole and McMahon (1999)
AP2	Luo *et al.* (2002)
Msx1, Msx2	Hu *et al.* (2001)
Id	Norton (2000)
Notch	Cornell and Eisen (2005)
Twist	Howe *et al.* (2003)
Hes1, Hes5	Kageyama *et al.* (2008a)
Maintain stem cell potential	
Foxd3	Hanna *et al.* (2002)
Sox10	Aoki *et al.* (2003)
Hes1	Kageyama *et al.* (2008a)
	Nakazaki *et al.* (2008)
	Ichi *et al.* (2010, 2011)
Transcriptional repressors	
Slug/Snail	Hemavathy *et al.* (2000)
Zic 1	Ebert *et al.* (2003)
Nbx	Kurata and Ueno (2003)
Msx1, Msx2	Catron *et al.* (1996)
Id2	Martinsen and Bronner-Fraser (1998)
Transcriptional activators	
Sox9	Scott *et al.* (2010)
Sox10	Kelsh (2006), Hong and Saint-Jeannet (2005)
Pax3	Maschhoff and Baldwin (2000)
c-Myc	Cole and McMahon (1999)
AP2	Williams and Tjian (1991), de Crozé *et al.* (2011)
Notch1	Kopan (2002)

1998; Meulemans and Bronner-Fraser, 2004; Monsoro-Burq *et al.*, 2003; Saint-Jeannet *et al.*, 1997). Neural crest specific transcription factors (see Table 8.1) have different functions, such as (a) stimulation of proliferation and prevention of differentiation, (b) maintenance of stem cells, (c) transcriptional repression, and (d) transcriptional activation.

Several GRN genes are modulated by maternal folate intake. These include *Tead2* (Kaneko *et al.*, 2007), *Hes1*, *Neurog2* (Ichi *et al.*, 2010, 2011), *Fgfr4* (Ichi *et al.*, 2011), *Notch* (Ichi *et al.*, 2011; Liu *et al.*, 2010), folate receptor—Folr1 (Zhu *et al.*, 2007), folate binding protein—Folbp1 (Tang *et al.*, 2005), the Wnt coreceptor—*Lrp6* (Carter *et al.*, 2005; Gray *et al.*, 2010), members of the Wnt family (Han *et al.*, 2009; Linask and Huhta, 2010), and members of the Shh signaling pathway (Gelineau-van Waes *et al.*, 2008), Additionally, microRNAs (miRNAs) are also regulated by folate (Ichi *et al.*, 2010; Shookhoff and Gallicano, 2010), suggesting that Dicer may be a target for folate action. Zehir *et al.* (2010) found that Dicer and miRNAs are required for survival of NC-derived tissues, playing a role in the prevention of apoptosis during differentiation. Nie *et al.* (2011) showed that Dicer activity in NCCs is essential for craniofacial organogenesis and pharyngeal arch artery morphogenesis.

V. Role of FA in Neural Crest Development

A. Cranial neural crest development

Several genes involved in the development of NC are responsive to maternal folate intake and resultant prevention of NTD.

1. Tead2

Interactions between neural ectoderm and epidermis initiate expression of early NC markers, including Slug, Snail, Pax3, and Wnt1 (Garcia-Castro and Bronner-Fraser, 1999; LaBonne and Bronner-Fraser, 1999). Tead2, a member of the Tead box family of transcription factors, binds a NC enhancer and activates Pax3, suggesting that Tead2 is an endogenous Pax3 activator (Milewski *et al.*, 2004), and an early NC marker. However, Tead2 is not required for *Pax3* expression (Kaneko *et al.*, 2007). Inactivation of *Tead2* in mice significantly increased the risk of anencephaly/exencephaly (Kaneko *et al.*, 2007); however, nullizygous embryos did not exhibit spina bifida, leading to the conclusion that Tead2 plays a role in neural tube closure independent of its putative role in *Pax3* regulation. Maternal intake of folate suppressed the risk of anencephaly/exencephaly in *Tead2* nullizygous females. Pifithrin-α, an inhibitor of p53-dependent apoptosis (Yan *et al.*, 2006), rescued NTDs in Pax3-deficient embryos (Pani *et al.*, 2002). Pifithrin-α also prevented Tead2-dependent exencephaly to the same extent as FA (Kaneko *et al.*, 2007), suggesting that TEAD2 regulation of neural tube closure may be related to inhibition of p53-dependent apoptosis, rather than induction of neural tube-specific gene expression.

2. Lrp6-Wnt coreceptor

NTDs in crooked tail (*Cd*) mice are preventable by maternal FA intake (Gray *et al.*, 2010). These mice have a gain-of-function mutation in low-density lipoprotein receptor-related protein 6 (*Lrp6*), a coreceptor for canonical WNT signaling. *Lrp6* encodes a single-pass transmembrane receptor and is a coreceptor with Frizzled in the canonical Wnt signaling pathway, which activates β-catenin-dependent TCF/LEF-regulated transcription (Schweizer and Varmus, 2003). *Cd* results from a single-nucleotide substitution in *Lrp6* (Carter *et al.*, 2005). The $Lrp6^{Cd}$ mutation replaces a highly conserved amino acid in the extracellular domain of LRP6 and interferes with the ability of dickkopf-1 (Dkk1) to inhibit canonical Wnt signaling, resulting in sustained elevation of cytosolic β-catenin levels in the presence of Wnt. Loss of *Lrp6* function produces several birth defects, including caudal axis truncation and limb deformities as well as exencephaly and spina bifida (Pinson *et al.*, 2000). Loss-of-function *Lrp6* mutations can also result in NTDs, a hypomorphic allele of *Lrp6*, caused by the ringelschwanz point mutation, is associated with spina bifida in otherwise viable homozygous pups (Kokubu *et al.*, 2004). FA supplementation rescues $Lrp6^{Cd/Cd}$ embryos by normalizing hyperactive Wnt signaling activity, whereas in *LRP6*-deficient embryos, FA further attenuates reduced Wnt activity, thereby compromising development (Gray *et al.*, 2010).

3. Folbp1-folic acid binding protein (Folbp1)

The developing embryo is entirely dependent on maternal folate to support high levels of proliferative activity. Folate receptors are crucial for transplacental transport (Henderson *et al.*, 1995), as well as embryonic folate utilization (Piedrahita *et al.*, 1999). Uptake of extracellular folate also depends on the reduced folate carrier (RFC1), an energy dependent, integral transmembrane protein with high affinity for folate, and folate binding protein (FolBp1) (Spiegelstein *et al.*, 2004). Folate enters cells via receptor-mediated endocytosis as 5-methyltetrahydrofolic acid, and then donates a 5-methyl group to Hcy during methionine synthesis. Folate receptors cycle back to the cell surface by endosomal trafficking (Birn *et al.*, 1993).

Folbp1 is bound to the plasma membrane by glycosyl–phosphatidyl–inositol (GPI) links (Lacey *et al.*, 1989). GPI-anchored proteins, such as FolBp1, are located in the plasma membrane in lipid rafts which often serve as platforms for signal transduction events and protein recycling (Miotti *et al.*, 2000). FolBp1 associates with cell surface proteins—megalin (Birn *et al.*, 2005), lyn (a src-family nonreceptor kinase), $G\alpha_{(i-3)}$ (Miotti *et al.*, 2000), Bat2d, and fibronectin type 3 domain (Fndc3a). Both *Bat2d* and *Fndc3a* mRNA are localized to the palatal shelves in the orofacial regions of E13.5 embryos (Pisano *et al.*, 2010). Studies suggest that Folbp1 expressed by NC cells in the developing embryo and its interaction with extracellular matrix proteins, and cell surface proteins may be important for NCSC migration and differentiation. Tang *et al.* (2005) suggested that a prominent role for Folbp1 is in regional regulation of apoptosis and cell proliferation that underlies aberrant neural tube and craniofacial defects.

4. Dicer and miRNA

Lack of miRNAs in NCCs leads to cell death and loss of NCC-derived craniofacial structures (Huang *et al.*, 2010; Nie *et al.*, 2011; Sheehy *et al.*, 2010; Zehir *et al.*, 2010). miRNAs target inhibitors of the Wnt signaling pathway (Liu *et al.*, 2011). Cranial and cardiac NCC migration or target colonization was not affected by Dicer disruption but subsequent development was disturbed (Nie *et al.*, 2011). In the head, loss of Dicer leads to a loss of NC-derived craniofacial bones while in the trunk, cells of the enteric, sensory and sympathetic nervous systems are lost during development. Loss of Dicer does not prevent initial NCC differentiation, but NC derivatives were lost due to apoptotic cell death (Zehir *et al.*, 2010). Thus Dicer and miRNAs have a central role in NCC survival, migration, and patterning in craniofacial and cardiovascular development. Defects in these factors lead to congenital neuro-craniofacial-cardiac defects (Schlesinger *et al.*, 2011).

miRNA misregulation may be linked to FA deficiency and NTDs (Shookhoff and Gallicano, 2010). Ichi *et al.* (2010) found increased expression of miR-138, miR-148a, miR-185, and miR-339-5p in folate deficient $Sp^{-/-}$ embryos. Maternal folate intake reversed this decrease. As a group these miRNAs targeted KDM6B expression and remodeled chromatin at *Hes1* and *Neurog2* promoters thus affecting maintenance and neurogenesis of NCSCs. miRNAs are predicted to be highly regulated by genomic methylation, which in turn is dependent on folate availability. It is interesting to note that at least half of the promoter regions for miRNAs are in close proximity to CpG islands and their methylation frequency is predicted to be at least an order of magnitude higher than that of protein-coding genes (Han *et al.*, 2011). Dietary regulation of miRNA may also be involved in cancer (Davis and Ross, 2008).

5. Trp53 or p53

Developmental processes and apoptosis are intertwined, for instance, deficiency of the p53 tumor suppressor protein impairs development. p53 null embryos display diverse craniofacial abnormalities, including exencephaly (Armstrong *et al.*, 1995; Sah *et al.*, 1995). Upregulation of p53-related apoptotic genes in NC progenitors leads to severe craniofacial defects (Jones *et al.*, 2008). Rinon *et al.* (2011) found that p53 coordinates cranial NCC growth and EMT/delamination processes by affecting cell cycle gene expression and proliferation at discrete developmental stages; disruption of these processes can lead to craniofacial defects.

Diabetes associated pregnancies have a higher probability of NTDs (Chappell *et al.*, 2009). Maternal diabetes inhibits *Pax3* expression. Insufficient Pax3 causes cardiac NC and neuroepithelial cell apoptosis, by a process dependent on p53. This provides a cellular explanation for cardiac outflow tract and NTDs induced by diabetic pregnancy. Additionally Pax3 inhibition of p53 dependent apoptosis is required for proper craniofacial development (Morgan *et al.*, 2008; Pani *et al.*, 2002). There is no direct evidence that folate changes p53 levels in NCSCs; however, FA supplementation significantly decreases glucose induced apoptosis. Hyperglycemia increases oxidative stress and p53 expression in embryos, leading to increased neuroepithelial apoptosis and NTDs (Loeken, 2006; Pani *et al.*, 2002). In *wild-type* (*WT*) embryos, inhibiting p53 neuroepithelial accumulation prevents cell death and allows neural tube closure, suggesting that p53 is a key regulatory gene for NTDs. FA supplementation of NSCs decreased apoptosis and nuclear p53 expression induced by high glucose, potentially via regulation of nuclear p53 expression (Jia *et al.*, 2008). Studies by Lotem *et al.* (1996) and Polyak *et al.* (1997) demonstrated that p53-induced apoptosis is mediated by ROS and that ROS-induced apoptosis was p53-dependent. Other studies found that FA reduced production of ROS and prevented DNA damage induced by increased oxidative stress (Huang *et al.*, 2004). Taken together, the data suggest that FA may protect NSCs from apoptosis induced by high glucose via inhibition of oxidative stress.

6. Notch

Notch signaling has two roles during cranial NC development: (i) establishment of the NC domain within the ectoderm via lateral induction and (ii) diversification of cell fates, for cells that arise from the NC via lateral inhibition. Specification of NC via lateral induction is brought about by Notch signaling, which regulates genes expressed by cranial NC through Bmp family members. Diversification of NC progeny via lateral inhibition results in inhibition of neurogenesis and promotion of proliferation (Cornell and Eisen, 2005; Kageyama *et al.*, 2008a,b). Supplementation of embryonic NSCs with FA stimulates dose-dependent NSC proliferation; increases mRNA and protein expression levels of *Notch1*, *Hes1*, and *Hes5*; and decreases *Mash1* expression (Liu *et al.*, 2010; Zhang *et al.*, 2008). Therefore, FA supplementation if given in the right chemical form, dose, and timing (Spiegelstein *et al.*, 2004) could positively affect NCSC maintenance by affecting *Hes1* activity via Notch activation, which inhibits neurogenesis and oligodendrocyte formation but promotes astrocyte formation from glial progenitors (Kageyama *et al.*, 2009; Louvi and Artavanis-Tsakonas, 2006). Thus, differentiation of neurons and maintenance of neural progenitor cells may be balanced due to Notch signaling.

B. Cardiac neural crest development

Cardiac NC development is impaired by folate deficiency. Boot *et al.* (2003) studied the behavior of neuroepithelial cells and cranial and cardiac NCCs *in vitro*. FA increased neuroepithelial cell outgrowth and NCC differentiation into nerve and smooth muscle cells. Addition of Hcy increased NCC outgrowth and migration from the neural tube and inhibited NCC differentiation. These findings suggested that NTDs caused by folate deficiency and hyperhomocysteinemia develop as a result of increased neuroepithelial to NCC transformation. Increased transformation leads to a shortage of neuroepithelial cells in the neural tube. Defects in orofacial and conotruncal development are explained by abnormal differentiation of NCCs in the presence of high Hcy concentrations, thus supporting involvement of FA and Hcy in the development in neural tube development and NC related heart formations.

Lack of folate increases Hcy levels which in turn inhibits cardiac NCC formation by inhibiting NC dispersal and decreasing the distance cells migrate from the neural tube (Tierney *et al.*, 2004). Lack of migration due to Hcy may result from increased intracellular Ca^{2+}, possibly due to generation of inositol triphosphate (Heidenreich *et al.*, 2008). Rosenquist *et al.* (2007) identified 65 gene transcripts altered by Hcy in cardiac NCCs. These included 19 genes involved in cell migration and adhesion, 13 gene involved in metabolism, 11 genes involved in DNA/RNA interaction, 10 genes involved in cell proliferation/apoptosis, and 6 genes involved in transporter/receptor related activities. Tang *et al.* (2004) showed that FA binding protein (*Folbp),* knockout mice died *in utero* with multiple malformations, including severe congenital heart defects. The beneficial effect of FA for congenital heart defects might be mediated via the the impact of Folbp on NCCs and by gene regulation of signaling pathways involved in the development of the pharyngeal arches and the secondary heart field (Zhu *et al.*, 2007).

The involvement of Pax3 in neural tube closure, cardiovascular development, and cardiac NC cell function cannot be overstated (Epstein, 1996; Epstein *et al.*, 2000; Gitler *et al.*, 2002; Jain *et al.*, 2011; Li *et al.*, 1999; Snider *et al.*, 2007). *Splotch* is a *Pax3* loss-of-function mouse mutant. Folate responsiveness in this mouse rescues genes regulated by Pax3; Including genes involved in migration, proliferation, and differentiation (Ichi *et al.*, 2010, 2011; Mayanil *et al.*, 2001). The role of miRNAs in cardiac development has been reviewed (Han *et al.*, 2011; Ono *et al.*, 2011; Thum *et al.*, 2008; Topkara and Mann, 2011). Studies have yet to examine if maternal folate intake changes miRNA levels, as part of a mechanism involved in rescuing heart defects.

1. Wnt signaling

Wnt–β-catenin signaling in the cardiogenic crescent is required for maintenance of undifferentiated cells. β-Catenin, a key downstream intermediary of the Wnt pathway, accumulates in the cytoplasm and subsequently translocates to the nucleus to activate target genes. Prior to specification, Wnt antagonists Crescent and Dickkopf-1 (Dkk-1) are expressed in the mesendoderm of the bilateral heart fields, this suppresses canonical Wnt signaling, resulting in decreased β-catenin levels. As a result, genes associated with the induction of cardiogenesis, such as *Hex* and *Isl1,* are upregulated. Hex is expressed in the prechordal plate and primary heart field, and Islet-1 is a marker for the secondary heart field (Cohen *et al.*, 2007; Foley and Mercola, 2004). Han *et al.* (2009) showed that lithium and Hcy repress and delay Hex and Islet-1 induction, by their ability to bypass extracellular Dkk-1 or Crescent antagonism. Therefore, both lithium and Hcy maintain Wnt–β-catenin activity at a time when it would normally be downregulated, thus producing cardiac anomalies. FA protects against the disruption of β-catenin signaling and ameliorates cardiac defects associated with Hcy or lithium exposure (Han *et al.*, 2009; Linask and Huhta, 2010).

Although maternal FA supplementation can be beneficial in terms of rescuing NTD and folate deficiency is deleterious to normal embryonic development, the optimal level of FA intake is highly dependent on genetic background (Gray and Ross, 2009; Gray *et al.*, 2010). For instance, a diet containing high FA levels (200 ppm FA chow) resulted in the greatest reduction in NTD rates in $Pax3^{Sp/Sp}$ mice; however, this diet also increased embryonic death (resorptions) sixfold (Wlodarczyk *et al.*, 2006). However, moderate dietary FA supplementation (10 ppm FA chow) produced NTD rescue in $Lrp6^{Cd/Cd}$ mice bearing a gain-of-function *Lrp6* allele, reduced NTD incidence, and enhanced embryonic survival. *In vitro* assays indicated that the level of FA in the media impacts the ability of recombinant Wnt3a to stimulate β-catenin stabilization and TCF/LEF dependent transcription (Gray *et al.*, 2010). Thus, FA metabolism impacts Wnt signaling as a function of mouse genotype, so that FA supplementation rescues NTD associated with the hyperactive $Lrp6^{Cd}$ mutation by attenuating Wnt signaling, while added FA increases embryopathy associated with *Lrp6* loss of function, by further suppressing reduced Wnt activity (Gray *et al.*, 2010). The data indicate that both the gene(s) and the nature of the mutation(s) are critical in defining individual risk and the optimal regimen for obtaining the best possible birth outcome. Thus, FA supplementation based on personalized genome readout should be the goal, in place of universal supplementation.

C. Caudal neural crest development

1. miRNAs

Pax3 loss-of-function mutant ($Sp^{-/-}$) embryos exhibit decreased NCSC proliferation potential and premature neurogenesis (Nakazaki *et al.*, 2008). In this folate responsive mouse model, maternal folate intake prior to conception reverses NCSC proliferation potential via epigenetic mechanisms (Ichi *et al.*, 2010). NCSCs from open caudal neural tubes of $Sp^{-/-}$ embryos exhibit increased H3K27 methylation and decreased expression of the H3K27 histone demethylase enzyme KDM6B, due to upregulation of KDM6B targeting miRNAs: miR-138, miR-148a, miR-185, and miR-339-5p. Maternal intake of folate downregulated these miRNAs and increased KDM6B expression in the caudal neural tube of rescued $Sp^{-/-}$ embryos. KDM6B, in turn, regulated H3K27 methylation marks on *Hes1* and *Neurog2* promoters, critical for stem cell proliferation and sensory neurogenesis, respectively.

2. Hes1 and Neurog2

Hes1 and *Neurog2* epigenomes in NCCs from *WT*, $Sp^{-/-}$, and FA-rescued $Sp^{-/-}$ embryos were examined at E10.5 and E12.5 (Ichi *et al.*, 2010). In E10.5 $Sp^{-/-}$ embryos, the Hes1 promoter primarily associated with a repressive chromatin mark, H3K27me2, whereas the *Neurog2* promoter primarily associated with an activating chromatin mark, H3K18Ac. *Hes1* pattern for *WT* was opposite that of $Sp^{-/-}$. The *Neurog2* promoter from *WT* embryos associated with H3K9Ac, H3K18Ac, and H3K27me. Association of transcriptionally active and inactive components of chromatin simultaneously on the *Neurog2* promoter suggests that stem cells are held transcriptionally inactive by H3K27me2 prior to differentiation. At E12.5, H3K27me2 continued to associate with the *Hes1* promoter, although at lower levels, minimal association was seen between H3K9Ac and H3K18Ac and *Hes1,* indicating a repressive chromatin structure. At the *Neurog2* promoter, $Sp^{-/-}$ embryos exhibited increased levels of H3K9 and H3K18 acetylation and KDM6B, compared to *WT*. The results suggested that premature neurogenesis observed in $Sp^{-/-}$ embryos may be due to increased transcriptional activity on Neurog2 at E10.5, coupled with the onset of a decrease in H3K27me2, and an increase in H3K27me3 at the *Hes1* promoter. FA treatment rescued levels of H3K27me2 and KDM6B, but not H3K9 or H3K18 acetylation, on the *Hes1* promoter in E10.5 $Sp^{-/-}$ neural tube progenitors. Overall, the data suggest that FA works via removal of methyl groups from H3K27me2 by increasing the levels of KDM6B associated with the *Hes1* promoter. FA-treated $Sp^{-/-}$ progenitor populations showed decreased H3K9Ac and H3K18Ac association with *Neurog2*. KDM6B also decreased on this promoter, however, increased H3K27me2 was not observed, possibly due to the presence of UTX at this developmental time point (Ichi *et al.*, 2010).

Studies examining neurosphere growth and proliferation from NCSCs isolated from *WT* and $Sp^{-/-}$ embryos support the conclusions indicated by epigenetic alterations at *Hes1* and *Neuog2* promoters. Caudal NCSCs isolated from *WT* embryos proliferate and form neurospheres in culture. These cells can also differentiate into neurons, oligodendrocytes, and astrocytes. Caudal NCSCs isolated from $Sp^{-/-}$ embryos exhibit minimal proliferating potential, generating very few neurospheres, even in the presence of EGF and bFGF. When FA is added to the $Sp^{-/-}$ culture media, cell proliferation, neurosphere growth, and differentiation potential is restored (Fig. 8.1).

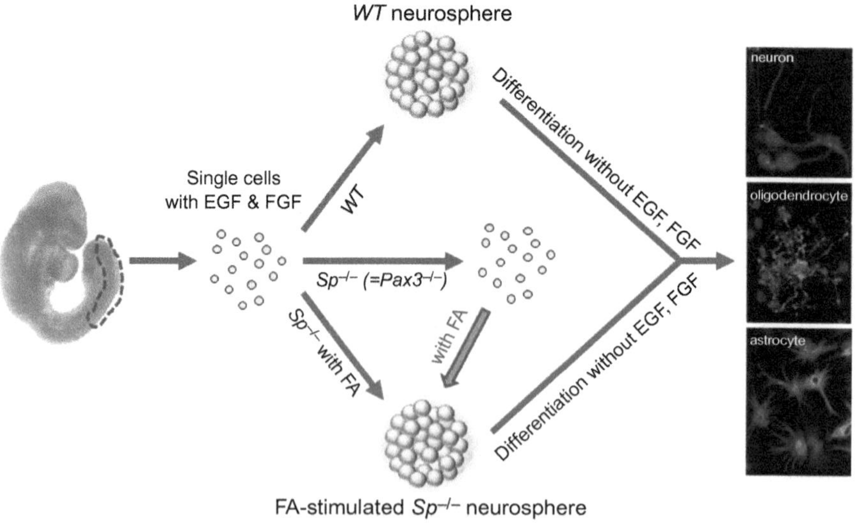

Figure 8.1 Caudal NCSCs isolated from WT embryos proliferate, generate neurospheres, and differentiate into cells of neuronal lineages. Caudal NCSCs isolated from $Sp^{-/-}$ embryos exhibit minimal proliferation and neurosphere formation. FA restores cell proliferation and differentiation potential in these cells (see explanation in the text). (For color version of this figure, the reader is referred to the Web version of this chapter.)

A hypothetical model for FA-mediated rescue of stem cell proliferation and neurogenesis is shown in Fig 8.2. In this model in $Sp^{-/-}$ embryos, several miRNAs are upregulated during early cell proliferation stages, some of these miRNAs target KDM6B. This results in decreased KDM6B expression and increased H3K27 methylation. The association between methylated H3K27 and the *Hes1* promoter results in (a) decreased *Hes1* expression and lowered stem cell proliferation. At the same time low association between methylated H3K27 and the *Neurog2* promoter, but high association between acetylated H3K9 and H3K18 and the *Neurog2* promoter results in (b) premature neurogenesis. FA-mediated rescue could be due to reversible downregulation of KDM6B targeting miRNAs, resulting in upregulation of KDM6B and subsequent lowering of H3K27 methylation. Demethylated H3K27 shows decreased binding to the *Hes1* promoter and increased binding to the *Neurog2* promoter, which results in (c) increased stem cell proliferation/maintenance and (d) normal neurogenesis.

3. Fgfr4

Fgfr4 is a Pax3 downstream target (Lagha *et al.*, 2008). FA rescues cell proliferation and upregulates fgfr4 mRNA and protein levels (Fig. 8.3), in NCSCs from $Sp^{-/-}$ embryos (Ichi *et al.*, 2011). Pax3 levels

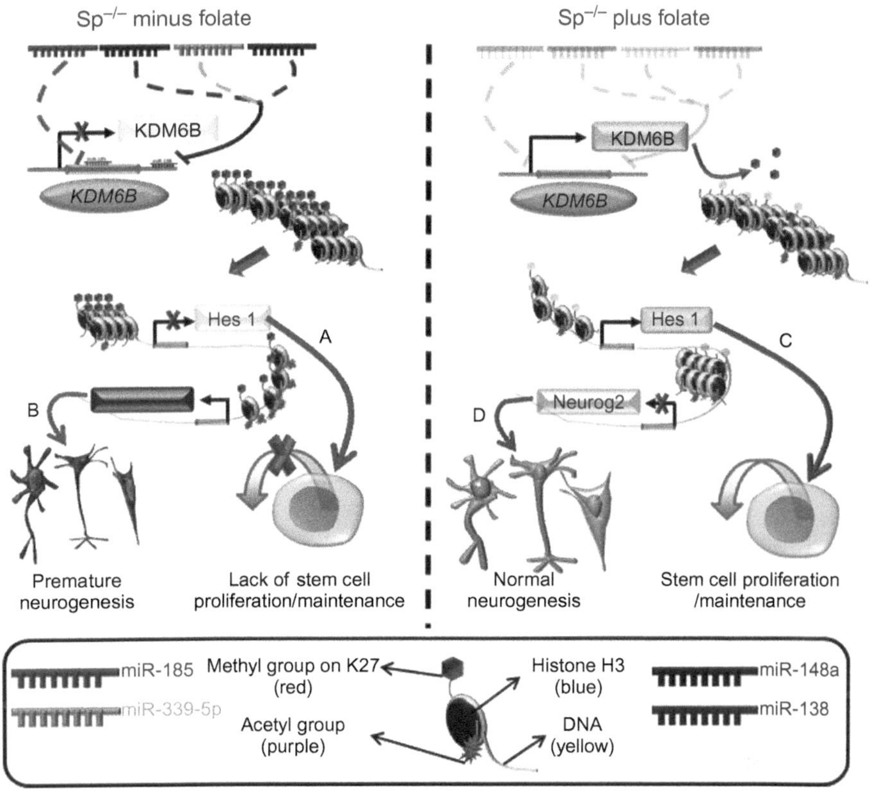

Figure 8.2 A hypothetical model for FA-mediated rescue of stem cell proliferation and neurogenesis (see explanation in the text). "This research was originally published in *Journal of Biological Chemistry* as Ichi *et al.* (2010). © The American Society for Biochemistry and Molecular Biology." (For color version of this figure, the reader is referred to the Web version of this chapter.)

in *WT* NCSCs are unaffected by FA treatment, suggesting that FA activates *Fgfr4* downstream of Pax3.

Interestingly, in this study when cells were allowed to differentiate, besides the presence of neural lineages, α-SMA positive smooth muscle cells differentiated from *WT* and $Sp^{-/-}$ *FA*-rescued cells (Ichi *et al.*, 2011). *FgfR4* expression is necessary for terminal muscle differentiation (Marics *et al.*, 2002). Disruption of FGFR4 signaling interrupts limb muscle formation (Yu *et al.*, 2004). *MyoD* induces ectopic *FgfR4* expression in the neural tube and in NCCs 1 day after electroporation. This observation indicates the existence of an unexpected positive feedback loop from *MyoD* to *FgfR4* during myogenesis (Delfini and Duprez, 2004). Appearance of α-SMA in FA-rescued $Sp^{-/-}$ neural stem cells may be due to FA-mediated upregulation of FGFR4, although this does not explain the α-SMA in *WT*.

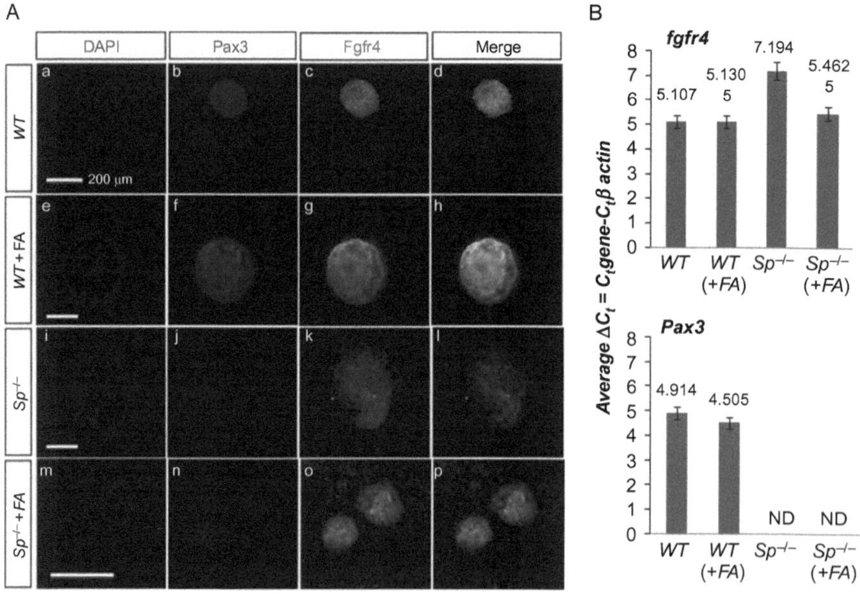

Figure 8.3 FA regulates *Fgfr4*, a Pax3 downstream target. (A) *WT* and $Sp^{-/-}$ stem cell colonies grown with and without FA, immunostained for Pax3 (b, f, j, n) and Fgfr4 (c, g, k, o) and counterstained with DAPI (a, e, I, m). Merged images are shown in (d, h, l, and p). (B) An approximately 2 cycle or fourfold increase in *fgfr4* message expression was observed in cells stimulated with FA as compared with nonstimulated $Sp^{-/-}$. A cycle threshold (C_t) is defined as the fractional cycle number at which reporter fluorescence passes a fixed threshold above base line. Samples with higher message expression will have a lower C_t value. This research is *in press* in *Stem Cells and Development*, as Ichi *et al.* (2011). (For color version of this figure, the reader is referred to the Web version of this chapter.)

4. Notch

Notch signaling is involved in cranial NC and caudal neural tube development. Notch pathway molecules are essential for maintenance, but not generation of mammalian NSCs (Hitoshi *et al.*, 2002). FA increases cell proliferation and Notch signaling in NSCs and alters components of the Notch signaling pathway including *Hes1* and *Hes5* (Liu *et al.*, 2010; Zhang *et al.*, 2008). NCSCs in FA-rescued $Sp^{-/-}$ neurospheres have increased Notch expression (Ichi *et al.*, 2011). Unpublished data from our lab show that FA increases expression of *Notch1, 2,* and *3 in* $Sp^{-/-}$ *and WT* embryos, suggesting that these molecules *are* not activated by Pax3. Activation may be via an alternative FA pathway.

Figure 8.4 shows a potential folate responsive gene regulatory network (GRN) in NCSCs. This network consists of growth factors, membrane receptors, transcription factors, and miRNA, histone modifications. Folate

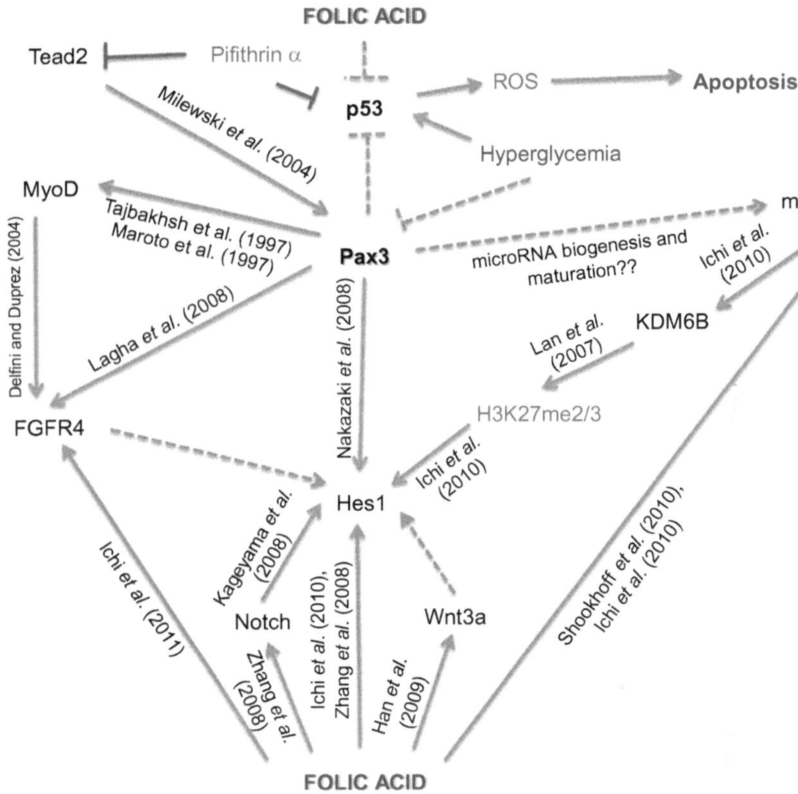

Figure 8.4 Potential folate responsive GRN in NCSCs comprised of growth factors, membrane receptors, transcription factors, microRNAs, and histone chromatin modifications. Solid lines represent interactions confirmed in the literature. Dashed lines represent potential interactions, which have yet to be investigated (see explanation in text). (For color version of this figure, the reader is referred to the Web version of this chapter.)

activates Pax3 downstream target genes, such as *FGFR4,* through yet undetermined mechanisms (Ichi *et al.*, 2011; Lagha *et al.*, 2008). *MyoD* is also a Pax3 downstream target; studies have not yet been done to determine if *MyoD* is regulated by FA. Expression of *Hes1* and *Neurog2* is regulated by FA (Ichi *et al.*, 2010; Nakazaki *et al.*, 2008), via downregulation of miRNAs, resulting in increased KDM6B expression and subsequent reduction of H3K27 methylation (Ichi *et al.*, 2010). *Hes1* is also a downstream target of Notch signaling. FA activates *Notch* (Liu *et al.*, 2010; Zhang *et al.*, 2008), including *Notch1, 2,* and *3* (Ichi *et al.*, 2011). FA may upregulate *Hes1* activity in $Sp^{-/-}$ FA-rescued neurospheres, by upregulating Notch expression. Wnt3a or FGFR4 signaling may also regulate *Hes1*; this has yet to be determined. Tead2 regulates Pax3 and is a folate responsive gene. Tead2 is inhibited by pifithrin-α, which also inhibits p53. FA downregulation of p53 needs to be examined further. FA regulates levels of miRNA in FA-rescued *Sp*$^{-/-}$ embryos, at this time it is not known if Pax3 regulates the biogenesis and maturation of these miRNAs.

VI. Folate Nonresponsive Genetic Mouse Models

One matter of utmost concern is to understand how to prevent NTD in folate nonresponsive individuals. Mouse models nonresponsive to FA (listed above) may help in this endeavor. The fact that these genetic mouse models do not respond to folate treatment, suggests that the genes involved are important, and may be a part of a yet undefined rescue mechanism(s) involving dorso-ventral patterning. It is therefore hypothesized that when these genes are non-functional or absent, then folate treatment can cause embryo lethality rather than rescue NTD. These genes must be studied to understand the complete picture of FA action on NCSCs, as they in conjunction with the FA-responsive genes form a neural tube defect rescue GRN.

VII. Conclusions and Future Directions

Although GRNs in cranial, cardiac, and caudal neural crest development have been elucidated, data are limited on how genes in these networks are affected by maternal FA intake. Future studies need to examine the impact of FA on these genes. Folate dose and timing also need to be optimized to ensure healthy birth outcomes and postnatal development bereft of complications, such as NTDs and possibly autism spectrum disorders. Birth outcomes in pregnant women on a folate rich diet are highly dependent upon genotypic variations in developmental pathways, for example, the Wnt or Notch signaling pathways. Thus, the prevailing paradigm of folate fortification in the same dose to everyone must be carefully reevaluated, while simultaneously continuing to provide FA supplementation to all pregnant women. Additionally, epigenomics of neurulation in the context of folate responsive and nonresponsive individuals must be examined. Such a study will have transformative potential on public health policies related to maternal nutrition and fetal development.

ACKNOWLEDGMENTS

This work was supported by the McLone Professorship Fund (C. S. M.), State of Illinois Excellence in Academic Medicine award (C. S. M.), a Grant from the Spastic Paralysis Research Foundation of Illinois-Eastern Iowa District of Kiwanis (C. S. M. and D. G. M.), and Spina Bifida Association and Chou Funds (C. S. M.). We would also like to thank Jose Hernandez, D. V. M. for the excellent working conditions for animal preparation and surgery, and William Goossens, manager of the Microscopy and Imaging Facility at Children's Memorial Research Center.

Disclosure statement. There is no conflict of interest.

REFERENCES

Abell, A. N., Rivera-Perez, J. A., Cuevas, B. D., Uhlik, M. T., Sather, S., Johnson, N. L., Minton, S. K., Lauder, J. M., Winter-Vann, A. M., Nakamura, K., Magnuson, T., Vaillancourt, R. R., *et al.* (2005). Ablation of MEKK4 kinase activity causes neurulation and skeletal patterning defects in the mouse embryo. *Mol. Cell. Biol.* **25,** 8948–8959.

Anderson, R. M., Stottmann, R. W., Choi, M., and Klingensmith, J. (2006). Endogenous bone morphogenetic protein antagonists regulate mammalian neural crest generation and survival. *Dev. Dyn.* **235,** 2507–2520.

Antony, A. C. (1996). Folate receptors. *Annu. Rev. Nutr.* **16,** 501–521.

Aoki, Y., Saint-Germain, N., Gyda, M., Magner-Fink, E., Lee, Y. H., Credidio, C., and Saint-Jeannet, J. P. (2003). Sox10 regulates the development of neural crest-derived melanocytes in Xenopus. *Dev. Biol.* **259,** 19–33.

Armstrong, J. F., Kaufman, M. H., Harrison, D. J., and Clarke, A. R. (1995). High-frequency developmental abnormalities in p53-deficient mice. *Curr. Biol.* **5,** 931–936.

Aruga, J., Tohmonda, T., Homma, S., and Mikoshiba, K. (2002). Zic1 promotes the expansion of dorsal neural progenitors in spinal cord by inhibiting neuronal differentiation. *Dev. Biol.* **244,** 329–341.

Barbera, J. P., Rodriguez, T. A., Greene, N. D., Weninger, W. J., Simeone, A., Copp, A. J., Beddington, R. S., and Dunwoodie, S. (2002). Folic acid prevents exencephaly in Cited2 deficient mice. *Hum. Mol. Genet.* **11,** 283–293.

Basch, M. L., and Bronner-Fraser, M. (2006). Neural crest inducing signals. *Adv. Exp. Med. Biol.* **589,** 24–31.

Beard, C. M., Panser, L. A., and Katusic, S. K. (2011). Is excess folic acid supplementation a risk factor for autism? *Med. Hypotheses* 10.1016/j.mehy.2011.03.013.

Berry, R. J., and Li, Z. (2002). Folic acid alone prevents neural tube defects: Evidence from the China study. *Epidemiology* **13,** 114–116.

Berry, R. J., Li, Z., Erickson, J. D., Li, S., Moore, C. A., Wang, H., Mulinare, J., Zhao, P., Wong, L. Y., Gindler, J., Hong, S. X., and Correa, A. (1999). Prevention of neural-tube defects with folic acid in China. China-U.S. Collaborative Project for Neural Tube Defect Prevention. *N. Engl. J. Med.* **341,** 1485–1490.

Betancur, P., Bronner-Fraser, M., and Sauka-Spengler, T. (2010). Assembling neural crest regulatory circuits into a gene regulatory network. *Annu. Rev. Cell Dev. Biol.* **26,** 581–603.

Birn, H., Selhub, J., and Christensen, E. I. (1993). Internalization and intracellular transport of folate-binding protein in rat kidney proximal tubule. *Am. J. Physiol.* **264**(2 Pt. 1), C302–C310.

Birn, H., Zhai, X., Holm, J., Hansen, S. I., Jacobsen, C., Christensen, E. I., and Moestrup, S. K. (2005). Megalin binds and mediates cellular internalization of folate binding protein. *FEBS J.* **272,** 4423–4430.

Bjorklund, N. K., and Gordon, R. (2006). A hypothesis linking low folate intake to neural tube defects due to failure of post-translation methylations of the cytoskeleton. *Int. J. Dev. Biol.* **50,** 135–141.

Boot, M. J., Steegers-Theunissen, R. P., Poelmann, R. E., Van Iperen, L., Lindemans, J., and Gittenberger-de Groot, A. C. (2003). Folic acid and homocysteine affect neural crest and neuroepithelial cell outgrowth and differentiation in vitro. *Dev. Dyn.* **227,** 301–308.

Boot, M. J., Gittenberger-de Groot, A. C., Poelmann, R. E., and Gourdie, R. G. (2006). Connexin43 levels are increased in mouse neural crest cells exposed to homocysteine. *Birth Defects Res. A Clin. Mol. Teratol.* **76,** 133–137.

Branum, A. M., and Lukacs, S. L. (2008). Food allergy among US children: Trends in prevalence and hospitalizations. *NCHS Data Brief* **10,** 1–8.

Brewster, R., Lee, J., Ruiz, I., and Altaba, A. (1998). Gli/Zic factors pattern the neural plate by defining domains of cell differentiation. *Nature* **393,** 579–583.

Buhr, G., and Bales, C. W. (2010). Nutritional supplements for older adults: Review and recommendations—Part II. *J. Nutr. Elder.* **29,** 42–71.

Calloni, G. W., Glavieux-Pardanaud, C., Le Douarin, N. M., and Dupin, E. (2007). Sonic Hedgehog promotes the development of multipotent neural crest progenitors endowed with both mesenchymal and neural potentials. *Proc. Natl. Acad. Sci. USA* **104,** 19879–19884.

Carter, M., Chen, X., Slowinska, B., Minnerath, S., Glickstein, S., Shi, L., Campagne, F., Weinstein, H., and Ross, M. E. (2005). Crooked tail (Cd) model of human folate-responsive neural tube defects is mutated in Wnt coreceptor lipoprotein receptor-related protein 6. *Proc. Natl. Acad. Sci. USA* **102,** 12843–12848.

Catron, K. M., Wang, H., Hu, G., Shen, M. M., and Abate-Shen, C. (1996). Comparison of MSX-1 and MSX-2 suggest a molecular basis for functional redundancy. *Mech. Dev.* **55,** 185–199.

Centers for Disease Control and Prevention (CDC) (2004). Asthma prevalence and control characteristics by race/ethnicity—United States, 2002. *MMWR Morb. Mortal. Wkly. Rep.* **53,** 145–148.

Chalepakis, G., Stoykova, A., Wijnholds, J., Tremblay, P., and Gruss, P. (1993). Pax: Gene regulators in the developing nervous system. *J. Neurobiol.* **24,** 1367–1384.

Chan, A., and Shea, T. B. (2007). Folate deprivation increases presenilin expression, gamma-secretase activity, and Abeta levels in murine brain: Potentiation by ApoE deficiency and alleviation by dietary S-adenosyl methionine. *J. Neurochem.* **102,** 753–760.

Chappell, J. H., Jr., Wang, X. D., and Loeken, M. R. (2009). Diabetes and apoptosis: Neural crest cells and neural tube. *Apoptosis* **14,** 1472–1483.

Cohen, E. D., Wang, Z., Lepore, J. J., Lu, M. M., Taketo, M. M., Epstein, D. J., and Morrisey, E. E. (2007). Wnt/beta-catenin signaling promotes expansion of Isl-1-positive cardiac progenitor cells through regulation of FGF signaling. *J. Clin. Invest.* **117,** 1794–1804.

Cole, M. D., and McMahon, S. B. (1999). The Myc oncoprotein: A critical evaluation of transactivation and target gene regulation. *Oncogene* **18,** 2916–2922.

Coppedè, F. (2010). One-carbon metabolism and Alzheimer's disease: Focus on epigenetics. *Curr. Genomics* **11,** 246–260.

Cornell, R. A., and Eisen, J. S. (2005). Notch in the pathway: The roles of Notch signaling in neural crest development. *Semin. Cell Dev. Biol.* **16,** 663–672.

Czeizel, A. E., and Dudas, I. (1992). Prevention of the first occurrence of neural-tube defects by periconceptional vitamin supplementation. *N. Engl. J. Med.* **327,** 1832–1835.

Davidson, G., and Niehrs, C. (2010). Emerging links between CDK cell cycle regulators and Wnt signaling. *Trends Cell Biol.* **20,** 453–460.

Davis, C. D., and Ross, S. A. (2008). Evidence for dietary regulation of microRNA expression in cancer cells. *Nutr. Rev.* **66,** 477–482.

De Castro, S. C., Leung, K. Y., Savery, D., Burren, K., Rozen, R., Copp, A. J., and Greene, N. D. (2010). Neural tube defects induced by folate deficiency in mutant curly tail (Grhl3) embryos are associated with alteration in folate one-carbon metabolism but are unlikely to result from diminished methylation. *Birth Defects Res. A Clin. Mol. Teratol.* **88,** 612–618.

de Crozé, N., Maczkowiak, F., and Monsoro-Burq, A. H. (2011). Reiterative AP2a activity controls sequential steps in the neural crest gene regulatory network. *Proc. Natl. Acad. Sci. USA* **108,** 155–160.

Delfini, M. C., and Duprez, D. (2004). Ectopic Myf5 or MyoD prevents the neuronal differentiation program in addition to inducing skeletal muscle differentiation, in the chick neural tube. *Development* **131,** 713–723.

Ebert, P. J., Timmer, J. R., Nakada, Y., Helms, A. W., Parab, P. B., Liu, Y., Hunsaker, T. L., and Johnson, J. E. (2003). Zic1 represses Math1 expression via interactions with the Math1 enhancer and modulation of Math1 autoregulation. *Development* **130,** 1949–1959.

Epeldegui, M., Pena-Melian, A., Varela-Moreiras, G., and Perez-Miguelsanz, J. (2002). Homocysteine modifies development of neurulation and dorsal root ganglia in chick embryos. *Teratology* **65,** 171–179.

Epstein, J. A. (1996). Pax3, neural crest and cardiovascular development. *Trends Cardiovasc. Med.* **6,** 255–260.

Epstein, J. A., Li, J., Lang, D., Chen, F., Brown, C. B., Jin, F., Lu, M. M., Thomas, M., Liu, E., Wessels, A., and Lo, C. W. (2000). Migration of cardiac neural crest cells in Splotch embryos. *Development* **127,** 1869–1878.

Ernest, S., Carter, M., Shao, H., Hosack, A., Lerner, N., Colmenares, C., Rosenblatt, D. S., Pao, Y. H., Ross, M. E., and Nadeau, J. H. (2006). Parallel changes in metabolite and expression profiles in crooked-tail mutant and folate-reduced wild-type mice. *Hum. Mol. Genet.* **15,** 3387–3393.

Essien, F. B., and Wannberg, S. L. (1993). Methionine but not folinic acid or vitamin B-12 alters the frequency of neural tube defects in Axd mutant mice. *J. Nutr.* **123,** 27–34.

Farah, A. (2009). The role of L-methylfolate in depressive disorders. *CNS Spectr.* **14**(1 Suppl. 2), 2–7.

Fava, M., and Mischoulon, D. (2009). Folate in depression: Efficacy, safety, differences in formulations, and clinical issues. *J. Clin. Psychiatry* **70**(Suppl. 5), 12–17.

Fleming, A., and Copp, A. J. (1998). Embryonic folate metabolism and mouse neural tube defects. *Science* **280,** 2107–2109.

Foley, A., and Mercola, M. (2004). Heart induction: Embryology to cardiomyocyteregeneration. *Trends Cardiovasc. Med.* **14,** 121–125.

Frey, L., and Hauser, W. A. (2003). Epidemiology of neural tube defects. *Epilepsia* **44** (Suppl. 3), 4–13.

Fuso, A., Seminara, L., Cavallaro, R. A., D'Anselmi, F., and Scarpa, S. (2005). S-adenosylmethionine/homocysteine cycle alterations modify DNA methylation status with consequent deregulation of PS1 and BACE and beta-amyloid production. *Mol. Cell. Neurosci.* **28,** 195–204.

Fuso, A., Cavallaro, R. A., Zampelli, A., D'Anselmi, F., Piscopo, P., Confaloni, A., and Scarpa, S. (2007). gamma-Secretase is differentially modulated by alterations of homocysteine cycle in neuroblastoma and glioblastoma cells. *J. Alzheimers Dis.* **11,** 275–290.

Fuso, A., Nicolia, V., Cavallaro, R. A., Ricceri, L., D'Anselmi, F., Coluccia, P., Calamandrei, G., and Scarpa, S. (2008). B-vitamin deprivation induces hyperhomocysteinemia and brain S-adenosylhomocysteine, depletes brain S-adenosylmethionine, and enhances PS1 and BACE expression and amyloid-beta deposition in mice. *Mol. Cell. Neurosci.* **37,** 731–746.

Fuso, A., Nicolia, V., Pasqualato, A., Fiorenza, M. T., Cavallaro, R. A., and Scarpa, S. (2011). Changes in Presenilin 1 gene methylation pattern in diet-induced B vitamin deficiency. *Neurobiol. Aging* **32,** 187–199.

Gammill, L. S., and Bronner-Fraser, M. (2003). Neural crest specification: Migrating into genomics. *Nat. Rev. Neurosci.* **4,** 795–805.

García-Castro, M., and Bronner-Fraser, M. (1999). Induction and differentiation of the neural crest. *Curr. Opin. Cell Biol.* **11,** 695–698.

García-Castro, M. I., Marcelle, C., and Bronner-Fraser, M. (2002). Ectodermal Wnt function as a neural crest inducer. *Science* **297,** 848–851.

Gelineau-van Waes, J., Maddox, J. R., Smith, L. M., Van Waes, M., Wilberding, J., Eudy, J. D., Bauer, L. K., and Finnell, R. H. (2008). Microarray analysis of E9.5 reduced folate carrier (RFC1; Slc19a1) knockout embryos reveals altered expression of genes in the cubilin-megalin multiligand endocytic receptor complex. *BMC Genomics* **9,** 156.

Gitler, A. D., Brown, C. B., Kochilas, L., Li, J., and Epstein, J. A. (2002). Neural crest migration and mouse models of congenital heart disease. *Cold Spring Harb. Symp. Quant. Biol.* **67,** 57–62.

Graham, A., Begbie, J., and McGonnell, I. (2004). Significance of the cranial neural crest. *Dev. Dyn.* **229,** 5–13.

Gray, J. D., and Ross, M. E. (2009). Mechanistic insights into folate supplementation from Crooked tail and other NTD-prone mutant mice. *Birth Defects Res. A Clin. Mol. Teratol.* **85,** 314–321.

Gray, J. D., Nakouzi, G., Slowinska-Castaldo, B., Dazard, J. E., Rao, J. S., Nadeau, J. H., and Ross, M. E. (2010). Functional interactions between the LRP6 WNT co-receptor and folate supplementation. *Hum. Mol. Genet.* **19,** 4560–4572.

Greene, N. D., Dunlevy, L. E., and Copp, A. J. (2003). Homocysteine is embryotoxic but does not cause neural tube defects in mouse embryos. *Anat. Embryol. (Berl)* **206,** 185–191.

Håberg, S. E., London, S. J., Stigum, H., Nafstad, P., and Nystad, W. (2009). Folic acid supplements in pregnancy and early childhood respiratory health. *Arch. Dis. Child.* **94,** 180–184.

Han, M., Serrano, M. C., Lastra-Vicente, R., Brinez, P., Acharya, G., Huhta, J. C., Chen, R., and Linask, K. K. (2009). Folate rescues lithium-, homocysteine- and Wnt3A-induced vertebrate cardiac anomalies. *Dis. Model. Mech.* **2,** 467–478.

Han, M., Toli, J., and Abdellatif, M. (2011). MicroRNAs in the cardiovascular system. *Curr. Opin. Cardiol.* **26,** 181–189.

Hanna, L. A., Foreman, R. K., Tarasenko, I. A., Kessler, D. S., and Labosky, P. A. (2002). Requirement for Foxd3 in maintaining pluripotent cells of the early mouse embryo. *Genes Dev.* **16,** 2650–2661.

Harris, M. J., and Juriloff, D. M. (2010). An update to the list of mouse mutants with neural tube closure defects and advances toward a complete genetic perspective of neural tube closure. *Birth Defects Res. A Clin. Mol. Teratol.* **88,** 653–669.

Heidenreich, D. J., Reedy, M. V., and Brauer, P. R. (2008). Homocysteine enhances cardiac neural crest cell attachment in vitro by increasing intracellular calcium levels. *Dev. Dyn.* **237,** 2117–2128.

Hemavathy, K., Ashraf, S. I., and Ip, Y. T. (2000). Snail/slug family of repressors: Slowly going into the fast lane of development and cancer. *Gene* **257,** 1–12.

Henderson, G. I., Perez, T., Schenker, S., Mackins, J., and Antony, A. C. (1995). Maternal-to-fetal transfer of 5-methyltetrahydrofolate by the perfused human placental cotyledon: Evidence for a concentrative role by placental folate receptors in fetal folate delivery. *J. Lab. Clin. Med.* **126,** 184–203.

Herrmann, W., and Obeid, R. (2011). Homocysteine: A biomarker in neurodegenerative diseases. *Clin. Chem. Lab. Med.* **49,** 435–441.

Hibbard, B. M., Hibbard, E. D., and Jeffcoate, T. N. (1965). Folic acid and reproduction. *Acta Obstet. Gynecol. Scand.* **44,** 375–400.

Hitoshi, S., Alexson, T., Tropepe, V., Donoviel, D., Elia, A. J., Nye, J. S., Conlon, R. A., Mak, T. W., Bernstein, A., and van der Kooy, D. (2002). Notch pathway molecules are essential for the maintenance, but not the generation, of mammalian neural stem cells. *Genes Dev.* **16,** 846–858.

Hofmann, M., Harris, M., Juriloff, D., and Boehm, T. (1998). Spontaneous mutations in SELH/Bc mice due to insertions of early transposons: Molecular characterization of null alleles at the nude and albino loci. *Genomics* **52,** 107–109.

Hollingsworth, J. W., Maruoka, S., Boon, K., Garantziotis, S., Li, Z., Tomfohr, J., Bailey, N., Potts, E. N., Whitehead, G., Brass, D. M., and Schwartz, D. A. (2008). In utero supplementation with methyl donors enhances allergic airway disease in mice. *J. Clin. Invest.* **118,** 3462–3469.

Hong, C. S., and Saint-Jeannet, J. P. (2005). Sox proteins and neural crest development. *Semin. Cell Dev. Biol.* **6,** 694–703.

Howe, L. R., Watanabe, O., Leonard, J., and Brown, A. M. (2003). Twist is up-regulated in response to Wnt1 and inhibits mouse mammary cell differentiation. *Cancer Res.* **63,** 1906–1913.

Hu, G., Lee, H., Price, S. M., Shen, M. M., and Abate-Shen, C. (2001). Msx homeobox genes inhibit differentiation through upregulation of cyclin D1. *Development* **128,** 2373–2384.

Huang, R. F., Yaong, H. C., Chen, S. C., and Lu, Y. F. (2004). In vitro folate supplementation alleviates oxidative stress, mitochondria-associated death signaling and apoptosis induced by 7-ketocholesterol. *Br. J. Nutr.* **92,** 887–894.

Huang, Z. P., Chen, J. F., Regan, J. N., Maguire, C. T., Tang, R. H., Dong, X. R., Majesky, M. W., and Wang, D. Z. (2010). Loss of microRNAs in neural crest leads to cardiovascular syndromes resembling human congenital heart defects. *Arterioscler. Thromb. Vasc. Biol.* **30,** 2575–2586.

Ichi, S., Costa, F. F., Bischof, J. M., Nakazaki, H., Shen, Y. W., Boshnjaku, V., Sharma, S., Mania-Farnell, B., McLone, D. G., Tomita, T., Soares, M. B., and Mayanil, C. S. (2010). Folic acid remodels chromatin on *Hes1* and *Neurog2* promoters during caudal neural tube development. *J. Biol. Chem.* **285,** 36922–36932.

Ichi, S., Nakazaki, H., Boshnjaku, V., Singh, R. M., Mania-Farnell, B., Xi, G., McLone, D. G., Tomita, T., and Mayanil, C. S. (2011). Fetal neural tube stem cells from Pax3 mutant mice proliferate differentiate and form synaptic connections when stimulated with folic acid. *Stem Cells Dev.* (in press).

Jain, R., Engleka, K. A., Rentschler, S. L., Manderfield, L. J., Li, L., Yuan, L., and Epstein, J. A. (2011). Cardiac neural crest orchestrates remodeling and functional maturation of mouse semilunar valves. *J. Clin. Invest.* **121,** 422–430.

James, S. J., Melnyk, S., Jernigan, S., Cleves, M. A., Halsted, C. H., Wong, D. H., Cutler, P., Bock, K., Boris, M., Bradstreet, J. J., Baker, S. M., and Gaylor, D. W. (2006). Metabolic endophenotype and related genotypes a.re associated with oxidative stress in children with autism. *Am. J. Med. Genet. B Neuropsychiatr. Genet.* **141B,** 947–956.

Jia, D. Y., Liu, H. J., Wang, F. W., Liu, S. M., Ling, E. A., Liu, K., and Hao, A. J. (2008). Folic acid supplementation affects apoptosis and differentiation of embryonic neural stem cells exposed to high glucose. *Neurosci. Lett.* **440,** 27–33.

Jones, N. C., Lynn, M. L., Gaudenz, K., Sakai, D., Aoto, K., Rey, J. P., Glynn, E. F., Ellington, L., Du, C., Dixon, J., Dixon, M. J., and Trainor, P. A. (2008). Prevention of the neurocristopathy Treacher Collins syndrome through inhibition of p53 function. *Nat. Med.* **14,** 125–133.

Jung, J. Y., Lee, H. S., Kang, D. G., Kim, N. S., Cha, M. H., Bang, O. S., Ryu, D. H., and Hwang, G. S. (2011). 1H-NMR-Based Metabolomics Study of Cerebral Infarction. *Stroke* **42,** 1282–1288.

Kageyama, R., Ohtsuka, T., and Kobayashi, T. (2008a). Roles of Hes genes in neural development. *Dev. Growth Differ.* **50**(Suppl. 1), S97–S103.

Kageyama, R., Ohtsuka, T., Shimojo, H., and Imayoshi, I. (2008b). Dynamic Notch signaling in neural progenitor cells and a revised view of lateral inhibition. *Nat. Neurosci.* **11,** 1247–1251.

Kageyama, R., Ohtsuka, T., Shimojo, H., and Imayoshi, I. (2009). Dynamic regulation of Notch signaling in neural progenitor cells. *Curr. Opin. Cell Biol.* **21,** 733–740.

Kaneko, K. J., Kohn, M. J., Liu, C., and DePamphilis, M. L. (2007). Transcription factor TEAD2 is involved in neural tube closure. *Genesis* **45,** 577–587.

Kapusta, L., Haagmans, M. L., Steegers, E. A., Cuypers, M. H., Blom, H. J., and Eskes, T. K. (1999). Congenital heart defects and maternal derangement of homocysteine metabolism. *J. Pediatr.* **135,** 773–774.

Kelsh, R. N. (2006). Sorting out Sox10 functions in neural crest development. *Bioessays* **28,** 788–798.

Kirby, M. L., and Waldo, K. L. (1995). Neural crest and cardiovascular patterning. *Circ. Res.* **77,** 211–215.

Klivenyi, P., and Vecsei, L. (2010). Novel therapeutic strategies in Parkinson's disease. *Eur. J. Clin. Pharmacol.* **66,** 119–125.

Knecht, A. K., and Bronner-Fraser, M. (2002). Induction of the neural crest: A multigene process. *Nat. Rev. Genet.* **3,** 453–456.

Kokubu, C., Heinzmann, U., Kokubu, T., Sakai, N., Kubota, T., Kawai, M., Wahl, M. B., Galceran, J., Grosschedl, R., Ozono, K., and Imai, K. (2004). Skeletal defects in ringelschwanz mutant mice reveal that Lrp6 is required for proper somitogenesis and osteogenesis. *Development* **131,** 5469–5480.

Kopan, R. (2002). Notch: A membrane-bound transcription factor. *J. Cell Sci.* **115,** 1095–1097.

Kronenberg, G., Colla, M., and Endres, M. (2009). Folic acid, neurodegenerative and neuropsychiatric disease. *Curr. Mol. Med.* **9,** 315–323.

Kurata, T., and Ueno, N. (2003). Xenopus Nbx, a novel NK-1 related gene essential for neural crest formation. *Dev. Biol.* **257,** 30–40.

LaBonne, C., and Bronner-Fraser, M. (1998). Neural crest induction in Xenopus: Evidence for a two-signal model. *Development* **125,** 2403–2414.

LaBonne, C., and Bronner-Fraser, M. (1999). Molecular mechanisms of neural crest formation. *Annu. Rev. Cell Dev. Biol.* **15,** 81–112.

Lacey, S. W., Sanders, J. M., Rothberg, K. G., Anderson, R. G., and Kamen, B. A. (1989). Complementary DNA for the folate binding protein correctly predicts anchoring to the membrane by glycosyl-phosphatidylinositol. *J. Clin. Invest.* **84,** 715–720.

Lagha, M., Kormish, J. D., Rocancourt, D., Manceau, M., Epstein, J. A., Zaret, K. S., Relaix, F., and Buckingham, M. E. (2008). Pax3 regulation of FGF signaling affects the progression of embryonic progenitor cells into the myogenic program. *Genes Dev.* **22,** 1828–1837.

Lakhwani, S., García-Sanz, P., and Vallejo, M. (2010). Alx3-deficient mice exhibit folic acid-resistant craniofacial midline and neural tube closure defects. *Dev. Biol.* **344,** 869–880.

Latacha, K. S., and Rosenquist, T. H. (2005). Homocysteine inhibits extra-embryonic vascular development in the avian embryo. *Dev. Dyn.* **234,** 323–331.

Le Douarin, N. M., and Teillet, M. A. (1973). The migration of neural crest cells to the wall of the digestive tract in avian embryo. *J. Embryol. Exp. Morphol.* **30,** 31–48.

Li, J., Liu, K. C., Jin, F., Lu, M. M., and Epstein, J. A. (1999). Transgenic rescue of congenital heart disease and spina bifida in Splotch mice. *Development* **126,** 2495–2503.

Lin, W., Zhang, Z., Srajer, G., Chen, Y. C., Huang, M., Phan, H. M., and Dent, S. Y. (2008). Proper expression of the Gcn5 histone acetyltransferase is required for neural tube closure in mouse embryos. *Dev. Dyn.* **237,** 928–940.

Linask, K. K., and Huhta, J. (2010). Folate protection from congenital heart defects linked with canonical Wnt signaling and epigenetics. *Curr. Opin. Pediatr.* **22,** 561–566.

Liu, H., Huang, G. W., Zhang, X. M., Ren, D. L., and X Wilson, J. (2010). Folic acid supplementation stimulates notch signaling and cell proliferation in embryonic neural stem cells. *J. Clin. Biochem. Nutr.* **47,** 174–180.

Liu, Y., Huang, T., Zhao, X., and Cheng, L. (2011). MicroRNAs modulate the Wnt signaling pathway through targeting its inhibitors. *Biochem. Biophys. Res. Commun.* **408,** 259–264.

Loeken, M. R. (2006). Advances in understanding the molecular causes of diabetes induced birth defects. *J. Soc. Gynecol. Investig.* **13,** 2–10.

Lotem, J., Peled-Kamar, M., Groner, Y., and Sachs, L. (1996). Cellular oxidative stress and the control of apoptosis by wild-type p53, cytotoxic compounds, and cytokines. *Proc. Natl. Acad. Sci. USA* **93,** 9166–9171.

Louvi, A., and Artavanis-Tsakonas, S. (2006). Notch signaling in vertebrate neural development. *Nat. Rev. Neurosci.* **7,** 93–102.

Luo, T., Matsuo-Takasaki, M., Thomas, M. L., Weeks, D. L., and Sargent, T. D. (2002). Transcription factor AP-2 is an essential and direct regulator of epidermal development in Xenopus. *Dev. Biol.* **245,** 136–144.

Maestro de las Casas, C., Epeldegui, M., Tudela, C., Varela-Moreiras, G., and Perez-Miguelsanz, J. (2003). High exogenous homocysteine modifies eye development in early chick embryos. *Birth Defects Res. A Clin. Mol. Teratol.* **67,** 35–40.

Marics, I., Padilla, F., Guillemot, J. F., Scaal, M., and Marcelle, C. (2002). FGFR4 signaling is a necessary step in limb muscle differentiation. *Development* **129,** 4559–4569.

Martignoni, E., Tassorelli, C., Nappi, G., Zangaglia, R., Pacchetti, C., and Blandini, F. (2007). Homocysteine and Parkinson's disease: A dangerous liaison? *J. Neurol. Sci.* **257,** 31–37.

Martinsen, B. J., and Bronner-Fraser, M. (1998). Neural crest specification regulated by the helix-loop-helix repressor Id2. *Science* **281,** 988–991.

Maschhoff, K. L., and Baldwin, H. S. (2000). Molecular determinants of neural crest migration. *Am. J. Med. Genet.* **97,** 280–288.

Mattson, M. P., and Shea, T. B. (2003). Folate and homocysteine metabolism in neural plasticity and neurodegenerative disorders. *Trends Neurosci.* **26,** 137–146.

Mayanil, C. S., George, D., Freilich, L., Miljan, E. J., Mania-Farnell, B., McLone, D. G., and Bremer, E. G. (2001). Microarray analysis detects novel Pax3 downstream target genes. *J. Biol. Chem.* **276,** 49299–49309.

Mayor, R., Guerrero, N., and Martínez, C. (1997). Role of FGF and noggin in neural crest induction. *Dev. Biol.* **189,** 1–12.

Meulemans, D., and Bronner-Fraser, M. (2004). Gene–regulatory interactions in neural crest evolution and development. *Dev. Cell* **7,** 291–299.

Milewski, R. C., Chi, N. C., Li, J., Brown, C., Lu, M. M., and Epstein, J. A. (2004). Identification of minimal enhancer elements sufficient for Pax3 expression in neural crest and implication of Tead2 as a regulator of Pax3. *Development* **131,** 829–837.

Miotti, S., Bagnoli, M., Tomassetti, A., Colnaghi, M. I., and Canevari, S. (2000). Interaction of folate receptor with signaling molecules lyn and G(alpha)(i-3) in detergent-resistant complexes from the ovary carcinoma cell line IGROV1. *J. Cell Sci.* **113,** 349–357.

Monsoro-Burq, A. H., Fletcher, R. B., and Harland, R. M. (2003). Neural crest induction by paraxial mesoderm in Xenopus embryos requires FGF signals. *Development* **130,** 3111–3124.

Morales, A. V., Barbas, J. A., and Nieto, M. A. (2005). How to become neural crest: From segregation to delamination. *Semin. Cell Dev. Biol.* **16,** 655–662.

Morgan, S. C., Lee, H. Y., Relaix, F., Sandell, L. L., Levorse, J. M., and Loeken, M. R. (2008). Cardiac outflow tract septation failure in Pax3-deficient embryos is due to p53-dependent regulation of migrating cardiac neural crest. *Mech. Dev.* **125,** 757–767.

Müller, T. (2008). Role of homocysteine in the treatment of Parkinson's disease. *Expert Rev. Neurother.* **8,** 957–967.

Murphy, M., and Bartlett, P. F. (1993). Molecular regulation of neural crest development. *Mol. Neurobiol.* **7,** 111–135.

Nakazaki, H., Reddy, A. C., Mania-Farnell, B. L., Shen, Y. W., Ichi, S., McCabe, C., George, D., McLone, D. G., Tomita, T., and Mayanil, C. S. (2008). Key basic helix-loop-helix transcription factor genes Hes1 and Ngn2 are regulated by Pax3 during mouse embryonic development. *Dev. Biol.* **316,** 510–523.

Nie, X., Wang, Q., and Jiao, K. (2011). Dicer activity in neural crest cells is essential for craniofacial organogenesis and pharyngeal arch artery morphogenesis. *Mech. Dev.* **128,** 200–207.

Northrup, H., and Volcik, K. A. (2000). Spina bifida and other neural tube defects. *Curr. Probl. Pediatr.* **30,** 313–332.

Norton, J. D. (2000). ID helix-loop-helix proteins in cell growth, differentiation and tumorigenesis. *J. Cell Sci.* **113,** 3897–3905.

Ono, K., Kuwabara, Y., and Han, J. (2011). MicroRNAs and cardiovascular diseases. *FEBS J.* 10.1111/j.1742-4658.2011.08090.x.

Pani, L., Horal, M., and Loeken, M. R. (2002). Rescue of neural tube defects in Pax-3-deficient embryos by p53 loss of function: Implications for Pax-3-dependent development and tumorigenesis. *Genes Dev.* **16,** 676–680.

Pezzini, A., Del Zotto, E., and Padovani, A. (2007). Homocysteine and cerebral ischemia: Pathogenic and therapeutical implications. *Curr. Med. Chem.* **14,** 249–263.

Piedrahita, J. A., Oetama, B., Bennett, G. D., van Waes, J., Kamen, B. A., Richardson, J., Lacey, S. W., Anderson, R. G., and Finnell, R. H. (1999). Mice lacking the folic acid-binding protein Folbp1 are defective in early embryonic development. *Nat. Genet.* **23,** 228–232.

Pinson, K. I., Brennan, J., Monkley, S., Avery, B. J., and Skarnes, W. C. (2000). An LDL-receptor-related protein mediates Wnt signalling in mice. *Nature* **407,** 535–538.

Pisano, M. M., Bhattacherjee, V., Wong, L., Finnell, R. H., and Greene, R. M. (2010). Novel folate binding protein-1 interactions in embryonic orofacial tissue. *Life Sci.* **86,** 275–280.

Pogribny, I. P., Basnakian, A. G., Miller, B. J., Lopatina, N. G., Poirier, L. A., and James, S. J. (1995). Breaks in genomic DNA and within the p53 gene are associated with hypomethylation in livers of folate/methyl-deficient rats. *Cancer Res.* **55,** 1894–1901Erratum in: Cancer Res 55, 2711.

Polyak, K., Xia, Y., Zweier, J. L., Kinzler, K. W., and Vogelstein, B. (1997). A model for p53-induced apoptosis. *Nature* **389,** 300–305.

Reeves, F. C., Burdge, G. C., Fredericks, W. J., Rauscher, F. J., and Lillycrop, K. A. (1999). Induction of antisense Pax-3 expression leads to the rapid morphological differentiation of neuronal cells and an altered response to the mitogenic growth factor bFGF. *J. Cell Sci.* **112,** 253–261.

Rinon, A., Molchadsky, A., Nathan, E., Yovel, G., Rotter, V., Sarig, R., and Tzahor, E. (2011). p53 coordinates cranial neural crest cell growth and epithelial-mesenchymal transition/delamination processes. *Development* **138,** 1827–1838.

Rodier, P. M., Ingram, J. L., Tisdale, B., and Croog, V. J. (1997). Linking etiologies in humans and animal models: Studies of autism. *Reprod. Toxicol.* **11,** 417–422.

Rogers, E. J. (2008). Has enhanced folate status during pregnancy altered natural selection and possibly Autism prevalence? A closer look at a possible link. *Med. Hypotheses* **71,** 406–410.

Rosenquist, T. H., Ratashak, S. A., and Selhub, J. (1996). Homocysteine induces congenital defects of the heart and neural tube: Effect of folic acid. *Proc. Natl. Acad. Sci. USA* **93,** 15227–15232.

Rosenquist, T. H., Bennett, G. D., Brauer, P. R., Stewart, M. L., Chaudoin, T. R., and Finnell, R. H. (2007). Microarray analysis of homocysteine-responsive genes in cardiac neural crest cells in vitro. *Dev. Dyn.* **236,** 1044–1054.

Sah, V. P., Attardi, L. D., Mulligan, G. J., Williams, B. O., Bronson, R. T., and Jacks, T. (1995). A subset of p53-deficient embryos exhibit exencephaly. *Nat. Genet.* **10,** 175–180.

Saint-Jeannet, J. P., He, X., Varmus, H. E., and Dawid, I. B. (1997). Regulation of dorsal fate in the neuraxis by Wnt-1 and Wnt-3a. *Proc. Natl. Acad. Sci. USA* **94,** 13713–13718.

Sasai, Y., and De Robertis, E. M. (1997). Ectodermal patterning in vertebrate embryos. *Dev. Biol.* **182,** 5–20.

Sauka-Spengler, T., and Bronner-Fraser, M. (2008). A gene regulatory network orchestrates neural crest formation. *Nat. Rev. Mol. Cell Biol.* **9,** 557–568.

Schlesinger, J., Schueler, M., Grunert, M., Fischer, J. J., Zhang, Q., Krueger, T., Lange, M., Tönjes, M., Dunkel, I., and Sperling, S. R. (2011). The cardiac transcription network modulated by Gata4 Mef2a Nkx2.5 Srf histone modifications and microRNAs. *PLoS Genet.* **7,** e1001313.

Schweizer, L., and Varmus, H. (2003). Wnt/Wingless signaling through beta-catenin requires the function of both LRP/Arrow and frizzled classes of receptors. *BMC Cell Biol.* **4,** 4.

Scott, J. M., and Weir, D. G. (1998). Folic acid, homocysteine and one-carbon metabolism: A review of the essential biochemistry. *J. Cardiovasc. Risk* **5,** 223–227.

Scott, C. E., Wynn, S. L., Sesay, A., Cruz, C., Cheung, M., Gomez Gaviro, M. V., Booth, S., Gao, B., Cheah, K. S., Lovell-Badge, R., and Briscoe, J. (2010). SOX9 induces and maintains neural stem cells. *Nat. Neurosci.* **13,** 1181–1189.

Selleck, M. A., and Bronner-Fraser, M. (1995). Origins of the avian neural crest: The role of neural plate-epidermal interactions. *Development* **121,** 525–538.

Shane, B. (1995). Folate chemistry and metabolism. *In* "Folate in Health and Disease," (L. B. Bailey, Ed.), pp. 1–22. Marcel Dekker, Inc., New York.

Sheehy, N. T., Cordes, K. R., White, M. P., Ivey, K. N., and Srivastava, D. (2010). The neural crest- enriched microRNA miR-452 regulates epithelial-mesenchymal signaling in the first pharyngeal arch. *Development* **137,** 4307–4316.

Shookhoff, J. M., and Gallicano, G. I. (2010). A new perspective on neural tube defects: Folic acid and microRNA misexpression. *Genesis* **48,** 282–294.

Smithells, R. W. (1985). Vitamins and neural tube defects. *Curr. Concepts Nutr.* **14,** 83–92.

Snider, P., Olaopa, M., Firulli, A. B., and Conway, S. J. (2007). Cardiovascular development and the colonizing cardiac neural crest lineage. *ScientificWorldJournal* **7,** 1090–1113.

Spiegelstein, O., Mitchell, L. E., Merriweather, M. Y., Wicker, N. J., Zhang, Q., Lammer, E. J., and Finnell, R. H. (2004). Embryonic development of folate binding protein-1 (Folbp1) knockout mice: Effects of the chemical form, dose, and timing of maternal folate supplementation. *Dev. Dyn.* **231,** 221–231.

Tamai, K., Semenov, M., Kato, Y., Spokony, R., Liu, C., Katsuyama, Y., Hess, F., Saint-Jeannet, J. P., and He, X. (2000). LDL-receptor-related proteins in Wnt signal transduction. *Nature* **407,** 530–535.

Tang, L. S., Wlodarczyk, B. J., Santillano, D. R., Miranda, R. C., and Finnell, R. H. (2004). Developmental consequences of abnormal folate transport during murine heart morphogenesis. *Birth Defects Res. A Clin. Mol. Teratol.* **70,** 449–458.

Tang, L. S., Santillano, D. R., Wlodarczyk, B. J., Miranda, R. C., and Finnell, R. H. (2005). Role of Folbp1 in the regional regulation of apoptosis and cell proliferation in the developing neural tube and craniofacies. *Am. J. Med. Genet. C Semin. Med. Genet.* **135C,** 48–58.

Thum, T., Catalucci, D., and Bauersachs, J. (2008). MicroRNAs: Novel regulators in cardiac development and disease. *Cardiovasc. Res.* **79,** 562–570.

Tierney, B. J., Ho, T., Reedy, M. V., and Brauer, P. R. (2004). Homocysteine inhibits cardiac neural crest cell formation and morphogenesis in vivo. *Dev. Dyn.* **229,** 63–73.

Ting, S. B., Wilanowski, T., Cerruti, L., Zhao, L. L., Cunningham, J. M., and Jane, S. M. (2003). The identification and characterization of human Sister-of-Mammalian Grainyhead (SOM) expands the grainyhead-like family of developmental transcription factors. *Biochem. J.* **370,** 953–962.

Topkara, V. K., and Mann, D. L. (2011). Role of MicroRNAs in Cardiac Remodeling and Heart Failure. *Cardiovasc. Drugs Ther.* **25,** 171–182.

van der Put, N. M., van Straaten, H. W., Trijbels, F. J., and Blom, H. J. (2001). Folate, homocysteine and neural tube defects: An overview. *Exp. Biol. Med. (Maywood)* **226,** 243–270.

Vanaerts, L. A., Blom, H. J., Deabreu, R. A., Trijbels, F. J., Eskes, T. K., Copius Peereboom-Stegeman, J. H., and Noordhoek, J. (1994). Prevention of neural tube defects by and toxicity of L-homocysteine in cultured postimplantation rat embryos. *Teratology* **50,** 348–360.

Verberne, M. E., Gittenberger-de Groot, A. C., van Iperen, L., and Poelmann, R. E. (2000). Distribution of different regions of cardiac neural crest in the extrinsic and the intrinsic cardiac nervous system. *Dev. Dyn.* **217,** 191–204.

Wainfan, E., and Poirier, L. A. (1992). Methyl groups in carcinogenesis: Effects on DNA methylation and gene expression. *Cancer Res.* **52**(7. Suppl), 2071s–2077s.

Wang, Q., Fang, W. H., Krupinski, J., Kumar, S., Slevin, M., and Kumar, P. (2008). Pax genes in embryogenesis and oncogenesis. *J. Cell. Mol. Med.* **12,** 2281–2294.

Washington, S. I., Byrd, N. A., Abu-Issa, R., Goddeeris, M. M., Anderson, R., Morris, J., Yamamura, K., Klingensmith, J., and Meyers, E. N. (2005). Sonic hedgehog is required for cardiac outflow tract and neural crest cell development. *Dev. Biol.* **283,** 357–372.

Wasson, G. R., McGlynn, A. P., McNulty, H., O'Reilly, S. L., McKelvey-Martin, V. J., McKerr, G., Strain, J. J., Scott, J., and Downes, C. S. (2006). Global DNA and p53 region-specific hypomethylation in human colonic cells is induced by folate depletion and reversed by folate supplementation. *J. Nutr.* **136,** 2748–2753.

Weinstein, D. C., and Hemmati-Brivanlou, A. (1999). Neural induction. *Annu. Rev. Cell Dev. Biol.* **15,** 411–433.

Wenstrom, K. D., Johanning, G. L., Johnston, K. E., and DuBard, M. (2001). Association of the C677T methylenetetrahydrofolate reductase mutation and elevated homocysteine levels with congenital cardiac malformations. *Am. J. Obstet. Gynecol.* **184,** 806–812.

Whitrow, M. J., Moore, V. M., Rumbold, A. R., and Davies, M. J. (2009). Effect of supplemental folic acid in pregnancy on childhood asthma: A prospective birth cohort study. *Am. J. Epidemiol.* **170,** 1486–1493.

Williams, T., and Tjian, R. (1991). Characterization of a dimerization motif in AP-2 and its function in heterologous DNA-binding proteins. *Science* **251,** 1067–1071.

Wlodarczyk, B. J., Tang, L. S., Triplett, A., Aleman, F., and Finnell, R. H. (2006). Spontaneous neural tube defects in splotch mice supplemented with selected micronutrients. *Toxicol. Appl. Pharmacol.* **213,** 55–63.

Wong, R. L., Wlodarczyk, B. J., Min, K. S., Scott, M. L., Kartiko, S., Yu, W., Merriweather, M. Y., Vogel, P., Zambrowicz, B. P., and Finnell, R. H. (2008). Mouse Fkbp8 activity is required to inhibit cell death and establish dorso-ventral patterning in the posterior neural tube. *Hum. Mol. Genet.* **17,** 587–601.

Yan, C. T., Kaushal, D., Murphy, M., Zhang, Y., Datta, A., Chen, C., Monroe, B., Mostoslavsky, G., Coakley, K., Gao, Y., Mills, K. D., Fazeli, A. P., *et al.* (2006). XRCC4 suppresses medulloblastomas with recurrent translocations in p53-deficient mice. *Proc. Natl. Acad. Sci. USA* **103,** 7378–7383.

Yu, S., Zheng, L., Trinh, D. K., Asa, S. L., and Ezzat, S. (2004). Distinct transcriptional control and action of fibroblast growth factor receptor 4 in differentiating skeletal muscle cells. *Lab. Invest.* **84,** 1571–1580.

Zehir, A., Hua, L. L., Maska, E. L., Morikawa, Y., and Cserjesi, P. (2010). Dicer is required for survival of differentiating neural crest cells. *Dev. Biol.* **340,** 459–467.

Zhang, X., Liu, H., Cong, G., Tian, Z., Ren, D., Wilson, J. X., and Huang, G. (2008). Effects of folate on notch signaling and cell proliferation in neural stem cells of neonatal rats in vitro. *J. Nutr. Sci. Vitaminol. (Tokyo)* **54,** 353–356.

Zhao, Q., Behringer, R. R., and de Crombrugghe, B. (1996). Prenatal folic acid treatment suppresses acrania and meroanencephaly in mice mutant for the Cart1 homeobox gene. *Nat. Genet.* **13,** 275–283.

Zhu, H., Wlodarczyk, B. J., Scott, M., Yu, W., Merriweather, M., Gelineau-van Waes, J., Schwartz, R. J., and Finnell, R. H. (2007). Cardiovascular abnormalities in Folr1 knockout mice and folate rescue. *Birth Defects Res. A Clin. Mol. Teratol.* **79,** 257–268.

Zohn, I. E., and Sarkar, A. A. (2008). Modeling neural tube defects in the mouse. *Curr. Top. Dev. Biol.* **84,** 1–35.

CHAPTER NINE

Modulation of the Generation of Dopaminergic Neurons from Human Neural Stem Cells by Bcl-X_L: Mechanisms of Action

Alberto Martínez-Serrano,*[,1] Claudia G. Castillo,[†,1] Elise T. Courtois,[‡,1] Elisa García-García,*[,1] *and* Isabel Liste*[,§,1]

Contents

I. Introduction 176
II. Sources of Human DAn for Cell Replacement in PD 178
  A. Fetal mesencephalic neurons/cells 178
  B. Stem cells 179
III. Epigenetic Cues and Genetic Manipulations to Improve hNSCs Differentiation Toward the A9 DA Phenotype 185
  A. Epigenetic cues help NSCs to differentiate into the A9 DA phenotype 185
  B. Genetic manipulations aimed at enhancing NSCs differentiation into the A9 DA phenotype 186
IV. Concluding Remarks 197
Acknowledgments 198
References 198

Abstract

Understanding the developmental mechanisms governing dopaminergic neuron generation and maintenance is crucial for the development of neuronal replacement therapeutic procedures, like in Parkinson's disease (PD), but also for research aimed at drug screening and pharmacology. In the present chapter, we review the present situation using stem cells of different origins (pluripotent

* Departamento de Biología Molecular and Centro de Biología Molecular "Severo Ochoa", Universidad Autónoma de Madrid - C.S.I.C. Cantoblanco, Madrid, Spain
† Departamento de Bioquímica, Facultad Medicina, Universidad de San Luis Potosí, San Luis Potosí, México
‡ Institute of Molecular and Cell Biology, A*STAR, Cancer Genetics group, Proteos, Singapore, Singapore
§ Área de Biología Celular y Desarrollo, Centro Nacional de Microbiología, Instituto de Salud Carlos III, Majadahonda, Madrid, Spain
1 All authors contributed equally to this manuscript.

Vitamins and Hormones, Volume 87
ISSN 0083-6729, DOI: 10.1016/B978-0-12-386015-6.00029-9

and multipotent) and summarize current manipulations of stem cells for the enhancement of dopaminergic neuron generation, focusing on the actions of Bcl-X_L. Bcl-X_L not only enhances dopaminergic neuron survival but also augments the expression of key developmental and maintenance genes, and, through the lengthening of the cell cycle early during differentiation, regulates cell fate decisions, producing a net enhancement of neurogenesis. The relevance of these findings is discussed in the context of basic neurogenesis and also for the development of efficient cell therapy in PD.

I. Introduction

The main pathology and motor symptoms in Parkinson's disease (PD) arise from the chronic and selective degeneration of mesencephalic dopaminergic neurons (DAn) of the A9 subgroup, located in the Substantia Nigra *pars compacta* (SNpc), which project to the striatum. The degeneration of these neurons results in an impaired function of the basal ganglia of the brain, causing impaired information flow, cortical activation, and the subsequent appearance of the cardinal motor symptoms (like tremor, rigidity, and hypokinesia). Current therapeutic options are limited, and efficacious only during short periods after treatment initiation. These include administration of L-DOPA, DA agonists, enzyme inhibitors (for instance of MAO-B), and surgery aimed at the stimulation of deep brain nuclei, like the thalamus, subthalamic nucleus and globus pallidus (Lew, 2007; Lindvall and Kokaia, 2009; Obeso *et al.*, 2010; Olanow and Tatton, 1999; Tzschentke and Schmidt, 2000). These strategies are effective only for limited periods of time, and aim at alleviating the symptoms of the disease, but not really to cure the disease, in the sense of retarding or halting neurodegeneration, stimulating regeneration, or replacing the lost DAn (Lindvall and Kokaia, 2009; Meissner *et al.*, 2011; Obeso *et al.*, 2010). In this scenario, the concept of treating the disease through the replacement of the lost DAn acquires special relevance. The rationale behind relies on decades of preclinical and clinical research studying the functional, therapeutic efficacy that the ectopic (striatal) placement of fresh fetal ventral mesencephalic tissue (as a source of DAn) shows in the experimentally or pathologically DA denervated/depleted striatum (Dunnett *et al.*, 2001; Lindvall and Björklund, 2004; Winkler *et al.*, 2005). However, the use of human fetal tissue raises several problems related to the standardization of cell preparations, tissue availability, clinical/gynecological logistics, and, in some countries, ethical or legal issues. These limitations pose significant hurdles for moving fresh tissue transplantation into a routine clinical practice, limiting its application to research in clinical trials. In this respect, many research groups has, over the last two decades, tried to find an alternative source of DA neuroblasts which could represent a safe and efficient, therapeutic cell replacement option (Brundin *et al.*, 2010; Hall *et al.*, 2007; Koch *et al.*, 2009a; Lindvall and Björklund, 2004; Lindvall and Kokaia, 2009, 2010; Lindvall *et al.*, 2004; Martínez-Serrano and Liste, 2010). The most intensively studied alternative cell sources are stem cells of different types, in particular embryonic stem cells (ESC), neural stem cells (NSC), and, more recently, the induced pluripotent stem cells (iPSC). Any potential DAn source must fulfill a series of criteria: (1) ability to proliferate extensively *in vitro* while retaining its properties in terms of phenotypical differentiation in a stable manner (at least to complete a Phase III clinical trial); (2) they must be absolutely safe (absence of tumor formation); (3) the implanted stem cells (or derivatives) should survive and differentiate only into neural cell types (neurons, astrocytes, and oligodendrocytes), and generate neurons typical of the ventral mesencephalon (A9 DAn, but also Glu and GABA neurons); (4) host immune reaction to the grafted cells should be minimal or at least controllable pharmacologically; (5) the resulting DAn should reinnervate the striatum, and release sufficient DA to stimulate its target neurons; (6) these DAn should functionally mature (electrophysiologically) and interact with the host circuitry in order to provide a regulated DA supply; and last, (7) they must exert relevant functional (behavioral) effects in tests designed to study Parkinson or parkinsonian (PD-like) symptoms. In the case of hemiparkinsonian rodents, for instance, the transplanted cells or derivatives should not only compensate drug-induced rotation, but also show an impact in spontaneous behavior, like in skilled paw use (paw-reaching or staircase tests) or the cylinder, rotarod, or rotating rod tests. So far, and to simplify, no stem cell source has been shown to match all of these criteria (see Table 9.1). The gold standard to which compare any stem cell or its derivatives remains being the fresh fetal VM tissue, which fulfills all criteria but for *in vitro* expansion (see below).

In this chapter we will concentrate mostly on VM derived NSCs and precursors, which are multipotent and do not pose the risk of tumor generation *in vivo*, but present the disadvantage of having a very limited proliferation potential, and the loss of their ability to generate neurons and DAn with time (passages) in culture (yielding NSC cultures inappropriate for DAn production). This loss of potency has been described for both rodent and human NSC cultures (Chung *et al.*, 2006; Hebsgaard *et al.*, 2009; Jensen and Parmar, 2006; Kim *et al.*, 2007, 2009; Maciaczyk *et al.*, 2008; Ostenfeld *et al.*, 2002; Roybon *et al.*, 2008; Yan *et al.*, 2001; reviewed in Brundin *et al.*, 2010; Lindvall and Kokaia, 2010; Martínez-Serrano and Liste, 2010).

Table 9.1 Overview of major requirements for safe and functional transplantation in PD

Parameter/cell type	Fresh VM	ESC	iPSC	NSC
In vitro expansion	No	Yes	Yes	Poor
Tumor formation *in vivo*	No	Yes	Yes	No
Only CNS derivatives *in vivo*	Yes	No	No	Yes
Host immune response	Moderate	Yes	No	Moderate
Striatal reinnervation, regulated DA supply	Yes	No (?)	No (?)	No (?)
In vivo electrophysiology	Yes	No (?)	No (?)	No (?)
Skilled motor behavior recovery	Yes	No (?)	No (?)	No (?)

II. Sources of Human DAn for Cell Replacement in PD

A. Fetal mesencephalic neurons/cells

Clinical cell replacement trials with intrastriatal transplantation of human fetal mesencephalic tissue, which is rich in postmitotic dopaminergic (DA) neuroblasts, have provided proof of principle that cell replacement therapy (CRT) can work in the human PD brain (Bjorklund and Dunnet, 2007). The DA neurons generated from the transplanted tissue reinnervate the denervated striatum and become functionally integrated, restoring striatal DA release and giving rise to clear symptomatic relief in some patients (Piccini *et al.*, 1999). However, the results have been variable, and the trials have revealed several limitations of the procedure for clinical routine practice that need to be improved in order to develop a viable CRT for PD. The main challenges are related to: (1) *Host tissue/environment.* Nowadays it is clear that the host immune system reacts to the cell transplant, specially through activated microglia and inflammation, affecting graft survival, functional improvement and probably helping the development of dyskinesias (Lindvall and Kokaia, 2009). Since autologous tissue cannot be used as donor for transplantation, long-term immunosuppressive treatment is needed to avoid graft rejection (Arenas, 2010; Piccini *et al.*, 2005). Another additional limitation is that the grafts may be affected by the pathology of the disease. In fact, evidences that PD pathology (Lewy bodies and α-synuclein aggregates) might propagate from host to graft have been recently reported (Li *et al.*, 2008b,c). The discussion remains related to the time the pathology will need to affect and deteriorate or kill the newly implanted neurons. (2) *Donor tissue and cell composition of the grafts.* Among the different DAn in the brain, it seems that the Substantia Nigra (A9) neuron subtype have the capacity to innervate and functionally activate their target structure, the dorsolateral region of the host striatum (Grealish *et al.*, 2010), better than preparations enriched in ventral tegmental (A10) DAn. Additionally, the dissection technique and cell cultivation/preservation before grafting are important for the survival of the DAn following grafting, as well as for the appearance of dyskinesias (related in part, to the serotonergic component in the grafts) (Carta *et al.*, 2008; Hagell *et al.*, 2002; Politis *et al.*, 2010). Other important challenges for fetal mesencephalic cells are related to ethical issues raised by the use of fetal tissue obtained from legal abortions, poor tissue availability (nowadays being aggravated since pharmacological abortion is replacing surgical methods in some countries), and poor standardization of the cell material, which contributes to a high variability in graft survival, and the degree of clinical benefit (Kokaia, Z., personal communication; Meyer *et al.*, 2010).

Due to these limitations, it is obvious that alternative sources of donor cells, different from fetal mesencephalic tissue, are needed.

B. Stem cells

Because of their capacity to expand indefinitely and give rise to large numbers of specialized cells, stem cells are nowadays considered the best candidates for DAn generation. However, so far, no scientifically based clinical trials with stem cell therapy have been performed in PD patients. After transplantation of stem-cell-derived DAn, the resultant cells must exhibit the properties of SNpc neurons in order to induce substantial benefit, which, in brief, are: (1) regulated DA release, and to achieve this the DAn need to be endowed with the molecular, morphological, and electrophysiological properties of SNpc A9 DAn; (2) reverse motor deficits in animal models of PD (optimally in rodents and monkeys); (3) reinnervate and become functionally integrated into the host neural circuitries without tumor formation (Lindvall and Kokaia, 2009).

1. Human embryonic stem cells (hESCs)

hESCs were first established by James Thomson in 1998 from the inner cell mass of the blastocyst (Thomson *et al.*, 1998). Because of their properties (self-renewal and pluripotency), these cells may constitute an optimal cell source for cell-replacement therapies. A prerequisite for clinical application of hESCs in PD is an efficient and strict differentiation of hESCs into midbrain DAn-like cells. Human DAn have been efficiently derived from ESC using a variety of methods, such as the use of the coculture with stromal cells, growth factors, secreted factors, transcription factors and morphogens, with some beneficial effects demonstrated after transplantation of these cells in animal models of PD (Arenas, 2010; Chambers *et al.*, 2009; Cho *et al.*, 2008; Koch *et al.*, 2009a,b; Malmersjo *et al.*, 2010).

Perrier *et al.* (2004) developed one of the most successful feeder-based protocols for DAn differentiation of hESCs. This protocol is based on cocultivation of hESCs with stromal cells and sequential treatment with morphogens and growth factors, including Shh, FGF8, FGF2, BDNF, GDNF, TGFB3, dcAMP, and Ascorbic Acid (AA). DA neurons obtained by this method express most of the typical markers and pharmacological responses of the midbrain DAn (Malmersjo *et al.*, 2010; Sacchetti *et al.*, 2009). However, these cells do not survive well after transplantation. More recently, a fast and efficient protocol based on blocking BMPs, and the use of Lefty, Activin, and TGFβ (using noggin and SB431542) for neural induction and DA differentiation has been described (Chambers *et al.*, 2009). These cells can be subsequently differentiated into DAn by sequential treatment with Shh, FGF8, BNDF, and AA, followed by GDNF, TGFB3, and dcAMP. So far, the *in vivo* function of the cells remains to be reported.

Although the above advancements encourage research aimed at using hESCs in CRT for PD, there are still several important problems that need a solution: (1) control of cell growth and differentiation. Teratoma formation or tissue overgrowths have been described in most of the studies using hESCs for grafting into the brain of PD animal models (Hwang, 2010; Roy *et al.*, 2006). The most likely explanation for the overgrowths is the persistence of undifferentiated cells or the presence of cells with chromosomal abnormalities inside the graft; (2) poor survival, low reinnervation potential and phenotypic instability of the transplanted cells (Cho *et al.*, 2008; Koch *et al.*, 2009b); (3) For clinical applications, it is crucial to eliminate any risk of contamination, including the risk of xenogenic contamination (from cell culture reagents or feeder layers). This implies that hESCs-derived DAn or their precursors must be generated and cultured under GLP/GMP (Good Laboratory and Manufacture Procedures) conditions in xenogen-free environment from the time of blastocyst isolation.

2. Human induced pluripotent stem (hiPS) cells

Human ESC-like pluripotent cell lines (hiPS, induced pluripotent stem cells) have been successfully established from adult human fibroblasts after overexpression of four transcription factors: Oct3/4, Sox2 and Klf4, and c-myc (or Nanog, Lin28) (Takahashi *et al.*, 2007; Yu *et al.*, 2007). These cells are similar to hESCs in morphology, gene expression profile and differentiation potential. The iPS cell technology offers new possibilities for biomedical research and clinical applications, as these cells could be used as an *in vitro* cellular model of PD, and for autologous transplantation (theoretically, no immunosuppressive therapy would be necessary). In addition, hiPS cells do not raise ethical concerns since they are derived from somatic cells, following routine tissue donation procedures.

In recent studies (Hargus *et al.*, 2010; Nguyen *et al.*, 2011; Soldner *et al.*, 2009) human iPS cells were isolated from patients suffering from sporadic or familial PD. These cells can be differentiated into DAn and, after grafting into the adult rodent brain, do not show signs of neurodegeneration and reduce motor asymmetry in parkinsonian rats (even though only few DAn reinnervate the host striatum) (Hargus *et al.*, 2010).

However, caution should be taken, since iPS cells are expected to face at least the same challenges as hESCs, plus additional problems related to the fact that they are patient derived (bearing the same genetic defects as the patient itself), the way they are generated (in most cases, genetically modified for the overexpression of cell growth-related transcription factors), and their *in vivo* performance.

In fact, recent studies have shown significant reprogramming variability, including structural chromosomal abnormalities, somatic memory and aberrant reprogramming of DNA methylation patterns in hiPS cells (Lister *et al.*, 2011).

Another important issue is the risk of tumor formation by iPS cells, which is even higher than that of ES cells, due to the reprogramming process. In addition to the overexpression of cell growth-related factors, most of the human iPS lines have been generated using viral methods, which can generate multiple chromosomal integrations and insertional mutagenesis.

3. Human neural stem cells (hNSCs)

VM DAn in both rodent and humans are generated from precursors located at the mesencephalic floor plate (FP) (Bonilla *et al.*, 2008; Ferri *et al.*, 2007; Hebsgaard *et al.*, 2009; Nakatani *et al.*, 2010; Nelander *et al.*, 2009; Ono *et al.*, 2007; Puelles *et al.*, 2004; reviewed in Abeilovich and Hammond, 2007; Alavian *et al.*, 2008; Ang, 2006; Gale and Li, 2008; Kim, 2011; Orme *et al.*, 2009; Prakash and Wurst, 2006; Puelles, 2007; Smidt and Burbach, 2007). Cultures of these precursors present a few but substantial limitations for their use as a stable source of human DAn. The cultures, in general, proliferate poorly, rapidly enter senescence, and evolve with time in culture, producing some DAn at early passages but mostly glia at later ones (Chung *et al.*, 2006; Kim *et al.*, 2007, 2009; Maciaczyk *et al.*, 2008; reviewed in Brundin *et al.*, 2010; Lindvall and Kokaia, 2010; Martínez-Serrano and Liste, 2010). To further complicate the scenario, it has been described that DAn originated from these expanded precursors are not born in culture, but from primary precursors still present at early passages (Kim *et al.*, 2009). As a consequence, in the field there is the uncertainty of whether or not these precursors of human DAn can be propagated in culture while retaining their properties. In practical terms, obtaining enough cells with homogeneous properties to conduct, for instance, the needed Phase I to III clinical trials, represents a formidable challenge. To facilitate basic and preclinical research, several groups, including ours, have explored the possibility of immortalizing the precursors. To this end, cell division-related genes have been engineered into the cells at early passage, as a means to immortalize them. The genes tested include *c-* and *v-myc*, resulting in the lines MESC2.10, NGC-407, ReNcell VM, vm c-mycER^{TAM}, and hVM1 (reported by Lotharius *et al.*, 2002; Paul *et al.*, 2007; Khan *et al.*, 2007; Donato *et al.*, 2007; Miljan *et al.*, 2009, and Villa *et al.*, 2009, respectively). The idea behind immortalization is to suspend or "freeze" the developmental program of the cells while conferring them unlimited expansion potential (for review see Martinez-Serrano and Bjorklund, 1997). The avian retroviral *v-myc* gene is, by far, the immortalizing gene explored in greater detail, providing consistent and reproducible immortalization. Avian *v-myc* was evolutionary originated by the rescue of endogenous cellular sequences by a lymphomyeloblastosis avian retrovirus. The gene incorporates exons II and III of the chicken *c-myc* joined to a portion of retroviral *gag* gene, and codes therefore for the fusion protein p110$^{gag\text{-}myc}$. Opposite to transformation, *v-myc* immortalizes the cells in a "conditional" way, in the sense that mitogen stimuli (normally EGF and/or FGF-2; but not serum, which induces differentiation) are still required for cell division to occur. Following mitogen withdrawal *in vitro*, or after *in vivo* implantation into the adult brain (environment devoid of mitogens), *v-myc* is downregulated (either physiologically by promoter silencing, or pharmacologically), and the cells cease division, and resume their differentiation program in the absence of overgrowth.

The properties of immortalized VM hNSC cell lines have been studied or reported to a variable degree. In most cases, not more than nestin expression (a marker of NSCs) and appearance of TH^{+} cells (TH, tyrosine hydroxylase, the rate limiting enzyme for catecholamine synthesis) after differentiation have been described (see references above). At present, it is not clear if the cells have not been studied in further detail, or if the results did not match expectations, as for MESC2.10 cells, in which TH expression is unstable *in vitro* and *in vivo* (Paul *et al.*, 2007). In contrast, the hVM1 cell line has been characterized in thorough detail (see Table 9.2). In brief, hVM1 are endowed with all the genuine properties of NSC that act as precursors of human A9 SNpc DAn. The molecular and functional properties of the differentiated progeny include the correct expression of multitude of markers, transcription factors needed for their generation and maintenance, and all genes required for functional maturation (Courtois *et al.*, 2010; Seiz *et al.*, submitted; Villa *et al.*, 2009). V-myc is profoundly downregulated with differentiation, and the cells exit the cell cycle as primary cells do (Villa *et al.*, 2009). Furthermore, the cells are functional in terms of calcium handling, electrophysiology, and DA synthesis and release (Krabbe *et al.*, 2009; Tønessen *et al.*, 2010). *In vivo*, the cells differentiate to the appropriate phenotypes, and result in the partial compensation of pharmacologically induced rotation (Courtois *et al.*, 2010). The long-term (>4 months post-grafting) survival, differentiation, and behavioral effects are being currently

Table 9.2 Summary of the properties of hVM1 cells and their differentiated progeny: Effects of Bcl-X_L

Parameter/property	Control and Bcl-X_L hVM1 cells	Changes in BCL-X_L HVM1 cells
Karyotype	SKY, G_banding = 46XX, stable up to passage 35[a,b]	
Expression of undifferentiation markers	Nestin, Vimentin, 3CB2, Sox2[c,d]	
Potency	Multipotent[c] → neurons (ß-III-tubulin, MAP2), astrocytes (GFAP), oligodendrocytes (NG2, MBP)	↑ ß-III-tubulin, ↑ Synapsin I, ↓ GFAP[a]
Downregulation of immature markers and genes with differentiation	Nestin, Vimentin, 3CB2, Ki-67, v-myc[b,c]	
Floor plate markers	Engrailed1, Otx2[a,d]	↑ Otx2[d]
Proneural genes	Mash1, NeuroD1, Neurogenin2, Msx1[a,c,d]	↑ Neurogenin2, ↑ Msx1[a,c,d]
Prodopaminergic and maintenance genes	En1, Lmx1b, Pitx3, Nurr1[a,c,d]	↑ En1, ↑ Lmx1b, ↑ Pitx3, ↑ Nurr1[a,d]
Cell death-related events with differentiation:		
Annexin-V binding		↓[a]
DNA fragmentation		↓[a]
Activated Caspase-3		↓[a]
Activated Caspase-9		↓[a]
AIF translocation		↓[a]
LDH release		↓[e]
Mitochondrial inner membrane potential		↑[a]
Maturation/function-related properties:		
Dopamine metabolism-related genes	TH, AADC, GTP-CH1[a,c,d]	↑ AADC, ↑ GTP-CH1[d]

(Continued)

Table 9.2 (*Continued*)

Parameter/property	Control and Bcl-X_L hVM1 cells	Changes in BCL-X_L HVM1 cells
Dopaminergic function-related genes	DAT, VMAT2, DRD2, GIRK2[a,d]	
Calcium binding proteins	Calretinin, Calbindin, no Parvalbumin[a,c,f]	↑ Calretinin[a]
Neurotransmitter receptors	GABA-B-R1, GABA-A-Rα3, GLUR2, NMDAR2A(GRIN2), CHRM3, CHRM5[f]	↑ All[f]
Neurotransmitter phenotypes generated	DA, GABA, Glu. Occasional Ser. No AcCho[a,d]	DA[a,d]
Dopamine synthesis and K^+-evoked release	Yes	↑[a]
Electrophysiological properties[g]	Resting membrane potential ≈ 60 mV Input resistance ≈ 2500 MΩ Inward rectification at hyperpolarization "Pace-maker" type activity Long (>2 s) duration of action potentials	
Calcium responses to . . .	High K^+, DA, Glu, GABA	↑ High K^+, ↑ DA
DAT activity	Yes (binding, 6-OH-DA and ^{3}H-DA transport)[d]	

[a] Courtois *et al.* (2010).
[b] Ramos, T., González-Lendínez, J., and Martínez-Serrano A, unpublished.
[c] Villa *et al.* (2009).
[d] Seiz *et al.* (submitted).
[e] Krabbe *et al.* (2009).
[f] Seiz E.G. (2010) Thesis manuscript, unpublished.
[g] Tønessen *et al.* (2010).

studied. The only drawback for the clinical application of the cells is that they have not been generated under GLP/GMP conditions, and that the insertion locus of the immortalizing gene in the genome is not known so far. For clinical application, the cells should be rederived and characterized, and at that time, the use of more sophisticated genetic engineering tools is envisaged.

III. Epigenetic Cues and Genetic Manipulations to Improve hNSCs Differentiation Toward the A9 DA Phenotype

A. Epigenetic cues help NSCs to differentiate into the A9 DA phenotype

As stated before, for stem cell-based replacement therapy in PD, differentiation of the stem cells into phenotypically stable functional DAn of the A9 SNpc phenotype is needed.

Theoretically, only NSCs isolated from the midbrain differentiate into functional, mature A9 DAn, since NSCs and restricted precursors become regionalized during embryonic development (meaning that they express region-specific developmental genes and markers). However, an important limitation of these cells is their low proliferation rate and the loss of their properties/potential with the time in culture. To avoid the latest, several extrinsic factors have been studied, and a few identified as important for the correct *in vitro* differentiation of stem/progenitor cells toward DAn.

Tissue samples from abortions at 6–10 weeks postfertilization have been used in most *in vitro* studies. In contrast to precursor cells derived, for example, from the forebrain, which are easily expanded using EGF/FGF-2 under 21% atmospheric oxygen, precursors from the mesencephalon need a reduced atmospheric oxygen of approximately 3–5% for long-term proliferation and differentiation into DAn (Storch *et al.*, 2001; Villa *et al.*, 2009).

In addition, the differentiation program of VM hNSCs can be influenced *in vitro* by addition of growth factors, agents that increase intracellular cyclic AMP or cytokines. In fact, several cocktails have been tested, with different success rate. Storch and collaborators (Storch *et al.*, 2001) demonstrated an enhanced induction of the DA phenotype by using a differentiating media containing cytokines (IL-1b, IL-11, and LIF) and GDNF (glial derived neurotrophic factor, that promotes the survival and morphological differentiation of DAn *in vivo*). Approximately 1% of the precursors became positive for TH, a marker for dopamine neurons. The neuronal progeny exhibited morphological and functional properties characteristic of DAn in culture.

Another factor combination promoting NSC differentiation toward the DA phenotype includes BDNF (brain derived neurotrophic factor), GDNF, DA, and Forskolin (a factor that increases intracellular cyclic AMP (cAMP) levels) (Riaz *et al.*, 2004). Using this protocol, around 12% of the *in vitro* differentiated hVM1 cells was TH positive and showed genuine properties of midbrain DA neurons (Villa *et al.*, 2009). A higher yield in DAn can be obtained by the use of a differentiation medium containing GDNF and dibutyryl-cAMP (db-cAMP; a factor mimicking the effects of cAMP) (Courtois *et al.*, 2010; Donato *et al.*, 2007; Lotharius *et al.*, 2002). However, the DA phenotype of the TH^+ neurons obtained from some VM hNSCs *in vivo* may be unstable (Paul *et al.*, 2007).

B. Genetic manipulations aimed at enhancing NSCs differentiation into the A9 DA phenotype

As noted above, hNSCs of VM origin generally show a poor capacity to generate DAn, and consistently loose this potential with passaging. This phenomenon was initially described in rodents, but it also affects both human primary and immortalized VM NSCs. Several attempts have been made to increase the neurogenic potential of NSCs, mostly through the expression of developmental transcription factors, and also other genes of the Bcl family, Bcl-X_L in particular. Transcription factors tested are described below, in the context of the interaction or crosstalk between elevated Bcl-X_L and developmental gene programs. Next, we will summarize current knowledge about Bcl-X_L actions on the differentiation of hNSCs.

1. Bcl-X_L effects on hNSCs fate decision

Bcl-X_L is the antiapoptotic member of the Bcl-2 family more expressed during CNS development, together with Mcl-1 protein. More specifically, Bcl-X_L is most expressed by postmitotic neurons while Mcl-1 is highly present in the Ventricular Zone (VZ) (Arbour *et al.*, 2008; Gonzalez-Garcia *et al.*, 1994; Krajewska *et al.*, 2002; Lindsten *et al.*, 2005). Bcl-X_L expression pattern and the fact that the knockout mouse dies at E13.5, when neurogenesis is peaking, makes Bcl-X_L a key factor for neuronal survival (see section below) (Motoyama *et al.*, 1995; Parsadanian *et al.*, 1998). In the context of neuronal DA survival, a tissue-specific knockout mouse where the Bcl-X gene (that generates the mRNAs of both Bcl-X_S and Bcl-X_L through alternative splicing) is deleted from TH-positive cells, presents a massive cell death of DAn, strengthening its antiapoptotic role in the DA neuronal population (Savitt *et al.*, 2005).

Due to this Bcl-X_L canonical and well-established survival role in the CNS, its forced expression in NSCs cultures constitutes a simple, clean experimental paradigm to learn how to increase the neuronal and, more specifically, the DA neuronal outcome after NSCs differentiation. This approach has been used in murine and human NSCs, and in both species the result was identical; the number of neurons substantially increased (Chang *et al.*, 2007; Courtois *et al.*, 2010; Krabbe *et al.*, 2009; Liste *et al.*, 2007). In the case of hNSCs, the experiment was performed both in forebrain and in VM hNSC and the neuronal enhancement was reproducible, indicating an absence of regional specificity for Bcl-X_L effects on neuronal differentiation. Different neuronal parameters were studied, and the neurons obtained from Bcl-X_L hNSCs were identical to those generated from naïve hNSCs, in terms of expression of neuronal markers such as Tuj1, Dcx and MAP2, and mature markers of synaptic function, like synapsin and synaptophysin.

When the time course of neuronal differentiation was studied, the neuron yield profile showed that neurons are generated at the same time in naïve and in Bcl-X_L hNSCs cultures. This suggested that the progenitor population already committed to become neurons was enlarged; indeed, BrdU incorporation experiments to quantify the size of the progenitor population showed that committed neuronal progenitors were more numerous in the presence of elevated Bcl-X_L levels (E. García-García, unpublished; Liste *et al.*, 2007). At the same time that neuron numbers increase after hNSCs differentiation, the number of cells expressing the glial marker GFAP decreased. These GFAP-positive cells remain vimentin positive but not nestin positive, leading to the conclusion that the forced expression of Bcl-X_L precludes the acquisition of a fully mature glial phenotype (Courtois *et al.*, 2010; E. García-García, unpublished; Liste *et al.*, 2007).

It is interesting to point out that the forced expression of BAX protein, an apoptotic member of the Bcl-2 family, has the opposite effect that the one of Bcl-X_L, that is, an increase in the number of cells positive for glial markers (GFAP and S100b) and a decrease the number of neurons, both in an *in vitro* and *in vivo* mouse NSCs models of differentiation (Chang *et al.*, 2007). Thus, not only the antiapoptotic members of the Bcl-2 family are important players in the fate decision choices taken by NSCs when generate their progeny.

Bcl-X_L is also expressed in the hematopoietic system during development. Its alternative restricted expression into certain cell fate lineages brought up the hypothesis that Bcl-X_L could have different roles in the different hematopoyetic cell fate lineages. To test this hypothesis, gain of function experiments were performed in hematopoyetic models, resulting in a similar result observed in neural differentiation: the overexpression of Bcl-X_L promotes erythroid fate while decreases the granulocytic fate (Haughn *et al.*, 2003). Thus, Bcl-X_L effects on fate decision are not only restricted to the CNS but it is also observed in the hematopoietic system.

Paying more attention to the differentiated neuronal phenotypes generated by hNSCs after increasing Bcl-X_L levels, a bias toward the DA phenotype was observed; other phenotypes, such as Glu and GABA, were increased proportionally to the total neuronal outcome (Liste *et al.*, 2004, 2007). In VM hNSCs, the newly TH-positive generated neurons were functional DAn of the SNpc type because they expressed the potassium channel GIRK2, and the transporters DAT and VMAT2, and were able to release dopamine to the extracellular medium after a KCl-induced depolarization. Moreover, when Bcl-X_L-modified hNSCs from midbrain are grafted into a rat striatum, a nonneurogenic area of the brain, they kept the property to produce more DAn in comparison to their control counterparts after 2 months of transplantation (Courtois *et al.*, 2010). The increase in functional DAn observed after the forced expression of Bcl-X_L has been achieved also in mouse and human ESCs. In the first case, partial behavioral recovery was observed in 6-OH-DA lesioned hemiparkinsonian rats (Shim *et al.*, 2004). The results obtained in hESC are discussed later on in this chapter (see section on genetic cascades). In relation to hNSCs, increased Bcl-X_L expression not only increases the numbers of DAn cells obtained from VM hNSCs, but also in the case of forebrain hNSCs; these DAn were not of the interneuron type—a neuronal population generated from progenitors residing in ventral areas of the forebrain—claiming for an specification action of Bcl-X_L toward the DA phenotype, and not an increase of an already existing neuronal population. As mentioned earlier, it is interesting to note that Bcl-X_L also promotes the DA phenotype when it is expressed in ESC cells, again supporting that Bcl-X_L could have an inducing role toward the DA phenotype.

2. Bcl-X_L effects on cell death

In the previous section we have described how members of the Bcl-2 family can modulate the final phenotypic outcome of the differentiating NSCs. With the aim of understanding the process involved in neuronal differentiation and the mechanisms by which Bcl-X_L alters the progeny in benefit of the neuronal DA phenotype, several studies have been performed.

The main known role of Bcl-X_L is to prevent apoptosis. Bcl-X_L, by localization, is at the crossroad of many apoptotic events. It has been described that Bcl-X_L plays crucial roles in the different cell death pathways that may occur in neural cells: the intrinsic and extrinsic apoptotic pathways. In the intrinsic pathway, mitochondria integrity is considered as the major apoptotic brake (Benn and Woolf, 2004), and Bcl-X_L can act both upstream (by directly interacting with pro-apoptotic Bcl-2 members BAX, BAK/BOK, and thus avoiding their mitochondrial translocation) or at the mitochondrial level by preventing the mitochondria inner membrane permeabilization and subsequent release of AIF (apoptosis inducer factor) or Cytochrome c. Moreover, Bcl-X_L was also identified as a regulator of the apoptosome and of several Caspases (-1, -3, and -9), that act downstream of the mitochondria, indicating its potent antiapoptotic function at different levels (Benn and Woolf, 2004).

It is then important to remember that Bcl-2, and Bcl-X_L in particular, play a fundamental role in the survival and cell death of NSCs, progenitors and differentiated cells during embryonic development, but also throughout the adult life. Indeed, studies in mice have described a concrete pattern of Bcl-2 family members' expression in the developing brain. It has been shown that Bcl-2, as well as Bcl-X_L, is abundantly expressed from early embryogenesis (Bcl-X_L is detectable as early as E8.5 in the mouse and continues being highly expressed until the end of the embryonic development). However Bcl-X_L, opposite to Bcl-2, remains the dominant form of the expressed Bcl-2 family antiapoptotic members in the CNS, also in postnatal and adult brain tissues (Gonzalez-Garcia *et al.*, 1994; Krajewska *et al.*, 2002). It is interesting to correlate this peak of expression with the developmental massive cell death that occurs during normal CNS formation, where about 85% of the neuronal cells are eliminated through programmed cell death (PCD). In this scenario, immature neurons are, by default, programmed to die, exception made of those cells that can find the trophic factors and signaling molecules that will inactivate the PCD scenario.

The concept of developmental cell death during normal brain development takes all its importance to explain the role of the different Bcl-2 family members in normal nervous system development. Bcl-X_L (but not Bcl-X_s) is expressed during embryonic brain development and remains highly expressed throughout adult life in certain brain areas, and the protein has been identified as crucial for the survival of postnatal CNS neurons (Parsadanian *et al.*, 1998). The generation of Bcl-X deficient mice commented above highlighted the importance of the expression of Bcl-X in neural progenitor cells, as well as in other progenitor cell types (hematopoietic stem cells). In another study using the fetal VM hNSC line hVM1, Bcl-X_L overexpression was shown to be necessary for the survival of the undifferentiated proliferating cells in an established culture, as well as during early differentiation, by reverting a caspase-independent apoptotic cell death (Courtois *et al.*, 2010).

Turning to *in vivo* models, Roth and collaborators, in an elegant study in mice, showed that Bcl-X_L was crucial to establish a balance between proapoptotic and antiapoptotic molecules during brain development (Bcl-X_L interacting with Caspase-9 and Bax in order to inhibit cell lost in the forming brain) (Zaidi *et al.*, 2001).

Once the brain neuronal network (the connectome) is successfully established in the developing brain, the neurons are to be maintained alive for the rest of the adult life. At this developmental time point, immature cells need to rapidly change from a "ready to die" state, to a "survivor" one, to assure the maintenance of the neurons of the CNS, but for a few neurogenic areas in the adult brain (Benn and Woolf, 2004; Cory and Adams, 2002).

When comparing the expression levels of the main antiapoptotic members in adult brain tissues, it is clear that Bcl-X_L is also predominantly expressed over Bcl-2 during adulthood, indicating thus an important role of this antiapoptotic factor not only during brain development but also in the adult tissue (Gonzalez-Garcia *et al.*, 1994).

Many additional studies have also addressed the question and reported the protective effect of Bcl-X_L in differentiated neuronal cells from multiple brain areas (Akchiche *et al.*, 2010; Zaidi *et al.*, 2001), and more specifically in DAn (Courtois *et al.*, 2010; Nair *et al.*, 2006; Seiz *et al.*, unpublished). Indeed, induced DAn cell death in a PD model (a proteasome inhibition model system) was shown to be due to an increase in phosphorylated levels of p53, which was shown to interact with mdm2 and Bcl-X_L, resulting in the induction of mitochondrial PCD cascade (Nair *et al.*, 2006). This study also included the analysis of human post mortem brain tissue of PD cases, which also showed the same increase in phosphorylated p53 levels, suggesting that the same mechanism could occur in the brain during the course of neurodegenerative disease (Nair *et al.*, 2006); p53 mediates nontranscriptional cell death in DAn by sequestering Bcl-X_L, results also demonstrated in the hematopoietic system, where p53 and Bcl-X_L collaborate in a cell death context in hematopoietic progenitors (Gu *et al.*, 2006). Altogether, these reports collectively indicate that Bcl-X_L expression is necessary for the survival of mature neurons in the adult CNS, and that the inactivation of this protein, mediated by an antiapoptotic stimulus, could be a crucial determinant for neurodegenerative disease progression.

In other studies, Bcl-X_L was shown to protect other cell types in the CNS, like oligodendrocytes (both the progenitors and the mature cells), which were protected from Activin and TGFß induced cell death. In this particular study, Bcl-X_L, because of its mitochondrial localization, was confirmed as conferring a crucial cell death inhibitory role, preserving mitochondrial integrity by inhibiting the release of Cytochrome c and AIF (Krantic *et al.*, 2007; Schulz *et al.*, 2008). These studies indicate that Bcl-X_L not only affects cell survival in the developing brain, but also plays a crucial and direct role in the protection of the brain components throughout adult life.

All of these studies argue in favor of Bcl-X_L as a key protein governing neuron survival in the two main periods of a neuron life: during development, where Bcl-X_L presence is needed to persist, or during adult life, where its expression protects against cell death stimuli.

3. Bcl-X_L effects on cell proliferation

The common Bcl-X_L effect discovered in many experiments, that is, an enhancement of neuron yield from NSCs, could be easily explained by the cell survival role of the protein. However, some of the collected data could not be explained in such a simple way. First, during differentiation of forebrain hNSCs there is not a substantial cell death, but rather marginal (around 5% of the total cells present), and the cell death rate did not change after elevating Bcl-X_L levels. Furthermore, pharmacological inhibition of caspases does not mimic the neurogenic effect of Bcl-X_L (Liste *et al.*, 2007). In the case of murine NSCs, mutation of the BAX/BAK binding domain of Bcl-X_L did not prevent or abolish its neurogenic effect (Chang *et al.*, 2007). All these results are consistent with a new, unknown action of Bcl-X_L, by which it is able to increase the number of neurons generated from NSCs, not related with its canonical survival effect.

Focusing on this Bcl-X_L action in different neuronal models, the same conclusion is reached: Bcl-X_L must have another role besides its well-known antiapoptotic action. Examples in the CNS are the increase in the synapse formation when Bcl-X_L is overexpressed in an *in vitro* model of hippocampal primary neurons (Li *et al.*, 2008a,b,c), and the enhancement in neurite growth and morphology in basal ganglion neurons (Kretz *et al.*, 2004). Furthermore, there are an increasing number of studies in which classical cell death-related proteins are being discovered to have other roles during development under physiological conditions. For instance, Caspase-3 has been involved in cell fate decision during development (Garrido and Kroemer, 2004; Kuranaga, 2011; Kuranaga and Miura, 2007).

Outside of the CNS, Bcl-X_L has been implicated in the control of cell cycle through yet poorly understood mechanisms, in which cell cycle inhibitors are involved (Janumyan *et al.*, 2008). Cell cycle and cell fate choice have been linked in many situations, for example, the forced exit of cell cycle induced by cdk inhibitors has the consequence of altering cell fate (Dehay and Kennedy, 2007; Doe, 2008). Furthermore, the length of the cell cycle has been correlated with the state of NSC, meaning that, as neurogenesis starts, the length of the cell cycle is progressively increasing (Salomoni and Calegari, 2010). This fact led to formulate the hypothesis of "the cell cycle length," that postulates that the solely increase of the length of the cell cycle promotes a neurogenic state. In more detail, the hypothesis proposes that neurogenic factors would have more time to exert their actions during critical phases of cell cycle, thus overcoming certain signaling threshold by extending the time they act. Experiments in which the cell cycle length has been artificially increased, for instance, pharmacologically, have resulted in an enlarged neuronal population, demonstrating the cell cycle length hypothesis in mouse *in vivo* and *in vitro* models (Calegari and Huttner, 2003)

We have recently discovered that the forced expression of Bcl-X_L induces the lengthening of the cell cycle in forebrain hNSCs *in vitro*, when the cells are taking cell fate decisions during differentiation (E. García-García *et al.*, unpublished). However, this cell cycle lengthening does not promote direct neurogenesis from the NSCs (an asymmetric division resulting in one stem cell and a neuron), rather it biases the NSCs for indirect neurogenesis, through intermediate progenitors (IPs). The result is an increase in IPs number, as demonstrated by the incorporation of BrdU later on during differentiation, as well as by the expression of the cell cycle inhibitor p27Kip1, which was lower in the presence of elevated Bcl-X_L levels. The study of the cell cycle inhibitors during the time of lengthened cell cycle aids to determine which protein(s) is responsible for the altered cell cycle control under high Bcl-X_L conditions. In this way, p21Cip1 and p53 were shown to be increased. Indeed, the knock down (by interference RNA) of p53 in Bcl-X_L overexpressing forebrain hNSCs prevented the neurogenic effect of the protein (E. García-García *et al.*, unpublished).

An increasing number of studies are reporting the important role of p53 in the control of the balance between differentiation and proliferation. In this line, it has been proved that p53 favors neurogenesis by controlling the expression of Id proteins, inhibitors of proneuronal genes (Aranha *et al.*, 2009). Despite the already reported interaction between Bcl-X_L and p53 in cell death models (Hagn *et al.*, 2010; Nair *et al.*, 2006), such a direct protein–protein binding was unlikely to occur due to the different subcellular localization of the two proteins in the forebrain hNSCs differentiation model, where cell death stimuli is lacking (E. García-García *et al.*, unpublished).

Another effect of Bcl-2 family members is the capacity to control calcium homeostasis. Firstly, this fact was related to the cell death decision-making role of the protein family, since a controlled release of calcium from internal stores could prevent calcium-related cell death (Li *et al.*, 2002, 2007; Pinton and Rizzuto, 2006; White *et al.*, 2005). However, recent studies have demonstrated that altering calcium homeostasis by Bcl-2 family proteins under more physiological conditions (absence of cell death stimulus) can exert other actions different from the canonical and cell death-related ones. In this way, Bax and Bak have been described to regulate T cell proliferation through the control of the release of calcium from (endoplasmic reticulum) ER stores. In the case of Bcl-X_L, the modulation of calcium homeostasis is linked with its ability to modulate signaling by the Ins3P receptor (REF; E. García-García *et al.*, unpublished; Jones *et al.*, 2007).

Back to the control of the cell cycle in the CNS, several groups have identified the importance of cytosolic calcium in the control of proliferation *versus* differentiation decision taken by NSCs (Lenos and Tsaniklidou, 2010). In fact, spontaneous calcium waves have been reported to increase in frequency and amplitude in the VZ of the developing mouse cortex when neurogenesis is peaking (Weissman *et al.*, 2004). As well, many experiments in which calcium homeostasis has been modulated by the addition of ATP, GABA, glutamate, acetylcholine, among others, have relevant effects on cell proliferation (Haydar *et al.*, 2000; Lipskaia *et al.*, 2009; Resende *et al.*, 2008; Ryu *et al.*, 2003).

The fact of the alteration of cell cycle length through the p53–p21 pathway when Bcl-X_L expression was forced, and the already published link between Bcl-X_L and calcium homeostasis lead to the hypothesis that Bcl-X_L could be modifying the function of the p53–p21 pathway through its ability to modify Ins3P receptor activity (E. García-García *et al.*, unpublished). Then, Bcl-X_L would modify the length of the cell cycle through the control of calcium homeostasis. Experiments testing this hypothesis have been performed in forebrain hNSCs and, for example, we have demonstrated that calcium chelation, that is, reducing and buffering cytosolic calcium, in Bcl-X_L overexpressing hNSCs prevents the neurogenic effect. Conversely, increasing the cytosolic calcium by Ins3P pathway stimulation in hNSCs promotes an increase in the number of IPs and in p53 levels, concomitantly enhancing neurogenesis (E. García-García *et al.*, unpublished).

Taken all of these results together, a clear picture is emerging where Bcl-X_L increases indirect neurogenesis from hNSCs by the modulation of calcium homeostasis and enlargement of the size of the IPs pool, instead of just through a direct and simple survival effect.

4. Interplay between Bcl-X_L and developmental gene cascades

The above-described enhancement of neurogenesis induced by Bcl-X_L, and more specifically toward the DA phenotype, is also detected at the molecular level. Indeed, genes controlling stemness and multipotency, neuronal fate choice, maturation, and phenotype specificity of NSCs are modulated in a concerted manner in order to increase the numbers of IPs, and promoting neuronal fate choice, maturation, and phenotype specification of the generated neurons (Guillemot, 2007; Nguyen *et al.*, 2006).

Studies in murine and forebrain human NSCs have shown that the overexpression of Bcl-X_L in hNSCs decreases the expression level of multipotency genes such as Hes5, Sox2, and Pax6 in an early temporal window during the differentiation (Chang *et al.*, 2007; E. García-García *et al.*, unpublished). These genes are expressed during development by noncommitted progenitor cells, so that a decrease in their expression correlates with a neuronal fate choice adopted (Guillemot, 2007). At the same time, transcription factors important for the generation of dividing IPs increase, such as Insm1 (Farkas *et al.*, 2008). Afterwards, the proneuronal gene Ngn2, member of the bHLH family, increases its expression level in a later temporal window during the process of differentiation. In murine NSCs models, Id proteins were also studied; their downregulation is needed for the neurogenic program to proceed, due to the inhibitory effects that they exert on the proneuronal bHLH proteins (Chang *et al.*, 2007; E. García-García *et al.*, unpublished). In relation to maturity, NeuroD1 was also enhanced, a well-known target of Ngn2 (Chang *et al.*, 2007; E. García-García *et al.*, unpublished; Guillemot, 2007)

However, it is not well understood yet if the increase in the expression of these genes is regulated by the ability of Bcl-X_L to control calcium homeostasis, or if it is just a consequence of the modulation of the systems controlling cell cycle length and progression, as the "cell cycle length hypothesis" would predict.

5. Other genetic manipulations of hNSCs aimed at enhancing neurogenesis and DAn generation

PD models offer a great opportunity for the development of new cell therapy strategies based on the generation of a highly specialized neuronal type, the A9/SNpc DAn, from a human NSC source. Thus, many laboratories have been concentrating their efforts in the identification of the stem cell of choice, developing cell culture procedures and optimizing them to enhance the yield in. As mentioned above, an important aspect in the development of these strategies is the generation of functional DAn, not just DAn, which, once transplanted into the lesioned or diseased brain, could supply for the lost DAergic functions. It has been shown that such event require the differentiation of the NSC toward A9 mesodiencephalic DAn-specific cell type (see for instance Grealish *et al.*, 2010). Indeed, not all the stem and NSCs sources can successfully generate such specific cell type endowed with the correct phenotype and functionality (Lindvall and Kokaia, 2010; Lindvall *et al.*, 2004; Martínez-Serrano and Liste, 2010; Studer *et al.*, 1998). Efforts have been oriented to the discovery and optimization of both epigenetic and genetic means of enhancing DAn generation from stem cells in general, and NSCs in particular. Epigentic cues were discussed early in this chapter. Below we summarize many attempts tried with transcriptions factors to achieve the goal.

In order to identify which transcription factor or protein (overexpressed or downregulated) could modulate NSC differentiation to favor the DAn phenotype, several crucial studies (of gain and loss of function) were conducted in mice to determine the genetic cascade(s) operating during embryonic development that lead to the generation of such neurons. An excellent review from Smidt and Burbach (2007) summarizes the different KO mice studies and consequent changes in the DAergic areas of the brain, highlighting the importance of different transcription factors (like Lmx1b, Lmx1a, Pitx3, Nurr1, Msx1, . . .) in the DAergic specification. A consequence of those mice studies was the attempts for the modification of various NSC cultures (from Zebrafish to rodent and human origin) by overexpressing those factors in order to augment the DAn yield. Those studies have mainly demonstrated that the genetic alteration of a single gene is not sufficient to obtain a pure DAn culture, although some could greatly enhance the DAn yield (as shown in Bcl-X_L modified hNSC cultures). Indeed, cell patterning and cooperation between different proneural/DAergic factors remain the most important clue (cf. Table 9.2, for the different effects exerted by Bcl-X_L forced expression on proneuronal and DAergic genes).

Among the genetic manipulation of NSC cultures, Ngn2 appears to be at a crucial point and its activation seems to be a major and necessary event, for the DAergic specification. Several genetic modifications of NSC cultures have been shown to directly interfere with Ngn2 upregulation, like Bcl-X_L or Msx1, event that was described as necessary at a specific, early developmental time for DA specification (Andersson *et al.*, 2006a,b; Kele *et al.*, 2006).

Other examples of genetic modification of *mouse* NSCs to enhance the DAergic specification are here briefly listed:

- Otx2 (Orthodenticle homeobox 2): Stem Cells can acquire neural progenitor characteristics to generate mesencephalic DAn by the overexpression of *Otx2* in mice caudal FP cells, conferring neurogenic activity and DAn capability of fate decision. (Ono *et al.*, 2007; Puelles *et al.*, 2004). Also, other studies have shown that overexpression of *Otx2* in mice caudal FP cells may confer neurogenic activity and fate decision oriented toward DAn. Thus, the repression of certain genes throughout the embryonic development seems to be necessary in order to repress other alternative fates that the NSCs could adopt during the time that neuronal induction lasts (*Otx2*, in particular, controls (by repressing) *Nkx2.2* and *Nkx6.1* gene expression in the ventral midbrain).
- bHLH transcription factors: Ngn2 and Mash1, which are expressed in the developing ventral mesencephalon, and act as proneuronal transcription factors; Nato3 (an FP-specific factor that represses *Hes1*; Hes1 promotes non neurogenic differentiation) (Ono *et al.*, 2010). Overexpression of *Ngn2* or other bHLH factors, by themselves, is not sufficient to increase the DAn number, but require the expression of the additional factor Nurr1 (Andersson *et al.*, 2007; Kim *et al.*, 2007).
- Lmx1a: (LIM homeobox transcription factor 1, alpha). The early expression of this factor in NSCs is of crucial importance for the correct expression of the pro-DAergic factors during the specification to DAergic progenitors, and also at the final steps of the DAn differentiation (Cai *et al.*, 2009).
- Pitx3 or Sox1: Sox1-GFP or Pitx3-GFP strategies were used as a selective marker for FACS sorting to enrich cell cultures in neuronal and DAergic progenitors and remove undesired cells (Aubert *et al.*, 2003; Hedlund *et al.*, 2008). Pitx3 seems to be an important player for the correct acquisition of the DA metabolism functions, maturation, and survival of the DAn.
- Wnts: Wnt1, a secreted glycoprotein has been shown to play a key role for the regulation of *Otx2* and *Nkx2.2* and the establishment of the field where DAn progenitors arise during embryonic development, but also during the later maturation period of DAn. The regulation of the

GSK3beta/ß-catenin pathway (by Wnt1, Wnt5a, and Wnt3a) is then actively involved in the DAn generation process (Castelo-Branco and Arenas, 2006; Castelo-Branco *et al.*, 2003, 2004; Tang *et al.*, 2010).
- Nurr1 and Foxa2: Both transcription factors are widely expressed during the embryonic development in the ventral midbrain. These two factors have been shown to cooperate for the induction of the transcription of pro-DAergic genes, cell cycle exit, DAn generation and survival (Lee *et al.*, 2010).
- Notch signaling: Notch has been determined to be an important player for the self-renewal of NSCs and the maintenance of stem cell pool. However, more recent studies conducted in the Zebrafish embryo (Mahler *et al.*, 2010) have shown that this key factor, and its ligands, Delta A&D, are also important players during the late specification and differentiation process of DAergic clusters. Such results, key factors with dual roles, indicate how complex is the regulation of the generation of functional DAn, and that genetic manipulation of the NSC has to be handle with great care.

As expected, the majority of the studies in the field have been conducted in mouse cells or tissue. Even if the mechanisms of the DAergic specification retains high homology between species, caution should be taken, being important to focus on the progress made understanding human NSC/precursor biology and generation of DAn in humans, as this is the material to be used in cell/gene therapy in a near (?) future in PD patients. Unfortunately, not as many reports as have described the manipulation of human ESC or NSC for the generation of DAn, when compared to rodent studies.

Herein, we present a brief summary of the different studies conducted in different sources of human cells, where genetic modifications were introduced in order to improve DAn differentiation.

Some laboratories have taken the option to use human embryonic stem (hESC) cells for the generation of a DAergic committed NSC population. Two interesting examples of hNSC genetic modification are Lmx1a and Bcl-X_L: In the case of Lmx1a Friling and collaborators showed that mouse ESC cells specifically increased DAn generation (over 80% of the neurons were TH positive), which indeed were functional as determined by electrophysiological recordings. In the case of human ESC, disappointingly, the effects were not so marked (50% of the neurons were DA), and function or performance following grafting was not reported (Friling *et al.*, 2009). The second example, Bcl-X_L, has been tested by Ko and collaborators in a study where Bcl-X_L and SHH were used to enhance DA neuron generation following a 1-month differentiation procedure that seems to result in the disappearance of undifferentiated cells *in vitro*, thus reducing the risk of tumor formation. DAn generation was improved in comparison to control hESC following grafting, and the transplants exerted moderate behavioral effects (50% improvement in amphetamine-induced rotations).

As described in a previous section, much progress has been done orienting hESCs to DAn, with variable success. In our case, we chose to use fetal hNSCs as a source ofDAn, and as previously described extensively, to modify those cells with the antiapoptotic family member Bcl-X_L to successfully improve the DAn yield.

There are few other examples of genetic modifications of hNSCs, reporting improvements of different magnitude, but generally low. Thus, bHLH proteins Ngn2 or Mash1 (*Ascl1* gene) did not enhance DA neuron generation from rat VM neurosphere cultures, and Nurr1 (alone or in combination with Ngn2 or Mash1) had only marginal effects (Kim *et al.*, 2007). When ASCL1 was tested in human VM neurosphere cultures, more neurons were obtained, but not fully mature DAn (Kim *et al.*, 2009). Interestingly, the authors studied if the TH-positive neurons present in differentiation rat and human VM neurosphere cultures were born in culture from their precursors (either stem or later precursors) or if the postmitotic precursors, or the neurons themselves, were surviving from the primary culture, being the latter true. The last example of genetic modification of VM precursors was reported by Roybon and collaborators, where Lmx1a, Msx1, Ngn2, and Pitx3 were overexpressed alone or in combinations in rat VM neurospheres (Roybon *et al.*, 2008). Notably, none of the genes or combinations enhanced differentiation into DAn. When the same genes were tested in the midbrain-derived progenitor cell line NGC-407, only Lmx1a enhanced differentiation toward a DA phenotype (threefold).

Using another system, Xiong *et al.* (2011) recently published that human umbilical cord mesenchymal stem cells (hUC-MSC) could be used, in combination with VEGF (human vascular endothelial growth factor) overexpression for cell therapy in a PD rat model.Such modification leads to some DAn differentiation enhancement of the transplanted cells, as well as partial functional recovery (drug-induced rotation) in hemiparkinsonian rats.

IV. Concluding Remarks

Stem cell therapy appears to be a promising therapeutic option in PD. Stem cells could overcome the problem of the limited availability of human fetal neural tissue, and the low survival rate of fetal DAn. However, there is still a lack of efficient and robust protocols for obtaining SNpc DAn from pluripotent SCs that would result in cell preparations free of immature cells. Thus, there is a current need for strategies that will improve ESCs differentiation, NSCs proliferation, and the proper differentiation of neuronal precursor cells into SNpc DAn, as well as the elimination of undifferentiated tumor-forming cells, and the further enhancement of the *in vivo* survival and integration of DAn following transplantation.

ACKNOWLEDGMENTS

We would like to thank the excellent technical assistance of Ignacio Tardieu de Chorro, Marta González Mella and Beatriz Moreno Moreno, and many fruitful discussions with members of AMS laboratory. Work at AMS laboratory was supported by grants from the EU (NMP-SL-2008-214706 EXCELL), MCyT (SAF2004-03405), MICINN (BIO2007-66807, PLE2009-0101, and SAF2010-17167), Carlos III Institute of Health grant RETICS TerCel (RD06/0010/0009), and La Caixa Foundation (BM05-22-0). The Center of Molecular Biology Severo Ochoa is also supported by an institutional grant from Foundation Ramón Areces.

The authors declare no conflict of interest.

REFERENCES

Abeilovich, A., and Hammond, R. (2007). Midbrain dopamine neuron differentiation: Factors and facts. *Dev. Biol.* **304,** 447–454.

Akchiche, N., Bossenmeyer-Pourie, C., Pourie, G., Koziel, V., Nedelec, E., Gueant, J. L., and Daval, J. L. (2010). Differentiation and neural integration of hippocampal neuronal progenitors: Signaling pathways sequentially involved. *Hippocampus* **20,** 949–961.

Alavian, K. N., Scholz, C., and Simon, H. H. (2008). Transcriptional regulation of mesencephalic dopaminergic neurons: The full circle of life and death. *Mov. Disord.* **23,** 319–328.

Andersson, E., Jensen, J. B., Parmar, M., Guillemot, F., and Bjorklund, A. (2006a). Development of the mesencephalic dopaminergic neuron system is compromised in the absence of neurogenin 2. *Development* **133,** 507–516.

Andersson, E., Tryggvason, U., Deng, Q., Friling, S., Alekseenko, Z., Robert, B., Perlmann, T., and Ericson, J. (2006b). Identification of intrinsic determinants of midbrain dopamine neurons. *Cell* **124,** 393–405.

Andersson, E. K., Irvin, D. K., Ahlsio, J., and Parmar, M. (2007). Ngn2 and Nurr1 act in synergy to induce midbrain dopaminergic neurons from expanded neural stem and progenitor cells. *Exp. Cell Res.* **313,** 1172–1180.

Ang, S. L. (2006). Transcriptional control of midbrain dopaminergic neuron development. *Development* **133,** 3499–3506.

Aranha, M. M., Sola, S., Low, W. C., Steer, C. J., and Rodrigues, C. M. (2009). Caspases and p53 modulate FOXO3A/Id1 signaling during mouse neural stem cell differentiation. *J. Cell. Biochem.* **107,** 748–758.

Arbour, N., Vanderluit, J. L., Le Grand, J. N., Jahani-Asl, A., Ruzhynsky, V. A., Cheung, E. C., Kelly, M. A., MacKenzie, A. E., Park, D. S., Opferman, J. T., and Slack, R. S. (2008). Mcl-1 is a key regulator of apoptosis during CNS development and after DNA damage. *J. Neurosci.* **28,** 6068–6078.

Arenas, E. (2010). Towards stem cell replacement therapies for Parkinson´s disease. *Biochem. Biophys. Res. Commun.* **21,** 152–156.

Aubert, J., Stavridis, M. P., Tweedie, S., O'Reilly, M., Vierlinger, K., Li, M., Ghazal, P., Pratt, T., Mason, J. O., Roy, D., and Smith, A. (2003). Screening for mammalian neural genes via fluorescence-activated cell sorter purification of neural precursors from Sox1-gfp knock-in mice. *Proc. Natl. Acad. Sci. USA* **100,** 11836–11841.

Benn, S. C., and Woolf, C. J. (2004). Adult neuron survival strategies—Slamming on the brakes. *Nat. Rev. Neurosci.* **5,** 686–700.

Bjorklund, A., and Dunnet, S. B. (2007). Dopamine neuron systems in the brain: An update. *Trends Neurosci.* **30,** 194–202.

Bonilla, S., Hall, A. C., Pinto, L., Attardo, A., Götz, M., Huttner, W. B., and Arenas, E. (2008). Identification of midbrain floor plate radial glia-like cells as dopaminergic progenitors. *Glia* **56,** 809–820.

Brundin, P., Barker, R. A., and Parmar, M. (2010). Neural grafting in Parkinson's disease: Problems and possibilities. *Prog. Brain Res.* **184,** 265–294.

Cai, J., Donaldson, A., Yang, M., German, M. S., Enikolopov, G., and Iacovitti, L. (2009). The role of Lmx1a in the differentiation of human embryonic stem cells into midbrain dopamine neurons in culture and after transplantation into a Parkinson's disease model. *Stem Cells* **27,** 220–229.

Calegari, F., and Huttner, W. B. (2003). An inhibition of cyclin-dependent kinases that lengthens, but does not arrest, neuroepithelial cell cycle induces premature neurogenesis. *J. Cell Sci.* **116,** 4947–4955.

Carta, M., Carlsson, T., Muñoz, A., Kirik, D., and Björklund, A. (2008). Serotonin-dopamine interaction in the induction and maintenance of L-DOPA-induced dyskinesias. *Prog. Brain Res.* **17,** 465–478.

Castelo-Branco, G., and Arenas, E. (2006). Function of Wnts in dopaminergic neuron development. *Neurodegener Dis* **3,** 5–11.

Castelo-Branco, G., Wagner, J., Rodriguez, F. J., Kele, J., Sousa, K., Rawal, N., Pasolli, H. A., Fuchs, E., Kitajewski, J., and Arenas, E. (2003). Differential regulation of midbrain dopaminergic neuron development by Wnt-1, Wnt-3a, and wnt-5a. *Proc. Natl. Acad. Sci. USA* **100,** 12747–12752.

Castelo-Branco, G., Rawal, N., and Arenas, E. (2004). GSK-3beta inhibition/beta-catenin stabilization in ventral midbrain precursors increases differentiation into dopamine neurons. *J. Cell Sci.* **117,** 5731–5737.

Chambers, S. M., *et al.* (2009). Highly efficient neural conversion of human ES and iPS cells by dual inhibition of SMAD signalling. *Nat. Biotechnol.* **27,** 275–280.

Chang, M. Y., Sun, W., Ochiai, W., Nakashima, K., Kim, S. Y., Park, C. H., Kang, J. S., Shim, J. W., Jo, A. Y., Kang, C. S., Lee, Y. S., Kim, J. S., *et al.* (2007). Bcl-XL/Bax proteins direct the fate of embryonic cortical precursor cells. *Mol. Cell. Biol.* **27,** 4293–4305.

Cho, M. S., *et al.* (2008). Highly efficient and large-scale generation of functional dopamine neurons from human embryonic stem cells. *Proc. Natl. Acad. Sci. USA* **105,** 3392–3397.

Chung, S., Shin, B. S., Hwang, M., Lardaro, T., Kang, U. J., Isacson, O., and Kim, K. S. (2006). Neural precursors derived from embryonic stem cells, but not those from fetal ventral mesencephalon, maintain the potential to differentiate into dopaminergic neurons after expansion *in vitro*. *Stem Cells* **24,** 1583–1593.

Cory, S., and Adams, J. M. (2002). The Bcl2 family: Regulators of the cellular life-or-death switch. *Nat. Rev. Cancer* **2,** 647–656.

Courtois, E. T., Castillo, C. G., Seiz, E. G., Ramos, M., Bueno, C., Liste, I., and Martinez-Serrano, A. (2010). *In vitro* and *in vivo* enhanced generation of human A9 dopamine neurons from neural stem cells by Bcl-XL. *J. Biol. Chem.* **285,** 9881–9897.

Dehay, C., and Kennedy, H. (2007). Cell-cycle control and cortical development. *Nat. Rev. Neurosci.* **8,** 438–450.

Doe, C. Q. (2008). Neural stem cells: Balancing self-renewal with differentiation. *Development* **135,** 1575–1587.

Donato, R., Miljan, E. A., Hines, S. J., Aouabdi, S., Pollock, K., Patel, S., Edwards, F. A., and Sinden, J. D. (2007). Differential development of neuronal physiological responsiveness in two human neural stem cell lines. *BMC Neurosci.* **8,** 36.

Dunnett, S. B., Björklund, A., and Lindvall, O. (2001). Cell therapy in Parkinson's disease—Stop or go? *Nat. Rev. Neurosci.* **2,** 365–369.

Farkas, L. M., Haffner, C., Giger, T., Khaitovich, P., Nowick, K., Birchmeier, C., Paabo, S., and Huttner, W. B. (2008). Insulinoma-associated 1 has a panneurogenic

role and promotes the generation and expansion of basal progenitors in the developing mouse neocortex. *Neuron* **60,** 40–55.
Ferri, A. L. M., Lin, W., Mavromatakis, Y. E., Wang, J. C., Sasaki, H., Whitsett, J. A., and Ang, S. L. (2007). Foxa1 and Foxa2 regulate multiple phases of midbrain dopaminergic neuron development in a dosage-dependent manner. *Development* **134,** 2761–2769.
Friling, S., Andersson, E., Thompson, L. H., Jönsson, M. E., Hebsgaard, J. B., Nanou, E., Alekseenko, Z., Marklund, U., Kjellander, S., Volakakis, N., Hovatta, O., El Manira, A., *et al.* (2009). Efficient production of mesencephalic dopamine neurons by Lmx1a expression in embryonic stem cells. *Proc. Natl. Acad. Sci. USA* **106**(18), 7613–7618.
Gale, E., and Li, M. (2008). Midbrain dopaminergic neuron fate specification: Of mice and embryonic stem cells. *Mol. Brain* **1,** 8.
Garrido, C., and Kroemer, G. (2004). Life's smile, death's grin: Vital functions of apoptosis-executing proteins. *Curr. Opin. Cell Biol.* **16,** 639–646.
Gonzalez-Garcia, M., Perez-Ballestero, R., Ding, L., Duan, L., Boise, L. H., Thompson, C. B., and Nunez, G. (1994). bcl-XL is the major bcl-x mRNA form expressed during murine development and its product localizes to mitochondria. *Development* **120,** 3033–3042.
Grealish, S., Jonsson, M. E., Li, M., Kirik, D., Bjorklund, A., and Thompson, L. H. (2010). The A9 dopamine neuron component in grafts of ventral mesencephalon is an important determinant for recovery of motor function in a rat model of Parkinson's disease. *Brain* **133,** 482–495.
Gu, Y., Siefring, J. E., Wang, L., Chae, H. D., Bailey, J. R., and Zheng, Y. (2006). Oncogenic Vav1 induces Rac-dependent apoptosis via inhibition of Bcl-2 family proteins and collaborates with p53 deficiency to promote hematopoietic progenitor cell proliferation. *Oncogene* **25,** 3963–3972.
Guillemot, F. (2007). Cell fate specification in the mammalian telencephalon. *Prog. Neurobiol.* **83,** 37–52.
Hagell, P., Piccin, P., Björklund, A., Brundin, P., Rehncrona, S., Widner, H., Crabb, L., Pavese, N., Oertel, W. H., Quinn, N., Brooks, D. J., and Lindvall, O. (2002). Dyskinesias following neural transplantation in Parkinson's disease. *Nat. Neurosci.* **5,** 627–628.
Hagn, F., Klein, C., Demmer, O., Marchenko, N., Vaseva, A., Moll, U. M., and Kessler, H. (2010). BclxL changes conformation upon binding to wild-type but not mutant p53 DNA binding domain. *J. Biol. Chem.* **285,** 3439–3450.
Hall, V., Li, J. Y., and Brundin, P. (2007). Restorative cell therapy for Parkinson's disease: A quest for the perfect cell. *Semin. Cell Dev. Biol.* **18,** 859–869.
Hargus, G., *et al.* (2010). Differentiated Parkinson patient-derived induced pluripotent stem cells grow in the adult rodent brain and reduce motor asymmetry in Parkinsonian rats. *Proc. Natl. Acad. Sci. USA* **107,** 15921–15926.
Haughn, L., Hawley, R. G., Morrison, D. K., von Boehmer, H., and Hockenbery, D. M. (2003). BCL-2 and BCL-XL restrict lineage choice during hematopoietic differentiation. *J. Biol. Chem.* **278,** 25158–25165.
Haydar, T. F., Wang, F., Schwartz, M. L., and Rakic, P. (2000). Differential modulation of proliferation in the neocortical ventricular and subventricular zones. *J. Neurosci.* **20,** 5764–5774.
Hebsgaard, J. B., Nelander, J., Sabelstrom, H., Jonsson, M. E., Stott, S., and Parmar, M. (2009). Dopamine neuron precursors within the developing human mesencephalon show radial glial characteristics. *Glia* **57,** 1648–1658.
Hedlund, E., Pruszak, J., Lardaro, T., Ludwig, W., Vinuela, A., Kim, K. S., and Isacson, O. (2008). Embryonic stem cell-derived Pitx3-enhanced green fluorescent protein midbrain dopamine neurons survive enrichment by fluorescence-activated cell sorting and function in an animal model of Parkinson's disease. *Stem Cells* **26,** 1526–1536.

Hwang, D. Y. (2010). Human ES and iPS cells as cell sources for the treatment of Parkinson's disease: Current state and problems. *J. Cell. Biochem.* **109,** 292–301.

Janumyan, Y., Cui, Q., Yan, L., Sansam, C. G., Valentin, M., and Yang, E. (2008). G0 function of BCL2 and BCL-xL requires BAX, BAK, and p27 phosphorylation by Mirk, revealing a novel role of BAX and BAK in quiescence regulation. *J. Biol. Chem.* **283,** 34108–34120.

Jensen, J. B., and Parmar, M. (2006). Strengths and limitations of the neurosphere culture system. *Mol. Neurobiol.* **34,** 153–161.

Jones, R. G., Bui, T., White, C., Madesh, M., Krawczyk, C. M., Lindsten, T., Hawkins, B. J., Kubek, S., Frauwirth, K. A., Wang, Y. L., Conway, S. J., Roderick, H. L., *et al.* (2007). The proapoptotic factors Bax and Bak regulate T Cell proliferation through control of endoplasmic reticulum Ca(2+) homeostasis. *Immunity* **27,** 268–280.

Kele, J., Simplicio, N., Ferri, A. L., Mira, H., Guillemot, F., Arenas, E., and Ang, S. L. (2006). Neurogenin 2 is required for the development of ventral midbrain dopaminergic neurons. *Development* **133,** 495–505.

Khan, Z., Akhtar, M., Asklund, T., Juliusson, B., Almqvist, P. M., and Ekström, T. J. (2007). HDAC inhibition amplifies gap junction communication in neural progenitors: Potential for cell-mediated enzyme prodrug therapy. *Exp. Cell Res.* **313,** 2958–2967.

Kim, H. J. (2011). Stem cell potential in Parkinson's disease and molecular factors for the generation of dopamineneurons. *Biochim. Biophys. Acta* **1812,** 1–11.

Kim, H. J., Sugimori, M., Nakafuku, M., and Svendsen, C. N. (2007). Control of neurogenesis and tyrosine hydroxylase expression in neural progenitor cells through bHLH proteins and Nurr1. *Exp. Neurol.* **203,** 394–405.

Kim, H. J., McMillan, E., Han, F., and Svendsen, C. N. (2009). Regionally specified human neural progenitor cells derived from the mesencephalon and forebrain undergo increased neurogenesis following overexpression of ASCL1. *Stem Cells* **27,** 390–398.

Koch, P., Opitz, T., Steinbeck, J. A., Ladewig, J., and Brustle, O. (2009a). A rosette-type, self-renewing human ES cell-derived neural stem cell with potential for *in vitro* instruction and synaptic integration. *Proc. Natl. Acad. Sci. USA* **106,** 3225–3230.

Koch, P., Kokaia, Z., Lindvall, O., and Brustle, O. (2009b). Emerging concepts in neural stem cell research: Autologous repair and cell-based disease modeling. *Lancet Neurol.* **8,** 819–829.

Krabbe, C., Courtois, E., Jensen, P., Jorgensen, J. R., Zimmer, J., Martinez-Serrano, A., and Meyer, M. (2009). Enhanced dopaminergic differentiation of human neural stem cells by synergistic effect of Bcl-xL and reduced oxygen tension. *J. Neurochem.* **110,** 1908–1920.

Krajewska, M., Mai, J. K., Zapata, J. M., Ashwell, K. W., Schendel, S. L., Reed, J. C., and Krajewski, S. (2002). Dynamics of expression of apoptosis-regulatory proteins Bid, Bcl-2, Bcl-X, Bax and Bak during development of murine nervous system. *Cell Death Differ.* **9,** 145–157.

Krantic, S., Mechawar, N., Reix, S., and Quirion, R. (2007). Apoptosis-inducing factor: A matter of neuron life and death. *Prog. Neurobiol.* **81,** 179–196.

Kretz, A., Kugler, S., Happold, C., Bahr, M., and Isenmann, S. (2004). Excess Bcl-XL increases the intrinsic growth potential of adult CNS neurons *in vitro*. *Mol. Cell. Neurosci.* **26,** 63–74.

Kuranaga, E. (2011). Caspase signaling in animal development. *Dev. Growth Differ.* **53,** 137–148.

Kuranaga, E., and Miura, M. (2007). Nonapoptotic functions of caspases: Caspases as regulatory molecules for immunity and cell-fate determination. *Trends Cell Biol.* **17,** 135–144.

Lee, H. S., Bae, E. J., Yi, S. H., Shim, J. W., Jo, A. Y., Kang, J. S., Yoon, E. H., Rhee, Y. H., Park, C. H., Koh, H. C., Kim, H. J., Choi, H. S., *et al.* (2010). Foxa2 and Nurr1 synergistically yield A9 nigral dopamine neurons exhibiting improved differentiation, function, and cell survival. *Stem Cells* **28,** 501–512.

Lenos, M. G., and Tsaniklidou, S. M. (2010). Calcium oscillations, G1 phase duration and neurogenesis timing. *Trends Cell Biol.* **20,** 577(author reply 578).

Lew, M. (2007). Overview of Parkinson's disease. *Pharmacotherapy* **27,** 155S–160S.

Li, C., Fox, C. J., Master, S. R., Bindokas, V. P., Chodosh, L. A., and Thompson, C. B. (2002). Bcl-X(L) affects Ca(2+) homeostasis by altering expression of inositol 1,4,5-trisphosphate receptors. *Proc. Natl. Acad. Sci. USA* **99,** 9830–9835.

Li, C., Wang, X., Vais, H., Thompson, C. B., Foskett, J. K., and White, C. (2007). Apoptosis regulation by Bcl-x(L) modulation of mammalian inositol 1,4,5-trisphosphate receptor channel isoform gating. *Proc. Natl. Acad. Sci. USA* **104,** 12565–12570.

Li, H., Chen, Y., Jones, A. F., Sanger, R. H., Collis, L. P., Flannery, R., McNay, E. C., Yu, T., Schwarzenbacher, R., Bossy, B., Bossy-Wetzel, E., Bennett, M. V., *et al.* (2008a). Bcl-xL induces Drp1-dependent synapse formation in cultured hippocampal neurons. *Proc. Natl. Acad. Sci. USA* **105,** 2169–2174.

Li, J. Y., *et al.* (2008b). Lewy bodies in grafted neurons in subjects with Parkinson's disease suggest host-to-graft disease propagation. *Nat. Med.* **14,** 501–503.

Li, J. Y., *et al.* (2008c). Critical issues of clinical human embryonic stem cell therapy for brain repair. *Trends Neurosci.* **31,** 146–153.

Lindsten, T., Zong, W. X., and Thompson, C. B. (2005). Defining the role of the Bcl-2 family of proteins in the nervous system. *Neuroscientist* **11,** 10–15.

Lindvall, O., and Björklund, A. (2004). Cell therapy in Parkinson's disease. *NeuroRx.* **1,** 382–393.

Lindvall, O., and Kokaia, Z. (2009). Prospects of stem cell therapy for replacing dopamine neurons in Parkinson´s disease. *Trends Pharmacol. Sci.* **30,** 260–267.

Lindvall, O., and Kokaia, Z. (2010). Stem cells in human neurodegenerative disorders- time for clinical translation? *J. Clin. Invest.* **120,** 29–40.

Lindvall, O., Kokaia, Z., and Martínez-Serrano, A. (2004). Stem cell therapy for human neurodegenerative disorders—How to make it work. *Nat. Med.* **10**(Suppl.), S42–S50.

Lipskaia, L., Hulot, J. S., and Lompre, A. M. (2009). Role of sarco/endoplasmic reticulum calcium content and calcium ATPase activity in the control of cell growth and proliferation. *Pflugers Arch.* **457,** 673–685.

Liste, I., Garcia-Garcia, E., and Martinez-Serrano, A. (2004). The generation of dopaminergic neurons by human neural stem cells is enhanced by Bcl-XL, both *in vitro* and *in vivo*. *J. Neurosci.* **24,** 10786–10795.

Liste, I., Garcia-Garcia, E., Bueno, C., and Martinez-Serrano, A. (2007). Bcl-XL modulates the differentiation of immortalized human neural stem cells. *Cell Death Differ.* **14,** 1880–1892.

Lister, R., *et al.* (2011). Hotspots of aberrant epigenomic reprogramming in human induced pluripotent stem cells. *Nature* **471,** 68–73.

Lotharius, J., Barg, S., Wiekop, P., Lundberg, C., Raymon, H. K., and Brundin, P. (2002). Effect of mutant alpha-synuclein on dopamine homeostasis in a new human mesencephalic cell line. *J. Biol. Chem.* **277,** 38884–38894.

Maciaczyk, J., Singec, I., Maciaczyk, D., and Nikkhah, G. (2008). Combined use of BDNF, ascorbic acid, low oxygen, and prolonged differentiation time generates tyrosine hydroxylase-expressing neurons after long-term *in vitro* expansion of human fetal midbrain precursor cells. *Exp. Neurol.* **213,** 354–362.

Mahler, J., Filippi, A., and Driever, W. (2010). DeltaA/DeltaD regulate multiple and temporally distinct phases of notch signaling during dopaminergic neurogenesis in zebrafish. *J. Neurosci.* **30,** 16621–16635.

Malmersjo, S., Liste, I., *et al.* (2010). Ca^{2+} and cAMP signaling in human embryonic stem cell-derived dopamine neurons. *Stem Cells Dev.* **19**(9), 1355–1364.

Martinez-Serrano, A., and Bjorklund, A. (1997). Immortalized neural progenitor cells for CNS gene transfer and repair. *Trends Neurosci.* **20,** 530–538.

Martínez-Serrano, A., and Liste, I. (2010). Recent progress and challenges for the use of stem cell derivatives in neuron replacement therapy for Parkinson's disease. *Future neurol.* **5,** 161–165.

Meissner, W. G., Frasier, M., Gasser, T., Coetz, C. G., Lozano, A., Piccini, P., Obeso, J. A., Rascol, O., Schapira, A., Voon, V., Weiner, D. M., Tison, F., *et al.* (2011). Priorities in Parkinson's disease research. *Nat. Rev. Drug Discov.* **10,** 377–393.

Meyer, A. K., Maisel, M., Hermann, A., Stirl, K., and Storch, A. (2010). Restorative approaches in Parkinson's disease: Which cell type wins the race? *J. Neurol. Sci.* **289,** 93–103.

Miljan, E. A., Hines, S. J., Pande, P., Corteling, R. L., Hicks, C., Zbarsky, V., Umachndran, M., Sowinski, P., Richardon, S., Tang, E., Wieruszew, M., Patel, S., *et al.* (2009). Implantation of c-mycERTAM immortalized human mesencephalic-derived clonal cell lines ameliorates behavior dysfunction in a rat model of Parkinson's disease. *Stem Cells Dev.* **18,** 307–319.

Motoyama, N., Wang, F., Roth, K. A., Sawa, H., Nakayama, K., Negishi, I., Senju, S., Zhang, Q., Fujii, S., *et al.* (1995). Massive cell death of immature hematopoietic cells and neurons in Bcl-x-deficient mice. *Science* **267,** 1506–1510.

Nair, V. D., McNaught, K. S., Gonzalez-Maeso, J., Sealfon, S. C., and Olanow, C. W. (2006). p53 mediates nontranscriptional cell death in dopaminergic cells in response to proteasome inhibition. *J. Biol. Chem.* **281,** 39550–39560.

Nakatani, T., Kumai, M., Mizuhara, E., Minaki, Y., and Ono, Y. (2010). Lmx1a and Lmx1b cooperate with Foxa2 to coordinate the specification of dopaminergic neurons and control floor plate cell differentiation in the developing mesencephalon. *Dev. Biol.* **339,** 101–113.

Nelander, J., Hebsgaard, J. B., and Parmar, M. (2009). Organization of the human embryonic ventral mesencephalon. *Gene Expr. Patterns* **9,** 555–561.

Nguyen, L., Besson, A., Roberts, J. M., and Guillemot, F. (2006). Coupling cell cycle exit, neuronal differentiation and migration in cortical neurogenesis. *Cell Cycle* **5,** 2314–2318.

Nguyen, H. N., *et al.* (2011). LRRK2 mutant iPSC-derived DA neurons demonstrate increased susceptibility to oxidative stress. *Cell Stem Cell* **8,** 267–280.

Obeso, J. A., Rodríguez-Oroz, M. C., Goetz, C. G., Marin, C., Kordower, J. H., Rodriguez, M., Hirsch, E. C., Farrer, M., Schapira, A. H. V., and Halliday, G. (2010). Missing pieces in the Parkinson's disease puzzle. *Nat. Med.* **16,** 563–661.

Olanow, C. W., and Tatton, W. G. (1999). Etiology and pathogenesis of Parkinson's disease. *Annu. Rev. Neurosci.* **22,** 123–144.

Ono, Y., Nakatani, T., Sakamoto, Y., Mizuhara, E., Minaki, Y., Kumai, M., Hamaguchi, A., Nishimura, M., Inoue, Y., Hayashi, H., Takahashi, J., and Imai, T. (2007). Differences in neurogenic potential in floor plate cells along an anteroposterior location: Midbrain dopaminergic neurons originate from mesencephalic floor plate cells. *Development* **134,** 3213–3225.

Ono, Y., Nakatani, T., Minaki, Y., and Kumai, M. (2010). The basic helix-loop-helix transcription factor Nato3 controls neurogenic activity in mesencephalic floor plate cells. *Development* **137,** 1897–1906.

Orme, R., Fricker-Gates, R. A., and Gates, M. A. (2009). Ontogeny of substantia nigra dopamine neurons. *J. Neural Transm. Suppl.* **73,** 3–18.

Ostenfeld, T., Joly, E., Tai, Y. T., Peters, A., Caldwell, M., Jauniaux, E., and Svendsen, C. N. (2002). Regional specification of rodent and human neurospheres. *Brain Res. Dev. Brain Res.* **134,** 43–55.

Parsadanian, A. S., Cheng, Y., Keller-Peck, C. R., Holtzman, D. M., and Snider, W. D. (1998). Bcl-xL is an antiapoptotic regulator for postnatal CNS neurons. *J. Neurosci.* **18,** 1009–1019.

Paul, G., Christophersen, N. S., Raymon, H., Kiaer, C., Smith, R., and Brundin, P. (2007). Tyrosine hydroxylase expression is unstable in a human immortalized mesencephalic cell line—studies in vitro and after intracerebral grafting in vivo. *Mol. Cell. Neurosci.* **34,** 390–399.

Perrier, A. L., Tabar, V., Barberi, T., Rubio, M. E., Bruses, J., Topf, N., Harrison, N. L., and Studer, L. (2004). Derivation of midbrain dopamine neurons from human embryonic stem cells. *Proc. Natl. Acad. Sci. USA* **101,** 12543–12548.
Piccini, P., Brooks, D. J., Björklund, A., Gunn, R. N., Grasby, P. M., Rimoldi, O., Brundin, P., Hagell, Rehncrona S., Widner, H., and Lindvall, O. (1999). Dopamine release from nigral transplants visualized *in vivo* in a Parkinson's patient. *Nat. Neurosci.* **2,** 1137–1140.
Piccini, P., Pavese, N., Hagell, P., Reimer, J., Björklund, A., Oertel, W. H., Quinn, N. P., Brooks, D. J., and Lindvall, O. (2005). Factors affecting the clinical outcome after neural transplantaion in Parkinson's disease. *Brain* **128,** 2977–2986.
Pinton, P., and Rizzuto, R. (2006). Bcl-2 and Ca2+ homeostasis in the endoplasmic reticulum. *Cell Death Differ.* **13,** 1409–1418.
Politis, M., Wu, K., Loane, C., Quinn, N. P., Brooks, D. J., Rehncrona, S., Bjorklund, A., Lindvall, O., and Piccini, P. (2010). Serotonergic neurons mediate dyskinesia side effects in Parkinson's disease patients with neural transplants. *Sci. Transl. Med.* **2,** 38ra46.
Prakash, N., and Wurst, W. (2006). Genetic netwrls controlling the development of midbrain dopaminergic neurons. *J. Physiol.* **575,** 403–410.
Puelles, E. (2007). Genetic control of basal midbrain development. *J. Neurosci. Res.* **85,** 3530–3534.
Puelles, E., Annino, A., Tuorto, F., Usiello, A., Acampora, D., Czerny, T., Brodski, C., Ang, S. L., Wurst, W., and Simeone, A. (2004). Otx2 regulates the extent, identity and fate of neuronal progenitor domains in the ventral midbrain. *Development* **131,** 2037–2048.
Resende, R. R., Gomes, K. N., Adhikari, A., Britto, L. R., and Ulrich, H. (2008). Mechanism of acetylcholine-induced calcium signaling during neuronal differentiation of P19 embryonal carcinoma cells *in vitro*. *Cell Calcium* **43,** 107–121.
Riaz, S. S., Theofilopoulos, S., Jauniaux, E., Stern, G. M., and Bradford, H. F. (2004). The differentiation potential of human foetal neuronal progenitor cells *in vitro*. *Brain Res. Dev. Brain Res.* **153,** 39–51.
Roy, N. S., *et al.* (2006). Functional engraftment of human ES cell-derived dopaminergic neurons enriched by coculture with telomerase-immortalized midbrain astrocytes. *Nat. Med.* **12,** 1259–1268.
Roybon, L., Hjalt, T., Christophersen, N. S., Li, J. Y., and Brundin, P. (2008). Effects on differentiation of embryonic ventral midbrain progenitors by Lmx1a, Msx1, Ngn2, and Pitx3. *J. Neurosci.* **28,** 3644–3656.
Ryu, J. K., Choi, H. B., Hatori, K., Heisel, R. L., Pelech, S. L., McLarnon, J. G., and Kim, S. U. (2003). Adenosine triphosphate induces proliferation of human neural stem cells: Role of calcium and p70 ribosomal protein S6 kinase. *J. Neurosci. Res.* **72,** 352–362.
Sacchetti, P., *et al.* (2009). Liver X receptors and oxysterols promote ventral midbrain neurogenesis in vivo and in human embryonic stem cells. *Cell Stem Cell* **5**(4), 409–419.
Salomoni, P., and Calegari, F. (2010). Cell cycle control of mammalian neural stem cells: Putting a speed limit on G1. *Trends Cell Biol.* **20,** 233–243.
Savitt, J. M., Jang, S. S., Mu, W., Dawson, V. L., and Dawson, T. M. (2005). Bcl-x is required for proper development of the mouse substantia nigra. *J. Neurosci.* **25,** 6721–6728.
Schulz, R., Vogel, T., Dressel, R., and Krieglstein, K. (2008). TGF-beta superfamily members, ActivinA and TGF-beta1, induce apoptosis in oligodendrocytes by different pathways. *Cell Tissue Res.* **334,** 327–338.
Shim, J. W., Koh, H. C., Chang, M. Y., Roh, E., Choi, C. Y., Oh, Y. J., Son, H., Lee, Y. S., Studer, L., and Lee, S. H. (2004). Enhanced *in vitro* midbrain dopamine neuron differentiation, dopaminergic function, neurite outgrowth, and 1-methyl-4-phenylpyridium resistance in mouse embryonic stem cells overexpressing Bcl-XL. *J. Neurosci.* **24,** 843–852.

Smidt, M. P., and Burbach, J. P. (2007). How to make a mesodiencephalic dopaminergic neuron. *Nat. Rev. Neurosci.* **8,** 21–32.

Soldner, F., Hockemeyer, D., Beard, C., Gao, Q., Bell, G. W., Cook, E. G., Hargus, G., Blak, A., Cooper, O., Mitalipova, M., Isacson, O., and Jaenisch, R. (2009). Parkinson's disease patient-derived induced pluripotent stem cells free of viral reprogramming factors. *Cell* **136,** 964–977.

Storch, A., Paul, G., Csete, M., Boehm, B. O., Carvey, P. M., Kupsch, A., and Schwarz, J. (2001). Long-term proliferation and dopaminergic differentiation of human mesencephalic neural precursor cells. *Exp. Neurol.* **170,** 317–325.

Studer, L., Tabar, V., and McKay, R. D. (1998). Transplantation of expanded mesencephalic precursors leads to recovery in parkinsonian rats. *Nat. Neurosci.* **1,** 290–295.

Takahashi, K., Tanabe, K., Ohnuki, M., Narita, M., Ichisaka, T., Tomoda, K., and Yamanaka, S. (2007). Induction of pluripotent stem cells from adult human fibroblasts by defined factors. *Cell* **131,** 861–872.

Tang, M., Villaescusa, J. C., Luo, S. X., Guitarte, C., Lei, S., Miyamoto, Y., Taketo, M. M., Arenas, E., and Huang, E. J. (2010). Interactions of Wnt/beta-catenin signaling and sonic hedgehog regulate the neurogenesis of ventral midbrain dopamine neurons. *J. Neurosci.* **30,** 9280–9291.

Thomson, J. A., Itskovitz-Eldor, J., Shapiro, S. S., Waknitz, M. A., Swiergiel, J. J., Marshall, V. S., and Jones, J. M. (1998). Embryonic stem cell lines derived from human blastocysts. *Science* **282,** 1145–1147.

Tønessen, J., Seiz, E. G., Ramos, M., Lindvall, O., Martínez-Serrano, A., and Kokaia, M. (2010). Functional properties of the human ventral mesecephalic neural stem cells line hVM1. *Exp. Neurol.* **223,** 653–656.

Tzschentke, T. M., and Schmidt, W. J. (2000). Functional relationship among medial prefrontal cortex, nucleus accumbens, and ventral tegmental area in locomotion and reward. *Crit. Rev. Neurobiol.* **14,** 131–142.

Villa, A., Liste, I., Courtois, E. T., Seiz, E. G., Ramos, M., Meyer, M., Juliusson, B., Kusk, P., and Martinez-Serrano, A. (2009). Generation and properties of a new human ventral mesencephalic neural stem cell line. *Exp. Cell Res.* **315,** 1860–1874.

Weissman, T. A., Riquelme, P. A., Ivic, L., Flint, A. C., and Kriegstein, A. R. (2004). Calcium waves propagate through radial glial cells and modulate proliferation in the developing neocortex. *Neuron* **43,** 647–661.

White, C., Li, C., Yang, J., Petrenko, N. B., Madesh, M., Thompson, C. B., and Foskett, J. K. (2005). The endoplasmic reticulum gateway to apoptosis by Bcl-X(L) modulation of the InsP3R. *Nat. Cell Biol.* **7,** 1021–1028.

Winkler, C., Kirirk, D., and Björklund, A. (2005). Cell transplantation in Parkinson's disease: How can we make it work? *Trends Neurosci.* **28,** 86–92.

Xiong, N., Zhang, Z., Huang, J., Chen, C., Jia, M., Xiong, J., Liu, X., Wang, F., Cao, X., Liang, Z., Sun, S., Lin, Z., *et al.* (2011). VEGF-expressing human umbilical cord mesenchymal stem cells, an improved therapy strategy for Parkinson's disease. *Gene Ther.* **18,** 394–402.

Yan, J., Studer, L., and McKay, R. D. G. (2001). Ascorbic acid increases de yield of dopaminergic neurons derived from basic fibroblast growth factor expanded mesencephalic precursors. *J. Neurochem.* **76,** 307–311.

Yu, J., *et al.* (2007). Induced pluripotent stem cell lines derived from human somatic cells. *Science* **318,** 1917–1920.

Zaidi, A. U., D'Sa-Eipper, C., Brenner, J., Kuida, K., Zheng, T. S., Flavell, R. A., Rakic, P., and Roth, K. A. (2001). Bcl-X(L)-caspase-9 interactions in the developing nervous system: Evidence for multiple death pathways. *J. Neurosci.* **21,** 169–175.

CHAPTER TEN

Glucocorticoid Hedgehog Agonists in Neurogenesis

Jiangbo Wang,* Larry S. Barak,† Robert A. Mook Jr.,* *and* Wei Chen*

Contents

I. Introduction 208
II. Select Glucocorticoids as Smoothened Agonists: Potential Effects for Neurogenesis 209
III. Mechanism of Action 210
IV. Structure–Activity Relationships (SAR) of Glucocorticoid Smoothened Agonists 211
V. Conclusion 212
References 214

Abstract

The process of neurogenesis in mammals, which is prolific and widespread at birth, gradually slows with aging and in humans becomes restricted to areas including the cerebellum and hippocampus. It has been reported that exposure to glucocorticoids can impair neurogenesis in both adults and children. Glucocorticoids are known to bind with high affinity to intracellular receptors. Glucocorticoid blood levels are normally regulated by environmental stresses, but because of their clinical utility, exogenous glucocorticoids are frequently administered in drug formulations. Consequently, concerns have arisen about the consequences of glucocorticoid use on neurogenesis and health, especially in the pediatric population. In this article, we will review recent findings that a select number of related glucocorticoids, halcinonide, fluticasone propionate, clobetasol propionate, and fluocinonide, also bind the hedgehog pathway receptor Smoothened. We will discuss their pharmacology and also a most surprising result; that this select group of compounds, which includes FDA approved drugs, unlike typical glucocorticoids such as dexamethasone, stimulate stem cell growth, and thus enhance neurogenesis. © 2011 Elsevier Inc.

* Department of Medicine, Duke University Medical Center, Durham, North Carolina, USA
† Department of Cell Biology, Duke University Medical Center, Durham, North Carolina, USA

Vitamins and Hormones, Volume 87
ISSN 0083-6729, DOI: 10.1016/B978-0-12-386015-6.00030-5

I. Introduction

The ability of stem cell neurons in the central nervous system of mammals to divide is not entirely lost with aging. It has recently been shown in humans that stem cell neuronal populations remain active in the hippocampus, suggesting that memory and behavior may remain plastic throughout adulthood and responsive to external stressors such as drugs (Schoenfeld and Gould, 2011). Mood disorders are prevalent in the general adult population and Lithium, a commonly used drug to treat depression, appears to enhance adult hippocampus neurogenesis by preventing GSK3β activity upregulation mediated by nuclear glucocorticoid receptors (GRs) (Boku *et al.*, 2010). The ability of glucocorticoids to inhibit neurogenesis has become a clinical concern in children where negative effects on brain development are apparent in infants exposed to dexamethasone (Heine and Rowitch, 2009), and the chronic use of steroids has increased in the pediatric population for diseases like asthma. In general, children exposed to glucocorticoids over extended periods have an increased association with cognitive deficits and a reduction in brain size (Damsted *et al.*, 2011).

Seven transmembrane receptors on stem cells regulate many of the growth and patterning pathways for organogenesis and tissue repair. For example, Frizzled receptors signal through GSK3β in the gastrointestinal tract to control levels of beta-catenin, whose elevation is associated with tumors (Doucas *et al.*, 2005). Similarly, abnormally elevated signaling of the seven transmembrane receptor Smoothened (Smo), a hedgehog pathway regulator normally suppressed by Patched, underlies development of medulloblastoma in the cerebellum (Corcoran and Scott, 2001). While endogenous Wnt agonists have been identified for Frizzled, no endogenous agonist has been discovered for Smo. A limited number of synthetic Smo agonists, however, have been discovered by screening compound libraries and testing their activity in the presence of the Smo antagonist cyclopamine (Chen *et al.*, 2002). For our program in stem cell and regenerative medicine, we developed a similar strategy to identify Smo ligands using a novel β-arrestin-GFP translocation assay (Wang *et al.*, 2010), but rather than screen combinatorial libraries we have concentrated our efforts on FDA approved drugs. In one such screen we surprisingly identified a restricted series of glucocorticoids with Smo activity (Wang *et al.*, 2010). Moreover, these select compounds, in contrast to glucocorticoids in general, stimulate rather than inhibit the growth of cerebellar neuronal progenitor cells.

II. Select Glucocorticoids as Smoothened Agonists: Potential Effects for Neurogenesis

The hedgehog signaling pathway, mediated by Smo, is critical for embryonic patterning, stem cell growth, and central nervous system development (Ingham and McMahon, 2001; Marti and Bovolenta, 2002; Palma *et al.*, 2005; Ruiz i Altaba *et al.*, 2002). Therapeutic activation of hedgehog signaling by Smo agonists has been proposed for restoring tissue function in the peripheral and central nervous systems for retinal injury, Parkinson's disease, spinal cord damage, and neuronal degeneration (Bambakidis *et al.*, 2009; Li *et al.*, 2008; Tsuboi and Shults, 2002; Wan *et al.*, 2007). The cerebellum contains one of the most abundant populations of neurons in the brain, derived from a stem cell population of cerebellar granular cell precursors (GCPs), and the *in vivo* expansion of these granule precursor cells is dependent on hedgehog/Smo pathway signaling (Wechsler-Reya and Scott, 1999). For the mouse cerebellum in the first postnatal week, purkinje cell secreted sonic hedgehog (Shh) ligand stimulates GCP proliferation in the outer layer of the cerebellar surface (external granular layer). GCPs then exit the proliferative phase of the cell cycle, differentiate into granule neurons, and migrate inward into the internal granule layer (Fig. 10.1).

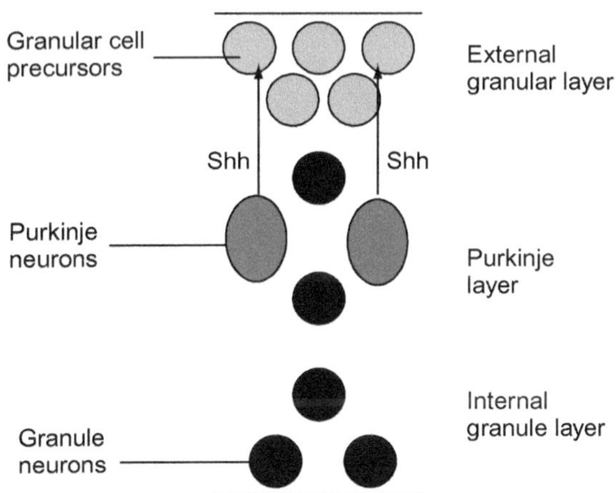

Figure 10.1 Primary neuronal granule cell precursor (GCP) proliferation in the cerebellum is hedgehog signaling dependent.

Using a high-throughput screen of cells containing Smo and β-arrestin-GFP reporter that binds activated Smo, we identified four glucocorticoids—halcinonide, fluticasone propionate, clobetasol propionate, and fluocinonide as Smo agonists (Wang *et al.*, 2010). Similar to some other known Smo agonists, SAG and purmorphamine, these glucocorticoids bind Smo, promote Smo internalization, and activate Gli. Consistent with Smo agonist activity, this select set of glucocorticoids stimulate the proliferation of GCPs alone and synergistically, and such stimulation was not affected by the glucocorticoid nuclear receptor antagonist Mifepristone (RU-486). In contrast, the GR agonist dexamethasone *inhibited* Shh-activated GCP proliferation. Additionally, GCP treatment with Shh, Purmorphamine, SAG, halcinonide, clobetasol propionate, and fluticasone propionate (but not the weak Smo agonist fluocinonide) inhibit Caspase-3 degradation and block GCP apoptosis by increasing endogenous cyclin D2 protein expression. No such growth enabling responses were observed in GCPs treated with cortisone, dexamethasone, prednisolone, and corticosterone.

As expected the four glucocorticoid Smo agonists and dexamethasone had similar effects in activating GR in a GR-GFP nuclear translocation assay, yet dexamethasone produced an opposite response to the glucocorticoid Smo agonists on GCP proliferation and apoptosis. Taken together these data suggest that the GCP proliferative response is due directly to activation of Smo and is independent of glucocorticoid nuclear receptor signaling. The ability of some glucocorticoid drugs, particularly FDA approved compounds, to activate hedgehog signaling has profound implications for pediatric therapeutics, stress-related disorders, and regenerative medicine. These drugs have well-known safety profiles and are well characterized, so that their application in new therapeutic areas can be evaluated much more expeditiously that unapproved compounds isolated in screens of compound libraries.

III. Mechanism of Action

Glucocorticoid steroids are a standard of care in many clinical settings including rheumatic disorders, asthma, inflammation, and cancer (Shimmer and Parker, 2006). An excellent summary of the medicinal chemistry of steroid hormones and therapeutically related compounds and a review of the X-ray crystal structures of steroidal compounds provide keen insights into the medicinal chemistry of steroidal glucocorticoids (Duax *et al.*, 1988; Proteau, 2004). Glucocorticoid steroids bind the GR, a member of the nuclear receptor superfamily (Fig. 10.2). Upon binding to GR, the cytoplasmic form of the receptor moves to the nucleus and activates the transcription and repression of multiple gene products involved in

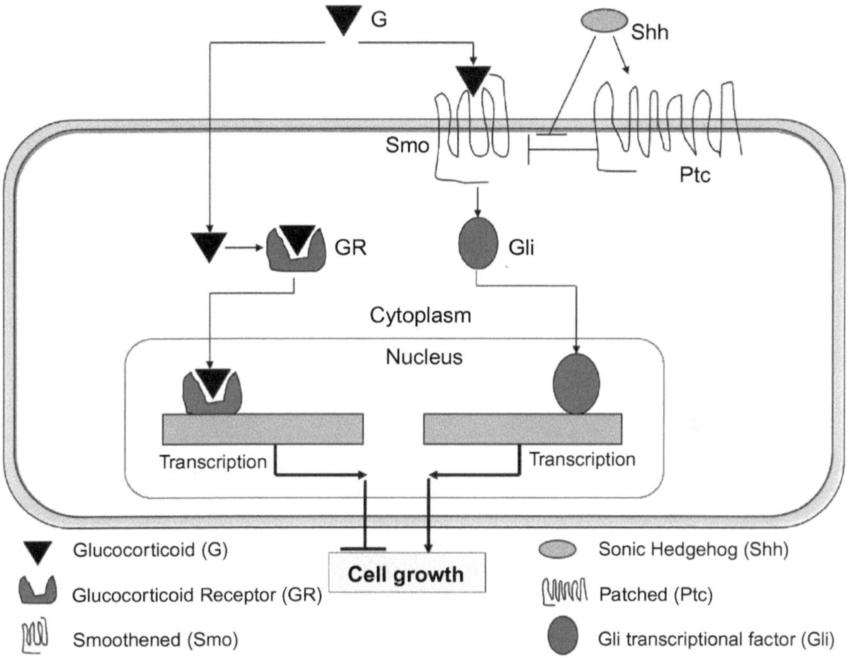

Figure 10.2 Conceptual model outlining cell signaling of select glucocorticoids through both the glucocorticoid and hedgehog/Smo pathway.

metabolism and inflammation. In particular, glucocorticoids such as dexamethasone acting through the GR receptor inhibit cell growth in GCPs. Independent of GR signaling, select glucocorticoids such as halcinonide bind to the Smo receptor, activate hedgehog signaling, and stimulate cell growth in GCPs (Fig. 10.2). While these select glucocorticoids can be both GR agonists and Smo agonists, the Smo agonist effects appear to be the more dominant in GCPs.

IV. Structure–Activity Relationships (SAR) of Glucocorticoid Smoothened Agonists

Subsequent screening of focused chemical libraries in the β-arrestin2-GFP translocation assay has increased the number of glucocorticoid derivatives identified with Smo agonist activity (Table 10.1). These efforts have helped define key SAR for hedgehog agonist activity while also identifying a number of structurally related glucocorticoid derivatives without Smo agonist activity (Table 10.2).

Table 10.1 Summary of glucocorticoid derivatives with Smo agonist activity in the β-arrestin2-GFP translocation assay

Compound	EC_{50} in β-arrestin2-GFP translocation assay (μM)
Fluticasone propionate[1]	0.099
Halcinonide[1]	1.1
Clobetasol propionate[1]	1.5
Fluocinonide[1]	>5
Fluticasone	0.1
Triamcinolone acetonide	0.4
Halobetasol propionate	0.8
Fluocinolone acetonide	1.5
Amcinonide	3

[1] (Wang *et al.*, 2010)

The most dominant trends noted for Smo agonist activity to date are the nature of the substituents at C-17, a hydroxyl group in the beta configuration at C-11, and a fluorine atom on the alpha face at C-9. Of these trends, a review of the inactive compounds suggests the most dominate feature for Smo agonist activity may be the nature of the substitution at the C-17 position in conjunction with a hydroxyl group in the beta position at C-11. Though each of the active Smo agonists contains a hydroxyl group in the beta configuration at C-11, there are a number of glucocorticoid derivatives with an 11-beta hydroxyl group that are inactive Smo agonists. These inactive Smo agonists often possess a C-17 substituent that is more hydrophilic, less branched and generally smaller. Some of these compounds also have a fluorine atom in the C-9 alpha position. Based on these observations, a working model was created to guide SAR exploration and to aid our understanding of the structural features of glucocorticoid derivatives that drive potency and activation of the hedgehog pathway (Fig. 10.3).

V. Conclusion

Although glucocorticoids for the most part are known to inhibit neurogenesis as nuclear receptor agonists, we identified a subgroup of them that bind Smo and are capable of stimulating neuron precursor cell proliferation. This unusual finding resulted from a high-throughput drug screening approach, was confirmed by secondary cell based assays, and formed the basis for a program of SAR analysis to characterize the chemical determinants underlying the binding. This subgroup of steroids presents us

Table 10.2 Smo agonist activity of representative glucocorticoid derivatives in the β-arrestin2-GFP translocation assay

Non cyclic series (NC)	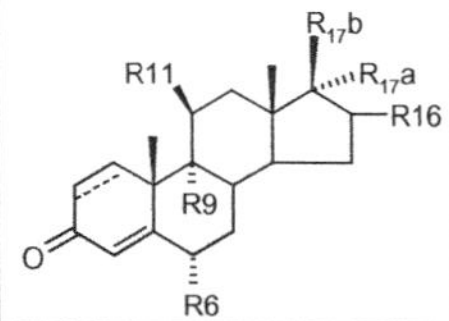					Cyclic series(C)	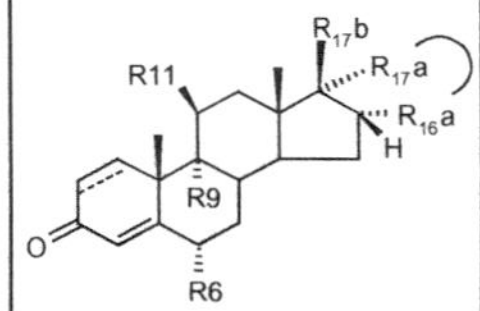		Activity in β-arrestin translocation assay at 12.5 uM active (A), inactive (IA)
Series	**Name**	**$D_{1,2}$**	**R6**	**R9**	**R11**	**R16**	**R17α**	**R17 β**	
NC	Fluticasone propionate	y	F	F	HO-b	CH_3-a	O–CO–CH_2–CH_3	CO–S–CH_2–F	**A**
C	Halcinonide	n	H	F	HO-b	–O–C$(CH_3)_2$–O–	Cyclized with R16	CO–CH_2–Cl	**A**
NC	Clobetasol propionate	y	H	F	HO-b	CH_3–b	O–CO–CH_2–CH_3	CO–CH_2–Cl	**A**
C	Fluocinonide	y	F	F	HO-b	–O–C$(CH_3)_2$–O–	Cyclized with R16	CO–CH_2–OCO–CH_3	**A**
NC	Fluticasone	y	F	F	HO-b	CH_3–a	OH	CO–S–CH_2–F	**A**
NC	Halobetasol propionate	y	F	F	HO-b	CH_3–	O–CO–CH_2–CH_3	CO–CH_2–Cl	**A**
C	Amcinonide	y	H	F	HO-b	–O–C(C_5H_{11})–O–	Cyclized with R16	CO–CH_2–O–CO–CH_3	**A**
C	Fluocinolone acetonide	y	F	F	HO-b	–O–C$(CH_3)_2$–O	Cyclized with R16	CO–CH_2–OH	**A**
C	Triamcinolone acetonide	y	H	F	HO-b	–O–C$(CH_3)_2$–O–	Cyclized with R16	CO–CH_2–OH	**A**
NC	Dexamethasone	y	H	F	HO-b	CH_3–a	OH	CO–CH_2–OH	IA
NC	Hydrocortisone-acetate	n	H	H	HO-b	H	OH	CO–CH_2–O–CO–CH_3	IA
NC	Dexamethasone 21 acetate	y	H	F	HO-b	CH_3–a	OH	CO–CH_2–O–CO–CH_3	IA
NC	Prednisolone	y	H	H	HO-b	H	OH	CO–CH_2–OH	IA
NC	Cortisone	n	H	H	O=	H	OH	CO–CH_2–OH	IA
NC	Fluticasone furoate	y	F	F	HO-b	CH_3–a	O-CO-(2-Furyl)	CO–S–CH_2–F	IA
NC	Flumethasone	y	F	F	HO-b	CH_3–a	OH	CO–CH_2–OH	IA
NC	Betamethasone 21 acetate	y	H	F	HO-b	CH_3–b	OH	CO–CH_2–O–CO–CH_3	IA
NC	Betamethasone	y	H	F	HO-b	CH_3–b	OH	CO–CH_2–OH	IA

$D_{1,2}$ (double bond between C1 and C2); y, yes; n, no.
Stereochemistry: a, alpha face; b, beta face of steroid

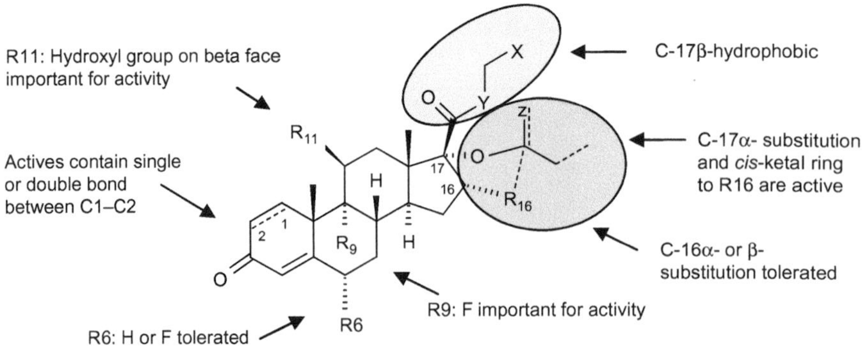

Figure 10.3 Working model of glucocorticoid Smo agonist activity. (For color version of this figure, the reader is referred to the Web version of this chapter.)

with an unprecedented opportunity to jump start clinical trials assessing the therapeutic benefits of potentiating hedgehog signaling, because these same glucocorticoid drugs have been used clinically by tens of millions and their safety profiles are well-known. Areas in which they may provide immediate benefit and could be tested include Parkinson's disease, spinal cord injury, myocardial repair post infarction, and wound healing.

REFERENCES

Bambakidis, N. C., Horn, E. M., Nakaji, P., Theodore, N., Bless, E., Dellovade, T., Ma, C., Wang, X., Preul, M. C., Coons, S. W., *et al.* (2009). Endogenous stem cell proliferation induced by intravenous hedgehog agonist administration after contusion in the adult rat spinal cord. *J. Neurosurg. Spine* **10,** 171–176.

Boku, S., Nakagawa, S., and Koyama, T. (2010). Glucocorticoids and lithium in adult hippocampal neurogenesis. *Vitam. Horm.* **82,** 421–431.

Chen, J. K., Taipale, J., Young, K. E., Maiti, T., and Beachy, P. A. (2002). Small molecule modulation of Smoothened activity. *Proc. Natl. Acad. Sci. USA* **99,** 14071–14076.

Corcoran, R. B., and Scott, M. P. (2001). A mouse model for medulloblastoma and basal cell nevus syndrome. *J. Neurooncol* **53,** 307–318.

Damsted, S. K., Born, A. P., Paulson, O. B., and Uldall, P. (2011). Exogenous glucocorticoids and adverse cerebral effects in children. *Eur. J. Paediatr. Neurol.* doi: 10.1016/j.ejpn.2011.05.002.

Doucas, H., Garcea, G., Neal, C. P., Manson, M. M., and Berry, D. P. (2005). Changes in the Wnt signalling pathway in gastrointestinal cancers and their prognostic significance. *Eur. J. Cancer* **41,** 365–379.

Duax, W. L., Griffin, J. F., Weeks, C. M., and Wawrzak, Z. (1988). The mechanism of action of steroid antagonists: Insights from crystallographic studies. *J. Steroid Biochem.* **31,** 481–492.

Heine, V. M., and Rowitch, D. H. (2009). Hedgehog signaling has a protective effect in glucocorticoid-induced mouse neonatal brain injury through an 11betaHSD2-dependent mechanism. *J. Clin. Invest.* **119,** 267–277.

Ingham, P. W., and McMahon, A. P. (2001). Hedgehog signaling in animal development: Paradigms and principles. *Genes Dev.* **15,** 3059–3087.

Li, X. J., Hu, B. Y., Jones, S. A., Zhang, Y. S., Lavaute, T., Du, Z. W., and Zhang, S. C. (2008). Directed differentiation of ventral spinal progenitors and motor neurons from human embryonic stem cells by small molecules. *Stem Cells* **26,** 886–893.

Marti, E., and Bovolenta, P. (2002). Sonic hedgehog in CNS development: One signal, multiple outputs. *Trends Neurosci.* **25,** 89–96.

Palma, V., Lim, D. A., Dahmane, N., Sanchez, P., Brionne, T. C., Herzberg, C. D., Gitton, Y., Carleton, A., Alvarez-Buylla, A., and Ruiz i Altaba, A. (2005). Sonic hedgehog controls stem cell behavior in the postnatal and adult brain. *Development* **132,** 335–344.

Proteau, P. J. (2004). Steroid hormones and therapeutically related compounds. *In* "Organic Medicinal Chemistry and Pharmaceutical Chemistry," (J. H. Block and M. John, Eds.), p. 767. Lippincott Williams and Wilkins, Baltimore.

Ruiz i Altaba, A., Palma, V., and Dahmane, N. (2002). Hedgehog-Gli signalling and the growth of the brain. *Nat. Rev. Neurosci.* **3,** 24–33.

Schoenfeld, T. J., and Gould, E. (2011). Stress, stress hormones, and adult neurogenesis. *Exp. Neurol.* doi: 10.1016/j.expneurol.2011.01.008.

Shimmer, B. P., and Parker, K. L. (2006). Adrenocorticotropic hormone; adrenocortical steroids and their synthetic analogs; Inhibitors of the synthesis and actions of adrenocorticoid hormones. *In* "Goodman and Gilman's: The Pharmacologic Basis of Therapeutics," (L. L. Brunton, J. S. Lazo, and K. L. Parker, Eds.), 11th edn., pp. 1587–1612. McGraw-Hill, New York.

Tsuboi, K., and Shults, C. W. (2002). Intrastriatal injection of sonic hedgehog reduces behavioral impairment in a rat model of Parkinson's disease. *Exp. Neurol.* **173,** 95–104.

Wan, J., Zheng, H., Xiao, H. L., She, Z. J., and Zhou, G. M. (2007). Sonic hedgehog promotes stem-cell potential of Muller glia in the mammalian retina. *Biochem. Biophys. Res. Commun.* **363,** 347–354.

Wang, J., Lu, J., Bond, M. C., Chen, M., Ren, X. R., Lyerly, H. K., Barak, L. S., and Chen, W. (2010). Identification of select glucocorticoids as Smoothened agonists: Potential utility for regenerative medicine. *Proc. Natl. Acad. Sci. USA* **107,** 9323–9328.

Wechsler-Reya, R. J., and Scott, M. P. (1999). Control of neuronal precursor proliferation in the cerebellum by Sonic Hedgehog. *Neuron* **22,** 103–114.

CHAPTER ELEVEN

Effect of Progesterone on Human Mesenchymal Stem Cells

Dobroslav S. Kyurkchiev,* Ekaterina Ivanova-Todorova,* *and* Stanimir D. Kyurkchiev†

Contents

I. Introduction 218
II. Biological Roles of Progesterone 218
  A. Progesterone in reproduction 218
  B. Progesterone in extra-reproductive tissues 219
  C. Receptors for progesterone 221
III. Mesenchymal Stem Cells 222
IV. Multipotent MSCs in Human Endometrium 224
V. Interaction Between Progesterone and MSCs 226
  A. Induction of HLA-G expression by progesterone 226
  B. Induction of progesterone-induced blocking factor (PIBF) expression 229
VI. Conclusions and Future Directions 231
Acknowledgments 231
References 231

Abstract

Progesterone is considered to be a major reproductive steroid hormone which supports the successful development of pregnancy. One of the basic targets of progesterone are stromal cells in endometrium which are triggered to undergo decidualization in preparation to accept the embryo. However, the endometrial stroma consists of different subpopulations of cells with varying characteristics and functions as one of the subpopulations are the endometrial mesenchymal stem cells (MSCs) which seem to be located both in the basal layer and in the functional layer of the endometrium. In all cases, these cells have the features of typical MSCs such as adherence and differentiation in multiple cell lineages. The endometrial MSCs are stimulated by progesterone to increase the expression and secretion of immunomodulatory proteins such as HLA-G and PIBF.

* Laboratory of Clinical Immunology, University Hospital "St. I.Rislki", Medical University Sofia, Sofia, Bulgaria
† Institute for Reproductive Health, Ob/Gyn Hospital "Dr. Shterev", Sofia, Bulgaria

Vitamins and Hormones, Volume 87
ISSN 0083-6729, DOI: 10.1016/B978-0-12-386015-6.00040-8

I. Introduction

Reproduction of mammalian animal species is a complex phenomenon which processes as a dynamic interplay between both molecular and cellular components from the mother and the fetus. Every stage of these interactions is governed by endocrine, immune, neurological, psychological, etc., factors which form a very complicated network. It seems quite impossible to cover all the richness of biologically effective players such as hormones, growth factors, prostaglandins, cytokines, endometrial cells, decidualization process, trophoblast cells, etc., as each of them is actively involved in development of a successful pregnancy.

However, progesterone has been considered as "the sentinel of pregnancy" since long ago and this popular saying truly reflects the importance of this steroid hormone with effect on numerous target cell populations such as lymphocytes, dendritic cells, endometrial cells, endothelial cells, etc. Recently it has been pointed out that progesterone actively modulates the functions of mesenchymal stem cells (MSCs) and particularly those cells located in human reproductive tract which are subjected to higher local concentrations of this hormone (Kyurkchiev *et al.*, 2010). Since the MSCs are involved in a number of different cellular interactions, it is of utmost interest to review published data about the effect of progesterone on these cells.

II. Biological Roles of Progesterone

A. Progesterone in reproduction

Progesterone is generally considered to be a reproduction-restricted hormone as the main targeted cells are those of ovary and endometrium. The main source of progesterone are the granulose cells and particularly luteinized granulosa cells forming corpus luteum in the postovulation luteal phase of the ovarian cycle. In case of developing pregnancy, the syncytiotrophoblast cells start secreting progesterone in large amount and after the ninth gestation week the placenta takes up the production of sufficient quantities of progesterone. The major target of progesterone action is the uterine endometrium, but in the early 1980s, it has been suggested that progesterone might have intraovarian effect although progesterone receptors (PRs) are not detected on granulosa cells (Rothchild, 1996). This concept was later developed by the assumption that progesterone actions in ovary might be mediated via PR located at the plasma membrane, membrane progestin receptor and a membrane receptor complex consisting of serpine 1 mRNA binding protein and PR membrane component 1.

Thus three signaling pathways might cause activation of granulosa cells resulting in different reactions in the ovary which have been extensively reviewed by Peluso (2006). Experimental studies both *in vivo* and *in vitro* have proved that progesterone inhibits the development of follicles in the ovary as there is a negative correlation between the plasma concentrations of progesterone and the growth of ovarian follicles (Hirshfield, 1984; Setty and Mills, 1987). In general, progesterone is considered to be involved in the control of follicular growth, ovulation, and luteinization. The inhibitory effect of progesterone on the proliferation of granulosa cells is not controlled by gonadotropin levels, and this strongly suggests that progesterone is directly acting on the granulosa cells (Peluso, 2006).

Progesterone is involved in preparation of the endometrium for possible implantation in a very specific process called decidualization which leads to the formation of decidua which is still a subject of some dispute. Most of the authors accept the concept that the decidua is a specifically transformed endometrium as the endometrial fibroblast-like stromal cells turn into epitheloid-like actively secreting decidual cells (Dunn *et al.*, 2003). This transformation takes place during the midluteal phase of the menstrual cycle when progesterone triggers the process in predominantly perivascularly located endometrial stromal cells, so in fact, the decidualization starts around the blood vessels. It is important to note this specific localization of the starting points of decidualization because they seem to coincide with the predilective localizations of the multipotent endometrial stromal cells (MESCs) as reported (Gargett, 2007). In humans, this transformation is unconditionally dependent on the presence of progesterone as reported by a number of research groups. However, the stimulating effect of progesterone seems to be complexed to the effects of some other growth factors and their corresponding receptors (Brosens *et al.*, 1999; Irwin *et al.*, 1989, 1991; Lockwood *et al.*, 2000; Piva *et al.*, 1996). It should be stressed out that in all these experiments the essential role of progesterone is clearly demonstrated. Thus it was shown that epidermal growth factor, basic fibroblast growth factor, insulin-like growth factor-I and II, or platelet-derived growth factor can stimulate the proliferation of human endometrial stromal cells only in the presence of progesterone in the culture (Irwin *et al.*, 1991).

B. Progesterone in extra-reproductive tissues

Mammary gland is another target organ for the effect of progesterone. It is well known that then mammary gland periodically undergoes significant changes during the menstrual cycle as the proliferation of the breast epithelia is with highest rate during the luteal phase. There is a well established correlation between the proliferation rate of the mammary gland epithelial cells and the progesterone levels. An increase of the DNA synthesis and mitoses in the epithelia of mammary gland are recorded during the luteal phase of the menstrual cycle which reaches a peak at the end of this phase and the switch-on apoptosis activity. It has been pointed out that the values for proliferative indices for the epithelial cells in breasts coincide with the levels of circulating progesterone in blood plasma (Navarrete *et al.*, 2005). However, it has been shown that a small portion of the epithelial cells express effective amounts of classical PRs, and these cells seem to have the basic characteristics of a "side population" as they are nondividing long-living population. The stimulating effect of progesterone is most probably achieved indirectly via triggering the expression of other growth factors (Ismail *et al.*, 2003; Rosen, 2003).

The list of extra-reproductive tissues which are targets for progesterone actions includes cardiovascular system, bones, respiratory systems, thymus, kidney, and adipose tissue (Graham and Clarke, 1997; Losel *et al.*, 2003). However, recently results from both experimental and clinical studies have shown that progesterone has various effects on the cells of the nervous system. Studies on the effect of progesterone applied for the purposes of hormone replacement therapy have shown that this hormone had promising neuroprotective and promyelinating effects. It is particularly interesting that in some cases age-dependent dysfunctions seemed to be influenced positively by progesterone (Schumacher *et al.*, 2007). Data from animal experiments show that progesterone should be considered as a neurosteroid which influences the neuronal differentiation in the fetus. Further on, experimental data suggest that progesterone seems to rebuild or at least protect the blood–brain barrier and reduce cerebral edema most probably by downregulating the process of inflammation taking place parallel to the brain injury (Singh, 2006). Recently its neuroprotective role is discovered in the studies of a number of research groups, and recent clinical trials indicated that progesterone treatment improves neurologic outcome after traumatic brain injury in adults (Wright *et al.*, 2007; Xiao *et al.*, 2008).

It is of particular interest the question about the effect of progesterone on the cells of the immune system because PRs have been shown to be expressed in peripheral blood monocytes, peripheral NK and T cells from pregnant women (Szekeres-Bartho *et al.*, 2001a). NK cells are of special interest because of their abundance in human endometrium during early pregnancy and because they differ in their phenotype in comparison to the NK cells in peripheral blood. Uterine NK cells (uNK cells) are positive for CD56 marker but negative for CD16, and the uNK cells do not have cytolytic activity (Dosiou and Guidice, 2001). The number of uNK cells is increased in the late luteal phase of the menstrual cycle which would suggest that a hormonal control is operational in that case. However, receptors for progesterone have not been detected neither at protein level nor at mRNA levels (Henderson *et al.*, 2003) and this would leave open the questions about the molecules involved in the hormonal control of the recruitment of uNK cells. It should be pointed out that progesterone exerts some modulating effect on the cytolytic activity of peripheral NK cells (Hansen *et al.*, 1992) and upregulates the expression of immunomodulating proteins by different cell types (to be discussed below).

C. Receptors for progesterone

Progesterone is a small hydrophobic molecule with steroid structure. Generally, it diffuses freely through the plasma membrane of all cells, however, in target cells progesterone binds to its specific receptors located in the cytoplasm. It is known that there are two protein isoforms of specific receptors (PR-A and PR-B) which are products of one gene (Li and O'Malley, 2003). Biochemically the difference between the two isoforms is that PR-A lacks a sequence about 42 amino acids at the N-terminus of the molecule. Recently, a third isoform of PR, PR-C, has been described which is a truncated molecule reportedly participating to the onset of labor (Condon *et al.*, 2006).

Experiments using the knockout approach in mice revealed that PR-A is both necessary and sufficient to trigger the hormone-dependent cell and tissue responses leading to successful pregnancy. Contrary to that, the PR-B isoform is required for the responses of other target organs such as mammary gland (Mulac-Jericevic *et al.*, 2000). Progesterone forms with its receptor a complex which is transported to the cell nucleus where the complex binds to progesterone responsive element, that is, specific DNA sequence in the promoter regions of some genes. The binding of the complex turns on/off the targeted gene, thus the complex progesterone/receptor acts a specific transcription factor. Changes of cell physiology resulting as a consequence of the activation of this transcription factors are termed genomic actions of the progesterone.

However, the actions of progesterone on target cells is not limited to genomic actions only but rapid responses taking place in seconds or minutes to progesterone stimulus have been described as well and termed "non-classic," "nongenomic," "extranuclear" (Falkenstein *et al.*, 1998; Losel *et al.*, 2003; Norman *et al.*, 2004). Different authors have proposed different criteria to discriminate genomic and nongenomic effects of progesterone such as: reactions detected minutes after stimulation of cells with progesterone and the best example of which is the acrosome reaction of human sperm (Meizel *et al.*, 1997). Moreover, the nongenomic actions of the progesterone are not blocked by antagonists of the classical PR as mifeprestone. These rapid actions are considered to be related to the expression of membrane PR as the progesterone receptor membrane component 1 (PRMC1) and the members of the progestin and adipoQ receptors family (PAQRs) are the most promising candidates (Gellersen *et al.*, 2009). PRMC1 is considered to be a single transmembrane protein with molecular mass ≈28 and ≈56 kDa which means to be expressed by variety of cell types. Depending on the tissue where it is expressed, PRMC1 can bind different ligands leading to different signaling pathways (Losel *et al.*, 2008).

It is obvious that a lot of research should be done before it would be possible to define precisely the cells expressing molecules which might be receptors for progesterone and the resulting reactions after the interaction between the receptor and the ligand which is progesterone. That makes it quite important to identify new populations of cells which might be targets of this multifaceted hormone.

III. Mesenchymal Stem Cells

By definition, the MSCs are multipotent stem cells present in both newborn and adult humans that can differentiate *in vitro* or *in vivo* into a variety of cell types such as osteoblasts, chondrocytes, adipocytes (Friedstein *et al.*, 1968; Owen, 1988). MSCs have been successfully isolated and characterized from a number of different human organs and/or tissues (bone marrow, adipose tissue, dental pulp, muscles, umbilical cord, amniotic fluid, etc.). MSCs are present in virtually every organ and/or tissue in the body in a dormant state and ready to start differentiating if needed, according to the opinion of some authors (Meirelles *et al.*, 2006; Minguelli *et al.*, 2001; Zuk *et al.*, 2002).

It is difficult to identify MSCs *in vivo* because it has been estimated that MSCs can be detected in bone marrow at a frequency between 1/10,000 and 1/100,000 bone marrow mononuclear cells (Castro-Malaspina *et al.*, 1980). So, that it is quite obvious that any studies on MSCs *in vivo* would be very challenging because there is no specific marker which can be used to identify these cells, and even some authors speculate certain proteins can be used as specific markers *in vivo*. A number of investigators have published that one or another marker can be used for isolation, but so far not a single one has been firmly established as useful (Simmons and Torok-Storb, 1991; Welm *et al.*, 2002; Xin *et al.*, 2005).

It has been proposed that Stro-1 antibody can be used as a marker for identification *in vivo* of MSCs in bone marrow (Simmons and Torok-Storb, 1991). Further studies have shown that osteoprogenitor cells in bone marrow are reacting positively with Stro-1 antibody and this marker can be used for their selection and enrichment prior to *in vitro* culture (Gronthos *et al.*, 1994). However, it is known that the surface marker recognized by the Stro-1 seems to be expressed at different cell populations in the bone marrow so that it is not very convincing that this approach could be useful.

Recently, it was reported that CD271 (low-affinity nerve growth factor receptor) is uniformly highly expressed by MSCs in bone marrow and lipoaspirate samples. A marker for human fibroblasts (D7-FIB) was used to select positively cells from human bone marrow by immunomagnetic procedure. Selected cells were characterized for their clonogenicity, immunophenotype, and capacity to differentiate as osteogenic, adipogenic, and chondrogenic lineages (Jones *et al.*, 2002). This finding has been supported by independent studies in other laboratories (Quirici *et al.*, 2002). CD271 is considered to be a promising candidate which is already commercially available for magnetic sorting of MSCs from bone marrow (Jones *et al.*, 2002).

Recently, it was reported that SSEA-4 (stage-specific embryonic antigen-4) which basically was used as a marker for human embryonic stem cells and the blastocyst stage of embryos could be used for identification of MSCs *in vivo*. The authors succeeded to isolate positive SSEA-4 cells from several samples of human bone marrow and these cells could adhere to solid phase *in vitro*, formed clones and could be differentiated as osteogenic, adipogenic, and chondrogenic lineages (Gang *et al.*, 2007).

Prospective isolation of homogeneous population of MSCs is a very attractive possibility which would enhance their further application in clinical practice but there is not a single, strictly specific surface marker which can be widely applied to this end. However, these stringent criteria would raise the question whether the MSCs populations are homogeneous *in vivo* by themselves.

Numerous studies have shown that cells isolated from different sources have both common features (clonogenicity, differentiation capacity, etc.) which would allow to be unitedly termed MSCs and differences in their activities. It has been shown that bone marrow derived MSCs at the initial stages of culture are morphologically heterogeneous. There is a population consisting of small spindle-like rapidly dividing cells and another type of cells which are flat polygonal slowly dividing cells (Colter *et al.*, 2001; Reyes *et al.*, 2002). Cells isolated from different sources and manifesting the basic common features of MSCs would have some specific characteristics too. Dental tissue derived stem cells seem to imitate the general behavior of the bone marrow derived MSCs but, however, the potency of their differentiation is rather different from the bone marrow derived MSCs (Huang *et al.*, 2009). Just to mention that MSCs isolated from adipose tissue have a stronger immunosuppressive activity as compared to bone marrow derived MSCs (Bochev *et al.*, 2008; Ivanova-Todorova *et al.*, 2009a,b,c). What is even more remarkable is that cells isolated from one niche represent a heterogeneous population. Heterogeneity at single-cell level was analyzed using cells isolated from human synovium. Among the 50 single-cell derived clones, there was great variety in the basic characteristics such as population doubling time, mean telomere lengths, chondrogenic and osteogenic differentiation, and no more than 30% of the cells could be differentiated as adipogenic (Karystinou *et al.*, 2009).

IV. Multipotent MSCs in Human Endometrium

It is known that the "niche" of the stem cells has utmost importance for their fate and activities so that the presence of stem cells in human endometrium is of particular interest because the cells are subjected to powerful hormonal control.

Human endometrium undergoes significant, both qualitative and quantitative, changes during the menstrual cycle of women in reproductive age. These changes are taking place under strict hormonal control and are due to both cell proliferation and differentiation. The ultimate aim of these processes is to prepare the uterus for the implantation and successful development of pregnancy.

Human endometrium consists of luminal and glandular epithelial cells, stromal fibroblasts, vascular smooth muscle cells, and endothelial cells forming basal and functional layer. A number of different lymphocytes and predominantly NK cells are recruited to the endometrium during the postovulatory phase of the cycle. The glandular epithelial layer is formed by epithelial cells, which are considered highly differentiated cells. As early as 1978, Prianishnikov proposed a model based on the hypothesis that endometrial epithelium comprises of populations of stem cells, capable of long-term self maintenance and producing differentiated epithelial cells, which can be divided according to the model, into three types: (i) oestradiol-sensitive cells, (ii) oestradiol- and progesterone-sensitive cells, and (iii) progesterone-sensitive cells (Prianishnikov, 1978).

Stromal cell component of human endometrium is an object of some discussions as one stream of discussion concerns the origin of the endometrial stromal cells. It has been reported that cells from donor origin (HLA-mismatched transplantation) have been detected in both epithelial glands and the stromal cells of endometrium of the recipient. A high level of chimerism, ranging from 0.2% up to 52%, was detected using RT-PCR technique and immunochemistry methods (Taylor, 2004). Animal experiments of the same group have demonstrated that cells derived from the bone marrow of male donors could be found in the uterine endometrium of female animals (Du and Taylor, 2007). The plausible explanation of these facts is the speculation that precursor cells are present in the bone marrow, which when recruited to the uterine microenvironment under the influence of the hormones present locally, can be transformed into stromal cells participating in physiological processes such as endometrium regeneration.

Other topics of discussion arise from the reports that some populations of stromal cells isolated from human decidua have the phenotype and the functions of antigen-presenting cells locally (Olivares *et al.*, 1997). Still, another unique population of endometrial stromal cells is the population of mesenchymal stem-like cells as reported by a number of research groups (Dimitrov *et al.*, 2008, 2010; Kato *et al.*, 2007; Schwab and Gargett, 2007; Schwab *et al.*, 2005, 2008). CD146 and platelet-derived growth factor receptor b (PDGF-Rb) are specific surface markers of the putative endometrial stromal stem cells. Immunocytochemical experiments demonstrated that specific membrane staining for CD146 is observed not only on endothelial cells of the arterioles but on perivascular stromal cells as well as PDGF-Rb positive cells localized perivascularly. These markers have been used in flow cytometry to isolate CD146+ PDFG-Rb+ endometrial stromal cells and test their potential to differentiate (Schwab and Gargett, 2007; Schwab *et al.*, 2008). Clonogenicity is considered to be a sign of the pluripotent character of a given cell population. The presence of small clonogenic subpopulations in single-cell suspensions isolated from the basal layer of human endometrium have been reported (Chan *et al.*, 2004). Using cell-cloning methods, it was shown that 0.22% of the epithelial and 1.25% of the stromal cells can form single colonies when seeded at concentrations of 500 cells/cm^2 for epithelial cells or 300 cells/cm^2 for stromal cells. A significantly higher cloning efficiency was recorded for the stromal cells (1.25 $\pm$ 0.18%) as compared with the epithelial cells (0.22 $\pm$ 0.07%). The most important feature of the stem cells generally is their potential for differentiation in different cell lineages. It has been shown that endometrial mesenchymal stem cells (EnMSCs) could be differentiated as osteogenic, adipogenic, myogenic, and chondrogenic cells (Schwab and Gargett, 2007). Further on, stromal cells isolated from early human deciduas could be induced to differentiate into endothelial-like cells, to enter the process of decidualization, and to secrete prolactin (Dimitrov *et al.*, 2010).

Still another discussion is going on and it's centered to the possible localization of the MSCs in endometrium. The first reports detected the EnMSCs in the basal layer and it's assumed that they reside there for long periods of time (Chan *et al.*, 2004), but more recent studies successfully isolated multipotent EnMSCs from menstrual blood (Meng *et al.*, 2007; Patel *et al.*, 2008) or from routinely collected endometrial biopsies (Schüring *et al.*, 2011). So, it is obvious that the stromal cells in endometrium are quite heterogenic and even more; we have observed morphological difference among endometrial stromal cells isolated from different sites of the uterine endometrium under hysteroscopic control (unpublished observations).

An important feature of the microenvironment occupied by the EnMSCs is the higher local concentrations of reproductive hormones and particularly progesterone. It has been shown that the progesterone can induce some changes in the production, expression, and secretion of some immunomodulatory proteins by EnMSCs and these data are discussed below.

V. Interaction Between Progesterone and MSCs

A. Induction of HLA-G expression by progesterone

The question about the immunomodulatory factors acting at the feto-maternal interface and their interactions leading to lack of rejecting immune response against the fetus have not been answered properly, yet. A number of factors have been identified and characterized which express predominantly immunosuppressive effect, and the progesterone is considered to be one of the most important factors in that context. However, the fact is that its effect by itself should not be related to its direct effect, only, but there are some indirect effects of progesterone which influence the expression of other immunomodulatory factors.

At the end of 1980s, a novel member of the human leukocyte antigen class Ib family (HLA class Ib) was characterized and labeled as HLA-G which showed a much lower degree of polymorphism and quite high restriction of tissue expression (Geraghty *et al.*, 1987; Le Bouteiller, 1994). Using specific monoclonal antibodies both cell surface and cytoplasmic forms of HLA-G were described. Human cytotrophoblast cells were the first targets for extensive studies on the expression and biological roles of HLA-G (Kovats *et al.*, 1990) then the expression of the HLA-G was detected in extravillous cytotrophoblasts, chorionic endothelial cells, and placental macrophages activated by interferon (Blaschitz *et al.*, 1997; Chu *et al.*, 1998; McMaster *et al.*, 1995). Further detailed investigations demonstrated the expression of HLA-G in certain populations of thymic epithelia (Mallet *et al.*, 1999), tumor cells (Wiendl *et al.*, 2002), and pathologically changed intestinal mucosa (Torres *et al.*, 2006). Recently, it was reported that HLA-G can be detected in very low amounts in decidual stromal cells cultures (Blanco *et al.*, 2008).

HLA-G is currently considered as the major factor to prevent the rejection of the semiallogenic fetus by the immune mechanisms of the mother (Le Bouteiller *et al.*, 1999). Numerous studies have proved that HLA-G suppresses the proliferation *in vitro* of CD4+ T lymphocytes (Bainbridge *et al.*, 2000), induces apoptosis in activated CD8+ T lymphocytes (Fournel *et al.*, 2000), and inhibits the cytotoxicity of the NK cells (Marchal-Bras-Goncalves *et al.*, 2001).

Quite expectedly the expression of HLA-G in different cell types is modulated by a number of agents such as viruses, cytokines, and hormones. It was reported that some neurotropic viruses (herpes simplex virus type 1 and rabies virus) stimulate the expression of HLA-G in human neural cells (Lafon *et al.*, 2005). It also was found that human immunodeficiency virus (HIV) significantly suppresses the surface expression of HLA-G by glioma cells and macrophages (Derrien *et al.*, 2004). Cytokines secreted by various cellular types are key players in the modulation of the HLA-G expression in its different isoforms. A good illustration of that statement is the fact that IL-10 could upregulate the HLA-G expression by cancer cells in primary cutaneous lymphomas and monocytes (Moreau *et al.*, 1999). It has been speculated that the increased expression of HLA-G in that case is a guarantee for the escape of the immune surveillance by cell clones with high HLA-G expression (Urosevic *et al.*, 2002). Another example is the reported data that when peripheral blood monocytes or macrophages which constitutively express rather low amounts of HLA-G have been treated with IFN-γ the cells show higher expression of HLA-G both at mRNA and protein level in a dose-dependent manner (Yang *et al.*, 1996).

There is some evidence that HLA-G expression is seemingly modulated by hormones and mainly by the reproductive steroid hormone progesterone. Sheshgiri *et al.* (2008) using immunoenzyme assay and flow cytometry analysis reported that human endothelial cells from heart vessels and smooth muscle cells do not express HLA-G, but however, after treatment of these cells with progesterone expression of HLA-G was detected. This fact would suggest that both surface and secreted forms of HLA-G were induced in progesterone-treated cells, and it has been reported that both primary cultures of cytotrophoblasts and JEG-3 chorioncarcinoma cells have increased the levels of HLA-G after being cultured in the presence of progesterone. The expression of HLA-G was shown to be dose and time dependent as the peak concentrations in cell lysates and culture supernatants were detected after 4–5 and 24 h of culture with progesterone (Yie *et al.*, 2006). In decidual stromal cell cultures, HLA-G expression is increased in the presence of IFN-γ and combination of progesterone with cAMP (Blanco *et al.*, 2008).

The data about the expression of HLA-G on the surface of multiple types of cells and its changes under the influence of various biologically active substances made us study the expression of HLA-G in MSCs and the effect of progesterone on this expression. Objects of these studies were MSCs derived from bone marrow (BM-MSCs), adipose tissue (AT-MSCs), and endometrium (EnMSCs). The presence of progesterone in cultures did not change either the phenotype or the growth kinetics of the cells compared to the control cell cultures. The cells of all types had their typical fibroblast-like morphology, proliferated adhering to the plastic and forming monolayer. Flow cytometry analysis of the cells did not reveal expression of HLA-G by the control cultures of the cells from the three sources. The presence of progesterone in physiological for pregnancy dozes in the cell cultures induced the expression of HLA-G as 7.9% of BM-MSCs were positively stained, 5.3% of EnMSCs were positive but the highest percentage of positive cells was recorded for the At-MSCs—11.4%. Experiments with BM-MSCs and EnMSCs showed a dose-dependent effect of the progesterone while such an effect was not detected when AT-MSCs were tested. Another significant difference in the behavior of the AT-MSCS is that the expression of HLA-G could be detected when higher doses of progesterone were used, that is, no dose-dependent effect was seen. So, it seems that MSCs derived from adipose tissue react in a different manner in comparison to the BM-MSCs and EnMSCs (Ivanova-Todorova *et al.*, 2009a).

In a parallel series of experiments by confocal microscopy, it was observed that the control MSCs cultures of the three different sources did not express HLA-G neither intracellular nor on the cell membrane. When the cells were cultured in the presence of progesterone, a specific perinuclear staining was observed. In EnMSC cultures stimulated with progesterone, higher numbers of positively stained cells were observed with higher intensity of staining when compared with the cells isolated from the other two sources. Further experiments using RT-PCR and western blot detected the presence of HLA-G at both nucleic acid and protein levels in EnMSCs cultured in the absence of progesterone. Some discrepancies of these results can be explained by the different sensitivity of the methods used (Mourdjeva *et al.*, 2009). Taken together, these results would suggest that HLA-G is expressed in the MSCs isolated from the three sources but its level of expression is significantly higher in EnMSCs which might be due to the fact that their source (human endometrium) is the site of higher concentrations of progesterone. When cultured *in vitro* in the presence of progesterone, the expression of HLA-G both in the cytoplasm and the cell surface is enhanced significantly. However, we couldn't detect secretion of HLA-G in the culture supernatant of the cells stimulated with progesterone, which might be due the fact that the ELISA kits that we have used were not sensitive and/or specific as needed.

Coming back to the specific features of AT-MSCs, it should be mentioned that in another series of experiments it has been shown that these cells have a higher immunosuppressive effect on immunoglobulin secretion by mitogen-stimulated B cells (Bochev *et al.*, 2008) and on the differentiation of human monocyte dendritic cells (Ivanova-Todorova *et al.*, 2009a).

It can be speculated that the progesterone effect on AT-MSCs might be due to common steroid effects while with BM-MSCs and EnMSCs the effect is strictly specific. Another speculation is that the BM-MSCs population might contain precursors which via circulation can reach the endometrium and differentiate as endometrial stromal cells as suggested earlier (Taylor, 2004). It is possible that these precursors are present in the population of the EnMSCs which are strongly influenced by the high local concentration of progesterone.

In conclusion, it should be pointed out that progesterone upregulates the expression by MSCs of HLA-G, one of the major players in maintenance of the immune balance between the mother and the fetus in developing pregnancy. An important finding is that the MSCs which have well documented immunomodulatory activity are newly detected targets of progesterone and are most probably involved in the immune interactions at the maternal fetal interface. It can be speculated that these cells contribute to both tissue remodeling and maintenance of immune tolerance during pregnancy (Ivanova-Todorova *et al.*, 2009b).

B. Induction of progesterone-induced blocking factor (PIBF) expression

PIBF was described initially by Szekeres-Bartho *et al.* (1985, 1989) as an active factor intimately involved in regulation of the immune response in pregnancy. It is a 34-kDa protein which inhibits the lymphocyte cytotoxicity, the prostaglandins secretion, and has an antiabortive effect. PIBF is firstly detected quite early after the implantation of successfully developing pregnancy (Szekeres-Bartho *et al.*, 1990). Receptors for PIBF are shown to be expressed by the majority of lymphocytes and it's secreted by Tγδ lymphocytes and decidual NK cells after stimulation with progesterone (Dosiou and Guidice, 2005; Szekeres-Bartho *et al.*, 2001a). Some authors claim that expression of PIBF receptors can be detected in CD8+ lymphocytes only during the pregnancy (van Kaer *et al.*, 1991). PIBF secretion by lymphocytes is triggered via the classic and/or membrane PRs. Allogeneic stimuli increase the numbers of PIBF receptors on the lymphocytes (Chek *et al.*, 1997; Szekeres-Bartho *et al.*, 1995).

The immunomodulatory effect of PIBF is quite variable but the total result is directed to preserving the pregnancy (Szekeres-Bartho *et al.*, 2001b). It has been established that PIBF inhibits significantly the NK cells activity through several mechanisms and namely: (i) strongly inhibits the liberation of perforin from the NK cells, (ii) blocks the liberation of arachidonic acid and thus limits the production and secretion of prostaglandins, and (iii) inhibits and lowers the secretion of IL-12 as a result of the diminished prostaglandin E2 secretion (Szekeres-Bartho *et al.*, 1985; van Kaer *et al.*, 1991). On the other hand, PIBF does not inhibit the IL-2 production and its concentration is significantly reduced in cases of immunomediated abortions (Szekeres-Bartho *et al.*, 1989). Detailed investigations have shown that PIBF enhanced the secretion of IL-3, IL-4, and IL-10 by spleen cells of mice when cultured *in vitro*, while the production of IFN gamma was not modified in comparison to controls (Szekeres-Bartho and Wegmann, 1996). Kozma *et al.* (2006) reported that the effect of PIBF is mediated via a novel IL-4 receptor which consists most probably of IL-4R chain and the PIBF receptor which is a GPI-anchored protein. The activation signal-induced phosphorylation and nuclear translocation of STAT-6 and inhibition of the STAT-4 pathway. Some studies report that at least in 89% of patients with abortion the concentration of PIBF is lower in comparison to normally developing pregnancy (Check *et al.*, 1996).

The presence of MSCs in human early deciduas raises the question about the expression of PIBF by EnMSCs. In our experiments, EnMSCs were isolated from 8 to 10 g.w. pregnancy and were cultured in the presence of progesterone for 60 h. Monoclonal antibody against PIBF produced in our laboratory which reacted positively against PIBF in immunoblot, immunofluorescence, and flow cytometry was applied in these experiments (Ivanova-Todorova *et al.*, 2008). A comparatively low percentage of EnMSCs (2.4%) were detected to express PIBF at their surface and stimulation with progesterone did not influence the expression. However, when confocal microscopy was used, it was found that a significant number of cells were positively stained as an intensive homogeneous intracellular staining was observed. Flow cytometry analysis showed that the percentage of intracellular positively stained cells reached as high as 90% of the cells in the cultures. However when compared with cells cultured in the presence of progesterone, the percentage of PIBF-positive cells dropped to 74% of the cells, and the confocal microscopy showed a lower perinuclear granulated staining (Ivanova-Todorova *et al.*, 2009c). These results would suggest that PIBF is constitutively expressed in EnMSCs and progesterone seems to decrease the intracytoplasmic PIBF content. The specific granular pattern of the positive staining in the presence of progesterone would strongly suggest that PIBF is secreted by the cells under these culture conditions. It can be speculated that EnMSCs may be considered as an alternative source of PIBF as initially they are acting as a depot for this protein and with the increase of the local concentration of progesterone its secretion would occur to express the immunosuppressive effect of PIBF on the locally recruited immune cells.

The results from our studies on the effect of progesterone on the MSCs convincingly show that its modulation of the expression of immunoregulatory proteins such as HLA-G and PIBF is just one of its effects in protection of pregnancy. Although there are significant differences in the biochemical structure, properties, and activities of the two proteins, there some analogies as far as the effect of progesterone on the MSCs are concerned. These are as follows:

- Both HLA-G and PIBF are immunomodulatory molecules expressed by MSCs and contributing to the development of a tolerant status allowing successful development of pregnancy.
- Minimal expression of the both molecules is observed at the MSCs surface, but the expression of HLA-G is strongly upregulated by progesterone while no changes were observed with PIBF expression.
- Both studied proteins are expressed intracellular in MSCs cytoplasm and their expression in the cells is strongly reduced in the presence of progesterone most probably due to their secretion from the cells.
- Unfortunately there are not, yet, well developed, standardized, and broadly accepted tests for detection and measurement of secreted forms of these proteins which would allow to compare their plasma concentrations in health and disease.

VI. Conclusions and Future Directions

Uterine endometrium and the stromal cells, in particular, are under the effective control of progesterone. Stromal cell compartment is represented by different subpopulations of stromal cells which differ in their phenotypic characteristics, location in the uterine wall, functions, and possibly origin. It can be speculated the endometrium itself is not uniform structure where all the effects of the hormonal control are simultaneously and equally executed but represent a mosaic of active sites where preferentially the physiological processes such as decidualization and possibly implantation will take place. It is quite possible that EnMSCs are the dominant cell type in such active sites, and under the influence of progesterone, they would upregulate expression and secretion of immunomodulatory proteins (HLA-G, PIBF) and thus will prepare the optimal conditions for embryo nidation and implantation. This hypothesis needs a lot of studies to be really supported by facts.

ACKNOWLEDGMENTS

This work was supported in part by Grant "Establishment of Center for research on problems of the reproductive health," "Development of the scientific infrastructure-2008," and Grant No DDVU 02/34 "Stimulating of scientific researches in medical high schools" Ministry of Education, Youth, and Science, Republic of Bulgaria.

REFERENCES

Bainbridge, D. R. J., Ellis, S. A., and Sargent, I. L. (2000). HLA-G suppresses proliferation of CD4+ T-lymphocytes. *J. Reprod. Immunol.* **48,** 17–26.

Blanco, O., Tirado, I., Munoz-Fernandez, R., Abadıa-Molina, C. A., Garcıa-Pacheco, M. J., Pena, J., and Olivares, E. G. (2008). Human decidual stromal cells express HLA-G effects of cytokines and decidualization. *Hum. Reprod.* **23,** 144–152.

Blaschitz, A., Lenfant, F., Mallet, V., Hartmann, M., Bensussan, A., Geraghty, D. E., Le Bouteiller, P., and Dohr, G. (1997). Endothelial cells in chorionic fetal vessels of first trimester placenta express HLA-G. *Eur. J. Immunol.* **27,** 3380–3386.

Bochev, I., Elmadjian, G., Kyurkchiev, D., Tzvetanov, L., Altankova, I., Tivchev, P., and Kyurkchiev, S. (2008). Mesenchymal stem cells from human bone marrow or adipose tissue differently modulate mitogen-stimulated B-cell immunoglobulin production in vitro. *Cell Biol. Int.* **32,** 384–393.

Brosens, J. J., Hayashi, N., and White, J. O. (1999). Progesterone receptor regulates decidual prolactin expression in differentiating human endometrial stromal cells. *Endocrinology* **140,** 4809–4820.

Castro-Malaspina, H., Gay, R. E., Resnick, G., Kapoor, N., Meyers, P., Chiarieri, D., McKenzie, S., Broxmeyer, H. H., and Moore, M. A. (1980). Characterisation of human bone marrow fibroblast colony-forming cells (CFU-F) and their progeny. *Blood* **56,** 289–301.

Chan, R. W. S., Schwab, K. E., and Garget, C. E. (2004). Clonogenicity of human endometrial epithelial and stromal cells. *Biol. Reproduction*. **70,** 1738–1750.
Check, J., Szekeres-Bartho, J., and O'Slaughnessy, A. (1996). Progesterone induced blocking factor seen in pregnancy lymphocytes soon after implantation. *Am. J. Reprod. Immunol.* **35,** 277–280.
Chek, J., Arwitz, M., Gross, J., Peymer, M., and Szekeres-Bartho, J. (1997). Lymphocyte immunotherapy (LI) increases serum levels of (PIBF). *Am. J. Reprod. Immunol.* **37,** 17–20.
Chu, W., Fant, M. E., Geraghty, D. E., and Hunt, J. S. (1998). Soluble HLA-G in human placentas: Synthesis in trophoblasts and interferon-g-activated macrophages but not placental fibroblasts. *Hum. Immunol.* **59,** 435–442.
Colter, D. C., Sekiya, I., and Prockop, D. J. (2001). Identification of a subpopulation of rapidly self-renewing and multipotential adult stem cells in colonies of human marrow stromal cells. *Proc. Natl. Acad. Sci. USA* **98**(14), 7841–7845.
Condon, J. C., Hardy, D. B., Kovaric, K., and Mendelson, C. R. (2006). Upregulation of the progesterone receptor (PR)-C isoform in laboring myometrium by activation of NF-{kappa}B may contribute to the onset of labor through inhibition of PR function. *Mol. Endocrinol.* **20,** 764–775.
Derrien, M., Pizzato, N., Dolcini, G., Menu, E., Chaouat, G., Lenfant, F., Barré-Sinoussi, F., and Le Bouteiller, P. (2004). Human immunodeficiency virus 1 down-regulates cell surface expression of the non-classical major histocompatibility class I molecule HLA-G1. *J. Gen. Virol.* **85,** 1945–1954.
Dimitrov, R., Timeva, T., Kyurkchiev, D., Stamenova, M., Shterev, A., Kostova, P., Zlatkov, V., Kehayov, I., and Kyurkchiev, S. (2008). Characterization of clonogenic stromal cells isolated from human endometrium. *Reproduction* **135,** 551–558.
Dimitrov, R., Kyurkchiev, D., Timeva, T., Yunakova, M., Stamenova, M., Shterev, A., and Kyurkchiev, S. (2010). First-trimester human decidua contains a population of mesenchymal stem cells. *Fertil. Steril.* **93,** 210–219.
Dosiou, C., and Guidice, L. C. (2001). Natural killer cells in pregnancy and reccurent pregnancy loss: Endocrine and immunologic perspectives. *Hum. Reprod.* **16,** 855–861.
Dosiou, C., and Guidice, L. C. (2005). NK cells in pregnancy and recurrent pregnancy loss: Endocrine and immunologic perspectives. *Endocr. Rev.* **26,** 44–62.
Du, H., and Taylor, H. S. (2007). Contribution of bone marrow-derived stem cells to endometrium and endometriosis. *Stem Cells* **25**(8), 2082–2086.
Dunn, C. L., Kelly, R. W., and Critchley, H. O. D. (2003). Decidualization of the human endometrial stromal cell: An enigmatic transformation. *Reprod. Biomed. Online* **7,** 151–161.
Falkenstein, E., Schmieding, K., Lange, A., Meyer, C., Gerdes, D., Welsch, U., and Wehling, M. (1998). Localization of a putative progesterone membrane binding protein in porcine hepatocytes. *Cell. Mol. Biol.* **44**(4), 571–578.
Fournel, S., Aguerre-Girr, M., Huc, X., Lenfant, F., Alam, A., Toubert, A., Bensussan, A., and Le Bouteiller, P. (2000). Soluble HLA-G1 triggers CD95 / CD95 ligand-mediated apoptosis in activated CD81 cells by interacting with CD81. *J. Immunol.* **164,** 6100–6104.
Friedstein, A. J., Petrakova, K. V., Kurolesova, A. I., and Frolova, G. P. (1968). Heterotopic transplants of bone marrow. Analysis of precursor cells for osteogenic and hematopoetic tissues. *Transplantation* **6,** 230–247.
Gang, E. J., Bosnakovski, D., Figueiredo, C. A., Visser, J. W., and Perlingeiro, R. C. (2007). SSEA-4 identifies mesenchymal stem cells from bone marrow. *Blood* **109,** 1743–1751.
Gargett, C. E. (2007). Uterine stem cells: What is the evidence? *Hum. Reprod. Update* **13,** 87–101.
Gellersen, B., Fernandes, M. S., and Brosens, J. J. (2009). Non- genomic progesterone actions in female reproduction. *Hum. Reprod. Update.* **15,** 119–138.

Geraghty, D. E., Koller, B. H., and Orr, H. T. (1987). A human majorhistocompatibility complex class I gene that encodes a protein with a shortened cytoplasmic segment. *Proc. Natl. Acad. Sci. USA* **84,** 9145–9149.

Graham, J. D., and Clarke, C. L. (1997). Physiological action of progesterone in target tissues. *Endocrinol Rev.* **18,** 502–519.

Gronthos, S., Graves, S. E., Ohta, S., and Simmons, P. J. (1994). The STRO-1+ fraction of adult human bone marrow contains the osteogenic precursors. *Blood* **84,** 4164–4173.

Hansen, P. J., and Skopets, B. (1992). Temporal relationship between progesterone and uterine lymphocyte-inhibitory activity in ewes. *Vet. Rec.* **131,** 371–372.

Henderson, T. A., Saunders, P. T., Moffet-King, A., Grrome, N. O., and Critchley, H. O. (2003). Steroid receptor expression in uterine natural killer cells. *J. Clin. Endocrinol. Metab.* **88,** 440–449.

Hirshfield, A. N. (1984). Stathmokinetic analysis of granulosa cell proliferation in antral follicles of cyclic rats. *Biol. Reprod.* **31,** 52–58.

Huang, G. T.-J., Gronthos, S., and Shi, S. (2009). Mesenchymal stem cells derived from dental tissues vs. those from other sources: Their biology and role in regenerative medicine. *J. Dent. Res.* **88,** 792–806.

Irwin, J. C., Kirk, D., King, R. J. B., Quigley, M. M., and Gwatkin, R. B. L. (1989). Hormonal regulation of human endometrial stromal cells in culture: And in vitro model for decidualization. *Fertil. Steril.* **52,** 761–768.

Irwin, J. C., Utian, W. H., and Eckert, R. L. (1991). Sex steroids and growth factors differentially regulate the growth and differentiation of cultured human endometrial stromal cells. *Endocrinology* **129,** 2385–2392.

Ismail, P. M., Amato, P., Soyal, S. M., DeMayo, F. J., Conneely, O. M., O'Malley, B. W., and Lydon, J. P. (2003). Progesterone involvement in breast development and tumorigenesis—As revealed by progesterone receptor "knockout" and "knockin" mouse models. *Steroids* **68,** 779–787.

Ivanova-Todorova, E., Kyurkchiev, D., Nalbanski, A., Timeva, T., Shterev, A., and Kyurkchiev, S. (2008). Production and characterization of a novel monoclonal antibody against progesterone-induced blocking factor (PIBF). *J. Reprod. Immunol.* **78,** 94–101.

Ivanova-Todorova, E., Ivan Bochev, I., Mourdjeva, M., Dimitrov, R., Bukarev, D., Kyurkchiev, S., Tivchev, P., Alrtankova, I., and Kyurkchiev, D. (2009a). Adipose tissue derived mesenchymal stem cells are more potent suppressors of dendritic cells differentiation copmpared with bone-marrow derived mesenchymal stem cells. *Immunol. Lett.* **126,** 37–42.

Ivanova-Todorova, E., Mourdjeva, M., Kyurkchiev, D., Bochev, I., Stoyanova, E., Dimitrov, R., Timeva, T., Yunakova, M., Bukarev, D., Shterev, A., Tivchev, P., and Kyurkchiev, S. (2009b). HLA-G expression is up-regulated by progesterone in mesenchymal stem cells. *Am. J. Reprod. Immunol.* **62,** 25–33.

Ivanova-Todorova, E., Kyurkchiev, D., Mourdjeva, M., Dimitreov, R., Stoyanova, E., Timeva, T., Yunakova, M., Shterev, A., and Kyurkchiev, S. (2009c). Pre-decidual multipotent stromal cells (PreDMSC) constitutively express progesterone induced blocking factor (PIBF). *C. R. Acad. Bulg. Sci.* **62**(12), 1567–1570.

Jones, R. A., Straszynski, L., Meredith, D. M., Markham, A. F., Jack, A., Emery, P., and McGonagle, D. (2002). Isolation and characterization of bone marrow multipotential mesenchymal progenitor cells. *Arthritis Rheum.* **46,** 3349–3360.

van Kaer, L., Wu, M., Ichikawa, Y., Ito, K., Bouneville, M., Ostrand-Rosenberg, S., Murphy, D. B., and Tonegawa, S. (1991). Recognition of MHC TL gene products by gd T cells. *Immunol Rev.* **120,** 89–115.

Karystinou, A., Dell'Accio, F., Kurth, T. B., Wackerhage, H., Khan, I. M., Archer, C. W., Jones, E. A., Mitsiadis, T. A., and De Bari, C. (2009). Distinct mesenchymal progenitor cell subsets in the adult human synovium. *Rheumatology* **48,** 1057–1064.

Kato, K., Yoshimoto, M., Keiji, Kato, Adachi1, S., Yamayoshi1, A., Arima, T., Asanoma, K., Kyo, S., Nakahata, T., and Wake, N. (2007). Characterization of side-population cells in human normal endometrium. *Hum. Reprod.* **22**(5), 1214–1223.

Kovats, S., Main, E. K., Librach, C., Stubblebine, M., Fisher, S. J., and DeMars, R. (1990). A class I antigen, HLA-G, expressed in human trophoblasts. *Science* **248,** 220–223.

Kozma, N., Halasz, M., Polgar, B., Poehlmann, T. G., Markert, U. R., Palkovics, T., Keszei, M., Par, G., Kiss, K., Szeberenyi, J., Grama, L., and Szekeres-Bartho, J. (2006). Progesterone-induced blocking factor ctivates STAT6 via binding to a novel IL-4 receptor. *J. Immunol.* **176,** 819–826.

Kyurkchiev, D., Ivanova-Todorova, E., and Kyurkchiev, S. (2010). New target cells of the immunomodulatory effects of progesterone. *Reprod. Biomed. Online* **21,** 304–311.

Lafon, M., Prehaud, C., Megret, F., Lafage, M., Mouillot, G., Roa, M., Moreau, P., Rouas-Freiss, N., and Carosella, E. D. (2005). Modulation of HLA-G expression in human neural cells after neurotropic viral infections. *J. Virol.* **79,** 15226–15237.

Le Bouteiller, P. (1994). HLA class I chromosomal region, genes, and products: Facts and questions. *Crit. Rev. Immunol.* **14,** 89–105.

Le Bouteiller, P., Solier, C., Pro'll, J., Aguerre-Girr, M., Fournel, S., and Lenfant, F. (1999). Mini symposium. The major histocompatibility complex in pregnancy: Part II. Placental HLA-G protein expression in vivo: Where and what for? *Hum. Reprod. Update* **5,** 223–233.

Li, X., and O'Malley, B. W. (2003). Unfolding the action of progesterone receptors. *J. Biol. Chem.* **278,** 39261–39264.

Lockwood, C. J., Krikun, G., Runic, R., Schwartz, L. B., Mesia, A. F., and Schatz, F. (2000). Progestin-epidermal growth factor regulation of tissue factor expression during decidualization of human endometrial stromal cells. *J. Clin. Endocrinol. Metab.* **85,** 297–301.

Losel, R. M., Falkenstein, E., Feuring, M., Schultz, A., Tillmann, H.-C., Rossol-Haseroth, K., and Wehling, M. (2003). Nongenomic steroid action: Controversies, questions, and answers. *Physiol. Rev.* **83,** 965–1016.

Losel, R. M., Besonga, D., Peluso, J. J., and Wehlinga, M. (2008). Progesterone receptor membrane component 1—Many tasks for a versatile protein. *Steroids* **73,** 929–934.

Mallet, V., Blaschitz, A., Crisa, L., Schmitt, C., Fournel, S., King, A., Loke, Y. W., Dohr, G., and Le Bouteiller, P. H. (1999). HLA-G in the human thymus: A subpopulation of medullary epithelial but not CD83F dendritic cells expresses HLA-G as a membrane-bound and soluble protein. *Int. Immunol.* **11,** 889–898.

Marchal-Bras-Goncalves, R., Rouas-Freiss, N., Connan, F., Choppin, J., Dausset, J., Carosella, E. D., Kirszenbaum, M., and Guillet, J.-G. (2001). A soluble HLA-G protein that inhibits natural killer cell-mediated cytotoxicity. *Transplant. Proc.* **33,** 2355–2359.

McMaster, M. T., Librach, C. L., Zhou, Y., Lim, K. H., Janatpour, M. J., DeMars, R., Kovats, S., Damsky, C., and Fisher, S. J. (1995). Human placental HLA-G expression is restricted to differentiated cytotrophoblasts. *J. Immunol.* **154,** 3771–3778.

Meirelles, L. D. S., Chagastelles, P. C., and Nardi, N. B. (2006). Mesenchymal stem cells reside in virtually all post-natal organs and tissues. *J. Cell Sci.* **119**(11), 2204–2213.

Meizel, S., Turner, K. O., and Nuccitelli, R. (1997). Progesterone triggers a wave of increased free calcium during the human sperm acrosome reaction. *Dev. Biol.* **182,** 67–75.

Meng, X., Ichim, T. E., Zhong, J., Rogers, A., Yin, Z., Jackson, J., Wang, H., Ge, W., Bogin, V., Chan, K. W., Thébaud, B., and Neil, H. R. (2007). Endometrial regenerative cells: A novel stem cell population. *J. Transl. Med.* **5,** 57–67.

Minguelli, J. J., Erices, A., and Conget, P. (2001). Mesenchymal stem cells. *Exp. Biol. Med.* **226,** 507–520.

Moreau, P., Adrian-Cabestre, F., Menier, C., Guiard, V., Gourand, L., Dausset, J., Carosella, E. D., and Paul, P. (1999). IL-10 selectively induces HLA-G expression in human trophoblasts and monocytes. *Int. Immunol.* **11,** 803–811.

Mourdjeva, M., Ivanova-Todorova, E., Kyurkchiev, D., Bochev, I., Stoyanova, E., Dimitrov, R., and Kyurkchiev, S. (2009). Progesterone enhances the HLA-G expression in mesenchymal stem cells. *C. R. Acad. Bulg. Sci.* **62**(12), 1567–1570.

Mulac-Jericevic, B., Mullinax, R. A., DeMayo, F. J., Lydon, J. P., and Conneely, O. M. (2000). Subgroup of reproductive functions of progesterone mediated by progesterone receptor-B isoform. *Science* **289,** 1751–1754.

Navarrete, M. A., Maier, C. M., Falzoni, R., Cuadros, L. G., Lima, G. R., Baracat, E. C., and Nazario, A. C. (2005). Assessment of the proliferative, apoptotic and cellular renovation indices of the human mammary epithelium during the follicular and luteal phases of the menstrual cycle. *Breast Cancer Res.* **7**(3), 306–313.

Norman, A. W., Mizwicki, M. T., and Norman, D. P. (2004). Steroid-hormone rapid actions, membrane receptors and a conformational ensemble model. *Nat. Rev. Drug Discov.* **3,** 27–41.

Olivares, E. G., Montes, M. J., Oliver, C., Galindo, J. A., and Riuz, C. (1997). Cultured human decidual stromal cells express B7–1 (CD80) and stimulate allogenic T cells. *Biology of Reproduction* **5709–615,** 6.

Owen, M. (1988). Marrow stromal cells. *J. Cell Sci. Suppl.* **10,** 63–76.

Patel, A. N., Park, E., Kuzman, M., Benetti, F., Silva, F. J., and Allickson, J. G. (2008). Multipotent menstrual blood stromal stem cells: isolation, characterization and differentiation. *Cell Transplantation.* **17,** 303–311.

Peluso, J. (2006). Multiplicity of progesterone's actions and receptors in the mammalian ovary. *Biol. Reprod.* **75,** 2–8.

Piva, M., Flieger, O., and Rider, V. (1996). Growth factor control of cultured rat uterine stromal cell proliferation is progesterone dependent. *Biol. Reprod.* **55,** 1333–1342.

Prianishnikov, V. A. (1978). On the concept of stem cell and a model of functional–morphological structure of the endometrium. *Contraception* **18,** 213–223.

Quirici, N., Soligo, D., Bossolasco, P., Servida, F., Lumini, C., and Deliliers, G. L. (2002). Isolation of bone marrow mesenchymal stem cells by anti-nerve growth factor receptor antibodies. *Exp. Hematol.* **30,** 783–791.

Reyes, M., Dudek, A., Jahagirdar, B., Koodie, L., Marker, P. H., and Verfaillie, C. M. (2002). Origin of endothelial progenitors in human postnatal bone marrow. *J. Clin. Invest.* **109**(3), 337–346.

Rosen, J. M. (2003). Hormone receptor patterning plays a critical role in normal lobuloalveolar development and breast cancer progression. *Breast Dis.* **18,** 3–9.

Rothchild, I. (1996). The corpus luteum revisited: Are the paradoxical effects of RU486 a clue to how progesterone stimulates its own secretion? *Biol. Reprod.* **55,** 1–4.

Schumacher, M., Guennoun, R., Ghoumari, A., Massaad, C., Robert, F., El-Etr, M., Akwa, Y., Rajkowski, K., and Baulieu, E. E. (2007). Novel perspectives for progesterone in hormone replacement therapy, with special reference to the nervous system. *Endocr. Rev.* **28,** 387–439.

Schüring, A. N., Schulte, N., Kelsch, R., Ropke, A., Kiesel, L., and Gotte, M. (2011). Characterization of endometrial mesenchymal stem-like cells obtained by endometrial biopsy during routine diagnostics. *Fertil. Steril.* **95,** 423–426.

Schwab, K. E., and Gargett, C. E. (2007). Co-expression of two perivascular cell markers isolates mesenchymal stem-like cells from human endometrium. *Hum. Reprod.* **22,** 2903–2911.

Schwab, K. E., Chan, R. W., and Gargett, C. E. (2005). Putative stem cell activity of human endometrial epithelial and stromal cells during the menstrual cycle. *Fertil. Steril.* **84,** 1124–1130.

Schwab, K. E., Hutchinson, P., and Gargett, C. E. (2008). Identification of surface markers for prospective isolation of human endometrial stromal colony-forming cells. *Hum. Reprod.* **23,** 934–943.

Setty, S. L., and Mills, T. M. (1987). The effects of progesterone on follicular growth in the rabbit ovary. *Biol. Reprod.* **36,** 1247–1252.

Sheshgiri, R., Rao, V., Tumiati, L. C., Xiao, R., Prodger, J. L., Badiwala, M., Librach, C., and Delgado, D. H. (2008). Progesterone induces human leukocyte antigen-G expression in vascular endothelial and smooth muscle cells. *Circulation* **118,** 58–64.

Simmons, P. J., and Torok-Storb, B. (1991). Identification of stromal cells precursors in human bone marrow by a novel monoclonal antibody STRO-1. *Blood* **78,** 55–62.

Singh, M. (2006). Progesterone-induced neuroprotection. *Endocrine* **29,** 271–274.

Szekeres-Bartho, J., and Wegmann, T. G. (1996). A progesterone-dependent immunomodulatory protein alters the Th1/Th2 balance. *J. Reprod. Immunol.* **31,** 81–95.

Szekeres-Bartho, J., Hadnagy, J., and Pacsa, A. S. (1985). The suppressive effect of progesterone on lymphocyte cytotoxicity: Unique progesterone sensitivity of pregnancy lymphocytes. *J. Reprod. Immunol.* **7,** 121–128.

Szekeres-Bartho, J., Autran, B., Debre, P., Andreu, G., Denver, L., and Chaouat, G. (1989). Immunoregulatory effects of a suppressor factor from healthy pregnant women's lymphocytes after progesterone induction. *Cell. Immunol.* **122,** 281–294.

Szekeres-Bartho, J., Szekeres, G. Y., Dedre, P., Autran, B., and Chaouat, G. (1990). Reactivity of lymphocytes to a progesterone receptor-specific monoclonal antibody. *Cell. Immunol.* **125,** 273–283.

Szekeres-Bartho, J., Faust, Zs, and Varga, P. (1995). The expression of a progesterone-induced immunomodulatory protein in pregnancy lymphocytes. *Am. J. Reprod. Immunol.* **34,** 342–348.

Szekeres-Bartho, J., Barakonyi, A., Miko, E., Polgar, B., and Palkovics, T. (2001a). The role of gamma/delta T cells in the feto-maternal relationship. *Semin. Immunol.* **13,** 229–233.

Szekeres-Bartho, J., Barakonyi, A., Par, G., Polgar, B., Palkovics, T., and Szereday, L. (2001b). Progesterone as an immunomodulatory molecule. *Int. Immunopharmacol.* **1,** 1037–1048.

Taylor, H. S. (2004). Endometrial cells derived from donor stem cells in bone marrow transplant recipients. *JAMA* **292,** 81–85.

Torres, M. I., Lopez-Casado, M. A., Luque, J., Peña, J., and Rıos, A. (2006). New advances in coeliac disease: Serum and intestinal expression of HLA-G. *Int. Immunol.* **18,** 713–718.

Urosevic, M., Willers, J., Mueller, B., Kempf, W., Burg, G., and Dummer, R. (2002). HLA-G protein up-regulation in primary cutaneous lymphomas is associated with interleukin-10 expression in large cell T-cell lymphomas and indolent B-cell lymphomas. *Blood* **99,** 609–617.

Welm, B. E., Tepera, S. B., Venezia, T., Graubert, T. A., Rosen, J. M., and Goodell, M. A. (2002). Sca-1 (pos) cells in the mouse mammary gland represent an enriched progenitor cell population. *Dev. Biol.* **245,** 42–56.

Wiendl, H., Mitsdoerffer, M., Hofmeister, V., Wischhusen, J., Bornemann, A., Meyermann, R., Weiss, E. H., Melms, A., and Weller, M. (2002). A functional role of HLA-G expression in human gliomas: An alternative strategy of immune escape. *J. Immunol.* **168,** 4772–4780.

Wright, D. W., Kellermann, A. L., Hertzberg, V. S., Clark, P. L., Frankel, M., Goldstein, F. C., Salomone, J. P., Dent, L. L., Harris, O. A., Ander, D. S., Lowery, D. W., Patel, M. M., *et al.* (2007). ProTECT: A randomized clinical trial of progesterone for acute traumatic brain injury. *Ann. Emerg. Med.* **49,** 391–402.

Xiao, G., Wei, J., Yan, W., Wang, W., and Lu, Z. (2008). Improved outcomes from the administration of progesterone for patients with acute severe traumatic brain injury: A randomized controlled trial. *Crit. Care* **12,** 153–163.

Xin, L., Lawson, D. A., and Witte, O. N. (2005). The Sca-1 cell surface marker enriches for a prostate-regenerating cell subpopulation that can initiate prostate tumorigenesis. *Proc. Natl. Acad. Sci. USA* **102,** 6942–6947.

Yang, Y., Chu, W., Geraghty, D. E., and Hunt, J. S. (1996). Expression of HLA-G in human mononuclear phagocytes and selective induction by IFN-gamma. *J. Immunol.* **156,** 4224–4231.

Yie, S., Li, L., Li, G., Xiao, R., and Librach, C. L. (2006). Progesterone enhances HLA-G gene expression in JEG-3 choriocarcinoma cells and human cytotrophoblasts in vitro. *Hum. Reprod.* **21,** 46–51.

Zuk, P. A., Zhu, M., Ashjian, P., De Ugarte, D. A., Huang, J. I., Mizuno, H., Alfonso, Z. C., Fraser, J. K., Benhaim, P., and Hedrick, M. H. (2002). Human adipose tissue is a source of multipotent stem cells. *Mol. Biol. Cell* **13,** 4279–4295.

CHAPTER TWELVE

Regulation of Muscle Stem Cells Activation: The Role of Growth Factors and Extracellular Matrix

Edyta Brzoska, Maria A. Ciemerych, Marta Przewozniak, *and* Małgorzata Zimowska

Contents

I. Cells Participating in Muscle Growth and Repair: Mechanisms of Activation 240
II. The Unique Ability of Skeletal Muscles to Regenerate 246
III. Muscle Stem Cells Activation: The Importance of Satellite Cell Niche 249
IV. The Interactions with ECM 252
V. Growth Factors Regulating Activation of Satellite Cells: The Case of HGF 255
VI. Other ECM-bound Growth Factors Regulating Myoblast Proliferation and Differentiation 258
VII. Concluding Remarks 260
Acknowledgments 261
References 261

Abstract

Vertebrate skeletal muscle is composed of organized multinucleate muscle fibers and also various subpopulations of cells localized in between. Some of them can be considered as the stem cells, however, few of them are able to follow myogenic program. First and most extensively studied so far, are the satellite cells that serve as tissue-specific precursors for muscle growth and repair. They are located between the basal membrane and the sarcolemma of adult muscle myofibers. They remain quiescent but can be activated in response to muscle damage resulted from mechanical injury, stretching, exercise, denervation, or progressing muscle dystrophy. Except the satellite cells also other stem cells could participate in muscle fibers reconstruction. Such cells as pericytes and mesangioblasts, muscle-derived stem cells, including so-called muscle side population, or CD133 expressing cells, were proved to be able to undergo

Department of Cytology, Institute of Zoology, Faculty of Biology, University of Warsaw, Warsaw, Poland

Vitamins and Hormones, Volume 87
ISSN 0083-6729, DOI: 10.1016/B978-0-12-386015-6.00031-7

myogenic differentiation in experiments involving their *in vitro* coculture with myoblasts or transplantation to injured skeletal muscle. In the current review, we will summarize stimuli influencing skeletal muscle stem cells activation, that is, growth factors which are secreted by muscle fibers, satellite cells, inflammatory cells, or released from basal lamina. We will also describe factors present within the skeletal muscle niche which interactions with stem cells lead to their activation, proliferation, asymmetric divisions, migration, and finally differentiation into myotubes, and then terminally differentiated myofibers.

I. Cells Participating in Muscle Growth and Repair: Mechanisms of Activation

Myogenesis occurs both during embryonic development and postnatal muscle growth (Buckingham *et al.*, 2003). Myogenic differentiation can also take place within the functional skeletal muscle of an adult organism. Such muscle retains a unique ability to reconstruct its structure in response to injury or disease (Moraczewski *et al.*, 2008; Wallace and McNally, 2009). In each case, that is, in developing, postnatally growing, or regenerating muscle, similar sequence of events occurs. Myogenic precursor cells are activated to start their proliferation and differentiation (Fig. 12.1). Finally, myoblasts derived from these cells fuse to form multinucleated myotubes and then, after innervation, muscle fibers (Rochlin *et al.*, 2010). It has to be noted, however, that during the embryonic myogenesis or during the regeneration, the activation of myogenic precursors and differentiation of myoblasts occurs within the evidently different environment. In the developing embryo, myoblasts forming skeletal muscles in the trunk and limbs originate from somites, which are the segments of paraxial mesoderm. In the adult muscle, myoblasts are the progeny of the unique population of muscle-specific precursor cells, that is, satellite cells (Bismuth and Relaix, 2010; Buckingham and Vincent, 2009). These cells serve as tissue-specific precursors prerequisite for muscle growth and repair (Kawiak *et al.*, 2006; Mauro, 1961; Zammit and Beauchamp, 2001). The somites-origin of satellite cells was shown (Gros *et al.*, 2005; Schienda *et al.*, 2006). Kassar-Duchossoy and collaborators documented that during mouse embryo development population of dermomyotome-origin cells expressing *Pax3* and *Pax7* but not muscle markers (myogenic regulatory factors, MRFs) give rise to satellite cells population in the trunk and limb muscles (Kassar-Duchossoy *et al.*, 2005).

In adult muscle, satellite cells are located between the basal membrane and the sarcolemma of myofibers, remain quiescent, and can be activated in response to muscle injury, denervation, stretching, exercise, or disease (Pallafacchina *et al.*, 2010). Within the skeletal muscle the satellite cell-derived myoblast could fuse with existing fibers or fuse with each other to form at first

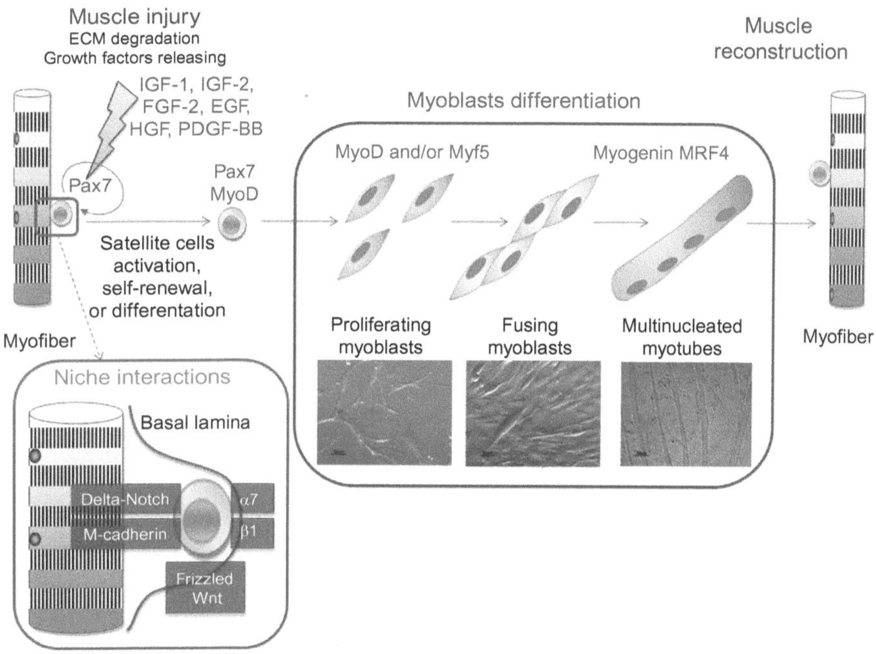

Figure 12.1 Factors controlling satellite cells activation and myoblast differentiation. In injured muscle, ECM (basal lamina) is degraded what leads to the release of growth factors, such as IGF-1, IGF-2, FGF-2, EGF, PDGF-BB, and HGF. Growth factors and cytokines might also originate from sources other than ECM, that is, myoblasts, myofibres, interstitial cell, vasculature, blood cells. Changes caused by the muscle injury lead to the satellite cells (muscle stem cells) activation and inflammatory cells mobilization. Pax7 expressing quiescent satellite cells interact with their niche and are characterized by asymmetric interactions with myofiber and basal lamina. These interactions secure both their quiescent state and ability to self-renew. When activated they start to proliferate and express myogenic regulatory factors *MyoD* and/or *Myf5*. Some of the cells downregulate *Pax7*, express *myogenin* and *MRF4*. These cells become myoblasts that differentiate, fuse, and form mulinucleated myotubes and then myofibers. Other subpopulations of cells downregulate *MyoD*, express *Pax7*, and proliferate to reestablish self-renewing satellite cells population.

new immature myotubes and then mature muscle fibers (Bischoff, 1975; Bischoff and Heintz, 1994; Lipton and Schultz, 1979). These cells can be also isolated, and thus, their differentiation could be followed under *in vitro* culture conditions (Fig. 12.1; Brzoska *et al.*, 2006a; Grabowska *et al.*, 2011). The number of satellite cells decreases during organism life, for example, 30–35% of the postnatal mouse myofibers nuclei belongs to the satellite cells, but in the adult mouse muscle they consist only 1–4% (Schultz, 1974). Aging satellite cells are characterized by their delayed response to activation and limited ability to proliferate and differentiate (Barani *et al.*, 2003; Bockhold *et al.*, 1998; Shefer *et al.*, 2006).

The population of satellite cells was shown to be heterogeneous in such organisms as mouse, rat, or human (Biressi and Rando, 2010; Collins *et al.*, 2005; Lindstrom *et al.*, 2010; Rossi *et al.*, 2010; Rouger *et al.*, 2004). However, almost all (more than 90%) of the quiescent satellite cells express *Pax7* (paired box proteins 7; Buckingham and Relaix, 2007; Olguin and Olwin, 2004; Seale *et al.*, 2000; Zammit *et al.*, 2006). Pax7 and its paralogue Pax3 are the members of the paired box transcription factors family (Relaix *et al.*, 2006) which regulate, either directly or indirectly, the expression of such important myogenic regulatory factors as MyoD (Buckingham *et al.*, 2006), Myf5 (Collins *et al.*, 2009; McKinnell *et al.*, 2008; Tajbakhsh *et al.*, 1996), and myogenin (Collins *et al.*, 2009). Moreover, in quiescent satellite cells Pax7 binds to the promoter of *Id3* (inhibitor of differentiation 3) gene, induces its expression, that is, prevents the premature differentiation (Kumar *et al.*, 2009). The same proportion of satellite cells, that is, approximately 90%, is characterized by the presence of Myf5 (Beauchamp *et al.*, 2000), muscle-specific adhesion proteins, such as syndecan-3 and -4 (Cornelison *et al.*, 2001; Tanaka *et al.*, 2009), integrin alpha7 (Kuang *et al.*, 2007), integrin beta1 (Kuang *et al.*, 2007), m-cadherin (Irintchev *et al.*, 1994; Wrobel *et al.*, 2007), VCAM1 (vascular cellular adhesion molecule-1) (Rosen *et al.*, 1992), CD34 (Beauchamp *et al.*, 2000), and c-met (Cornelison and Wold, 1997). Satellite cells also synthesize CXCR4 (Cerletti *et al.*, 2008) which is a receptor of SDF-1 (stromal derived factor-1), that is, cytokine involved in regulation of stem cells trafficking (Kucia *et al.*, 2004; Libura *et al.*, 2002; Pituch-Noworolska *et al.*, 2003). The main source of SDF-1 is bone marrow, but it is also produced within the injured tissues like skeletal muscles, heart, or liver. Importantly, its gradient is responsible for the directed migration of hematopoietic stem cells (HSCs), mesenchymal stem cells (MSCs), and tissue-specific stem cells to the site of the injured tissue (Kucia *et al.*, 2006), and also to the site of inflammation.

Any given stem cell, including satellite cells, retains capability to self-renew. This crucial feature secures maintaining of the satellite cells pool necessary for muscle repair and growth. Pax7 transcription factor was suggested to be a major player in satellite cell self-renewal. The role of *Pax7* gene in this process was tested by the generation of *Pax7*-null mice. These mice were characterized either by the complete absence of satellite cells (Seale *et al.*, 2000) or by dramatic decrease in their number (Oustanina *et al.*, 2004; Relaix *et al.*, 2005). Moreover, even if in the *Pax7*-null newborn mice muscle satellite cells were present, their number noticeably decreased during knock-out animal life span comparing to wild-type one (Oustanina *et al.*, 2004; Relaix *et al.*, 2005). What important, Pax7 was also shown to play a role in human satellite cells renewal (Pawlikowski *et al.*, 2009). Notably, *Pax7* is expressed by the majority of quiescent, but also in rat and mouse activated satellite cells and proliferating myoblasts (Brzoska *et al.*, 2009; Zammit *et al.*, 2006). *Pax3*, however, was shown to be expressed only by a small subpopulation of quiescent and also activated satellite cells of some muscles, but in diaphragm, the majority of Pax7-positive satellite cells also coexpress *Pax3* (Buckingham and Relaix, 2007; Montarras *et al.*, 2005; Relaix *et al.*, 2006). Zammit and coworkers showed that induction of *Pax7* expression in *Pax*-7-null C2C12 myoblasts leads to the increase of MyoD-positive cells number (Zammit *et al.*, 2006). On the other hand, overexpression of *Pax7* downregulates *MyoD*, prevents *myogenin* induction, and promotes cell cycle exit (Olguin and Olwin, 2004). However, all activated satellite cells, and derived from them proliferating myoblasts, coexpress the transcription factors Pax7 and MyoD, as it was documented in experiments involving *in vitro* culture of rat or mouse satellite cells on isolated muscle fibers (Yablonka-Reuveni and Rivera, 1994; Zammit *et al.*, 2002). Also among the activated satellite cells different subpopulations emerge. Those are: the cells that downregulate *Pax7*, express *MyoD*, and complete myogenic program, and the cells that maintain *Pax7* expression, downregulate *MyoD*, and then withdraw from the cell cycle to restore satellite cells population (Zammit *et al.*, 2004b). However, it was also reported that satellite cells obtained from *Pax7*-null mice proliferated and differentiated efficiently (Mitchell *et al.*, 2010). It was documented that Pax7 inhibited MyoD function and that myogenin negatively regulated Pax7 (Olguin *et al.*, 2007). Inhibition of *Pax7* expression could be executed by the member of TGF-beta protein family engaged in inhibition of myoblasts differentiation, that is, myostatin. Myostatin was also shown to have impact on *Pax7* expression via ERK1/2 (extracellular signal-regulated kinase-1/2) signaling, and thus, to be involved in the control of the satellite cell self-renewal (McFarlane *et al.*, 2008). Also other signaling pathways were suggested to be involved in the regulation of Pax7 levels. As it was reported by Perez-Ruiz and coworkers, constitutive retroviral-driven expression of beta-catenin results in increase in the number of satellite cells, expressing *Pax7* but not *MyoD*, that were able to undergo self-renewing (Perez-Ruiz *et al.*, 2008). Thus, authors concluded that beta-catenin signaling directs these cells toward the self-renewal pathway. Yet, another factor involved in the process of self-renewal is Sprouty1, an inhibitor of the signaling pathways depending on the activation of receptor tyrosine kinases (RTKs). Sprouty1 is highly expressed in quiescent satellite cells in intact adult muscle, but in proliferating satellite cells, that is, after muscle injury, its expression is downregulated and resurrected again in satellite cells "returning" to quiescent state (Shea *et al.*, 2010). Thus, Sprouty1 is required for the return to quiescence during the self-renewing of Pax7-expressing satellite cells (Abou-Khalil and Brack, 2010). Therefore, it can be concluded that Pax7 is required for the maintenance and survival of satellite cells. Pax7 influence on satellite cell differentiation is still unclear; however, differentiation of satellite cell-derived myoblasts is associated with the decrease in *Pax7* expression and regulated by precisely and timely controlled expression of MRFs.

Once satellite cells become activated, they downregulate the expression of *Pax7* and induce the synthesis of so-called early myogenic regulatory factors (early MRFs) such as MyoD and Myf5 (Cooper *et al.*, 1999; Yablonka-Reuveni *et al.*, 2008; Zammit *et al.*, 2004a). Next, the terminal differentiation of myoblasts is governed by so-called late MRFs, like myogenin and MRF4 (Parker *et al.*, 2003). MRFs contain bHLH (basic helix-loop-helix) domain which enables dimerization with regulatory factors, that is, E-proteins. These heterodimers bind to the DNA consensus sequence CANNTG and regulate the expression of proteins crucial for proper functioning of skeletal muscle fiber, such as myosin heavy chains (MyHC), troponin, and muscle creatinine kinase (Blais *et al.*, 2005; Charge and Rudnicki, 2004; Yokoyama and Asahara, 2011).

The growing body of evidence suggests that the processes of satellite cells quiescence, self-renewal, as well as myoblast differentiation are also controlled by the microRNAs molecules, which are a class of about 22-nucleotide noncoding RNAs posttranscriptionaly regulating gene expression. Among those involved in the restriction of myoblast differentiation are miR-1 and miR-206 (Chen *et al.*, 2010). Chen and coworkers cultured satellite cells isolated from transgenic mice in which genome exons 22 and 23 of the *Dicer* gene (RNase III endonuclease responsible for miRNA maturation) had been flanked by loxP sequences. Infection of $\text{Dicer}_{\text{flox/flox}}$ satellite cells with viruses encoding Cre recombinase led to the deletion of *Dicer* and depletion of all mature, functional microRNAs (Chen *et al.*, 2010). *Dicer*-null satellite cells retained their undifferentiated character and did not follow myogenic program, as it was showed by very low expression of the MyHCs. Moreover, miR-1 and miR-206 overexpression leads to the decrease in the level of Pax7 protein (Chen *et al.*, 2010). Recently, it was reported that miR-206 and miR-486 are induced by MyoD and as a result are upregulated during the differentiation of mouse C2C12 myoblasts (Dey *et al.*, 2011). Moreover, transfection of C2C12 with vectors encoding miR-206 or miR-486 increased myogenin and MyHC expression. It was also showed that miR-206 and miR-486 improved myoblast differentiation by inhibition of Pax7 expression (Dey *et al.*, 2011).

Except the satellite cells, few populations of other stem cells are able to undergo myogenic differentiation. The most important source of such stem cells is obviously skeletal muscle, but also its associated tissues and blood. Among muscle-associated cells one can list so-called muscle side population (SP) cells that are characterized by the expression of syndecan-4, stem cell marker ABCG2—a member of the ATP binding cassette (ABC) transporters, and also Sca1 (stem cell antigen-1) marker, which belongs to the *Ly-6* gene superfamily that transcripts are markers of lymphocyte activation and differentiation. These cells engraft injured muscle with higher efficiency than that achieved as a result of the transplantation of whole satellite cells population (Tanaka *et al.*, 2009). Recently, Mitchell and coworkers reported muscle-resident stem cells present within the interstitium, so-called PICs (PW1+/Pax7-interstitial cells), that expressed stress mediator PW1 (protein that participates in the TNFα–NFκB signaling cascade) and did not express *Pax7* (Mitchell *et al.*, 2010). PW1 expression was observed in early embryonic mesoderm, and it was shown to be a marker of primary myoblasts and myogenic cell lines. It was also detected in the satellite and interstitial cells of adult muscles (Nicolas *et al.*, 2005; Relaix *et al.*, 1996). PICs were shown to follow myogenic program, that is, express *MyoD*, *Pax7*, and *MyHC*. Importantly, they formed multinucleated myotubes when cultured *in vitro* and participated in muscle regeneration after transplantation into injured tibialis anterior of nude mouse (Mitchell *et al.*, 2010).

Myogenic potential was also attributed to the stem cells connected with blood vessels, that is, pericytes or mesangioblasts, as well as CD133-positive cells. Pericytes isolated from human postnatal skeletal muscle and transplanted into skeletal muscles of scid/mdx mice, that is, dystrophic mice which are characterized by immunodeficiency allowing for xenogenic transplantation, contributed to formation of numerous muscle fibers expressing human dystrophin (Dellavalle *et al.*, 2007). Also mesangioblasts were shown to effectively engraft injured muscle of dystrophic mice and dogs, and to participate in the formation of 80% new myofibers (Galvez *et al.*, 2006; Guttinger *et al.*, 2006; Morosetti *et al.*, 2010; Sampaolesi *et al.*, 2003, 2006). CD133-expressing cells follow myogenic program when cocultured with myoblasts or improve muscle regeneration in mdx mice. (Gavina *et al.*, 2006; Shi *et al.*, 2009; Torrente *et al.*, 2004). Other cells that can serve as myogenic progenitors are MSCs which differentiation into skeletal muscle myoblasts and muscle fibers was observed in *in vitro* culture (Gang *et al.*, 2008; Wakitani *et al.*, 1995), and also *in vivo*, after engraftment into the scid/mdx mice muscle in which they were able to participate in the formation of as many as 60% (De Bari *et al.*, 2003) or 30–40% new myofibers (Dezawa *et al.*, 2005). Next, the myogenic potential, however limited, was proved for multipotent adult stem cells (MASCs) (Schulze *et al.*, 2005), adipose-derived stem cells (Di Rocco *et al.*, 2006), neural stem cells (Galli *et al.*, 2000), and also the hematopoietic- and bone marrow-derived stem cells (LaBarge and Blau, 2002). The HSCs were able to follow myogenic program *in vitro* and *in vivo*, but their ability to participate in the skeletal muscle regeneration was very low, as they were able to participate in the creation of as few as 1–3.5% of newly formed myofibers (Abedi *et al.*, 2007; Camargo *et al.*, 2003; Corbel *et al.*, 2003; Dreyfus *et al.*, 2004; LaBarge and Blau, 2002; Sherwood *et al.*, 2004). Further, also stem cells that come from an exogenous source, that is, human umbilical cord blood cells, or mesenchymal stromal cells present within the human umbilical cord, were tested according to their potential to differentiate into myoblasts (Brzoska *et al.*, 2006b), but similar to HSCs, they did not efficiently participate in the skeletal muscle regeneration. Another promising source of exogenous stem cells are embryonic stem (ES) cells (Barberi *et al.*, 2005, 2007; Chang *et al.*, 2009; Kamochi *et al.*, 2006; Maltsev *et al.*, 1993; Zheng *et al.*, 2006) and induced pluripotent stem (iPS) cells (Mizuno *et al.*, 2010). Unfortunately, in order to efficiently differentiate into skeletal muscle tissue, these cells require additional "support," such as chemical treatment (Stavropoulos *et al.*, 2009), sorting of mesodermal progenitor cells (Sakurai *et al.*, 2008), or overexpression of myogenic regulatory factors such as *MyoD* (Dekel *et al.*, 1992), *Pax3* (Darabi *et al.*, 2008a), or *Pax7* (Darabi *et al.*, 2011). For example, when *Pax3* was overexpressed in differentiating mouse ES cells, resulting myoblasts were able to form 20% new myofibers in regenerating muscle (Darabi *et al.*, 2008a).

The search for the "replacement" stem cells is one of the crucial strategies in the therapy of diseases such as muscular dystrophies, that is, conditions which might be associated with the loss of the satellite cells population and resulting in the inability to regenerate (Cerletti *et al.*, 2008; Torrente *et al.*, 2003). However, the outcome is far from being satisfactory (Darabi *et al.*, 2008b; Farini *et al.*, 2009; Tedesco *et al.*, 2010; Wallace and McNally, 2009). In order to introduce them into the clinic, one has to fully understand the sophisticated and multifaceted process associated with the injury-induced activation and differentiation of the stem cells present within the skeletal muscle.

II. The Unique Ability of Skeletal Muscles to Regenerate

Regeneration of injured skeletal muscle occurs in two overlapping stages: myolysis and reconstruction (Bassaglia and Gautron, 1995). Myolysis encompasses degeneration of muscle fibers associated with the acute inflammation. Reconstruction includes activation of satellite cells and differentiation of resulting myoblasts leading to the formation of new muscle fibers (Garrett, 1990; Kasemkijwattana *et al.*, 1998). Moreover, it is associated with the remodeling of the existing and synthesis of the new extracellular matrix (ECM) components (Stauber *et al.*, 1990). The skeletal muscle regeneration has been extensively experimentally tested in various mammalian species, such as mouse, rat, dog, bovine, and also humans, what allowed to state that in each case this process occurs with different, species-characteristic dynamics. In rat, the first steps of skeletal muscle regeneration, that is, degeneration and accompanied inflammation, take place within the first few days after injury (up to fifth day), whereas formation of new myofibers usually begins between 7 to 10 days after injury. The regeneration process peaks at two and then decreases within 3–4 weeks after injury (Bassaglia and Gautron, 1995). In the mouse, the first signs of reconstruction, that is, the presence of myotubes and so-called early myofibers with centrally located myonuclei, are observed on fifth day after injury (e.g., Brzoska *et al.*, 2006b).

In injured muscle, the integrity of the myofiber plasma membrane, that is, sarcolemma, and basal lamina is destroyed, leading to the increase of extracellular calcium (Belcastro *et al.*, 1998). Disintegration of myofibers is initiated by the proteolysis of the structural proteins, such as myosin, actin, desmin, nebulin, titin, alpha actinin, and other that build the muscle fiber. The degradation of muscle proteins results from the action of cytosolic rather than lysosomal proteases (Furuno and Goldberg, 1986; Ueda *et al.*, 1998). Among the enzymes that play the major role in the protein degradation induced by muscle injury are calcium-activated neutral cysteine proteinases, referred to as calpains (Moraczewski *et al.*, 1996; Murachi, 1984; Sultan *et al.*, 2000; Zimowska *et al.*, 2001). These enzymes are involved in the initiation of myofibrillar and cytoskeletal protein breakdown, and activity of mili-calpain and micro-calpain leads to the extensive degradation of such muscle structural proteins as desmin, nebulin, titin, alpha actinin, troponin, and tropomyosin, resulting in devastation of muscle architecture (Huang and Forsberg, 1998). Such proteins as myosin and actin were also shown to be also degraded via ubiquitin-dependent manner, that is, in proteasomes (Taillandier *et al.*, 2004). Simultaneously, with the initiation of massive proteolysis the early phase of myolysis involves the inflammation manifested by the migration of inflammatory cells to the site of the injury and phagocytosis of cellular debris (Smith *et al.*, 2008; Tidball, 1995). Recruitment of inflammatory cells is a complex process. It involves proinflammatory mediators, such as interleukin-1 and -6 (IL-1, IL-6), tumor necrosis factor-alpha (TNF-alpha), histamine, platelet-activated factor (PAF), and prostaglandins. These factors can enhance the production of specific neutrophil and macrophage chemoattractants, such as macrophage inflammatory protein-2 (MIP-2), the rodent analog of interleukin-8 (IL-8), macrophage chemoattractant protein-1 (MCP-1), and MIP-1 which are released by blood cells (Cote *et al.*, 2008). Among the cells that produce such attractants, and are involved in the control of tissue injury and repair, are mast cells. They are present within various tissues, such as connective tissue, but also localized within the skeletal muscle, and can secrete a wide variety of mediators causing inflammatory responses, for example, IL-1beta, IL-6, IL-8, and TNF-alpha (Biedermann *et al.*, 2000). They were also shown to modulate the postinjury accumulation of neutrophils (Cote *et al.*, 2008). As a result, neutrophil numbers increase within 2 h postinjury and remain above control levels at least for next 48 h. So, the neutrophils are the cells that are the first to invade the injury site and release cytokines that attract and activate other types of inflammatory cells. Neutrophils also produce cytolytic and cytotoxic molecules, that is, hypochlorous acid, reactive oxygen species, that have a negative impact on damaged muscle (Tidball, 1995). Moreover, they also release matrix metalloproteinases (MMPs), that is, zinc-dependent enzymes responsible for the degradation of ECM components (see Section IV). *In vitro* studies showed that human blood neutrophils added to the human myoblasts caused their damage by inducing membrane rupture and increased myotubes' endocytosis (Pizza *et al.*, 2001). Thus, neutrophils, except being involved in phagocytosis of degenerating muscle fibers, can also damage previously uninjured ones and can be blamed as a "factors" that exacerbate skeletal muscle damage after injury. Moreover, it is also possible that neurophils can negatively impact not only at myoblasts and myotubes but also at the satellite cells present within the injured muscle (Pizza *et al.*, 2001).

As it was mentioned previously, the appearance of neutrophils precedes the infiltration of the macrophages (Cote *et al.*, 2008). Two populations of macrophages appear to perform multiple functions at various times in the inflammatory process: $ED1^+$ and $ED2^+$. The $ED1^+$ macrophages appear within necrotic fibers as early as 1 day after neutrophil invasion and are activated by the proinflammatory cytokines, such as TNF-alpha and IL-1beta. They were shown to be involved in phagocytosis and removal of cellular debris of injured muscle (Prisk and Huard, 2003). $ED1^+$ cells are followed by an influx of $ED2^+$ macrophages. These second population of macrophages do not enter necrotic tissues but appear during the latter stages of inflammation. In rat muscle their elevated number is observed from 24 to 48 h after injury (McLennan, 1996). These macrophages do not phagocytose degenerating skeletal muscle fibers but rather produce growth factors such as fibroblast growth factor (FGF), insulin-like growth factor (IGF-1), and transforming growth factor-beta1 (TGF-beta1). As result, they promote muscle regeneration by stimulation of myoblast proliferation (Massimino *et al.*, 1997; Smith *et al.*, 2008; St Pierre and Tidball, 1994). Consequently, suppression of their infiltration results in the impairment of skeletal muscle regeneration as it was shown in mice (Segawa *et al.*, 2008). Thus, macrophages are involved in both phases of skeletal muscle regeneration: first, inflammation and cleansing of necrosis and then myogenic differentiation and tissue repair (Chazaud *et al.*, 2009).

As it was presented in the first part of this review, the cells that are involved in the reconstruction of the skeletal muscle are satellite cells. Upon muscle injury, they become activated and give rise to myoblasts that proliferate and differentiate to form muscle fibers. These fibers restore, at least partially, architecture and function of regenerated muscle. Under normal conditions, the regenerated muscle is morphologically and functionally indistinguishable from undamaged muscle. The reconstruction of muscle depends on proper innervation of the newly formed muscle fibers (Pinter *et al.*, 2003). Deregulation of this process may result in the abnormal muscle architecture and also in the development of fibrosis, that is, accumulation of ECM components, such as collagen and fibronectin (Kaariainen *et al.*, 2000; Miller *et al.*, 1999). Excessive production of fibrotic tissue can affect regeneration of skeletal muscle. Since various muscle types are differently prone to the excessive fibrosis, the general characteristic of skeletal muscles, as a tissue that is characterized by great ability to regenerate, does not apply to all known muscle types.

Mature muscles are a mosaic of fast and slow fibers (Keller *et al.*, 2000; Pallafacchina *et al.*, 2002). For example, extensor digitorum longus (edl) can serve as an example of fast-twitch and soleus as slow-twitch muscle. Mechanical injury of each of these muscles results in the different progression of the repair process. In case of slow-twitch soleus muscle, regeneration is often inefficient and hindered by the development of fibrosis, that is, scar tissue, which results from the accumulation of ECM components, such as collagen and fibronectin. In rat soleus muscle, the formation of scar tissue begins usually between the second and third week postinjury and is the end product of the muscle repair process (Bassaglia and Gautron, 1995). Next, the number of satellite cells present within the slow-twitch muscle was shown to be higher than in the fast-twitch one what also can result in the differences in the process of muscle regeneration (Collins *et al.*, 2005; Zammit *et al.*, 2002). Interestingly, these cells also varied in their behavior during *in vitro* culture. Rat myoblast resulting from satellite cells isolated from slow-twitch muscle soleus, plated at high density on gelatin-coated dishes, proliferated more vigorously but less efficiently fused into myotubes than those isolated from fast-twitched edl muscle (Lagord *et al.*, 1998). The delayed differentiation of soleus-derived cells was proved by the lower expression of mRNA encoding muscular creatine kinase enzyme. In contrast, our own observations documented that *in vitro* cultured rat edl myoblasts differentiate at a lower rate than soleus myoblasts (Brzoska *et al.*, 2009). The background of the observed differences is not clear. Soleus- and edl-derived myoblasts also differed in *Pax7* expression, that is, at late stages of differentiation edl-derived myoblasts expressed higher levels of *Pax7* then soleus ones (Brzoska *et al.*, 2009). Last but not least, edl and soleus differed in MMP expression and activity observed during muscle regeneration (Zimowska *et al.*, 2008) (see Section IV). Importantly, it was also suggested that the properties of satellite cells in their myofiber niche could depend on their origin (Zammit *et al.*, 2004b). Summarizing, the behavior of satellite cells isolated from soleus and edl can serve as another example of differences existing between fast and slow muscles.

III. Muscle Stem Cells Activation: The Importance of Satellite Cell Niche

One of the most crucial components regulating the properties and impacting at the fate of satellite cells is their niche (Gopinath and Rando, 2008). Under normal physiological conditions, interactions with the niche control satellite cells quiescence and self-renewal. In injured muscles, as it was presented above, niche changes rapidly, and thus, the satellite cells activation and further differentiation of resulting myoblasts becomes possible.

What can be described as satellite cells niche is a highly asymmetrical environment restricted by myofibre sarcolemma, which is characterized by the presence of such proteins as m-cadherin and Delta–Notch ligand as well as the basal lamina which is composed of type IV collagen, laminin, fibronectin, entacin, and other proteoglycans and glycoproteins (Fig. 12.1; Sanes, 2003). Moreover, satellite cells localized within the niche react not only to the physical interactions with myofiber and basal lamina components but are also exposed to various growth factors or cytokines released from ECM or secreted by the myofibre, the microvasculature, that is, endothelial cells and periendothelial cells, the blood cells and interstitial cells localized in the vicinity of the skeletal muscle (Christov *et al.*, 2007).

As it was mentioned above, the asymmetry of the satellite cells, and also of their niche, is manifested by the specific localization of such adhesion proteins as m-cadherin or integrin alpha7. M-cadherin localized on sarcolemma of myofiber interacts with m-cadherin detected at this satellite cell surface (apical side) that interacts with myofiber (Cornelison and Wold, 1997; Irintchev *et al.*, 1994). Integrin alpha7 subunit, which is also considered as a marker of satellite cells, or integrin beta1 subunit is localized at this surface of the satellite cell (basal side) that communicates with the basal lamina and form linkage between cytoskeleton and laminin (Burkin and Kaufman, 1999; Li *et al.*, 2003). Such asymmetry may also impact at the plane of cell divisions.

As it is generally accepted, asymmetric stem cell division leads to the generation of two types of cells, that is, the stem cell (satellite stem cells) and the one which will start to differentiate (satellite myogenic precursor cells). Growing body of evidence suggests that asymmetric interactions of satellite cells with their niche result in their unequal division. As it was described by Kuang and coworkers, about 10% of the adult satellite cells expressing *Pax7* and lacking Myf5 (in brief described as Pax7+/Myf5−) represent a stem cell population. These cells undergo apical-basal oriented divisions resulting in the generation of basally located Pax7+/Myf5− (satellite stem cells) and an apically positioned cells that express *Myf 5*, that is, Pax7+/Myf5+ cells (satellite myogenic cells) (Kuang *et al.*, 2007). Isolation of both subpopulations of cells and their transplantation into muscle showed that Pax7+/Myf5+ cells differentiated into myofibers and Pax7+/Myf5− cells contributed to the satellite cell reservoir (Kuang *et al.*, 2007).

The asymmetric satellite cells division is regulated by the Numb which inhibits Notch signaling. The myofiber express Delta-like protein, that is, ligand which binds and activates Notch receptor positioned at the apical side of satellite cell. In adult mice, the satellite cells express Notch1, Notch2, and Notch3 that interact with Notch ligands Delta-like1 and Jagged1 (Conboy and Rando, 2002; Fukada *et al.*, 2007). Activation of Notch leads to the cleavage of its intracellular domain, that is, NICD (Notch intracellular domain) by γ-secretase. NICD is then translocated into the nucleus where binds transcription factor CBF1/suppressor of hairless/Lag-1 (CSL), known also as RBPJ, and activates expression of target genes. Activation of Notch in mouse satellite cells leads to their activation, proliferation, but inhibits their differentiation (Conboy and Rando, 2002; Conboy *et al.*, 2003). Notably, the satellite cell niche is unoccupied in the mice in which muscle fibers both alleles of *RBPJ* were conditionally disrupted (Vasyutina *et al.*, 2007). As it was mentioned above, Notch signaling is negatively regulated by Numb (Chapman *et al.*, 2006) which plays a crucial role in the satellite cells activation and self-renewal (Conboy and Rando, 2002). Numb is asymmetrically localized in cytoplasm of dividing mouse satellite cells, and consequently it is involved in the regulation of asymmetric cell divisions (Conboy and Rando, 2002). Chapman and coworkers documented that proliferating C2C12 myoblasts which receive high levels of Numb down-regulate Notch signaling. These cells to which none or low amounts of Numb were segregated continue Notch signaling (Chapman *et al.*, 2006). Thus, cells that retain Numb expression follow differentiation program, in opposite to these ones that lose Numb (Chapman *et al.*, 2006; Conboy *et al.*, 2003; Shinin *et al.*, 2006).

Notch signaling pathway was also shown to be deeply immersed in network of other signaling pathways. Among them is Wnt, and its receptor Frizzled, pathway. Wnt proteins compose large family of secreted signaling molecules that act, among others, throughout the stabilization of beta-catenin that avoids degradation, becomes translocated into the nucleus, and controls transcription (Willert and Jones, 2006). Among the Wnts impacting at myoblasts is Wnt3a that promotes their differentiation via beta-catenin-dependent mechanism (Pansters *et al.*, 2011). Crosstalk between Notch and Wnt signaling pathways, through GSK3beta (glycogen synthase kinase-3), provides the differentiation signal (Brack *et al.*, 2008). On the other hand, Wnt7a when bound to Frizzled7 stimulates symmetric satellite stem cell division (from 30% to 67%), what allows the increase of the satellite stem cell pool (Le Grand *et al.*, 2009). By contrast, the treatment of satellite cells isolated from mouse edl muscle with Wnt3a did not change the portion of symmetric and asymmetric divisions, and Wnt7a treatment did not influence satellite cell-derived myoblast differentiation.

Notch signaling is also regulated by its interactions with syndecan-3 that is transmembrane heparan sulfate proteoglycan (HSPG) expressed in satellite cells (Cornelison *et al.*, 2004; Fuentealba *et al.*, 1999; Pisconti *et al.*, 2010). It serves as a link that via its extracellular domain connects ECM proteins and growth factors with cytoskeleton and via its cyto-plasmic domain impacts on the intracellular signaling molecules. Pisconti and coworkers showed that the absence of syndecan-3 disrupt Notch signaling, and in consequence, the satellite cells stay activated, fail to return to quiescence, and delay terminal differentiation affecting the skeletal muscle regeneration (Pisconti *et al.*, 2010).

The changes in the properties of satellite cells niche in aging muscle appear to be the main factor responsible for the impair in their regeneration ability (Collins *et al.*, 2007; Conboy and Rando, 2005). In aged muscles, lower expression of Delta causes decrease in the proliferation of satellite cells, dependently of Notch signaling (Conboy *et al.*, 2003). Deregulation of functioning of aging satellite cells is also manifested by the fact that adult satellite cells, but not embryonic or prenatal myogenic precursors, require neither Pax7 nor Pax3 for myogenic differentiation during muscle regeneration (Lepper *et al.*, 2009). Moreover, limited satellite cells proliferation and myoblasts differentiation observed in aging muscles is connected with the decreased "efficiency" of various signaling pathways crucial for these processes such as Wnt, TGF, or IGF signaling (Barton-Davis *et al.*, 1998; Carlson *et al.*, 2008, 2009).

As it was already mentioned the niche, and in particular basal lamina, except being involved in the control of satellite cell self-renewal and activation, is also a reservoir for inactive forms of growth factors that are released after myofiber damage and induce satellite cells activation (Boonen and Post, 2008; Cornelison, 2008; Ten Broek *et al.*, 2010). Besides, the growth factors, and also cytokines, might originate from various sources other than ECM (Taipale and Keski-Oja, 1997), including the myofibre and interstitial cells, the vasculature (Christov *et al.*, 2007), and the blood cells (Marino *et al.*, 2011; Tidball, 2005).

IV. The Interactions with ECM

The ECM is a structure formed by macromolecules capable of self-assembly, mainly via noncovalent bonds. It is composed predominantly of collagens, noncollagenous glycoproteins, and proteoglycans. In muscle, ECM provides a basic mechanical support for fibers, vessels, and nerves, and its composition and plasticity is crucial not only for the satellite cell differentiation but also for retaining their undifferentiating character (Gilbert *et al.*, 2010). During skeletal muscle injury and following repair components of ECM, that is, collagens, laminins, fibronectin, entactin/nidogen, and HSPGs, are degraded and then resynthesized by fibroblasts and muscle cells (Mott and Werb, 2004). The notion that processes that lead to the ECM remodeling are necessary for the proper skeletal muscle repair is supported by the results of multiple studies. For example, inhibition of collagen synthesis blocked the differentiation of cultured L6 rat skeletal myoblasts (Nandan *et al.*, 1990) or prevented myotube formation by differentiating C2C12 mouse myoblasts (Saitoh *et al.*, 1992). Both, inhibition of proteoglycan synthesis by sodium chlorate or blocking the integrin by beta-D-xyloside and RGDS peptides, inhibited fusion and differentiation of these cells (Osses and Brandan, 2002). Interestingly, not only the presence but also the quality of the ECM proteins was shown to impact at the myoblast behavior. Thus, decorin-, but not fibronectin-modified collagen I stimulated myoblast migration *in vitro* (Goetsch *et al.*, 2011).

ECM is also an important reservoir for growth factors and cytokines modulating not only their activity but also turnover. By interacting with matrix molecules, the growth factors may become sequestered from their receptors, may become activated, for example, by proteolytic processing, or they may be presented to the cells in a manner significantly altering their bioactivity. For example, the ability of myoblasts to response to TGF-beta is directly modulated by the expression of decorin, a chondroitin/dermatan sulfate proteoglycan. Reduction of decorin expression or of decorin availability results in a decreased responsiveness to TGF-beta (Riquelme *et al.*, 2001). Thus, during the skeletal muscle injury and regeneration, the proper sequence of degradation and synthesis of ECM components, and also the proper levels of specific growth factors acting at the subsequent stages of repair, can either promote or inhibit the differentiation of myoblasts and muscle reconstruction.

As it was mentioned above, the injury of the skeletal muscle initiates the proteolysis of the protein components of the muscle fibers and also ECM proteins. Next, fibroblasts, leukocytes, and activated myoblast present within the injured tissue start to resynthesize ECM. Fibronectin is the first to be produced, as early as 1–5 h after muscle injury (Gulati *et al.*, 1982). Then, the fibronectin network is associated with collagen, which is synthesized by fibroblasts and forms the scaffold for the adhering and aggregating platelets, thus, provides the anchorage for the migrating fibroblast and satellite cells (Gulati *et al.*, 1984; Maley *et al.*, 1995). At the beginning of tissue repair, the fibronectin synthesis occurs simultaneously with the production of type III collagen, which at that stage of muscle reconstruction is the predominant form of collagen. Later, the production of type I collagen significantly increases (Lehto and Jarvinen, 1985; Salonen *et al.*, 1985). An important phase of muscle regeneration involves synthesis of glycosaminoglycans (GAGs) which form a hydrophilic matrix securing the suitable environment for ECM remodeling and muscle healing (Lehto *et al.*, 1985). GAGs moiety composition changes during the tissue repair. Initially, the predominantly synthesized GAG is hyaluronic acid, then its production drops, but the increase in the levels of chondroitin sulfate and dermatan sulfate is observed (Shum and Chau, 1996). Finally, when proliferation of myoblasts reaches a plateau, the synthesis of heparan sulfate is initiated. An increased expression of HSPGs: syndecan-3, glypican, and perlecan was found in murine skeletal muscle regeneration induced by barium chloride injection (Casar *et al.*, 2004). Such changes in the GAGs compositions result in the modification of growth factors, cytokines, and chemokines binding and activity.

Except the growth factors also the enzymes capable to modify the muscle environment were shown to be stored within ECM. Among them are MMPs that play the pivotal role in the ECM remodeling during muscle regeneration. MMPs family includes more than 20 identified members. They are responsible for the degradation of native forms of collagen types I, II, and III and also denatured collagen, laminin, and other ECM proteins (Lewis *et al.*, 2001; Mott and Werb, 2004; Sternlicht and Werb, 2001). The MMPs are assembled as so-called zymogens, that is, inactive endopeptidases (Okada *et al.*, 1988). To become active, they need to be processed by other proteolytic enzymes, such as serine proteases: furin or plasmin. Moreover, MMPs' activity is efficiently controlled by tissue inhibitors of MMP (TIMPs). Four TIMPs, that is, TIMP-1, -2, -3, and -4, inhibit MMPs' activity by binding to their catalytic sites (Visse and Nagase, 2003). Except TIMPs the broad-spectrum inhibitor of proteases, that is, alpha 2-macroglobulin, is able to arrest MMPs' action. Such enzymes as MMP-2, MMP-9, and MMP-14 can be also inhibited by the membrane-bound glycoprotein RECK (reversion including cysteine-rich protein with Kazal motifs; Parks *et al.*, 2004). MMPs are synthesized by many different types of cells, such as myoblast and also muscle fibers, fibroblasts, and neutrophils. They can be either anchored to the cell surface or secreted into extracellular space (Sternlicht and Werb, 2001). Although several MMPs exist within the skeletal muscle tissue, two of them, that is, MMP-9 and MMP-2, are thought to play an important role during skeletal muscle response to injury (Kherif *et al.*, 1998; Zimowska *et al.*, 2009). In the inactive form they are present in the intact muscle, and their activation is associated with the inflammatory-induced changes related to inflammatory myopathies of the skeletal muscle (Kieseier *et al.*, 2001; Schoser *et al.*, 2002). Importantly, increased expression and activation of both MMP isoforms can be caused by muscle injury, such as induced by cardiotoxin injection or occurring in mdx mice, and is associated with the regeneration (Bani *et al.*, 2008; Kherif *et al.*, 1999). Importantly, the levels of MMP-9 and -2 expression and activation depend on the phase of muscle regeneration and also on the muscle type, that is, fast- versus slow-twitch muscle (Zimowska *et al.*, 2008). In soleus muscle elevated levels of MMP-9 were observed during the myolysis and reconstruction phase, while regenerating edl muscles exhibited decreased MMP-9 activity during myolysis and increased MMP-2 activity at the reconstruction phase (Zimowska *et al.*, 2008). Thus, at various stages of the soleus and edl muscles' regeneration different MMPs are involved in ECM remodeling. Moreover, it can be linked to previously mentioned differences in the ability of these muscles to regenerate (see Section II). Extensive fibrosis in soleus muscle can result, at least partially, from the insufficient degradation of ECM components and can be reversed for example by the treatments blocking the growth factors involved in fibrosis development (see Section VI).

The role of MMP during muscle repair is not only restricted to the proteolysis of ECM structural proteins, that is, collagen. MMPs can directly or indirectly affect the activity of various growth factors that function during the inflammation and repair processes. For example, the binding of TGF-beta to decorin inactivates this growth factor, and its activation requires proteolytic release from ECM (Casar *et al.*, 2004). Also, FGF binding to syndecan-1 and glypican is essential for the function of FGF during muscle development and repair, since they affect the proliferation and differentiation of turkey myoblasts (Velleman *et al.*, 2007). Therefore, surrounding ECM plays an important role in growth control and differentiation. It acts not only as a scaffold for the cells but also as a reservoir of growth factors and cytokines, regulating their activation status and turnover. Disorganization of the ECM affects the proper progression of the myogenic program, independently of myogenin expression (Osses and Brandan, 2002). *In vivo*, muscular injuries that destroy the muscle basal lamina generally present a poorer functional recovery than injuries that minimally disrupt its integrity and orientation (Sanes, 2003). Thus, skeletal muscle differentiation and regeneration are regulated by interactions between exogenous growth factor-activated signaling cascades and endogenous muscle-specific transcriptional programs. Various growth factors are thought to play a role in different stages of muscle regeneration. Among them are the members of the different growth factor families, such as HGF/SF (hepatocyte growth factor/scatter factor), TGFs, IGFs, FGFs, regulating muscle regeneration.

V. Growth Factors Regulating Activation of Satellite Cells: The Case of HGF

Together with cell–ECM interactions, controlled secretion of growth factors plays a key role during embryonic myogenesis, postnatal growth, and regeneration of skeletal muscles. Part of growth factors can be stored in ECM where they are bound to proteoglycans (Chu *et al.*, 2004; Forsten-Williams *et al.*, 2008; Goerges and Nugent, 2003) and can be released during ECM remodeling mediated by MMPs that accompanies muscle regeneration (Kurisaki *et al.*, 2003; Levi *et al.*, 1996). In regenerating skeletal muscles, growth factors are secreted by proliferating and differentiating myoblasts and also cells infiltrating damaged tissue, such as neutrophils, macrophages, T-cells, and platelets (Arsic *et al.*, 2004; Deasy *et al.*, 2009; Fielding *et al.*, 1993; Lu *et al.*, 2011; Shimokado *et al.*, 1985; Ten Broek *et al.*, 2010; Tidball, 2005). The action of growth factors depends on their concentration, type of receptors that are used, and also interplay between various signaling pathways activated.

Satellite cells proliferation is controlled by variety of growth factors, such as IGF-1 and IGF-2, FGF-2, EGF (epidermal growth factor), PDGF-BB (platelet-derived growth factor-BB), and LIF (leukemia inhibitory factor; for review, see Ciemerych *et al.*, 2011). However, to this point only HGF/SF was shown to induce satellite cells exit from quiescent state, that is, their transition from G0 to G1 phase (Bischoff, 1986; Tatsumi *et al.*, 1998). Other growth factors such as FGF-2, IGF-1 and 2, PDGF-BB, TGF-beta1 and 2 were shown to be unable to activate *in vitro* cultured rat satellite cells (Johnson and Allen, 1995). However, they were able to promote their proliferation when those cells were already activated (Chakravarthy *et al.*, 2000; Doumit *et al.*, 1993; Schabort *et al.*, 2009; Sheehan and Allen, 1999; Tanaka *et al.*, 2008). Among the growth factors that regulate myoblast differentiation are IGF-2, TNF-alpha, and TGF-beta (Carter *et al.*, 2009; Chen *et al.*, 2005; Rosenthal *et al.*, 1991; Schabort *et al.*, 2009).

HGF/SF, as it was previously mentioned, plays a crucial role in satellite cells activation. It was first purified from rat hepatocytes by Nakamura and coworkers (Nakamura *et al.*, 1986). Similarly to other growth factors, also HGF/SF can be stored in ECM where it is bound to HSPGs and released upon muscle injury, during ECM remodeling by MMPs (Yamada *et al.*, 2008). Subsequently, HGF/SF binds to c-met receptor and tyrosine kinase residues of c-met intracellular domains become autophosphorylated, and thus, able to transduce the signal (Allen *et al.*, 1995; Basilico *et al.*, 2008). In satellite cells, this interaction was shown to activate ERK1 and ERK2 MAP (mitogen-activated protein) kinase pathway or to induce phosphatidylinositol 3-kinase (PI3K) pathway (Halevy and Cantley, 2004; Leshem *et al.*, 2002; Tatsumi *et al.*, 1998). Activation of ERK1 and ERK2 leads to their translocation into nucleus and results in the transcription of such factors as cyclin D1, which is a crucial cell cycle regulator (Coolican *et al.*, 1997). Therefore, levels of phosphorylated, that is, activated, ERK1 and ERK2 proteins can be considered as a marker of myogenic cells responsiveness to growth factors (McFarland and Pesall, 2008), and their activation is necessary for cell proliferation (Chambard *et al.*, 2007). In certain types of myoblast cell lines, such as murine i28 cells, HGF/SF-mediated phosphorylation of ERK1 and ERK2 kinases is abolished by PI3K inhibitor Wortmannin suggesting that activation of MAP kinase pathway may require PI3K activity (Halevy and Cantley, 2004).

Characterization of HGF/SF role in the satellite cells activation is based on the observation made by Bischoff in 1986. He showed that crushed muscle extract (CME) contains a factor which induces activation of quiescent satellite cells associated with isolated muscle fibers, but not of other cells, for example, muscle fibroblasts (Bischoff, 1986). However, he was unable to identify the proteins present in CME which are responsible for satellite cells activation (Bischoff, 1986, 1990). Also, other authors verified that factors present in extracts from murine muscles may stimulate cell cycle reentry of satellite cells (Mezzogiorno *et al.*, 1993). Growth factor responsible for satellite cells activation was finally identified by Tatsumi and coworkers in 1998 and described as HGF/SF (Tatsumi *et al.*, 1998). Analysis of CME revealed that it contains HGF/SF protein (Tatsumi *et al.*, 1998). Further, this growth factor is localized in ECM surrounding uninjured muscle fibers, and c-met receptor is synthesized by satellite cells (Tatsumi *et al.*, 1998). Injection of HGF/SF to rat tibialis anterior muscle increases the number of DNA-replicating cells, indicating that this growth factor impacts on cellular proliferation. Tatsumi and colleagues also established that immunoneutralization of HGF/SF activity decreases the number of proliferating myoblasts (Tatsumi *et al.*, 1998). Presented results are consistent with those obtained by others indicating that blocking of HGF/SF with an antibody inhibits proliferation of *in vitro* cultured rat myoblasts (Sheehan *et al.*, 2000). Further, this growth factor was shown to activate satellite cells in skeletal muscles of patients suffering from polymyositis/dermatomyositis (PM/DM) that manifests by chronic inflammation and weakness of skeletal muscles (Sugiura *et al.*, 2010). In mdx mice, in satellite cells present within the regions of muscle repair, HGF/SF colocalizes with c-met receptor (Tatsumi *et al.*, 1998). Expression of mRNA encoding HGF/SF was also shown to be upregulated in regenerating rat and bovine skeletal muscles (Hayashi *et al.*, 2004; Jennische *et al.*, 1993). Importantly, level of HGF/SF expression is proportional to the extent of muscle injury, that is, to the number of activated satellite cells that are required for regeneration (Cornelison and Wold, 1997). Moreover, HGF/SF is also released upon contractions in exercised human skeletal muscles (O'Reilly *et al.*, 2008). It has been shown that not all c-met-positive satellite cells respond equally to HGF/SF, serving as additional proof of the heterogeneity of satellite cells population (Wozniak *et al.*, 2003, 2005).

Both elevated expression and colocalization of HGF/SF and c-met are considered as an earliest markers of satellite cells activation (Tatsumi *et al.*, 1998). Although HGF/SF levels are increased in regenerating muscles, ectopic administration of this growth factor to regenerating murine tibialis anterior may also inhibit muscle regeneration (Miller *et al.*, 2000). At the time of injury, exogenous administration of HGF/SF (6.25 ng and 50 ng per muscle) increases the number of activated satellite cells, but when muscle exposure to this growth factor is prolonged to 3 days after injury, regeneration is inhibited in dose-dependent manner. This is manifested by decreased number and area of regenerated muscle fibers when higher doses (50 ng per muscle) of this growth factor are administrated into muscles (Miller *et al.*, 2000).

In vitro studies in that rat and bovine satellite cells were cultured in presence of HGF/SF confirm results obtained from analyses of skeletal muscle regeneration. They prove that cell proliferation is stimulated by this growth factor proportionally to increased concentrations of HGF/SF (0.1–30 ng/ml). Moreover, HGF/SF-treated satellite cells enter cell cycle earlier than control cells and express higher levels of cyclin D1 and PCNA (proliferating cell nuclear antigen)(Allen *et al.*, 1995; Hayashi *et al.*, 2004), and concomitantly were shown to contain hyperphosphorylated, that is, inactive form of crucial cell cycle suppressor, that is, retinoblastoma protein (pRB) (Leshem and Halevy, 2002). Increased expression of cyclin D1 leads to activation of CDK4/6 (cyclin-dependent kinase 4 and 6) which phosphorylates and inactivates pRB, and as a result allows the cell cycle reentry. Moreover, HGF/SF action on satellite cells is accompanied by decreased expression of CDK inhibitor $p27^{KIP1}$, again allowing cells to proliferate (Leshem and Halevy, 2002). Interestingly, Yamada and coworkers showed that HGF/SF has a dual impact on satellite cells cultured *in vitro* depending on its dose. High concentration of this growth factor (500 ng/ml) inhibited their proliferation, and induced expression of myostatin which, as shown by other authors, leads to cyclin D1 degradation in murine C2C12 myoblasts (Yamada *et al.*, 2010; Yang *et al.*, 2007). High concentrations of this growth factor stimulate expression of another CDK inhibitor, that is, $p21^{CIP1}$, thus, induce withdrawal from the cell cycle and reestablish quiescence of satellite cells. Addition of anti-myostatin antibody to the satellite cells *in vitro* cultured in the presence of high concentrations (500 ng/ml) of HGF/SF inhibited $p21^{CIP1}$ expression and prevented cell cycle arrest (Yamada *et al.*, 2010). Together, those results clearly show that low concentrations of HGF/SF activate satellite cells, but its high concentrations inhibit proliferation of those cells.

At present HGF/SF remains the only characterized growth factor that triggers satellite cells activation and that is indispensable for efficient skeletal muscle regeneration. One has to remember that few markers of satellite cells activation that are known to this point include, as it was described previously, colocalization of HGF/SF and c-met receptor, expression of c-fos and c-jun genes, that appear in satellite cells 3–6 h after muscle injury, expression of cyclin D1, which is a general marker of cell cycle reentry, and other factors such as Myf5 (Allen *et al.*, 1995; Beauchamp *et al.*, 2000; Kami *et al.*, 1995; Tatsumi *et al.*, 1998; Wozniak *et al.*, 2005). Activated satellite cells are also distinguished from the quiescent ones by the appearance of such proteins like PCNA associated with DNA replication, not transition from G0 to G1. Therefore, determination of all growth factors that may activate quiescent satellite cells requires identification of markers that are directly engaged in satellite cells exit from quiescence.

VI. Other ECM-bound Growth Factors Regulating Myoblast Proliferation and Differentiation

Except HGF/SF which was shown to be the only factor impacting at the satellite cells activation also multiple other ECM-stored growth factors that have varying affinities for heparin and heparin-like molecules impact at the myoblast proliferation and differentiation. Among them are the members of FGF family. Several of these factors are expressed in the developing skeletal muscle, such as FGF-1, FGF-2 (Doumit *et al.*, 1993; Moore *et al.*, 1991). FGF-1 and FGF-2 receptors are present on quiescent satellite cells and thus, may be involved in the activation of those cells (Kastner *et al.*, 2000). Moreover, FGFs are secreted in injured skeletal muscles, and their pro-proliferative potential is at least equal to this of HGF/SF (Johnson and Allen, 1995; Tanaka *et al.*, 2008; Yablonka-Reuveni *et al.*, 1999). FGF-6 expression was shown to be upregulated during muscle regeneration, strongly suggesting its participation in muscle repair (Floss *et al.*, 1997). The single injection of FGF-6 in soleus muscle accelerates its regeneration in adult mice (Armand *et al.*, 2003). FGF-6 improves muscle regeneration by stimulating differentiation but not proliferation of satellite cells and/or myoblasts (Sakuma *et al.*, 2000). Moreover, FGF-6 is also involved in the maintenance of the reserve pool of satellite cells in skeletal muscle (Israeli *et al.*, 2004).

Other factors that impact at almost all stages of muscle regeneration are IGFs-1 and -2. IGF bioavailability is modulated by IGF receptor and by six IGF binding proteins (IGFBPs) that attach IGFs to ECM proteins (Clemmons *et al.*, 1993). *In vitro* studies showed that IGF-1 and -2 promote myoblast proliferation and differentiation. Treatment of *in vitro* cultured primary human skeletal muscle cells with IGF-1 stimulates their proliferation and fusion (Czifra *et al.*, 2006). It also stimulates terminal differentiation of myoblasts by inducing myogenin expression (Florini *et al.*, 1991). As far as the role of IGF-2 is concerned, it has been shown that it stimulates the proliferation and differentiation of *in vitro* cultured L6 myoblasts (Bach *et al.*, 1995). It was also documented that IGF-2-initiated signaling targets transcriptional coregulators that are essential cofactors for MyoD, and thus, inhibition of IGF-2 synthesis impaired the transcriptional activity of this transcription factor (Wilson and Rotwein, 2006).

Analyses of regenerating muscles showed that *IGF-1* and *-2* expression become upregulated after injury. Moreover, both factors promote reinnervation of reconstructed muscle fibers (Caroni and Grandes, 1990; Vergani *et al.*, 1998). IGF-1 appears to induce muscle hypertrophy, that is, increase in the size of myofibres, during muscle regeneration (Adams and McCue, 1998; Bark *et al.*, 1998) and improves regeneration via promoting the cell survival (Lawlor and Rotwein, 2000). IGF-2 was shown to play variable roles throughout the skeletal muscle regeneration. The local administration of IGF-2 to regenerating skeletal muscle, at the myolysis phase, results in delayed myoblast proliferation and differentiation, however, the continuous administration of IGF-2 resulted in the enhancement of muscle repair (Kirk *et al.*, 2003).

In contrast to IGF-1 and IGF-2 that can promote muscle differentiation *in vitro* and can enhance muscle repair *in vivo*, members of the TGF-beta superfamily prominently inhibit myoblast differentiation and muscle regeneration. The TGF-betas are essential regulators of cell proliferation and also ECM synthesis (Massague *et al.*, 1986). In addition, TGF-beta disrupts an IGF-2-stimulated signaling pathway that is necessary for muscle differentiation *in vitro* (Gardner *et al.*, 2011). TGF-beta regulates cellular processes by binding to three high-affinity cell surface receptors: TGF-beta type I (TGF-betaRI), type II (TGF-betaRII), and type III (TGF-betaRIII, also known as betaglycan). Activity of TGF-beta is also controlled by decorin and biglycan that are proteoglycans binding TGF-beta in ECM.

Studies involving *in vitro* cultured Sol8 myoblasts revealed that TGF-beta1 action leads to the downregulation of the expression of MRFs, such as *MyoD*, and completely inhibited expression of *MyHC* (Allegra *et al.*, 2004). Analyses of mechanically injured, regenerating rat muscles revealed that the fibrosis, evident in slow-twitch soleus muscle, was accompanied by elevated TGFbeta1 expression (Zimowska *et al.*, 2009). Injection of antibodies against TGF-beta1 or TGF-betaRI into regenerating soleus or edl muscles resulted in the acceleration and improvement of the repair process. Also the treatment of *in vitro* cultured myoblasts isolated either from soleus or edl muscles with same antibodies increased proliferation rate and also caused precocious fusion into myotubes (Zimowska *et al.*, 2009). Thus, the muscle environment has a profound effect on the regenerative capacity of the resident muscle precursor cells. Control of growth factors is necessary to proper regeneration of injured muscle and myoblast differentiation, and manipulations of their levels and activities may have a beneficiary impact on the progress of muscle regeneration.

VII. Concluding Remarks

The exceptional ability of skeletal muscles to efficiently regenerate relies on the presence of satellite cells. These cells remain under the constant influence of their niche, that is, environment delineated by the myofibre sarcolemma and the basal lamina. If niche remains stable, it is able to sustain satellite cell self-renewal necessary to secure proper functioning of the skeletal muscle. However, upon the muscle injury this environment rapidly changes leading to satellite cells activation, proliferation, and differentiation of resulting myoblasts. Activation of satellite cells results from the degradation of muscle fibers and ECM (basal lamina) and release of the variety of growth factors and cytokines, such as IGF-1, IGF-2, FGF-2, EGF, PDGF-BB, HGF/SF, SDF-1. These factors are either accumulated in ECM or secreted by myoblasts, interstitial cells, vasculature, or blood cells. Such changing environment is also able to "attract" other stem cells originating from muscle itself, its associated cells and blood. When introduced into the muscle, such cells are able to follow myogenic program. Thus, rapidly and dynamically changing environment accompanies at first the injured muscle degradation and then its reconstruction. Full understanding of all the processes involved in the skeletal muscle regeneration, the role of ECM components, and identification of cells that can participate in this process is crucial for the development of therapies improving the repair of skeletal muscles that have been severely injured or are affected by the muscular disorders.

ACKNOWLEDGMENTS

The study was supported by The Polpharma Scientific Foundation and Medicine (grant number: 2/VII/2008) and the Ministry of Science and Higher Education (grant numbers N N301 304037 and N N301 422438). We would like to thank to all the members of the Department of Cytology for their constant support and helpful discussions. We are grateful to Karolina Archacka for her very careful reading of this manuscript and helpful comments. The authors regret the omission of any relevant literature, which happened either because of space constraint or oversight.

REFERENCES

Abedi, M., Foster, B. M., Wood, K. D., Colvin, G. A., McLean, S. D., Johnson, K. W., and Greer, D. A. (2007). Haematopoietic stem cells participate in muscle regeneration. *Br. J. Haematol.* **138,** 792–801.

Abou-Khalil, R., and Brack, A. S. (2010). Muscle stem cells and reversible quiescence: The role of sprouty. *Cell Cycle* **9,** 2575–2580.

Adams, G. R., and McCue, S. A. (1998). Localized infusion of IGF-I results in skeletal muscle hypertrophy in rats. *J. Appl. Physiol.* **84,** 1716–1722.

Allegra, S., Li, J. Y., Saez, J. M., and Langlois, D. (2004). Terminal differentiation of Sol 8 myoblasts is retarded by a transforming growth factor-beta autocrine regulatory loop. *Biochem. J.* **381,** 429–436.

Allen, R. E., Sheehan, S. M., Taylor, R. G., Kendall, T. L., and Rice, G. M. (1995). Hepatocyte growth factor activates quiescent skeletal muscle satellite cells in vitro. *J. Cell. Physiol.* **165,** 307–312.

Armand, A. S., Launay, T., Pariset, C., Della Gaspera, B., Charbonnier, F., and Chanoine, C. (2003). Injection of FGF6 accelerates regeneration of the soleus muscle in adult mice. *Biochim. Biophys. Acta* **1642,** 97–105.

Arsic, N., Zacchigna, S., Zentilin, L., Ramirez-Correa, G., Pattarini, L., Salvi, A., Sinagra, G., and Giacca, M. (2004). Vascular endothelial growth factor stimulates skeletal muscle regeneration in vivo. *Mol. Ther.* **10,** 844–854.

Bach, L. A., Salemi, R., and Leeding, K. S. (1995). Roles of insulin-like growth factor (IGF) receptors and IGF-binding proteins in IGF-II-induced proliferation and differentiation of L6A1 rat myoblasts. *Endocrinology* **136,** 5061–5069.

Bani, C., Lagrota-Candido, J., Pinheiro, D. F., Leite, P. E., Salimena, M. C., Henriques-Pons, A., and Quirico-Santos, T. (2008). Pattern of metalloprotease activity and myofiber regeneration in skeletal muscles of mdx mice. *Muscle Nerve* **37,** 583–592.

Barani, A. E., Durieux, A. C., Sabido, O., and Freyssenet, D. (2003). Age-related changes in the mitotic and metabolic characteristics of muscle-derived cells. *J. Appl. Physiol.* **95,** 2089–2098.

Barberi, T., Willis, L. M., Socci, N. D., and Studer, L. (2005). Derivation of multipotent mesenchymal precursors from human embryonic stem cells. *PLoS Med.* **2,** e161.

Barberi, T., Bradbury, M., Dincer, Z., Panagiotakos, G., Socci, N. D., and Studer, L. (2007). Derivation of engraftable skeletal myoblasts from human embryonic stem cells. *Nat. Med.* **13,** 642–648.

Bark, T. H., McNurlan, M. A., Lang, C. H., and Garlick, P. J. (1998). Increased protein synthesis after acute IGF-I or insulin infusion is localized to muscle in mice. *Am. J. Physiol.* **275,** E118–E123.

Barton-Davis, E. R., Shoturma, D. I., Musaro, A., Rosenthal, N., and Sweeney, H. L. (1998). Viral mediated expression of insulin-like growth factor I blocks the aging-related loss of skeletal muscle function. *Proc. Natl. Acad. Sci. USA* **95,** 15603–15607.

Basilico, C., Arnesano, A., Galluzzo, M., Comoglio, P. M., and Michieli, P. (2008). A high affinity hepatocyte growth factor-binding site in the immunoglobulin-like region of Met. *J. Biol. Chem.* **283,** 21267–21277.

Bassaglia, Y., and Gautron, J. (1995). Fast and slow rat muscles degenerate and regenerate differently after whole crush injury. *J. Muscle Res. Cell Motil.* **16,** 420–429.

Beauchamp, J. R., Heslop, L., Yu, D. S., Tajbakhsh, S., Kelly, R. G., Wernig, A., Buckingham, M. E., Partridge, T. A., and Zammit, P. S. (2000). Expression of CD34 and Myf5 defines the majority of quiescent adult skeletal muscle satellite cells. *J. Cell Biol.* **151,** 1221–1234.

Belcastro, A. N., Shewchuk, L. D., and Raj, D. A. (1998). Exercise-induced muscle injury: A calpain hypothesis. *Mol. Cell. Biochem.* **179,** 135–145.

Biedermann, T., Kneilling, M., Mailhammer, R., Maier, K., Sander, C. A., Kollias, G., Kunkel, S. L., Hultner, L., and Rocken, M. (2000). Mast cells control neutrophil recruitment during T cell-mediated delayed-type hypersensitivity reactions through tumor necrosis factor and macrophage inflammatory protein 2. *J. Exp. Med.* **192,** 1441–1452.

Biressi, S., and Rando, T. A. (2010). Heterogeneity in the muscle satellite cell population. *Semin. Cell Dev. Biol.* **21,** 845–854.

Bischoff, R. (1975). Regeneration of single skeletal muscle fibers in vitro. *Anat. Rec.* **182,** 215–235.

Bischoff, R. (1986). A satellite cell mitogen from crushed adult muscle. *Dev. Biol.* **115,** 140–147.

Bischoff, R. (1990). Interaction between satellite cells and skeletal muscle fibers. *Development* **109,** 943–952.

Bischoff, R., and Heintz, C. (1994). Enhancement of skeletal muscle regeneration. *Dev. Dyn.* **201,** 41–54.

Bismuth, K., and Relaix, F. (2010). Genetic regulation of skeletal muscle development. *Exp. Cell Res.* **316,** 3081–3086.

Blais, A., Tsikitis, M., Acosta-Alvear, D., Sharan, R., Kluger, Y., and Dynlacht, B. D. (2005). An initial blueprint for myogenic differentiation. *Genes Dev.* **19,** 553–569.

Bockhold, K. J., Rosenblatt, J. D., and Partridge, T. A. (1998). Aging normal and dystrophic mouse muscle: Analysis of myogenicity in cultures of living single fibers. *Muscle Nerve* **21,** 173–183.

Boonen, K. J., and Post, M. J. (2008). The muscle stem cell niche: Regulation of satellite cells during regeneration. *Tissue Eng. Part B Rev.* **14,** 419–431.

Brack, A. S., Conboy, I. M., Conboy, M. J., Shen, J., and Rando, T. A. (2008). A temporal switch from notch to Wnt signaling in muscle stem cells is necessary for normal adult myogenesis. *Cell Stem Cell* **2,** 50–59.

Brzoska, E., Bello, V., Darribere, T., and Moraczewski, J. (2006a). Integrin alpha3 subunit participates in myoblast adhesion and fusion in vitro. *Differentiation* **74,** 105–118.

Brzoska, E., Grabowska, I., Hoser, G., Streminska, W., Wasilewska, D., Machaj, E. K., Pojda, Z., Moraczewski, J., and Kawiak, J. (2006b). Participation of stem cells from human cord blood in skeletal muscle regeneration of SCID mice. *Exp. Hematol.* **34,** 1262–1270.

Brzoska, E., Przewozniak, M., Grabowska, I., Janczyk-Ilach, K., and Moraczewski, J. (2009). Pax3 and Pax7 expression during myoblast differentiation in vitro and fast and slow muscle regeneration in vivo. *Cell Biol. Int.* **33,** 483–492.

Buckingham, M., and Relaix, F. (2007). The role of Pax genes in the development of tissues and organs: Pax3 and Pax7 regulate muscle progenitor cell functions. *Annu. Rev. Cell Dev. Biol.* **23,** 645–673.

Buckingham, M., and Vincent, S. D. (2009). Distinct and dynamic myogenic populations in the vertebrate embryo. *Curr. Opin. Genet. Dev.* **19,** 444–453.

Buckingham, M., Bajard, L., Chang, T., Daubas, P., Hadchouel, J., Meilhac, S., Montarras, D., Rocancourt, D., and Relaix, F. (2003). The formation of skeletal muscle: From somite to limb. *J. Anat.* **202,** 59–68.

Buckingham, M., Bajard, L., Daubas, P., Esner, M., Lagha, M., Relaix, F., and Rocancourt, D. (2006). Myogenic progenitor cells in the mouse embryo are marked by the expression of Pax3/7 genes that regulate their survival and myogenic potential. *Anat. Embryol. (Berl)* **211**(Suppl. 1), 51–56.

Burkin, D. J., and Kaufman, S. J. (1999). The alpha7beta1 integrin in muscle development and disease. *Cell Tissue Res.* **296,** 183–190.

Camargo, F. D., Green, R., Capetanaki, Y., Jackson, K. A., and Goodell, M. A. (2003). Single hematopoietic stem cells generate skeletal muscle through myeloid intermediates. *Nat. Med.* **9,** 1520–1527.

Carlson, M. E., Hsu, M., and Conboy, I. M. (2008). Imbalance between pSmad3 and Notch induces CDK inhibitors in old muscle stem cells. *Nature* **454,** 528–532.

Carlson, M. E., Conboy, M. J., Hsu, M., Barchas, L., Jeong, J., Agrawal, A., Mikels, A. J., Agrawal, S., Schaffer, D. V., and Conboy, I. M. (2009). Relative roles of TGF-beta1 and Wnt in the systemic regulation and aging of satellite cell responses. *Aging Cell* **8,** 676–689.

Caroni, P., and Grandes, P. (1990). Nerve sprouting in innervated adult skeletal muscle induced by exposure to elevated levels of insulin-like growth factors. *J. Cell Biol.* **110,** 1307–1317.

Carter, E. J., Cosgrove, R. A., Gonzalez, I., Eisemann, J. H., Lovett, F. A., Cobb, L. J., and Pell, J. M. (2009). MEK5 and ERK5 are mediators of the pro-myogenic actions of IGF-2. *J. Cell Sci.* **122,** 3104–3112.

Casar, J. C., McKechnie, B. A., Fallon, J. R., Young, M. F., and Brandan, E. (2004). Transient up-regulation of biglycan during skeletal muscle regeneration: Delayed fiber growth along with decorin increase in biglycan-deficient mice. *Dev. Biol.* **268,** 358–371.

Cerletti, M., Jurga, S., Witczak, C. A., Hirshman, M. F., Shadrach, J. L., Goodyear, L. J., and Wagers, A. J. (2008). Highly efficient, functional engraftment of skeletal muscle stem cells in dystrophic muscles. *Cell* **134,** 37–47.

Chakravarthy, M. V., Abraha, T. W., Schwartz, R. J., Fiorotto, M. L., and Booth, F. W. (2000). Insulin-like growth factor-I extends in vitro replicative life span of skeletal muscle satellite cells by enhancing G1/S cell cycle progression via the activation of phosphatidylinositol 3′-kinase/Akt signaling pathway. *J. Biol. Chem.* **275,** 35942–35952.

Chambard, J. C., Lefloch, R., Pouyssegur, J., and Lenormand, P. (2007). ERK implication in cell cycle regulation. *Biochim. Biophys. Acta* **1773,** 1299–1310.

Chang, H., Yoshimoto, M., Umeda, K., Iwasa, T., Mizuno, Y., Fukada, S., Yamamoto, H., Motohashi, N., Miyagoe-Suzuki, Y., Takeda, S., Heike, T., and Nakahata, T. (2009). Generation of transplantable, functional satellite-like cells from mouse embryonic stem cells. *FASEB J.* **23,** 1907–1919.

Chapman, G., Liu, L., Sahlgren, C., Dahlqvist, C., and Lendahl, U. (2006). High levels of Notch signaling down-regulate Numb and Numblike. *J. Cell Biol.* **175,** 535–540.

Charge, S. B., and Rudnicki, M. A. (2004). Cellular and molecular regulation of muscle regeneration. *Physiol. Rev.* **84,** 209–238.

Chazaud, B., Brigitte, M., Yacoub-Youssef, H., Arnold, L., Gherardi, R., Sonnet, C., Lafuste, P., and Chretien, F. (2009). Dual and beneficial roles of macrophages during skeletal muscle regeneration. *Exerc. Sport Sci. Rev.* **37,** 18–22.
Chen, S. E., Gerken, E., Zhang, Y., Zhan, M., Mohan, R. K., Li, A. S., Reid, M. B., and Li, Y. P. (2005). Role of TNF-{alpha} signaling in regeneration of cardiotoxin-injured muscle. *Am. J. Physiol. Cell Physiol.* **289,** C1179–C1187.
Chen, J. F., Tao, Y., Li, J., Deng, Z., Yan, Z., Xiao, X., and Wang, D. Z. (2010). microRNA-1 and microRNA-206 regulate skeletal muscle satellite cell proliferation and differentiation by repressing Pax7. *J. Cell Biol.* **190,** 867–879.
Christov, C., Chretien, F., Abou-Khalil, R., Bassez, G., Vallet, G., Authier, F. J., Bassaglia, Y., Shinin, V., Tajbakhsh, S., Chazaud, B., and Gherardi, R. K. (2007). Muscle satellite cells and endothelial cells: Close neighbors and privileged partners. *Mol. Biol. Cell* **18,** 1397–1409.
Chu, C. L., Buczek-Thomas, J. A., and Nugent, M. A. (2004). Heparan sulphate proteoglycans modulate fibroblast growth factor-2 binding through a lipid raft-mediated mechanism. *Biochem. J.* **379,** 331–341.
Ciemerych, M. A., Archacka, K., Grabowska, I., and Przewozniak, M. (2011). Cell cycle regulation during proliferation and differentiation of mammalian muscle precursor cells. *In* "Cell Cycle in Development, Results and Problems in Cell Differentiation," (J. Z. Kubiak, Ed.), Vol. 53, pp. 498–507. Springer-Verlag, Berlin/Heidelberg.
Clemmons, D. R., Jones, J. I., Busby, W. H., and Wright, G. (1993). Role of insulin-like growth factor binding proteins in modifying IGF actions. *Ann. N. Y. Acad. Sci.* **692,** 10–21.
Collins, C. A., Olsen, I., Zammit, P. S., Heslop, L., Petrie, A., Partridge, T. A., and Morgan, J. E. (2005). Stem cell function, self-renewal, and behavioral heterogeneity of cells from the adult muscle satellite cell niche. *Cell* **122,** 289–301.
Collins, C. A., Zammit, P. S., Ruiz, A. P., Morgan, J. E., and Partridge, T. A. (2007). A population of myogenic stem cells that survives skeletal muscle aging. *Stem Cells* **25,** 885–894.
Collins, C. A., Gnocchi, V. F., White, R. B., Boldrin, L., Perez-Ruiz, A., Relaix, F., Morgan, J. E., and Zammit, P. S. (2009). Integrated functions of Pax3 and Pax7 in the regulation of proliferation, cell size and myogenic differentiation. *PLoS One* **4,** e4475.
Conboy, I. M., and Rando, T. A. (2002). The regulation of Notch signaling controls satellite cell activation and cell fate determination in postnatal myogenesis. *Dev. Cell* **3,** 397–409.
Conboy, I. M., and Rando, T. A. (2005). Aging, stem cells and tissue regeneration: Lessons from muscle. *Cell Cycle* **4,** 407–410.
Conboy, I. M., Conboy, M. J., Smythe, G. M., and Rando, T. A. (2003). Notch-mediated restoration of regenerative potential to aged muscle. *Science* **302,** 1575–1577.
Coolican, S. A., Samuel, D. S., Ewton, D. Z., McWade, F. J., and Florini, J. R. (1997). The mitogenic and myogenic actions of insulin-like growth factors utilize distinct signaling pathways. *J. Biol. Chem.* **272,** 6653–6662.
Cooper, R. N., Tajbakhsh, S., Mouly, V., Cossu, G., Buckingham, M., and Butler-Browne, G. S. (1999). In vivo satellite cell activation via Myf5 and MyoD in regenerating mouse skeletal muscle. *J. Cell Sci.* **112**(Pt 17), 2895–2901.
Corbel, S. Y., Lee, A., Yi, L., Duenas, J., Brazelton, T. R., Blau, H. M., and Rossi, F. M. (2003). Contribution of hematopoietic stem cells to skeletal muscle. *Nat. Med.* **9,** 1528–1532.
Cornelison, D. D. (2008). Context matters: In vivo and in vitro influences on muscle satellite cell activity. *J. Cell. Biochem.* **105,** 663–669.
Cornelison, D. D., and Wold, B. J. (1997). Single-cell analysis of regulatory gene expression in quiescent and activated mouse skeletal muscle satellite cells. *Dev. Biol.* **191,** 270–283.
Cornelison, D. D., Filla, M. S., Stanley, H. M., Rapraeger, A. C., and Olwin, B. B. (2001). Syndecan-3 and syndecan-4 specifically mark skeletal muscle satellite cells and are implicated in satellite cell maintenance and muscle regeneration. *Dev. Biol.* **239,** 79–94.

Cornelison, D. D., Wilcox-Adelman, S. A., Goetinck, P. F., Rauvala, H., Rapraeger, A. C., and Olwin, B. B. (2004). Essential and separable roles for Syndecan-3 and Syndecan-4 in skeletal muscle development and regeneration. *Genes Dev.* **18,** 2231–2236.

Cote, C. K., DiMezzo, T. L., Banks, D. J., France, B., Bradley, K. A., and Welkos, S. L. (2008). Early interactions between fully virulent Bacillus anthracis and macrophages that influence the balance between spore clearance and development of a lethal infection. *Microbes Infect.* **10,** 613–619.

Czifra, G., Toth, I. B., Marincsak, R., Juhasz, I., Kovacs, I., Acs, P., Kovacs, L., Blumberg, P. M., and Biro, T. (2006). Insulin-like growth factor-I-coupled mitogenic signaling in primary cultured human skeletal muscle cells and in C2C12 myoblasts. A central role of protein kinase Cdelta. *Cell. Signal.* **18,** 1461–1472.

Darabi, R., Gehlbach, K., Bachoo, R. M., Kamath, S., Osawa, M., Kamm, K. E., Kyba, M., and Perlingeiro, R. C. (2008a). Functional skeletal muscle regeneration from differentiating embryonic stem cells. *Nat. Med.* **14,** 134–143.

Darabi, R., Santos, F. N., and Perlingeiro, R. C. (2008b). The therapeutic potential of embryonic and adult stem cells for skeletal muscle regeneration. *Stem Cell Rev.* **4,** 217–225.

Darabi, R., Pan, W., Bosnakovski, D., Baik, J., Kyba, M., and Perlingeiro, R. C. (2011). Functional myogenic engraftment from mouse iPS cells. *Stem Cell Rev.* [Epub ahead of print]. Doi: 10.1007/s12015-011-9258-2.

De Bari, C., Dell'Accio, F., Vandenabeele, F., Vermeesch, J. R., Raymackers, J. M., and Luyten, F. P. (2003). Skeletal muscle repair by adult human mesenchymal stem cells from synovial membrane. *J. Cell Biol.* **160,** 909–918.

Deasy, B. M., Feduska, J. M., Payne, T. R., Li, Y., Ambrosio, F., and Huard, J. (2009). Effect of VEGF on the regenerative capacity of muscle stem cells in dystrophic skeletal muscle. *Mol. Ther.* **17,** 1788–1798.

Dekel, I., Magal, Y., Pearson-White, S., Emerson, C. P., and Shani, M. (1992). Conditional conversion of ES cells to skeletal muscle by an exogenous MyoD1 gene. *New Biol.* **4,** 217–224.

Dellavalle, A., Sampaolesi, M., Tonlorenzi, R., Tagliafico, E., Sacchetti, B., Perani, L., Innocenzi, A., Galvez, B. G., Messina, G., Morosetti, R., Li, S., Belicchi, M., *et al.* (2007). Pericytes of human skeletal muscle are myogenic precursors distinct from satellite cells. *Nat. Cell Biol.* **9,** 255–267.

Dey, B. K., Gagan, J., and Dutta, A. (2011). miR-206 and -486 induce myoblast differentiation by downregulating Pax7. *Mol. Cell. Biol.* **31,** 203–214.

Dezawa, M., Ishikawa, H., Itokazu, Y., Yoshihara, T., Hoshino, M., Takeda, S., Ide, C., and Nabeshima, Y. (2005). Bone marrow stromal cells generate muscle cells and repair muscle degeneration. *Science* **309,** 314–317.

Di Rocco, G., Iachininoto, M. G., Tritarelli, A., Straino, S., Zacheo, A., Germani, A., Crea, F., and Capogrossi, M. C. (2006). Myogenic potential of adipose-tissue-derived cells. *J. Cell Sci.* **119,** 2945–2952.

Doumit, M. E., Cook, D. R., and Merkel, R. A. (1993). Fibroblast growth factor, epidermal growth factor, insulin-like growth factors, and platelet-derived growth factor-BB stimulate proliferation of clonally derived porcine myogenic satellite cells. *J. Cell. Physiol.* **157,** 326–332.

Dreyfus, P. A., Chretien, F., Chazaud, B., Kirova, Y., Caramelle, P., Garcia, L., Butler-Browne, G., and Gherardi, R. K. (2004). Adult bone marrow-derived stem cells in muscle connective tissue and satellite cell niches. *Am. J. Pathol.* **164,** 773–779.

Farini, A., Razini, P., Erratico, S., Torrente, Y., and Meregalli, M. (2009). Cell based therapy for Duchenne muscular dystrophy. *J. Cell. Physiol.* **221,** 526–534.

Fielding, R. A., Manfredi, T. J., Ding, W., Fiatarone, M. A., Evans, W. J., and Cannon, J. G. (1993). Acute phase response in exercise. III. Neutrophil and IL-1 beta accumulation in skeletal muscle. *Am. J. Physiol.* **265,** R166–R172.

Florini, J. R., Ewton, D. Z., and Roof, S. L. (1991). Insulin-like growth factor-I stimulates terminal myogenic differentiation by induction of myogenin gene expression. *Mol. Endocrinol.* **5,** 718–724.

Floss, T., Arnold, H. H., and Braun, T. (1997). A role for FGF-6 in skeletal muscle regeneration. *Genes Dev.* **11,** 2040–2051.

Forsten-Williams, K., Chu, C. L., Fannon, M., Buczek-Thomas, J. A., and Nugent, M. A. (2008). Control of growth factor networks by heparan sulfate proteoglycans. *Ann. Biomed. Eng.* **36,** 2134–2148.

Fuentealba, L., Carey, D. J., and Brandan, E. (1999). Antisense inhibition of syndecan-3 expression during skeletal muscle differentiation accelerates myogenesis through a basic fibroblast growth factor-dependent mechanism. *J. Biol. Chem.* **274,** 37876–37884.

Fukada, S., Uezumi, A., Ikemoto, M., Masuda, S., Segawa, M., Tanimura, N., Yamamoto, H., Miyagoe-Suzuki, Y., and Takeda, S. (2007). Molecular signature of quiescent satellite cells in adult skeletal muscle. *Stem Cells* **25,** 2448–2459.

Furuno, K., and Goldberg, A. L. (1986). The activation of protein degradation in muscle by Ca2+ or muscle injury does not involve a lysosomal mechanism. *Biochem. J.* **237,** 859–864.

Galli, R., Borello, U., Gritti, A., Minasi, M. G., Bjornson, C., Coletta, M., Mora, M., De Angelis, M. G., Fiocco, R., Cossu, G., and Vescovi, A. L. (2000). Skeletal myogenic potential of human and mouse neural stem cells. *Nat. Neurosci.* **3,** 986–991.

Galvez, B. G., Sampaolesi, M., Brunelli, S., Covarello, D., Gavina, M., Rossi, B., Constantin, G., Torrente, Y., and Cossu, G. (2006). Complete repair of dystrophic skeletal muscle by mesoangioblasts with enhanced migration ability. *J. Cell Biol.* **174,** 231–243.

Gang, E. J., Bosnakovski, D., Simsek, T., To, K., and Perlingeiro, R. C. (2008). Pax3 activation promotes the differentiation of mesenchymal stem cells toward the myogenic lineage. *Exp. Cell Res.* **314,** 1721–1733.

Gardner, S., Alzhanov, D., Knollman, P., Kuninger, D., and Rotwein, P. (2011). TGF-beta inhibits muscle differentiation by blocking autocrine signaling pathways initiated by IGF-II. *Mol. Endocrinol.* **25,** 128–137.

Garrett, W. E., Jr. (1990). Muscle strain injuries: Clinical and basic aspects. *Med. Sci. Sports Exerc.* **22,** 436–443.

Gavina, M., Belicchi, M., Rossi, B., Ottoboni, L., Colombo, F., Meregalli, M., Battistelli, M., Forzenigo, L., Biondetti, P., Pisati, F., Parolini, D., Farini, A., *et al.* (2006). VCAM-1 expression on dystrophic muscle vessels has a critical role in the recruitment of human blood-derived CD133+ stem cells after intra-arterial transplantation. *Blood* **108,** 2857–2866.

Gilbert, P. M., Havenstrite, K. L., Magnusson, K. E., Sacco, A., Leonardi, N. A., Kraft, P., Nguyen, N. K., Thrun, S., Lutolf, M. P., and Blau, H. M. (2010). Substrate elasticity regulates skeletal muscle stem cell self-renewal in culture. *Science* **329,** 1078–1081.

Goerges, A. L., and Nugent, M. A. (2003). Regulation of vascular endothelial growth factor binding and activity by extracellular pH. *J. Biol. Chem.* **278,** 19518–19525.

Goetsch, K. P., Kallmeyer, K., and Niesler, C. U. (2011). Decorin modulates collagen I-stimulated, but not fibronectin-stimulated, migration of C2C12 myoblasts. *Matrix Biol.* **30,** 109–117.

Gopinath, S. D., and Rando, T. A. (2008). Stem cell review series: Aging of the skeletal muscle stem cell niche. *Aging Cell* **7,** 590–598.

Grabowska, I., Szeliga, A., Moraczewski, J., Czaplicka, I., and Brzoska, E. (2011). Comparison of satellite cell-derived myoblasts and C2C12 differentiation in two- and three-dimensional cultures: Changes in adhesion protein expression. *Cell Biol. Int.* **35,** 125–133.

Gros, J., Manceau, M., Thome, V., and Marcelle, C. (2005). A common somitic origin for embryonic muscle progenitors and satellite cells. *Nature* **435,** 954–958.

Gulati, A. K., Reddi, A. H., and Zalewski, A. A. (1982). Distribution of fibronectin in normal and regenerating skeletal muscle. *Anat. Rec.* **204,** 175–183.

Gulati, A. K., Reddi, A. H., and Zalewski, A. A. (1984). Changes in the extracellular matrix components fibronectin and laminin during immune rejection of skeletal muscle. *Anat. Rec.* **209,** 21–27.

Guttinger, M., Tafi, E., Battaglia, M., Coletta, M., and Cossu, G. (2006). Allogeneic mesoangioblasts give rise to alpha-sarcoglycan expressing fibers when transplanted into dystrophic mice. *Exp. Cell Res.* **312,** 3872–3879.

Halevy, O., and Cantley, L. C. (2004). Differential regulation of the phosphoinositide 3-kinase and MAP kinase pathways by hepatocyte growth factor vs. insulin-like growth factor-I in myogenic cells. *Exp. Cell Res.* **297,** 224–234.

Hayashi, S., Aso, H., Watanabe, K., Nara, H., Rose, M. T., Ohwada, S., and Yamaguchi, T. (2004). Sequence of IGF-I, IGF-II, and HGF expression in regenerating skeletal muscle. *Histochem. Cell Biol.* **122,** 427–434.

Huang, J., and Forsberg, N. E. (1998). Role of calpain in skeletal-muscle protein degradation. *Proc. Natl. Acad. Sci. USA* **95,** 12100–12105.

Irintchev, A., Zeschnigk, M., Starzinski-Powitz, A., and Wernig, A. (1994). Expression pattern of M-cadherin in normal, denervated, and regenerating mouse muscles. *Dev. Dyn.* **199,** 326–337.

Israeli, D., Benchaouir, R., Ziaei, S., Rameau, P., Gruszczynski, C., Peltekian, E., Danos, O., and Garcia, L. (2004). FGF6 mediated expansion of a resident subset of cells with SP phenotype in the C2C12 myogenic line. *J. Cell. Physiol.* **201,** 409–419.

Jennische, E., Ekberg, S., and Matejka, G. L. (1993). Expression of hepatocyte growth factor in growing and regenerating rat skeletal muscle. *Am. J. Physiol.* **265,** C122–C128.

Johnson, S. E., and Allen, R. E. (1995). Activation of skeletal muscle satellite cells and the role of fibroblast growth factor receptors. *Exp. Cell Res.* **219,** 449–453.

Kaariainen, M., Jarvinen, T., Jarvinen, M., Rantanen, J., and Kalimo, H. (2000). Relation between myofibers and connective tissue during muscle injury repair. *Scand. J. Med. Sci. Sports* **10,** 332–337.

Kami, K., Noguchi, K., and Senba, E. (1995). Localization of myogenin, c-fos, c-jun, and muscle-specific gene mRNAs in regenerating rat skeletal muscle. *Cell Tissue Res.* **280,** 11–19.

Kamochi, H., Kurokawa, M. S., Yoshikawa, H., Ueda, Y., Masuda, C., Takada, E., Watanabe, K., Sakakibara, M., Natuki, Y., Kimura, K., Beppu, M., Aoki, H., *et al.* (2006). Transplantation of myocyte precursors derived from embryonic stem cells transfected with IGFII gene in a mouse model of muscle injury. *Transplantation* **82,** 516–526.

Kasemkijwattana, C., Menetrey, J., Somogyl, G., Moreland, M. S., Fu, F. H., Buranapanitkit, B., Watkins, S. C., and Huard, J. (1998). Development of approaches to improve the healing following muscle contusion. *Cell Transplant.* **7,** 585–598.

Kassar-Duchossoy, L., Giacone, E., Gayraud-Morel, B., Jory, A., Gomes, D., and Tajbakhsh, S. (2005). Pax3/Pax7 mark a novel population of primitive myogenic cells during development. *Genes Dev.* **19,** 1426–1431.

Kastner, S., Elias, M. C., Rivera, A. J., and Yablonka-Reuveni, Z. (2000). Gene expression patterns of the fibroblast growth factors and their receptors during myogenesis of rat satellite cells. *J. Histochem. Cytochem.* **48,** 1079–1096.

Kawiak, J., Brzoska, E., Grabowska, I., Hoser, G., Streminska, W., Wasilewska, D., Machaj, E. K., Pojda, Z., and Moraczewski, J. (2006). Contribution of stem cells to skeletal muscle regeneration. *Folia Histochem. Cytobiol.* **44,** 75–79.

Keller, A., Demeurie, J., Merkulova, T., Geraud, G., Cywiner-Golenzer, C., Lucas, M., and Chatelet, F. P. (2000). Fibre-type distribution and subcellular localisation of alpha and beta enolase in mouse striated muscle. *Biol. Cell* **92,** 527–535.

Kherif, S., Dehaupas, M., Lafuma, C., Fardeau, M., and Alameddine, H. S. (1998). Matrix metalloproteinases MMP-2 and MMP-9 in denervated muscle and injured nerve. *Neuropathol. Appl. Neurobiol.* **24,** 309–319.
Kherif, S., Lafuma, C., Dehaupas, M., Lachkar, S., Fournier, J. G., Verdiere-Sahuque, M., Fardeau, M., and Alameddine, H. S. (1999). Expression of matrix metalloproteinases 2 and 9 in regenerating skeletal muscle: A study in experimentally injured and mdx muscles. *Dev. Biol.* **205,** 158–170.
Kieseier, B. C., Schneider, C., Clements, J. M., Gearing, A. J., Gold, R., Toyka, K. V., and Hartung, H. P. (2001). Expression of specific matrix metalloproteinases in inflammatory myopathies. *Brain* **124,** 341–351.
Kirk, S. P., Oldham, J. M., Jeanplong, F., and Bass, J. J. (2003). Insulin-like growth factor-II delays early but enhances late regeneration of skeletal muscle. *J. Histochem. Cytochem.* **51,** 1611–1620.
Kuang, S., Kuroda, K., Le Grand, F., and Rudnicki, M. A. (2007). Asymmetric self-renewal and commitment of satellite stem cells in muscle. *Cell* **129,** 999–1010.
Kucia, M., Jankowski, K., Reca, R., Wysoczynski, M., Bandura, L., Allendorf, D. J., Zhang, J., Ratajczak, J., and Ratajczak, M. Z. (2004). CXCR4-SDF-1 signalling, locomotion, chemotaxis and adhesion. *J. Mol. Histol.* **35,** 233–245.
Kucia, M., Wojakowski, W., Reca, R., Machalinski, B., Gozdzik, J., Majka, M., Baran, J., Ratajczak, J., and Ratajczak, M. Z. (2006). The migration of bone marrow-derived non-hematopoietic tissue-committed stem cells is regulated in an SDF-1-, HGF-, and LIF-dependent manner. *Arch. Immunol. Ther. Exp. (Warsz)* **54,** 121–135.
Kumar, D., Shadrach, J. L., Wagers, A. J., and Lassar, A. B. (2009). Id3 is a direct transcriptional target of Pax7 in quiescent satellite cells. *Mol. Biol. Cell* **20,** 3170–3177.
Kurisaki, T., Masuda, A., Sudo, K., Sakagami, J., Higashiyama, S., Matsuda, Y., Nagabukuro, A., Tsuji, A., Nabeshima, Y., Asano, M., Iwakura, Y., and Sehara-Fujisawa, A. (2003). Phenotypic analysis of Meltrin alpha (ADAM12)-deficient mice: Involvement of Meltrin alpha in adipogenesis and myogenesis. *Mol. Cell. Biol.* **23,** 55–61.
LaBarge, M. A., and Blau, H. M. (2002). Biological progression from adult bone marrow to mononucleate muscle stem cell to multinucleate muscle fiber in response to injury. *Cell* **111,** 589–601.
Lagord, C., Soulet, L., Bonavaud, S., Bassaglia, Y., Rey, C., Barlovatz-Meimon, G., Gautron, J., and Martelly, I. (1998). Differential myogenicity of satellite cells isolated from extensor digitorum longus (EDL) and soleus rat muscles revealed in vitro. *Cell Tissue Res.* **291,** 455–468.
Lawlor, M. A., and Rotwein, P. (2000). Insulin-like growth factor-mediated muscle cell survival: Central roles for Akt and cyclin-dependent kinase inhibitor p21. *Mol. Cell. Biol.* **20,** 8983–8995.
Le Grand, F., Jones, A. E., Seale, V., Scime, A., and Rudnicki, M. A. (2009). Wnt7a activates the planar cell polarity pathway to drive the symmetric expansion of satellite stem cells. *Cell Stem Cell* **4,** 535–547.
Lehto, M., and Jarvinen, M. (1985). Collagen and glycosaminoglycan synthesis of injured gastrocnemius muscle in rat. *Eur. Surg. Res.* **17,** 179–185.
Lehto, M., Sims, T. J., and Bailey, A. J. (1985). Skeletal muscle injury—Molecular changes in the collagen during healing. *Res. Exp. Med. (Berl)* **185,** 95–106.
Lepper, C., Conway, S. J., and Fan, C. M. (2009). Adult satellite cells and embryonic muscle progenitors have distinct genetic requirements. *Nature* **460,** 627–631.
Leshem, Y., and Halevy, O. (2002). Phosphorylation of pRb is required for HGF-induced muscle cell proliferation and is p27kip1-dependent. *J. Cell. Physiol.* **191,** 173–182.
Leshem, Y., Gitelman, I., Ponzetto, C., and Halevy, O. (2002). Preferential binding of Grb2 or phosphatidylinositol 3-kinase to the met receptor has opposite effects on HGF-induced myoblast proliferation. *Exp. Cell Res.* **274,** 288–298.

Levi, E., Fridman, R., Miao, H. Q., Ma, Y. S., Yayon, A., and Vlodavsky, I. (1996). Matrix metalloproteinase 2 releases active soluble ectodomain of fibroblast growth factor receptor 1. *Proc. Natl. Acad. Sci. USA* **93,** 7069–7074.

Lewis, M. P., Machell, J. R., Hunt, N. P., Sinanan, A. C., and Tippett, H. L. (2001). The extracellular matrix of muscle—Implications for manipulation of the craniofacial musculature. *Eur. J. Oral Sci.* **109,** 209–221.

Li, J., Rao, H., Burkin, D., Kaufman, S. J., and Wu, C. (2003). The muscle integrin binding protein (MIBP) interacts with alpha7beta1 integrin and regulates cell adhesion and laminin matrix deposition. *Dev. Biol.* **261,** 209–219.

Libura, J., Drukala, J., Majka, M., Tomescu, O., Navenot, J. M., Kucia, M., Marquez, L., Peiper, S. C., Barr, F. G., Janowska-Wieczorek, A., and Ratajczak, M. Z. (2002). CXCR4-SDF-1 signaling is active in rhabdomyosarcoma cells and regulates locomotion, chemotaxis, and adhesion. *Blood* **100,** 2597–2606.

Lindstrom, M., Pedrosa-Domellof, F., and Thornell, L. E. (2010). Satellite cell heterogeneity with respect to expression of MyoD, myogenin, Dlk1 and c-Met in human skeletal muscle: Application to a cohort of power lifters and sedentary men. *Histochem. Cell Biol.* **134,** 371–385.

Lipton, B. H., and Schultz, E. (1979). Developmental fate of skeletal muscle satellite cells. *Science* **205,** 1292–1294.

Lu, H., Huang, D., Saederup, N., Charo, I. F., Ransohoff, R. M., and Zhou, L. (2011). Macrophages recruited via CCR2 produce insulin-like growth factor-1 to repair acute skeletal muscle injury. *FASEB J.* **25,** 358–369.

Maley, M. A., Davies, M. J., and Grounds, M. D. (1995). Extracellular matrix, growth factors, genetics: Their influence on cell proliferation and myotube formation in primary cultures of adult mouse skeletal muscle. *Exp. Cell Res.* **219,** 169–179.

Maltsev, V. A., Rohwedel, J., Hescheler, J., and Wobus, A. M. (1993). Embryonic stem cells differentiate in vitro into cardiomyocytes representing sinusnodal, atrial and ventricular cell types. *Mech. Dev.* **44,** 41–50.

Marino, M., Scuderi, F., Provenzano, C., and Bartoccioni, E. (2011). Skeletal muscle cells: From local inflammatory response to active immunity. *Gene Ther.* **18,** 109–116.

Massague, J., Cheifetz, S., Endo, T., and Nadal-Ginard, B. (1986). Type beta transforming growth factor is an inhibitor of myogenic differentiation. *Proc. Natl. Acad. Sci. USA* **83,** 8206–8210.

Massimino, M. L., Rapizzi, E., Cantini, M., Libera, L. D., Mazzoleni, F., Arslan, P., and Carraro, U. (1997). ED2+ macrophages increase selectively myoblast proliferation in muscle cultures. *Biochem. Biophys. Res. Commun.* **235,** 754–759.

Mauro, A. (1961). Satellite cell of skeletal muscle fibers. *J. Biophys. Biochem. Cytol.* **9,** 493–495.

McFarland, D. C., and Pesall, J. E. (2008). Phospho-MAPK as a marker of myogenic satellite cell responsiveness to growth factors. *Comp. Biochem. Physiol. B Biochem. Mol. Biol.* **149,** 463–467.

McFarlane, C., Hennebry, A., Thomas, M., Plummer, E., Ling, N., Sharma, M., and Kambadur, R. (2008). Myostatin signals through Pax7 to regulate satellite cell self-renewal. *Exp. Cell Res.* **314,** 317–329.

McKinnell, I. W., Ishibashi, J., Le Grand, F., Punch, V. G., Addicks, G. C., Greenblatt, J. F., Dilworth, F. J., and Rudnicki, M. A. (2008). Pax7 activates myogenic genes by recruitment of a histone methyltransferase complex. *Nat. Cell Biol.* **10,** 77–84.

McLennan, I. S. (1996). Degenerating and regenerating skeletal muscles contain several subpopulations of macrophages with distinct spatial and temporal distributions. *J. Anat.* **188**(Pt 1), 17–28.

Mezzogiorno, A., Coletta, M., Zani, B. M., Cossu, G., and Molinaro, M. (1993). Paracrine stimulation of senescent satellite cell proliferation by factors released by muscle or myotubes from young mice. *Mech. Ageing Dev.* **70,** 35–44.

Miller, G. R., Smith, C. A., and Stauber, W. T. (1999). Determination of fibrosis from cryostat sections using high performance liquid chromatography: Skeletal muscle. *Histochem. J.* **31,** 89–94.

Miller, K. J., Thaloor, D., Matteson, S., and Pavlath, G. K. (2000). Hepatocyte growth factor affects satellite cell activation and differentiation in regenerating skeletal muscle. *Am. J. Physiol. Cell Physiol.* **278,** C174–C181.

Mitchell, K. J., Pannerec, A., Cadot, B., Parlakian, A., Besson, V., Gomes, E. R., Marazzi, G., and Sassoon, D. A. (2010). Identification and characterization of a non-satellite cell muscle resident progenitor during postnatal development. *Nat. Cell Biol.* **12,** 257–266.

Mizuno, Y., Chang, H., Umeda, K., Niwa, A., Iwasa, T., Awaya, T., Fukada, S., Yamamoto, H., Yamanaka, S., Nakahata, T., and Heike, T. (2010). Generation of skeletal muscle stem/progenitor cells from murine induced pluripotent stem cells. *FASEB J.* **24,** 2245–2253.

Montarras, D., Morgan, J., Collins, C., Relaix, F., Zaffran, S., Cumano, A., Partridge, T., and Buckingham, M. (2005). Direct isolation of satellite cells for skeletal muscle regeneration. *Science* **309,** 2064–2067.

Moore, J. W., Dionne, C., Jaye, M., and Swain, J. L. (1991). The mRNAs encoding acidic FGF, basic FGF and FGF receptor are coordinately downregulated during myogenic differentiation. *Development* **111,** 741–748.

Moraczewski, J., Piekarska, E., Bonavaud, S., Wosinska, K., Chazaud, B., and Barlovatz-Meimon, G. (1996). Differential intracellular distribution and activities of mu- and m-calpains during the differentiation of human myogenic cells in culture. *C. R. Acad. Sci. III* **319,** 681–686.

Moraczewski, J., Archacka, K., Brzoska, E., Ciemerych, M. A., Grabowska, I., Janczyk-Ilach, K., Streminska, W., and Zimowska, M. (2008). From planarians to mammals—The many faces of regeneration. *Int. J. Dev. Biol.* **52,** 219–227.

Morosetti, R., Gidaro, T., Broccolini, A., Gliubizzi, C., Sancricca, C., Tonali, P. A., Ricci, E., and Mirabella, M. (2010). Mesoangioblasts from Facioscapulohumeral Muscular Dystrophy display in vivo a variable myogenic ability predictable by their in vitro behavior. *Cell Transplant* [Epub ahead of print]. Doi: 10.3727/096368910X546571.

Mott, J. D., and Werb, Z. (2004). Regulation of matrix biology by matrix metalloproteinases. *Curr. Opin. Cell Biol.* **16,** 558–564.

Murachi, T. (1984). Calcium-dependent proteinases and specific inhibitors: Calpain and calpastatin. *Biochem. Soc. Symp.* **49,** 149–167.

Nakamura, T., Teramoto, H., and Ichihara, A. (1986). Purification and characterization of a growth factor from rat platelets for mature parenchymal hepatocytes in primary cultures. *Proc. Natl. Acad. Sci. USA* **83,** 6489–6493.

Nandan, D., Clarke, E. P., Ball, E. H., and Sanwal, B. D. (1990). Ethyl-3,4-dihydroxybenzoate inhibits myoblast differentiation: Evidence for an essential role of collagen. *J. Cell Biol.* **110,** 1673–1679.

Nicolas, N., Marazzi, G., Kelley, K., and Sassoon, D. (2005). Embryonic deregulation of muscle stress signaling pathways leads to altered postnatal stem cell behavior and a failure in postnatal muscle growth. *Dev. Biol.* **281,** 171–183.

Okada, Y., Harris, E. D., Jr., and Nagase, H. (1988). The precursor of a metalloendopeptidase from human rheumatoid synovial fibroblasts. Purification and mechanisms of activation by endopeptidases and 4-aminophenylmercuric acetate. *Biochem. J.* **254,** 731–741.

Olguin, H. C., and Olwin, B. B. (2004). Pax-7 up-regulation inhibits myogenesis and cell cycle progression in satellite cells: A potential mechanism for self-renewal. *Dev. Biol.* **275,** 375–388.

Olguin, H. C., Yang, Z., Tapscott, S. J., and Olwin, B. B. (2007). Reciprocal inhibition between Pax7 and muscle regulatory factors modulates myogenic cell fate determination. *J. Cell Biol.* **177,** 769–779.

O'Reilly, C., McKay, B., Phillips, S., Tarnopolsky, M., and Parise, G. (2008). Hepatocyte growth factor (HGF) and the satellite cell response following muscle lengthening contractions in humans. *Muscle Nerve* **38,** 1434–1442.

Osses, N., and Brandan, E. (2002). ECM is required for skeletal muscle differentiation independently of muscle regulatory factor expression. *Am. J. Physiol. Cell Physiol.* **282,** C383–C394.

Oustanina, S., Hause, G., and Braun, T. (2004). Pax7 directs postnatal renewal and propagation of myogenic satellite cells but not their specification. *EMBO J.* **23,** 3430–3439.

Pallafacchina, G., Calabria, E., Serrano, A. L., Kalhovde, J. M., and Schiaffino, S. (2002). A protein kinase B-dependent and rapamycin-sensitive pathway controls skeletal muscle growth but not fiber type specification. *Proc. Natl. Acad. Sci. USA* **99,** 9213–9218.

Pallafacchina, G., Francois, S., Regnault, B., Czarny, B., Dive, V., Cumano, A., Montarras, D., and Buckingham, M. (2010). An adult tissue-specific stem cell in its niche: A gene profiling analysis of in vivo quiescent and activated muscle satellite cells. *Stem Cell Res.* **4,** 77–91.

Pansters, N. A., van der Velden, J. L., Kelders, M. C., Laeremans, H., Schols, A. M., and Langen, R. C. (2011). Segregation of myoblast fusion and muscle-specific gene expression by distinct ligand-dependent inactivation of GSK-3beta. *Cell. Mol. Life Sci.* **68,** 523–535.

Parker, M. H., Seale, P., and Rudnicki, M. A. (2003). Looking back to the embryo: Defining transcriptional networks in adult myogenesis. *Nat. Rev. Genet.* **4,** 497–507.

Parks, W. C., Wilson, C. L., and Lopez-Boado, Y. S. (2004). Matrix metalloproteinases as modulators of inflammation and innate immunity. *Nat. Rev. Immunol.* **4,** 617–629.

Pawlikowski, B., Lee, L., Zuo, J., and Kramer, R. H. (2009). Analysis of human muscle stem cells reveals a differentiation-resistant progenitor cell population expressing Pax7 capable of self-renewal. *Dev. Dyn.* **238,** 138–149.

Perez-Ruiz, A., Ono, Y., Gnocchi, V. F., and Zammit, P. S. (2008). beta-Catenin promotes self-renewal of skeletal-muscle satellite cells. *J. Cell Sci.* **121,** 1373–1382.

Pinter, S., Mendler, L., and Dux, L. (2003). Neural impacts on the regeneration of skeletal muscles. *Acta Biochim. Pol.* **50,** 1229–1237.

Pisconti, A., Cornelison, D. D., Olguin, H. C., Antwine, T. L., and Olwin, B. B. (2010). Syndecan-3 and Notch cooperate in regulating adult myogenesis. *J. Cell Biol.* **190,** 427–441.

Pituch-Noworolska, A., Majka, M., Janowska-Wieczorek, A., Baj-Krzyworzeka, M., Urbanowicz, B., Malec, E., and Ratajczak, M. Z. (2003). Circulating CXCR4-positive stem/progenitor cells compete for SDF-1-positive niches in bone marrow, muscle and neural tissues: An alternative hypothesis to stem cell plasticity. *Folia Histochem. Cytobiol.* **41,** 13–21.

Pizza, F. X., McLoughlin, T. J., McGregor, S. J., Calomeni, E. P., and Gunning, W. T. (2001). Neutrophils injure cultured skeletal myotubes. *Am. J. Physiol. Cell Physiol.* **281,** C335–C341.

Prisk, V., and Huard, J. (2003). Muscle injuries and repair: The role of prostaglandins and inflammation. *Histol. Histopathol.* **18,** 1243–1256.

Relaix, F., Weng, X., Marazzi, G., Yang, E., Copeland, N., Jenkins, N., Spence, S. E., and Sassoon, D. (1996). Pw1, a novel zinc finger gene implicated in the myogenic and neuronal lineages. *Dev. Biol.* **177,** 383–396.

Relaix, F., Rocancourt, D., Mansouri, A., and Buckingham, M. (2005). A Pax3/Pax7-dependent population of skeletal muscle progenitor cells. *Nature* **435,** 948–953.

Relaix, F., Montarras, D., Zaffran, S., Gayraud-Morel, B., Rocancourt, D., Tajbakhsh, S., Mansouri, A., Cumano, A., and Buckingham, M. (2006). Pax3 and Pax7 have distinct and overlapping functions in adult muscle progenitor cells. *J. Cell Biol.* **172,** 91–102.

Riquelme, C., Larrain, J., Schonherr, E., Henriquez, J. P., Kresse, H., and Brandan, E. (2001). Antisense inhibition of decorin expression in myoblasts decreases cell responsiveness to transforming growth factor beta and accelerates skeletal muscle differentiation. *J. Biol. Chem.* **276,** 3589–3596.

Rochlin, K., Yu, S., Roy, S., and Baylies, M. K. (2010). Myoblast fusion: When it takes more to make one. *Dev. Biol.* **341,** 66–83.

Rosen, G. D., Sanes, J. R., LaChance, R., Cunningham, J. M., Roman, J., and Dean, D. C. (1992). Roles for the integrin VLA-4 and its counter receptor VCAM-1 in myogenesis. *Cell* **69,** 1107–1119.

Rosenthal, S. M., Brunetti, A., Brown, E. J., Mamula, P. W., and Goldfine, I. D. (1991). Regulation of insulin-like growth factor (IGF) I receptor expression during muscle cell differentiation. Potential autocrine role of IGF-II. *J. Clin. Invest.* **87,** 1212–1219.

Rossi, C. A., Pozzobon, M., Ditadi, A., Archacka, K., Gastaldello, A., Sanna, M., Franzin, C., Malerba, A., Milan, G., Cananzi, M., Schiaffino, S., Campanella, M., *et al.* (2010). Clonal characterization of rat muscle satellite cells: Proliferation, metabolism and differentiation define an intrinsic heterogeneity. *PLoS One* **5,** e8523.

Rouger, K., Brault, M., Daval, N., Leroux, I., Guigand, L., Lesoeur, J., Fernandez, B., and Cherel, Y. (2004). Muscle satellite cell heterogeneity: In vitro and in vivo evidences for populations that fuse differently. *Cell Tissue Res.* **317,** 319–326.

Saitoh, O., Periasamy, M., Kan, M., and Matsuda, R. (1992). cis-4-Hydroxy-L-proline and ethyl-3,4-dihydroxybenzoate prevent myogenesis of C2C12 muscle cells and block MyoD1 and myogenin expression. *Exp. Cell Res.* **200,** 70–76.

Sakuma, K., Watanabe, K., Sano, M., Uramoto, I., and Totsuka, T. (2000). Differential adaptation of growth and differentiation factor 8/myostatin, fibroblast growth factor 6 and leukemia inhibitory factor in overloaded, regenerating and denervated rat muscles. *Biochim. Biophys. Acta* **1497,** 77–88.

Sakurai, H., Okawa, Y., Inami, Y., Nishio, N., and Isobe, K. (2008). Paraxial mesodermal progenitors derived from mouse embryonic stem cells contribute to muscle regeneration via differentiation into muscle satellite cells. *Stem Cells* **26,** 1865–1873.

Salonen, V., Lehto, M., Kalimo, M., Penttinen, R., and Aro, H. (1985). Changes in intramuscular collagen and fibronectin in denervation atrophy. *Muscle Nerve* **8,** 125–131.

Sampaolesi, M., Torrente, Y., Innocenzi, A., Tonlorenzi, R., D'Antona, G., Pellegrino, M. A., Barresi, R., Bresolin, N., De Angelis, M. G., Campbell, K. P., Bottinelli, R., and Cossu, G. (2003). Cell therapy of alpha-sarcoglycan null dystrophic mice through intra-arterial delivery of mesoangioblasts. *Science* **301,** 487–492.

Sampaolesi, M., Blot, S., D'Antona, G., Granger, N., Tonlorenzi, R., Innocenzi, A., Mognol, P., Thibaud, J. L., Galvez, B. G., Barthelemy, I., Perani, L., Mantero, S., *et al.* (2006). Mesoangioblast stem cells ameliorate muscle function in dystrophic dogs. *Nature* **444,** 574–579.

Sanes, J. R. (2003). The basement membrane/basal lamina of skeletal muscle. *J. Biol. Chem.* **278,** 12601–12604.

Schabort, E. J., van der Merwe, M., Loos, B., Moore, F. P., and Niesler, C. U. (2009). TGF-beta's delay skeletal muscle progenitor cell differentiation in an isoform-independent manner. *Exp. Cell Res.* **315,** 373–384.

Schienda, J., Engleka, K. A., Jun, S., Hansen, M. S., Epstein, J. A., Tabin, C. J., Kunkel, L. M., and Kardon, G. (2006). Somitic origin of limb muscle satellite and side population cells. *Proc. Natl. Acad. Sci. USA* **103,** 945–950.

Schoser, B. G., Blottner, D., and Stuerenburg, H. J. (2002). Matrix metalloproteinases in inflammatory myopathies: Enhanced immunoreactivity near atrophic myofibers. *Acta Neurol. Scand.* **105,** 309–313.

Schultz, E. (1974). A quantitative study of the satellite cell population in postnatal mouse lumbrical muscle. *Anat. Rec.* **180,** 589–595.

Schulze, M., Belema-Bedada, F., Technau, A., and Braun, T. (2005). Mesenchymal stem cells are recruited to striated muscle by NFAT/IL-4-mediated cell fusion. *Genes Dev.* **19,** 1787–1798.

Seale, P., Sabourin, L. A., Girgis-Gabardo, A., Mansouri, A., Gruss, P., and Rudnicki, M. A. (2000). Pax7 is required for the specification of myogenic satellite cells. *Cell* **102,** 777–786.

Segawa, M., Fukada, S., Yamamoto, Y., Yahagi, H., Kanematsu, M., Sato, M., Ito, T., Uezumi, A., Hayashi, S., Miyagoe-Suzuki, Y., Takeda, S., Tsujikawa, K., *et al.* (2008). Suppression of macrophage functions impairs skeletal muscle regeneration with severe fibrosis. *Exp. Cell Res.* **314,** 3232–3244.

Shea, K. L., Xiang, W., LaPorta, V. S., Licht, J. D., Keller, C., Basson, M. A., and Brack, A. S. (2010). Sprouty1 regulates reversible quiescence of a self-renewing adult muscle stem cell pool during regeneration. *Cell Stem Cell* **6,** 117–129.

Sheehan, S. M., and Allen, R. E. (1999). Skeletal muscle satellite cell proliferation in response to members of the fibroblast growth factor family and hepatocyte growth factor. *J. Cell. Physiol.* **181,** 499–506.

Sheehan, S. M., Tatsumi, R., Temm-Grove, C. J., and Allen, R. E. (2000). HGF is an autocrine growth factor for skeletal muscle satellite cells in vitro. *Muscle Nerve* **23,** 239–245.

Shefer, G., Van de Mark, D. P., Richardson, J. B., and Yablonka-Reuveni, Z. (2006). Satellite-cell pool size does matter: Defining the myogenic potency of aging skeletal muscle. *Dev. Biol.* **294,** 50–66.

Sherwood, R. I., Christensen, J. L., Weissman, I. L., and Wagers, A. J. (2004). Determinants of skeletal muscle contributions from circulating cells, bone marrow cells, and hematopoietic stem cells. *Stem Cells* **22,** 1292–1304.

Shi, M., Ishikawa, M., Kamei, N., Nakasa, T., Adachi, N., Deie, M., Asahara, T., and Ochi, M. (2009). Acceleration of skeletal muscle regeneration in a rat skeletal muscle injury model by local injection of human peripheral blood-derived CD133-positive cells. *Stem Cells* **27,** 949–960.

Shimokado, K., Raines, E. W., Madtes, D. K., Barrett, T. B., Benditt, E. P., and Ross, R. (1985). A significant part of macrophage-derived growth factor consists of at least two forms of PDGF. *Cell* **43,** 277–286.

Shinin, V., Gayraud-Morel, B., Gomes, D., and Tajbakhsh, S. (2006). Asymmetric division and cosegregation of template DNA strands in adult muscle satellite cells. *Nat. Cell Biol.* **8,** 677–687.

Shum, D. K., and Chau, C. H. (1996). Changes in glycosaminoglycans during regeneration of post-crush sciatic nerves of adult guinea pigs. *J. Neurosci. Res.* **46,** 465–476.

Smith, C., Halvorsen, B., Otterdal, K., Waehre, T., Yndestad, A., Fevang, B., Sandberg, W. J., Breland, U. M., Froland, S. S., Oie, E., Gullestad, L., Damas, J. K., *et al.* (2008). High levels and inflammatory effects of soluble CXC ligand 16 (CXCL16) in coronary artery disease: Down-regulatory effects of statins. *Cardiovasc. Res.* **79,** 195–203.

St Pierre, B. A., and Tidball, J. G. (1994). Differential response of macrophage subpopulations to soleus muscle reloading after rat hindlimb suspension. *J. Appl. Physiol.* **77,** 290–297.

Stauber, W. T., Fritz, V. K., and Dahlmann, B. (1990). Extracellular matrix changes following blunt trauma to rat skeletal muscles. *Exp. Mol. Pathol.* **52,** 69–86.

Stavropoulos, M. E., Mengarelli, I., and Barberi, T. (2009). Differentiation of multipotent mesenchymal precursors and skeletal myoblasts from human embryonic stem cells. *Curr. Protoc. Stem Cell Biol.* (Chapter 1, Unit 1 F 8.). 9:1F.8.1-1F.8.10.

Sternlicht, M. D., and Werb, Z. (2001). How matrix metalloproteinases regulate cell behavior. *Annu. Rev. Cell Dev. Biol.* **17,** 463–516.

Sugiura, T., Kawaguchi, Y., Soejima, M., Katsumata, Y., Gono, T., Baba, S., Kawamoto, M., Murakawa, Y., Yamanaka, H., and Hara, M. (2010). Increased HGF and c-Met in muscle tissues of polymyositis and dermatomyositis patients: Beneficial roles of HGF in muscle regeneration. *Clin. Immunol.* **136,** 387–399.

Sultan, K. R., Dittrich, B. T., and Pette, D. (2000). Calpain activity in fast, slow, transforming, and regenerating skeletal muscles of rat. *Am. J. Physiol. Cell Physiol.* **279,** C639–C647.

Taillandier, D., Combaret, L., Pouch, M. N., Samuels, S. E., Bechet, D., and Attaix, D. (2004). The role of ubiquitin-proteasome-dependent proteolysis in the remodelling of skeletal muscle. *Proc. Nutr. Soc.* **63,** 357–361.

Taipale, J., and Keski-Oja, J. (1997). Growth factors in the extracellular matrix. *FASEB J.* **11,** 51–59.

Tajbakhsh, S., Rocancourt, D., and Buckingham, M. (1996). Muscle progenitor cells failing to respond to positional cues adopt non-myogenic fates in myf-5 null mice. *Nature* **384,** 266–270.

Tanaka, Y., Yamaguchi, A., Fujikawa, T., Sakuma, K., Morita, I., and Ishii, K. (2008). Expression of mRNA for specific fibroblast growth factors associates with that of the myogenic markers MyoD and proliferating cell nuclear antigen in regenerating and overloaded rat plantaris muscle. *Acta Physiol (Oxf.)* **194,** 149–159.

Tanaka, K. K., Hall, J. K., Troy, A. A., Cornelison, D. D., Majka, S. M., and Olwin, B. B. (2009). Syndecan-4-expressing muscle progenitor cells in the SP engraft as satellite cells during muscle regeneration. *Cell Stem Cell* **4,** 217–225.

Tatsumi, R., Anderson, J. E., Nevoret, C. J., Halevy, O., and Allen, R. E. (1998). HGF/SF is present in normal adult skeletal muscle and is capable of activating satellite cells. *Dev. Biol.* **194,** 114–128.

Tedesco, F. S., Dellavalle, A., Diaz-Manera, J., Messina, G., and Cossu, G. (2010). Repairing skeletal muscle: Regenerative potential of skeletal muscle stem cells. *J. Clin. Invest.* **120,** 11–19.

Ten Broek, R. W., Grefte, S., and Von den Hoff, J. W. (2010). Regulatory factors and cell populations involved in skeletal muscle regeneration. *J. Cell. Physiol.* **224,** 7–16.

Tidball, J. G. (1995). Inflammatory cell response to acute muscle injury. *Med. Sci. Sports Exerc.* **27,** 1022–1032.

Tidball, J. G. (2005). Inflammatory processes in muscle injury and repair. *Am. J. Physiol. Regul. Integr. Comp. Physiol.* **288,** R345–R353.

Torrente, Y., Camirand, G., Pisati, F., Belicchi, M., Rossi, B., Colombo, F., El Fahime, M., Caron, N. J., Issekutz, A. C., Constantin, G., Tremblay, J. P., and Bresolin, N. (2003). Identification of a putative pathway for the muscle homing of stem cells in a muscular dystrophy model. *J. Cell Biol.* **162,** 511–520.

Torrente, Y., Belicchi, M., Sampaolesi, M., Pisati, F., Meregalli, M., D'Antona, G., Tonlorenzi, R., Porretti, L., Gavina, M., Mamchaoui, K., Pellegrino, M. A., Furling, D., *et al.* (2004). Human circulating AC133(+) stem cells restore dystrophin expression and ameliorate function in dystrophic skeletal muscle. *J. Clin. Invest.* **114,** 182–195.

Ueda, Y., Wang, M. C., Ou, B. R., Huang, J., Elce, J., Tanaka, K., Ichihara, A., and Forsberg, N. E. (1998). Evidence for the participation of the proteasome and calpain in early phases of muscle cell differentiation. *Int. J. Biochem. Cell Biol.* **30,** 679–694.

Vasyutina, E., Lenhard, D. C., Wende, H., Erdmann, B., Epstein, J. A., and Birchmeier, C. (2007). RBP-J (Rbpsuh) is essential to maintain muscle progenitor cells and to generate satellite cells. *Proc. Natl. Acad. Sci. USA* **104,** 4443–4448.

Velleman, S. G., Coy, C. S., and McFarland, D. C. (2007). Effect of syndecan-1, syndecan-4, and glypican-1 on turkey muscle satellite cell proliferation, differentiation, and responsiveness to fibroblast growth factor 2. *Poult. Sci.* **86,** 1406–1413.

Vergani, L., Di Giulio, A. M., Losa, M., Rossoni, G., Muller, E. E., and Gorio, A. (1998). Systemic administration of insulin-like growth factor decreases motor neuron cell death and promotes muscle reinnervation. *J. Neurosci. Res.* **54,** 840–847.

Visse, R., and Nagase, H. (2003). Matrix metalloproteinases and tissue inhibitors of metalloproteinases: Structure, function, and biochemistry. *Circ. Res.* **92,** 827–839.

Wakitani, S., Saito, T., and Caplan, A. I. (1995). Myogenic cells derived from rat bone marrow mesenchymal stem cells exposed to 5-azacytidine. *Muscle Nerve* **18,** 1417–1426.

Wallace, G. Q., and McNally, E. M. (2009). Mechanisms of muscle degeneration, regeneration, and repair in the muscular dystrophies. *Annu. Rev. Physiol.* **71,** 37–57.

Willert, K., and Jones, K. A. (2006). Wnt signaling: Is the party in the nucleus? *Genes Dev.* **20,** 1394–1404.

Wilson, E. M., and Rotwein, P. (2006). Control of MyoD function during initiation of muscle differentiation by an autocrine signaling pathway activated by insulin-like growth factor-II. *J. Biol. Chem.* **281,** 29962–29971.

Wozniak, A. C., Pilipowicz, O., Yablonka-Reuveni, Z., Greenway, S., Craven, S., Scott, E., and Anderson, J. E. (2003). C-Met expression and mechanical activation of satellite cells on cultured muscle fibers. *J. Histochem. Cytochem.* **51,** 1437–1445.

Wozniak, A. C., Kong, J., Bock, E., Pilipowicz, O., and Anderson, J. E. (2005). Signaling satellite-cell activation in skeletal muscle: Markers, models, stretch, and potential alternate pathways. *Muscle Nerve* **31,** 283–300.

Wrobel, E., Brzoska, E., and Moraczewski, J. (2007). M-cadherin and beta-catenin participate in differentiation of rat satellite cells. *Eur. J. Cell Biol.* **86,** 99–109.

Yablonka-Reuveni, Z., and Rivera, A. J. (1994). Temporal expression of regulatory and structural muscle proteins during myogenesis of satellite cells on isolated adult rat fibers. *Dev. Biol.* **164,** 588–603.

Yablonka-Reuveni, Z., Seger, R., and Rivera, A. J. (1999). Fibroblast growth factor promotes recruitment of skeletal muscle satellite cells in young and old rats. *J. Histochem. Cytochem.* **47,** 23–42.

Yablonka-Reuveni, Z., Day, K., Vine, A., and Shefer, G. (2008). Defining the transcriptional signature of skeletal muscle stem cells. *J. Anim. Sci.* **86,** E207–E216.

Yamada, M., Sankoda, Y., Tatsumi, R., Mizunoya, W., Ikeuchi, Y., Sunagawa, K., and Allen, R. E. (2008). Matrix metalloproteinase-2 mediates stretch-induced activation of skeletal muscle satellite cells in a nitric oxide-dependent manner. *Int. J. Biochem. Cell Biol.* **40,** 2183–2191.

Yamada, M., Tatsumi, R., Yamanouchi, K., Hosoyama, T., Shiratsuchi, S., Sato, A., Mizunoya, W., Ikeuchi, Y., Furuse, M., and Allen, R. E. (2010). High concentrations of HGF inhibit skeletal muscle satellite cell proliferation in vitro by inducing expression of myostatin: A possible mechanism for reestablishing satellite cell quiescence in vivo. *Am. J. Physiol. Cell Physiol.* **298,** C465–C476.

Yang, W., Zhang, Y., Li, Y., Wu, Z., and Zhu, D. (2007). Myostatin induces cyclin D1 degradation to cause cell cycle arrest through a phosphatidylinositol 3-kinase/AKT/GSK-3 beta pathway and is antagonized by insulin-like growth factor 1. *J. Biol. Chem.* **282,** 3799–3808.

Yokoyama, S., and Asahara, H. (2011). The myogenic transcriptional network. *Cell. Mol. Life Sci.* **68**(11), 1843–1849.

Zammit, P., and Beauchamp, J. (2001). The skeletal muscle satellite cell: Stem cell or son of stem cell? *Differentiation* **68,** 193–204.

Zammit, P. S., Heslop, L., Hudon, V., Rosenblatt, J. D., Tajbakhsh, S., Buckingham, M. E., Beauchamp, J. R., and Partridge, T. A. (2002). Kinetics of myoblast proliferation show that resident satellite cells are competent to fully regenerate skeletal muscle fibers. *Exp. Cell Res.* **281,** 39–49.

Zammit, P. S., Carvajal, J. J., Golding, J. P., Morgan, J. E., Summerbell, D., Zolnerciks, J., Partridge, T. A., Rigby, P. W., and Beauchamp, J. R. (2004a). Myf5 expression in satellite cells and spindles in adult muscle is controlled by separate genetic elements. *Dev. Biol.* **273,** 454–465.

Zammit, P. S., Golding, J. P., Nagata, Y., Hudon, V., Partridge, T. A., and Beauchamp, J. R. (2004b). Muscle satellite cells adopt divergent fates: A mechanism for self-renewal? *J. Cell Biol.* **166,** 347–357.

Zammit, P. S., Relaix, F., Nagata, Y., Perez Ruiz, A., Collins, C. A., Partridge, T. A., and Beauchamp, J. R. (2006). Pax7 and myogenic progression in skeletal muscle satellite cells. *J. Cell Sci.* **119,** 1824–1832.

Zheng, J. K., Wang, Y., Karandikar, A., Wang, Q., Gai, H., Liu, A. L., Peng, C., and Sheng, H. Z. (2006). Skeletal myogenesis by human embryonic stem cells. *Cell Res.* **16,** 713–722.

Zimowska, M., Szczepankowska, D., Streminska, W., Papy, D., Tournaire, M. C., Gautron, J., Barritault, D., Moraczewski, J., and Martelly, I. (2001). Heparan sulfate mimetics modulate calpain activity during rat Soleus muscle regeneration. *J. Cell. Physiol.* **188,** 178–187.

Zimowska, M., Brzoska, E., Swierczynska, M., Streminska, W., and Moraczewski, J. (2008). Distinct patterns of MMP-9 and MMP-2 activity in slow and fast twitch skeletal muscle regeneration in vivo. *Int. J. Dev. Biol.* **52,** 307–314.

Zimowska, M., Duchesnay, A., Dragun, P., Oberbek, A., Moraczewski, J., and Martelly, I. (2009). Immunoneutralization of TGFbeta1 mproves skeletal muscle regeneration: Effects on myoblast differentiation and glycosaminoglycan content. *Int. J. Cell Biol.* **2009,** 659372.

CHAPTER THIRTEEN

Thymosins and Muscle Regeneration

Takahiko Hara

Contents

I. Introduction 278
II. Basic Properties of Thymosins 278
A. Classification of thymosins 278
B. $T\beta_4$ gene and mRNA expression 279
C. Structure and posttranslational processing of the $T\beta_4$ protein 280
D. β-thymosins in other species 280
III. Physiological Activities of $T\beta_4$ 281
A. Intracellular activities of $T\beta_4$ 281
B. Extracellular activities of $T\beta_4$ 281
IV. Roles of $T\beta_4$ in Skin Tissue Regeneration 282
V. Roles of $T\beta_4$ in Heart Regeneration 282
VI. Roles of $T\beta_4$ in Skeletal Muscle Regeneration 283
A. Induction upon muscle injury 283
B. Recruitment of myoblasts 283
C. Therapeutic applications in muscular dystrophy 284
VII. Signaling Mechanism Involved in the Chemotactic Activity of $T\beta_4$ 285
VIII. Concluding Remarks 285
References 286

Abstract

Thymosins are a family of highly conserved small peptides originally isolated from calf thymus. One representative member of the family is thymosin-β_4 ($T\beta_4$), a major G-actin-sequestering peptide present in many tissues. In the last decade, various studies have uncovered several important functions for $T\beta_4$ related to the regeneration of injured tissues including skin and heart. In particular, $T\beta_4$ promotes endothelial cell migration via the activation of Akt2 kinase at the leading edge of the cell. In the case of skeletal muscle injury, increased levels of $T\beta_4$ are produced by muscle fibers and surrounding immune cells. Satellite cell-derived myoblasts and myocytes are chemoattracted by $T\beta_4$, which facilitates skeletal muscle regeneration. Recently, it was reported that $T\beta_4$ interacts physically with F_1-F_o ATP synthase on the plasma membrane to increase the local concentration of ATP, which stimulates the $P2X_4$ purinergic

Stem Cell Project, Tokyo Metropolitan Institute of Medical Science, Setagaya-ku, Tokyo, Japan

Vitamins and Hormones, Volume 87
ISSN 0083-6729, DOI: 10.1016/B978-0-12-386015-6.00032-9

receptor to elicit a migratory response from endothelial cells. Thus, it is clear that $T\beta_4$ is an important chemotactic factor involved in stem/progenitor cell-mediated tissue regeneration.

I. Introduction

Thymosins are a group of small peptides with molecular weights of 1000–15,000Da, originally isolated from the thymus gland (Goldstein, 2007; Goldstein *et al.*, 2005), although it is now known that the major thymosins [thymosin-α_1, thymosin-β_4 ($T\beta_4$), thymosin-β_{10}, thymosin-β_{15}] are present in a variety of mammalian tissues. Since thymosins modulate cell migration, angiogenesis and immune responses (Mannherz and Hannappel, 2009; Sosne *et al.*, 2010), they have been categorized as biological response modifiers, and accumulating evidence has demonstrated that $T\beta_4$, in particular, is a key regulator of tissue regeneration. This chapter provides an overview of beta thymosin research and describes recent findings about the signaling mechanism of $T\beta_4$ and its role in skeletal muscle regeneration.

II. Basic Properties of Thymosins

A. Classification of thymosins

Thymosins are divided into three groups based on their isoelectric points: α (<pH 5.0), β (pH 5.0–7.0), and γ (>pH 7.0). Several protein species (thymosin-α_1, thymosin-β_1, $T\beta_4$, thymosin-β_{10}, thymosin-β_{15}) have been characterized molecularly from among the 40 to 60 small polypeptides present in thymosin fraction 5 (TF5), biochemically isolated by A. L. Goldstein's laboratory in 1966 (Goldstein *et al.*, 1966). Thymosin-α_1 is a potent immunomodulatory peptide (Goldstein and Badamchian, 2004) and is used in patients infected with the hepatitis-B or -C virus as an immune adjuvant, and in melanoma patients in combination with chemotherapy. Thymosin-β_1 turned out to be ubiquitin (Schlesinger *et al.*, 1975), while $T\beta_4$ is the most important thymosin with regard to tissue regeneration (see below), and thymosin-β_{10} and thymosin-β_{15} are highly homologous to $T\beta_4$ (Fig. 13.1). Three of these β-thymosins have both unique and overlapping functions; thymosin-β_{10} is the predominant form of β-thymosin in embryonic brain (Lugo *et al.*, 1991) and binds to E-tropomodulin (an actin filament-capping protein) to regulate cell viability (Rho *et al.*, 2004), while thymosin-β_{10} either promotes or inhibits the malignancy of cancer cells depending on the experimental conditions (Hall *et al.*, 1990; Lee *et al.*, 2005). Finally, as is the case for $T\beta_4$, thymosin-β_{15} is abundantly expressed

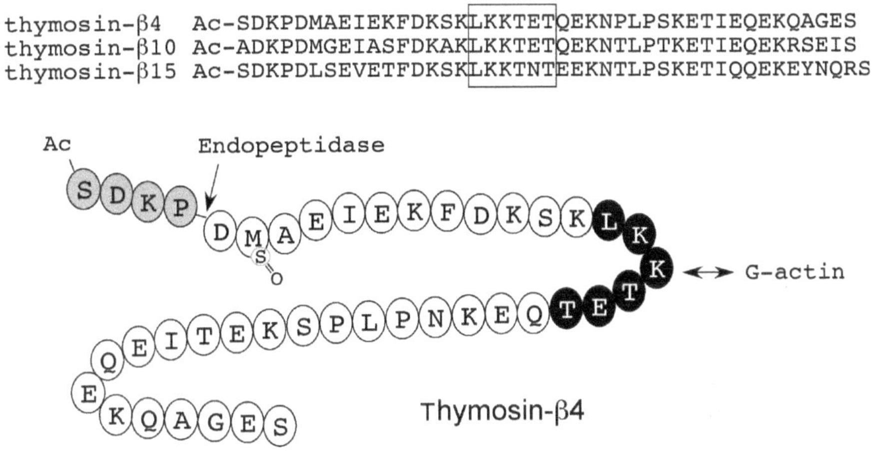

Figure 13.1 Primary structures of representative β-thymosins. (Upper) Amino acid sequence comparison of human thymosin-β_4, thymosin-β_{10}, and thymosin-β_{15}. (Lower) Schematic representation showing posttranslational modification sites, position of Ac-SDKP peptide (amino acid residues shown in gray), and actin-binding region (amino acid residues shown in black) of thymosin-β_4.

in prostate cancer and promotes cell migration and angiogenesis (Bao *et al.*, 1996). In addition, the levels of human thymosin-β_{15} in urine have shown promise as a diagnostic marker for prostate cancer (Hutchinson *et al.*, 2005).

B. Tβ_4 gene and mRNA expression

In humans, the Tβ_4 gene *TMSB4X* is localized to the X chromosome at Xq21.3–q22 (Clauss *et al.*, 1991). The Tβ_4 cDNA open reading frame contains an initial methionine codon followed by a codon for the N-terminal serine and, although cells secrete a certain amount of Tβ_4, there is no hydrophobic signal sequence. The initial methionine residue of the nascent Tβ_4 polypeptide is removed and the N-terminal serine residue is often acetylated in the cells.

The expression of Tβ_4 mRNA is cell cycle dependent and is highest at the G_0/G_1 transition and during S-phase (Schobitz *et al.*, 1991), and changes in the expression of Tβ_4 appear to be related to cell differentiation. It has been reported that hepatocyte growth factor, nerve growth factor or fibroblast growth factor (FGF) can increase the level of Tβ_4 mRNA (Leonard *et al.*, 1987; Oh *et al.*, 2002) and, in addition, interferon treatment augments the transcription of the Tβ_4 gene (Friedman *et al.*, 1984). It has also been shown that increased Tβ_4 expression in cancer cells promotes metastasis, possibly by increasing cell mobility.

C. Structure and posttranslational processing of the $T\beta_4$ protein

$T\beta_4$ is a small, acidic (pI 5.1) polypeptide of 4964 Da composed of 43 amino acid residues and is the most abundant actin-sequestering protein (~0.3 mM) in human platelets. The N-terminal half of $T\beta_4$ exhibits a strong similarity to Wasp Homology Domain 2 (WH_2) (Paunola *et al.*, 2002), a small, widespread actin-binding motif. Since $T\beta_4$ binds to G-actin through the LKKTET sequence in the WH_2 module (Fig. 13.1), it is possible that $T\beta_4$ evolved from the addition of C-terminal sequence to the ancestral WH_2 module (Dominguez, 2007). Interestingly, NMR studies revealed that $T\beta_4$ is unstructured in aqueous solution (Zarbock *et al.*, 1990) and folds on binding to G-actin (Domanski *et al.*, 2004). In addition, $T\beta_4$ forms different secondary structures upon interaction with other proteins. Such multitasking by partner promiscuity has been referred to as "moonlighting" (Jeffery, 1999), a property that undoubtedly facilitates the multiple functions of $T\beta_4$ (Jeffery, 2003).

The methionine residue at position six of $T\beta_4$ is easily oxidized (Fig. 13.1) and, in monocytes, the resultant $T\beta_4$ sulfoxide inhibits the chemotactic response of neutrophils (Young *et al.*, 1999), whereas in myoblasts, both $T\beta_4$ and $T\beta_4$ sulfoxide stimulate the myoblast migratory response (Tokura *et al.*, 2011).

The peptide *N*-acetyl-Ser-Asp-Lys-Pro (Ac-SDKP) is released from the N-terminus of the $T\beta_4$ protein by the action of an unidentified endopeptidase (Grillon *et al.*, 1990; Fig. 13.1). The Ac-SDKP peptide is abundant in serum and other body fluids (Azizi *et al.*, 1997) and is known to inhibit the proliferation of hematopoietic stem cells (Bonnet *et al.*, 1993), stimulate endothelial cell migration (Wang *et al.*, 2004), and decrease inflammation (Yang *et al.*, 2004). Extracellular concentrations of Ac-SDKP are regulated through degradation by an angiotensin-converting enzyme (Rieger *et al.*, 1993).

D. β-thymosins in other species

The primary amino acid sequences of $T\beta_4$ and $T\beta_{10}$ are identical between human and mouse, indicating a high level of conservation between these molecules. Orthologues of β-thymosin have been found from mammals to echinoderms but are not present in yeast or prokaryotes. Tetrapods express three monomeric β-thymosins, which are orthologues of human $T\beta_4$, thymosin-β_{10}, and thymosin-β_{15} (Safer and Chowrashi, 1997). Bony fish express an additional copy of a $T\beta_4$ orthologue in addition to three β-thymosins (Edwards, 2010). In the fruit fly and nematodes, larger proteins containing repeated β-thymosin-like sequences have been discovered (Manuel *et al.*, 2000).

III. Physiological Activities of Tβ$_4$

A. Intracellular activities of Tβ$_4$

As mentioned above, Tβ$_4$ is present at high concentrations in a wide variety of mammalian cell types and forms a 1:1 complex with G-actin (Safer *et al.*, 1991). In cells, G-actin monomers continuously polymerize to create F-actin, which forms cellular microfilaments with other actin-binding proteins; however, the formation of the G-actin-Tβ$_4$ complex prevents the polymerization of G-actin and subsequent microfilament formation. Due to its profusion in the cytosol and its ability to bind G-actin but not F-actin, Tβ$_4$ is regarded as the principal actin-sequestering protein.

The release of G-actin monomers from Tβ$_4$ occurs as part of the mechanism that drives actin polymerization and is essential for maintaining cell shape and motility. At the leading edge of the cell, the association of the actin-binding protein profilin with G-actin liberates Tβ$_4$, which subsequently binds to integrin-linked kinase (ILK) in the lamellipodia (Fan *et al.*, 2009). Tβ$_4$-ILK complexes then activate Akt2 kinase, leading to the production of metalloproteinase-2, which is involved in matrix degradation. Thus, Tβ$_4$ acts as an actin buffer, preventing its spontaneous polymerization into actin filaments and supplying a pool of actin monomers when the cell requires filaments.

B. Extracellular activities of Tβ$_4$

In addition to its various intracellular roles, Tβ$_4$ shows a remarkably diverse range of effects when present in the fluid surrounding animal tissues. A substantial amount of Tβ$_4$ is released from cells to act as a paracrine factor in support of angiogenesis, wound healing, hair growth, and cardiac regeneration. It is particularly interesting that Tβ$_4$ induces the expression of the vascular endothelial growth factor (VEGF) gene by stabilizing hypoxia-inducible factor 1 alpha (HIF1α) protein (Jo *et al.*, 2010). Tβ$_4$ also enhances the viability of endothelial progenitor cells; however, it is considered unlikely that Tβ$_4$ exerts all these effects through the intracellular sequestration of G-actin.

It has been reported that Tβ$_4$ stimulates the chemotaxis of endothelial cells (Malinda *et al.*, 1997), keratinocytes (Philp *et al.*, 2004) and cardiomyocytes (Smart *et al.*, 2007). In contrast, Tβ$_4$ inhibits the chemoattraction of neutrophils to sites of infection (Young *et al.*, 1999) and it is interesting that Tβ$_4$ sulfoxide has an even greater ability to inhibit the migration of neutrophils than Tβ$_4$. It promoted the dispersion of neutrophils from a focus, inhibited their response to fMet-Leu-Phe, and lowered their adhesion to endothelial cells. Tβ$_4$ sulfoxide is an active factor in the culture fluid of cells after treatment with steroid hormone, suggesting that it could be a steroid-inducible anti-inflammatory factor (Young *et al.*, 1999).

Some of the multiple activities of Tβ_4, unrelated to actin, may be mediated by Ac-SDKP, best known as an inhibitor of the proliferation of hematopoietic stem cells in bone marrow (Liu *et al.*, 2003). Ac-SDKP stimulates endothelial cell migration and differentiation and enhances the secretion of metalloproteinase-1. In addition, Ac-SDKP induces angiogenesis in rats (Liu *et al.*, 2003) and reduces cardiac fibrosis in rats with myocardial infarction (Yang *et al.*, 2004).

IV. Roles of Tβ_4 in Skin Tissue Regeneration

Tβ_4 promotes wound healing when it is added either directly to the surface of dermal wounds or administered intraperitoneally (Malinda *et al.*, 1999). The largest concentrations of Tβ_4 are produced by platelets and leukocytes, both of which are the first cells to encounter a wound (Hannappel and van Kampen, 1987). Tβ_4 not only stimulates the migration of endothelial cells and keratinocytes (Malinda *et al.*, 1997) but also increases collagen deposition and wound contraction (Blain *et al.*, 2002). Tβ_4 also induces the production of metalloproteinases and inhibits the mobilization of inflammatory cells. Moreover, Tβ_4 enhances stem cell migration from the bulge region of the hair follicle (Philp *et al.*, 2004) and, therefore, is very effective in accelerating hair growth. All of these biological activities of Tβ_4 appear to contribute to its positive impact on the regeneration of injured skin.

The N-terminal residues of Tβ_4 are responsible for its anti-inflammatory action, while cell migration and angiogenesis are promoted via its central actin-binding domain (LKKTET) (Philp *et al.*, 2003). When synthetic Tβ_4 was administered intraperitoneally in pregnant mice, it was incorporated into the developing fetus to upregulate the endogenous expression of the Tβ_4 gene (Smart *et al.*, 2010). Therefore, it appears that an autoregulatory loop involving extracellular and intracellular Tβ_4 supports its strong regenerative capacity.

V. Roles of Tβ_4 in Heart Regeneration

It has been demonstrated that Tβ_4 and its cleavage product are important mediators of cardiac regeneration (Bock-Marquette *et al.*, 2004; Hinkel *et al.*, 2008; Smart *et al.*, 2007). Tβ_4 mRNA was found to be elevated during hypoxia, and Tβ_4 was shown to associate physically with Particularly Interesting New Cys-His protein (PINCH) and ILK to activate the survival kinase Akt in cells. Thus, it appears that Tβ_4 promotes the survival and repair of cardiomyocytes after cardiac injury through the PINCH/ILK signaling pathway, in addition to stimulating the migration of cardiomyocytes and endothelial cells. $T\beta_4$ also decreases infarct size, presumably due to the attenuated inflammation. Intriguingly, a very recent report demonstrated that the basic helix-loop-helix (bHLH) transcription factor, Hand1, is essential for yolk sac vasculogenesis and embryo survival via transcriptional activation of the $T\beta_4$ gene in mouse embryos (Smart *et al.*, 2010). Thus, $T\beta_4$ plays critical roles in cardiac development and cardioprotection in the adult.

VI. Roles of $T\beta_4$ in Skeletal Muscle Regeneration

A. Induction upon muscle injury

Damaged skeletal muscles are repaired everyday to support normal locomotor activity. During muscle regeneration, various cytokines and secretary proteins, including FGF and insulin-like growth factor-2 (IGF-2), are produced by regenerating muscle fibers and surrounding immune cells (Charge and Rudnicki, 2004). These factors coordinately support the recovery of functional muscle fibers by stimulating the cell cycle entry of satellite cells (muscle stem cells) and by the recruitment and differentiation of myoblasts and myotube formation and fusion (Collins *et al.*, 2005).

It has been reported that the expression of $T\beta_4$ mRNA is upregulated in the skeletal muscles of dystrophin-deficient *mdx* mice (Tseng *et al.*, 2002; Turk *et al.*, 2005) and in injured porcine muscles (Ferre *et al.*, 2007) compared with their respective healthy controls. In addition, mRNA levels of $T\beta_4$ and thymosin-β_{10} are increased after cardiotoxin-induced skeletal muscle damage in mice (Tokura *et al.*, 2011), and the major producers of these molecules are regenerating muscle fibers and inflammatory hematopoietic cells. Large amounts of macrophages and T cells infiltrate the degenerated areas of skeletal muscle after cardiotoxin injection, and these immune cells are known to play an important role in the regeneration process by secreting a number of cytokines, including $T\beta_4$ and thymosin-β_{10}. Although it remains unclear how $T\beta_4$ and thymosin-β_{10} are secreted from muscle and immune cells, it is possible that they are provided, in part, directly from the disrupted muscle tissues at the beginning of the repair process.

B. Recruitment of myoblasts

$T\beta_4$ is known to chemoattract human umbilical vein endothelial cells (Malinda *et al.*, 1997). Similarly, $T\beta_4$ and $T\beta_4$ sulfoxide exhibit a chemoattractant effect on C2C12 cells (immortalized murine myoblast cells) (Tokura *et al.*, 2011). This is also true for skeletal muscle-derived primary myoblasts and myocytes in mice, and it has been shown that Tβ_4 sulfoxide is more potent than Tβ_4 in promoting the migration of C2C12 cells in the scratch wound assay.

Tβ_4 and thymosin-β_{10} mRNAs are expressed in proliferating myoblasts; however, much larger amounts of these mRNAs are produced from differentiated C2C12-derived myotubes to generate a local gradient of these proteins (Tokura *et al.*, 2011). Presumably, newly formed myofibers, as well as inflammatory hematopoietic cells, produce larger amounts of Tβ_4 and thymosin-β_{10} in a time- and space-dependent manner, thereby facilitating the mobilization of myoblasts and myocytes into the regenerating skeletal muscles (Fig. 13.2).

C. Therapeutic applications in muscular dystrophy

Since Tβ_4 is one of the factors involved in the regeneration of skeletal muscle, it is potentially useful for the treatment of muscular dystrophy. A recent report demonstrated that Tβ_4-treated *mdx* mice showed a significant increase in the number of regenerating muscle fibers compared with untreated mice (Spurney *et al.*, 2010). Since ILK and Akt are activated in Tβ_4-stimulated C2C12 myoblast cells, as in cardiac myocytes, Tβ_4 may enhance the survival of myoblasts via the PINCH-ILK-Akt pathway. Additionally, a reduction in

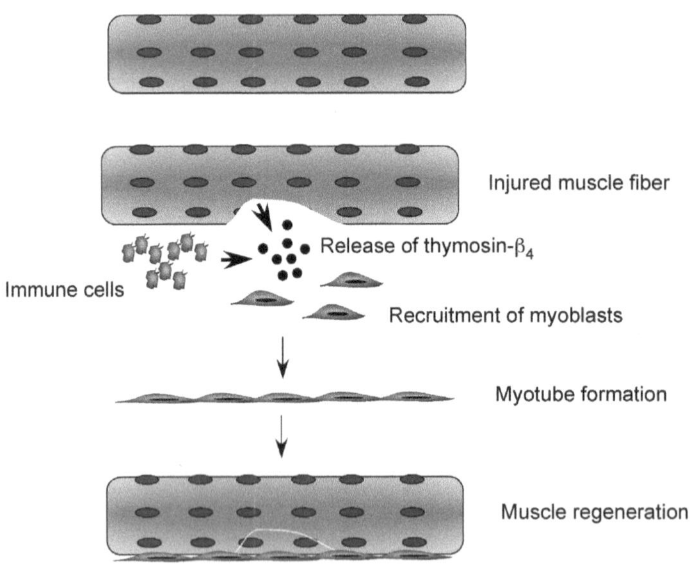

Figure 13.2 Mode of action of thymosin-β4 in skeletal muscle regeneration. Summary of the mechanism underlying the involvement of thymosin-β_4 in the regeneration of skeletal muscle.

NF-κB activity induced by Tβ$_4$ might play a role in the antiapoptotic effects seen in skeletal muscles (Baudy *et al.*, 2009). However, the impaired skeletal muscle strength and fibrosis seen in *mdx* mice were not ameliorated after chronic administration of Tβ$_4$ protein (Spurney *et al.*, 2010). Even though Tβ$_4$ alone may not be sufficient for the functional recovery of muscle, it is worth testing Tβ$_4$ in combination with other cytokines and drugs in various animal models of muscular dystrophy.

VII. Signaling Mechanism Involved in the Chemotactic Activity of Tβ$_4$

Despite the accumulating evidence for multiple biological roles played by Tβ$_4$, it remains unclear which molecules act as Tβ$_4$ receptors. Studies in the past have indicated that, intracellularly, Tβ$_4$ binds to G-actin (Safer *et al.*, 1991), PINCH-1/ILK (Bock-Marquette *et al.*, 2004), hMLH1 (Brieger *et al.*, 2007), Ku80 (Bednarek *et al.*, 2008) and stabilin-2 (Lee *et al.*, 2008); proteins that are important partners in the cell-migration and anti-inflammatory activities of Tβ$_4$. The physical association of Tβ$_4$ with various different protein types might be explained by its moonlighting property, as mentioned above.

The extracellular action of Tβ$_4$ must be mediated by cell surface receptors, and recent studies have proposed two candidate molecules: adenosine receptors (Huang *et al.*, 2006) and F_1-F_0 ATP synthase (Freeman *et al.*, 2011). Although both groups employed a proteomics approach, the F_1-F_0 ATP synthase study revealed the existence of a unique signal transduction mechanism with ATP as a second messenger (Fig. 13.3). In this model, Tβ$_4$ binds to F_1-F_0 ATP synthase on the plasma membrane to inhibit its ATP hydrolysis activity, thereby increasing the cell surface ATP concentration. In fact, Tβ$_4$ and the mitochondrial ATP synthase inhibitor IF1 share a structural similarity in the long alpha-helical region (Freeman *et al.*, 2011). Increased ATP levels predominantly stimulated the purinergic receptor $P2X_4$ to elicit a migratory response from human umbilical vein endothelial cells. The use of antibodies against F_1-F_0 ATP synthase or the siRNA-mediated knockdown of *$P2X_4$* resulted in the reduction of Tβ$_4$-induced cell migration. Future investigations with genetic models would help to clarify whether F_1-F_0 ATP synthase is involved in the various biological activities of Tβ$_4$.

VIII. Concluding Remarks

Recent studies with both cell cultures and animal experiments have demonstrated clearly that the administration of Tβ$_4$ can promote the migration of endothelial cells and various types of stem/progenitor cells, and

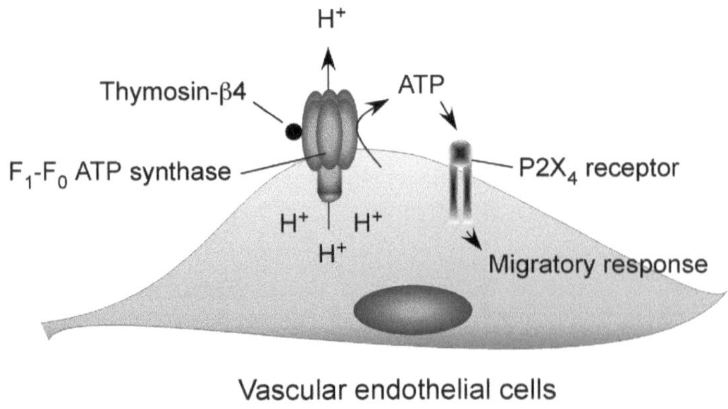

Figure 13.3 Cell surface F_1-F_0 ATP synthase as a thymosin-β_4 receptor. As was recently demonstrated by Freeman *et al.* (2011), F_1-F_0 ATP synthase mediates the thymosin-β_4-induced migration of endothelial cells. This figure depicts the signal transduction mechanism involved in this process.

stimulate the formation of blood vessels and improve the survival of mature cardiac cells. In addition, Tβ_4 attenuates leukocyte recruitment and the subsequent production of inflammatory cytokines. Such multiple biological activities exhibited by a single molecule provide the impetus for on-going world-wide clinical trials of Tβ_4 and its derivatives. As Tβ_4 exerts its effect in a manner distinct from typical cytokines, future investigations of the Tβ_4 receptors and downstream signaling events would provide valuable insight into their possible therapeutic applications. Moreover, this fascinating multitasking small peptide may help to shed light on the mechanism linking the dynamism of intracellular actin to the cellular responses to external stimuli that occur throughout the human body.

REFERENCES

Azizi, M., Ezan, E., Nicolet, L., Grognet, J. M., and Menard, J. (1997). High plasma level of *N*-acetyl-seryl-aspartyl-lysyl-proline: A new marker of chronic angiotensin-converting enzyme inhibition. *Hypertension* **30,** 1015–1019.

Bao, L., Loda, M., Janmey, P. A., Stewart, R., Anand-Apte, B., and Zetter, B. R. (1996). Thymosin beta 15: A novel regulator of tumor cell motility upregulated in metastatic prostate cancer. *Nat. Med.* **2,** 1322–1328.

Baudy, A. R., Saxena, N., Gordish, H., Hoffman, E. P., and Nagaraju, K. (2009). A robust in vitro screening assay to identify NF-kappaB inhibitors for inflammatory muscle diseases. *Int. Immunopharmacol.* **9,** 1209–1214.

Bednarek, R., Boncela, J., Smolarczyk, K., Cierniewska-Cieslak, A., Wyroba, E., and Cierniewski, C. S. (2008). Ku80 as a novel receptor for thymosin beta4 that mediates its intracellular activity different from G-actin sequestering. *J. Biol. Chem.* **283,** 1534–1544.

Blain, E. J., Mason, D. J., and Duance, V. C. (2002). The effect of thymosin beta4 on articular cartilage chondrocyte matrix metalloproteinase expression. *Biochem. Soc. Trans.* **30,** 879–882.

Bock-Marquette, I., Saxena, A., White, M. D., Dimaio, J. M., and Srivastava, D. (2004). Thymosin beta4 activates integrin-linked kinase and promotes cardiac cell migration, survival and cardiac repair. *Nature* **432,** 466–472.

Bonnet, D., Lemoine, F. M., Pontvert-Delucq, S., Baillou, C., Najman, A., and Guigon, M. (1993). Direct and reversible inhibitory effect of the tetrapeptide acetyl-*N*-Ser-Asp-Lys-Pro (Seraspenide) on the growth of human CD34+ subpopulations in response to growth factors. *Blood* **82,** 3307–3314.

Brieger, A., Plotz, G., Zeuzem, S., and Trojan, J. (2007). Thymosin beta 4 expression and nuclear transport are regulated by hMLH1. *Biochem. Biophys. Res. Commun.* **364,** 731–736.

Charge, S. B., and Rudnicki, M. A. (2004). Cellular and molecular regulation of muscle regeneration. *Physiol. Rev.* **84,** 209–238.

Clauss, I. M., Wathelet, M. G., Szpirer, J., Islam, M. Q., Levan, G., Szpirer, C., and Huez, G. A. (1991). Human thymosin-beta 4/6-26 gene is part of a multigene family composed of seven members located on seven different chromosomes. *Genomics* **9,** 174–180.

Collins, C. A., Olsen, I., Zammit, P. S., Heslop, L., Petrie, A., Partridge, T. A., and Morgan, J. E. (2005). Stem cell function, self-renewal, and behavioral heterogeneity of cells from the adult muscle satellite cell niche. *Cell* **122,** 289–301.

Domanski, M., Hertzog, M., Coutant, J., Gutsche-Perelroizen, I., Bontems, F., Carlier, M. F., Guittet, E., and van Heijenoort, C. (2004). Coupling of folding and binding of thymosin beta4 upon interaction with monomeric actin monitored by nuclear magnetic resonance. *J. Biol. Chem.* **279,** 23637–23645.

Dominguez, R. (2007). The beta-thymosin/WH2 fold: Multifunctionality and structure. *Ann. N. Y. Acad. Sci.* **1112,** 86–94.

Edwards, J. (2010). Vertebrate beta-thymosins: Conserved synteny reveals the relationship between those of bony fish and of land vertebrates. *FEBS Lett.* **584,** 1047–1053.

Fan, Y., Gong, Y., Ghosh, P. K., Graham, L. M., and Fox, P. L. (2009). Spatial coordination of actin polymerization and ILK-Akt2 activity during endothelial cell migration. *Dev. Cell* **16,** 661–674.

Ferre, P. J., Liaubet, L., Concordet, D., SanCristobal, M., Uro-Coste, E., Tosser-Klopp, G., Bonnet, A., Toutain, P. L., Hatey, F., and Lefebvre, H. P. (2007). Longitudinal analysis of gene expression in porcine skeletal muscle after post-injection local injury. *Pharm. Res.* **24,** 1480–1489.

Freeman, K. W., Bowman, B. R., and Zetter, B. R. (2011). Regenerative protein thymosin beta-4 is a novel regulator of purinergic signaling. *FASEB J.* **25,** 907–915.

Friedman, R. L., Manly, S. P., McMahon, M., Kerr, I. M., and Stark, G. R. (1984). Transcriptional and posttranscriptional regulation of interferon-induced gene expression in human cells. *Cell* **38,** 745–755.

Goldstein, A. L. (2007). History of the discovery of the thymosins. *Ann. N. Y. Acad. Sci.* **1112,** 1–13.

Goldstein, A. L., and Badamchian, M. (2004). Thymosins: Chemistry and biological properties in health and disease. *Expert Opin. Biol. Ther.* **4,** 559–573.

Goldstein, A. L., Slater, F. D., and White, A. (1966). Preparation, assay, and partial purification of a thymic lymphocytopoietic factor (thymosin). *Proc. Natl. Acad. Sci. USA* **56,** 1010–1017.

Goldstein, A. L., Hannappel, E., and Kleinman, H. K. (2005). Thymosin beta4: Actin-sequestering protein moonlights to repair injured tissues. *Trends Mol. Med.* **11,** 421–429.

Grillon, C., Rieger, K., Bakala, J., Schott, D., Morgat, J. L., Hannappel, E., Voelter, W., and Lenfant, M. (1990). Involvement of thymosin beta 4 and endoproteinase Asp-N in the biosynthesis of the tetrapeptide AcSerAspLysPro a regulator of the hematopoietic system. *FEBS Lett.* **274,** 30–34.

Hall, A. K., Hempstead, J., and Morgan, J. I. (1990). Thymosin beta 10 levels in developing human brain and its regulation by retinoic acid in the HTB-10 neuroblastoma. *Brain Res. Mol. Brain Res.* **8,** 129–135.

Hannappel, E., and van Kampen, M. (1987). Determination of thymosin beta 4 in human blood cells and serum. *J. Chromatogr.* **397,** 279–285.

Hinkel, R., El-Aouni, C., Olson, T., Horstkotte, J., Mayer, S., Muller, S., Willhauck, M., Spitzweg, C., Gildehaus, F. J., Munzing, W., Hannappel, E., *et al.* (2008). Thymosin beta4 is an essential paracrine factor of embryonic endothelial progenitor cell-mediated cardioprotection. *Circulation* **117,** 2232–2240.

Huang, C. M., Wang, C. C., Barnes, S., and Elmets, C. A. (2006). In vivo detection of secreted proteins from wounded skin using capillary ultrafiltration probes and mass spectrometric proteomics. *Proteomics* **6,** 5805–5814.

Hutchinson, L. M., Chang, E. L., Becker, C. M., Shih, M. C., Brice, M., DeWolf, W. C., Gaston, S. M., and Zetter, B. R. (2005). Use of thymosin beta15 as a urinary biomarker in human prostate cancer. *Prostate* **64,** 116–127.

Jeffery, C. J. (1999). Moonlighting proteins. *Trends Biochem. Sci.* **24,** 8–11.

Jeffery, C. J. (2003). Multifunctional proteins: Examples of gene sharing. *Ann. Med.* **35,** 28–35.

Jo, J. O., Kim, S. R., Bae, M. K., Kang, Y. J., Ock, M. S., Kleinman, H. K., and Cha, H. J. (2010). Thymosin beta4 induces the expression of vascular endothelial growth factor (VEGF) in a hypoxia-inducible factor (HIF)-1alpha-dependent manner. *Biochim. Biophys. Acta* **1803,** 1244–1251.

Lee, S. H., Son, M. J., Oh, S. H., Rho, S. B., Park, K., Kim, Y. J., Park, M. S., and Lee, J. H. (2005). Thymosin beta10 inhibits angiogenesis and tumor growth by interfering with Ras function. *Cancer Res.* **65,** 137–148.

Lee, S. J., So, I. S., Park, S. Y., and Kim, I. S. (2008). Thymosin beta4 is involved in stabilin-2-mediated apoptotic cell engulfment. *FEBS Lett.* **582,** 2161–2166.

Leonard, D. G., Ziff, E. B., and Greene, L. A. (1987). Identification and characterization of mRNAs regulated by nerve growth factor in PC12 cells. *Mol. Cell. Biol.* **7,** 3156–3167.

Liu, J. M., Lawrence, F., Kovacevic, M., Bignon, J., Papadimitriou, E., Lallemand, J. Y., Katsoris, P., Potier, P., Fromes, Y., and Wdzieczak-Bakala, J. (2003). The tetrapeptide AcSDKP, an inhibitor of primitive hematopoietic cell proliferation, induces angiogenesis in vitro and in vivo. *Blood* **101,** 3014–3020.

Lugo, D. I., Chen, S. C., Hall, A. K., Ziai, R., Hempstead, J. L., and Morgan, J. I. (1991). Developmental regulation of beta-thymosins in the rat central nervous system. *J. Neurochem.* **56,** 457–461.

Malinda, K. M., Goldstein, A. L., and Kleinman, H. K. (1997). Thymosin beta 4 stimulates directional migration of human umbilical vein endothelial cells. *FASEB J.* **11,** 474–481.

Malinda, K. M., Sidhu, G. S., Mani, H., Banaudha, K., Maheshwari, R. K., Goldstein, A. L., and Kleinman, H. K. (1999). Thymosin beta4 accelerates wound healing. *J. Invest. Dermatol.* **113,** 364–368.

Mannherz, H. G., and Hannappel, E. (2009). The beta-thymosins: Intracellular and extracellular activities of a versatile actin binding protein family. *Cell Motil. Cytoskeleton* **66,** 839–851.

Manuel, M., Kruse, M., Muller, W. E., and Le Parco, Y. (2000). The comparison of beta-thymosin homologues among metazoa supports an arthropod-nematode clade. *J. Mol. Evol.* **51,** 378–381.

Oh, I. S., So, S. S., Jahng, K. Y., and Kim, H. G. (2002). Hepatocyte growth factor upregulates thymosin beta4 in human umbilical vein endothelial cells. *Biochem. Biophys. Res. Commun.* **296,** 401–405.

Paunola, E., Mattila, P. K., and Lappalainen, P. (2002). WH2 domain: A small, versatile adapter for actin monomers. *FEBS Lett.* **513,** 92–97.

Philp, D., Huff, T., Gho, Y. S., Hannappel, E., and Kleinman, H. K. (2003). The actin binding site on thymosin beta4 promotes angiogenesis. *FASEB J.* **17,** 2103–2105.

Philp, D., Nguyen, M., Scheremeta, B., St-Surin, S., Villa, A. M., Orgel, A., Kleinman, H. K., and Elkin, M. (2004). Thymosin beta4 increases hair growth by activation of hair follicle stem cells. *FASEB J.* **18,** 385–387.

Rho, S. B., Chun, T., Lee, S. H., Park, K., and Lee, J. H. (2004). The interaction between E-tropomodulin and thymosin beta-10 rescues tumor cells from thymosin beta-10 mediated apoptosis by restoring actin architecture. *FEBS Lett.* **557,** 57–63.

Rieger, K. J., Saez-Servent, N., Papet, M. P., Wdzieczak-Bakala, J., Morgat, J. L., Thierry, J., Voelter, W., and Lenfant, M. (1993). Involvement of human plasma angiotensin I-converting enzyme in the degradation of the haemoregulatory peptide N-acetyl-seryl-aspartyl-lysyl-proline. *Biochem. J.* **296,** 373–378.

Safer, D., and Chowrashi, P. K. (1997). Beta-thymosins from marine invertebrates: Primary structure and interaction with actin. *Cell Motil. Cytoskeleton* **38,** 163–171.

Safer, D., Elzinga, M., and Nachmias, V. T. (1991). Thymosin beta 4 and Fx, an actin-sequestering peptide, are indistinguishable. *J. Biol. Chem.* **266,** 4029–4032.

Schlesinger, D. H., Goldstein, G., and Niall, H. D. (1975). The complete amino acid sequence of ubiquitin, an adenylate cyclase stimulating polypeptide probably universal in living cells. *Biochemistry* **14,** 2214–2218.

Schobitz, B., Netzker, R., Hannappel, E., and Brand, K. (1991). Cell-cycle-regulated expression of thymosin beta 4 in thymocytes. *Eur. J. Biochem.* **199,** 257–262.

Smart, N., Risebro, C. A., Melville, A. A., Moses, K., Schwartz, R. J., Chien, K. R., and Riley, P. R. (2007). Thymosin beta4 induces adult epicardial progenitor mobilization and neovascularization. *Nature* **445,** 177–182.

Smart, N., Dube, K. N., and Riley, P. R. (2010). Identification of Thymosin beta4 as an effector of Hand1-mediated vascular development. *Nat. Commun.* **1,** 46.

Sosne, G., Qiu, P., Goldstein, A. L., and Wheater, M. (2010). Biological activities of thymosin beta4 defined by active sites in short peptide sequences. *FASEB J.* **24,** 2144–2151.

Spurney, C. F., Cha, H.-J., Sali, A., Pandey, G. S., Pistilli, E., Guerron, A. D., Gordish-Dressman, H., Hoffman, E. P., and Nagaraju, K. (2010). Evaluation of skeletal and cardiac muscle function after administration of thymosin beta-4 in the dystrophin deficient mouse. *PLoS One* **5,** e8976.

Tokura, Y., Nakayama, Y., Fukada, S., Nara, N., Yamamoto, H., Matsuda, R., and Hara, T. (2011). Muscle injury-induced thymosin beta4 acts as a chemoattractant for myoblasts. *J. Biochem.* **149,** 43–48.

Tseng, B. S., Zhao, P., Pattison, J. S., Gordon, S. E., Granchelli, J. A., Madsen, R. W., Folk, L. C., Hoffman, E. P., and Booth, F. W. (2002). Regenerated mdx mouse skeletal muscle shows differential mRNA expression. *J. Appl. Physiol.* **93,** 537–545.

Turk, R., Sterrenburg, E., de Meijer, E. J., van Ommen, G. J., den Dunnen, J. T., and T Hoen, P. A. (2005). Muscle regeneration in dystrophin-deficient mdx mice studied by gene expression profiling. *BMC Genomics* **6,** 98.

Wang, D., Carretero, O. A., Yang, X. Y., Rhaleb, N. E., Liu, Y. H., Liao, T. D., and Yang, X. P. (2004). N-acetyl-seryl-aspartyl-lysyl-proline stimulates angiogenesis in vitro and in vivo. *Am. J. Physiol. Heart Circ. Physiol.* **287,** H2099–H2105.

Yang, F., Yang, X. P., Liu, Y. H., Xu, J., Cingolani, O., Rhaleb, N. E., and Carretero, O. A. (2004). Ac-SDKP reverses inflammation and fibrosis in rats with heart failure after myocardial infarction. *Hypertension* **43,** 229–236.
Young, J. D., Lawrence, A. J., MacLean, A. G., Leung, B. P., McInnes, I. B., Canas, B., Pappin, D. J., and Stevenson, R. D. (1999). Thymosin beta 4 sulfoxide is an anti-inflammatory agent generated by monocytes in the presence of glucocorticoids. *Nat. Med.* **5,** 1424–1427.
Zarbock, J., Oschkinat, H., Hannappel, E., Kalbacher, H., Voelter, W., and Holak, T. A. (1990). Solution conformation of thymosin beta 4: A nuclear magnetic resonance and simulated annealing study. *Biochemistry* **29,** 7814–7821.

CHAPTER FOURTEEN

MicroRNAs and Mesenchymal Stem Cells

Federica Collino,* Stefania Bruno,*,† Maria Chiara Deregibus,* Ciro Tetta,†,‡ *and* Giovanni Camussi*

Contents

I. Introduction 292
II. MicroRNAs 293
III. Mesenchymal Stem Cells 296
IV. miRNAs and Stem Cells 296
  A. miRNA profile in MSCs 298
  B. Comparative expression profiles of miRNAs among human MSCs derived from different tissues 298
V. Role of miRNAs in MSC Differentiation 299
  A. miRNAs and mesodermal lineage differentiation of MSCs 299
  B. miRNAs and MSC plasticity 305
VI. The Role of miRNAs in Cell-to-Cell Communication 306
  A. Microvesicles 307
  B. miRNA transfer mediated by MVs 308
  C. Biological effects of MVs 310
VII. Conclusions 312
Acknowledgments 312
References 312

Abstract

In the adult body, mesenchymal stem cells (MSCs) represent a population with self-renewal ability and multipotent differentiation capabilities. The phenotype of these cells is modulated by a dynamic interplay of signals within a defined microenvironment. Recent studies indicate that microRNAs (miRNAs) act as regulatory signals for maintaining of stemless, self-renewal, and differentiation in embryonic and adult stem cells. miRNAs are noncoding RNAs with pleiotropic effect dependent on posttranscriptional regulation of gene expression. In the

* Department of Internal Medicine and Center for Molecular Biotechnology University of Torino, Torino, Italy
† SiS-Ter S.p.A., Palazzo Pignano, Crema, Italy
‡ Fresenius Medical Care, Bad Homburg, Germany

Vitamins and Hormones, Volume 87
ISSN 0083-6729, DOI: 10.1016/B978-0-12-386015-6.00033-0

stem cell biology, miRNAs by repressing translation of specific mRNAs, may determine the fate of these cells. The characterization of miRNAs present in MSCs may be relevant not only as signature of the cell type but also for the understanding of their biological activities. Recent studies indicate also that the exchange of miRNAs between neighboring cells is an integral part of MSC communication with tissue-injured cells. The transport of miRNAs within biological fluids is guaranteed by microvesicles (MVs) that after release from the cell of origin may enter into a target cell delivering their cargo. MVs may allow a bidirectional exchange of miRNAs between injured cells and MSCs. The exchange of genetic information may on one hand, reprogram the phenotype of MSCs, to acquire features of the injured tissues. On the other hand, MVs derived from stem cells may activate regenerative programs in cells survived to injury. The study of miRNAs, their biological function, and their transfer opens a new dimension on the fate and behavior of MSCs and on their potential application in regenerative medicine.

I. Introduction

Stem cells represent a population with unlimited self-renewal and multilineage differentiation capabilities that plays an essential role in organogenesis during the embryonic development and in the growth, homeostasis, and repair of many tissues in the adult.

The embryonic stem cells (ESCs) are pluripotent cells derived from the inner cell mass of blastocysts that have the capacity to maintain an undifferentiated state indefinitely, express typical stem cell markers, possess high levels of telomerase activity, and have the potential to differentiate in all cell types of the organism both *in vitro* and *in vivo*.

Adult stem cells (ASCs) subsist in most tissues and organs of mammalian organisms (Mimeault and Batra, 2008). However, the most abundant source of ASCs is the bone marrow. Bone marrow contains two main stem cell populations, the hematopoietic stem cells, which are committed to differentiate into blood cells (erythrocytes, thrombocytes, and leukocytes) and the mesenchymal stem cells (MSCs), which originate from connective tissue lineages, including bone, fat, cartilage, and muscle (Jiang *et al.*, 2002). In addition, MSCs, that localize in perivascular areas, provide stromal support for the hematopoietic stem cells in the bone marrow (Noort *et al.*, 2002) and play a role in maintaining these cells quiescent by inhibiting proliferation and differentiation (Glennie *et al.*, 2005). Moreover, resident adult stem/progenitor cells are present in several tissues and preferentially generate differentiated cells of their tissue of origin, suggesting a role in the physiological turnover and in tissue repair. Several studies indicate also a cross lineage differentiation potential of bone marrow and tissue-resident MSCs that could be involved in the repair of different organs and tissues (Phinney and Prockop, 2007). In this context, microRNAs (miRNAs) are candidates for the modulation of MSC phenotype and for the crosstalk of these cells with tissue-resident cells.

II. MicroRNAs

The miRNAs are a class of small noncoding RNAs of 19–23 nucleotides that are derived from a 70-nucleotide precursor (Lakshmipathy and Hart, 2008). They are known to modulate gene expression by translational inhibition or by promoting the degradation of target mRNAs (Krol *et al.*, 2010). miRNAs were initially found in worms and flies (Lagos-Quintana *et al.*, 2001; Lau *et al.*, 2001; Lee and Ambros, 2001) and subsequently studied also in plants, green algae, viruses, and mammalians (Griffiths-Jones *et al.*, 2008). miRNAs, by exploiting numerous biological functions, act as key regulators of many processes ranging from immune response (Bak and Mikkelsen, 2010; Gantier, 2010; Pedersen *et al.*, 2007; Zhang and Bevan, 2010), apoptosis, and cellular metabolism to cellular proliferation and differentiation (Ambros, 2004; Bartel, 2004). Moreover, specific miRNAs have been associated to organogenesis, particularly to the control of neural development, hematopoiesis, cardiovascular development, and angiogenesis (Gangaraju and Lin, 2009; Zhang, 2008).

In mammals, the biogenesis of miRNAs and their action machinery have been characterized in detail (Carthew and Sontheimer, 2009; Ding *et al.*, 2009; Huntzinger and Izaurralde, 2011; Winter *et al.*, 2009; Fig. 14.1). Primary miRNA (pri-miRNA) transcripts are processed into a 70-nucleotide stem loop precursor miRNA (pre-miRNA) by the Drosha–DGCR8 microprocessor complex in the nucleus. In particular, the domain binding partner protein (dsRBD) of DGCR8 functions as a molecular anchor to correctly locate the DROSHA catalytic site on the pri-miRNA stem sequence, to achieve the cleavage reaction leading to pre-miRNA formation (Han *et al.*, 2006; Fig. 14.1). The second processing step to create mature miRNAs is performed by the ribonuclease type III, Dicer, after the pre-miRNAs export from the nucleus to the cytoplasm. In the cytoplasm, Dicer cleaves preferentially the terminal loop of the precursor to generate mature miRNA duplex of approximately 22 nucleotide long (Kim, 2005). Some miRNAs, called mirtrons, are generated by a noncanonical pathway from introns with a mechanism that bypass Drosha requirement. In this pathway, the early processing step is done by spliceosome and a debranching enzyme that generate a short hairpin, ready for further processing by Dicer (Krol *et al.*, 2010; Fig. 14.1). Intron-derived miRNAs are also cleaved by Dicer, with the support of TRBP protein, in the cytoplasm and generate a approximately 22 nucleotides miRNA/miRNA* duplex.

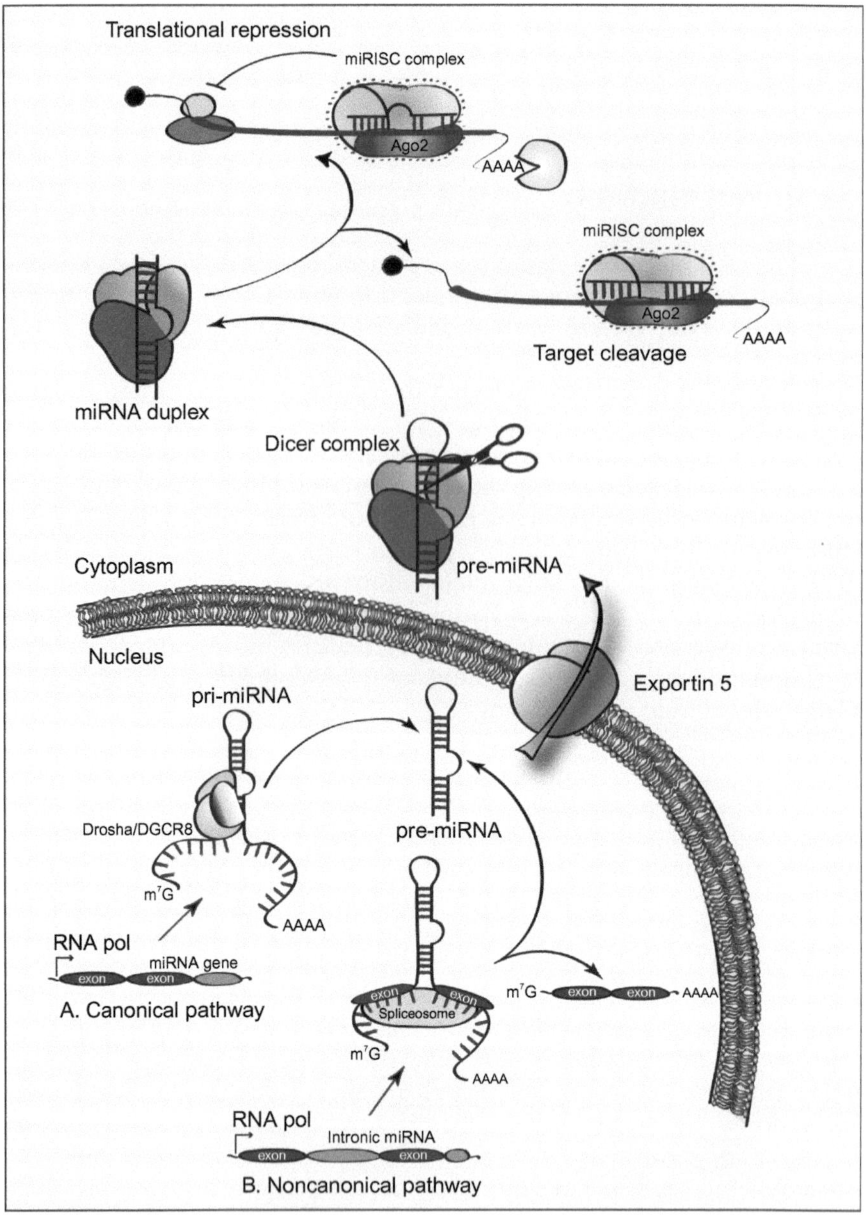

Figure 14.1 Schematic representation of miRNA biogenesis. miRNAs may be generated by canonical and noncanonical pathways. The canonical pathway involves the transcription of miRNA genes into a pre-miRNA precursor after proteolytic cleavage of a pri-miRNA by Drosha/DGCR8 complex. In the noncanonical pathway, pre-miRNA is generated from small introns (mintrons) by alternative splicing and debranching with no need of Drosha/DGCR8 processing. In both cases, the pre-miRNA is then translocated in the cytoplasm by exportin 5 and after cleavage by Dicer complex,

After Dicer cleavage, only one of the two strands of the duplex, derived from both canonical and noncanonical pathways is then loaded into the miRNA-induced silencing complexes (miRISCs), through its interaction with one of the member of the Argonaute (Ago) family. Although mammalian genomes encode four classes of Ago proteins, Ago2 protein is the only one capable to directly cleave the complementary target mRNAs (Liu *et al.*, 2009) and for this reason is commonly called the "sole vertebrate Slicer Argonaute" (Yang and Lai, 2010). The formed miRISC may then silence the expression of target genes predominantly through a posttranscriptional repression. The silence of specific targets is dependent on a base-pairing interaction between the incorporated miRNA in miRISC and the target. The fate of targets is dependent on their partially or fully complementary sequence with the loaded miRNA (Carthew and Sontheimer, 2009; Huntzinger and Izaurralde, 2011; Krol *et al.*, 2010). In particular, a perfect interaction may induce the target cleavage by the miRISC component Ago2. More commonly, base pairing is imperfect resulting in translational repression of the target gene (Fig. 14.1).

In the cytoplasm, mRNAs can be accumulated in foci known as processing bodies (P-bodies) (Krol *et al.*, 2010). P-bodies are bifunctional as they can act as storage or as degradation site of mRNAs. P-bodies are enriched of components of the miRISC and of miRNAs and their subsistence depends on the presence of Ago and GW182 proteins of the miRISC and on a functional miRNA machinery (Eulalio *et al.*, 2007; Pauley *et al.*, 2006). Stress granules are another site of dynamic aggregates of untranslated mRNAs within the cytoplasm, generated in response to cell stress. mRNAs within stress granules associate with a subset of translation-initiating factors, the 40S ribosomal subunit and several RNA-binding proteins. Stress granules are considered aggregates of mRNAs stalled in a process of transient inhibition of translation. Stress granules and P-bodies are frequently located adjacent to each other suggesting that mRNA may move between the two compartments. As P-bodies, stress granules are enriched of miRISC proteins and may accumulate artificial miRNA mimics, synthetic reporter mRNAs, and components of the miRNA machinery (Krol *et al.*, 2010). Whether this reflects a functional role in mRNA silencing by miRNAs or only an enrichment of miRISCs, as result of a passive dragging of mRNA-associated miRISCs within these structures, remains to be determined (Krol *et al.*, 2010).

About 40–50% of mammalian mRNAs are thought to be regulated by miRNAs at the translational level. Based on their potential to target hundreds of different mRNAs, miRNAs may finely coordinate the transcriptome expression in the cell. The functional consequence of miRNA

generates a mature miRNA duplex. One of the two strands of the duplex is loaded within the miRISC and based on the complementary of base pairs may induce translational repression and deadenylation or cleavage of the target. (For color version of this figure, the reader is referred to the Web version of this chapter.)

action implicates a tight regulation of cellular development and differentiation by a complex and dynamic mechanism (Krol *et al.*, 2010). Therefore, alterations of miRNA pattern can lead to pathogenic consequences (Dykxhoorn, 2010; Guil and Esteller, 2009).

III. Mesenchymal Stem Cells

MSCs or multipotent stromal cells are characterized as plastic adherent cells with fibroblastoid aspect, with the capacity of extensive proliferation and self-renewal *in vitro*. Besides bone marrow, MSC-like cells have been isolated from adult skeletal muscle, adipose tissue, synovium, the circulatory system, dental pulp, lung (da Silva Meirelles *et al.*, 2006; Phinney and Prockop, 2007), liver (Herrera *et al.*, 2006), and glomeruli (Bruno *et al.*, 2009a,b). However, specific identification markers for isolation of MSCs and for their identification in tissues are still missing. The *Mesenchymal and Tissue Stem Cell Committee of the International Society for Cellular Therapy* proposed the minimal criteria to define human MSCs (da Silva Meirelles *et al.*, 2008; Dominici *et al.*, 2006). These criteria include the expression of surface receptors, such as CD105, CD73, CD29, and CD90 and the absence of hematopoietic markers including CD45, CD34, CD14, or CD11b, CD79a, or CD19 and HLA-DR surface molecules (Dominici *et al.*, 2006). The minimal criteria to define human MSCs include also osteo-, chondro-, and adipogenic differentiation capabilities.

During the past decades, MSCs have generated a great deal of interest in many clinical setting, in particular, MSCs were shown to ameliorate tissue damage in response to injury and disease. MSCs, particularly those derived from bone marrow, have demonstrated efficacy as therapeutic vectors in animal models of lung injury (Ortiz *et al.*, 2003), kidney disease (Herrera *et al.*, 2004; Morigi *et al.*, 2004), diabetes (Lee *et al.*, 2006), graft versus host disease (Ringdén *et al.*, 2006), myocardial infarction (Minguell and Erices, 2006), and various neurological disorders (Phinney and Isakova, 2005). Moreover, MSCs have a low immunogenicity and possess immunoregulatory properties including the ability to increase *in vitro* the percentage of T regulatory cells, when cocultured with T-lymphocyte and to inhibit the proliferative response of antigen-specific memory T-cells (Crop *et al.*, 2009).

IV. miRNAs and Stem Cells

Several studies indicate that miRNAs are strongly involved in stem cell maintenance and differentiation (Gangaraju and Lin, 2009). The role of miRNAs in stem cell biology emerged from studies showing a specific gene signature related to stemness. In addition, small group of genes were found to selectively associate with different stem cell populations (Gangaraju and Lin, 2009). Based on these considerations, researchers focused the attention on the epigenetic regulation responsible for differential gene expression among stem cells of different origin. The comparison of miRNA signatures among pluripotent, multipotent, and somatic stem cells revealed specific miRNAs involved in the regulation of pluripotent and differentiation genes (Aranda *et al.*, 2009). In particular, hierarchical cluster analysis of 250 mature human miRNAs (from Sanger database), in different stem cell lines (pluripotent ESCs versus different sources of multipotent ASCs) classified the cell lines tested into two groups, one of which included the ESCs and the other all the different ASCs tested (Aranda *et al.*, 2009). ESCs differ from ASCs for 64 miRNAs, 34 of which are upregulated and 30 downregulated.

At variance, the two ASC populations studied, namely the adult progenitors and the MSCs, differ only for the expression of few miRNAs. miR-143, miR-129, and miR-199b were found to be downregulated and miR-204 and miR-424 were upregulated in adult progenitors in respect to MSC (Aranda *et al.*, 2009). Moreover, the association between the transcriptome data and miRNA analyzes demonstrated an inverse correlation between the expression of genes involved in stem cell phenotype and miRNA expression (Assou *et al.*, 2007).

In mammals, so far only few miRNAs have been assigned a functional role in developmental processes: miR-181a promotes the differentiation of hematopoietic stem cells into B-cells in mice (Chen *et al.*, 2004). miR-181a targets the homeobox protein Hox-A11 which is involved in mammalian myoblast differentiation (Naguibneva *et al.*, 2006). Another miRNA involved in the developmental processes is miR-196a that regulates several *Hox* genes (Yekta *et al.*, 2004). miR-134, which is considered brain specific, regulates the development of dendritic spine (Schratt *et al.*, 2006). miR-143 has been involved in the differentiation of preadipocytes into mature adipocytes (Esau *et al.*, 2004) and miR-1, miR-133, and miR-206 in myogenesis (Kim *et al.*, 2006; Rao *et al.*, 2006).

On the other hand, a large body of studies focused on the role of miRNAs in ESC biology. In particular, miRNA screening in human and mouse ESC revealed that they express a unique set of miRNAs that are downregulated after ESC differentiation, suggesting a role for miRNAs in the maintenance of pluripotency (Kanellopoulou *et al.*, 2005; Morin *et al.*, 2008; Murchison *et al.*, 2005; Tang *et al.*, 2006).

Functional studies relayed on experiments based on the generation of ESC deficient for components involved in miRNA biogenesis. It was found that mouse ESC knockout for Dicer (Kanellopoulou *et al.*, 2005) or for DGCR8 (Wang *et al.*, 2007) fail to undergo differentiation in embryo bodies, supporting the role of miRNAs in self-renewal. In addition, mice lacking Ago2 show defects in neural development (Liu *et al.*, 2004).

All these results support the role of miRNAs in epigenetic regulation of stemness together with promoter DNA methylations and histone modifications (Aranda *et al.*, 2009).

A. miRNA profile in MSCs

Only few studies have so far addressed the miRNA profile on ASCs. Most of the studies focus on MSCs, searching for a miRNA signature that may allow to define and track this cell population. Moreover, the identification of specific pathways involved in MSC proliferation and differentiation has provided functional information that could be useful to develop correct strategy to culture MSC for cell therapy applications (Lakshmipathy and Hart, 2008). The importance of miRNA as epigenetic regulators of MSC behavior and function has been recently recognized (Guo *et al.*, 2011). Most of the miRNAs highly expressed in human ESCs (Lakshmipathy *et al.*, 2007) are absent in MSCs. In particular, miRNAs related to pluripotency in ESCs, such as miR302a, b, c, and d, and miR200c are not expressed by MSCs (Goff *et al.*, 2008). On the contrary, mature let-7 family miRNAs, not present in ESCs, are expressed by MSCs and are involved in the osteogenic differentiation induced by platelet-derived growth factor (Goff *et al.*, 2008).

B. Comparative expression profiles of miRNAs among human MSCs derived from different tissues

As described above, cells with MSC properties have been isolated from different tissues. These populations are heterogeneous and under certain conditions show pluripotent-like properties (Aquino *et al.*, 2010). In the past years, researchers explored the transcriptomes of MSCs derived from different organs, to identify the core feature of all MSCs and the individual signature gene expression profile of each MSC population. Tsai *et al.* compared the transcriptome expression of MSCs derived from amniotic fluid, amniotic membrane, cord blood, and bone marrow (Tsai *et al.*, 2007). They demonstrated that adult MSCs share a core signature profile characteristic of all MSCs and distinct from that of various fetal organs. Moreover, MSCs from different organs exhibit a specific signature expression profile (Tsai *et al.*, 2007). The same approach has been used to identify miRNA signature of MSCs from different origin (Collino *et al.*, 2010; Wang *et al.*, 2010). Wang *et al.* compared the miRNA expression profiles of human adipose tissue-derived MSCs (hADSCs) from breast, face, and abdominal adipose tissue (Wang *et al.*, 2010). All the different hADSC populations express extremely low levels of the ESC- and tissue (liver, heart, muscle, pancreas, placenta, and testis)-specific miRNAs (Wang *et al.*, 2010). Moreover, the miRNA expression profile of hADSCs from abdominal adipose tissue is extremely similar to that of hADSCs from breast, but not to that of hADSCs from face (Wang *et al.*, 2010).

Comparison of the miRNA expression profiles on bone marrow-derived MSCs and on hADSCs shows that the only difference is the expression of miR-424, which is downregulated in bone marrow-derived MSCs (Aranda *et al.*, 2009). We recently analyzed the miRNA expression profile of 365 mature miRNAs, from human bone marrow-derived MSCs and human liver stem cells (HLSCs) (Collino *et al.*, 2010). HLSCs are a population of MSCs showing a partial hepatic commitment previously isolated and characterized by Herrera *et al.* (2006). Clustering similarity analysis between these two cell types shows that human bone marrow-derived MSCs and HLSCs share 128 miRNAs, 98 of which clustered in the same expression group, suggesting a common origin of the two populations (Collino *et al.*, 2010). Moreover, we showed four miRNAs selectively expressed by HLSCs (miR-7, -95, -204, and -650) and eight by MSCs (miR-196b, -196a, -615, -501, -449, -17-3p, -497, and-486). Among miRNAs expressed by both cell types, we found that 32 miRNAs were upregulated in MSCs compared to HLSCs and three miRNAs in HLSCs in respect to MSCs (Collino *et al.*, 2010). These data indicate that MSCs from different organs share a common pattern of miRNAs, characteristic of MSC lineage. In the mean time, few miRNAs are specific for the tissue origin of MSCs, suggesting an influence of the niche in their expression.

V. Role of miRNAs in MSC Differentiation

The observation that specific populations of miRNAs are differentially expressed during MSC differentiation leads to investigate the role of miRNAs in this process (Fig. 14.2).

A. miRNAs and mesodermal lineage differentiation of MSCs

1. MSCs and osteogenic differentiation

The identification of miRNA pattern on MSC undergoing osteogenic differentiation has been fully described in several studies. In particular, it has been evaluated whether they play a role in the autonomous differentiation of MSC toward osteoblasts during successive culture expansion (Krampera *et al.*, 2005). Indeed, in physiological conditions, MSCs differentiate into osteoblasts by genetic and epigenetic mechanisms (Hu *et al.*, 2010).

Bone morphogenetic proteins (BMPs) are potent morphogens that activate transcriptional programs involved in lineage determination of multiple cell types (Plikus *et al.*, 2008; Rendl *et al.*, 2008). BMPs are also involved in the commitment of MSCs (Lee *et al.*, 2000). BMP2, in particular, has been

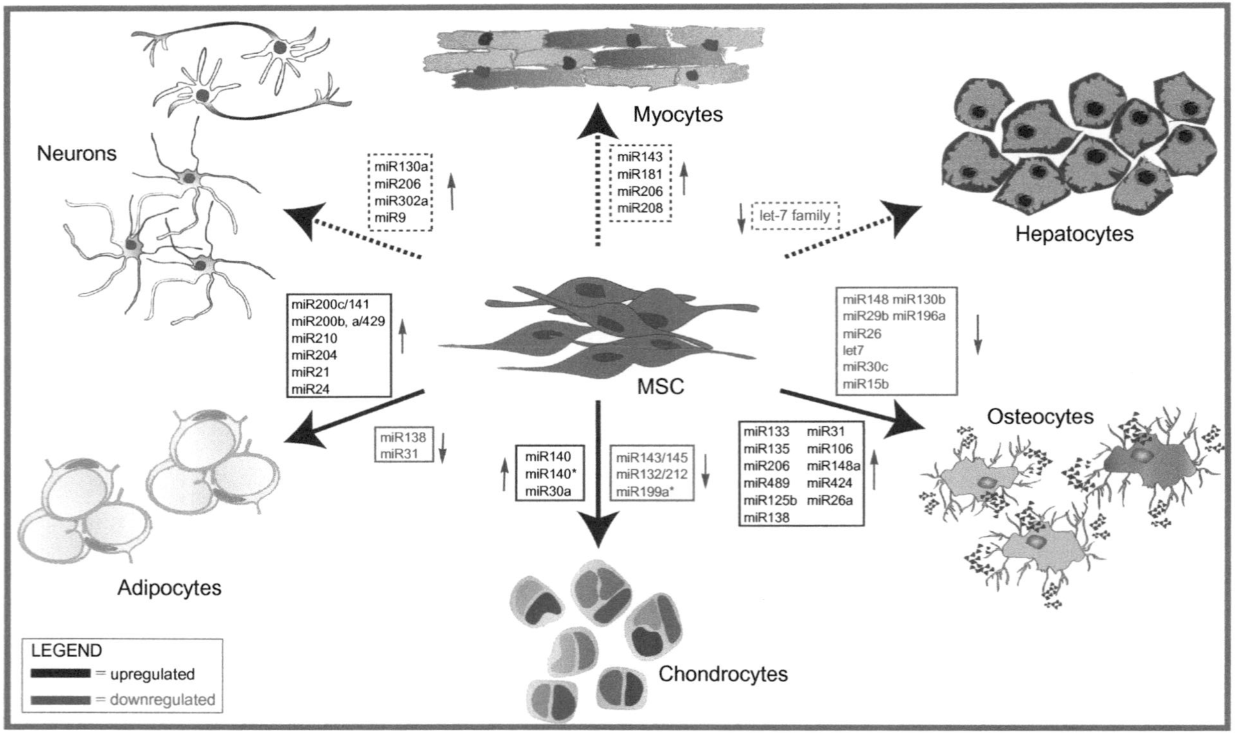

Figure 14.2 Role of miRNAs that regulate differentiation of MSCs in mesodermal and nonmesodermal tissues. Red-labeled miRNAs are downregulated (↓), whereas black-labeled miRNAs are upregulated (↑) during the process of differentiation. (See Color Insert.)

described to sustain osteoblastogenesis (Li *et al.*, 2008). By profiling miRNAs during BMP2-induced osteogenesis of mouse premyoblast mesenchymal cells, Li *et al.* demonstrated that 25 miRNAs result significantly regulated during the differentiation process. In particular, 22 of these miRNAs are downregulated. Target prediction analysis showed a tight correlation between modulated miRNAs and multiple osteogenic signaling pathways including that of Wnt, BMP, and FGF, and transcriptional regulators of osteogenesis. miR-133 was predicted to target osteogenic transcription factor (Runx2), whereas miR-135 was predicted to target many components of BMP signaling such as Smad5 protein (Li *et al.*, 2008).

Other studies on mouse premyoblast mesenchymal cells before and after BMP-2 treatment demonstrated that miR-206, previously viewed as a muscle-specific miRNA, is another potential key regulator of the osteogenic process (Inose *et al.*, 2009). miR-206 is expressed in osteoblasts, and its expression decreased during differentiation into osteocytes. Overexpression of miR-206 in osteoblasts inhibits their differentiation, whereas knockdown of miR-206 expression promotes the differentiation into ostecytes. In silico analysis has identified connexin 43 (Cx43), a major gap junction protein in osteoblasts, as a target of miR-206 (Inose *et al.*, 2009). Schoolmeesters *et al.*, through the screening of a collection of miRNA inhibitors, found that three miRNAs, namely miR-148b, -27a, and -489, are capable to regulate osteogenic cell fate of human MSCs (Schoolmeesters *et al.*, 2009). The specificity of these miRNAs for the osteogenic process is supported by experiments showing that their inhibition do not affect adipogenesis. Moreover, cells transfected with miR-489 inhibitor or miR-148b mimic alone or in combination demonstrated a drastic increase of different markers of early osteogenic differentiation, such as osteopontin and alkaline phosphatase activity (Schoolmeesters *et al.*, 2009). Further studies indicate an important role of miR-125b in the control of osteogenic differentiation of ST2 mouse MSCs by regulating cell proliferation (Mizuno *et al.*, 2008). The expression of miR-125b is attenuated in osteoblastic-differentiated ST2 cells induced by BMP-4, in respect to undifferentiated ST2 cells. Moreover, the transfection of exogenous miR-125b inhibits osteoblastic differentiation after BMP-4 treatment (Mizuno *et al.*, 2008). Li *et al.* recently described the potential involvement of miRNAs during the differentiation of MC3T3 preosteoblast cells (Li *et al.*, 2009). Among miRNAs differentially expressed during the osteoblastic differentiation, miR-29, let-7, and miR-26 families are upregulated. Interestingly, they may target genes involved in collagen protein accumulation during the mineralization stage. Further, Li *et al.* demonstrated that miR-29b promotes osteogenesis by downregulating known inhibitors of osteoblast differentiation, such as HDAC4, TGFbeta3, ACVR2A, CTNNBIP1, and DUSP2 proteins through binding to their 3′-untranslated region (UTR) sequences (Li *et al.*, 2009). These researchers proposed that this mechanism may prevent fibrosis and facilitates mineral deposition.

Recently, Eskildsen *et al.* showed that miR-138 may modulate osteogenic differentiation of human bone marrow-derived MSCs (Eskildsen *et al.*, 2011). miRNA array profiling revealed that miR-138 is downregulated during osteoblast differentiation of hMSCs. Overexpression of miR-138 inhibits *in vitro* osteoblast differentiation, whereas its inhibition promotes the expression of osteoblast-specific genes, alkaline phosphatase activity, and matrix mineralization. Target prediction analysis and experimental validation by luciferase 3′-UTR reporter assay indicate that the focal adhesion kinase, which is involved in osteoblast differentiation, is a target of miR-138 (Eskildsen *et al.*, 2011). All these results suggest that the pharmacological administration of antimiR-138 could represent a therapeutic strategy for enhancing bone formation *in vivo*. Gao *et al.* recently identified miR-31, -106a, -148a, and -424 as miRNAs that are under-expressed in osteo-differentiated bone marrow MSCs (Gao *et al.*, 2011). The putative targets of these miRNAs, predicted by bioinformatic analysis, include RUNX2, CBFB, and BMPs which are involved in bone formation. Further, the putative targets of miRNAs overexpressed in osteo-differentiated bone marrow MSCs (miR-30c, -15b, and -130b), are the CD29 MSC maker, the stemness-maintaining factor, and genes related to cell differentiation (Gao *et al.*, 2011).

Similar studies were performed on human adipose tissue-derived mesenchymal stem cells (hADSCs) which are able to generate bone *in vivo* (Hicok *et al.*, 2004; Zuk *et al.*, 2001). Luzi *et al.* demonstrated that miR-26a, which is upregulated during the late stages of hADSC osteogenic differentiation, modulates expression of SMAD1 (Luzi *et al.*, 2008), predicted as its potential target by bioinformatic approaches (Rajwsky, 2006). The inhibition of miR-26a increased protein levels of SMAD1 transcription factor enhancing osteoblast differentiation. Another group reported that miR-196a, which is upregulated during osteogenic hADSC differentiation, inhibits proliferation and enhances the differentiation process by binding to specific target sequences harbored in the 3′-UTR of *HOXC8* mRNA (Kim *et al.*, 2009a,b). The obvious importance of miRNAs for osteoblastic differentiation and bone formation suggests that the manipulation of miRNA expression may provide therapeutic opportunities for the treatment of bone related diseases (Hu *et al.*, 2010).

2. MSCs and adipogenic differentiation

The complete understanding of the molecular mechanisms underlying the differentiation of MSCs during adipogenesis could provide new insights into the pathogenesis of a number of diseases, such as obesity and diabetes (Yang *et al.*, 2011b). Several high-throughput studies have identified hundreds of miRNAs that are differentially expressed during the formation of metabolic tissues or in related pathophysiological states (Alexander *et al.*, 2011). Interestingly, adipose tissue is increasingly being recognized as an important mediator of energy distribution through the body, not only for its role as energy storage tissue but also for its influence on body metabolic functions, through cytokine release (Alexander *et al.*, 2011). Mature adipocyte formation is made up by two important steps: determination and terminal differentiation of preadipocytes. The first step involves the commitment of MSCs to become preadipocytes, losing their ability to differentiate in other cell lineages. Numerous studies described the important role of TGF-β, WNT, and BMP signaling in the determination of MSC fate. In particular, the balance between these signaling factors may direct MSCs to differentiate in adipocytes, chondrocytes, or osteocytes. In this context, miRNAs play a critical role in regulating these pathways. Previous studies demonstrated that the Wnt signaling potently inhibits the differentiation of ST2 mouse MSCs into adipocytes. Kennell *et al.* recently showed that the mammalian homologues of miR-8 (miR-200c/141 and miR-200b,a/429) may promote adipogenesis through the inhibition of the Wnt signaling in these cells (Kennell *et al.*, 2008). The generation of stable cell lines expressing the miR-200c/141 or miR-200b,a/429 clusters induces the ST2 cells to differentiate into adipocytes. Moreover, the expression of both miRNA clusters increases adipogenesis (Kennell *et al.*, 2008). Another miRNA identified by high-throughput microarray data, the miR-210, has an important role in promoting adipogenesis, supporting the terminal differentiation of preadipocytes. In particular, the overexpression of miR-210 on 3T3-L1 preadipocyte cells promotes the lipids formation by repressing WNT signaling by targeting Tcf7l2 (Qin *et al.*, 2010). In human and mouse bone marrow-derived MSCs, miR-204 was shown to be involved in their differentiation fate. In fact, overexpression of miR-204 inhibited osteoblast differentiation and promoted adipocyte differentiation. On the contrary, the inhibition of miR-204 promotes osteogenesis and prejudices adipocyte formation (Jian Huang *et al.*, 2010). miR-204/211 may act as important negative modulators of Runx2, a master regulator of osteoblast differentiation (Jian Huang *et al.*, 2010). Similarly, studies on adipogenic differentiation of human hADSCs identified miR-138 and miR-21 as factors involved in adipogenesis. miR-21, which is transiently upregulated during the adipogenic differentiation of hADSCs, promotes adipogenesis by binding to target sequences in the 3′-UTR of TGFBR2, protein of the TGF-β pathway (Kim *et al.*, 2009a,b). miR-138, which is significantly downregulated during adipogenesis, may inhibit the differentiation of hADSCs on adipocytes, through the downregulation of key adipogenic transcription factors, such as cytidine–cytidine–adenosine–adenosine–thymidine (CCAAT) enhancer binding protein alpha, peroxisome proliferator-activated receptor gamma 2, and several other adipogenic marker genes, including fatty acid binding protein 4 and lipoprotein lipase. In addition, luciferase reporter assays demonstrated that miR-138 directly targets EID-1, a nuclear receptor coregulator, implying that the negative role of miR-138 in the adipocyte differentiation of hAD-MSCs is at least partially mediated via repression of EID-1 transcript (Yang and Lai, 2010). Sun *et al.* demonstrated the upregulation of miR-24-1 and miR-31 during BMP-2-induced adipocyte differentiation of C3H10T1/2 murine MSC line (Sun *et al.*, 2009). The combinatory treatment of C3H10T1/2 cells with both miR-24 and BMP-2 induces adipogenesis more efficiently than BMP-2 alone. The opposite effect is observed when cells are treated with both miR-31 and BMP-2 (Sun *et al.*, 2009). Interestingly, overexpression of miR-24 may affect also cell cycle progression, through the enhancement of BMP-2-induced G1 arrest. In contrast, miR-31 overexpression results in an accumulation of cells at the G2-M transition. These data sustain the role of miR-24 in supporting the commitment of MSCs to adipocyte lineage and growth arrest. On the contrary, miR-31 has the opposite effect (Sun *et al.*, 2009).

3. MSCs and chondrogenic differentiation

miRNAs regulate also the cartilage homeostasis (Kobayashi *et al.*, 2011), and changes in miRNA expression and function have been associated with the development and progress of diseases affecting articular cartilage (Stanczyk *et al.*, 2008). The *in vivo* skeletal formation is mimicked *in vitro* by chondrogenesis assay of MSC differentiation (García-Álvarez *et al.*, 2011). Recently, Han *et al.*, using microarray technique, explored the expression of miRNAs in the *in vitro* model of MSC-induced chondrogenesis (Han *et al.*, 2010), showing the upregulation of four miRNAs (miR-130b, -152, -28, and -26b). Predicted targets of these miRNAs include several genes associated with collagen proteins such as *COL2A1*, *COL4A1*, and *COL6A1* (Han *et al.*, 2010). These genes encode their correspondent type collagen, and their mutation usually result in pathologic skeletal disorders, involving especially chondrocytes. Interestingly, IGF-1 (insulin-like growth factor I), a target predicted by miR-26b, promotes the differentiation of MSCs into chondrocytes in combination with TGF-β (Longobardi *et al.*, 2006). The identification of miRNAs expressed in articular chondrocytes and their changes in osteoarthritic cartilage may be relevant to complete the understanding of the osteoarthritis pathogenesis. Miyaki *et al.* compared miRNA expression between chondrocytes and MSCs showing that these cells differ mainly for the expression of miR-140 (Miyaki *et al.*, 2009). Normal human articular cartilage expressed miR-140 and this was significantly reduced in osteoarthritis tissue. During the chondrogenic differention of MSCs, miR-140 expression increases simultaneously with the expression of *Sox9* and *Col2a1* which are chondrocyte-specific transcripts. This finding suggests that miR-140 is a marker and possibly a regulator of chondrogenic differentiation (Miyaki *et al.*, 2009). miRNA microarray profile of mouse MSCs, at four different stages of TGF-β3-induced chondrogenic differentiation, shows that eight miRNAs are significantly upregulated and five miRNAs are downregulated (Yang *et al.*, 2011b). Interestingly, the both miR-143/145 and miR-132/212 clusters are downregulated during chondrocyte differentiation. Bioinformatic analysis identifies four predicted target genes, namely *Sox6*, *ACVR1B*, *Runx2*, and *ADAMTS5*, transcripts known to be involved in chondrogenesis (Yang *et al.*, 2011a). Yang *et al.* showed that the expression of miR-140* and miR-30a is upregulated whereas their predicted target mRNAs (*ADAMTS5*, *Runx2*) are downregulated. At variance, two downregulated miRNA clusters (miR-132/212, miR-143/145) are associated with increased expression of their corresponding predicted target mRNAs (*Sox6*, *ACVR1B*). Among the predicted target of the above mentioned miRNAs, Yang *et al.* identified a group of SMAD family members, previously known to be involved in signaling pathways associated with chondrogenesis (Fisher *et al.*, 2006). Moreover, miRNA screening on BMP2-treated C3H10T1/2 murine MSC identified miR-199a* as a novel miRNA directly implicated in the early stage of chondrogenic differentiation process (Lin *et al.*, 2009). miR-199a* directly inhibits early chondrogenesis by targeting and suppressing the expression of Smad1, a key downstream mediator of BMP signaling and a major regulator of bone and cartilage development (Lin *et al.*, 2009). Taken together, these results indicate that mesodermal cell-type differentiation of MSCs is regulated by specific sets of miRNAs and that different phenotypes are associated with the differential expression of these miRNAs.

B. miRNAs and MSC plasticity

Besides the ability of MSCs to differentiate in cells of all the mesenchymal lineage, it has been suggested that these cells may contribute to the repair of tissue of nonmesodermal origin. This property defined as plasticity has been related to the ability of MSCs to undergo transdifferentiation into tissue-resident cells. The mechanism involved in plasticity remains still controversial (Aquino *et al.*, 2010). *In vitro* studies demonstrated that MSCs can differentiate into different cell lineages, such as myocytes, hepatocytes, and neurons (Guo *et al.*, 2011; Fig. 14.2).

The involvement of miRNAs in myocyte differentiation of MSCs has been extensively studied. Shan *et al.* demonstrated that cardiomyocyte-like cells differentiated from human MSCs, by 5-azacytidine treatment or by an indirect coculture with neonatal rat myocytes, start to express representative heart-specific pri-miRNAs (Shan *et al.*, 2007). In particular, miR-143 and -181 were induced in human MSCs by 5-azacytidine, and miR-143, -181, -206, and -208 were induced by indirect coculture with neonatal rat myocytes (Shan *et al.*, 2007). Interestingly, miR-181 has been previously described as strongly upregulated during embryo differentiation. Moreover, miR-181 participates in establishing the muscle phenotype, through the downregulation of the homeobox protein Hox-A11, a repressor of the differentiation process (Naguibneva *et al.*, 2006).

Despite the differentiation of MSCs into neurons is highly controversial, several studies addressed the potential involvement of miRNAs in the generation of neuronal-like cells. Greco *et al.* compared the miRNAs expression profile of human bone marrow MSCs and MSC-derived neuronal-like cells after 12 days of culture in neuronal induction medium (Greco and Rameshwar, 2007). They identified 16 different miRNAs that are upregulated after differentiation. In particular, MSC-derived neuronal-like cells show significantly elevated levels of miR-130a, miR-206, and miR-302a compared with undifferentiated bone marrow MSCs. Moreover, the inhibition of miR-130a and miR-206 in MSC-derived neuronal-like cells causes significant increase in the synaptic transcript Tac1 mRNA translation, as indicated by high levels of its encoded protein, neurotransmitter substance P (Greco and Rameshwar, 2007). Other miRNAs identified in undifferentiated bone marrow MSCs, such as miR-125a, miR-125b, and miR-23b, are implicated in neural differentiation (Greco and Rameshwar, 2007). Jing *et al.* recently demonstrated that miRNA-9 could promote mouse bone marrow MSC neuronal differentiation, through the perturbation of the Notch signaling (Jing *et al.*, 2011). Moreover, they demonstrated that the expression of the neural cell-specific marker, microtubule-associated protein 2, increases during the differentiation process, whereas the expression of Notch-1 decreases (Jing *et al.*, 2011).

The involvement of miRNA in the hepatic commitment of MSCs is not completely understood. Koh *et al.* recently showed that human ESC-derived MSCs express high level of let-7 family of miRNAs. Combining different network prediction algorithms, they identified the hepatic nuclear factor 4 alpha transcript (*HNF4A*) as indirectly regulated by the let-7 family of miRNAs (Koh *et al.*, 2010). Moreover, it was observed that in undifferentiated MSCs, where the let-7 family of miRNAs is highly expressed, the expression of *HNF4A* is very low. In contrast, the HepG2 hepatoma tumor cells expressing high level of *HNF4A* show a very low expression of let-7 miRNAs. Since HNF transcription factors are upregulated in hepatocytes derived from ADSCs (Yamamoto *et al.*, 2008), it has been suggested that let-7 miRNAs regulate *HNF4A* levels during hepatic differentiation of MSCs (Koh *et al.*, 2010).

Finally, miRNA bioarrays show that undifferentiated MSCs contain miRNAs, such as miR-222, miR-221, and miR-214, that are involved in angiogenenic, myogenic, and hematopoietic differentiation (Greco and Rameshwar, 2007).

VI. The Role of miRNAs in Cell-to-Cell Communication

In vitro studies on MSC transdifferentiation, however, were not confirmed *in vivo*. Despite contribution of MSCs in the repair of tissue injury has been consistently shown in different organs, such as heart, liver, and kidney, evidence for transdifferentiation does not justify the biological effects (Gnecchi *et al.*, 2008; Tögel *et al.*, 2005). Recent studies demonstrate that the beneficial effects observed after treatment of different experimental diseases with exogenous MSC administration are associated only with a transient recruitment of MSCs at the site of injury with a minimal incorporation within the injured tissue, suggesting that MSCs may provide a paracrine/endocrine support to the repair (Humphreys *et al.*, 2008). Experiments of administration of MSC-conditioned medium demonstrate that factors released from MSCs mimic their beneficial effects (Bi, 2007; Gnecchi *et al.*, 2008). A paracrine/endocrine mechanism has been also implicated in the immunosuppressive and anti-inflammatory properties of MSCs. A complete characterization of the paracrine/endocrine factors and their pathways would be critical for a better understanding of the MSC regenerative potential and for the identification of new therapeutic strategies. Recent studies suggest that transfer of miRNAs between stem and differentiated cells may account, at least in part, for such paracrine/endocrine action. In this context, microvesicles (MVs) released from MSCs may act as a vehicle for such transfer.

A. Microvesicles

MVs include a heterogeneous population of vesicles released as exosomes from the endosomal compartment or as shedding vesicles from the cell surface. After their release, MVs may remain in the extracellular space in proximity of the cell of origin or may enter the biological fluids. Therefore, MVs not only operate as an integral component of the intercellular microenvironment but they may also transfer information at distances (Camussi *et al.*, 2011). Indeed, MVs display a broad spectrum of biological activities instrumental in cell-to-cell communication. Once released from a given cell type, MVs may interact through specific receptor–ligand with target cells and transfer various bioactive molecules including membrane receptors, proteins, and organelles (Camussi *et al.*, 2011). Attention has been recently paid on the capacity of MVs to induce epigenetic changes in target cells by delivering specific subsets of mRNAs and miRNAs. This transfer of genetic information has been recognized as a universal mechanism used by different cell types to influence behavior of other cells. The ability of MVs to reprogram recipient cells was first established by Ratajczak *et al.* (2006) that demonstrated that MVs derived from murine ESCs may reprogram adult haematopoietic stem/progenitor cells inducing an upregulation of early pluripotent hematopoietic markers and phosphorylation of MAPKp42/44 and Akt. This biologic effect was ascribed to a horizontal transfer of mRNAs mediated by MVs. Several subsequent studies indicate that mRNAs delivered by MVs can be translated into the corresponding proteins by the target cells (Bruno *et al.*, 2009a,b; Deregibus *et al.*, 2010; Herrera *et al.*, 2010; Valadi *et al.*, 2007). These studies suggest the potential of different stem cell population to reprogram differentiated cells during tissue regeneration and repair.

B. miRNA transfer mediated by MVs

Besides mRNA, MVs may transfer miRNAs. Valadi *et al.* (2007) described that exosomes derived from mast cells carry specific subsets of miRNAs. In addition, MVs derived from blood cells were shown to contain miRNAs associated with different cell functions such as differentiation, metabolic pathways, and modulation of immune response (Hunter *et al.*, 2008). Indeed, MVs may allow detection of miRNAs in the circulation. In fact, miRNAs in plasma would be degraded by RNase if not protected by the membrane envelope provided by MVs.

The finding that ESCs release an abundant amount of MVs that contain miRNAs suggests a role of MVs in signaling within the stem cell niche (Yuan *et al.*, 2009). Only few studies have so far defined the miRNA content in MVs derived from MSCs. Chen *et al.* described that the conditioned medium of human MSCs derived from ESCs contains "microparticles" enriched in pre-miRNAs (Chen *et al.*, 2010). They detected in the conditioned medium 60 miRNAs, 45 of which are also present within the cells. Two members of the most highly conserved and developmentally important human let-7 family, let-7b and let-7d, were only detected in the conditioned medium and not in human ESC-derived MSCs (Chen *et al.*, 2010), suggesting a mechanism of sorting. Recently, Koh *et al.* demonstrated that the conditioned medium of human MSCs derived from ESCs contains MVs that shuttled specific miRNAs, including the let-7 family miRNAs (Koh *et al.*, 2010). The let-7 family miRNAs, as described above, downregulate *HNF4A*, a physiological factor that in the liver activates *apoB* gene expression. Elevated *apoB* gene expression has been correlated with an increased risk of atherosclerosis and coronary heart diseases. Moreover, *HNF4A* is also involved in the regulation of serum lipid levels. Since human MSC-conditioned medium has been shown to reduce infarct size in patients with acute myocardial infarction (Timmers *et al.*, 2007), Koh *et al.* speculated that MSC-derived MVs carrying let-7 family of miRNAs (Koh *et al.*, 2010) may account for the effect of the conditioned medium. Moreover, they identified three miRNAs, namely miR-21, -133, and -30, shuttled by MVs which are known to be involved in cardiac remodeling after myocardial infarction (Duisters *et al.*, 2009; Thum *et al.*, 2008).

We recently demonstrated that MVs released from adult human MSCs contain selected pattern of mature miRNAs that can be transferred to target cells (Collino *et al.*, 2010; Fig. 14.3). Comparison of miRNA profile among human bone marrow MSCs, liver HLSCs and

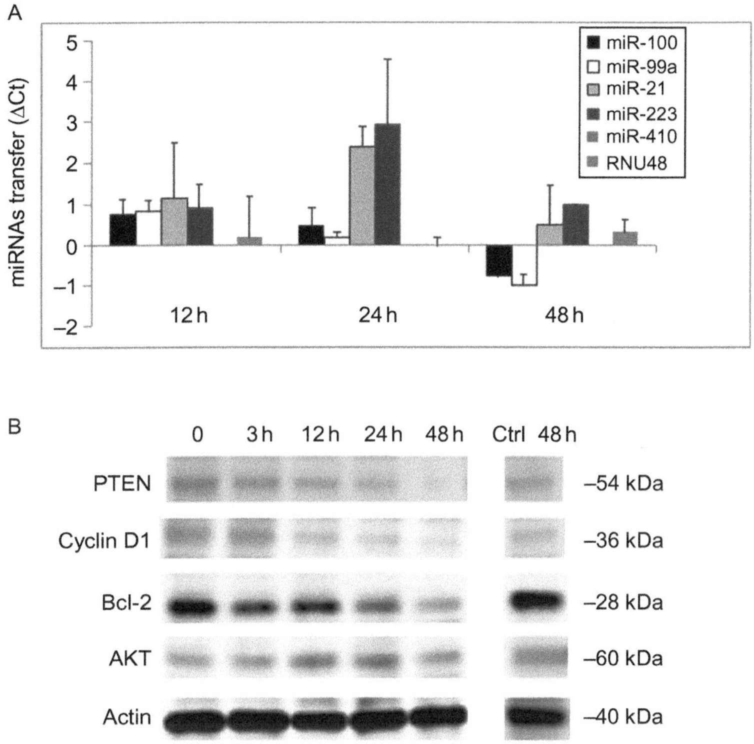

Figure 14.3 miRNA transfer and MV uptake by murine tubular epithelial cells. (A) The transfer of selected miRNAs was evaluated by qRT-PCR in tubular epithelial cells incubated with MVs derived from MSCs in the presence of an inhibitor of transcription. (B) Proteins targeted by the enriched miRNAs present in MVs were downregulated indicating a functional activity of transferred miRNAs. Reproduced from Collino *et al.* (2010). (For color version of this figure, the reader is referred to the Web version of this chapter.)

their released MVs showed that MVs contain a pattern of miRNAs shared with the cells of origin (91 detected miRNAs, of which 41 clustered in same expression group). Moreover, we observed that MVs expressed a group of miRNAs described as "selectively" expressed by MSCs, such as miR-103-1, -140, -143-5p, and -340. These data imply that miRNA patterns may be used as signatures to define the cellular origin of MVs. On the other hand, selected miRNAs were detectable only in the released MVs but absent in their cell of origin (miR-223, miR-142-3p, and miR-451 in MVs from HLSCs and miR-223, miR-564, and miR-451 in MVs from MSCs). Our data and Chen's study support the theory of a specific and organized package of miRNAs in MVs before their secretion (Chen *et al.*, 2010; Collino *et al.*, 2010).

Gene ontology analysis shows that the high expressed miRNAs in cells and MVs could be involved in multiorgan development, cell survival, and differentiation (Collino *et al.*, 2010). Moreover, we found that miRNAs highly expressed within MVs are transferred to murine tubular epithelial cells after MV internalization. miRNAs shuttled by MVs are functional because they are able to downregulate proteins such as Bcl-2, cyclin-D1, and PTEN, known to be targeted by the transferred miRNAs (Fig. 14.3). The mechanism of mRNA and miRNA compartmentalization within MVs has not been completely clarified. Since MVs contain stress granule specific ribonucleoproteins that are known to be involved in the traffic of RNAs within the cell and Ago2 protein of the miRISC, we speculate on the possible role of such proteins in the package of miRNAs in MVs and in their release (Collino *et al.*, 2010). A recent study by Kosaka *et al.* (2010) adds further inside on the secretory mechanisms and the intercellular transfer of miRNAs via MVs. They demonstrated that the MV secretion is dependent on ceramide-dependent secretory machinery rather than on the endosomal sorting complex required for transport (ESCRT) (Kosaka *et al.*, 2010).

C. Biological effects of MVs

It has been suggested that MVs mediate a bidirectional exchange of genetic information between stem and injured cells (Camussi *et al.*, 2010). Transcripts and miRNAs delivered from injured cells may reprogram the phenotype of stem cells to acquire specific features of the tissue. On the other hand, MVs released from stem cells may induce dedifferentiation of cells survived to injury with a cell cycle re-entry and tissue self-repair by delivering stem cell-specific mRNAs and miRNAs.

Several studies indicate that bone marrow-derived stem cells have the capacity to produce nonmarrow cells in many injured tissues after engraftment and may contribute to tissue repair (Dooner *et al.*, 2008; Quesenberry *et al.*, 2007). The underlying mechanism leading to stem cell differentiation after interaction with injured cells is only partially known. Transdifferentiation has been implicated in bone marrow stem cell plasticity in some but not in others studies (Deregibus *et al.*, 2010). Quesenberry and Aliotta (2008) suggested that differentiation occurs in response to specific signals derived from injured cells and identified in MVs, the carriers of genetic information (Fig. 14.4) able to change the phenotype of bone marrow-derived stem cells. In line with this hypothesis, Aliotta *et al.* demonstrated that murine bone marrow cells express genes for lung-specific proteins such as Clara cell-specific protein, surfactant B, and surfactant C after incubation with the injured lung-conditioned medium and that this activity resided in MVs released in the cell supernatant (Aliotta *et al.*, 2010). MVs derived from injured lung contain high levels of a lung-specific mRNA and transfer this

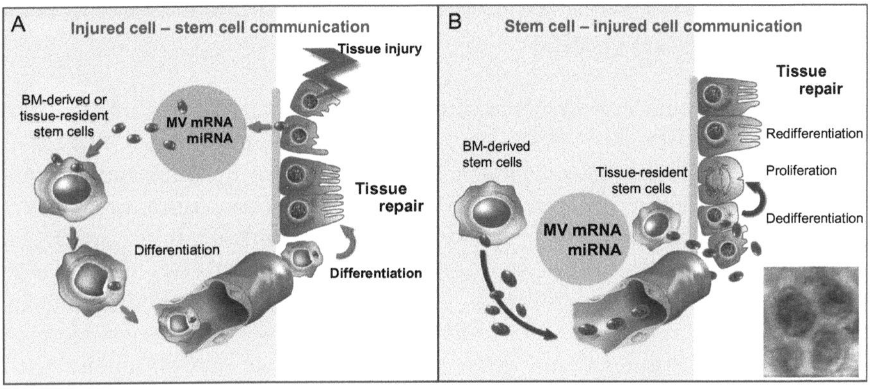

Figure 14.4 Schematic representation of MV-mediated mechanism of tissue repair. The MV-mediated exchange of genetic information may be bidirectional. (A) MVs derived from injured cells may transfer to stem cell-specific subsets of mRNAs and miRNAs that induce a stem cell differentiation with a contribution to the repair of the injured tissue. (B) MV-mediated transfer of mRNAs, miRNAs, or proteins from bone marrow-derived or tissue-resident stem cells may reprogram the phenotype of mature cells by inducing dedifferentiation of cells which survived injury and their reentry into the cell cycle, with repopulation of the injured tissue and redifferentiation into mature tissue cells. Inset: micrograph of transmission electron microscopy of MVs released from human MSCs. The image shows small vesicles of approximately 120 nm in diameter (reproduced with permission from Deregibus *et al.*, 2010). (For color version of this figure, the reader is referred to the Web version of this chapter.)

mRNA to bone marrow cells. These experiments suggest that MVs derived from injured tissue mediate bone marrow cell phenotype changes during physiological tissue repair.

On the other hand, MVs released from MSCs may account for the paracrine/endocrine action of these cells and for their beneficial effect in favoring the repair of tissue injury without the need of MSC differentiation. Indeed, MVs derived from MSCs may reprogram cells survived to injury inducing their dedifferentiation and reentry to cell cycle with activation of tissue regenerative programs (Fig. 14.4). This possibility was proved by administration of MVs released from MSCs *ex vivo* in different experimental models of renal and liver injury (Bruno *et al.*, 2009a,b; Gatti *et al.*, 2011; Herrera *et al.*, 2010). MVs were found to be able to enter in the renal tubular epithelial cells and in the hepatocytes delivering their mRNA and miRNA cargo. The tissue regenerative program activated by MVs is abrogated by the inactivation of RNA content of MVs, suggesting that the biological effects are related to the delivery of selected RNAs. This was confirmed by experiments showing *in vitro* and *in vivo* translation into proteins of reporter mRNAs and by experiments showing the transfer of biological active miRNAs (Collino *et al.*, 2010; Deregibus *et al.*, 2007).

VII. Conclusions

It is becoming increasingly evident that miRNAs play a fundamental role in the development of the organism. In stem cell biology, miRNAs have a crucial role in self-renewal, in maintenance of stemness, and in cell differentiation. Therefore, the understanding of the fine-tuning of miRNA generation, function, and targets may provide insight both in physiological and pathological conditions. In particular, functional studies should take into account the complex interplay of different miRNAs in modulating the same pathways and the action of individual miRNAs on different targets. In addition, recent studies demonstrate that some targets may modulate with an autoregulatory feedback loop the expression of their correspondent miRNAs in a sort of complex crosstalk (Krol *et al.*, 2010). The exchange of miRNAs between different cells is emerging as a mechanism of cell-to-cell communication. In this context, MVs act as vehicle for miRNA exchange to neighbor cells. MVs released from MSCs may transfer genetic information to injured cells activated program of tissue repair. The prediction of defined functions of miRNAs shuttled by MVs derived from MSCs indicate that they may deliver information important for the regulation of development, differentiation, and cell survival.

Acknowledgments

This material is based on work supported by Associazione Italiana per la Ricerca sul Cancro (AIRC) CT 8912, Ministero Università e Ricerca Scientifica, PRIN and by Grants from Regione Piemonte, Piattaforme Biotecnologiche, Progetto PiSTEM, Sis-Ter, and Fresenius Medical Care.

References

Alexander, R., Lodish, H., and Sun, L. (2011). MicroRNAs in adipogenesis and as therapeutic targets for obesity. *Expert Opin. Ther. Targets* **15**(5), 623–636.

Aliotta, J. M., Pereira, M., Johnson, K. W., de Paz, N., Dooner, M. S., Puente, N., Ayala, C., Brilliant, K., Berz, D., Lee, D., Ramratnam, B., McMillan, P. N., *et al.* (2010). Microvesicle entry into marrow cells mediates tissue-specific changes in mRNA by direct delivery of mRNA and induction of transcription. *Exp. Hematol.* **38,** 233–245.

Ambros, V. (2004). The functions of animal microRNAs. *Nature* **431,** 350–355.

Aquino, J. B., Bolontrade, M. F., García, M. G., Podhajcer, O. L., and Mazzolini, G. (2010). Mesenchymal stem cells as therapeutic tools and gene carriers in liver fibrosis and hepatocellular carcinoma. *Gene Ther.* **17**(6), 692–708.

Aranda, P., Agirre, X., Ballestar, E., Andreu, E. J., Román-Gómez, J., Prieto, I., Martín-Subero, J. I., Cigudosa, J. C., Siebert, R., Esteller, M., and Prosper, F. (2009).

Epigenetic signatures associated with different levels of differentiation potential in human stem cells. *PLoS One* **4**(11), e7809.
Assou, S., Le Carrour, T., Tondeur, S., Strom, S., Gabelle, A., *et al.* (2007). A meta analysis of human embryonic stem cells transcriptome integrated into a web-based expression atlas. *Stem Cells* **25,** 961–973.
Bak, R. O., and Mikkelsen, J. G. (2010). Regulation of cytokines by small RNAs during skin inflammation. *J. Biomed. Sci.* **17**(1), 53.
Bartel, D. P. (2004). MicroRNAs: Genomics, biogenesis, mechanism, and function. *Cell* **116,** 281–297.
Bruno, S., Bussolati, B., Grange, C., Collino, F., di Cantogno, L. V., Herrera, M. B., Biancone, L., Tetta, C., Segoloni, G., and Camussi, G. (2009a). Isolation and characterization of resident mesenchymal stem cells in human glomeruli. *Stem Cells Dev.* **18,** 867–880.
Bruno, S., Grange, C., Deregibus, M. C., Calogero, R. A., Saviozzi, S., Collino, F., Morando, L., Busca, A., Falda, M., Bussolati, B., Tetta, C., and Camussi, G. (2009b). Mesenchymal stem cell-derived microvesicles protect against acute tubular injury. *J. Am. Soc. Nephrol.* **20**(5), 1053–1067.
Camussi, G., Deregibus, M. C., Bruno, S., Cantaluppi, V., and Biancone, L. (2010). Exosomes/microvesicles as a mechanism of cell-to-cell communication. *Kidney Int.* **78** (9), 838–848.
Camussi, G., Deregibus, M. C., Bruno, S., Grange, C., Fonsato, V., and Tetta, C. (2011). Exosome/microvesicle-mediated epigenetic reprogramming of cells. *Am. J. Cancer. Res.* **1**(1), 98–110.
Carthew, R. W., and Sontheimer, E. J. (2009). Origins and mechanisms of miRNAs and siRNAs. *Cell* **136**(4), 642–655.
Chen, C. Z., Li, L., Lodish, H. F., and Bartel, D. P. (2004). MicroRNAs modulate hematopoietic lineage differentiation. *Science* **303,** 83–86.
Chen, T. S., Lai, R. C., Lee, M. M., Choo, A. B., Lee, C. N., *et al.* (2010). Mesenchymal stem cell secretes microparticles enriched in pre-microRNAs. *Nucleic Acids Res.* **38,** 215–224.
Collino, F., Deregibus, M. C., Bruno, S., Sterpone, L., Aghemo, G., Viltono, L., Tetta, C., and Camussi, G. (2010). Microvesicles derived from adult human bone marrow and tissue specific mesenchymal stem cells shuttle selected pattern of miRNAs. *PLoS One* **5** (7), e11803.
Crop, M., Baan, C., Weimar, W., and Hoogduijn, M. (2009). Potential of mesenchymal stem cells as immune therapy in solid organ transplantation. *Transpl. Int.* **22,** 365–376.
da Silva Meirelles, L., Chagastelles, P. C., and Nardi, N. B. (2006). Mesenchymal stem cells reside in virtually all post-natal organs and tissues. *J. Cell Sci.* **119,** 2204–2213.
da Silva Meirelles, L., Caplan, A. I., and Nardi, N. B. (2008). In search of the in vivo identity of mesenchymal stem cells. *Stem Cells* **26,** 2287–2299.
Deregibus, M. C., Cantaluppi, V., Calogero, R., Lo Iacono, M., Tetta, C., Biancone, L., Bruno, S., Bussolati, B., and Camussi, G. (2007). Endothelial progenitor cell derived microvesicles activate an angiogenic program in endothelial cells by a horizontal transfer of mRNA. *Blood* **110,** 2440–2448.
Deregibus, M. C., Tetta, C., and Camussi, G. (2010). The dynamic stem cell microenvironment is orchestrated by microvesicle-mediated transfer of genetic information. *Histol. Histopathol.* **25**(3), 397–404.
Ding, X. C., Weiler, J., and Grosshans, H. (2009). Regulating the regulators: Mechanisms controlling the maturation of microRNAs. *Trends Biotechnol.* **27,** 27–36.
Dominici, M., Le Blanc, K., Mueller, I., Slaper-Cortenbach, I., Marini, F., Krause, D., Deans, R., Keating, A., Dj, Prockop, and Horwitz, E. (2006). Minimal criteria for defining multipotent mesenchymal stromal cells. The International Society for Cellular Therapy position statement. *Cytotherapy* **8,** 315–3177.

Dooner, G. J., Colvin, G. A., Dooner, M. S., Johnson, K. W., and Quesenberry, P. J. (2008). Gene expression fluctuations in murine hematopoietic stem cells with cell cycle progression. *J. Cell. Physiol.* **214,** 786–795.

Duisters, R. F., Tijsen, A. J., Schroen, B., Leenders, J. J., Lentink, V., van der Made, I., Herias, V., van Leeuwen, R. E., Schellings, M. W., Barenbrug, P., Maessen, J. G., Heymans, S., *et al.* (2009). miR-133 and miR-30 regulate connective tissue growth factor: Implications for a role of microRNAs in myocardial matrix remodeling. *Circ. Res.* **104**(2), 170–178.

Dykxhoorn, D. M. (2010). MicroRNAs and metastasis: Little RNAs go a long way. *Cancer Res.* **70**(16), 6401–6406.

Esau, C., Kang, X., Peralta, E., Hanson, E., Marcusson, E. G., Ravichandran, L. V., Sun, Y., Koo, S., Perera, R. J., Jain, R., Dean, N. M., Freier, S. M., *et al.* (2004). MicroRNA-143 regulates adipocyte differentiation. *J. Biol. Chem.* **279,** 52361–52365.

Eskildsen, T., Taipaleenmäki, H., Stenvang, J., Abdallah, B. M., Ditzel, N., Nossent, A. Y., Bak, M., Kauppinen, S., and Kassem, M. (2011). MicroRNA-138 regulates osteogenic differentiation of human stromal (mesenchymal) stem cells in vivo. *Proc. Natl. Acad. Sci. USA* **108**(15), 6139–6144.

Eulalio, A., Behm-Ansmant, I., Schweizer, D., and Izaurralde, E. (2007). P-body formation is a consequence, not the cause, of RNA-mediated gene silencing. *Mol. Cell. Biol.* **27**(11), 3970–3981.

Fisher, M. C., Li, Y., Seghatoleslami, M. R., Dealy, C. N., and Kosher, R. A. (2006). Heparan sulfate proteoglycans including syndecan-3 modulate BMP activity during limb cartilage differentiation. *Matrix Biol.* **25,** 27–39.

Gangaraju, V. K., and Lin, H. (2009). MicroRNAs: Key regulators of stem cells. *Nat. Rev. Mol. Cell Biol.* **10**(2), 116–125.

Gantier, M. P. (2010). New perspectives in MicroRNA regulation of innate immunity. *J. Interferon Cytokine Res.* **30**(5), 283–289.

Gao, J., Yang, T., Han, J., Yan, K., Qiu, X., Zhou, Y., Fan, Q., and Ma, B. (2011). MicroRNA expression during osteogenic differentiation of human multipotent mesenchymal stromal cells from bone marrow. *J. Cell. Biochem.* **112**(7), 1844–1856.

García-Álvarez, F., Alegre-Aguarón, E., Desportes, P., Royo-Cañas, M., Castiella, T., Larrad, L., and Martínez-Lorenzo, M. J. (2011). Chondrogenic differentiation in femoral bone marrow-derived mesenchymal cells (MSC) from elderly patients suffering osteoarthritis or femoral fracture. *Arch. Gerontol. Geriatr.* **52**(2), 239–242.

Gatti, S., Bruno, S., Deregibus, M. C., Sordi, A., Cantaluppi, V., Tetta, C., and Camussi, G. (2011). Microvesicles derived from human adult mesenchymal stem cells protect against ischaemia-reperfusion-induced acute and chronic kidney injury. *Nephrol. Dial. Transplant.* **26**(5), 1474–1483.

Glennie, S., Soeiro, I., Dyson, P. J., Lam, E. W., and Dazzi, F. (2005). Bone marrow mesenchymal stem cells induce division arrest anergy of activated T cells. *Blood* **105,** 2821–2827.

Gnecchi, M., Zhang, Z., Ni, A., and Dzau, V. J. (2008). Paracrine mechanisms in adult stem cell signaling and therapy. *Circ. Res.* **103,** 1204–1219.

Goff, L. A., Boucher, S., Ricupero, C. L., Fenstermacher, S., Swerdel, M., Chase, L. G., Adams, C. C., Chesnut, J., Lakshmipathy, U., and Hart, R. P. (2008). Differentiating human multipotent mesenchymal stromal cells regulate microRNAs: Prediction of microRNA regulation by PDGF during osteogenesis. *Exp. Hematol.* **36**(10), 1354–1369.

Greco, S. J., and Rameshwar, P. (2007). MicroRNAs regulate synthesis of the neurotransmitter substance P in human mesenchymal stem cell-derived neuronal cells. *Proc. Natl. Acad. Sci. USA* **104**(39), 15484–15489.

Griffiths-Jones, S., Saini, H. K., Dongen, S. V., and Enright, A. J. (2008). miR Base: Tools for microRNA genomics. *Nucleic Acids Res.* **36,** 154–158.

Guil, S., and Esteller, M. (2009). DNA methylomes, histone codes and miRNAs: Tying it all together. *Int. J. Biochem. Cell Biol.* **41**(1), 87–95.

Guo, L., Zhao, R. C., and Wu, Y. (2011). The role of microRNAs in self-renewal and differentiation of mesenchymal stem cells. *Exp. Hematol.* **39**(6), 608–616.

Han, Y., Lee, K. H., Yeom, J. W., Nam, I., Heo, J. K., Rhee, S. Y., Sohn, Y., Cho, B. T. Zhang, and Kim, V. N. (2006). Molecular basis for the recognition of primary microRNAs by the Drosha-DGCR8 complex. *Cell* **125,** 887–901.

Han, J., Yang, T., Gao, J., Wu, J., Qiu, X., Fan, Q., and Ma, B. (2010). Specific microRNA expression during chondrogenesis of human mesenchymal stem cells. *Int. J. Mol. Med.* **25** (3), 377–384.

Herrera, M. B., Bussolati, B., Bruno, S., Fonsato, V., Romanazzi, G. M., and Camussi, G. (2004). Mesenchymal stem cells contribute to the renal repair of acute tubular epithelial injury. *Int. J. Mol. Med.* **14,** 1035–1041.

Herrera, M. B., Bruno, S., Buttiglieri, S., Tetta, C., Gatti, S., Deregibus, M. C., Bussolati, B., and Camussi, G. (2006). Isolation and characterization of a stem cell population from adult human liver. *Stem Cells* **24,** 2840–2850.

Herrera, M. B., Fonsato, V., Gatti, S., Deregibus, S., Sordi, A., Cantarella, D., Calogero, R., Bussolati, B., Tetta, C., and Camussi, G. (2010). Human liver stem cell-derived microvesicles accelerate hepatic regeneration in hepatectomized rats. *J. Cell. Mol. Med.* **14**(6B), 1605–1618.

Hicok, K. C., Du Laney, T. V., Zhou, Y. S., Halvorsen, Y. D., Hitt, D. C., Cooper, L. F., and Gimble, J. M. (2004). Human adipose-derived adult stem cells produce osteoid in vivo. *Tissue Eng.* **10,** 371–380.

Hu, R., Li, H., Liu, W., Yang, L., Tan, Y. F., and Luo, X. H. (2010). Targeting miRNAs in osteoblasts differentiation and bone formation. *Expert Opin. Ther. Targets* **14**(10), 1109–1120.

Humphreys, B. D., Valerius, M. T., Kobayashi, A., Mugford, J. W., Soeung, S., Duffield, J. S., McMahon, A. P., and Bonventre, J. V. (2008). Intrinsic epithelial cells repair the kidney after injury. *Cell Stem Cell* **2,** 284–291.

Hunter, M. P., Ismail, N., Zhang, X., Aguda, B. D., Lee, E. J., Yu, L., Xiao, T., Schafer, J., Lee, M. L., Schmittgen, T. D., Nana-Sinkam, S. P., Jarjoura, D., *et al.* (2008). Detection of microRNA expression in human peripheral blood microvesicles. *PLoS One* **5**(3), e3694.

Huntzinger, E., and Izaurralde, E. (2011). Gene silencing by microRNAs: Contributions of translational repression and mRNA decay. *Nat. Rev. Genet.* **12**(2), 99–110.

Inose, H., Ochi, H., Kimura, A., Fujita, K., Xu, R., Sato, S., Iwasaki, M., Sunamura, S., Takeuchi, Y., Fukumoto, S., Saito, K., Nakamura, T., *et al.* (2009). microRNA regulatory mechanism of osteoblast differentiation. *Proc. Natl. Acad. Sci. USA* **106**(49), 20794–20799.

Jian Huang, A., Lan Zhao, B., Lianping Xing, B., and Chen, D. (2010). MicroRNA-204 regulates Runx2 protein expression and mesenchymal progenitor cell differentiation. *Stem Cells* **28**(2), 357–364.

Jiang, Y., Jahagirdar, B. N., Reinhardt, R. L., Schwartz, R. E., Keene, C. D., Ortiz-Gonzalez, X. R., Reyes, M., Lenvik, T., Lund, T., Blackstad, M., Du, J., Aldrich, S., *et al.* (2002). Pluripotency of mesenchymal stem cells derived from adult marrow. *Nature* **418,** 41–49.

Jing, L., Jia, Y., Lu, J., Han, R., Li, J., Wang, S., Peng, T., and Jia, Y. (2011). MicroRNA-9 promotes differentiation of mouse bone mesenchymal stem cells into neurons by Notch signaling. *Neuroreport* **22**(5), 206–211.

Kanellopoulou, C., Muljo, S. A., Kung, A. L., Ganesan, S., Drapkin, R., Jenuwein, T., Livingston, D. M., and Rajewsky, K. (2005). Dicer-deficient mouse embryonic stem cells are defective in differentiation and centromeric silencing. *Genes Dev.* **19,** 489–501.

Kennell, J. A., Gerin, I., MacDougald, O. A., and Cadigan, K. M. (2008). The microRNA miR-8 is a conserved negative regulator of Wnt signaling. *Proc. Natl. Acad. Sci. USA* **105** (40), 15417–15422.

Kim, V. N. (2005). MicroRNA biogenesis: Coordinated cropping and dicing. *Nat. Rev. Mol. Cell Biol.* **6,** 376–385.

Kim, H. K., Lee, Y. S., Sivaprasad, U., Malhotra, A., and Dutta, A. (2006). Muscle-specific microRNA miR-206 promotes muscle differentiation. *J. Cell Biol.* **28,** 677–687.

Kim, Y. J., Hwang, S. J., Bae, Y. C., and Jung, J. S. (2009a). MiR-21 regulates adipogenic differentiation through the modulation of TGF-beta signaling in mesenchymal stem cells derived from human adipose tissue. *Stem Cells* **27**(12), 3093–3102.

Kim, Y. J., Bae, S. W., Yu, S. S., Bae, Y. C., and Jung, J. S. (2009b). miR-196a regulates proliferation and osteogenic differentiation in mesenchymal stem cells derived from human adipose tissue. *J. Bone Miner. Res.* **24,** 816–825.

Kobayashi, T., Lu, J., Cobb, B. S., Rodda, S. J., McMahon, A. P., Schipani, E., Merkenschlager, M., and Kronenberg, H. M. (2008). Dicer-dependent pathways regulate chondrocyte proliferation and differentiation. *Proc. Natl. Acad. Sci. USA* **105**(6), 1949–1954.

Koh, W., Sheng, C. T., Tan, B., Lee, Q. Y., Kuznetsov, V., Kiang, L. S., and Tanavde, V. (2010). Analysis of deep sequencing microRNA expression profile from human embryonic stem cells derived mesenchymal stem cells reveals possible role of let-7 microRNA family in downstream targeting of hepatic nuclear factor 4 alpha. *BMC Genomics* **11**(Suppl 1), S6.

Kosaka, N., Iguchi, H., Yoshioka, Y., Takeshita, F., Matsuki, Y., and Ochiya, T. (2010). Secretory mechanisms and intercellular transfer of microRNAs in living cells. *J. Biol. Chem.* **285,** 17442–17452.

Krampera, M., Pasini, A., Rigo, A., Scupoli, M. T., Tecchio, C., Malpeli, G., Scarpa, A., Dazzi, F., Pizzolo, G., and Vinante, F. (2005). HB-EGF/HER-1 signaling in bone marrow mesenchymal stem cells: Inducing cell expansion and reversibly preventing multilineage differentiation. *Blood* **106,** 59–66.

Krol, J., Loedige, I., and Filipowicz, W. (2010). The widespread regulation of microRNA biogenesis, function and decay. *Nat. Rev. Genet.* **11,** 597–610.

Lagos-Quintana, M., Rauhut, R., Lendeckel, W., and Tuschl, T. (2001). Identification of novel genes coding for small expressed RNAs. *Science* **294**(5543), 853–858.

Lakshmipathy, U., and Hart, R. P. (2008). Concise review: MicroRNA expression in multipotent mesenchymal stromal cells. *Stem Cells* **26**(2), 356–363.

Lakshmipathy, U., Love, B., Goff, L. A., Jörnsten, R., Graichen, R., Hart, R. P., and Chesnut, J. D. (2007). Micro RNA expression pattern of undifferentiated and differentiated human embryonic stem cells. *Stem Cells Dev.* **16**(6), 1003–1016.

Lau, N. C., Lim, L. P., Weinstein, E. G., and Bartel, D. P. (2001). An abundant class of tiny RNAs with probable regulatory roles in Caenorhabditis elegans. *Science* **294**(5543), 858–862.

Lee, R. C., and Ambros, V. (2001). An extensive class of small RNAs in Caenorhabditis elegans. *Science* **294**(5543), 862–864.

Lee, K. S., Kim, H. J., Li, Q. L., Chi, X. Z., Ueta, C., Komori, T., Wozney, J. M., Kim, E. G., Choi, J. Y., Ryoo, H. M., and Bae, S. C. (2000). Runx2 is a common target of transforming growth factor beta1 and bone morphogenetic protein 2, and cooperation between runx2 and smad5 induces osteoblast-specific gene expression in the pluripotent mesenchymal precursor cell line C2C12. *Mol. Cell. Biol.* **20,** 8783–8792.

Lee, R. H., Seo, M. J., Reger, R. L., Spees, J. L., Pulin, A. A., Olson, S. D., and Prockop, D. J. (2006). Multipotent stromal cells from human marrow home to and promote repair of pancreatic islets and renal glomeruli in diabetic pancreatic islets and renal glomeruli in diabetic NOD/scid mice. *Proc. Natl. Acad. Sci. USA* **103,** 17438–17443.

Li, Z., Hassan, M. Q., Volinia, S., van Wijnen, A. J., Stein, J. L., Croce, C. M., Lian, J. B., and Stein, G. S. (2008). A microRNA signature for a BMP2-induced osteoblast lineage commitment program. *Proc. Natl. Acad. Sci. USA* **105**(37), 13906–13911.

Li, Z., Hassan, M. Q., Jafferji, M., Aqeilan, R. I., Garzon, R., Croce, C. M., van Wijnen, A. J., Stein, J. L., Stein, G. S., and Lian, J. B. (2009). Biological functions of miR-29b contribute to positive regulation of osteoblast differentiation. *J. Biol. Chem.* **284**(23), 15676–15684.

Lin, E. A., Kong, L., Bai, X. H., Luan, Y., and Liu, C. J. (2009). miR-199a, a bone morphogenic protein 2-responsive MicroRNA, regulates chondrogenesis via direct targeting to Smad1. *J. Biol. Chem.* **284**(17), 11326–11335.

Liu, J., Carmell, M. A., Rivas, F. V., Marsden, C. G., Thomson, J. M., Song, J. J., Hammond, S. M., Joshua-Tor, L., Hannon, G. J., *et al.* (2004). Argonaute2 is the catalytic engine of mammalian RNAi. *Science* **305,** 1437–1441.

Liu, S. P., Fu, R. H., Yu, H. H., Li, K. W., Tsai, C. H., Shyu, W. C., and Lin, S. Z. (2009). MicroRNAs regulation modulated self-renewal and lineage differentiation of stem cells. *Cell Transplant.* **18**(9), 1039–1045.

Longobardi, L., O'Rear, L., Aakula, S., Johnstone, B., Shimer, K., Chytil, A., Horton, W. A., Moses, H. L., and Spagnoli, A. (2006). Effect of IGF-I in the chondrogenesis of bone marrow mesenchymal stem cells in the presence or absence of TGF-beta signaling. *J. Bone Miner. Res.* **21,** 626–636.

Luzi, E., Marini, F., Sala, S. C., Tognarini, I., Galli, G., and Brandi, M. L. (2008). Osteogenic differentiation of human adipose tissue-derived stem cells is modulated by the miR-26a targeting of the SMAD1 transcription factor. *J. Bone Miner. Res.* **23**(2), 287–295.

Mimeault, M., and Batra, S. K. (2008). Recent progress on tissue-resident adult stem cell biology and their therapeutic implications. *Stem Cell Rev.* **4,** 27–49.

Minguell, J. J., and Erices, A. (2006). Mesenchymal stem cells and the treatment of cardiac disease. *Exp. Biol. Med. (Maywood)* **231,** 39–49.

Miyaki, S., Nakasa, T., Otsuki, S., Grogan, S. P., Higashiyama, R., Inoue, A., Kato, Y., Sato, T., Lotz, M. K., and Asahara, H. (2009). MicroRNA-140 is expressed in differentiated human articular chondrocytes and modulates interleukin-1 responses. *Arthritis Rheum.* **60**(9), 2723–2730.

Mizuno, Y., Yagi, K., Tokuzawa, Y., Kanesaki-Yatsuka, Y., Suda, T., Katagiri, T., Fukuda, T., Maruyama, M., Okuda, A., Amemiya, T., Kondoh, Y., Tashiro, H., *et al.* (2008). miR-125b inhibits osteoblastic differentiation by down-regulation of cell proliferation. *Biochem. Biophys. Res. Commun.* **368,** 267–272.

Morigi, M., Imberti, B., Zoja, C., Corna, D., Tomasoni, S., Abbate, M., Rottoli, D., Angioletti, S., Benigni, A., Perico, N., Alison, M., and Remuzzi, G. (2004). Mesenchymal stem cells are renotropic, helping to repair the kidney and improve function in acute renal failure. *J. Am. Soc. Nephrol.* **15,** 1794–1804.

Morin, R. D., O'Connor, M. D., Griffith, M., Kuchenbauer, F., Delaney, A., Prabhu, A. L., Zhao, Y., McDonald, H., Zeng, T., Hirst, M., Eaves, C. J., and Marra, M. A. (2008). Application of massively parallel sequencing to microRNA profiling and discovery in human embryonic stem cells. *Genome Res.* **18,** 610–621.

Murchison, E. P., Partridge, J. F., Tam, O. H., Cheloufi, S., and Hannon, G. J. (2005). Characterization of Dicer deficient murine embryonic stem cells. *Proc. Natl. Acad. Sci. USA* **102,** 12135–12140.

Naguibneva, I., Ameyar-Zazoua, M., Polesskaya, A., Ait-Si-Ali, S., Groisman, R., Souidi, M., Cuvellier, S., and Harel-Bellan, A. (2006). The microRNA miR-181 targets the homeobox protein Hox-A11during mammalian myoblast differentiation. *Nat. Cell Biol.* **8,** 278–284.

Noort, W. A., Kruisselbrink, A. B., In't Anker, P. S., Kruger, M., Van Bezooijen, R. L., De Paus, R. A., Heemskerk, M. H., Löwik, C. W., Falkenburg, J. H., Willemze, R., and Fibbe, W. E. (2002). Mesenchymal stem cells promote engraftment of human umbilical cord blood-derived CD34(+) cells in NOD/SCID mice. *Exp. Hematol.* **30,** 870–878.

Ortiz, L. A., Gambelli, F., McBride, C., Gaupp, D., Baddoo, M., Kaminski, N., and Phinney, D. G. (2003). Mesenchymal stem cell engraftment in lung is enhanced in response to bleomycin exposure and ameliorates its fibrotic effects. *Proc. Natl. Acad. Sci. USA* **100,** 8407–8411.

Pauley, K. M., Eystathioy, T., Jakymiw, A., Hamel, J. C., Fritzler, M. J., and Chan, E. K. (2006). Formation of GW bodies is a consequence of microRNA genesis. *EMBO Rep.* **7** (9), 904–910.

Pedersen, I. M., Cheng, G., Wieland, S., Volinia, S., Croce, C. M., Chisari, F. V., and David, M. (2007). Interferon modulation of cellular microRNAs as an antiviral mechanism. *Nature* **449**(7164), 919–922.

Phinney, D. G., and Isakova, I. (2005). Plasticity and therapeutic potential of mesenchymal stem cells in the nervous system. *Curr. Pharm. Des.* **11,** 1255–1265.

Phinney, D. G., and Prockop, D. J. (2007). Concise review: Mesenchymal stem/multipotent stromal cells: The state of transdifferentiation and modes of tissue repair—Current views. *Stem Cells* **25**(11), 2896–2902.

Plikus, M. V., Mayer, J. A., de la Cruz, D., Baker, R. E., Maini, P. K., Maxson, R., and Chuong, C. M. (2008). Cyclic dermal BMP signalling regulates stem cell activation during hair regeneration. *Nature* **451,** 340–344.

Qin, L., Chen, Y., Niu, Y., Chen, W., Wang, Q., Xiao, S., Li, A., Xie, Y., Li, J., Zhao, X., He, Z., and Mo, D. (2010). A deep investigation into the adipogenesis mechanism: Profile of microRNAs regulating adipogenesis by modulating the canonical Wnt/beta-catenin signaling pathway. *BMC Genomics* **11,** 320–330.

Quesenberry, P. J., and Aliotta, J. M. (2008). The paradoxical dynamism of marrow stem cells: Considerations of stem cells, niches, and microvesicles. *Stem Cell Rev.* **4,** 137–147.

Quesenberry, P. J., Colvin, G., Dooner, G., Dooner, M., Aliotta, J. M., and Johnson, K. (2007). The stem cell continuum: Cell cycle, injury, and phenotype lability. *Ann. N.Y. Acad. Sci.* **1106,** 20–29.

Rajwsky, N. (2006). microRNA target predictions in animals. *Nat. Genet.* **38,** 8–13.

Rao, P. K., Kumar, R. M., Farkhondeh, M., Baskerville, S., and Lodish, H. F. (2006). Myogenic factors that regulate expression of musclespecific microRNAs. *Proc. Natl. Acad. Sci. USA* **6,** 8721–8726.

Ratajczak, J., Miekus, K., Kucia, M., Zhang, J., Reca, R., Dvorak, P., and Ratajczak, M. Z. (2006). Embryonic stem cell-derived microvesicles reprogram hematopoietic progenitors: Evidence for horizontal transfer of mRNA and protein delivery. *Leukemia* **20**(5), 847–856.

Rendl, M., Polak, L., and Fuchs, E. (2008). BMP signaling in dermal papilla cells is required for their hair follicle-inductive properties. *Genes Dev.* **22,** 543–557.

Ringdén, O., Uzunel, M., Rasmusson, I., Remberger, M., Sundberg, B., Lönnies, H., Marschall, H. U., Dlugosz, A., Szakos, A., Hassan, Z., Omazic, B., Aschan, J., *et al.* (2006). Mesenchymal stem cells for the treatment of therapy-resistant graft-versus-host disease. *Transplantation* **81,** 1390–1397.

Schoolmeesters, A., Eklund, T., Leake, D., Vermeulen, A., Smith, Q., Force Aldred, S., and Fedorov, Y. (2009). Functional profiling reveals critical role for miRNA in differentiation of human mesenchymal stem cells. *PLoS One* **4**(5), e5605.

Schratt, G. M., Tuebing, F., Nigh, E. A., Kane, C. G., Sabatini, M. E., Kiebler, M., and Greenberg, M. E. (2006). A brain-specific microRNA regulates dendritic spine development. *Nature* **19,** 283–289.

Shan, Z. X., Lin, Q. X., Yu, X. Y., Deng, C. Y., Li, X. H., Zhang, X. C., Liu, X. Y., and Fu, Y. H. (2007). MicroRNAs can be expressed in cardiomyocyte-like cells differentiated from human mesenchymal stem cells. *Nan Fang Yi Ke Da Xue Xue Bao* **27**(12), 1813–1816.

Stanczyk, J., Pedrioli, D. M., Brentano, F., Sanchez-Pernaute, O., Kolling, C., Gay, R. E., Detmar, M., Gay, S., and Kyburz, D. (2008). Altered expression of MicroRNA in synovial fibroblasts and synovial tissue in rheumatoid arthritis. *Arthritis Rheum.* **58**(4), 1001–1009.

Sun, F., Wang, J., Pan, Q., Yu, Y., Zhang, Y., Wan, Y., Wang, J., Li, X., and Hong, A. (2009). Characterization of function and regulation of miR-24-1 and miR-31. *Biochem. Biophys. Res. Commun.* **380**(3), 660–665.

Tang, F., Hajkova, P., Barton, S. C., Tang, F., Hajkova, P., Barton, S. C., Lao, K., Surani, M. A., *et al.* (2006). MicroRNA expression profiling of single whole embryonic stem cells. *Nucleic Acids Res.* **34**(2), e9.

Thum, T., Gross, C., Fiedler, J., Fischer, T., Kissler, S., Bussen, M., Galuppo, P., Just, S., Rottbauer, W., Frantz, S., Castoldi, M., Soutschek, J., *et al.* (2008). MicroRNA-21 contributes to myocardial disease by stimulating MAP kinase signalling in fibroblasts. *Nature* **456**(7224), 980–984.

Timmers, L., Lim, S. K., Arslan, F., Armstrong, J. S., Hoefer, I. E., Doevendans, P. A., Piek, J. J., El Oakley, R. M., Choo, A., Lee, C. N., Pasterkamp, G., and de Kleijn, D. P. (2007). Reduction of myocardial infarct size by human mesenchymal stem cell conditioned medium. *Stem Cell Res.* **1**(2), 129–137.

Tögel, F., Hu, Z., Weiss, K., Isaac, J., Lange, C., and Westenfelder, C. (2005). Administered mesenchymal stem cells protect against ischemic acute renal failure through differentiation-independent mechanisms. *Am. J. Physiol. Renal Physiol.* **289**(1), 31–42.

Tsai, M. S., Hwang, S. M., Chen, K. D., Lee, Y. S., Hsu, L. W., Chang, Y. J., Wang, C. N., Peng, H. H., Chang, Y. L., Chao, A. S., Chang, S. D., Lee, K. D., *et al.* (2007). Functional network analysis of the transcriptomes of mesenchymal stem cells derived from amniotic fluid, amniotic membrane, cord blood, and bone marrow. *Stem Cells* **25** (10), 2511–2523.

Valadi, H., Ekstrom, K., Bossios, A., Sjostrand, M., Lee, J. J., *et al.* (2007). Exosome mediated transfer of mRNAs and microRNAs is a novel mechanism of genetic exchange between cells. *Nat. Cell Biol.* **9,** 654–659.

Wang, Y., Medvid, R., Melton, C., Jaenisch, R., and Blelloch, R. (2007). DGCR8 is essential for microRNA biogenesis and silencing of embryonic stem cell self-renewal. *Nat. Genet.* **39,** 380–385.

Wang, K. H., Kao, A. P., Singh, S., Yu, S. L., Kao, L. P., Tsai, Z. Y., Lin, S. D., and Li, S. S. (2010). Comparative expression profiles of mRNAs and microRNAs among human mesenchymal stem cells derived from breast, face, and abdominal adipose tissues. *Kaohsiung J. Med. Sci.* **26**(3), 113–122.

Winter, J., Jung, S., Keller, S., Gregory, R. I., and Diederichs, S. (2009). Many roads to maturity: MicroRNA biogenesis pathways and their regulation. *Nat. Cell Biol.* **11,** 228–234.

Yamamoto, Y., Banas, A., Murata, S., Ishikawa, M., Lim, C. R., Teratani, T., Hatada, I., Matsubara, K., Kato, T., and Ochiya, T. (2008). A comparative analysis of the transcriptome and signal pathways in hepatic differentiation of human adipose mesenchymal stem cells. *FEBS J.* **275**(6), 1260–1273.

Yang, J. S., and Lai, E. C. (2010). Dicer-independent, Ago2-mediated microRNA biogenesis in vertebrates. *Cell Cycle* **9**(22), 4455–4460.

Yang, B., Guo, H., Zhang, Y., Dong, S., and Ying, D. (2011a). The microRNA expression profiles of mouse mesenchymal stem cell during chondrogenic differentiation. *BMB Rep.* **44**(1), 28–33.

Yang, Z., Bian, C., Zhou, H., Huang, S., Wang, S., Liao, L., and Zhao, R. C. (2011b). MicroRNA hsa-miR-138 inhibits adipogenic differentiation of human adipose tissue-derived mesenchymal stem cells through adenovirus EID-1. *Stem Cells Dev.* **20**(2), 259–267.

Yekta, S., Shih, I. H., and Bartel, D. P. (2004). MicroRNA-directed cleavage of HOXB8 mRNA. *Science* **304,** 594–596.

Yuan, A., Farber, E. L., Rapoport, A. L., Tejada, D., Deniskin, R., Akhmedov, N. B., and Farber, D. B. (2009). Transfer of microRNAs by embryonic stem cell microvesicles. *PLoS One* **4,** e4722.

Zhang, C. (2008). MicroRNAs: Role in cardiovascular biology and disease. *Clin. Sci. (Lond.)* **114**(12), 699–706.

Zhang, N., and Bevan, M. J. (2010). Dicer controls CD8+ T-cell activation, migration, and survival. *Proc. Natl. Acad. Sci. USA* **107**(50), 21629–21634.

Zuk, P. A., Zhu, M., Mizuno, H., Huang, J., Futrell, J. W., Katz, A. J., Benhaim, P., Lorenz, H. P., and Hedrick, M. H. (2001). Multilineage cells from human adipose tissue: Implications for cell-based therapies. *Tissue Eng.* **7,** 211–228.

CHAPTER FIFTEEN

MicroRNA and Vascular Smooth Muscle Cells

Changqing Xie, Jifeng Zhang, *and* Y. Eugene Chen

Contents

I. Introduction 322
II. miRNA Biogenesis and Mechanism 323
III. miRNA and VSMC Differentiation 326
IV. miRNA and VSMC Phenotypic Switch 327
V. miRNA and VSMC Neointima Hyperplasia 330
VI. miRNA and VSMC-Related Diseases 331
  A. Atherosclerosis 331
  B. Hypertension 331
  C. Aneurysm 332
  D. Diabetic vascular complications 333
VII. Prospective Application of miRNAs as Therapeutics for Vascular Disease 333
Acknowledgments 334
References 334

Abstract

Vascular smooth muscle cells (VSMCs) exhibit extraordinary plasticity during postnatal development. Vascular injury initiates VSMC phenotypic switch from the contractile to proliferative phenotype, which plays a central role in vascular lesion formation and diverse vascular diseases. MicroRNAs (miRNAs) regulate gene expression posttranscriptionally by either degrading target mRNAs or repressing their translation. Emerging evidence has revealed miRNAs are critical regulators in VSMC differentiation from stem cells, phenotypic switch, and various vascular pathogenesis. Here, we review recent advances regarding functions of specific miRNAs in vasculature and discuss possible mechanisms by which miRNAs affect VSMC biology.

Cardiovascular Center, Department of Internal Medicine, University of Michigan Medical Center, Ann Arbor, Michigan, USA

Vitamins and Hormones, Volume 87
ISSN 0083-6729, DOI: 10.1016/B978-0-12-386015-6.00034-2

I. Introduction

Vascular diseases such as atherosclerosis, restenosis, hypertension, and aneurysm are among the leading causes of morbidity and mortality in developed countries (Murray and Lopez, 1997). Differentiation, migration, proliferation, and apoptosis of vascular smooth muscle cells (VSMCs) and endothelial cells (ECs) are responsible for the development and vascular disease. Specifically, VSMCs exhibit a plasticity in which a process of VSMC dedifferentiation occurs in the context of vascular injury and lesion formation. Alterations in VSMC normal functions and phenotypic modulation play a major role in the pathogenesis of vascular diseases (Owens *et al.*, 2004).

Understanding normal VSMC differentiation and phenotypic switch mechanisms will contribute to novel cell-based therapies for aberrant VSMC differentiation-related diseases. VSMC differentiation and phenotypic plasticity is tightly regulated through a complex, multilayered, synergistic combination ofDNA-binding transcription factors (e.g., SRF (serum response factor)), accessory cofactors for the DNA-binding proteins (e.g., myocardin (MyoCD)/Elk-1 (ETS domain-containing protein 1)/KLF4 (Kruppel like factor 4)), the direct interaction of DNA and transcription factor complexes (e.g., the SRF/CArG interaction), and DNA/histone modifications present within promoters chromatin (e.g., H4 acetylation at CArG boxes) (McDonald and Owens, 2007). Strikingly, nearly all VSMC-restricted contractile protein genes and many other genes important for migration, proliferation, and extracellular matrix production, contain evolutionarily conserved CArG box DNA sequences within their promoters that are required for VSMC transcription *in vivo* (Miano, 2003; Sun *et al.*, 2006). SRF binding to CArG boxes activates expression of VSMC differentiation marker genes in cooperation with its cofactors, MyoCD and Elk-1, for adaptation to varying microenvironments (Buchwalter *et al.*, 2004; Wang *et al.*, 2004; Yoshida and Owens, 2005; Yoshida *et al.*, 2007; Zhou *et al.*, 2005). MyoCD is appreciated as the most potent SRF transcriptional coactivator and it functions to physically associate with SRF and form a ternary complex with CArG box DNA (Spiegelman and Heinrich, 2004; Wang and Olson, 2004; Wang *et al.*, 2001). On the other hand, KLF4 is a binding factor for the transforming growth factor beta (TGF-β) control element and it potently represses VSMC differentiation marker genes by inhibiting MyoCD expression, thus preventing SRF/MyoCD activity (Liu *et al.*, 2003, 2005).

Histone modifications have an enormous influence on the control of VSMC gene expression (Cao *et al.*, 2005; Manabe and Owens, 2001; McDonald *et al.*, 2006; Nie *et al.*, 2003; Qiu and Li, 2002). Other than histone methylation (McDonald *et al.*, 2006), histone acetylation is one of the important categories of histone modification that regulates multiple cellular functions, including chromatin remodeling and transcriptional regulation (Bhaumik *et al.*, 2007; Davie and Spencer, 1999; Langley *et al.*, 2005). Levels of histone acetylation depend on the activities of histone acetylases (HATs) and histone deacetylases (HDACs), which add or remove acetyl groups from protein substrates, respectively (Sun *et al.*, 2003). During development, VSMC-restricted sets of histone acetylation/deacetylation occur within the CArG box chromatin of VSMC genes affecting chromatin-binding properties of SRF. For example, H4 acetylation opens up the chromatin structure facilitating SRF access to CArG box DNA (McDonald *et al.*, 2006). On the other hand, KLF4 can recruit histone H4 deacetylase activity to VSMC genes releasing the association of SRF to methylated histones and CArG box chromatin for the repression of VSMC gene expression (Kawai-Kowase *et al.*, 2009; McDonald *et al.*, 2006; Yoshida *et al.*, 2008a,b).To date, it has been well-documented that HDAC-1 to -6 inhibit VSMC differentiation by repressing SRF activation (Cao *et al.*, 2005; Paroni *et al.*, 2007; Qiu and Li, 2002; Yoshida *et al.*, 2007).

Recent work identifies a new regulatory layer, that of microRNA (miRNA)-mediated mechanisms, superimposed to the VSMC transcriptional components to further fine-tune protein levels via posttranscriptional regulation, in response to varying environmental and metabolic cues resulting in changes of vascular cell differentiation, migration, proliferation, and apoptosis (Cheng *et al.*, 2009; Ji *et al.*, 2007; Kuehbacher *et al.*, 2007; Liu *et al.*, 2009; Poliseno *et al.*, 2006; Suarez *et al.*, 2007; Table 15.1 and Fig. 15.1). The current review chapter is intended to summarize recent research progress regarding the roles of miRNAs in vascular biology and vascular disease.

II. miRNA Biogenesis and Mechanism

miRNA is a class of highly conserved, single-stranded, noncoding small RNAs that control cellular function by either degrading mRNAs or inhibiting their translation. miRNAs are produced from longer primary RNA precursors (pri-miRNAs) containing stem-loop structures that are transcribed from genomes by RNA polymerase II and cleaved in the nucleus by the complex of the RNase III enzyme Drosha and its partner DGCR8/Pasha to form approximately 70-nucleotide pre-miRNAs (Bartel, 2004; Denli *et al.*, 2004; Gregory *et al.*, 2004; Han *et al.*, 2004; Krol *et al.*, 2010; Lee *et al.*, 2003). Pre-miRNAs are transported into the cytoplasm by Exportin-5 and subsequently processed by the nuclease Dicer into the 20- to 24-nucleotide mature miRNA (Bohnsack *et al.*, 2004; Hutvagner *et al.*, 2001; Lund *et al.*, 2004; Yi *et al.*, 2003). It has been established that mature miRNAs associate with a nuclease complex coupled to target mRNAs for the purpose of mediating mRNA silencing through their degradation through Argonaute-catalyzed mRNA cleavage and translational repression (Hammond *et al.*, 2000, 2001;

Table 15.1 MicroRNA:target pairs identified and their function in VSMCs

miRNA	Targets	Functions	Reference
miR-1	KLF4	(+) differentiation	(Xie *et al.*, 2011)
miR-1	Pim1	(–) proliferation	(Chen *et al.*, 2011; Jiang *et al.*, 2010)
miR-10a	HDAC4	(+) differentiation	(Huang *et al.*, 2010)
miR-125b	Suv39h1	epigenetics and (–) inflammation	(Villeneuve *et al.*, 2010)
miR-143/145	CaMKII-δ, ACE	(–) contraction	(Boettger *et al.*, 2009; Cordes *et al.*, 2009)
miR-143/145	KLF4, MyoCD, ELK1	(+) differentiation, (–) proliferation	(Cordes *et al.*, 2009)
miR-145	KLF5	(+) differentiation, (–) proliferation	(Cheng *et al.*, 2009)
miR-143/145	KLF4, KLF5, SSH1, SRGAP1/2, ADD3	(+) differentiation, (–) proliferation	(Xin *et al.*, 2009)
miR-146a	KLF4	(+) proliferation	(Sun *et al.*, 2011)
miR-21	PTEN, BCL-2, PDCD4	(+) proliferation, (–) apoptosis, (–) differentiation,	(Davis *et al.*, 2008a; Ji *et al.*, 2007; Lin *et al.*, 2009)
miR-221/222	p27, p57	(–) differentiation, (+) proliferation	(Liu *et al.*, 2009)
miR-221	c-KIT, p27	(–) differentiation, (+) proliferation	(Davis *et al.*, 2008b)
miR-24	TRB3	(–) proliferation	(Chan *et al.*, 2010)
miR-26a	SMAD1, SMAD4	(–) differentiation, (+) proliferation, (–) apoptosis, (+) migration	(Leeper *et al.*, 2011)
miR-30a	Ca(v)1.2	(–) vascular tone	(Rhee *et al.*, 2009)

Hutvagner and Zamore, 2002; Liu *et al.*, 2004; Rana, 2007), which have been proven to be involved into widespread cellular functions, such as differentiation, proliferation, migration, and apoptosis (Anderson *et al.*, 2006; Chien, 2007; Kim *et al.*, 2006; McCarthy and Esser, 2007; Naguibneva *et al.*, 2006; Nakajima *et al.*, 2006; Rana, 2007; Schickel *et al.*, 2008; van Rooij *et al.*, 2007; Yang *et al.*, 2007; Zhao *et al.*, 2005, 2007).

Though well established in plants that the main mechanism for miRNA-mediated repression involves mRNA degradation, in mammalian cells, the predominant mechanism is less clear. Initially, miRNAs were thought to repress protein output with little or no influence on mRNA levels (Jones-Rhoades *et al.*, 2006; Olsen and Ambros, 1999). Nonetheless, several mRNA-array experiments showed that miRNAs also decrease the levels of many targeted mRNAs (Giraldez *et al.*, 2006; Krutzfeldt *et al.*, 2005; Lim *et al.*, 2005; Rehwinkel *et al.*, 2006). A recent study used mRNA arrays to monitor effects on both mRNA levels and mRNA ribosome density and occupancy, thereby providing a more sensitive analysis of changes in mRNA utilization and bypassing the need to compare protein and mRNA (Hendrickson *et al.*, 2009). This array study supports the mRNA-destabilization scenario but, since it examines the response to an ectopically introduced miRNA, it still leaves open the question of whether endogenous miRNA-target interactions might impart additional translational repression. A sensitive method based on deep sequencing of ribosome-protected

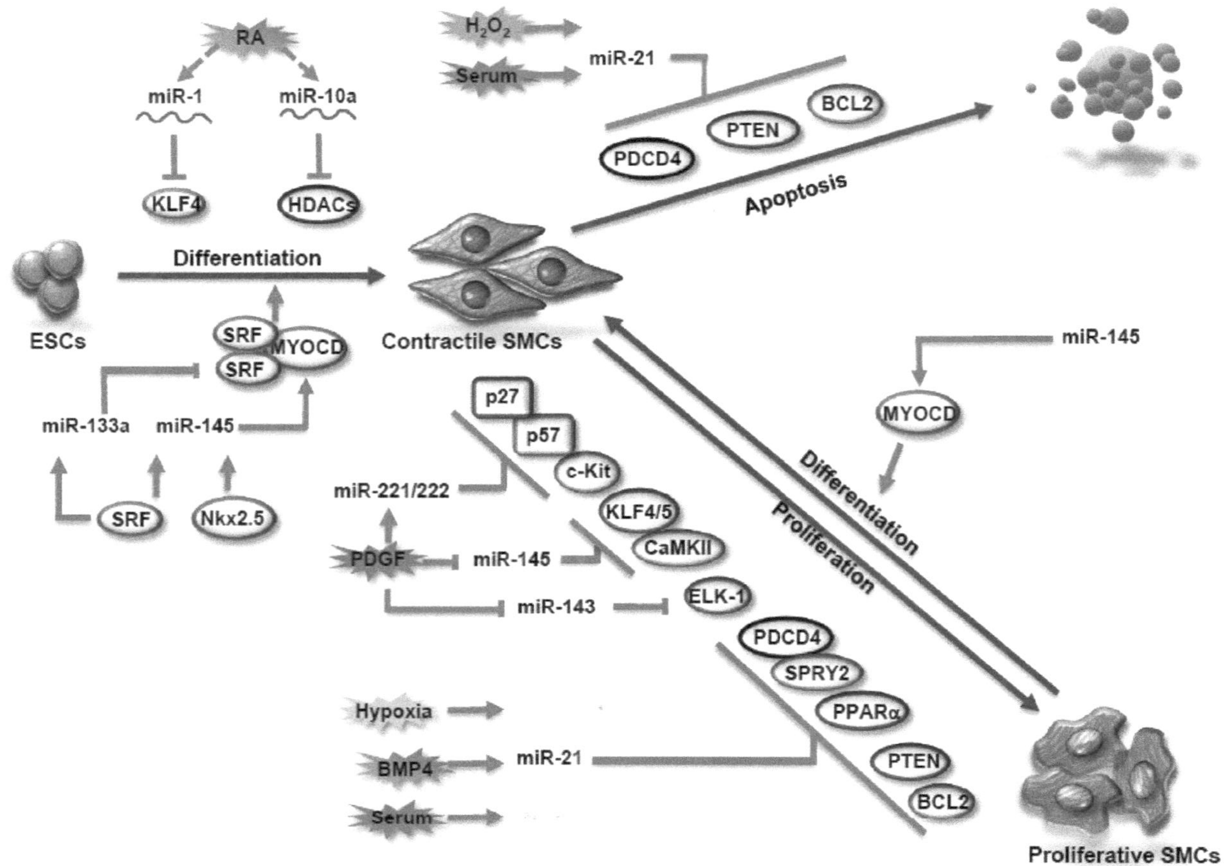

Figure 15.1 Schematic diagram of microRNA:targets identified in VSMCs. KLF-4, Kruppel like factor 4; HDACs, histone deacetylases; miR, microRNA; Nkx2.5, NK2 transcription factor related, locus 5; MYOCD, myocardin; SRF, serum response factor; ELK1, ETS domain-containing protein 1; PPARα, peroxisome proliferator-activated receptor alpha; PTEN, phosphatase and tensin homolog; BCL2, B-cell CLL/lymphoma 2; CaMKII, calcium/calmodulin-dependent protein kinase II; PDCD4, programmed cell death 4; SPRY2, sprouty homolog 2 (Drosophila); BMP4, bone morphogenetic protein 4; PDGF, platelet-derived growth factor; BMP4, bone morphogenetic protein 4. (For color version of this figure, the reader is referred to the Web version of this chapter.)

mRNA fragments (RPFs) was used for analyzing the effects of miRNAs (e.g., miR-1, miR-155, and miR-223) on their targets in human and mouse cells (Guo *et al.*, 2010). The miRNAs did not seem to repress target mRNAs that were expressed at a low level more strongly than targets expressed at higher levels, suggesting that the repressive effects of miRNAs are independent of target expression levels. For both ectopic and endogenous miRNA regulatory interactions, lowered mRNA levels account for most ($\geq$84%) of the decreased protein production (Guo *et al.*, 2010), which is greater than previous estimates using array analyzes. Although the detailed underlying mechanism by which miRNAs destabilize their mRNA targets is largely unclear in the mammalian system, several studies appear to indicate that decreases in the levels of target mRNAs are associated with poly(A)-tail shortening, leading to a model in which miRNAs cause mRNA deadenylation, which, in turn, promotes decapping and more rapid degradation through standard mRNA-turnover processes (Behm-Ansmant *et al.*, 2006; Eulalio *et al.*, 2009; Giraldez *et al.*, 2006; Wu *et al.*, 2006).

III. miRNA and VSMC Differentiation

With a recently established model for *in vitro* VSMC differentiation from embryonic stem cells (ES cells), we found that during the process from ES cells differentiation to VSMCs, a number of miRNAs showed dynamic changes (Huang *et al.*, 2010; Xie *et al.*, 2011). Among those, miR-1 and miR-10a showed a tendency to be upregulated and blockade of miR-1 and miR-10a respectively repressed VMSC differentiation, evidenced by reduced percentage of VSMC differentiation. Further, individual duplexes between these miRNAs and two potential targets relevant to differentiation were identified, that is, miR-1:KLF4 and miR-10a:HDAC4 (Huang *et al.*, 2010; Xie *et al.*, 2011). Interestingly, miR-10a expression was regulated by NF-κB in the setting of SMC differentiation from ES cells (Huang *et al.*, 2010). These results suggest that miRNAs play a critical role in regulating VSMC differentiation from ES cells *in vitro*.

The miR-143/145 cluster has been shown to regulate SMC differentiation from fibroblasts and neural crest cells (Cordes *et al.*, 2009). Deletion of miR-145 impaired the conversion from fibroblasts to VSMCs induced by MyoCD overexpression and resulted in reduced expression of components of the SMC contractile apparatus. Overexpression of miR-145 enhanced neural crest stem cells differentiation into the VSMC lineage. Further, the number of contractile VSMCs significantly decreased and the number of synthetic VSMCs remarkably increased in the aorta and the femoral artery in miR-143/miR-145 knockout mice, simultaneously, while the number of noncontracting, proliferating precursors increased (Boettger *et al.*, 2009; Elia *et al.*, 2009; Xin *et al.*, 2009). Nevertheless, VSMCs within the miR-143/145 mutant artery showed a significant inhibition of the expression of SMC-specific differentiation markers (Boettger *et al.*, 2009; Elia *et al.*, 2009). Expression of the miR-143/145 cluster responsible for regulating VSMC differentiation was controlled by SRF, Nkx2.5, and MyoCD which bind to the promoter of the miR-143 and miR-145 cluster (Cordes *et al.*, 2009; Xin *et al.*, 2009).

More direct evidence to show that miRNAs regulate fundamental differentiation of VSMCs *in vivo* came from the study of the mouse conditional knockout of the rate-limiting enzyme Dicer in VSMCs of blood vessels (Albinsson *et al.*, 2010). Deletion of Dicer in VSMCs causes late embryonic lethality at E16–17. Loss of VSMC Dicer results in dilated, thin-walled blood vessels, which may be due to reduction in cellular proliferation. Moreover, the mutant VSMCs exhibit loss of contractile apparatuses and the ensuing impaired contractility which was partly rescued by over-expression of miR-145 or MyoCD. This study further suggested Dicer-dependent miRNAs are important for VSMC development and function by regulating differentiation (Albinsson *et al.*, 2010).

Taken together, these results suggest that miRNAs have critical roles for the differentiation of VSMCs from stem cells/precursors both *in vitro* and *in vivo*.

IV. miRNA and VSMC Phenotypic Switch

Under normal physiologic conditions, VSMCs stay quiescent, form layers within the vessel wall and control blood flow by contracting or relaxing in response to external stimuli. However, these contractile VSMCs will regain proliferation capability under various environmental conditions. Over time, unchecked proliferation can lead to pathologic changes in the vascular wall. The switch between the proliferative and the contractile phenotypes is governed by a complex regulatory protein network. Intriguingly, the intrinsic coordination between transcription factors and miRNAs in regulating VSMC phenotypic switch is just starting to be revealed in its biologic beauty.

In cultured VSMCs, platelet-derived growth factor (PDGF)-BB treatment is a well-known pro-proliferative trigger to initiate VSMC phenotypic switch from contractile to proliferative. miR-221 and miR-222 expression is transcriptionally induced by PDGF-BB (Davis *et al.*, 2008b; Liu *et al.*, 2009). The knockdown of miR-221 enhanced the expression of VSMCs specific markers and significantly abrogated the PDGF-BB-dependent induction of VSMCs migration and proliferation, which are fundamental characteristics of the proliferative state of VSMCs (Davis *et al.*, 2008b), which suggests that miR-221 is essential for the PDGF-mediated stimulation of VSMCs phenotypic switching. On the contrary, transfection of exogenous miR-221 into VSMCs reduced the expression of VSMCs markers and remarkably improved VSMCs migration and proliferation. Moreover, the effect by miR-221 on the VSMC phenotype was mediated by repressing c-Kit and MyoCD to impair SMC-specific gene expression and repressing p27Kip1 to promotes cell proliferation in VSMCs, respectively (Davis *et al.*, 2008b).

Recent findings from several groups indicate that VSMCs undergo phenotypic switch in response to the knockdown of miR-143/145 (Boettger *et al.*, 2009; Cordes *et al.*, 2009). miR-143/145 is enriched in mature VSMCs where the expression of miR-143/145 can be directly activated by SRF–MyoCD interaction (Cordes *et al.*, 2009). *In vitro* overexpression of miR-145 or miR-143 was sufficient to promote the VSMC contractile state by enhancing expression of the contractile apparatus and inhibit proliferation of cultured VSMCs (Cheng *et al.*, 2009; Cordes *et al.*, 2009). On the other hand, contractile abilities in response to vasopressive stimuli were dramatically impaired in the absence of the miR-143/145 in VSMCs, which may result from the reduction of the VSMC contractile apparatus (Boettger *et al.*, 2009; Elia *et al.*, 2009). Isolated VSMCs from miR-143/145 knockout mice have enhanced migratory activity in response to PDGF-BB, compared to wild type VSMCs (Boettger *et al.*, 2009; Elia *et al.*, 2009). Further studies revealed that miR-143/145 targets Elk1, MyoCD, KLF4, KLF5, CamkIIδ, and angiotensin-converting enzyme (ACE) in VSMCs (Boettger *et al.*, 2009; Cheng *et al.*, 2009; Cordes *et al.*, 2009), indicating that miR-143/145 targets a broad network of transcription factors and structural and membrane-bound proteins in VSMCs in order to regulate contractility and proliferation.

The TGF-β and bone morphogenetic protein (BMP) family of trophic factors contribute to VSMC phenotypic switch by promoting VSMC contractility due to enhanced expression of the contractile apparatus (Lagna *et al.*, 2007; Owens, 1995). Treatment of VSMC with TGFβ and BMP4 resulted in change in the expression of miRs, including miR-21, through promoting the processing of primary transcripts of miRNA into precursor miRNA by regulating DROSHA complex (Davis *et al.*, 2008a). The study further identified that the induction of a contractile phenotype in VSMCs by TGFβ and BMPs was mediated by miR-21: programmed cell death 4 (PDCD4) binding duplex (Davis *et al.*, 2008a). Conversely, the TGFβ signaling pathway was indicated to be under the control of miRNAs. Recent evidence showed that inhibition of miR-26a increased gene expression of mothers against decapentaplegic homolog SMAD-1 and SMAD-4, while overexpression inhibited SMAD-1. Simultaneously, the inhibition of miR-26a promoted VSMC apoptosis and phenotypic switch to a contractile status while inhibiting proliferation and migration (Leeper *et al.*, 2011).

Intriguingly, miR-24 is one of the miRNAs involved into both PDGF-BB and TGFβ signaling pathways, which respectively represent the distinguishing trigger of VSMC proliferation and differentiation (Chan *et al.*, 2010). PDGF-BB induced miR-24 expression, which subsequently led to Tribbles-like protein-3 (TRB3) downregulation by a posttranscriptional effect. As the result, the expression of SMAD proteins were reduced, indicating the decrease in BMP and TGFβ signaling, promoting VSMC proliferation. Inhibition of miR-24 enhanced Trb3 expression and impaired VSMC proliferation activity. Thus, this study provides a simple and elegant molecular basis for the antagonism between the PDGF and TGFβ pathways, and its effect on the control of the VSMC phenotype (Chan *et al.*, 2010).

Beyond regulating cardiomyocyte differentiation and function, recently, miR-1 has been shown to contribute to the regulation of VSMC phenotype controlled by MyoCD, which induces miR-1 expression in VSMC (Chen *et al.*, 2011; Jiang *et al.*, 2010). Surprisingly, the contractility of VSMCs enhanced by MyoCD was remarkably inhibited when exogenous miR-1 was introduced into VSMCs *in vitro* through suppressing the expression of contractile proteins, such as smooth muscle alpha actin (SM α-actin) and SM22alpha and impairing the actin cytoskeletal organization (Jiang *et al.*, 2010). Yet, the introduction of miR-1 into VSMCs dramatically inhibited cell proliferation, which was mediated by repression of the miR-1 target Pim-1 (a proto-oncogene encoding a serine/threonine protein kinase) in VSMCs (Chen *et al.*, 2011; Jiang *et al.*, 2010). These findings clearly underscore the importance of contextual effects on miRNAs function and the need to further investigate their roles in physiologically relevant scenarios.

Previous studies have demonstrated that KLF4 potently represses the expression of multiple VSMC specific genes by repressing MyoCD levels during phenotypic switch of VSMC in response to TGF-β or PDGF-BB (Kawai-Kowase *et al.*, 2009; King *et al.*, 2003; Liu *et al.*, 2003, 2005; Pidkovka *et al.*, 2007; Yoshida *et al.*, 2008a,b; Zeng *et al.*, 2006). Recent data has further demonstrated that KLF4 competed with KLF5 to bind to and regulate the miR-146a promoter, and that KLF4 and KLF5 exerted opposing effects on that promoter. Meanwhile, miR-146a promotes VSMC proliferation *in vitro*. Interestingly, KLF4 was one of the targets of miR-146a-dependent regulation of VSMC proliferation *in vitro*, evidenced by enhanced expression of KLF4 upon silencing of miR-146a and decreased KLF4 levels by over-expression of miR-146a. Thus, miR-146a and KLF4 formed a negative regulatory loop to control VSMC phenotypic switch (Sun *et al.*, 2011).

Reactive oxygen species are potent stimuli involved in vascular pathogenesis, including atherosclerosis and restenosis after angioplasty (Cordes *et al.*, 2009). Altered expression of miRNAs has been shown in VSMCs upon treatment with hydrogen peroxide (Lin *et al.*, 2009), including miR-21 that was highly enriched. Further, hydrogen peroxide-induced VSMC apoptosis and death were potentiated by miR-21 inhibitor and repressed by pre-miR-21. The protective effect of miR-21 on VSMCs was mediated by targeting PDCD4 (Lin *et al.*, 2009). Nevertheless, miR-21 enhanced VSMC proliferation and repressed cell apoptosis through inhibiting expression of phosphatase and tensin homolog (PTEN) and B-cell CLL/lymphoma

2 (Bcl-2) posttranscriptionally (Ji *et al.*, 2007). Thus, miR-21 might represent a pro-proliferation factor in vascular diseases related to vascular injury.

Collectively, these observations indicate that the VSMC phenotypic switch is under the control of a complex network comprising transcription factors and miRNAs, which results in vascular modification in response to a wide spectrum of stimuli.

V. miRNA and VSMC Neointima Hyperplasia

Neointimal hyperplasia has broadly been considered as the main process of vascular repair in response to various injuries. Importantly, multiple studies have demonstrated that neointimal progression is a serial scenario of molecular and cellular events. Undoubtedly, VSMC phenotypic switch is one of these critical scenarios. The miRNAs mentioned above, as powerful regulators of VSMC specific gene expression, are also involved in the modulation of the phenotypic switch happening in intimal thickening after vascular injury.

Microarray analysis indicated that miR-21(Ji *et al.*, 2007), miR-221/222 (Liu *et al.*, 2009), and miR-143/145 (Boettger *et al.*, 2009; Cheng *et al.*, 2009; Cordes *et al.*, 2009; Elia *et al.*, 2009) are highly expressed in normal murine arteries. In response to vascular injury, expression profiles of miRNAs experiences dynamic changes (Ji *et al.*, 2007). Among those, miR-21 (Ji *et al.*, 2007) and miR-221/222 (Liu *et al.*, 2009) were dramatically increased and miR-145 downregulated (Cheng *et al.*, 2009). In addition, in the rat carotid artery balloon-injury model, neointimal hyperplasia of injured artery was significantly suppressed by downregulation of miR-21 (Ji *et al.*, 2007), miR-221/222 (Liu *et al.*, 2009), miR-146a (Sun *et al.*, 2011) and was accompanied by inhibition of VSMC proliferation. On the other hand, restoration of miR-145 in balloon-injured arteries via miR-145 overexpression using adenovirus-mediated gene transfer remarkably inhibited neointimal growth concurrent with upregulation of VSMC differentiation markers (Cheng *et al.*, 2009).

In a different vascular injury model, miR-143/145 expression was significantly decreased in the thickened vessels of the ligated carotid arteries when compared with contralateral control arteries, (Cordes *et al.*, 2009). As they aged, up to 18-month-old, miR-143/145 mutant mice showed a significant increase in spontaneous neointimal formation with large amounts of VSMCs and macrophages and deposits of amorphous collagen I in the femoral arteries (Boettger *et al.*, 2009). However, this appears to be contrary to another report indicated that neointimal lesion development after ligation was significantly impaired in mice for either single miR-143/145 deletion or the double miRNAs deletion, due to reduced VSMC migration (Xin *et al.*, 2009).

This result may be explained partially by either different vascular injury models (mouse vs. rat; young vs. old mice) or different genetic modification approaches (locally virus mediated vs. systemically modified). Taken together, these studies highlight that miRNAs play pivotal roles in the control of VSMC phenotypic switch and the response of the vascular wall to injury.

VI. miRNA and VSMC-Related Diseases

It is well established that various vascular diseases are caused by change in the VSMC phenotype in response to a spectrum of environmental stimuli. Consequently, miRNAs are also emerging as critical regulators during the pathogenesis of those vascular diseases.

A. Atherosclerosis

Atherosclerosis involves sequential processes including endothelial dysfunction, inflammation, cell proliferation, and matrix alteration. VMSC phenotypic switch contribute to the pathogenesis of atherosclerosis and are linked to all those processes (Dzau *et al.*, 2002). Meanwhile, it is well known that neointima formation is a major event during the pathogenesis of atherosclerosis. Thus, the miRNAs, which are known to interfere with neointima formation, could be predicted to be important players in controlling the atherosclerotic process (Boettger *et al.*, 2009; Cheng *et al.*, 2009; Cordes *et al.*, 2009; Elia *et al.*, 2009; Ji *et al.*, 2007; Liu *et al.*, 2009; Sun *et al.*, 2011). As shown in the literature, transcripts of miR-145 were indeed decreased to nearly undetectable levels in mouse arteriosclerotic lesions, whether spontaneous as the animals aged (Boettger *et al.*, 2009) or induced by a Western diet (Cordes *et al.*, 2009). The aorta of ApoE KO mice had markedly decreased constitutive levels of miR-143/145, and their expression was further reduced when fed a high-fat diet (HFD) (Elia *et al.*, 2009). These data strongly indicates a role for miRNA in the pathogenesis of atherosclerosis.

B. Hypertension

Hypertension is a multifactorial trait resulting from the combination of environmental and genetic factors. Several polymorphisms of candidate genes that are linked to hypertensive phenotypes have been identified (Benetos *et al.*, 1996; Bonnardeaux *et al.*, 1994; Davies *et al.*, 1994). Genetic variations of pre-miRNAs, mature miRNAs, and their target genes determined by gene polymorphisms and single-nucleotide polymorphisms have been proposed to be related to many human diseases including hypertension

(Mishra *et al.*, 2008; Sethupathy and Collins, 2008; Wang *et al.*, 2010). Thus, for instance, it has been demonstrated that miR-155 is complementary with a specific sequence in the 3′ UTR of angiotensin II receptor type 1 (AT1R) gene harboring the A1166C polymorphism (Martin *et al.*, 2007). Recent data showed that the interplay between miR-155 expression, +1166C polymorphism, and AT1R protein expression may play a role in the regulation of blood pressure (Ceolotto *et al.*, 2011).

Fascinatingly, miR-143/145 knockout mice showed a significant reduction in systolic blood pressure due to reduced vascular tone, resulting from decreased contractile ability of the vessels (Boettger *et al.*, 2009; Xin *et al.*, 2009). Combined genomics/proteomics and transcriptional analyzes identified multiple potential miR-143/145 targets including ACE. In this regard, it is noteworthy that angiotensin (Ang) II, a product of ACE-mediated cleavage of Ang I, is bound by the AT1R on the surface of VSMCs, promoting vasoconstriction and alteration of VSMC phenotype (Schieffer *et al.*, 2000); this provides one potential explanation for selective targeting of the vascular system in these mutant mice (Table 15.1).

In different rodent models of hypertension, vascular voltage-gated L-type calcium channel [Ca(L)] current and vascular tone is increased because of increased expression of the noncardiac form of the Ca(L) [Ca(v)1.2] (Rhee *et al.*, 2009). A modified miR-30a-based short hairpin RNA driven by the cytomegalovirus promoter reduced endogenous Ca(v)1.2 expression by 61% and decreased the Ca(L) current carried by barium by 47%. This molecular intervention *in vivo* may provide a novel long-term vascular-specific gene therapy for hypertension (Rhee *et al.*, 2009). Additionally, differential expression of miRNAs, including miR-155 and miR-208, has been revealed in the aorta of spontaneously hypertensive rats (SHR) (Xu *et al.*, 2008). In this study, miR-155 level was significantly lower in aorta of SHR and miR-208 expression in aorta was negatively correlated with blood pressure and age, indicating that these two miRNAs play a role in the pathogenesis of hypertension.

C. Aneurysm

Abdominal aortic aneurysm (AAA) is defined as a permanent segmental dilatation of the abdominal aorta. Despite improvement in the surgical treatment and perioperative care, AAA is among the 15 leading causes of death in the USA (Hallett, 2000). The expression of miR-143/145 was suppressed by stress in experimental models, either acute (transverse aortic constriction) or chronic stress (ApoE knockout mouse) (Elia *et al.*, 2009). Interestingly, the same change in miR-143/145 expression in the human aorta has been identified in human aneurysm samples (Elia *et al.*, 2009). In addition, miR-26a was found to be downregulated in either the elastase or the AngII-ApoE$^{-/-}$ mouse models of AAA formation, coinciding temporally with decreased SM α-actin staining and vessel dilation. Such outcome may be due to the alteration of the TGF-β pathway signaling (Leeper *et al.*, 2011) and suggests a possible novel treatment approach for AAA in the future.

D. Diabetic vascular complications

Diabetes is associated with increased risk for cardiovascular diseases associated with vascular inflammation and atherosclerosis (Devaraj *et al.*, 2010). The relationship between diabetic vascular complications and miRNAs remains scantly explored. Recent evidence shows that miR-125b impaired the expression of chromatin histone H3 lysine-9 methyltransferase Suv39h1 and resulted in the modification of the epigenetic marker histone H3 lysine-9 trimethylation (H3K9me3) and reduced expression of inflammatory genes in VSMCs isolated from diabetic mice (Villeneuve *et al.*, 2010). These results demonstrate a novel line of evidence in the epigenetic regulation of inflammatory genes in VSMCs of diabetic mice through the regulation of Suv39h1 and subsequent changes in the pathogenesis of diabetic vascular complications (Villeneuve *et al.*, 2010).

VII. Prospective Application of miRNAs as Therapeutics for Vascular Disease

The initial promising results outlined here have revealed that multiple miRNAs play critical roles on the development of vascular diseases partially via regulating VSMC phenotype and function through their corresponding target genes. Several *in vivo* preclinical animal studies have demonstrated the exciting therapeutic potential and results in the context of models of vascular disease (Cheng *et al.*, 2009; Cordes *et al.*, 2009; Ji *et al.*, 2007; Liu *et al.*, 2009). Thus, targeted miRNA therapeutics has been anticipated as the potential candidate treatment option via elegantly restoring the miRNAs reduced in diseases and, on the other hand, inhibiting miRNAs upregulated in the pathological state and opens a new dimension in the treatment of VSMC-related vascular diseases.

However, we have to acknowledge that this therapeutic strategy remains in the early stages of development and is mostly limited primarily to *in vitro* and rodent models of diseases. Essential challenges regarding miRNA-targeted therapy in the future clinic application remain to be solved. First, the main challenge is highlighted by the very nature of miRNAs, in that multiple target genes of single miRNAs might induce unexpected side effects and result in other vascular disease-related cellular events (Fluiter *et al.*, 2009). Second, the detailed contextual cellular and molecular mechanisms of tissue specific effects of miRNAs should be broadly studied in all categories of human vascular diseases. In that regard, either the miRNA expression or that of the potential targets could be altered in a same cell type in response to different stimuli or, alternatively, available targets to a given miRNA may vary in different tissue contexts. Thirdly, modes of delivery have to be explored and developed in a contextual basis as well. Adequate tissue specific targeting, vascular disease specific targeting or, even personalized and individual tailoring of the delivery approach may eventually become a key obstacle for expediting clinical application. Several therapeutic approaches and modes of delivery have recently been reviewed and deserve to be explored and validated in a systematic fashion in preclinical and clinical settings (Montgomery and van Rooij, 2010; Pan *et al.*, 2010) with attention to feasibility, safety, and specificity of these approaches.

ACKNOWLEDGMENTS

The work in our laboratory was partially supported by National Institutes of Health (HL068878, HL089544, and HL105114), J. Z. and C. X. are supported by American Heart Association Scientific Development Grant 0835237N and 09SDG2260023, respectively. Y. E. C. is an established investigator of American Heart Association (0840025N).

REFERENCES

Albinsson, S., *et al.* (2010). MicroRNAs are necessary for vascular smooth muscle growth, differentiation, and function. *Arterioscler. Thromb. Vasc. Biol.* **30,** 1118–1126.

Anderson, C., *et al.* (2006). MIR-206 regulates connexin43 expression during skeletal muscle development. *Nucleic Acids Res.* **34,** 5863–5871.

Bartel, D. P. (2004). MicroRNAs: Genomics, biogenesis, mechanism, and function. *Cell* **116,** 281–297.

Behm-Ansmant, I., *et al.* (2006). mRNA degradation by miRNAs and GW182 requires both CCR4:NOT deadenylase and DCP1:DCP2 decapping complexes. *Genes Dev.* **20,** 1885–1898.

Benetos, A., *et al.* (1996). Influence of angiotensin-converting enzyme and angiotensin II type 1 receptor gene polymorphisms on aortic stiffness in normotensive and hypertensive patients. *Circulation* **94,** 698–703.

Bhaumik, S. R., *et al.* (2007). Covalent modifications of histones during development and disease pathogenesis. *Nat. Struct. Mol. Biol.* **14,** 1008–1016.

Boettger, T., *et al.* (2009). Acquisition of the contractile phenotype by murine arterial smooth muscle cells depends on the Mir143/145 gene cluster. *J. Clin. Invest.* **119,** 2634–2647.

Bohnsack, M. T., *et al.* (2004). Exportin 5 is a RanGTP-dependent dsRNA-binding protein that mediates nuclear export of pre-miRNAs. *RNA* **10,** 185–191.

Bonnardeaux, A., *et al.* (1994). Angiotensin II type 1 receptor gene polymorphisms in human essential hypertension. *Hypertension* **24,** 63–69.

Buchwalter, G., *et al.* (2004). Ets ternary complex transcription factors. *Gene* **324,** 1–14.

Cao, D., *et al.* (2005). Modulation of smooth muscle gene expression by association of histone acetyltransferases and deacetylases with myocardin. *Mol. Cell. Biol.* **25,** 364–376.

Ceolotto, G., *et al.* (2011). Interplay between miR-155, AT1R A1166C polymorphism, and AT1R expression in young untreated hypertensives. *Am. J. Hypertens.* **24,** 241–246.

Chan, M. C., *et al.* (2010). Molecular basis for antagonism between PDGF and the TGFbeta family of signalling pathways by control of miR-24 expression. *EMBO J.* **29,** 559–573.

Chen, J., *et al.* (2011). Induction of MicroRNA-1 by myocardin in smooth muscle cells inhibits cell proliferation. *Arterioscler. Thromb. Vasc. Biol.* **31,** 368–375.

Cheng, Y., *et al.* (2009). MicroRNA-145, a novel smooth muscle cell phenotypic marker and modulator, controls vascular neointimal lesion formation. *Circ. Res.* **105,** 158–166.

Chien, K. R. (2007). Molecular medicine: MicroRNAs and the tell-tale heart. *Nature* **447,** 389–390.

Cordes, K. R., *et al.* (2009). miR-145 and miR-143 regulate smooth muscle cell fate and plasticity. *Nature* **460,** 705–710.

Davie, J. R., and Spencer, V. A. (1999). Control of histone modifications. *J. Cell. Biochem. Suppl.* **32–33,** 141–148.

Davies, E., *et al.* (1994). Angiotensin II (type-1) receptor locus: CA repeat polymorphism and genetic mapping. *Hum. Mol. Genet.* **3,** 838.

Davis, B. N., *et al.* (2008a). SMAD proteins control DROSHA-mediated microRNA maturation. *Nature* **454,** 56–61.

Davis, B. N., *et al.* (2008b). Induction of microRNA-221 by platelet-derived growth factor signaling is critical for modulation of vascular smooth muscle phenotype. *J. Biol. Chem.* **284,** 3728–3738.

Denli, A. M., *et al.* (2004). Processing of primary microRNAs by the microprocessor complex. *Nature* **432,** 231–235.

Devaraj, S., *et al.* (2010). Diabetes is a proinflammatory state: A translational perspective. *Expert Rev. Endocrinol. Metab.* **5,** 19–28.

Dzau, V. J., *et al.* (2002). Vascular proliferation and atherosclerosis: New perspectives and therapeutic strategies. *Nat. Med.* **8,** 1249–1256.

Elia, L., *et al.* (2009). The knockout of miR-143 and -145 alters smooth muscle cell maintenance and vascular homeostasis in mice: Correlates with human disease. *Cell Death Differ.* **16,** 1590–1598.

Eulalio, A., *et al.* (2009). Deadenylation is a widespread effect of miRNA regulation. *RNA* **15,** 21–32.

Fluiter, K., *et al.* (2009). The therapeutic potential of LNA-modified siRNAs: Reduction of off-target effects by chemical modification of the siRNA sequence. *Methods Mol. Biol.* **487,** 189–203.

Giraldez, A. J., *et al.* (2006). Zebrafish MiR-430 promotes deadenylation and clearance of maternal mRNAs. *Science* **312,** 75–79.

Gregory, R. I., *et al.* (2004). The microprocessor complex mediates the genesis of microRNAs. *Nature* **432,** 235–240.

Guo, H., *et al.* (2010). Mammalian microRNAs predominantly act to decrease target mRNA levels. *Nature* **466,** 835–840.

Hallett, J. W., Jr. (2000). Management of abdominal aortic aneurysms. *Mayo Clin. Proc.* **75,** 395–399.

Hammond, S. M., *et al.* (2000). An RNA-directed nuclease mediates post-transcriptional gene silencing in Drosophila cells. *Nature* **404,** 293–296.

Hammond, S. M., *et al.* (2001). Argonaute2, a link between genetic and biochemical analyses of RNAi. *Science* **293,** 1146–1150.

Han, J., *et al.* (2004). The Drosha-DGCR8 complex in primary microRNA processing. *Genes Dev.* **18,** 3016–3027.

Hendrickson, D. G., *et al.* (2009). Concordant regulation of translation and mRNA abundance for hundreds of targets of a human microRNA. *PLoS Biol.* **7,** e1000238.

Huang, H., *et al.* (2010). miR-10a contributes to retinoid acid-induced smooth muscle cell differentiation. *J. Biol. Chem.* **285,** 9383–9389.

Hutvagner, G., *et al.* (2001). A cellular function for the RNA-interference enzyme Dicer in the maturation of the let-7 small temporal RNA. *Science* **293,** 834–838.
Hutvagner, G., and Zamore, P. D. (2002). A microRNA in a multiple-turnover RNAi enzyme complex. *Science* **297,** 2056–2060.
Ji, R., *et al.* (2007). microRNA expression signature and antisense-mediated depletion reveal an essential role of microRNA in vascular neointimal lesion formation. *Circ. Res.* **100,** 1579–1588.
Jiang, Y., *et al.* (2010). microRNA-1 inhibits myocardin-induced contractility of human vascular smooth muscle cells. *J. Cell. Physiol.* **225,** 506–511.
Jones-Rhoades, M. W., *et al.* (2006). microRNAS and their regulatory roles in plants. *Annu. Rev. Plant Biol.* **57,** 19–53.
Kawai-Kowase, K., *et al.* (2009). PIAS1 mediates TGFbeta-induced SM alpha-actin gene expression through inhibition of KLF4 function-expression by protein sumoylation. *Arterioscler. Thromb. Vasc. Biol.* **29,** 99–106.
Kim, H. K., *et al.* (2006). Muscle-specific microRNA miR-206 promotes muscle differentiation. *J. Cell Biol.* **174,** 677–687.
King, K. E., *et al.* (2003). Kruppel like factor 4 (KLF4/GKLF) is a target of bone morphogenetic proteins and transforming growth factor beta 1 in the regulation of vascular smooth muscle cell phenotype. *J. Biol. Chem.* **278,** 11661–11669.
Krol, J., *et al.* (2010). The widespread regulation of microRNA biogenesis, function and decay. *Nat. Rev. Genet.* **11,** 597–610.
Krutzfeldt, J., *et al.* (2005). Silencing of microRNAs in vivo with "antagomirs". *Nature* **438,** 685–689.
Kuehbacher, A., *et al.* (2007). Role of Dicer and Drosha for endothelial microRNA expression and angiogenesis. *Circ. Res.* **101,** 59–68.
Lagna, G., *et al.* (2007). Control of phenotypic plasticity of smooth muscle cells by bone morphogenetic protein signaling through the myocardin-related transcription factors. *J. Biol. Chem.* **282,** 37244–37255.
Langley, B., *et al.* (2005). Remodeling chromatin and stress resistance in the central nervous system: Histone deacetylase inhibitors as novel and broadly effective neuroprotective agents. *Curr. Drug Targets CNS Neurol. Disord.* **4,** 41–50.
Lee, Y., *et al.* (2003). The nuclear RNase III Drosha initiates microRNA processing. *Nature* **425,** 415–419.
Leeper, N. J., *et al.* (2011). microRNA-26a is a novel regulator of vascular smooth muscle cell function. *J. Cell. Physiol.* **226,** 1035–1043.
Lim, L. P., *et al.* (2005). Microarray analysis shows that some microRNAs downregulate large numbers of target mRNAs. *Nature* **433,** 769–773.
Lin, Y., *et al.* (2009). Involvements of microRNAs in hydrogen peroxide-mediated gene regulation and cellular injury response in vascular smooth muscle cells. *J. Biol. Chem.* **284,** 7903–7913.
Liu, J., *et al.* (2004). Argonaute2 is the catalytic engine of mammalian RNAi. *Science* **305,** 1437–1441.
Liu, X., *et al.* (2009). A necessary role of miR-221 and miR-222 in vascular smooth muscle cell proliferation and neointimal hyperplasia. *Circ. Res.* **104,** 476–487.
Liu, Y., *et al.* (2005). Kruppel-like factor 4 abrogates myocardin-induced activation of smooth muscle gene expression. *J. Biol. Chem.* **280,** 9719–9727.
Liu, Y., *et al.* (2003). A transforming growth factor-beta control element required for SM alpha-actin expression in vivo also partially mediates GKLF-dependent transcriptional repression. *J. Biol. Chem.* **278,** 48004–48011.
Lund, E., *et al.* (2004). Nuclear export of microRNA precursors. *Science* **303,** 95–98.
Manabe, I., and Owens, G. K. (2001). Recruitment of serum response factor and hyperacetylation of histones at smooth muscle-specific regulatory regions during differentiation

of a novel P19-derived in vitro smooth muscle differentiation system. *Circ. Res.* **88,** 1127–1134.

Martin, M. M., *et al.* (2007). The human angiotensin II type 1 receptor +1166 A/C polymorphism attenuates microRNA-155 binding. *J. Biol. Chem.* **282,** 24262–24269.

McCarthy, J. J., and Esser, K. A. (2007). microRNA-1 and microRNA-133a expression are decreased during skeletal muscle hypertrophy. *J. Appl. Physiol.* **102,** 306–313.

McDonald, O. G., and Owens, G. K. (2007). Programming smooth muscle plasticity with chromatin dynamics. *Circ. Res.* **100,** 1428–1441.

McDonald, O. G., *et al.* (2006). Control of SRF binding to CArG box chromatin regulates smooth muscle gene expression in vivo. *J. Clin. Invest.* **116,** 36–48.

Miano, J. M. (2003). Serum response factor: Toggling between disparate programs of gene expression. *J. Mol. Cell. Cardiol.* **35,** 577–593.

Mishra, P. J., *et al.* (2008). MiRSNPs or MiR-polymorphisms, new players in microRNA mediated regulation of the cell: Introducing microRNA pharmacogenomics. *Cell Cycle* **7,** 853–858.

Montgomery, R. L., and van Rooij, E. (2010). microRNA regulation as a therapeutic strategy for cardiovascular disease. *Curr. Drug Targets* **11,** 936–942.

Murray, C. J., and Lopez, A. D. (1997). Global mortality, disability, and the contribution of risk factors: Global burden of disease study. *Lancet* **349,** 1436–1442.

Naguibneva, I., *et al.* (2006). The microRNA miR-181 targets the homeobox protein Hox-A11 during mammalian myoblast differentiation. *Nat. Cell Biol.* **8,** 278–284.

Nakajima, N., *et al.* (2006). microRNA-1 facilitates skeletal myogenic differentiation without affecting osteoblastic and adipogenic differentiation. *Biochem. Biophys. Res. Commun.* **350,** 1006–1012.

Nie, M., *et al.* (2003). Transcriptional regulation of cyclooxygenase 2 by bradykinin and interleukin-1beta in human airway smooth muscle cells: Involvement of different promoter elements, transcription factors, and histone h4 acetylation. *Mol. Cell. Biol.* **23,** 9233–9244.

Olsen, P. H., and Ambros, V. (1999). The lin-4 regulatory RNA controls developmental timing in Caenorhabditis elegans by blocking LIN-14 protein synthesis after the initiation of translation. *Dev. Biol.* **216,** 671–680.

Owens, G. K. (1995). Regulation of differentiation of vascular smooth muscle cells. *Physiol. Rev.* **75,** 487–517.

Owens, G. K., *et al.* (2004). Molecular regulation of vascular smooth muscle cell differentiation in development and disease. *Physiol. Rev.* **84,** 767–801.

Pan, Q., *et al.* (2010). microRNA 21: Response to hormonal therapies and regulatory function in leiomyoma, transformed leiomyoma and leiomyosarcoma cells. *Mol. Hum. Reprod.* **16,** 215–227.

Paroni, G., *et al.* (2007). Dephosphorylation and caspase processing generate distinct nuclear pools of histone deacetylase 4. *Mol. Cell. Biol.* **27,** 6718–6732.

Pidkovka, N. A., *et al.* (2007). Oxidized phospholipids induce phenotypic switching of vascular smooth muscle cells in vivo and in vitro. *Circ. Res.* **101,** 792–801.

Poliseno, L., *et al.* (2006). microRNAs modulate the angiogenic properties of HUVECs. *Blood* **108,** 3068–3071.

Qiu, P., and Li, L. (2002). Histone acetylation and recruitment of serum responsive factor and CREB-binding protein onto SM22 promoter during SM22 gene expression. *Circ. Res.* **90,** 858–865.

Rana, T. M. (2007). Illuminating the silence: Understanding the structure and function of small RNAs. *Nat. Rev. Mol. Cell Biol.* **8,** 23–36.

Rehwinkel, J., *et al.* (2006). Genome-wide analysis of mRNAs regulated by Drosha and Argonaute proteins in Drosophila melanogaster. *Mol. Cell. Biol.* **26,** 2965–2975.

Rhee, S. W., *et al.* (2009). Vascular smooth muscle-specific knockdown of the noncardiac form of the L-type calcium channel by microRNA-based short hairpin RNA as a potential antihypertensive therapy. *J. Pharmacol. Exp. Ther.* **329,** 775–782.
Schickel, R., *et al.* (2008). microRNAs: Key players in the immune system, differentiation, tumorigenesis and cell death. *Oncogene* **27,** 5959–5974.
Schieffer, B., *et al.* (2000). Expression of angiotensin II and interleukin 6 in human coronary atherosclerotic plaques: Potential implications for inflammation and plaque instability. *Circulation* **101,** 1372–1378.
Sethupathy, P., and Collins, F. S. (2008). microRNA target site polymorphisms and human disease. *Trends Genet.* **24,** 489–497.
Spiegelman, B. M., and Heinrich, R. (2004). Biological control through regulated transcriptional coactivators. *Cell* **119,** 157–167.
Suarez, Y., *et al.* (2007). Dicer dependent microRNAs regulate gene expression and functions in human endothelial cells. *Circ. Res.* **100,** 1164–1173.
Sun, J. M., *et al.* (2003). Measurement of histone acetyltransferase and histone deacetylase activities and kinetics of histone acetylation. *Methods* **31,** 12–23.
Sun, Q., *et al.* (2006). Defining the mammalian CArGome. *Genome Res.* **16,** 197–207.
Sun, S. G., *et al.* (2011). miR-146a and Kruppel-like factor 4 form a feedback loop to participate in vascular smooth muscle cell proliferation. *EMBO Rep.* **12,** 56–62.
van Rooij, E., *et al.* (2007). Control of stress-dependent cardiac growth and gene expression by a microRNA. *Science* **316,** 575–579.
Villeneuve, L. M., *et al.* (2010). Enhanced levels of microRNA-125b in vascular smooth muscle cells of diabetic db/db mice lead to increased inflammatory gene expression by targeting the histone methyltransferase Suv39h1. *Diabetes* **59,** 2904–2915.
Wang, D., *et al.* (2001). Activation of cardiac gene expression by myocardin, a transcriptional cofactor for serum response factor. *Cell* **105,** 851–862.
Wang, D. Z., and Olson, E. N. (2004). Control of smooth muscle development by the myocardin family of transcriptional coactivators. *Curr. Opin. Genet. Dev.* **14,** 558–566.
Wang, M., *et al.* (2010). Common genetic variants in pre-microRNAs are associated with risk of coal workers' pneumoconiosis. *J. Hum. Genet.* **55,** 13–17.
Wang, Z., *et al.* (2004). Myocardin and ternary complex factors compete for SRF to control smooth muscle gene expression. *Nature* **428,** 185–189.
Wu, L., *et al.* (2006). microRNAs direct rapid deadenylation of mRNA. *Proc. Natl. Acad. Sci. USA* **103,** 4034–4039.
Xie, C., *et al.* (2011). microRNA-1 regulates smooth muscle cell differentiation by repressing Kruppel-like factor 4. *Stem Cells Dev.* **20,** 205–210.
Xin, M., *et al.* (2009). microRNAs miR-143 and miR-145 modulate cytoskeletal dynamics and responsiveness of smooth muscle cells to injury. *Genes Dev.* **23,** 2166–2178.
Xu, C. C., *et al.* (2008). Differential expression of microRNAs in the aorta of spontaneously hypertensive rats. *Sheng Li Xue Bao* **60,** 553–560.
Yang, B., *et al.* (2007). The muscle-specific microRNA miR-1 regulates cardiac arrhythmogenic potential by targeting GJA1 and KCNJ2. *Nat. Med.* **13,** 486–491.
Yi, R., *et al.* (2003). Exportin-5 mediates the nuclear export of pre-microRNAs and short hairpin RNAs. *Genes Dev.* **17,** 3011–3016.
Yoshida, T., *et al.* (2008a). Kruppel-like factor 4, Elk-1, and histone deacetylases cooperatively suppress smooth muscle cell differentiation markers in response to oxidized phospholipids. *Am. J. Physiol. Cell Physiol.* **295,** C1175–C1182.
Yoshida, T., *et al.* (2007). Platelet-derived growth factor-BB represses smooth muscle cell marker genes via changes in binding of MKL factors and histone deacetylases to their promoters. *Am. J. Physiol. Cell Physiol.* **292,** C886–C895.

Yoshida, T., *et al.* (2008b). Conditional deletion of Kruppel-like factor 4 delays down-regulation of smooth muscle cell differentiation markers but accelerates neointimal formation following vascular injury. *Circ. Res.* **102,** 1548–1557.
Yoshida, T., and Owens, G. K. (2005). Molecular determinants of vascular smooth muscle cell diversity. *Circ. Res.* **96,** 280–291.
Zeng, Y., *et al.* (2006). Regulation of cGMP-dependent protein kinase expression by Rho and Kruppel-like transcription factor-4. *J. Biol. Chem.* **281,** 16951–16961.
Zhao, Y., *et al.* (2007). Dysregulation of cardiogenesis, cardiac conduction, and cell cycle in mice lacking miRNA-1-2. *Cell* **129,** 303–317.
Zhao, Y., *et al.* (2005). Serum response factor regulates a muscle-specific microRNA that targets Hand2 during cardiogenesis. *Nature* **436,** 214–220.
Zhou, J., *et al.* (2005). Smooth muscle-specific genes are differentially sensitive to inhibition by Elk-1. *Mol. Cell. Biol.* **25,** 9874–9885.

CHAPTER SIXTEEN

Transforming Growth Factor-Beta Superfamily in Mouse Embryonic Stem Cell Self-Renewal

Katherine E. Galvin-Burgess* *and* Jay L. Vivian*

Contents

I. Introduction	342
II. Overview of TGF-Beta-Related Signaling	343
A. Ligands and transmembrane receptors	343
B. Smad proteins	345
C. Additional signaling modulators	346
III. ES Cell Regulation by the BMP Pathway	347
A. Roles in pluripotency	347
B. Roles in differentiation	349
IV. ES Cell Regulation by the Nodal Pathway	350
A. Roles in pluripotency	350
B. Roles in differentiation	353
V. Interactions of TGF-Beta Signaling with Other Pluripotency Pathways	354
VI. TGF-Beta Signaling in Other Pluripotent Cells	354
A. Human ES cells	354
B. Epiblast stem cells	356
C. Induced pluripotent stem cells	357
VII. Conclusions and Future Directions	358
Acknowledgment	359
References	359

Abstract

Embryonic stem (ES) cells are pluripotent cells that maintain the capability of undifferentiated self-renewal in culture. As mouse ES cells have the capacity to give rise to all the tissues of the body, they are an excellent developmental biology model system and a model for regenerative therapies. The extracellular cues and the intracellular signaling cascades that regulate ES cell self-renewal

* Department of Pathology and Laboratory Medicine, Institute for Reproductive Health and Regenerative Medicine, University of Kansas Medical Center, Kansas City, USA

Vitamins and Hormones, Volume 87
ISSN 0083-6729, DOI: 10.1016/B978-0-12-386015-6.00035-4

and cell-fate choices are complex and actively studied. Many developmental signaling pathways regulate the ES cell phenotype, and their intracellular programs interact to modulate the gene networks controlling ES cell pluripotency. This review focuses on the current understanding and outstanding questions of the roles of the transforming growth factor-beta-related signaling pathways in regulating pluripotency and differentiation of mouse ES cells. The complex dichotomic roles of bone morphogenetic protein signaling in maintaining the undifferentiated state and also inducing specific cell fates will be reviewed. The emerging roles of Nodal signaling in ES cell self-renewal will also be discussed.

I. Introduction

Embryonic stem (ES) cells are derived from the inner cell mass of the blastocyst (Evans and Kaufman, 1981; Martin, 1981). As the inner cell mass gives rise to the epiblast and subsequently to all three germ layers of the embryo, ES cells maintain this property of pluripotency and can differentiate into most cell fates. The capability of ES cells to have sustained pluripotent growth during long-term *in vitro* culture produces a valuable system for both studying early embryonic development and investigating key signaling pathways controlling pluripotency and differentiation. The capacity for an ES cell to remain undifferentiated and germline competent is a function of the intrinsic transcriptional machinery that directs ES cell self-renewal. Transcriptional regulatory networks are being uncovered in ES cells, including essential DNA-binding proteins such as Oct4, Sox2, and Nanog (Orkin *et al.*, 2008). The self-renewal, pluripotency, and differentiative capacity of ES cells are regulated by numerous extracellular stimuli of the culture conditions as well as the autocrine signaling produced by the ES cells themselves. The downstream molecular mechanisms by which different signaling pathways affect stem cell pluripotency and interact with other intracellular pathways are an active and growing area of developmental biology and stem cell research.

The transforming growth factor-beta (TGF-beta)-related signaling pathways have complex roles in regulating the pluripotency and cell fate of ES cells. The TGF-beta pathway is composed to two main branches, bone morphogenetic proteins (BMP) and Activin/Nodal/TGF-beta subfamilies. While BMP signaling is a well-studied pluripotency pathway in mouse ES cells, the role of Activin/Nodal/TGF-beta signaling is less clear and appears to affect pluripotency indirectly. This chapter will focus on the current understanding of the roles of TGF-beta-related signaling in mouse ES cell biology.

II. Overview of TGF-Beta-Related Signaling

Work from many models systems has constructed a canonical model for the molecular cascades that direct TGF-beta-related signaling from the extracellular ligand to the nucleus to regulate changes in gene expression. This literature is extensive and cannot be comprehensively detailed here; this chapter will provide a basic overview of the signaling pathways and components most relevant to mouse ES cell renewal and differentiation.

A. Ligands and transmembrane receptors

The TGF-beta-related ligands consist of over 30 cytokines. All ligands exhibit a related amino-terminal signaling peptide for secretion, with the functional peptide residing in the carboxy-terminal region. Prior to secretion, the signaling ligand is activated when furin or proprotein convertase 6 proteolytically cleaves the precursor protein to form a carboxy-terminal mature protein dimer (Cui *et al.*, 1998). The cytokines with signaling activity are categorized into two major arms, the BMP ligands and Activin/Nodal/TGF-beta ligands, canonically distinguished by the intracellular signaling pathways that mediate the signaling (see below; Fig. 16.1). Depending on the culture conditions, ES cells are exposed to a variety of TGF-beta ligands, particularly from the complex cocktail of factors present in serum-supplemented media. Additionally, several TGF-beta ligands are highly expressed by ES cells themselves, suggesting potential autocrine modes of signaling. Microarray analysis of mouse ES cells demonstrated that the ligands *BMP1*, *BMP2*, *BMP4*, and *BMP8b* are expressed in ES cells (Galvin *et al.*, 2010). Of the Activin/Nodal/TGF-beta ligands, *Nodal* has the highest mRNA expression in mouse ES cells grown under serum-free conditions. Additionally, two distinct inhibitory TGF-beta-related ligands, the *Lefty* factors and *GDF3* (growth differentiation factor-3), are highly expressed by ES cells.

The capacity of ES cells to respond to different ligands is predicated by the expression of the transmembrane receptors. The TGF-beta-related ligands bind to single-pass type I and type II transmembrane serine/threonine kinase receptors, consisting of an extracellular ligand-binding domain and an intracellular kinase signaling motif. Upon ligand–receptor oligomerization, type II receptors phosphorylate the glycine and serine-rich domain of the type I receptors. Each signaling arm has its own set of serine/threonine kinase type I receptors. The type II Activin receptors have a broad specificity and bind to Activin, Nodal, and BMP (Kitisin *et al.*, 2007). Type II receptors, including BMP type II receptor (BmprII) and Activin type IIA and IIB receptors (ActrIIA, ActrIIB), are constitutively active and expressed in mouse ES cells.

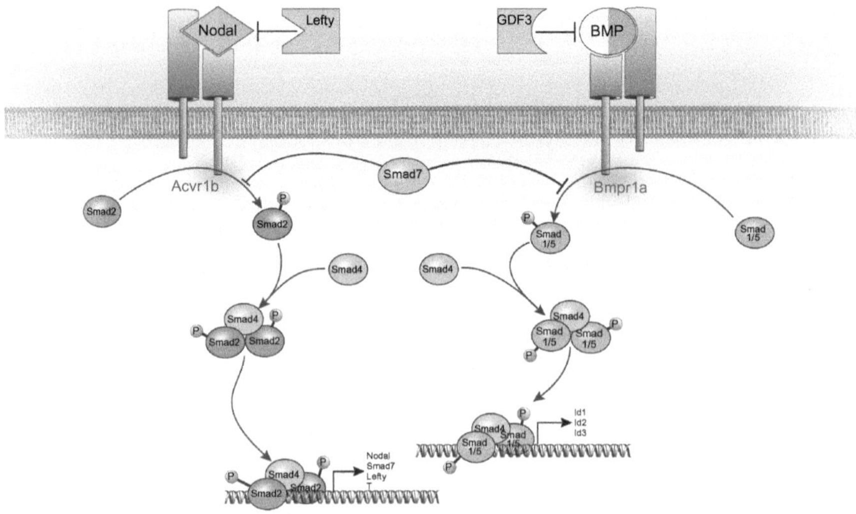

Figure 16.1 Canonical TGF-beta-related signaling in mouse ES cells. BMP and Nodal ligands signal through distinct sets of transmembrane serine/threonine kinase receptors. Upon receptor activation, receptor-regulated Smad proteins are phosphorylated. BMP signaling induces phosphorylation of Smad1/5, while stimulation with Nodal phosphorylates Smad2. Activated receptor-regulated Smad proteins form oligomeric complexes with co-Smad4 and enter the nucleus. In cohort with other transcription factors, this complex binds to DNA and enhances or represses specific target genes. BMP signaling induces transcription of the *Id* genes, which are known as ES cell self-renewal factors. Target genes of Nodal signaling include feedback inhibitors. Enhanced expression of Smad7 feeds back to inhibit both TGF-beta-related signaling pathways. The signaling pathways may also be inhibited extracellularly; GDF3 blocks BMP signaling, and the Lefty factors antagonize Nodal signaling. (For color version of this figure, the reader is referred to the Web version of this chapter.)

Phosphorylation of the type I receptor by the type II receptor converts its kinase domain to an active conformation (Miyazono *et al.*, 2010). Intracellular signaling specificity is largely determined through the type I kinase receptors (Shi and Massague, 2003). The type I receptors that mediate BMP signaling include Activin receptor type Ia (Actr1a or Alk2), BMP receptor type Ia (Bmpr1a or Alk3), and BMP receptor type Ib (Bmpr1b or Alk6). Activin, Nodal, and TGF-beta ligands signal via Activin receptor type-Ib (Acvr1b or ALK4), TGF-beta receptor type I (Tgfbr1 or ALK5), and Activin receptor type-Ic (Acvr1c or ALK7). Of the type I receptors, Bmpr1a, Acvr1b, and Tgfbr1 are most highly expressed in mouse ES cells (Galvin *et al.*, 2010).

The BMP and TGF-beta subfamilies also differ in their modes of the ligand–receptor interaction (Shi and Massague, 2003). BMP ligands preferentially bind to type I receptors and subsequently recruit the type II receptors to the ligand–receptor complex. In contrast, TGF-beta ligands exhibit a high affinity for the type II receptors and very low affinity for isolated type I receptors. After the TGF-beta ligand and type II receptor kinase bind, the type I receptor is incorporated in the complex. The differences in binding affinities between the TGF-beta-related ligands and their signaling receptors allow oligomerization to be controlled in a precise manner, which consequently results in the activation of distinct cell-type-specific signaling cascades (Sieber *et al.*, 2009).

B. Smad proteins

Following receptor activation, canonical TGF-beta-related signaling is transmitted through the intracellular Smad proteins via type I-mediated phosphorylation of receptor-regulated Smads (R-Smad) at their C-terminal SSXS-motif. The Smad proteins exhibit extensive homology, harboring a C-terminal Mad-homology 2 (MH2) domain which functions in receptor recognition, nuclear import, and Smad oligomerization (Wrana, 2000). The N-terminal of the Smad proteins contain a highly conserved Mad-homology 1 (MH1) domain that carries out numerous protein–protein interactions and also binds DNA. Between the MH1 and MH2 domains is a proline-rich linker region which contains regulatory phosphorylation sites and protein-binding domains (Moustakas and Heldin, 2009). BMP signaling utilizes R-Smads including Smad1, Smad5, and Smad8. In mouse ES cells, *Smad1,* and to a lesser extent *Smad5*, are expressed at higher transcript levels than *Smad8* (Galvin *et al.*, 2010). In contrast, Activin, TGF-beta, and Nodal signaling is transmitted by Smad2 and Smad3. ES cells express higher levels of *Smad2* than *Smad3* (Dunn *et al.*, 2004), suggesting that Nodal signaling is preferentially mediated by Smad2 in mouse ES cells.

Phosphorylated R-Smads homotrimerize or form heteromeric complexes with Smad4, the common-mediator Smad (co-Smad; Shi and Massague, 2003). Active R-Smad/Smad4 complexes translocate to the nucleus, interact with additional transcription factors, bind to DNA, and modulate the expression of downstream target genes. The Smad factors bind to DNA with relatively low affinity and limited sequence complexity; thus, much of the specificity of localization relies on interactions with other nuclear cofactors to regulate transcription (Wrana, 2000). Identified DNA-binding partners include forkhead activin signal transducer (FAST), CBP/p300, MSG1, and others. Smad transcriptional complexes may also recruit transcriptional repressors to negatively regulate transcription of specific genes. The cell-type-specific responses to signaling are regulated in part by the transcriptional protein partners that mediate transcriptional activation or repression (Shi and Massague, 2003). The protein partners of the Smad proteins are numerous and highly cell-type-dependent, thus allowing for a complexity of downstream responses. In undifferentiated ES cells, binding sites for Smad1, Smad2, and Smad4 have been localized to the promoter regions of numerous developmental regulators (Fei *et al.*, 2010a,b; Lopez-Rovira *et al.*, 2002), suggesting that both BMP and Nodal-related signaling directly regulate gene regulation and, by inference, participate in ES cell pluripotency and cell-fate determination.

Smad6 and Smad7 are members of a third class of Smad proteins, the inhibitory Smads (I-Smad) that antagonize TGF-beta-related signaling. The I-Smads act as major negative regulators forming autoinhibitory feedback loops and mediating cross talk with other signaling pathways (Park, 2005). Smad6 and Smad7 differ in the antagonistic effects they have on the different arms of the TGF-beta superfamily. While Smad6 preferentially inhibits BMP signaling, Smad7 inhibits signaling by TGF-beta, Activin, Nodal, and BMP (Park, 2005). Smad7 inhibits signaling through multiple potential mechanisms. These proteins may compete with R-Smads for binding to activated type I receptor, degrade activated type 1 receptors via E3-ubiquitin ligases, and/or dephosphorylate activated type I receptors (ten Dijke and Hill, 2004). ES cells express both *Smad6* and *Smad7* (Galvin *et al.*, 2010); however, only *Smad7* increases expression in response to Activin stimulation, while decreasing expression following Smad2 inhibition. The modulation of Smad7 expression succeeding changes in Smad2 activity plays an important role in the cross talk between Nodal and BMP activity in ES cells (Galvin *et al.*, 2010; Park, 2005; see below).

C. Additional signaling modulators

The canonical TGF-beta-related signaling is regulated through numerous extrinsic and intrinsic mechanisms. Both intracellular and extracellular modulators of BMP signaling are present in mouse ES cells, and their roles in stem cell biology are beginning to be explored. A wide variety of extracellular inhibitors binds to the free ligand and/or competes for receptor occupation sites, resulting in decreased signal propagation. Nodal itself can inhibit BMP signaling by heterodimerization of the ligands (Yeo and Whitman, 2001). Similarly, a relatively unstudied member of the TGF-beta family, GDF3 also appears to act as a bifunctional protein (Levine and Brivanlou, 2006a), inhibiting BMP signaling (Levine and Brivanlou, 2006b) while also working as a Nodal-like ligand (Chen *et al.*, 2006). The coreceptor Cripto is required for the majority of Nodal signaling (Constam, 2009). However, Cripto also inhibits Activin signaling by acting as a noncompetitive antagonist (Kelber *et al.*, 2008). Additionally, extracellular antagonists that were first demonstrated to inhibit Activin signaling have also been shown to antagonize the BMP pathway. For example, the inhibitor Follistatin, which is expressed in mouse ES cells, inhibits both Activin and BMP activity (Iemura *et al.*, 1998; Pierre *et al.*, 2005). Lefty proteins are structurally related to the TGF-beta-related ligands and are extracellular inhibitors to Nodal signaling. The Lefty factors (Lefty1 and Lefty2) are highly expressed in human and mouse ES cells, and their expression is controlled by multiple stem-cell pluripotency pathways (Tabibzadeh and Hemmati-Brivanlou, 2006). Analysis of pluripotent cell lines suggests that Lefty may also inhibit TGF-beta signaling (Ulloa and Tabibzadeh, 2001). However, the specificity and putative antagonistic activities of the Lefty proteins in ES cells is not resolved.

III. ES Cell Regulation by the BMP Pathway

BMPs were originally identified for their ability to induce bone formation (Urist, 1965); however, a large body of work has shown that these secreted factors play an active role in the prenatal and postnatal development of nearly all tissues and organ systems, including the nervous system, somites, lung, kidney, skin, muscle, gonads, and the hematopoietic system. Moreover, genetic and developmental studies have shown that BMP signaling plays a vital role in body plan development and pattern formation (Hogan, 1996). Recent work paradoxically demonstrates that the BMPs are important in ES cell proliferation, self-renewal, and differentiation (Watabe and Miyazono, 2009; Xu *et al.*, 2008). In this review, we will focus on the activities and downstream targets of the BMP–Smad1/5 signaling involved in maintaining ES cell pluripotency and controlling cell-fate choices.

A. Roles in pluripotency

To maintain undifferentiated ES cells, standard cell culture conditions consist of a co-culture of ES cells with mitotically inactivated embryonic fibroblast cells in a media supplemented with serum. The feeder cells are known to secrete factors vital in maintaining the proliferative and pluripotent state of stem cells. One known essential secreted factor is leukemia inhibitory factor (LIF; Williams *et al.*, 1988). Signal transducer and activation of transcription 3 (Stat3) is activated by LIF stimulation (Niwa *et al.*, 1998). While activation of Stat3 is necessary and sufficient for ES cell self-renewal in media containing serum, neither LIF nor activated Stat3 signaling can maintain ES cells in serum-free medium (Liu *et al.*, 2007). These results indicate that other signaling pathways provided by serum are important for maintaining pluripotency.

By examining other signaling pathways essential for pluripotency, Ying and colleagues found that BMP4, in combination with LIF, maintains ES cell pluripotency and blocks neural differentiation in serum-free media (Ying *et al.*, 2003). Similarly, BMP2 or the BMP relative growth differentiation factor-6 (GDF6) also support ES cell self-renewal in the presence of LIF. BMP stimulation of ES cells induces phosphorylation of Smad1/5 and expression of known BMP target genes (Galvin *et al.*, 2010; Ying *et al.*, 2003). Inhibition of BMP signaling also affects ES cell pluripotency. GDF3, a secreted TGF-beta family member, binds to BMP4 and blocks downstream signaling. ES cells harboring a genetrap at the *Gdf3* locus demonstrate a normal, undifferentiated morphology and pluripotent gene expression in the absence of LIF (Levine and Brivanlou, 2006b). These results suggest an important role for BMP-related signaling in stem cell self-renewal.

To gain insight into the mechanisms by which BMP signaling maintains pluripotency, downstream targets have been identified. *Id* (inhibitors of differentiation) genes are prominent targets of BMP–Smad signaling in a variety of cell types, including undifferentiated ES cells (Hollnagel *et al.*, 1999). The Id family of transcriptional regulators contains four members, Id1, Id2, Id3, and Id4, and they play important roles during early development, cell-fate determination, and embryonic and somatic stem cell biology (Lasorella *et al.*, 2001; Sikder *et al.*, 2003). These helix-loop-helix proteins negatively control the function of basic helix-loop-helix transcription factors by heterodimerizing with and sequestering pools of transiently diffusing transcription factors to prevent their reassociation with DNA (O'Toole *et al.*, 2003). The *Id* genes are direct targets of BMP signaling via Smad1 and Smad4, which bind at *cis*-acting regulatory elements (Lopez-Rovira *et al.*, 2002).

Austin Smith's group addressed the relevance of the Id factors in BMP-mediated ES cell pluripotency (Ying *et al.*, 2003). The addition of serum or recombinant BMP to media increases *Id* expression, consistent with BMP or serum-dependent regulation of these factors. Constitutive expression of Id1 liberates ES cells from BMP or serum dependence for self-renewal, demonstrating that the *Id* genes are the primary downstream targets of BMP signaling in maintaining the undifferentiated state. The presence of LIF is still required for Id expression to block ES differentiation, consistent with the observations of BMP signaling activity. Other factors present in serum or secreted by the feeder layer may also induce the expression of *Id* factors independent of the BMP signaling and the Smad factors (Benezra, 2001). This observation may explain why ES cells can be derived from Smad4-deficient blastocysts (Sirard *et al.*, 1998) and why knockdown of Smad1 or Smad4 expression does not impair self-renewal ability of ES cells (Fei *et al.*, 2010a). Importantly, the detailed molecular mechanisms by which the propluripotency activity of BMP–Id axis acts to attenuate prodifferentiative pathways remain to be elucidated. In particular, it remains to be determined how the Id factors, presumptively by binding to specific protein targets, may function to enhance self-renewal in ES cells.

Zhao and colleagues have proposed an additional, potentially Smad-independent mechanism for the function of BMP signaling in ES pluripotency (Qi *et al.*, 2004). In this study, the authors suggest that BMP signaling also acts to inhibit extracellular receptor kinase (ERK) and p38 mitogen-activated protein kinase (p38 MAPK) pathways. The mechanism of this activity is as-yet undefined; the authors suggest a possible "noncanonical" signaling pathway not utilizing Smad signaling. This work found that pharmacological inhibitors of ERK and p38 MAPKs mimicked the effect of BMP4 on ES cells. Subsequent work with additional pharmacological inhibitors confirmed that blocking phosphorylation of ERK allows mouse ES cells to be maintained in the absence of LIF and BMP4 (Ying *et al.*, 2008). When pERK signaling is strongly inhibited, concurrent inhibition of the glycogen synthase kinase 3 pathway increased the overall viability of ES cells in serum-free conditions. Strikingly, simultaneous inhibition of these two pathways is sufficient to allow for the derivation of germline-competent ES cells from blastocysts in serum-free and feeder-free media from the mouse and rat (Buehr *et al.*, 2008). This work establishes a "ground state" model for ES cell self-renewal. This model suggests that pluripotency signals such as BMP–Smad–Id and LIF–Stat3 signaling function in self-renewal by blocking differentiating-inductive signals such as ERK signaling. The capacity for maintaining self-renewal can thus be mimicked by inhibiting downstream signal transduction pathways such as FGF–ERK signaling. The molecular mechanisms by which these pathways communicate with one another remain to be determined.

A variety of genetic studies implicate BMP signaling in important and complex roles in pluripotency of the early embryo. Low levels of *BMP4* transcript are detected in embryonic day 6.5 embryos, and targeted disruption of the BMP4 locus produces homozygous mutant embryos which die between embryonic day 6.5 and 9.5 (Winnier *et al.*, 1995). Most BMP4 homozygous mutant embryos do not proceed beyond the egg cylinder stage. Additionally, targeted inactivation of BMP type II receptor *BmprII* or type I receptor *Bmpr1a* leads to death of the mutant embryos prior to embryonic day 9.5 (Beppu *et al.*, 2000; Mishina *et al.*, 1995). Strong *in vivo* evidence of the importance of BMP signaling in pluripotency comes from analysis of the *Bmpr1a* mutant mice. Crosses of mice with only one copy of the BMP type I receptor ($Bmpr1a^{+/-}$) produced wild type and heterozygous ES cell lines but failed to establish $Bmpr1a^{-/-}$ ES cell lines (Qi *et al.*, 2004). These *in vivo* data are consistent with a central role for BMP signaling in the establishment of pluripotent cell lineages.

B. Roles in differentiation

BMP signaling has functions in a plethora of early developmental events, including gastrulation, germ-cell-fate determination, establishment of the left–right asymmetry, mesoderm formation, and neural patterning (Kishigami and Mishina, 2005; Zhao, 2003). Given the diverse roles of

BMP signaling in development, a variety of studies have sought to use ES cell differentiation as an *in vitro* mechanistic model. Using a variety of ES differentiation techniques, recombinant BMP4 has been shown to negatively regulate neural induction by ES cells (Finley *et al.*, 1999; Kawasaki *et al.*, 2000; Ying *et al.*, 2003), while inhibition of BMP signaling by Noggin increases neural stem cell differentiation (Tropepe *et al.*, 2001). Overexpression of Id factors also blocks neural cell fates of ES cells, consistent with the notion that the Id factors remain important downstream targets of BMP signaling to modulate differentiation. Further work has demonstrated that BMP4 treatment also inhibits cardiomyocyte differentiation (Hao *et al.*, 2008; Taha *et al.*, 2007).

While BMP signaling has been shown to block neural and cardiac lineages, it also promotes alternative germ layer cell fates. The timecourse of BMP activity likely influences the cell-fate decisions of differentiating ES cells. In ES cells, BMP4 induces cells of the mesodermal lineages, including hematopoietic precursors (Finley *et al.*, 1999; Johansson and Wiles, 1995). Knockdown of Smad1 and Smad4 significantly reduces expression of endoderm markers in embryoid bodies (Fei *et al.*, 2010a). Moreover, embryoid bodies from loss-of-function GDF3 genetraps, a TGF-beta family member which inhibits BMP4 signaling, do not form mesoderm or endoderm (Levine and Brivanlou, 2006b). These data point to a function for BMP signaling in enhancing the differentiation of ES cells toward certain "ventral" mesodermal fates, while actively inhibiting neural and cardiac differentiation.

These findings indicate that under distinct environmental conditions, BMP signaling has dramatically different functions to either stimulate ES cell self-renewal or direct specific differentiation pathways. However, the mechanism of the dichotomy of BMP signaling in ES cell pluripotency and differentiation remains to be fully elucidated. Given the complex transcriptional partners of the Smad proteins, the changes in gene expression during differentiation almost certainly influence the downstream targets of BMP signaling. Further work is needed to discern the intracellular signaling and molecular targets of BMP signaling in ES cells in these distinct contexts.

IV. ES Cell Regulation by the Nodal Pathway

A. Roles in pluripotency

Recent work has been undertaken to define the roles of Smad2 signaling in ES cells. ES cells have an active Smad2 signaling axis which can be manipulated *in vitro* through recombinant ligands and pharmacological inhibitors (Watabe and Miyazono, 2009). Importantly, this signaling activity is present even under serum-free conditions (Galvin *et al.*, 2010), strongly suggesting that Smad2 activation in undifferentiated ES cells is via an autocrine signaling mechanism. Of the Activin/Nodal/TGF-beta ligands, only *Nodal* has high mRNA expression in mouse ES cells (Galvin *et al.*, 2010). Knockdown of Nodal significantly decreases the expression of Smad2 gene targets; thus, mouse ES cells primarily activate Smad2 signaling through autocrine Nodal secretion. However, as recombinant Activin produces stronger Smad2 phosphorylation at lower concentrations than commercially available recombinant Nodal (Ogawa *et al.*, 2007), Activin is often used experimentally to stimulate the Smad2 pathway in ES cells.

Pharmacological agents have been employed to understand the functions of Nodal–Smad2 signaling in ES cells. The small molecule inhibitor SB431542 inhibits phosphorylation of Smad2 by the type I receptor kinases. James and colleagues used blastocyst outgrowth cultures to investigate the requirement for Smad2 activation in the establishment of the stem cell compartment. In control cultures, blastocysts attached and grew $Oct4^+$ outgrowths; however, blastocysts cultured with SB431542 have outgrowths that displayed a complete loss of Oct4 staining. Thus, an active Smad2 signaling pathway is required to maintain the $Oct4^+$ compartment of the inner cell mass *in vitro*.

In contrast, disruption of Smad2 signaling in established mouse ES cell lines does not grossly affect self-renewal (Fei *et al.*, 2010b; Galvin *et al.*, 2010; James *et al.*, 2005). No significant changes in Oct4 protein expression were noted in established ES cells after 3 days of inhibition of Smad2 activation by SB341542 (James *et al.*, 2005). However, treatment with SB431542 slows the growth of mouse ES cells and also reduces the spontaneous differentiation on the edge of ES cell colonies (Galvin *et al.*, 2010; Ogawa *et al.*, 2007). These data suggest an essential function for Smad2 signaling for the entry to stemness, but that Smad2 signaling has more subtle functions in modulating the established stem cell phenotype.

Two independent studies have demonstrated that inhibition of Smad2 phosphorylation indirectly induces the expression of the *Id* genes (Galvin *et al.*, 2010; Ogawa *et al.*, 2007). Decreased Smad2 activity reduces the expression of *Smad7*, a direct target of Smad2 (Guzman-Ayala *et al.*, 2009); the reduced Smad7 levels feed back to enhance BMP signaling activity (Galvin *et al.*, 2010). Thus, through indirectly enhancing the BMP pathway, inhibition of Nodal signaling enhances *Id* expression. ES cells overexpressing Smad7 fail to induce *Id1* following Smad2 inhibition, demonstrating that the cross talk via this inhibitory Smad was necessary for pathway interactions (Galvin *et al.*, 2010). Thus, in an environment with decreased Smad2 activity, mouse ES cells may have enhanced self-renewal through indirect enhancement of BMP signaling. These observations indicate that autocrine Nodal signaling activity in ES cells modulates the BMP pluripotency signaling pathways. More research is necessary to determine the exact relationship between Smad2 signaling and mouse ES cell pluripotency.

The regulation of BMP signaling by Nodal signaling suggests indirect functions of Nodal signaling in self-renewal through feedback regulators and modulation of additional signaling pathways.

Analyses of other Smad2-regulated genes in undifferentiated ES cells reveal only a small subset of genes with increased expression in response to Smad2 activation and decreased expression in response to Smad2 inhibition. The most highly responsive genes to modulation of Smad2 activity are the *Lefty* factors, inhibitory members of the TGF-beta family (Galvin *et al.*, 2010; Guzman-Ayala *et al.*, 2009). Both *Lefty1* and *Lefty2* increase expression over threefold with Activin stimulation and drastically decrease expression following SB431542 inhibition of Smad2 phosphorylation (Galvin *et al.*, 2010). Lefty factors were first identified as morphogenic factors that are asymmetrically expressed during gastrulation (Meno *et al.*, 1997, 1998). Recent work has identified more complex expression of the Lefty factors, including high expression in both mouse and human ES cells (Tabibzadeh and Hemmati-Brivanlou, 2006), as well as the early preimplantation embryo (Takaoka *et al.*, 2006). The expression of *Lefty1* in ES cells is additionally regulated by core pluripotency genes, including Oct4, Klf4, and Sox2 (Nakatake *et al.*, 2006). Further, ES cells overexpressing Stat3, the main downstream target of LIF signaling, significantly downregulate the expression of *Lefty2* (Cinelli *et al.*, 2008). These data point to a complex regulation of the Lefty factors in undifferentiated stem cells.

Inhibition of Nodal–Smad2 signaling also largely eliminates *Brachyury* expression (Galvin *et al.*, 2010). While Brachyury has historically been used as a marker of mesoderm tissues, it is also expressed in most undifferentiated ES cell lines. Nodal–Smad2 signaling regulates *Brachyury* transcription through recruitment of the histone demethylase Jmjd3 by Smad2 (Dahle *et al.*, 2010). During development, Nodal signaling regulates gene expression by counteracting the effects of the Polycomb-mediated repression of target genes. Recent work shows that Brachyury and Stat3 bind to an enhancer sequence of *Nanog* locus and enhance its transcription in mouse ES cells (Suzuki *et al.*, 2006). Nanog is a significant transcription factor involved in the self-renewal of undifferentiated ES cells. The binding of Brachyury to the *Nanog* locus suggests a role for Brachyury, and by inference Nodal–Smad2 signaling, in ES cell self-renewal.

At the blastocyst stage, *Smad2*, *Smad3*, and *Smad4* transcripts are expressed in the embryo (Dunn *et al.*, 2004). From the four-cell stage of the early embryo through formation of the blastocyst, Smad2 is phosphorylated and localized to the nucleus (James *et al.*, 2005). *Smad2* transcripts are present at approximately twofold higher levels than *Smad3* in ES cells but are coexpressed at approximately equivalent levels by embryonic day 8.5 (Dunn *et al.*, 2004). Genetic depletion of *Smad2* and *Smad3* causes severely limited development with deficits that closely resemble *Smad4*-deficient embryos. Additionally, Smad2/Smad3 homozygous mutants lack *Oct4* expression, a distinguishing marker of pluripotent cells. These data are consistent with functions for Smad2/3/4 signaling in early development, including the lineages giving rise to ES cells.

B. Roles in differentiation

Activin/Nodal signaling plays a clear role in the development of mesoderm and endoderm tissues (Schier, 2003; Tiedemann *et al.*, 2001; Tremblay *et al.*, 2000). Such an induction of mesoderm and endoderm cell lineages is also demonstrated in ES cells grown in culture. Exposure of embryoid bodies to serum or recombinant Activin enhances differentiation to definitive endodermal cells (Kubo *et al.*, 2004). Moreover, expression of Nodal induces the specification of ES cells into definitive endoderm and mesoderm derivatives using an inducible ES cell differentiation model (Takenaga *et al.*, 2007).

In mouse ES cells, knockdown of Smad2 or pharmacological inhibition of Smad2 phosphorylation significantly reduces expression of mesodermal and endodermal markers (Fei *et al.*, 2010b; Galvin *et al.*, 2010). To determine underlying mechanisms in Smad2-mediated differentiation events, Smad2-associated genes were identified by genome-wide chromatin immunoprecipitation-chip analysis (Fei *et al.*, 2010b). Genes involved in metabolic and developmental processes were enriched populations of Smad2-bound factors. *Tapbp*, which encodes the transmembrane glycoprotein tapasin, was identified as a direct, downstream target of Activin/Nodal–Smad2 signaling. *Tapbp* knockdown significantly attenuates Activin-mediated differentiation of mesoderm and endoderm cell fates and was thus demonstrated to be an essential player in Activin-mediated mesendoderm differentiation in ES cells.

While inhibition of Smad2 signaling hinders differentiation of the mesoderm and endoderm cell lineages, it surprisingly appears to enhance differentiation of trophectoderm lineages (Galvin *et al.*, 2010). Under normal culture conditions, mouse ES cells will not readily differentiate to the trophectoderm lineages, and ES cells will not contribute to the trophectoderm lineages when injected into blastocysts. However, using Wnt or BMP4 stimulation, mouse ES cells can be triggered to express genes involved in trophoblast differentiation (Hayashi *et al.*, 2010; He *et al.*, 2008). Under pluripotent, defined-media conditions, inhibition of Smad2 activation significantly enhances *Ascl2* and *Gata2* expression, markers of some differentiated trophectoderm lineages (Galvin *et al.*, 2010). The increased expression of trophectoderm markers generally requires long-term Smad2 inhibition, suggesting that its signaling mechanism is likely indirect and involves other pathways. Thus Activin/Nodal–Smad2 signaling functions in a cascade of differentiation events that while promoting endodermal and mesodermal cell fates also appear to endogenously inhibit the propensity of ES cells to differentiate down a trophectoderm pathway.

V. Interactions of TGF-Beta Signaling with Other Pluripotency Pathways

While cross talk occurs between the BMP and Nodal pathways in ES cells, these pathways also interact with other signaling pathways. As discussed previously, BMP signaling may support ES cell self-renewal via inhibition of both ERK and p38 MAPK pathways (Qi *et al.*, 2004). Additionally, BMP signaling has been shown to interact with the Wnt pathway. In both human and mouse ES cells, activation of the canonical Wnt pathway is sufficient to maintain pluripotency (Sato *et al.*, 2004). Stimulation of ES cells with BMP4 increases RNA and protein expression of Wnt1 and activates beta-catenin (Lee *et al.*, 2009). BMP4 signaling also induces Akt phosphorylation. The BMP4-induced Wnt pathway activity is attenuated by phosphatidyl inositol 3-kinase (PI3K) siRNA and an Akt inhibitor. Moreover, Wnt signaling also affects BMP activity. Knockdown of Wnt1 inhibits BMP target genes and ES cell proliferation. Overall, BMP4 signaling interactions with PI3K/Akt and Wnt1/beta-catenin contribute to the maintenance of pluripotency and proliferation in mouse ES cells (Fig. 16.2).

Investigation of Nodal signaling interactions is less studied beyond its previously discussed function in modulating the BMP pathway (Galvin *et al.*, 2010). The majority of gene targets modulated by changes in Nodal activity appears to feedback and affect the TGF-beta signaling pathway itself (Smad7, Lefty, Nodal, Cripto, SnoN, Ski, and Bambi; Galvin *et al.*, 2010; Guzman-Ayala *et al.*, 2009). Gene expression analyses of ES cells with activated or inhibited Nodal signaling demonstrate changes in expression of ligands for Fgf (*Fgf15* and *Fgf8*) and Notch (*Notch3* and *Notch4*) signaling pathways (Galvin *et al.*, 2010; Guzman-Ayala *et al.*, 2009), suggesting potential roles for Nodal signaling in regulating these pathways. However, further investigation is needed to determine if Nodal signaling cooperates with these or other signal transduction pathways to regulate mouse ES cell pluripotency.

VI. TGF-Beta Signaling in Other Pluripotent Cells

A. Human ES cells

Mouse and human ES cells have distinct self-renewal mechanisms. Unlike mouse ES cells, which can utilize the combination of LIF and BMP signaling to maintain stemness, human ES cells utilize alternative signaling pathways to retain their pluripotency (Daheron *et al.*, 2004). Activin signaling through Smad2 phosphorylation maintains human ES cells in their

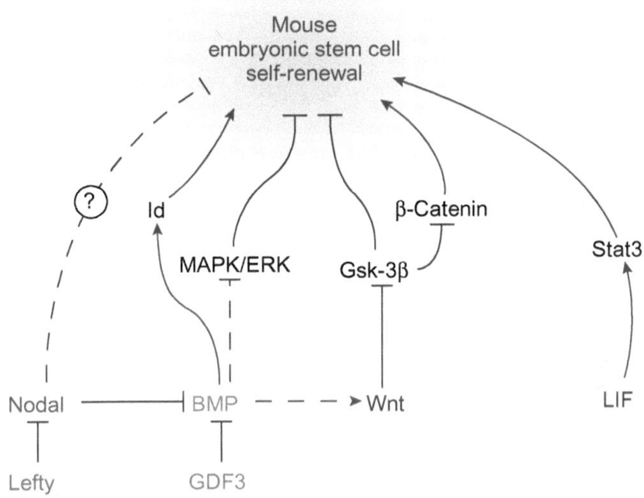

Figure 16.2 Signaling networks that regulate mouse ES cell self-renewal. The BMP pathway can affect mouse ES cell pluripotency through multiple signaling pathways. BMP signaling can be antagonized by GDF3, a TGF-beta-related family member which regulates both pluripotency and cell-fate determination in mouse ES cells (Levine and Brivanlou, 2006b). BMP stimulation induces expression of the Id factors, which in combination with LIF maintain ES cells pluripotency (Ying *et al.*, 2003). BMP signaling also inhibits extracellular receptor kinase (ERK) and mitogen-activated protein kinase (MAPK) pathways to enhance pluripotency (Qi *et al.*, 2004; Ying *et al.*, 2008). Last, stimulation with recombinant BMP activates Wnt signaling to induce ES cell pluripotency (Lee *et al.*, 2009). As inhibition of Nodal signaling enhances phosphorylation of Smad1/5, endogenous Nodal activity likely works to diminish BMP signaling (Galvin *et al.*, 2010). Thus, Nodal signaling may either regulate high levels of intracellular programs of BMP signaling or decrease ES cell pluripotency. Signal transduction stimulated by Nodal is inhibited by the Lefty1 and Lefty2, two of the most highly expressed inhibitors in mouse and human ES cells. Dotted lines and "?" between pathways denote an undefined signaling mechanism. (For color version of this figure, the reader is referred to the Web version of this chapter.)

undifferentiated state and sustains the expression of pluripotency genes (Beattie *et al.*, 2005; James *et al.*, 2005). Recombinant Activin or Nodal stimulation of human ES cells induces higher protein expression of Oct4 and Nanog, while inhibition of Smad2 signaling via Lefty or Follistatin decreases protein expression of these pluripotency markers (Xiao *et al.*, 2006). Analyses of human ES cells grown on Matrigel demonstrate that Activin is necessary and sufficient for the maintenance of self-renewal and pluripotency under long-term feeder-free and serum-free growth conditions (Xiao *et al.*, 2006). Correspondingly, inhibition of Activin/Nodal signaling by Follistatin, Lefty, or Smad2 pharmacological inhibition induces human ES cell differentiation (Vallier *et al.*, 2005). A more detailed analysis of multiple pluripotency markers on human ES cells lines by fluorescent-activated cell sorting demonstrated that FGF signaling cooperates with Smad2 activation to keep cells pluripotent over prolonged periods of time in the absence of feeders, Matrigel, or serum (Vallier *et al.*, 2005). Thus, while Smad2 signaling is not required to maintain self-renewal of mouse ES cells, it is essential in human ES cells. These differences may be attributed to either species-specific differences or the embryonic cell type which human ES cells may resemble (Ginis *et al.*, 2004; Rao, 2004); in particular, human ES cells resemble mouse epiblast stem cells (see below).

In human ES cells, suppression of BMP signaling with stimulation of basic FGF signaling maintains undifferentiated proliferation of human ES cells, while activation of BMP signaling induces differentiation (Chaturvedi *et al.*, 2009; Xu *et al.*, 2005). In undifferentiated human ES cells, the Activin/Nodal–Smad2 is active, whereas during differentiation, Smad2 signaling is decreased and Smad1/5 signaling is activated (James *et al.*, 2005). Long-term stimulation with BMP4 induces human ES cells to express trophoblast markers and secrete placental hormones (Xu *et al.*, 2002), and short-term BMP4 treatment initiates mesoderm induction (Zhang *et al.*, 2008).

Recent work has focused on the molecular mechanisms underlying the intracellular networks in human ES cells that retain pluripotency. A transcriptional interaction exists between the self-renewal pathways; Activin controls expression of *Wnt3*, *bFGF*, and *FGF8* (Xiao *et al.*, 2006), and FGF signaling can correspondingly direct expression of TGF-beta family members (Greber *et al.*, 2007). Analyses of vital pluripotency transcription factors revealed that *Nanog* expression is induced by TGF-beta signaling (Xu *et al.*, 2008). Upon TGF-beta stimulation, Smad2/3 binds to the Nanog promoter in undifferentiated human ES cells. Putative Smad-binding sites were identified, which when mutated reduce the response of Nanog to modulation of TGF-beta signaling. Nanog also participates with Smad2 in a negative-feedback loop to inhibit differentiation (Vallier *et al.*, 2009). As Nanog overexpression induces Oct4 expression in the absence of FGF and TGF-beta signaling (Xu *et al.*, 2008), further understanding of the regulation of Nanog may divulge other signaling pathways involved in human ES cell pluripotency.

B. Epiblast stem cells

Whereas ES cells are derived from the inner cell mass of a preimplantation blastocyst, other stem cells of embryonic origin can also be established from tissues derived from the postimplantation embryo. The epiblast, which is the postimplantation layer of cells derived from the inner cell mass, gives rise to the embryo proper. Mouse epiblast stem cells can be established from late epiblast layers of an embryonic day 5.75 pregastrulation stage embryo (Brons *et al.*, 2007; Tesar *et al.*, 2007). Epiblast stem cells can be grown for over 40 passages and express Oct4, Nanog, and SSEA-1 pluripotency markers. In contrast to mouse ES cells which rely on LIF and BMP signaling to maintain pluripotency, mouse epiblast stem cells depend on Activin/Nodal and FGF signaling, conditions similar to human ES cell culture conditions. In epiblast stem cells, pharmacological inhibition of Activin/Nodal signaling hinders pluripotency and promotes differentiation (Tesar *et al.*, 2007). Similar culture conditions have been noted for epiblast stem cells derived from pig tissue (Alberio *et al.*, 2010).

While ES cell colonies have a round morphology, epiblast stem cells grow with a flat, compact morphology (Brons *et al.*, 2007; Tesar *et al.*, 2007). Moreover, ES cells and epiblast stem cells maintain gene expression patterns more reminiscent of the preimplantation inner cell mass and postimplantation epiblast, respectively. After *in vitro* differentiation, epiblast stem cells are capable of giving rise to tissues derived from each of the three germ layers, and injections of epiblast stem cells to immunodeficient mice formed teratomas containing a wide variety of tissues. However, only sparse chimerism is noted when testing the integration of epiblast stem cells into preimplantation-stage mouse embryos and germline transmission is not demonstrated. Yet, treatment of epiblast stem cells with BMP4 induces expression of primordial germ-cell makers *in vitro* (Tesar *et al.*, 2007).

As epiblast stem cells share culture conditions that also support human ES cell self-renewal, it has been suggested that human ES cells may be closer to cells of a postimplantation/pregastrulation stage of mammalian development. Similar to human ES cells, Smad2/3 signaling promotes self-renewal via *Nanog* expression in mouse epiblast stem cells (Greber *et al.*, 2010). The putative Smad-binding sites are not well conserved on the *Nanog* promoter between human and mouse. However, one functional binding site is present on the mouse *Nanog* promoter several basepairs upstream of the known Sox–Oct bipartite motif, and this site binds phospho-Smad2/3 antibodies and actively responds to changes in Smad signaling. Mouse epiblast stem cells represent a later developmental stage of ES cells and demonstrate that the signaling systems which regulate stem cell self-renewal can vary between pluripotent populations with distinct temporal origins.

C. Induced pluripotent stem cells

Nuclear transplantation reprograms differentiated cells to ES cell-like states. In recent years, pioneering work by Yamanaka and colleagues demonstrated that combined expression of different transcription factors is sufficient to reprogram differentiated fibroblasts to pluripotent stem cells (Takahashi and Yamanaka, 2006). These cells, termed induced pluripotent stem (iPS) cells, express ES cell markers, exhibit ES cell morphology, differentiate into tissues of all three germ layers, and generate germline-competent chimeras (Takahashi and Yamanaka, 2006; Wernig *et al.*, 2007). Among others, the factors used to transform fibroblasts to iPS cells include Oct4, Sox2, c-Myc, and Klf4. iPS cells have also been generated from human fibroblasts (Park *et al.*, 2008) and thus raise the possibility that patient-specific stem cells could be made from adult human cells and be of great therapeutic use (Rossant, 2007).

Current work is moving away from virally introduced reprogramming factors (which potentially hold a degree of safety and efficiency obstacles) and moving toward reprogramming methods utilizing small molecules (which enhances the therapeutic potential of iPS cells; Seifinejad *et al.*, 2010). These studies are also defining the process of reprogramming. A number of studies have found that inhibition of Smad2 signaling increases the efficiency of iPS cell generation (Ichida *et al.*, 2009; Li *et al.*, 2009; Lin *et al.*, 2009; Maherali and Hochedlinger, 2009; Yu *et al.*, 2011). In human and rat iPS cells, inhibition of the MEK, GSK3-beta, and Activin/Nodal signaling chemically selects for undifferentiated iPS cells (Li *et al.*, 2009). Inhibition of Smad2 activity has been specifically noted to inhibit spontaneous differentiation and increase the clonal expansion efficiency in rat iPS cells. In mouse iPS cells, Smad2 inhibition strikingly increases the number of reprogrammed iPS cell colonies while treatment with TGF-beta eliminates colony formation (Maherali and Hochedlinger, 2009).

While inhibition of Activin/Nodal signaling enhances iPS cell generation, it can also act as a replacement factor for some of the reprogramming factors. Inhibition of Smad2 signaling bypasses the requirement for exogenous c-Myc or Sox2, but not both, in the generation of iPS cells from mouse fibroblasts (Maherali and Hochedlinger, 2009). As the Smad2 inhibitor did not induce markedly higher levels of cMyc or Sox2, the authors suggest that the inhibition may act on a pathway that circumvents the need for the individual reprogramming factors. Work by Ichida and colleagues demonstrated that inhibition of TGF-beta signaling replaces the requirement for Sox2 by inducing the expression of *Nanog* (Ichida *et al.*, 2009); after 48 h of TGF-beta inhibition, *Nanog* expression is increased over 10-fold. Correspondingly, knockdown of *Nanog* reduces the reprogramming frequency of iPS cells by 50-fold. In part, inhibition of TGF-beta signaling enhances iPS cell generation through induction of the pluripotency transcription factor *Nanog*, which likely serves to repress differentiation signals. The molecular mechanisms by which Smad2 signaling modulates the reprogramming of somatic cells to pluripotent stem cells remain to be elucidated.

VII. Conclusions and Future Directions

The TGF-beta superfamily elicits responses from a wide-ranging variety of cells, from kidney cells to neurons to cancer cells to stem cells. Understanding the intrinsic factors that create a unique pluripotent environment in ES cells is essential to discerning how BMP and Nodal signals influence that pluripotent state. While BMP induces expression of the *Id* factors and acts to inhibit differentiating-inductive signals (Ying *et al.*, 2003, 2008), its effect on ES cell pluripotency is also dependent on LIF–Stat3 stimulation. Thus, BMP functions in concert with other signaling pathways to induce factors necessary for self-renewal and to inhibit factors that would lead the cell to differentiate. Inhibition of Nodal signaling enhances BMP signaling (Galvin *et al.*, 2010), suggesting a role for autocrine Nodal signaling in modulating pluripotency. As the pluripotent role of Nodal varies dramatically between mouse ES cells and mouse epiblast stem cells, this disparate finding proposes that the intracellular atmosphere may change quickly between the inner cell mass and epiblast stages of development to modulate Smad2 signaling functions.

Future studies focusing on understanding the interactions of the multiple pathways simultaneously stimulating ES cells must also be coupled with studies investigating the pluripotent nature of stem cells. While many ES cells are considered to be "undifferentiated," these populations are actually composed of heterogeneous populations of functionally distinct cell types (Canham *et al.*, 2010; Chambers *et al.*, 2007). A more detailed analysis of the complex signaling pathways will be needed to elucidate the important downstream targets and signaling interactions present in mixed cell populations. Comparisons of BMP or Nodal signaling modulations between highly pluripotent populations of ES to less pluripotent ES cells will allow discovery of signaling responses that may be lost in the heterogeneous populations that are currently studied.

ACKNOWLEDGMENT

We wish to acknowledge Stanton Fernald for his work on the illustrations.

REFERENCES

Alberio, R., Croxall, N., and Allegrucci, C. (2010). Pig epiblast stem cells depend on activin/nodal signaling for pluripotency and self-renewal. *Stem Cells Dev.* **19,** 1627–1636.

Beattie, G. M., Lopez, A. D., Bucay, N., Hinton, A., Firpo, M. T., King, C. C., and Hayek, A. (2005). Activin A maintains pluripotency of human embryonic stem cells in the absence of feeder layers. *Stem Cells* **23,** 489–495.

Benezra, R. (2001). Role of Id proteins in embryonic and tumor angiogenesis. *Trends Cardiovasc. Med.* **11,** 237–241.

Beppu, H., Kawabata, M., Hamamoto, T., Chytil, A., Minowa, O., Noda, T., and Miyazono, K. (2000). BMP type II receptor is required for gastrulation and early development of mouse embryos. *Dev. Biol.* **221,** 249–258.

Brons, I. G., Smithers, L. E., Trotter, M. W., Rugg-Gunn, P., Sun, B., de Sousa, Chuva, Lopes, S. M., Howlett, S. K., Clarkson, A., Ahrlund-Richter, L., Pedersen, R. A., and Vallier, L. (2007). Derivation of pluripotent epiblast stem cells from mammalian embryos. *Nature* **448,** 191–195.

Buehr, M., Meek, S., Blair, K., Yang, J., Ure, J., Silva, J., McLay, R., Hall, J., Ying, Q. L., and Smith, A. (2008). Capture of authentic embryonic stem cells from rat blastocysts. *Cell* **135,** 1287–1298.

Canham, M. A., Sharov, A. A., Ko, M. S., and Brickman, J. M. (2010). Functional heterogeneity of embryonic stem cells revealed through translational amplification of an early endodermal transcript. *PLoS Biol.* **8,** e1000379.

Chambers, I., Silva, J., Colby, D., Nichols, J., Nijmeijer, B., Robertson, M., Vrana, J., Jones, K., Grotewold, L., and Smith, A. (2007). Nanog safeguards pluripotency and mediates germline development. *Nature* **450,** 1230–1234.

Chaturvedi, G., Simone, P. D., Ain, R., Soares, M. J., and Wolfe, M. W. (2009). Noggin maintains pluripotency of human embryonic stem cells grown on Matrigel. *Cell Prolif.* **42,** 425–433.

Chen, C., Ware, S. M., Sato, A., Houston-Hawkins, D. E., Habas, R., Matzuk, M. M., Shen, M. M., and Brown, C. W. (2006). The Vg1-related protein Gdf3 acts in a Nodal signaling pathway in the pre-gastrulation mouse embryo. *Development* **133,** 319–329.

Cinelli, P., Casanova, E. A., Uhlig, S., Lochmatter, P., Matsuda, T., Yokota, T., Rulicke, T., Ledermann, B., and Burki, K. (2008). Expression profiling in transgenic FVB/N embryonic stem cells overexpressing STAT3. *BMC Dev. Biol.* **8,** 57.

Constam, D. B. (2009). Riding shotgun: A dual role for the epidermal growth factor-Cripto/FRL-1/Cryptic protein Cripto in Nodal trafficking. *Traffic* **10,** 783–791.

Cui, Y., Jean, F., Thomas, G., and Christian, J. L. (1998). BMP-4 is proteolytically activated by furin and/or PC6 during vertebrate embryonic development. *EMBO* **17,** 4735–4743.

Daheron, L., Opitz, S. L., Zaehres, H., Lensch, M. W., Andrews, P. W., Itskovitz-Eldor, J., and Daley, G. Q. (2004). LIF/STAT3 signaling fails to maintain self-renewal of human embryonic stem cells. *Stem Cells* **22,** 770–778.

Dahle, O., Kumar, A., and Kuehn, M. R. (2010). Nodal signaling recruits the histone demethylase Jmjd3 to counteract polycomb-mediated repression at target genes. *Sci. Signal.* **3,** ra48.

Dunn, N. R., Vincent, S. D., Oxburgh, L., Robertson, E. J., and Bikoff, E. K. (2004). Combinatorial activities of Smad2 and Smad3 regulate mesoderm formation and patterning in the mouse embryo. *Development* **131,** 1717–1728.

Evans, M. J., and Kaufman, M. H. (1981). Establishment in culture of pluripotential cells from mouse embryos. *Nature* **292,** 154–156.

Fei, T., Xia, K., Li, Z., Zhou, B., Zhu, S., Chen, H., Zhang, J., Chen, Z., Xiao, H., Han, J. D., and Chen, Y. G. (2010a). Genome-wide mapping of SMAD target genes reveals the role of BMP signaling in embryonic stem cell fate determination. *Genome Res.* **20,** 36–44.

Fei, T., Zhu, S., Xia, K., Zhang, J., Li, Z., Han, J. D., and Chen, Y. G. (2010b). Smad2 mediates Activin/Nodal signaling in mesendoderm differentiation of mouse embryonic stem cells. *Cell Res.* **20,** 1306–1318.

Finley, M. F., Devata, S., and Huettner, J. E. (1999). BMP-4 inhibits neural differentiation of murine embryonic stem cells. *J. Neurobiol.* **40,** 271–287.

Galvin, K. E., Travis, E. D., Yee, D., Magnuson, T., and Vivian, J. L. (2010). Nodal signaling regulates the bone morphogenic protein pluripotency pathway in mouse embryonic stem cells. *J. Biol. Chem.* **285,** 19747–19756.

Ginis, I., Luo, Y., Miura, T., Thies, S., Brandenberger, R., Gerecht-Nir, S., Amit, M., Hoke, A., Carpenter, M. K., Itskovitz-Eldor, J., and Rao, M. S. (2004). Differences between human and mouse embryonic stem cells. *Dev. Biol.* **269,** 360–380.

Greber, B., Lehrach, H., and Adjaye, J. (2007). Fibroblast growth factor 2 modulates transforming growth factor beta signaling in mouse embryonic fibroblasts and human ESCs (hESCs) to support hESC self-renewal. *Stem Cells* **25,** 455–464.

Greber, B., Wu, G., Bernemann, C., Joo, J. Y., Han, D. W., Ko, K., Tapia, N., Sabour, D., Sterneckert, J., Tesar, P., and Scholer, H. R. (2010). Conserved and divergent roles of FGF signaling in mouse epiblast stem cells and human embryonic stem cells. *Cell Stem Cell* **6,** 215–226.

Guzman-Ayala, M., Lee, K. L., Mavrakis, K. J., Goggolidou, P., Norris, D. P., and Episkopou, V. (2009). Graded Smad2/3 activation is converted directly into levels of target gene expression in embryonic stem cells. *PLoS One* **4,** e4268.

Hao, J., Daleo, M. A., Murphy, C. K., Yu, P. B., Ho, J. N., Hu, J., Peterson, R. T., Hatzopoulos, A. K., and Hong, C. C. (2008). Dorsomorphin, a selective small molecule inhibitor of BMP signaling, promotes cardiomyogenesis in embryonic stem cells. *PLoS One* **3,** e2904.

Hayashi, Y., Furue, M. K., Tanaka, S., Hirose, M., Wakisaka, N., Danno, H., Ohnuma, K., Oeda, S., Aihara, Y., Shiota, K., Ogura, A., Ishiura, S., *et al.* (2010). BMP4 induction of trophoblast from mouse embryonic stem cells in defined culture conditions on laminin. *In Vitro Cell. Dev. Biol. Anim.* **46**(5), 416–430.

He, S., Pant, D., Schiffmacher, A., Meece, A., and Keefer, C. L. (2008). Lymphoid enhancer factor 1-mediated Wnt signaling promotes the initiation of trophoblast lineage differentiation in mouse embryonic stem cells. *Stem Cells* **26,** 842–849.

Hogan, B. L. (1996). Bone morphogenetic proteins: Multifunctional regulators of vertebrate development. *Genes Dev.* **10,** 1580–1594.

Hollnagel, A., Oehlmann, V., Heymer, J., Ruther, U., and Nordheim, A. (1999). Id genes are direct targets of bone morphogenetic protein induction in embryonic stem cells. *J. Biol. Chem.* **274,** 19838–19845.

Ichida, J. K., Blanchard, J., Lam, K., Son, E. Y., Chung, J. E., Egli, D., Loh, K. M., Carter, A. C., Di Giorgio, F. P., Koszka, K., Huangfu, D., Akutsu, H., *et al.* (2009). A small-molecule inhibitor of tgf-Beta signaling replaces sox2 in reprogramming by inducing nanog. *Cell Stem Cell* **5,** 491–503.

Iemura, S., Yamamoto, T. S., Takagi, C., Uchiyama, H., Natsume, T., Shimasaki, S., Sugino, H., and Ueno, N. (1998). Direct binding of follistatin to a complex of bone-morphogenetic protein and its receptor inhibits ventral and epidermal cell fates in early Xenopus embryo. *Proc. Natl. Acad. Sci. USA* **95,** 9337–9342.

James, D., Levine, A. J., Besser, D., and Hemmati-Brivanlou, A. (2005). TGFbeta/activin/nodal signaling is necessary for the maintenance of pluripotency in human embryonic stem cells. *Development* **132,** 1273–1282.

Johansson, B. M., and Wiles, M. V. (1995). Evidence for involvement of activin A and bone morphogenetic protein 4 in mammalian mesoderm and hematopoietic development. *Mol. Cell. Biol.* **15,** 141–151.

Kawasaki, H., Mizuseki, K., Nishikawa, S., Kaneko, S., Kuwana, Y., Nakanishi, S., Nishikawa, S. I., and Sasai, Y. (2000). Induction of midbrain dopaminergic neurons from ES cells by stromal cell-derived inducing activity. *Neuron* **28,** 31–40.

Kelber, J. A., Shani, G., Booker, E. C., Vale, W. W., and Gray, P. C. (2008). Cripto is a noncompetitive activin antagonist that forms analogous signaling complexes with activin and nodal. *J. Biol. Chem.* **283,** 4490–4500.

Kishigami, S., and Mishina, Y. (2005). BMP signaling and early embryonic patterning. *Cytokine Growth Factor Rev.* **16,** 265–278.

Kitisin, K., Saha, T., Blake, T., Golestaneh, N., Deng, M., Kim, C., Tang, Y., Shetty, K., Mishra, B., and Mishra, L. (2007). Tgf-Beta signaling in development. *Sci. STKE* **2007,** cm1.

Kubo, A., Shinozaki, K., Shannon, J. M., Kouskoff, V., Kennedy, M., Woo, S., Fehling, H. J., and Keller, G. (2004). Development of definitive endoderm from embryonic stem cells in culture. *Development* **131,** 1651–1662.

Lasorella, A., Uo, T., and Iavarone, A. (2001). Id proteins at the cross-road of development and cancer. *Oncogene* **20,** 8326–8333.

Lee, M. Y., Lim, H. W., Lee, S. H., and Han, H. J. (2009). Smad, PI3K/Akt, and Wnt-dependent signaling pathways are involved in BMP-4-induced ESC self-renewal. *Stem Cells* **27,** 1858–1868.

Levine, A. J., and Brivanlou, A. H. (2006a). GDF3 at the crossroads of TGF-beta signaling. *Cell Cycle* **5,** 1069–1073.

Levine, A. J., and Brivanlou, A. H. (2006b). GDF3, a BMP inhibitor, regulates cell fate in stem cells and early embryos. *Development* **133,** 209–216.

Li, W., Wei, W., Zhu, S., Zhu, J., Shi, Y., Lin, T., Hao, E., Hayek, A., Deng, H., and Ding, S. (2009). Generation of rat and human induced pluripotent stem cells by combining genetic reprogramming and chemical inhibitors. *Cell Stem Cell* **4,** 16–19.

Lin, T., Ambasudhan, R., Yuan, X., Li, W., Hilcove, S., Abujarour, R., Lin, X., Hahm, H. S., Hao, E., Hayek, A., and Ding, S. (2009). A chemical platform for improved induction of human iPSCs. *Nat. Methods* **6,** 805–808.

Liu, N., Lu, M., Tian, X., and Han, Z. (2007). Molecular mechanisms involved in self-renewal and pluripotency of embryonic stem cells. *J. Cell. Physiol.* **211,** 279–286.

Lopez-Rovira, T., Chalaux, E., Massague, J., Rosa, J. L., and Ventura, F. (2002). Direct binding of Smad1 and Smad4 to two distinct motifs mediates bone morphogenetic protein-specific transcriptional activation of Id1 gene. *J. Biol. Chem.* **277,** 3176–3185.

Maherali, N., and Hochedlinger, K. (2009). Tgfbeta signal inhibition cooperates in the induction of iPSCs and replaces Sox2 and cMyc. *Curr. Biol.* **19,** 1718–1723.

Martin, G. R. (1981). Isolation of a pluripotent cell line from early mouse embryos cultured in medium conditioned by teratocarcinoma stem cells. *Proc. Natl. Acad. Sci. USA* **78,** 7634–7638.

Meno, C., Ito, Y., Saijoh, Y., Matsuda, Y., Tashiro, K., Kuhara, S., and Hamada, H. (1997). Two closely-related left-right asymmetrically expressed genes, lefty-1 and lefty-2: Their distinct expression domains, chromosomal linkage and direct neuralizing activity in Xenopus embryos. *Genes Cells* **2,** 513–524.

Meno, C., Shimono, A., Saijoh, Y., Yashiro, K., Mochida, K., Ohishi, S., Noji, S., Kondoh, H., and Hamada, H. (1998). Lefty-1 is required for left-right determination as a regulator of lefty-2 and nodal. *Cell* **94,** 287–297.

Mishina, Y., Suzuki, A., Ueno, N., and Behringer, R. R. (1995). Bmpr encodes a type I bone morphogenetic protein receptor that is essential for gastrulation during mouse embryogenesis. *Genes Dev.* **9,** 3027–3037.

Miyazono, K., Kamiya, Y., and Morikawa, M. (2010). Bone morphogenetic protein receptors and signal transduction. *J. Biochem.* **147,** 35–51.

Moustakas, A., and Heldin, C. H. (2009). The regulation of TGFbeta signal transduction. *Development* **136,** 3699–3714.

Nakatake, Y., Fukui, N., Iwamatsu, Y., Masui, S., Takahashi, K., Yagi, R., Yagi, K., Miyazaki, J., Matoba, R., Ko, M. S., and Niwa, H. (2006). Klf4 cooperates with Oct3/4 and Sox2 to activate the Lefty1 core promoter in embryonic stem cells. *Mol. Cell. Biol.* **26,** 7772–7782.

Niwa, H., Burdon, T., Chambers, I., and Smith, A. (1998). Self-renewal of pluripotent embryonic stem cells is mediated via activation of STAT3. *Genes Dev.* **12,** 2048–2060.

Ogawa, K., Saito, A., Matsui, H., Suzuki, H., Ohtsuka, S., Shimosato, D., Morishita, Y., Watabe, T., Niwa, H., and Miyazono, K. (2007). Activin-Nodal signaling is involved in propagation of mouse embryonic stem cells. *J. Cell Sci.* **120,** 55–65.

Orkin, S. H., Wang, J., Kim, J., Chu, J., Rao, S., Theunissen, T. W., Shen, X., and Levasseur, D. N. (2008). The transcriptional network controlling pluripotency in ES cells. *Cold Spring Harb. Symp. Quant. Biol.* **73,** 195–202.

O'Toole, P. J., Inoue, T., Emerson, L., Morrison, I. E., Mackie, A. R., Cherry, R. J., and Norton, J. D. (2003). Id proteins negatively regulate basic helix-loop-helix transcription factor function by disrupting subnuclear compartmentalization. *J. Biol. Chem.* **278,** 45770–45776.

Park, S. H. (2005). Fine tuning and cross-talking of TGF-beta signal by inhibitory Smads. *J. Biochem. Mol. Biol.* **38,** 9–16.

Park, I. H., Zhao, R., West, J. A., Yabuuchi, A., Huo, H., Ince, T. A., Lerou, P. H., Lensch, M. W., and Daley, G. Q. (2008). Reprogramming of human somatic cells to pluripotency with defined factors. *Nature* **451,** 141–146.

Pierre, A., Pisselet, C., Monget, P., Monniaux, D., and Fabre, S. (2005). Testing the antagonistic effect of follistatin on BMP family members in ovine granulosa cells. *Reprod. Nutr. Dev.* **45,** 419–425.

Qi, X., Li, T. G., Hao, J., Hu, J., Wang, J., Simmons, H., Miura, S., Mishina, Y., and Zhao, G. Q. (2004). BMP4 supports self-renewal of embryonic stem cells by inhibiting mitogen-activated protein kinase pathways. *Proc. Natl. Acad. Sci. USA* **101,** 6027–6032.

Rao, M. (2004). Conserved and divergent paths that regulate self-renewal in mouse and human embryonic stem cells. *Dev. Biol.* **275,** 269–286.

Rossant, J. (2007). Stem cells: The magic brew. *Nature* **448,** 260–262.

Sato, N., Meijer, L., Skaltsounis, L., Greengard, P., and Brivanlou, A. H. (2004). Maintenance of pluripotency in human and mouse embryonic stem cells through activation of Wnt signaling by a pharmacological GSK-3-specific inhibitor. *Nat. Med.* **10,** 55–63.

Schier, A. F. (2003). Nodal signaling in vertebrate development. *Annu. Rev. Cell Dev. Biol.* **19,** 589–621.

Seifinejad, A., Tabebordbar, M., Baharvand, H., Boyer, L. A., and Hosseini Salekdeh, G. (2010). Progress and promise towards safe induced pluripotent stem cells for therapy. *Stem Cell Rev.* **6**(2), 297–306.

Shi, Y., and Massague, J. (2003). Mechanisms of TGF-beta signaling from cell membrane to the nucleus. *Cell* **113,** 685–700.

Sieber, C., Kopf, J., Hiepen, C., and Knaus, P. (2009). Recent advances in BMP receptor signaling. *Cytokine Growth Factor Rev.* **20,** 343–355.

Sikder, H. A., Devlin, M. K., Dunlap, S., Ryu, B., and Alani, R. M. (2003). Id proteins in cell growth and tumorigenesis. *Cancer Cell* **3,** 525–530.

Sirard, C., de la Pompa, J. L., Elia, A., Itie, A., Mirtsos, C., Cheung, A., Hahn, S., Wakeham, A., Schwartz, L., Kern, S. E., Rossant, J., and Mak, T. W. (1998). The tumor suppressor gene Smad4/Dpc4 is required for gastrulation and later for anterior development of the mouse embryo. *Genes Dev.* **12,** 107–119.

Suzuki, A., Raya, A., Kawakami, Y., Morita, M., Matsui, T., Nakashima, K., Gage, F. H., Rodriguez-Esteban, C., and Izpisua Belmonte, J. C. (2006). Nanog binds to Smad1 and blocks bone morphogenetic protein-induced differentiation of embryonic stem cells. *Proc. Natl. Acad. Sci. USA* **103,** 10294–10299.

Tabibzadeh, S., and Hemmati-Brivanlou, A. (2006). Lefty at the crossroads of "stemness" and differentiative events. *Stem Cells* **24,** 1998–2006.

Taha, M. F., Valojerdi, M. R., and Mowla, S. J. (2007). Effect of bone morphogenetic protein-4 (BMP-4) on cardiomyocyte differentiation from mouse embryonic stem cell. *Int. J. Cardiol.* **120,** 92–101.

Takahashi, K., and Yamanaka, S. (2006). Induction of pluripotent stem cells from mouse embryonic and adult fibroblast cultures by defined factors. *Cell* **126,** 663–676.

Takaoka, K., Yamamoto, M., Shiratori, H., Meno, C., Rossant, J., Saijoh, Y., and Hamada, H. (2006). The mouse embryo autonomously acquires anterior-posterior polarity at implantation. *Dev. Cell* **10,** 451–459.

Takenaga, M., Fukumoto, M., and Hori, Y. (2007). Regulated Nodal signaling promotes differentiation of the definitive endoderm and mesoderm from ES cells. *J. Cell Sci.* **120,** 2078–2090.

ten Dijke, P., and Hill, C. S. (2004). New insights into TGF-beta-Smad signalling. *Trends Biochem. Sci.* **29,** 265–273.

Tesar, P. J., Chenoweth, J. G., Brook, F. A., Davies, T. J., Evans, E. P., Mack, D. L., Gardner, R. L., and McKay, R. D. (2007). New cell lines from mouse epiblast share defining features with human embryonic stem cells. *Nature* **448,** 196–199.

Tiedemann, H., Asashima, M., Grunz, H., and Knochel, W. (2001). Pluripotent cells (stem cells) and their determination and differentiation in early vertebrate embryogenesis. *Dev. Growth Differ.* **43,** 469–502.

Tremblay, K. D., Hoodless, P. A., Bikoff, E. K., and Robertson, E. J. (2000). Formation of the definitive endoderm in mouse is a Smad2-dependent process. *Development* **127,** 3079–3090.

Tropepe, V., Hitoshi, S., Sirard, C., Mak, T. W., Rossant, J., and van der Kooy, D. (2001). Direct neural fate specification from embryonic stem cells: A primitive mammalian neural stem cell stage acquired through a default mechanism. *Neuron* **30,** 65–78.

Ulloa, L., and Tabibzadeh, S. (2001). Lefty inhibits receptor-regulated Smad phosphorylation induced by the activated transforming growth factor-beta receptor. *J. Biol. Chem.* **276,** 21397–21404.

Urist, M. R. (1965). Bone: Formation by autoinduction. *Science* **150,** 893–899.

Vallier, L., Alexander, M., and Pedersen, R. A. (2005). Activin/Nodal and FGF pathways cooperate to maintain pluripotency of human embryonic stem cells. *J. Cell Sci.* **118,** 4495–4509.

Vallier, L., Mendjan, S., Brown, S., Chng, Z., Teo, A., Smithers, L. E., Trotter, M. W., Cho, C. H., Martinez, A., Rugg-Gunn, P., Brons, G., and Pedersen, R. A. (2009). Activin/Nodal signalling maintains pluripotency by controlling Nanog expression. *Development* **136,** 1339–1349.

Watabe, T., and Miyazono, K. (2009). Roles of TGF-beta family signaling in stem cell renewal and differentiation. *Cell Res.* **19,** 103–115.

Wernig, M., Meissner, A., Foreman, R., Brambrink, T., Ku, M., Hochedlinger, K., Bernstein, B. E., and Jaenisch, R. (2007). In vitro reprogramming of fibroblasts into a pluripotent ES-cell-like state. *Nature* **448,** 318–324.

Williams, R. L., Hilton, D. J., Pease, S., Willson, T. A., Stewart, C. L., Gearing, D. P., Wagner, E. F., Metcalf, D., Nicola, N. A., and Gough, N. M. (1988). Myeloid leukaemia inhibitory factor maintains the developmental potential of embryonic stem cells. *Nature* **336,** 684–687.

Winnier, G., Blessing, M., Labosky, P. A., and Hogan, B. L. (1995). Bone morphogenetic protein-4 is required for mesoderm formation and patterning in the mouse. *Genes Dev.* **9,** 2105–2116.

Wrana, J. L. (2000). Regulation of Smad activity. *Cell* **100,** 189–192.

Xiao, L., Yuan, X., and Sharkis, S. J. (2006). Activin A maintains self-renewal and regulates fibroblast growth factor, Wnt, and bone morphogenic protein pathways in human embryonic stem cells. *Stem Cells* **24,** 1476–1486.

Xu, R. H., Chen, X., Li, D. S., Li, R., Addicks, G. C., Glennon, C., Zwaka, T. P., and Thomson, J. A. (2002). BMP4 initiates human embryonic stem cell differentiation to trophoblast. *Nat. Biotechnol.* **20,** 1261–1264.

Xu, R. H., Peck, R. M., Li, D. S., Feng, X., Ludwig, T., and Thomson, J. A. (2005). Basic FGF and suppression of BMP signaling sustain undifferentiated proliferation of human ES cells. *Nat. Methods* **2,** 185–190.

Xu, R. H., Sampsell-Barron, T. L., Gu, F., Root, S., Peck, R. M., Pan, G., Yu, J., Antosiewicz-Bourget, J., Tian, S., Stewart, R., and Thomson, J. A. (2008). NANOG is a direct target of TGFbeta/activin-mediated SMAD signaling in human ESCs. *Cell Stem Cell* **3,** 196–206.

Yeo, C., and Whitman, M. (2001). Nodal signals to Smads through Cripto-dependent and Cripto-independent mechanisms. *Mol. Cell* **7,** 949–957.

Ying, Q. L., Nichols, J., Chambers, I., and Smith, A. (2003). BMP induction of Id proteins suppresses differentiation and sustains embryonic stem cell self-renewal in collaboration with STAT3. *Cell* **115,** 281–292.

Ying, Q. L., Wray, J., Nichols, J., Batlle-Morera, L., Doble, B., Woodgett, J., Cohen, P., and Smith, A. (2008). The ground state of embryonic stem cell self-renewal. *Nature* **453,** 519–523.

Yu, J., Chau, K. F., Vodyanik, M. A., Jiang, J., and Jiang, Y. (2011). Efficient feeder-free episomal reprogramming with small molecules. *PLoS One* **6,** e17557.

Zhang, P., Li, J., Tan, Z., Wang, C., Liu, T., Chen, L., Yong, J., Jiang, W., Sun, X., Du, L., Ding, M., and Deng, H. (2008). Short-term BMP-4 treatment initiates mesoderm induction in human embryonic stem cells. *Blood* **111,** 1933–1941.

Zhao, G. Q. (2003). Consequences of knocking out BMP signaling in the mouse. *Genesis* **35,** 43–56.

CHAPTER SEVENTEEN

The Biology of HIFα Proteins in Cell Differentiation and Disease

Inmaculada Royo González,* Victoria Moreno-Manzano,† Francisco Javier Rodríguez-Jimenez,† Pilar Sepúlveda,†,‡ *and* José-María Sánchez-Puelles*,§

Contents

I. Introduction 368
II. HIF and Cancer 369
  A. Genetic instability 369
  B. HIFs in cancer proliferation and metastasis 369
III. HIF and Stem 370
  A. HIFs and cancer stem cells 371
IV. HIF and Neurodegenerative Diseases 372
V. HIFs in Cardiac Ischemic Diseases 373
VI. HIFs as Target 374
VII. Conclusion 375
References 375

Abstract

The biology of the α subunits of the hypoxia-inducible factors (HIFα) has expanded in the past years from their original role in angiogenesis to their nowadays position in the self-renewal and differentiation of stem cells. Hypoxia is a physiological condition in different tissues—including tumors—and, may cause stem cells in the onset of genomic instability, this last associated in the scientific literature with the acquisition of a malignant phenotypes. HIFα proteins have been the subjects of excellent scientific contributions in the past years, providing new paradigms in the biology of these key proteins and their pivotal role in cell homeostasis. Over other therapeutic implications, the relevance of studies focused on the etiology of tumor-initiating cells and the characterization of the mechanisms

* Molecular Pharmacology Unit, Molecular Medicine Department, Centro de Investigaciones Biológicas, Consejo Superior de Investigaciones Científicas, Madrid, Spain
† MGM Integrated Solutions, Madrid, Spain
‡ Fundación Valenciana Centro de Investigación Príncipe Felipe, Valencia, Spain
§ Fundación para la Investigación Hospital La Fe, Valencia, Spain

Vitamins and Hormones, Volume 87
ISSN 0083-6729, DOI: 10.1016/B978-0-12-386015-6.00036-6

that could lead to their malignancy, is gaining significance in the health areas of cancer and regenerative medicine.

I. Introduction

Hypoxia-inducible transcription factors (HIFs) have been the subject of numerous research studies, as they are the key regulators of cell reaction to the lack of cell oxygen and are widely referred to in the context of pathological processes of cancer, inflammation, cardiovascular, and neurodegenerative diseases and, in general, all the angiogenic pathologies (reviewed in Bertout *et al.*, 2008; Lisy and Peet, 2008; Majmundar *et al.*, 2010; Pouysségur *et al.*, 2006; Rankin and Giaccia, 2008; Semenza, 2003, 2010). Oxygen depletion triggers cell changes in gene expression driven by HIF proteins and their signaling pathway. HIF is a heterodimer consisting of an oxygen-regulated α subunit (1α and 2α) and a constitutively expressed β subunit, also known as aryl hydrocarbon receptor nuclear translocator (ARNT). HIF proteins are members of the bHLH-PAS family and bind to canonical DNA sequences (hypoxia-regulated elements or HREs) in the promoters or enhancers of target genes. Despite the existing similarities, α subunits trigger overlapping and specific genes and are therefore involved in different molecular pathways with different physiological consequences for the cells. Briefly, HIF1α but not HIF2α induces genes involved in the glycolysis process, whereas HIF2α regulates the angiogenic route, even in the absence of hypoxia. A third α subunit has been identified, HIF3α, that lacks the transactivation domain and may function as an inhibitor of HIF1α and HIF2α. There is a great lack of knowledge regarding HIF3α since most publications in the past decades provides insights for the other two α isoforms. Further proofs of the different biological contributions of the HIFα proteins were obtained from studies of mutant phenotypes in mice embryos (Bertout *et al.*, 2008). Regulation of HIF activity is mainly exerted at protein level, due to the hydroxylation of the key proline residues present in the oxygen-dependent degradation domain (ODD) of the α subunits. Under aerobic conditions, hydroxylation of the proline residues by the prolyl-hydroxylases (PHDs, also known as EGLNs) (del Peso *et al.*, 2003) in the ODD permits the polyubiquitination and rapid degradation of the HIFα subunits through an E3 ubiquitin ligase complex containing the von Hippel–Lindau protein, complex which targets this protein to the 26S proteasome. Depletion of oxygen prevents destruction, leading to stabilized a proteins that interact with ARNT in the nucleus, recognize the HRE sequences in the DNA and activate the transcription mediated by the p300/CBP complexes leading to overlapping and specific target genes (Hu *et al.*, 2003; Raval *et al.*, 2005).

II. HIF AND CANCER

A. Genetic instability

Hypoxia is responsible for diminished DNA repair and, therefore, high mutagenesis (Yuan and Glazer, 1998; Yuan *et al.*, 2000) as well as for enrichment of mismatch repair (MMR)-deficient cells with augmented microsatellite instability (MSI) (Kondo *et al.*, 2001). The role of MMR deficiency in the pathogenesis of brain tumors remains controversial (Eckert *et al.*, 2007) but is considered a hallmark of nonpolyposis colorectal cancer (Fishel *et al.*, 1993) and is also present in a significant proportion of other cancer types (Aaltonen *et al.*, 1993; Geary *et al.*, 2008). In addition to this, there is a wide body of research concerning the epigenetic modifications that occur in tumor cells. However, little is known about the effects that altered epigenetic events may produce on the expression of DNA repair genes and the genomic integrity of stem cells under hypoxic conditions. We recently reported for the first time in the stem field, the repression of the MMR system in murine and human, adult, and stem cells under hypoxia (Rodríguez-Jiménez *et al.*, 2008). We have shown that HIF1α is not directly involved in the repression of MMR in as it is described for cancer cells. Our results showed that MMR transcriptional downregulation in hypoxia is associated with hypoacetylated and hypermethylated histone H3 that impair SP1 binding in their promoter regions. The deregulated MMR system, caused by hypoxia, may contribute to generating genomic instability in stem cell niches, which may lead to malignant transformation into cancer stem cells. *In vitro* hypoxic preconditioning of embryonic stem cells has been put forward as a strategy for promoting cell survival and functional benefits after transplantation into the ischemic rat brain or for increasing efficacy of human endothelial progenitor cells for therapeutic neovascularization (Prado-Lopez *et al.*, 2010). Therefore, monitoring the status of DNA repair machinery in hypoxic conditions may contribute to better present and future use of cell therapies as promising nanomedicines.

B. HIFs in cancer proliferation and metastasis

HIFα proteins are associated with poor prognosis in a variety of cancer diseases (Majmundar *et al.*, 2010; Semenza, 2010) from which solid tumors constitute the vast majority of cancer burden. There is distinctive behavior for α isoforms in prognosis, whereas 1α is associated with lower mortality in certain cancers (e.g., Non-Small-Cell Lung Carcinoma (NSCLC) and neuroblastoma, among others) HIF2α expression in these diseases is concomitant with negative prognosis (Bertout *et al.*, 2008; Majmundar *et al.*, 2010). Although there are recent results that may indicate this greater oncogenic properties of HIF2α

(Franovic *et al.*, 2009), it is not clear whether the lack of this isoform is associated with malignancy and promote tumor growth or, a more complex scenario for HIF2α that deserve further investigation, especially considering the controversy of data published in lung tumorogenesis (Franovic, 2009; Kim *et al.*, 2009a; Mazumdar *et al.*, 2010).

Results published so far show that HIF1α is much related with cancer metastasis. Expression of HIF1α implies multiple changes including, but not restricted to, the loss of E-Cadherin, the induction of extracellular matrix degradation enzymes like the lysil oxidase (LOX), the regulation of *TWIST1* transcription, and also favoring the nuclear localization of SNAIL1 in a vascular endothelial growth factor (VEGF) dependent manner in prostate cancer (see Lu and Kang, 2010; Mazumdar *et al.*, 2010; Rankin and Giaccia 2008; Semenza, 2003). Hence, the role of HIFα proteins in cancer progression leading to poor prognosis in many cancer types seems to be largely evidenced (Semenza, 2010).

III. HIF AND STEM

Hypoxia is associated with the undifferentiated status of stem cells and the role of HIFα proteins in the maintenance of multipotency is relatively modern. The real mechanisms by which HIF pathway interacts with others to keep stemness are still largely unknown, albeit the abundant publications in the past few years (reviewed in Keith and Simon, 2007; Lisy and Peet, 2008; Simon and Keith, 2008). HIFα proteins affect the self-renewal and the differentiation process of stem cells by specific regulation of relevant genes and the key transcription factors involved in these processes. HIF1α interacts with Notch1 to maintain undifferentiated cell state, whereas HIF2α binds to the marker of the undifferentiated state *Oct*4 promoter, inducing its expression and transcriptional activity (Boyer *et al.*, 2005; Patel and Simon, 2008; Rankin *et al.*, 2008). *Sox*2 exerts the control of pluripotency by direct modulation of *Oct*4 levels in embryonic stem cells of mice (Masui *et al.*, 2007). The experiments of Takahashi and coworkers showed that somatic cells can be reprogrammed into pluripotent stem cells by transduction of four defined transcription factors—c-*Myc*, *Klf*4, *Sox*2, and *Oct*4 (Takahashi and Yasmanaka, 2006). Although further studies concerning the key factors that regulate the differentiation process are required, inactivation of the signaling that maintains pluripotency in hypoxic tissues may promote cell differentiation and could be a therapeutic strategy in cancer and regenerative medicine. Keith and Simon elegantly hypothesized that *Sox*2 and *Klf*4 might also be HIF targets (Keith and Simon, 2007). However, no experimental data had been supported this conjecture. To further explore whether HIFα proteins control *Sox*2 expression, we first searched *in silico* for putative HRE binding sequences within the promoter regions of rat and human *Sox2* genes. ChIP data and the use of shRNAi in experiments leading to HIF2α-knockdown cells strongly suggested that *Sox2* is a direct target of the HIF2α isoform (McCord *et al.*, 2009; Moreno-Manzano *et al.*, 2009a). Complementary information that reinforces the role of HIF2α in the direct control of *Sox2* was provided by the ChIP experiments carried out in the presence of FM19G11, a specific HIFα inhibitor (Moreno-Manzano *et al.*, 2009b). All the above clearly strengthen the hierarchy of HIF2α in the control of two of the key genetic factors that govern pluripotency. Recent contribution of Anokye-Danso and coworkers illustrates one step forward in the cell reprogramming area. By using miRNA that shows to be a direct target of *Oct4* and *Sox2*, the authors lead Induced-pluripotent-stem cells (IPS) state in which *Oct4* gene expression is activated and cooperates with Hdac-mediated pathways without the requirement of exogenous transcription factors (Anokye-Danso *et al.*, 2011).

A. HIFs and cancer stem cells

It does exist a debate whether the cancer stem cells are responsible of the functional heterogeneity of the tumor and constitute the "initiating cells," being much resistant to chemio and radio therapies, thus having a phenomenal clinical relevance (Clarke and Fuller, 2006; Polyak and Hahn, 2006; reviewed in Visvader and Lindeman, 2008). Although it may be that not all tumors might respond to a single theory, since there are clear evidences that pointed out the clonal origin for certain tumor diseases such as melanomas (Quintana *et al.*, 2008; Visvader and Lindeman, 2008). Although the wide number of publications and crescent interest and efforts ongoing in this research, at present there is not a single marker profile that lead to the undoubtly identification of a cancer stem cell population for colon and intestinal cancers, nor for breast and other cancer diseases. It is worthy for the topic of the present report to mention that it does exist scientific literature that associate the acquisition of malignant phenotype by certain stem cells markers and hypoxia, but still limited, although it has been reported examples from nononcogenic patients (Ono *et al.*, 2007; Rubio *et al.*, 2008). Methodological procedures are crucial to determine the percentages of cells from the tumor side populations—cells endowed with stem-like properties—that are required to originate oncogenic diseases in mice xenograft models (Quintana *et al.*, 2008). A target gene of HIF2α, the already mentioned *Oct4*, a key factor associated with undifferentiated cells, in cooperation with c-*Myc*, has a role in promoting proliferation of progenitor and/or stem cells, thereby undoubtedly contributing to tumor growth (Majmundar *et al.*, 2010; Patel and Simon, 2008). Also NOTCH interaction with HIF1α implicates bad prognosis in colon and renal cell carcinoma and recent literature seems to indicate the specific function of HIF2α in breast cancer and glioblastoma cells (Keith and Simon, 2007; McCord *et al.*, 2009; Polyak and Hahn, 2006; Simon and Keith, 2008).

IV. HIF and Neurodegenerative Diseases

Neurons and the whole central nervous system function and viability require a constant oxygen and glucose supply. Neurons have developed high and specific mechanism of hypoxia/ischemic adaptive response for a proper neuroprotection and therefore giving an important role to HIFs as master regulators of the survival signaling. Though oxygen and/or HIF-dependent regulation, the receptor mediated phosphorylation cascades provides a prevailing control for the adaptive induced circumstances. Growth factors and cytokines, including the neuronal growth factor (NGF), epidermal growth factor (EGF), platelet derived growth factor (PDGF), tumor necrosis factor (TNF) alpha, or insulin-like growth factor IGF-1 and/or -2, represent the major enhancers for HIF transcriptional activity under normoxic conditions (Bilton and Booker, 2003). Cerebral stroke is the main representative example of ischemic damage in the brain, which implies adaptive responses activation on a hypoxic ambient. Protein accumulation of the main two α isoforms rapidly occur. Interestingly, strong evidence supports the differential role on gene regulation of the two isoforms. Even do concomitant induction of adaptive mechanism such as glycolytic metabolism and angiogenesis, support the basal neuroprotective for the total HIF activity. VEGF induced by HIF1α supports survival of neural precursor cells after stroke (Harms *et al.*, 2010) and HIF2α, with a prominent expression on the carotid body (Tian *et al.*, 1998) keeps redox homeostasis. Ying-Jie *et al.* recently showed a marked influence of HIF2α on the mitochondrial electron transport chain complex I, resulting in decreased expression of major cytosolic and mitochondrial antioxidant enzymes on apnea (Peng *et al.*, 2011).

In neurodegenerative disorders such as Alzehimer's, Parkinson's, Huntington's diseases, or amiotrophic lateral sclerosis (ALS), all associated with toxic accumulations inducing a neuroinflammatory stress response that finally relates to the progressive neuronal degeneration, HIF1α expression is related with a neuroprotective function. A strong contribution on the regulatory neuroprotection effect has been shown for VEGF-mediated expression through HIF1α protein activation (Lambrechts *et al.*, 2003). On Alzehimer's disease the β-amyloid peptide accumulation is related to an increased glucose metabolism, which, in turn, is concomitant with HIFα elevated protein levels and activity, again with a general neuroprotective role against the generated oxidative stress (Soucek *et al.*, 2003). ALS, a progressive motorneuron degeneration disease, also implies a deregulation on the oxygen metabolism with accumulative oxidative damage due to superoxide dismutase mutation (*SOD*1) (Rakhit *et al.*, 2002).

V. HIFs in Cardiac Ischemic Diseases

Myocardial infarction is one of the main causes of mortality and morbidity in developed countries. Trombotic occlusion of an epicardial coronary artery triggers several pathological processes leading to impaired cardiac function. In the irrigation zone affected by the occlusion, there is a massive cardiomyocyte death by apoptotic and necrotic mechanisms (Palojoki *et al.*, 2001). As a response, ischemic tissue elevates levels of HIFα, a set of transcription factors that regulate the expression of nearly 200 genes that can affect the cellular adaptative responses to stress (Loor and Schumacker, 2008). Equally than in the central nervous system, HIFα proteins play a cardio protective role. Inducible nitric oxide synthase (iNOS) and cyclooxygenase-2 (COX-2) are gene targets of HIF, both of which have been associated with enhanced resistance to ischemia (Bolli *et al.*, 2002; Kido *et al.*, 2005). Other genes regulated by HIFα include VEGF, erythropoietin, angiopoietin, placental growth factor, and platelet-derived growth factor (Patel *et al.*, 2009; Pichiule *et al.*, 2004; Wang and Semenza, 1993). Some of these angiogenic factors recruit subsets of proangiogenic hematopoietic cells. These, in turn, may also release new angiogenic factors and contribute to the repair by paracrine effects (Akita *et al.*, 2003)

Therapies based on the expression of HIF1α induce neoangiogenesis in the ischemic tissue. Cell lines transfected with HIF1α showed increased survival and connectivity with adjacent cells. The preconditioning with hypoxia of endothelial precursor cells increased the ability to repair ischemic limbs through the activation of the angiogenic program (Akita *et al.*, 2003). The intracardiac injection of HIF1α naked DNA reduced the size of infarction and increased neovascularization in the area (Shyu *et al.*, 2002). In this context, preconditioning of mesenchymal stem cells (MSCs) with hypoxia induced the prosurvival Akt pathway and increased their motility and therapeutic potential in a model of hind limb ischemia (Rosova *et al.*, 2008). Recently, the role of HIF1a in MSC survival has been described (Kim *et al.*, 2009b). In addition, culture of H9 ESC line in hypoxic conditions promoted endothelial cell differentiation that improved their therapeutic potential in a model of myocardial infarction (Prado-Lopez *et al.*, 2010).

Considering the impact of HIFs in the adaptative response to hypoxia the ability to manipulate HIF activity will have an impact in the treatment of cardiac ischemic diseases.

VI. HIFs as Target

HIFα proteins, a hallmark of different tumor types, were the desired strategy of many drug discovery efforts, but most inhibitors did not comply with the pharmacological properties or toxicity profiles required for approval of the drug by the regulatory agencies. Given the central role that HIF-driven transcription factor activity has in compensating for loss of oxygen, the control of this activity offers a potent tool for treating a wide range of hypoxia-related pathologies and physiological disorders. The Nobel Prize Judah Folkman already predicted the use of angiogenesis therapies in the control of obesity, vascular malformations, Chron disease, and others. Antiangiogenic molecules are already in the market and the discovery efforts have also been in line of using this "organizing principle" defined by Folkman, for developing therapeutic approaches for various diseases (Folkman, 2007). Over the past 20 years, major drug discovery efforts have gone into the search for HIFα inhibitors for use in new drugs. Though a wide range of diverse molecules have been found to inhibit the HIF pathway, these molecules often have other actions that indirectly cause lower level of both HIF1,2α protein levels (Folkman, 2007; Moreno-Manzano *et al.*, 2009a; Rodríguez-Jiménez *et al.*, 2010; Semenza, 2010). A good example is the prolyl hydroxylase inhibitors of the PHD enzymes that constitute the most advanced ones in clinical trials (Majmundar, 2010). None of the reported HIFα inhibitors have met the pharmacokinetic requirements for human therapeutic use. However, as elegantly pointed out very recently but Majmundar *et al.*, HIF-targeted therapies must meet HIFα repression as sufficient for their intended biological effects. These authors have also suggested that the crescent distinctive biological roles that the HIFα subunits are showing in disease, fully justifies the rational for selection of highly specific inhibitors for each isoform (Majmundar, 2010). Although limited efforts has been done in this direction, an elegant academicals approach in the Structural Biology area, showed the structural differences of HIF2α to approach the design of novel specific inhibitors (Key *et al.*, 2009; Scheuermanna *et al.*, 2009).

Considering the strong HIFα interactions with cell proliferation and differentiation pathways specially its downstream targets (Keith and Simon, 2007; Patel and Simon, 2008), there is room to approach novel HIFα inhibitors for intervention in cancer and degenerative diseases. The previous results of our group showed the discovery of FM19G11, a novel, selective, and potent low-molecular-weight inhibitor of HIF1,2α proteins (PCTES2009/000135; Moreno-Manzano *et al.*, 2009a; Rodríguez-Jiménez *et al.*, 2010) now under Preclinical Development under stringent regulatory rules. It represses the target genes of the two α subunits in different tumor cell lines and in adult and embryonic stem cell models, from rodents and humans, respectively. FM19G11 inhibits at nanomolar range the transcriptional and protein expression of *Oct4* and *Sox2*, both in a neural stem cell from rats—a model reported for spinal cord regeneration (Moreno-Manzano *et al.*, 2009b) and in human embryonic stem cells (hESC). Further, *Nanog* and *Tgfα* inhibition were also shown by RT-PCR and protein analysis in the presence of FM19G11. Experiments using shRNAi showed a significant depletion of *Sox2* protein only in the case of HIF2α silencing, but not in HIF1α mediate dablation. Therefore, our results with this novel inhibitor that controls the expression of *Sox2*, *Oct4*, *Nanog*, and *Tgfα* in murine and human stem cell and triggers cell differentiation leading to oligodendrocyte population (Moreno-Manzano *et al.*, 2009a). To date, this drug provides the first evidence to demonstrate that *Sox2* is a direct target of HIF2α but not of HIF1α, and that cell reprogramming is feasible using a single small molecule with a safe toxicity profile. The results with FM19G11 in cell differentiation clearly position HIFα and its downstream gene targets, as attractive pharmacological approaches in the design of modern cancer and regenerative drug-based therapies in human diseases. In addition, the activity data regarding efficacy and toxicity together with the robust protection allows considering this novel family of HIFα inhibitors as promising candidates for future clinical use.

VII. Conclusion

As already anticipated by Judah Folkman, new strategies for drug discovery research demonstrate that HIFα inhibitors may not only be restricted to the area of angiogenic pathologies, but nowadays also entering in other seemingly unrelated diseases. This last includes the design of novel therapeutic interventions in the promising field of regenerative medicine, providing new tools for the treatments of degenerative diseases and/or in the preconditioning of the stem cells used for cell transplantation therapies.

References

Aaltonen, L. A., Peltomäki, P., Leach, F. S., Sistonen, P., Pylkkänen, L., Mecklin, J. P., Järvinen, H., Powell, S. M., Jen, J., Hamilton, S. R., *et al.* (1993). Clues to the pathogenesis of familial colorectal cancer. *Science* **260,** 812–816.

Akita, T., Murohara, T., Ikeda, H., Sasaki, K., Shimada, T., Egami, K., and Imaizumi, T. (2003). Hypoxic preconditioning augments efficacy of human endothelial progenitor cells for therapeutic neovascularization. *Lab. Invest.* **83,** 65–73.

Anokye-Danso, F., Trivedi, C. M., Juhr, D., Gupta, M., Cui, Z., Tian, Y., Zhang, Y., Yang, W., Gruber, P. J., Epstein, J. A., and Morrisey, E. E. (2011). Highly efficient miRNA-mediated reprogramming of mouse and human somatic cells to pluripotency. *Cell Stem Cell* **8,** 376–388.

Bertout, J. A., Patel, S. A., and Simon, M. C. (2008). The impact of O_2 availability on human cancer. *Nat. Rev. Cancer* **8,** 967–975.

Bilton, R. L., and Booker, G. W. (2003). The subtle side to hypoxia inducible factor (HIFalpha) regulation. *Eur. J. Biochem.* **270,** 791–798.

Bolli, R., Shinmura, K., Tang, X. L., Kodani, E., Xuan, Y. T., Guo, Y., and Dawn, B. (2002). Discovery of a new function of cyclooxygenase (COX)-2: COX-2 is a cardioprotective protein that alleviates ischemia/reperfusion injury and mediates the late phase of preconditioning. *Cardiovasc. Res.* **55,** 506–519.

Boyer, L. A., Lee, T. I., Cole, M. F., Johnstone, S. E., Levine, S. S., Zucker, J. P., Guenther, M. G., Kumar, R. M., Murray, H. L., Jenner, R. G., Gifford, D. K., Melton, D. A., *et al.* (2005). Core transcriptional regulatory circuitry in human embryonic stem cells. *Cell* **122,** 947–956.

Clarke, M. F., and Fuller, M. (2006). Stem cells and cancer: Two faces of eve. *Cell* **124,** 1111–1115.

Del Peso, L., Castellanos, M. C., Temes, E., Martin-Puig, S., Cuevas, Y., Olmos, G., and Landazuri, M. O. (2003). The von Hippel Lindau/hypoxia-inducible factor (HIF) pathway regulates the transcription of the HIF-proline hydroxylase genes in response to low oxygen. *J. Biol. Chem.* **278,** 48690–48695.

Eckert, A., Kloor, M., Giersch, A., Ahmadi, R., Herold-Mende, C., Hampl, J. A., Heppner, F. L., Zoubaa, S., Holinski-Feder, E., Pietsch, T., Wiestler, O. D., von Knebel Doeberitz, M., *et al.* (2007). Microsatellite instability in pediatric and adult high-grade gliomas. *Brain Pathol.* **17,** 146–150.

Fishel, R., Lescoe, M. K., Rao, M. R., Copeland, N. G., Jenkins, N. A., Garber, J., Kane, M., and Kolodner, R. (1993). The human mutator gene homolog *MSH2* and its association with hereditary nonpolyposis colon cancer. *Cell* **75,** 1027–1038.

Folkman, J. (2007). Angiogenesis: An organizing principle for drug discovery? *Nat. Rev. Drug Discov.* **6,** 273–286.

Franovic, A., Holterman, C. E., Payette, J., and Lee, S. (2009). Human cancers converge at the HIF-2a oncogenic axis. *Proc. Natl. Acad. Sci. USA* **106,** 21306–21311.

Geary, J., Sasieni, P., Houlston, R., Izatt, L., Eeles, R., Payne, S. J., Fisher, S., and Hodgson, S. V. (2008). Gene-related cancer spectrum in families with hereditary nonpolyposis colorectal cancer (HNPCC). *Fam. Cancer* **7,** 163–172.

Harms, K. M., Li, L., and Cunningham, L. A. (2010). Murine neural stem/progenitor cells protect neurons against ischemia by HIF-1alpha-regulated VEGF signaling. *PLoS One* **5,** e9767.

Hu, C. J., Wang, L. Y., Chodosh, L. A., Keith, B., and Simon, M. C. (2003). Differential roles of hypoxia-inducible factor 1 (HIF-1) and HIF-2 in hypoxic gene regulation. *Mol. Cell. Biol.* **23,** 9361–9374.

Keith, B., and Simon, M. C. (2007). Hypoxia-inducible factors, stem cells, and cancer. *Cell* **129,** 465–472.

Key, J., Scheuermann, T. H., Anderson, P. C., Daggett, V., and Gardner, K. H. (2009). Principles of ligand binding within a completely buried cavity in HIF2r PAS-B. *J. Am. Chem. Soc.* **131,** 17647–17654.

Kido, M., Du, L., Sullivan, C. C., Li, X., Deutsch, R., Jamieson, S. W., and Thistlethwaite, P. A. (2005). Hypoxia-inducible factor 1-alpha reduces infarction and attenuates progression of cardiac dysfunction after myocardial infarction in the mouse. *J. Am. Coll. Cardiol.* **46,** 2116–2124.

Kim, H. W., Haider, H. K., Jiang, S., and Ashraf, M. (2009a). **Ischemic preconditioning augments survival of stem cells via miR-210 expression by targeting caspase-8-associated protein 2. *J. Biol. Chem.* **284,** 33161–33168.

Kim, W. Y., Perera, S., Zhou, B., Carretero, J., Yeh, J. J., Heathcote, S. A., Jackson, A. L., Nikolinakos, P., Ospina, B., Naumov, G., Brandstetter, K. A., Weigman, V. J., *et al.* (2009b). HIF2α cooperates with RAS to promote lung tumorigenesis in mice. *J. Clin. Invest.* **119**(8), 2160–2170.

Kondo, A., Safaei, R., Mishima, M., Niedner, H., Lin, X., and Howell, S. B. (2001). Hypoxia-induced enrichment and mutagenesis of cells that have lost DNA mismatch repair. *Cancer Res.* **61,** 7603–7607.

Lambrechts, D., Storkebaum, E., Morimoto, M., Del-Favero, J., Desmet, F., Marklund, S. L., Wyns, S., Thijs, V., Andersson, J., van Marion, I., Al-Chalabi, A., Bornes, S., *et al.* (2003). VEGF is a modifier of amyotrophic lateral sclerosis in mice and humans and protects motoneurons against ischemic death. *Nat. Genet.* **34,** 383–394.

Lisy, K., and Peet, D. J. (2008). Turn me on: Regulating HIF transcriptional activity. *Cell Death Differ.* **15,** 642–649.

Loor, G., and Schumacker, P. T. (2008). Role of hypoxia-inducible factor in cell survival during myocardial ischemia-reperfusion. *Cell Death Differ.* **15,** 686–690.

Lu, X., and Kang, Y. (2010). Hypoxia and hypoxia-inducible factors: master regulators of metastasis. *Clin. Cancer Res.* **16,** 5928–5935.

Majmundar, A. J., Wong, W. J., and Simon, C. S. (2010). Hypoxia-inducible factors and the response to hypoxic stress. *Mol. Cell* **40,** 294–309.

Masui, S., Nakatake, Y., Toyooka, Y., Shimosato, D., Yagi, R., Takahashi, K., Okochi, H., Okuda, A., Matoba, R., Sharov, A. A., Ko, M. S., and Niwa, H. (2007). Pluripotency governed by Sox2 via regulation of Oct3/4 expression in mouse embryonic stem cells. *Nat. Cell Biol.* **9**(6), 625–635.

Mazumdar, J., Hickey, M. M., Pant, D. K., Durham, A. C., Sweet-Cordero, A., Vachani, A., Jacks, T., Chodosh, L. A., Kissil, J. L., Simon, M. C., and Keith, B. (2010). HIF-2 alpha deletion promotes Kras-driven lung tumor development. *Natl. Acad. Sci. USA* **107,** 14182–14187.

McCord, A. M., Jamal, M., Shankavaram, U. T., Lang, F. F., Camphausen, K., and Tofilon, P. J. (2009). Physiologic oxygen concentration enhances the stem-like properties of CD133+human glioblastoma cells in vitro. *Mol. Cancer Res.* **7**(4), 489–497.

Moreno-Manzano, V., Rodríguez-Jiménez, F. J., Aceña-Bonilla, J. L., Fustero-Lardíes, S., Erceg, S., Dopazo, J., Montaner, D., Stojkovic, M., and Sánchez-Puelles, J. M. (2009a). FM19G11, a new HIF modulator, affect stem cell differentiation status. *J. Biol. Chem.* **285**(2), 1333–1342.

Moreno-Manzano, V., Rodríguez-Jiménez, F. J., García-Roselló, M., Laínez, S., Erceg, S., Calvo, M. T., Ronaghi, M., Lloret, M., Planells-Cases, R., Sánchez-Puelles, J. M., and Stojkovic, M. (2009b). Activated spinal cord ependymal stem cells rescue neurological function. *Stem Cells* **27,** 733–743.

Ono, M., Maruyama, T., Masuda, H., Kajitani, T., Nagashima, T., Arase, T., Ito, M., Ohta, K., Uchida, H., Asada, H., Yoshimura, Y., Okano, H., *et al.* (2007). Side population in human uterine myometrium displays phenotypic and functional characteristics of myometrial stem cells. *PNAS* **104**(47), 18700–18705.

Palojoki, E., Saraste, A., Eriksson, A., Pulkki, K., Kallajoki, M., Voipio-Pulkki, L. M., and Tikkanen, I. (2001). Cardiomyocyte apoptosis and ventricular remodeling after myocardial infarction in rats. *Am. J. Physiol. Heart Circ. Physiol.* **280,** H2726–H2731.

Patel, S. A., and Simon, M. C. (2008). Biology of hypoxia-inducible factor-2alpha in development and disease. *Cell Death Differ.* **15,** 628–634.

Patel, N., Gonsalves, C. S., Yang, M., Malik, P., and Kalra, V. K. (2009). Placenta growth factor induces 5-lipoxygenase-activating protein to increase leukotriene formation in sickle cell disease. *Blood* **113,** 1129–1138.

Peng, Y. J., Nanduri, J., Khan, S. A., Yuan, G., Wang, N., Kinsman, B., Vaddi, D. R., Kumar, G. K., Garcia, J. A., Semenza, G. L., and Prabhakar, N. R. (2011). Hypoxia-inducible factor 2alpha (HIF-2alpha) heterozygous-null mice exhibit exaggerated carotid body sensitivity to hypoxia, breathing instability, and hypertension. *Proc. Natl. Acad. Sci. USA* **108,** 3065–3070.

Pichiule, P., Chavez, J. C., and LaManna, J. C. (2004). Hypoxic regulation of angiopoietin-2 expression in endothelial cells. *J. Biol. Chem.* **279,** 12171–12180.

Polyak, K., and Hahn, W. C. (2006). Roots and stems: Stem cells in cancer. *Nat. Med.* **12,** 296–300.

Pouysségur, J., Dayan, F., and Mazure, N. M. (2006). Hypoxia signalling in cancer and approaches to enforce tumour regression. *Nature* **441,** 437–443.

Prado-Lopez, S., Conesa, A., Arminan, A., Martinez-Losa, M., Escobedo-Lucea, C., Gandia, C., Tarazona, S., Melguizo, D., Blesa, D., Montaner, D., Sanz-Gonzalez, S., Sepulveda, P., *et al.* (2010). Hypoxia promotes efficient differentiation of human embryonic stem cells to functional endothelium. *Stem Cells* **28**(3), 407–418.

Quintana, E., Shackleton, M., Sabel, M. S., Fullen, D. R., Johnson, T. M., and Morrison, S. J. (2008). Efficient tumour formation by single human melanoma cells. *Nature* **456,** 593–599.

Rakhit, R., Cunningham, P., Furtos-Matei, A., Dahan, S., Qi, X. F., Crow, J. P., Cashman, N. R., Kondejewski, L. H., and Chakrabartty, A. (2002). Oxidation-induced misfolding and aggregation of superoxide dismutase and its implications for amyotrophic lateral sclerosis. *J. Biol. Chem.* **277,** 47551–47556.

Rankin, E. B., and Giaccia, A. J. (2008). The role of hypoxia-inducible factors in tumorigenesis. *Cell Death Differ.* **15,** 678–685.

Rankin, E. B., Rha, J., Unger, T. L., Wu, C. H., Shutt, H. P., Johnson, R. S., Simon, M. C., Keith, B., and Haase, V. H. (2008). Hypoxia-inducible factor-2 regulates vascular tumorigenesis in mice. *Oncogene* **27,** 5354–5358.

Raval, R. R., Lau, K. W., Tran, M. G., Sowter, H. M., Mandriota, S. J., Li, J. L., Pugh, C. W., Maxwell, P. H., Harris, A. L., and Ratcliffe, P. J. (2005). Contrasting properties of hypoxia-inducible factor 1 (HIF-1) and HIF-2 in von Hippel-Lindau-associated renal cell carcinoma. *Mol. Cell. Biol.* **25,** 5675–5686.

Rodríguez-Jiménez, F. J., Moreno-Manzano, V., Lucas-Dominguez, R., and Sánchez-Puelles, J. M. (2008). Hypoxia causes downregulation of mismatch repair system and genomic instability in stem cells. *Stem Cells* **26,** 2052–2062.

Rodríguez-Jiménez, F. J., Moreno-Manzano, V., Mateos-Gregorio, P., Royo, I., Erceg, S., Murguia, J. R., and Sánchez-Puelles, J. M. (2010). FM19G11 A new modulator of hif that links mTOR activation with the DNA damage checkpoint pathways. *Cell Cycle* **9** (14), 2803–2813.

Rosova, I., Dao, M., Capoccia, B., Link, D., and Nolta, J. A. (2008). Hypoxic preconditioning results in increased motility and improved therapeutic potential of human mesenchymal stem cells. *Stem Cells* **26,** 2173–2182.

Rubio, D., Garcia, S., Paz, M. F., De la Cueva, T., Lopez-Fernandez, L. A., Lloyd, A. C., Garcia-Castro, J., and Bernad, A. (2008). Molecular characterization of spontaneous mesenchymal stem cell transformation. *PLoS One* **3**(1), e1398.

Scheuermanna, T. H., Tomchicka, D. R., Machiusa, M., Guoa, Y., Bruicka, R. K., and Gardnera, K. H. (2009). Artificial ligand binding within the HIF2_ PAS-B domain of the HIF2 transcription factor. *PNAS* **106**(2), 450–455.

Semenza, G. L. (2003). Targeting HIF-1 for cancer therapy. *Nat. Rev. Cancer* **3,** 721–732.

Semenza, G. L. (2010). Defining the role of hypoxia-inducible factor 1 in cancer biology and therapeutics. *Oncogene* **29,** 625–634.

Shyu, K. G., Wang, M. T., Wang, B. W., Chang, C. C., Leu, J. G., Kuan, P., and Chang, H. (2002). Intramyocardial injection of naked DNA encoding HIF-1alpha/VP16 hybrid to enhance angiogenesis in an acute myocardial infarction model in the rat. *Cardiovasc. Res.* **54,** 576–583.

Simon, M. C., and Keith, B. (2008). The role of oxygen availability in embryonic development and stem cell function. *Nat. Rev. Mol. Cell Biol.* **9,** 285–296.

Soucek, T., Cumming, R., Dargusch, R., Maher, P., and Schubert, D. (2003). The regulation of glucose metabolism by HIF-1 mediates a neuroprotective response to amyloid beta peptide. *Neuron* **39,** 43–56.

Takahashi, K., and Yasmanaka, S. Y. (2006). Induction of pluripotent stem cells from mouse embryonic and adult fibroblast cultures by defined factors. *Cell* **126,** 663–676.

Tian, H., Hammer, R. E., Matsumoto, A. M., Russell, D. W., and McKnight, S. L. (1998). The hypoxia-responsive transcription factor EPAS1 is essential for catecholamine homeostasis and protection against heart failure during embryonic development. *Genes Dev.* **12,** 3320–3324.

Visvader, J. E., and Lindeman, G. (2008). Cancer stem cells in solid tumors: Accumulating evidence and unresolved questions. *Nat. Rev. Cancer* **8,** 755–768.

Wang, G. L., and Semenza, G. L. (1993). Desferrioxamine induces erythropoietin gene expression and hypoxia-inducible factor 1 DNA-binding activity: Implications for models of hypoxia signal transduction. *Blood* **82,** 3610–3615.

Yuan, J., and Glazer, P. M. (1998). Mutagenesis induced by the tumor microenvironment. *Mutat. Res.* **400,** 439–446.

Yuan, J., Narayanan, L., Rockwell, S., and Glazer, P. M. (2000). Diminished DNA repair and elevated mutagenesis in mammalian cells exposed to hypoxia and low pH. *Cancer Res.* **60,** 4372–4376.

CHAPTER EIGHTEEN

Regulatory Role of Klf5 in Early Mouse Development and in Embryonic Stem Cells

Silvia Parisi *and* Tommaso Russo

Contents

I. Klf5: A Krüppel-Like Transcription Factor 382
II. Multiple Functions of Klf5 382
III. Klfs in Reprogramming 383
IV. Gene Expression Network for the Maintenance of ESC Pluripotent State 384
V. Klf5 Function in ESCs 385
VI. Klf5 Role in Early Embryonic Development 387
VII. Klf5 Targets 390
VIII. Klf5 Connection to the Core Pluripotency Network 391
IX. Conclusions 392
Acknowledgments 392
References 393

Abstract

Pluripotency and self-renewal of embryonic stem cells (ESCs) are maintained by regulatory mechanisms, based on a sophisticated network of transcription factors. Recently, a growing body of evidence has indicated that Klf5, a transcription factor highly expressed in mouse ESCs and during the early phases of mouse development, plays a crucial role in maintaining ESC self-renewal and pluripotency, in governing ESC fate decisions and proper development of blastocyst *in vivo*. Indeed, Klf5-null mice show developmental defects at blastocyst stage, due to the defective establishment of the inner cell mass. Moreover, Klf5 knockdown in ESCs results in the loss of undifferentiated phenotype, whereas its ectopic expression is sufficient to maintain ESC in the undifferentiated state, even in the absence of LIF. Finally, it has been recently reported that Klf5 activates the expression of self-renewal-promoting genes and, simultaneously, it inhibits the expression of differentiation-related genes. Here, we discuss the functional

Dipartimento di Biochimica e Biotecnologie Mediche, Università di Napoli "Federico II," CEINGE Biotecnologie Avanzate, European School of Molecular Medicine (SEMM), Naples, Italy

Vitamins and Hormones, Volume 87
ISSN 0083-6729, DOI: 10.1016/B978-0-12-386015-6.00037-8

role of Klf5 in the control on ESC self-renewal and pluripotency and its integration in the core transcriptional network governing ESC state.

I. Klf5: A Krüppel-Like Transcription Factor

Klf5 belongs to the Krüppel-like factor family that in mammals includes 17 members identified so far. Klf5, also known as intestinal-enriched Krüppel-like factor (IKLF), was originally identified through screening of a human placenta cDNA library (Sogawa *et al.*, 1993) as basic transcription element-binding protein 2 (BTEB2). The 457 aa of the mouse Klf5 protein contain three contiguous zinc finger elements in the carboxyl terminal region and undergo posttranslational modifications such as acetylation, phosphorylation, and SUMOylation, that have been reported to modulate Klf5 biological activities (Guo *et al.*, 2009a; Oishi *et al.*, 2008; Zhang and Teng, 2003).

Klf5 is widely expressed during embryogenesis and in the adult. During mouse gastrulation, Klf5 is expressed in the anterior primitive streak (7 dpc) and then its expression becomes more restricted to the endoderm of midgut and hindgut (8.5 dpc) (Moore-Scott *et al.*, 2007). At 10.5 dpc, Klf5 is highly expressed in primitive gut, and at 15.5 dpc, Klf5 transcripts are abundant in developing gut, skin, and meninges. Later, in 16.5 dpc embryos, Klf5 expression is high in both intestine and skin as well as in the meninges, the bronchiolar epithelium of the lung, the epithelial lining of the trachea, and the surface epithelium of the tongue (Ohnishi *et al.*, 2000). In adult mice, Klf5 is mainly expressed in stomach, small intestine, and colon with lower expression in skin, lung, uterus, placenta, and testis (Conkright *et al.*, 1999).

II. Multiple Functions of Klf5

The expression pattern of Klf5 suggests that it is mainly localized in the proliferative compartment rather than in differentiated cells. Indeed, high expression of Klf5 is detected in basal layer of epidermis and at the base of the intestine crypts where epithelial proliferating cells are localized (McConnell *et al.*, 2007; Ohnishi *et al.*, 2000). This expression of Klf5 in proliferating compartment rather than in differentiated cells indicated a possible role as a pro-proliferative factor. Indeed, it has been demonstrated that Klf5 constitutive expression in NIH3T3 resulted in accelerated cell growth and, at the same time, in serum- and anchorage-independent growth of the cells (Sun *et al.*, 2001). Moreover, the pro-proliferative effect of Klf5 has been demonstrated also in nontransformed cell lines from intestine and skin (Bateman *et al.*, 2004; Guo *et al.*, 2009a;

Nandan *et al.*, 2008). Many groups showed that Klf5 promotes cell proliferation by accelerating G1/S and G2/M cell cycle transition (Chen *et al.*, 2006; Nandan *et al.*, 2008). However, a recent study suggests that Klf5 may have opposite roles in the control of proliferation, acting as pro-proliferative or antiproliferative regulator as a consequence of a TGF-β-mediated acetylation (Guo *et al.*, 2009b). Moreover, Klf5 is also able to inhibit proliferation in several cancer cell lines (Bateman *et al.*, 2004; Chen *et al.*, 2003; Yang *et al.*, 2005), indicating that this factor may have a context dependent role, thus acting as tumor suppressor or as oncogene (Dong and Chen, 2009). Besides the role in cell proliferation, Klf5 is involved in many biological processes such as cell cycle, apoptosis, migration, and differentiation. Indeed, Klf5 has an antiapoptotic function by acting both in a p53-dependent and -independent manner (Zhao *et al.*, 2008; Zhu *et al.*, 2006). Finally, Klf5 has also a role in promoting epithelial cell migration, an important process in normal epithelial homeostasis in gut and skin (Yang *et al.*, 2008).

III. Klfs in Reprogramming

Very recently, Klf5 was demonstrated to promote reprogramming. Pluripotent stem cells, such as embryonic stem cells (ESCs), have the ability to self-renew indefinitely in culture and to differentiate into a broad spectrum of derivatives of all three embryonic germ layers (pluripotency). The potential of ESCs to give raise to specialized and functionally active cell types and the possibility to expand them in a virtually unlimited manner have captured the attention of scientists as a possible novel source of cell populations for therapeutic purposes, such as transplantation and tissue regeneration. However, ethical issues hamper the use of ESCs in regenerative medicine. In contrast, the reprogramming of somatic cells toward pluripotent stem cells overcomes ethical issues and represents an attractive and efficient alternative to ESCs. The first report of the induction of pluripotent stem cells from mouse embryonic and adult fibroblasts was based on the expression of four transcription factors: Klf4, Sox2, Oct3/4, and c-Myc (Takahashi and Yamanaka, 2006). Later, in an attempt to identify additional factors able to generate iPS cells, the above mentioned factors were substituted for their respective homologs (Nakagawa *et al.*, 2008). In this study, the ability to give raise to Nanog-GFP-positive colonies was evaluated upon overexpression of Sox2, c-Myc, Oct3/4 and alternatively Klf2, Klf4, or Klf5. This analysis showed that Klf2 is able to fully substitute Klf4, whereas Klf5 can induce reprogramming, but with lower efficiency compared to Klf4 (Nakagawa *et al.*, 2008).

IV. Gene Expression Network for the Maintenance of ESC Pluripotent State

Recently, the key role of Klf5 in the maintenance of ESC stemness has emerged. Self-renewal of ESCs hinges on many extrinsic and intrinsic signals. The main extracellular signal that sustains ESC self-renewal and pluripotency is the activation of leukemia inhibitory factor (LIF) pathway. LIF, by binding to the LIF receptor, signals the activation of the transcription factor STAT3 that translocates into the nucleus and activates a variety of downstream genes, including Sall4, Myc, and Klf4 (Cartwright *et al.*, 2005; Niwa *et al.*, 2009). STAT3 suppression results in ESC differentiation (Burdon *et al.*, 2002; Niwa *et al.*, 1998), whereas constitutive activation of STAT3 is sufficient to maintain ESCs in the undifferentiated state, even in the absence of LIF (Matsuda *et al.*, 1999). Together with LIF–STAT3 pathway, the pluripotency of ESCs is modulated by BMP4 and Nodal/Activin-TGFβ family members. Indeed, it has been shown that BMP4, through activation of intracellular receptors SMADs, induces the expression of Id genes, which suppress ESC differentiation in combination with LIF signaling (Ying *et al.*, 2003). In further studies, BMP4 was suggested to specifically inhibit neural differentiation, rather than to promote self-renewal (Fei *et al.*, 2010). Moreover, we have recently found that the suppression of BMP4 signaling results in the upregulation of Nodal/Activin pathway along with maintenance of ESC undifferentiated state even in differentiation-promoting conditions (Aloia *et al.*, 2010).

Another extracellular signal that is involved in pluripotency maintenance is Wnt. Indeed, Wnt proteins contribute to maintain ESC pluripotency by activating transcription factor 3 (Tcf3) that collaborates with the pivotal transcription factors Oct3/4, Sox2, and Nanog (Hao *et al.*, 2006; Ogawa *et al.*, 2006; Sato *et al.*, 2004).

All the extracellular signals required for stemness maintenance converge on a complex transcriptional network. Further than the already mentioned STAT3, three transcription factors have been considered for several years the crucial players in the stemness regulatory circuit: Oct3/4, Sox2, and Nanog. Oct3/4, also known as Pou5f1, is expressed in pluripotent cell lineages, such as ICM and germ cells *in vivo* and undifferentiated ESC *in vitro*, where it quickly disappears when the cells undergo differentiation (Okamoto *et al.*, 1990; Rosner *et al.*, 1990). Both suppression and overexpression of Oct3/4 in ESCs causes differentiation (Niwa *et al.*, 2000), indicating that a tight control of Oct3/4 expression levels is required to maintain self-renewal.

Sox2 is expressed in pluripotent cells and in the nervous system (Avilion *et al.*, 2003). Sox2 deletion in ESCs causes differentiation with a phenotype similar to that observed upon Oct3/4 deletion (Masui *et al.*, 2007). On the other hand, Sox2 overexpressing cells are biased toward neural differentiation (Kopp *et al.*, 2008; Zhao *et al.*, 2004). Oct3/4 and Sox2 bound the DNA cooperatively (Ambrosetti *et al.*, 2000) at regulatory elements present in their own promoters and in many genes that are specifically expressed in pluripotent stem cells such as Nanog, REST, and Lefty (Matoba *et al.*, 2006).

Nanog was discovered at the same time by two independent groups for its specific expression in ESCs (Mitsui *et al.*, 2003) and its ability to maintain pluripotency in the absence of LIF (Chambers *et al.*, 2003). Nanog-null ESCs can be established and they maintain pluripotency although with an increasing tendency to differentiate (Chambers *et al.*, 2007).

V. Klf5 Function in ESCs

Recently, a role for Klf2, Klf4, and Klf5 genes in the pluripotency control has been highlighted. Jiang *et al.* (2008), on the basis of Klfs ability to induce cell reprogramming, investigated the expression profile of all known Klfs in undifferentiated and differentiated ESCs. They reported that only Klf2, Klf4, and Klf5 seem to have an ESC-specific expression with high levels of transcripts in undifferentiated ESCs that dramatically decrease when differentiation occurs. Thus, to elucidate the role of these three Klfs in ESC pluripotency, they performed single knockdown (KD) by RNA interference of each gene. No evident impairment of ESC undifferentiated state was observed when morphology and alkaline phosphatase (AP) expression were analyzed (Jiang *et al.*, 2008). However, triple KD of Klf2, Klf4, and Klf5 resulted in a dramatic decrease of AP expression supported by an evident appearance of differentiated cells. These results prompted Jiang *et al.* to conclude that only the triple KD of these Klfs led to ESC differentiation and consequently loss of stemness, indicating that these genes have redundant function in ESCs. However, more recently several papers reported that each factor plays a unique role in the maintenance of ESC undifferentiated state. Indeed, we showed that single KD of Klf5 induces an evident impairment of ESC undifferentiated state (Parisi *et al.*, 2008): upon Klf5 KD, we found a significant reduction of AP staining supported by an evident decrease in Oct3/4, Nanog, and Sox2 expression. Moreover, we showed that this loss of undifferentiated state is due to the onset of differentiation program biased toward mesoderm and trophoblast fates (Parisi *et al.*, 2008). Later, Ema *et al.* (2008) demonstrated that ESCs couldn't be established *in vitro* from inner cell mass (ICM) of Klf5-null blastocysts. Moreover, these authors generated Klf5 KO ESCs *in vitro* by targeting the wild type allele in Klf5 heterozygous cells and observed that these cells have a differentiated morphology and exhibited reduced Nanog expression after few passages in culture. They also found that Klf5-null ESCs showed an increased expression of primitive ectoderm, mesodermal, and trophoblast markers (Ema *et al.*, 2008). All these data indicated that suppression of Klf5 impairs the maintenance of ESC undifferentiated state by driving the cells toward an inappropriate differentiation. In support of the nonredundant function of Klfs, a recent paper demonstrated also that Klf4 KD led to a loss of the ESC undifferentiated state accompanied by an upregulation of genes associated with a differentiation state such as Brachyury and Cdx2 (Zhang *et al.*, 2010). Thus, all these recent results indicate that the single KD of Klf4 and Klf5 is enough to induce ESC differentiation (Parisi *et al.*, 2008; Zhang *et al.*, 2010). Discrepancy in KD experimental conditions may justify these differences. In fact, the single KD of Klf4 and Klf5 was performed by stably expressing the specific shRNAs resulting in a lasting suppression of Klf function. Short term KD of one Klfs may be compensated by the others. Instead, if the KD is persistent the redundant effect of other Klfs may be insufficient to maintain ESC identity. Moreover, the effects on AP decrease, observed upon Klf2, Klf4, or Klf5 single KD could be masked in the study of Jiang *et al.* by a too high cell density. Finally, also in the conditions of Jiang *et al.*, the single KD of Klf5 is able to significantly reduce the levels of some stemness genes such as Nanog, Tbx3, and Nr5a2 (Jiang *et al.*, 2008). Thus, it may be that KD of single Klf can be detrimental for ESC self-renewal but the triple inactivation of Klfs is required to trigger the extensive differentiation observed by Jiang *et al.*

The key role of Klf5 in the maintenance of ESC potential is also proved by its ability to sustain ESC stemness in the absence of LIF. Indeed, we have demonstrated that sustained expression of Klf5 in ESCs cultured in the absence of LIF or even in culture conditions enhancing differentiation is able to maintain expression of stemness markers such as AP, Oct3/4, and Nanog (Parisi *et al.*, 2008). Moreover, in agreement with this observation, Klf5 ectopic expression results in a block of differentiation (Parisi *et al.*, unpublished data). Similar results were also obtained by Ema *et al.* that showed that ESCs overexpressing Klf5 were resistant to differentiation stimuli, and also after a prolonged culture in the absence of LIF, these cells were able to generate chimeric embryos upon injection into a host blastocyst (Ema *et al.*, 2008). Klf4 and Klf2 were also described to be able to overcome the withdrawal of LIF by maintaining ESC undifferentiated state (Hall *et al.*, 2009; Zhang *et al.*, 2010). Although the effects of Klf2 suppression were not reported, a hierarchical relationship in the ability of these three Klfs to support self-renewal in the absence of LIF was suggested, with Klf2 as the most potent (Hall *et al.*, 2009). This finding is in agreement with the previous observation that Klf2 and Klf4 are more efficient of Klf5 in the reprogramming of somatic cells into iPS cells (Nakagawa *et al.*, 2008).

It has been suggested that one of the possible contributes of Klf5 in the maintenance of ESC potential may be the proliferation (Ema *et al.*, 2008). Although Ema *et al.* reported that Klf5 KO ESCs have a reduced proliferation rate, the same authors claimed that upon Klf5 KD no effects on proliferation were detected in agreement with our results (Ema *et al.*, 2008; Parisi *et al.*, 2008). Moreover, we did not find significant proliferative effects after short term Klf5 overexpression. However, we found that Klf5 over-expressing cells maintain their proliferative potential upon the exposure to differentiation-inducing conditions (Parisi *et al.*, unpublished data), in agreement with the observation that Klf5overexpressing cells showed an increased number of cells in S phase compared to the control (Ema *et al.*, 2008).

VI. Klf5 Role in Early Embryonic Development

The expression of Klf5 in the very early stages of mouse embryogenesis was recently elucidated. Klf5 expression can be easily detected at early cleavage stages (8 cell embryos) before compaction (morula stage), when Oct3/4 protein is uniformly present in all the cells of the embryo, and thus it is coexpressed with Klf5. At the early blastocyst stage, Klf5 expression is present both in the inner and outer cell layer (Ema *et al.*, 2008; Lin *et al.*, 2010; Parisi *et al.*, 2008), colocalizing alternatively with Nanog and Oct3/4 or Cdx2 respectively. By the expanded blastocyst stage, Klf5 appears to be reduced in epiblast cells, still positive for Nanog, whereas its expression seems to be specifically localized in primitive endoderm cells, originating from ICM (GATA4-positive cells) and trophoblast cells (Fig. 18.1). During gastrulation (7 dpc), as mentioned above, Klf5 localized in the anterior primitive streak (Ema *et al.*, 2008; Moore-Scott *et al.*, 2007).

Klf5+/− mice were viable and fertile. However, they showed reduced response to injury and angiogenesis as well as defects in adipocyte differentiation and skeletal growth retardation during perinatal period (Oishi *et al.*, 2005; Shindo *et al.*, 2002; Shinoda *et al.*, 2008). Klf5-null mice were initially reported to exhibit embryonic lethality before E8.5 (Shindo *et al.*, 2002). Later, it has been showed that Klf5 KO embryos fail to implant (Ema *et al.*, 2008), due a defective development in earlier stages of development. Indeed, although Klf5 KO embryos were morphologically not distinguishable from controls at 2.5 dpc (Lin *et al.*, 2010), with a roughly normal expression of Oct3/4, they do not show detectable levels of the trophoblast marker Cdx2 (Lin *et al.*, 2010). At blastocyst stage (3.5 dpc), the expression of the pluripotency markers Oct3/4 and Nanog was spatially restricted to the ICM of wt and heterozygous blastocysts, while both markers appeared scattered in Klf5 KO collapsed blastocysts (Fig. 18.2). Moreover, in agreement with other observations (Ema *et al.*, 2008), we found that the expression of the trophoectoderm marker Cdx2 is dramatically reduced and barely detectable in mutant blastocysts (Fig. 18.2). Although in some cases partial cavitation was found, the mutant embryos were rarely fully expanded

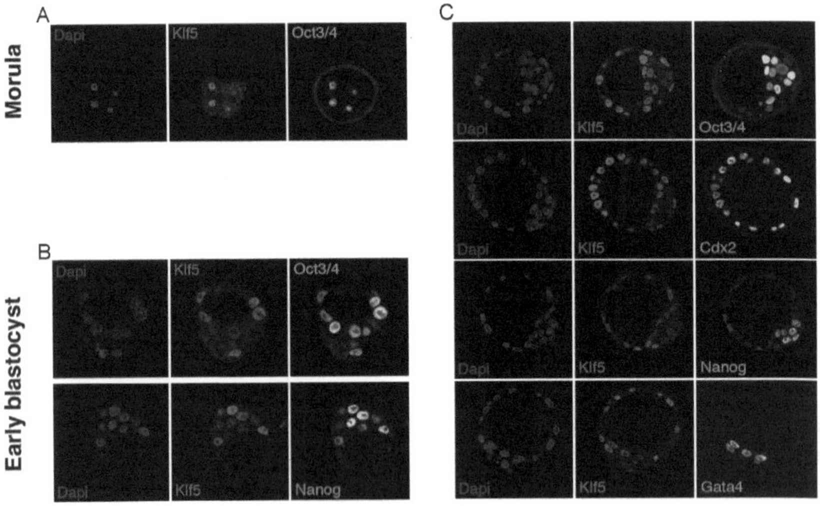

Figure 18.1 Klf5 expression during preimplantation mouse development. Immunostaining analysis of Klf5 expression pattern during the earliest phases of mouse development. Colocalization with markers of ICM (Oct3/4 and Nanog), throphoblast (Cdx2), primitive endoderm (GATA4) is shown. (A) Morula stage, (B) early blastocyst, (C) hatched blastocyst. (For color version of this figure, the reader is referred to the Web version of this chapter.)

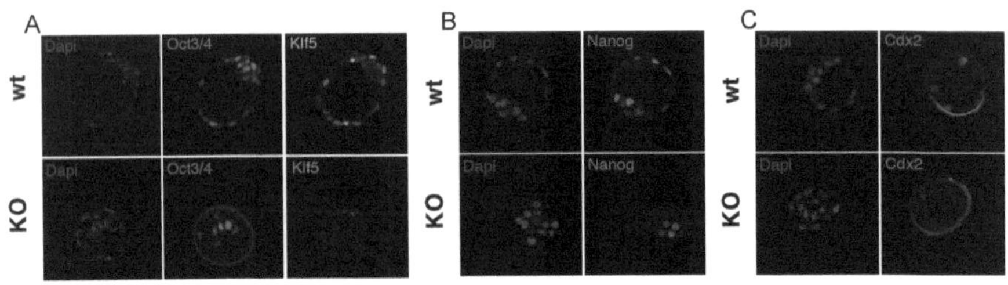

Figure 18.2 Phenotype of Klf5-null blastocysts. The expression of Oct3/4 (A) and Nanog (B) as markers of ICM and Cdx2 for trophoblast (C) was analyzed by means of immunostaining in wt and Klf5-null embryos. (For color version of this figure, the reader is referred to the Web version of this chapter.)

in contrast to the majority of wild type or heterozygous embryos (Lin *et al.*, 2010). Thus, Klf5 deficiency prevents formation of a blastocelic cavity, resulting in the appearance of abnormal morulae at 3.5 dpc. As a result of these developmental defects, Klf5 KO embryos failed to give rise to ICM cells *in vitro*. Indeed, whereas wt and heterozygous embryos hatched from the zona pellucida and attached onto the culture dish at 24 h in culture, mutant embryos sometimes shed their zona pellucida but never attached to the dish, rapidly collapsed, and finally degenerated inside the zona pellucida after 48 h (Fig. 18.3). In culture conditions permissive for the outgrowth of trophoblast cells (i.e., in the presence of Fgf4 and heparin), mutant embryos were unable to attach to the plate and to develop properly compared to the controls (Fig. 18.3). Only when the zonae pellucidae were removed before culturing the embryos, mutant blastocysts were able to outgrow, but they gave raise to cells defective for Cdx2 expression (Lin *et al.*, 2010). Taken together, these observations imply that Klf5 KO embryos remain viable until 3.5 dpc but probably do not progress during development. The impaired blastocyst development is likely due to a defect in proper signals that allow the specification of trophoblast versus ICM fate during the transition from morula to blastocyst stages (Lin *et al.*, 2010). Moreover, in the subsequent stage, when epiblast and primitive endoderm cells have to segregate in the embryo, Klf5 is required to control the relative number of cells contributing to these two lineages (Lin *et al.*, 2010).

Although Klf2 and Klf4 are induced during early embryogenesis (Jiang *et al.*, 2008), only Klf5 exhibits a very early embryonic lethality (Ema *et al.*, 2008). Indeed, Klf2 KO mice show embryonic lethality at E12.5–E14.5 due to intraembryonic and intra-amniotic hemorrhage (Kuo *et al.*, 1997). Klf4 KO mice result in perinatal lethality due to defective formation of skin epithelial barrier and globet cell differentiation (Katz *et al.*, 2002; Segre *et al.*, 1999). The phenotype of these KO mice clearly demonstrates that *in vivo*

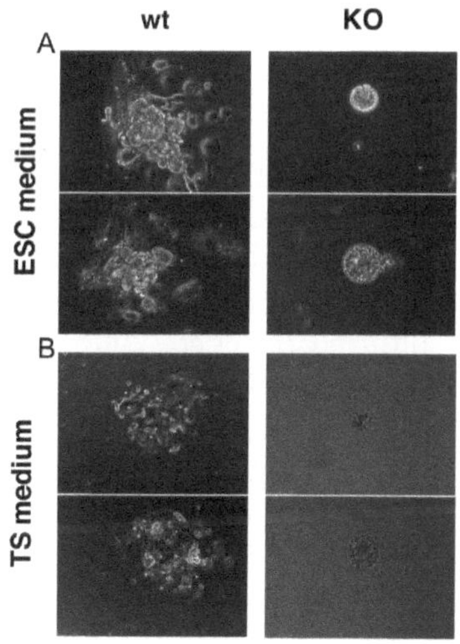

Figure 18.3 Blastocyst outgrowth. Phase contrast images of outgrowth cultures from wild type and Klf5-null blastocysts. Freshly isolated blastocysts were grown in ESC medium (A) or trophoblast stem cells (TS) medium (B) to allow the outgrowth of ESCs or trophoblast cells, respectively.

there is no redundancy among the three Klfs, given that Klf2 and Klf4 are unable to overcome the absence of Klf5 during the earliest phases of mouse development. The molecular basis for this functional differences can be the presence of Klf5-specific cofactors during preimplantation development and/or a unique set of transcriptional targets for Klf5.

VII. Klf5 Targets

Jiang *et al.* (2008), by means of ChIP-on-chip analysis, reported that Klf2, Klf4, and Klf5 share many common target sites. We have recently performed a detailed analysis of Klf5 target genes by microarray in combination with chromatin IP coupled with massively parallel short-tag-based sequencing (ChIP-seq). Microarray analysis showed that Klf5 could act both as activator or repressor by activating at least 88 targets and repressing 300 genes (Parisi *et al.*, 2010). Among the genes controlled by Klf5 were found genes that play a crucial role in ESCs such as Tcl1 (Ivanova *et al.*, 2006; Matoba *et al.*, 2006), BMP4 (Ying *et al.*, 2003), Nr0b1 (Khalfallah *et al.*, 2009) indicating that Klf5 may be required in ESCs to maintain pluripotency by activating expression of these self-renewal promoting genes and simultaneously by inhibiting expression of differentiation-promoting genes such as TGFβ2 (Singla and Sun, 2005), Otx2, Pitx2 (Ivanova *et al.*, 2006), GDNF (Morizane *et al.*, 2009). Notably, although Nanog and Oct3/4 have been described as Klf5 targets (Jiang, *et al.*, 2008; Parisi *et al.*, 2008), they were not found to change significantly in the microarray analysis (Parisi *et al.*, 2010). This can be explained by a functional redundancy among the Klfs in activating components of the core pluripotency network. This is also consistent with the observations of Ema and coworkers: (i) Nanog and Oct3/4 are still expressed in Klf5 KO embryos, and thus their transcription could not be solely dependent on Klf5, (ii) during the first passages of Klf5 KO ESCs, Nanog expression is comparable to wild type cells, while it is altered after prolonged culture (Ema *et al.*, 2008). Thus, it may be that the binding to Nanog and Oct3/4 promoters can be a characteristic of Klf4 and/or Klf2 rather than of Klf5. The matching of microarray data with ChIP-seq results allowed identifying at least 161 genes that can be directly regulated by Klf5 (Parisi *et al.*, 2010). Among the primary targets of Klf5, there is Tcl1 that was already described to be downstream of Oct3/4 and Klf5 (Ema *et al.*, 2008; Matoba *et al.*, 2006). Moreover, 8 out of 23 Klf5 direct targets were shown, by means of RNA interference, to contribute to the maintenance of ESC undifferentiated state. Among these Klf5 targets, we found the kinase Hck already described to be involved in gp130-mediated signaling of LIF, since introduction of constitutively activated Hck alleviates the requirement of ESCs for LIF to remain undifferentiated (Ernst *et al.*, 1994). Thus, Klf5 seems to regulate genes that contribute to the maintenance of ESC undifferentiated state. Moreover, it has been demonstrated that Klf5 acts in the maintenance of ESC state through suppression of genes related to differentiation such as Serpine1. Indeed, ectopic expression of Serpine1 resulted in an impairment of ESC phenotype by promoting the appearance of differentiation markers even in presence of LIF (Parisi *et al.*, 2010). This phenotype mimics the effect of Klf5 KD as well as Klf5 KO ESCs (Ema *et al.*, 2008; Parisi *et al.*, 2008). We have found that 50% of 53 tested genes are specifically regulated by Klf5 and 90% are regulated differentially by Klf5 and Klf4. Interestingly, four of Klf5 targets that were shown to influence the maintenance of ESC state (i.e., Igfbp3, Niban, Perp, and Serpine1) resulted to be regulated by Klf5 and not by Klf2 and Klf4 (Parisi *et al.*, 2010), indicating that the control of ESC undifferentiated state by Klf5 is carried out at least in part through nonredundant targets of Klfs.

VIII. Klf5 Connection to the Core Pluripotency Network

Klf5, together with Klf2 and Klf4, has been implicated in the regulation of a signaling network controlling ESC self-renewal and pluripotency. This network is composed of several intrinsic and extrinsic signals that all converge on the key transcription factors STAT3, Oct3/4, Sox2, and Nanog (Masui, 2010; Niwa, 2007). Klfs seem to be strictly interconnected with these factors both by controlling them and being controlled by them. Indeed, it has been demonstrated that Klf4 and Klf5 genes bear binding sites for Nanog in their promoter regions (Chen *et al.*, 2008; Loh *et al.*, 2006; Parisi *et al.*, 2008), and Nanog KD induces an evident decrease of Klf4 and Klf5 expression levels although Nanog overexpression does not significantly change Klf4 and Klf5 expression (Bourillot *et al.*, 2009; Parisi *et al.*, 2008). Moreover, STAT3 activation, upon LIF treatment of ESCs, induces an increase in the expression of Klf4 and Klf5, whereas Klf2 showed no induction by LIF (Bourillot *et al.*, 2009; Hall *et al.*, 2009). Moreover, null STAT3 ESCs stimulated with LIF retain full expression of Klf2, but these cells show reduced expression of both Klf4 and Klf5 (Hall *et al.*, 2009), indicating that STAT3 controls Klf4 and Klf5 but not Klf2. However, it has been demonstrated that upon expression of a mutant form of LIF receptor, that can signal only through STAT3 and not through alternative signaling (i.e., MAPK or PI3K), LIF stimulation induces Klf4 but not Klf5 upregulation (Hall *et al.*, 2009), indicating that Klf5 is controlled by LIF in a STAT3-dependent manner but involving additional factors.

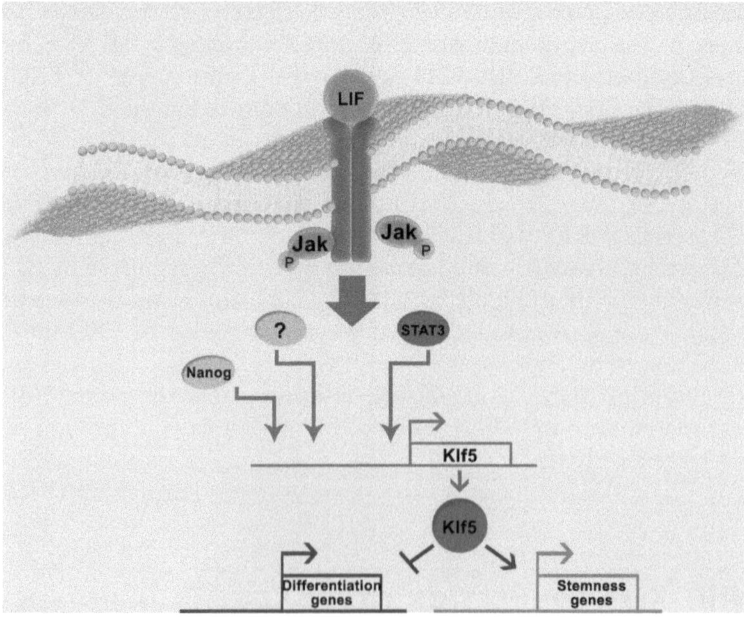

Figure 18.4 Model of Klf5 integration into the core transcriptional network controlling ESC state. (For color version of this figure, the reader is referred to the Web version of this chapter.)

IX. Conclusions

Klf5, together with Klf2 and Klf4, represent key regulators of ESC self-renewal and pluripotency. Klf5 is integrated in ESC core transcriptional network being controlled by Nanog and STAT3. On the other hand, Klf5 controls ESC state by regulating, alone or together with Klf2 and Klf4, the expression of master genes, that is, Oct3/4, Nanog, and Sox2, and genes contributing to maintenance of ESC undifferentiated state, such as Hck, Perp, Bcam, etc. Moreover, at the same time, Klf5 seems to be required in ESCs to maintain pluripotency by preventing the activation of lineage-specific differentiation programs (i.e., mesoderm and trophoblast) through the suppression of differentiation-related genes (Fig. 18.4).

ACKNOWLEDGMENTS

We thank Ryozo Nagai and Ichiro Manabe for providing us with Klf5-null mice. We thank Mrs. Danila Scarpa for precious assistance. This work was supported by grants from EC-Sirocco Consortium, Associazione Italiana Ricerca sul Cancro (AIRC), and Italian Ministry of Health.

REFERENCES

Aloia, L., Parisi, S., Fusco, L., Pastore, L., and Russo, T. (2010). Differentiation of embryonic stem cells 1 (Dies1) is a component of bone morphogenetic protein 4 (BMP4) signaling pathway required for proper differentiation of mouse embryonic stem cells. *J. Biol. Chem.* **285,** 7776–7783.

Ambrosetti, D. C., Scholer, H. R., Dailey, L., and Basilico, C. (2000). Modulation of the activity of multiple transcriptional activation domains by the DNA binding domains mediates the synergistic action of Sox2 and Oct-3 on the fibroblast growth factor-4 enhancer. *J. Biol. Chem.* **275,** 23387–23397.

Avilion, A. A., Nicolis, S. K., Pevny, L. H., Perez, L., Vivian, N., and Lovell-Badge, R. (2003). Multipotent cell lineages in early mouse development depend on SOX2 function. *Genes Dev.* **17,** 126–140.

Bateman, N. W., Tan, D., Pestell, R. G., Black, J. D., and Black, A. R. (2004). Intestinal tumor progression is associated with altered function of KLF5. *J. Biol. Chem.* **279,** 12093–12101.

Bourillot, P. Y., Aksoy, I., Schreibe, V., Wianny, F., Schulz, H., Hummel, O., Hubner, N., and Savatier, P. (2009). Novel STAT3 target genes exert distinct roles in the inhibition of mesoderm and endoderm differentiation in cooperation with Nanog. *Stem Cells* **27,** 1760–1771.

Burdon, T., Smith, A., and Savatier, P. (2002). Signaling, cell cycle and pluripotency in embryonic stem cells. *Trends Cell Biol.* **12,** 432–438.

Cartwright, P., McLean, C., Sheppard, A., Rivett, D., Jones, K., and Dalton, S. (2005). LIF/STAT3 controls ES cell self-renewal and pluripotency by a Myc-dependent mechanism. *Development* **132,** 885–896.

Chambers, I., Colby, D., Robertson, M., Nichols, J., Lee, S., Tweedie, S., and Smith, A. (2003). Functional expression cloning of Nanog, a pluripotency sustaining factor in embryonic stem cells. *Cell* **113,** 643–655.

Chambers, I., Silva, J., Colby, D., Nichols, J., Nijmeijer, B., Robertson, M., Vrana, J., Jones, K., Grotewold, L., and Smith, A. (2007). Nanog safeguards pluripotency and mediates germline development. *Nature* **450,** 1230–1234.

Chen, C., Bhalala, H. V., Vessella, R. L., and Dong, J. T. (2003). KLF5 is frequently deleted and down-regulated but rarely mutated in prostate cancer. *Prostate* **55,** 81–88.

Chen, C., Benjamin, M. S., Sun, X., Otto, K. B., Guo, P., Dong, X. Y., Bao, Y., Zhou, Z., Cheng, X., Simons, J. W., and Dong, J. T. (2006). KLF5 promotes cell proliferation and tumorigenesis through gene regulation in the TSU-Pr1 human bladder cancer cell line. *Int. J. Cancer* **118,** 1346–1355.

Chen, X., Xu, H., Yuan, P., Fang, F., Huss, M., Vega, V. B., Wong, E., Orlov, Y. L., Zhang, W., Jiang, J., Loh, Y. H., Yeo, H. C., *et al.* (2008). Integration of external signaling pathways with the core transcriptional network in embryonic stem cells. *Cell* **133,** 1106–1117.

Conkright, M. D., Wani, M. A., Anderson, K. P., and Lingrel, J. B. (1999). A gene encoding an intestinal-enriched member of the Krüppel-like factor family expressed in intestinal epithelial cells. *Nucleic Acids Res.* **27,** 1263–1270.

Dong, J. T., and Chen, C. (2009). Essential role of KLF5 transcription factor in cell proliferation and differentiation and its implications for human diseases. *Cell. Mol. Life Sci.* **66,** 2691–2706.

Ema, M., Mori, D., Niwa, H., Hasegawa, Y., Yamanaka, Y., Hitoshi, S., Mimura, J., Kawabe, Y., Hosoya, T., Morita, M., Shimosato, D., Uchida, K., *et al.* (2008). Kruppel-like factor 5 is essential for blastocyst development and the normal self-renewal of mouse ESCs. *Cell Stem Cell* **3,** 555–567.

Ernst, M., Gearing, D. P., and Dunn, A. R. (1994). Functional and biochemical association of Hck with the LIF/IL-6 receptor signal transducing subunit gp130 in embryonic stem cells. *EMBO J.* **13,** 1574–1584.

Fei, T., Xia, K., Li, Z., Zhou, B., Zhu, S., Chen, H., Zhang, J., Chen, Z., Xiao, H., Han, J. D., and Chen, Y. G. (2010). Genome-wide mapping of SMAD target genes reveals the role of BMP signaling in embryonic stem cell fate determination. *Genome Res.* **20,** 36–44.

Guo, P., Dong, X. Y., Zhang, X., Zhao, K. W., Sun, X., Li, Q., and Dong, J. T. (2009a). Pro-proliferative factor KLF5 becomes anti-proliferative in epithelial homeostasis upon signaling-mediated modification. *J. Biol. Chem.* **284,** 6071–6078.

Guo, P., Zhao, K. W., Dong, X. Y., Sun, X., and Don, J.-T. (2009b). Acetylation of KLF5 alters the assembly of p15 transcription factors in TGFb-mediated induction in epithelial cells. *J. Biol. Chem.* **284**(27), 18184–18193.

Hall, J., Guo, G., Wray, J., Eyres, I., Nichols, J., Grotewold, L., Morfopoulou, S., Humphreys, P., Mansfield, W., Walker, R., Tomlinson, S., and Smith, A. (2009). Oct4 and LIF/Stat3 additively induce Kruppel factors to sustain embryonic stem cell self-renewal. *Cell Stem Cell* **5,** 597–609.

Hao, J., Li, T. G., Qi, X., Zhao, D. F., and Zhao, G. Q. (2006). WNT/beta-catenin pathway up-regulates Stat3 and converges on LIF to prevent differentiation of mouse embryonic stem cells. *Dev. Biol.* **290,** 81–91.

Ivanova, N., Dobrin, R., Lu, R., Kotenko, I., Levorse, J., DeCoste, C., Schafer, X., Lun, Y., and Lemischka, I. R. (2006). Dissecting self-renewal in stem cells with RNA interference. *Nature* **442,** 533–538.

Jiang, J., Chan, Y. S., Loh, Y. H., Cai, J., Tong, G. Q., Lim, C. A., Robson, P., Zhong, S., and Ng, H. H. (2008). A core Klf circuitry regulates self-renewal of embryonic stem cells. *Nat. Cell Biol.* **10,** 353–360.

Katz, J. P., Perreault, N., Goldstein, B. G., Lee, C. S., Labosky, P. A., Yang, V. W., and Kaestner, K. H. (2002). The zinc-finger transcription factor Klf4 is required for terminal differentiation of goblet cells in the colon. *Development* **129,** 2619–2628.

Khalfallah, O., Rouleau, M., Barbry, P., Bardoni, B., and Lalli, E. (2009). Dax-1 knockdown in mouse embryonic stem cells induces loss of pluripotency and multilineage differentiation. *Stem Cells* **27,** 1529–1537.

Kopp, J. L., Ormsbee, B. D., Desler, M., and Rizzino, A. (2008). Small increases in the level of Sox2 trigger the differentiation of mouse embryonic stem cells. *Stem Cells* **26,** 903–911.

Kuo, C. T., Veselits, M. L., Barton, K. P., Lu, M. M., Clendenin, C., and Leiden, J. M. (1997). The LKLF transcription factor is required for normal tunica media formation and blood vessel stabilization during murine embryogenesis. *Genes Dev.* **11,** 2996–3006.

Lin, S. C., Wani, M. A., Whitsett, J. A., and Wells, J. M. (2010). Klf5 regulates lineage formation in the pre-implantation mouse embryo. *Development* **137,** 3953–3963.

Loh, Y. H., Wu, Q., Chew, J. L., Vega, V. B., Zhang, W., Chen, X., Bourque, G., George, J., Leong, B., Liu, J., Wong, K. Y., Sung, K. W., *et al.* (2006). The Oct4 and Nanog transcription network regulates pluripotency in mouse embryonic stem cells. *Nat. Genet.* **38,** 431–440.

Masui, S. (2010). Pluripotency maintenance mechanism of embryonic stem cells and reprogramming. *Int. J. Hematol.* **91,** 360–372.

Masui, S., Nakatake, Y., Toyooka, Y., Shimosato, D., Yagi, R., Takahashi, K., Okochi, H., Okuda, A., Matoba, R., Sharov, A. A., Ko, M. S. H., and Niwa, H. (2007). Pluripotency governed by Sox2 via regulation of Oct3/4 expression in mouse embryonic stem cells. *Nat. Cell Biol.* **9,** 625–635.

Matoba, R., Niwa, H., Masui, S., Ohtsuka, S., Carter, M. G., Sharov, A. A., and Ko, M. S. H. (2006). Dissecting Oct3/4-regulated gene networks in embryonic stem cells by expression profiling. *PLoS One* **1,** e26.

Matsuda, T., Nakamura, T., Nakao, K., Arai, T., Katsuki, M., Heike, T., and Yokota, T. (1999). STAT3 activation is sufficient to maintain an undifferentiated state of mouse embryonic stem cells. *EMBO J.* **18,** 4261–4269.

McConnell, B. B., Ghaleb, A. M., Nandan, M. O., and Yang, V. W. (2007). The diverse functions of Krüppel-like factors 4 and 5 in epithelial biology and pathobiology. *Bioessays* **29,** 549–557.

Mitsui, K., Tokuzawa, Y., Itoh, H., Segawa, K., Murakami, M., Takahashi, K., Maruyama, M., Maeda, M., and Yamanaka, S. (2003). The homeoprotein Nanog is required for maintenance of pluripotency in mouse epiblast and ES cells. *Cell* **113,** 631–642.

Moore-Scott, B. A., Opoka, R., Lin, S. C., Kordich, J. J., and Wells, J. M. (2007). Identification of molecular markers that are expressed in discrete anterior-posterior domains of the endoderm from the gastrula stage to mid-gestation. *Dev. Dyn.* **236,** 1997–2003.

Morizane, R., Monkawa, T., and Itoh, H. (2009). Differentiation of murine embryonic stem and induced pluripotent stem cells to renal lineage in vitro. *Biochem. Biophys. Res. Commun.* **390,** 1334–1339.

Nakagawa, M., Koyanagi, M., Tanabe, K., Takahashi, K., Ichisaka, T., Aoi, T., Okita, K., Mochiduki, Y., Takizawa, N., and Yamanaka, S. (2008). Generation of induced pluripotent stem cells without Myc from mouse and human fibroblasts. *Nat. Biotechnol.* **26,** 101–106.

Nandan, M. O., McConnell, B. B., Ghaleb, A. M., Bialkowska, A. B., Sheng, H., Shao, J., Babbin, B. A., Robine, S., and Yang, V. W. (2008). Krüppel-like factor 5 mediates cellular transformation during oncogenic KRAS-induced intestinal tumorigenesis. *Gastroenterology* **134,** 120–130.

Niwa, H. (2007). How is pluripotency determined and maintained? *Development* **134,** 635–646.

Niwa, H., Burdon, T., Chambers, I., and Smith, A. (1998). Self-renewal of pluripotent embryonic stem cells is mediated via activation of STAT3. *Genes Dev.* **12,** 2048–2060.

Niwa, H., Miyazaki, J., and Smith, A. G. (2000). Quantitative expression of Oct 3/4 defines differentiation, dedifferentiation or self-renewal of ES cells. *Nat. Genet.* **24,** 372–376.

Niwa, H., Ogawa, K., Shimosato, D., and Adachi, K. (2009). A parallel circuit of LIF signaling pathways maintains pluripotency of mouse ES cells. *Nature* **460,** 118–122.

Ogawa, K., Nishinakamura, R., Iwamatsu, Y., Shimosato, D., and Niwa, H. (2006). Synergistic action of Wnt and LIF in maintaining pluripotency of mouse ES cells. *Biochem. Biophys. Res. Commun.* **343,** 159–166.

Ohnishi, S., Laub, F., Matsumoto, N., Asaka, M., Ramirez, F., Yoshida, T., and Terada, M. (2000). Developmental expression of the mouse gene coding for the Krüppel-like transcription factor KLF5. *Dev. Dyn.* **217,** 421–429.

Oishi, Y., Manabe, I., Tobe, K., Tsushima, K., Shindo, T., Fujiu, K., Nishimura, G., Maemura, K., Yamauchi, T., Kubota, N., Suzuki, R., Kitamura, T., *et al.* (2005). Krüppel-like transcription factor KLF5 is a key regulator of adipocyte differentiation. *Cell Metab.* **1,** 27–39.

Oishi, Y., Manabe, I., Tobe, K., Ohsugi, M., Kubota, T., Fujiu, K., Maemura, K., Kubota, N., Kadowaki, T., and Nagai, R. (2008). SUMOylation of Krüppel-like transcription factor 5 acts as a molecular switch in transcriptional programs of lipid metabolism involving PPAR-delta. *Nat. Med.* **14,** 656–666.

Okamoto, K., Okazawa, H., Okuda, A., Sakai, M., Muramatsu, M., and Hamada, H. (1990). A novel octamer binding transcription factor is differentially expressed in mouse embryonic cells. *Cell* **60,** 461–472.

Parisi, S., Passaro, F., Aloia, L., Manabe, I., Nagai, R., Pastore, L., and Russo, T. (2008). Klf5 is involved in self-renewal of mouse embryonic stem cells. *J. Cell Sci.* **121,** 2629–2634.

Parisi, S., Cozzuto, L., Tarantino, C., Passaro, F., Ciriello, S., Aloia, L., Antonini, D., De Simone, V., Pastore, L., and Russo, T. (2010). Direct targets of Klf5 transcription factor contribute to the maintenance of mouse embryonic stem cell undifferentiated state. *BMC Biol.* **8,** 128.

Rosner, M. H., Vigano, M. A., Ozato, K., Timmons, P. M., Poirier, F., Rigby, P. W., and Staudt, L. M. (1990). A POU-domain transcription factor in early stem cells and germ cells of the mammalian embryo. *Nature* **345,** 686–692.

Sato, N., Meijer, L., Skaltsounis, L., Greengard, P., and Brivanlou, A. H. (2004). Maintenance of pluripotency in human and mouse embryonic stem cells through activation of Wnt signaling by a pharmacological GSK-3-specific inhibitor. *Nat. Med.* **10,** 55–63.

Segre, J. A., Bauer, C., and Fuchs, E. (1999). Klf4 is a transcription factor required for establishing the barrier function of the skin. *Nat. Genet.* **22,** 356–360.

Shindo, T., Manabe, I., Fukushima, Y., Tobe, K., Aizawa, K., Miyamoto, S., Kawai-Kowase, K., Moriyama, N., Imai, Y., Kawakami, H., Nishimatsu, H., Ishikawa, T., *et al.* (2002). Kruppel-like zincfinger transcription factor KLF5/BTEB2 is a target for angiotensin II signaling and an essential regulator of cardiovascular remodeling. *Nat. Med.* **8,** 856–863.

Shinoda, Y., Ogata, N., Higashikawa, A., Manabe, I., Shindo, T., Yamada, T., Kugimiya, F., Ikeda, T., Kawamura, N., Kawasaki, Y., Tsushima, K., Takeda, N., *et al.* (2008). Kruppel-like factor 5 causes cartilage degradation through transactivation of matrix metalloproteinase 9. *J. Biol. Chem.* **283,** 24682–24689.

Singla, D. K., and Sun, B. (2005). Transforming growth factor-beta2 enhances differentiation of cardiac myocytes from embryonic stem cells. *Biochem. Biophys. Res. Commun.* **332,** 135–141.

Sogawa, K., Imataka, H., Yamasaki, Y., Kusume, H., Abe, H., and FujiiKuriyama, Y. (1993). cDNA cloning and transcriptional properties of a novel GC box-binding protein, BTEB2. *Nucleic Acids Res.* **21,** 1527–1532.

Sun, R., Chen, X., and Yang, V. W. (2001). Intestinal-enriched Krüppel-like factor (Krüppel-like factor 5) is a positive regulator of cellular proliferation. *J. Biol. Chem.* **276,** 6897–6900.

Takahashi, K., and Yamanaka, S. (2006). Induction of pluripotent stem cells from mouse embryonic and adult fibroblast cultures by defined factors. *Cell* **126,** 663–676.

Yang, Y., Goldstein, B. G., Chao, H. H., and Katz, J. P. (2005). KLF4 and KLF5 regulate proliferation, apoptosis and invasion in esophageal cancer cells. *Cancer Biol. Ther.* **4,** 1216–1221.

Yang, Y., Tetreault, M. P., Yermolina, Y. A., Goldstein, B. G., and Katz, J. P. (2008). Kruppel-like factor 5 controls keratinocyte migration via the integrin-linked kinase. *J. Biol. Chem.* **283,** 18812–18820.

Ying, Q. L., Nichols, J., Chambers, I., and Smith, A. (2003). BMP induction of Id proteins suppresses differentiation and sustains embryonic stem cell self-renewal in collaboration with STAT3. *Cell* **115,** 281–292.

Zhang, Z., and Teng, C. T. (2003). Phosphorylation of Krüppel-like factor 5 (KLF5/IKLF) at the CBP interaction region enhances its transactivation function. *Nucleic Acids Res.* **31,** 2196–2208.

Zhang, P., Andrianakos, R., Yang, Y., Liu, C., and Lu, W. (2010). Kruppel-like factor 4 (Klf4) prevents embryonic stem (ES) cell differentiation by regulating Nanog gene expression. *J. Biol. Chem.* **285,** 9180–9189.

Zhao, S., Nichols, J., Smith, A. G., and Li, M. (2004). SoxB transcription factors specify neuroectodermal lineage choice in ES cells. *Mol. Cell. Neurosci.* **27,** 332–342.

Zhao, Y., Hamza, M. S., Leong, H. S., Lim, C. B., Pan, Y. F., Cheung, E., Soo, K. C., and Iyer, N. G. (2008). Kruppel-like factor 5 modulates p53-independent apoptosis through Pim1 survival kinase in cancer cells. *Oncogene* **27,** 1–8.

Zhu, N., Gu, L., Findley, H. W., Chen, C., Dong, J. T., Yang, L., and Zhou, M. (2006). KLF5 interacts with P53 in regulating survivin expression in acute lymphoblastic leukemia. *J. Biol. Chem.* **281,** 14711–14718.

CHAPTER NINETEEN

Bam and Bgcn in *Drosophila* Germline Stem Cell Differentiation

Sreejith Perinthottathil *and* Changsoo Kim[1]

Contents

I. Introduction 400
II. Bam Repression of Stem Cell Maintenance Factors 402
A. Bam and Bgcn are a necessary and sufficient differentiation promoting factor 402
B. Bam repression of Nos translation 403
C. Bam repression of Pum function at the protein level 404
D. Bam repression of E-cadherin translation 405
E. Bam repression of germline Piwi expression 405
F. Bam repression of microRNAs via Mei-P26 406
G. Bam repression of Dpp signaling 407
H. Brat, Fu, and Dsmurf repression of Dpp signaling 409
III. Conclusion 409
Acknowledgments 412
References 413

Abstract

The female *Drosophila* reproductive organ, the ovary, has provided researchers with an incisive genetic system with which principle regulation of stem cell maintenance and differentiation has been delineated. An environmental niche regulates a stem cell's asymmetric self-renewal division that produces a daughter stem cell and a differentiated daughter cell, which further differentiate into eggs. A number of extrinsic and intrinsic factors have been identified that are required either for stem cell maintenance or differentiation. Bam/Bgcn complex plays a pivotal role in promoting stem cell differentiation. Recent papers suggest that Bam/Bgcn complex regulates translation of important maintenance factors and is also involved in the regulation of microRNA-dependent translational repression. Here, we focus on Bam and Bgcn repression of stem cell maintenance factors in the differentiation of germline stem cells (GSCs). © 2011 Elsevier Inc.

Hormone Research Center, School of Biological Sciences and Technology, Chonnam National University, Gwangju, South Korea
[1] Corresponding author.

Vitamins and Hormones, Volume 87
ISSN 0083-6729, DOI: 10.1016/B978-0-12-386015-6.00038-X

I. Introduction

Germline stem cells (GSCs) are a source of continuous production of gametes, sperms, and eggs, with the ability to produce a differentiated daughter cell and a daughter stem cell through asymmetric self-renewal cell division. Molecular mechanisms governing stem cell maintenance and differentiation have been unearthed by studying the Drosophila ovary, in which eggs are continuously produced (Fig. 19.1; Lin, 2002; Spradling *et al.*, 2001). GSCs are reliably recognizable from differentiated daughter cells, cystoblasts (CBs) and mitotically dividing cysts. GSCs are anteriorly located, and close to niche cap cells, and possess a single round fusome (also termed a spectrosome) at the anterior region of the GSCs (Fig. 19.1). A fusome is a membranous organelle which can be readily labeled with antibodies against spectrins and Hu-li tai shao (Hts), a Drosophila adducin-like protein (Fig. 19.2; de Cuevas *et al.*, 1996; Lin *et al.*, 1994; Yue and Spradling, 1992).

Two or three GSCs are attached to surrounding stromal niche cells at the anterior end of the tubular structure, the germaria, of the ovarioles (Fig. 19.1). The somatic niche cells provide special microenvironments

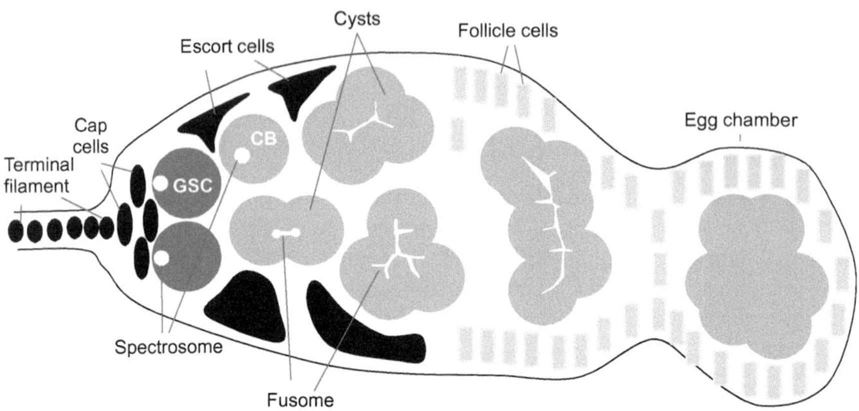

Figure 19.1 Schematic drawing of a germarium. Two to three germline stem cells (GSCs) reside at the anterior tip in association with somatic cap cells. A GSC undergoes asymmetric self-renewal division, producing a daughter GSC that retains association with cap cells, and a cystoblast (CB) displaced one cell away from cap cells (Lin, 2002). The CB undergoes four more incomplete divisions to produce 16-cell germline cysts that are interconnected by ring canals. One of these cells becomes an oocyte and the others become nurse cells. The stem cells and CBs contain a spectrosome, a subcellular organelle enriched in membrane skeletal proteins (Lin *et al.*, 1994). The spectrosome changes its shape to a branched structure called the fusome in the dividing cysts. The 16-cells are surrounded by one layer of follicle cells, giving rise to an egg chamber and budding off from the germaria.

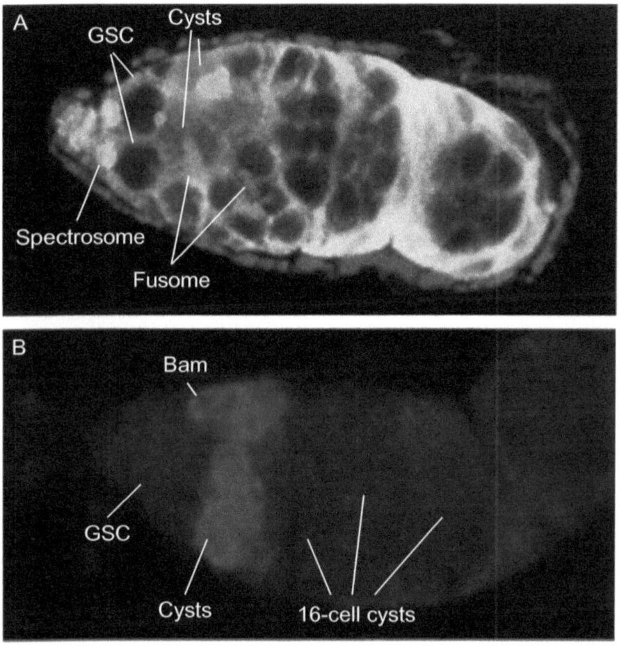

Figure 19.2 Bam expression in dividing cysts. (A) A confocal image of a germarium carrying a GFP reporter bearing 3′-UTR of Bam (Pek *et al.*, 2009) that was double stained with anti-GFP (green) and 1B1 (red) (Lin *et al.*, 1994). 1B1 stains spectrosomes and fusomes. GFP (green) marks GSCs and dividing cysts. (B) Wild-type germarium stained with Bam-C antibody (Ohlstein and McKearin, 1997). Bam is not detected in the GSCs and late 16-cell cysts, but highly expressed in the dividing cysts (red). (See Color Insert.)

that regulate the asymmetric cell division of stem cells (Spradling *et al.*, 2001; Xie and Spradling, 2000). A daughter cell remains in contact with somatic cap cells through DE-cadherin-mediated adhesion (Song *et al.*, 2002) retaining stem cell properties, while the distant cell, which is one cell away from the niche cap cells, becomes a differentiating daughter cell (the CB cell), that activates its own differentiation program. While stem cells can divide indefinitely, the cystoblast undergoes exactly four rounds of synchronous divisions with incomplete cytokinesis, to generate cysts of 2, 4, 8, and 16 that are interconnected through intercellular structures termed ring canals and fusomes (Lin, 2002). Of the 16, one cell becomes an oocyte, while other 15 cells become nurse cells.

Extrinsic and intrinsic factors have been identified through isolation of sterile mutants affecting GSC self-renewal and differentiation (Xie *et al.*, 2005). One group of mutants generated ovarian "tumors" full of undifferentiated GSCs-like cells. Ovarian tumors are formed by a block of the progression into differentiation of stem cells by mutations in *bag-of-marbles* (*bam*) (McKearin and Spradling, 1990), *benign gonial cell neoplasm* (*bgcn*) (Ohlstein *et al.*, 2000), and *mei-P26* (Page *et al.*, 2000). These groups of genes represent differentiating-promoting factors, which are required for stem cell differentiation. The mutations of other groups led to depletion of germ cells in the germarium, resulting in rudimentary germarium which is caused by precociously differentiated stem cells that left the germaria. These groups include *pumilio* (*pum*), *nanos* (*nos*) (Lin and Spradling, 1997), *piwi* (Cox *et al.*, 2000), and *fs(1)Yb* (King and Lin, 1999). These groups represent stem cell maintenance factors that are required for stem cell maintenance. One of the important mechanisms in stem cell maintenance is *decapentaplegic* (*dpp*) signaling, which was identified from the overexpression phenotype of *dpp* which led to ovarian "tumors" phenotypes (Xie and Spradling, 1998). Molecular interplay between these factors has been the focus of intense research, revealing the principles governing stem cell maintenance and differentiation. Emergent themes are that maintenance factors repress differentiation-promoting factors in the GSCs, while differentiation factors repress maintenance factors in the CBs. In this review, we will focus on Bam and Bgcn which downregulate stem cell maintenance factors in the CBs to promote stem cell differentiation.

II. Bam Repression of Stem Cell Maintenance Factors

A. Bam and Bgcn are a necessary and sufficient differentiation promoting factor

bam mutant flies were identified through mutagenesis as being female sterile, producing no eggs (McKearin and Spradling, 1990). *bam* mutant ovaries contain a few small abnormal egg chambers filled with hundreds of undifferentiated single cells. The "tumorous" ovarian phenotypes suggest that Bam requires transition of GSCs into differentiation. Molecular cloning revealed that Bam is a novel protein with no apparent motifs except for a consensus for PEST sequence, which is related to protein instability (Rogers *et al.*, 1986). Bam transcripts are detected in CBs and very early cystocytes (McKearin and Spradling, 1990), but absent in GSCs (Fig. 19.2). Bam protein is localized to the cytoplasm (Bam-C) and fusomes (Bam-F). Bam-C accumulation parallels the bam transcript. Interestingly, Bam-F is detectable in all germline cells, including GSC (McKearin and Ohlstein, 1995). *bgcn* mutations resulted in "tumorous" ovaries, with phenotypes similar to *bam* mutations (Lavoie *et al.*, 1999). Bgcn is distantly related to the DExH-box ATP-dependent RNA helicase, lacking motifs for ATP binding and unwinding (Ohlstein *et al.*, 2000). Bgcn mRNA expression is very similar to bam except that it is additionally found in the GSC where cytoplasmic Bam (Bam-C) protein is not detectable (Ohlstein *et al.*, 2000).

Several observations strongly suggest that Bam and Bgcn proteins may form a protein complex. First, Bam-F staining is absent in *bgcn* mutant germaria (Lavoie *et al.*, 1999). Second, Bgcn acts as a dominant enhancer of the *bam* mutant phenotype (Ohlstein *et al.*, 2000). Third, heat-induced ectopic Bam expression eliminated stem cells only in the presence of Bgcn function (Lavoie *et al.*, 1999; Ohlstein and McKearin, 1997). In accordance with genetic interactions, three groups have reported that Bam physically interacts with Bgcn to form a protein complex (Kim *et al.*, 2010; Li *et al.*, 2009; Shen *et al.*, 2009). Protein expression pattern and loss-of-function mutant phenotypes suggest that both Bam and Bgcn are required in the CBs for promoting their differentiation. Ectopic Bam expression in the GSCs triggered differentiation of GSCs, resulting in empty germaria, indicating that Bam and Bgcn are principle factors promoting differentiation of stem cells (Chen and McKearin, 2003a; Ohlstein and McKearin, 1997). What is the molecular mechanism of the Bam/Bgcn complex in promoting GSC differentiation? Recent evidence suggests that the Bam/Bgcn complex downregulates important stem cell maintenance factors, including Pumilio (Pum)/Nanos (Nos), E-cadherin, eIF4F, germline Piwi, microRNAs, and Dpp signaling.

B. Bam repression of Nos translation

pum and *nos* were originally isolated from mutants with affected embryo segmentation (Lehmann and Nusslein-Volhard, 1987, 1991; Macdonald, 1992; Struhl, 1989) and were subsequently shown to be required for stem cell maintenance (Forbes and Lehmann, 1998; Lin and Spradling, 1997; Parisi and Lin, 1999). Embryo patterning requires a ternary complex involving Pum, Nos, and Brain tumor (Brat) to repress Hunchback (Hb) translation (Sonoda and Wharton, 1999, 2001; Wharton *et al.*, 1998). The Hb 3′-UTR contains a specific sequence known as the *N*anos *R*esponsive *E*lement (NRE) to which the Pum protein binds and recruits Nos and Brat proteins (Murata and Wharton, 1995; Sonoda and Wharton, 2001; Wharton and Struhl, 1991). Since Pum and Nos are both expressed in the GSCs, it was tempting to speculate that Pum and Nos repress factors required for differentiation in the GSCs (Forbes and Lehmann, 1998; Parisi and Lin, 1999; Wang and Lin, 2004; Wang *et al.*, 1994). Identifying Pum/Nos targets in the GSCs has been an important focus of research.

Since Pum/Nos is a stem cell maintenance factor, its activity should be downregulated in the CBs. Bam is a differentiation-promoting factor so that it is possible that Bam could downregulate Pum function in the CBs. This possibility was examined by genetic epistasis analysis. In the *pum bam* double mutant germaria all GSCs went into differentiation which resembled phenotypes of *pum* mutant germaria (Chen and McKearin, 2005; Szakmary *et al.*, 2005). This result suggests that Bam, a differentiation-promoting factor, is not required for differentiation that is triggered in the absence of Pum. In other words, Bam's main role in promotion of stem cell differentiation is to block the Pum/Nos function that prevents differentiation. Consistent with this hypothesis, reduction of one copy of Bam partially rescued the hypomorphic *nos* mutant ovaries (Li *et al.*, 2009). The question is how Bam downregulates Pum/Nos function.

Li *et al.* (2009) demonstrated that Bam downregulates Nos translation via the 3′-UTR of Nos transcripts. Nos protein accumulation is completely reciprocal to Bam protein accumulation; Nos protein is high in GSCs, and declines in the CB, and cystocytes of two-, four-, and eight-cell cysts, and then increases again to high levels in 16-cell cysts, whereas Bam protein is absent in GSCs, but high in CBs and dividing cystocyts, and then declines in 16-cell cysts (Li *et al.*, 2009; Ohlstein and McKearin, 1997; Verrotti and Wharton, 2000; Wang and Lin, 2004; Wang *et al.*, 1994). The Nos protein decline is not due to transcriptional repression since a green fluorescence protein (GFP) reporter fused to the Nos promoter is present throughout the germarium (Li *et al.*, 2009). In contrast, a GFP reporter bearing Nos 3′-UTR recapitulated Nos protein accumulation, suggesting that the Nos 3′-UTR is responsible for Nos protein accumulation (Li *et al.*, 2009). Ectopic Bam expression in GSCs and 16-cell cysts reduced the levels of Nos protein containing the Nos 3′-UTR. Thus Bam acts via the Nos 3′-UTR to repress Nos accumulation. Of note, tethering Bam to the 3′-UTR of a reporter caused repression of translation of the reporter; importantly this repression does not require Bgcn (Shen *et al.*, 2009), raising the possibility that Bam acts as a translational repressor and Bgcn's role is to recruit Bam to the 3′-UTR of its target mRNAs. Bam physically and genetically interacts with eIF4A, a component of the translation initiation complex eIF4F, which promotes translation initiation (Gingras *et al.*, 1999). So a plausible hypothesis could be that Bam is recruited to the 3′-UTR in conjunction with Bgcn inhibiting eIF4A interacting with other translational initiation complexes, thus inhibiting translation of the transcript. Currently, it is not known whether or not Bam/Bgcn directly binds to the Nos 3′-UTR.

C. Bam repression of Pum function at the protein level

Genetic epistasis analysis indicated that Bam antagonizes Pum/Nos function (Chen and McKearin, 2005; Szakmary *et al.*, 2005), prompting us to investigate direct inhibition of Pum/Nos function by Bam. We examined whether Bam physically interacts with Pum/Nos. We exploited diverse tools such as yeast two- and trihybrid assays, protein fragment complementation assays, as well as pull-down assays. Different approaches showed similar results that Bam physically interacts with Pum, and the interaction is strongly augmented by the presence of Bgcn (Kim *et al.*, 2010). There is no direct interaction observed between Pum and Bgcn. We discovered a quaternary complex including Bam, Bgcn, Pum, and Nos in which Bam interacts with the N-terminal region of Pum, while Nos interacts with the C-terminal Puf domain. In HEK293 cells, Bam binding to Pum is responsible for Bam inhibition of Pum translational repression. Thus, in the CBs, it could be possible that Bam binding to Pum will reduce the ability of Pum/Nos function at the protein level, so that translational repression of differentiation genes is derepressed.

D. Bam repression of E-cadherin translation

E-cadherin encoded by *shotgun* (*shg*) mediates cell adhesion between GSC and cap cells (Song *et al.*, 2002). This anchorage to the cap cells is essential for self-renewal of GSCs as judged by the GSCs loss when E-cadherin is absent in the GSCs (Song *et al.*, 2002). *bam* mutant GSCs express more E-cadherin than wild-type GSCs (Jin *et al.*, 2008). Conversely, ectopic Bam expression in the GSCs downregulated E-cadherin accumulation between cap cells and GSCs (Jin *et al.*, 2008). These data suggest that E-cadherin protein level depends on Bam protein level. In S2 cells, coexpression of Bam and Bgcn reduced expression of a reporter bearing the Shg (E-cadherin) 3′-UTR (Shen *et al.*, 2009), suggesting that Bam/Bgcn acts via the 3′-UTR of E-cadherin. It remains to be determined if Bam/Bgcn directly binds the 3′-UTR of E-cadherin.

E. Bam repression of germline Piwi expression

Female sterile mutant screening identified *piwi* (*P*-element *i*nduced *wi*mpy testis) (Cox *et al.*, 1998; Lin and Spradling, 1997). The cloned *piwi* gene encodes a novel protein conserved in all metazoans (Cox *et al.*, 2000). Piwi proteins bind a small noncoding RNAs called piRNAs (Piwi-associated RNA) and Piwi-piRNA complex executes epigenetic regulation (Lin and Yin, 2008). Piwi is expressed in apical somatic cells (niche Piwi) (Szakmary *et al.*, 2005) and also expressed in GSCs (germline Piwi) (Cox *et al.*, 2000; Szakmary *et al.*, 2005). *piwi* mutant ovarioles contain germaria that are depleted of germline cells, suggesting that Piwi is required for GSC maintenance (Cox *et al.*, 1998; Lin and Spradling, 1997). This Piwi's role depends on somatic but not germline Piwi (Cox *et al.*, 2000; Szakmary *et al.*, 2005).

bam and *piwi* double mutant ovarioles are filled with hundreds of undifferentiated germ cells, phenotypes similar to that of *bam* mutant ovaries (Chen and McKearin, 2005; Szakmary *et al.*, 2005), suggesting that Piwi's role in stem cell maintenance is achieved through antagonizing Bam function. Consistent with this notion, overexpression of Piwi in apical somatic cells increases the number of GSC-like cells which lack Bam transcriptional expression (Cox *et al.*, 2000). Conversely, Bam transcription is derepressed in the GSCs when Piwi is removed (Chen and McKearin, 2005), indicating that Bam transcriptional silencing in the GSCs requires niche Piwi function. Since Dpp signaling is responsible for repression of Bam transcription, it is possible that Piwi functions through augmenting Dpp signaling.

Dsmurf is a HECT domain-containing ubiquitin E3 ligase that has been shown to downregulate Dpp signaling in somatic cells (Liang *et al.*, 2003; Podos *et al.*, 2001). One possibility was that Piwi could inhibit Dsmurf to augment Dpp signaling in the GSCs. Ovarioles from *Dsmurf* mutant increased the number of GSCs that are active in Dpp signaling (Casanueva and Ferguson, 2004), as ectopic Piwi expression does (King *et al.*, 2001). *piwi Dsmurf* double mutant females resulted in supernumerary GSCs similar to that observed for the *Dsmurf* mutant, suggesting that Piwi acts by antagonizing Dsmurf function (Chen and McKearin, 2005). In addition, *piwi Dsmurf* double mutant GSC-like cells did not express Bam protein, suggesting that *piwi Dsmurf* cells augmented Dpp signaling and Piwi's role is to repress Dsmurf (Chen and McKearin, 2005). In the *bam* mutant germaria, Dpp signaling is active only in cells close to the niche. Of note, in the *Dsmurf bam* double mutant germaria, Dpp signaling is active in most germ cells, indicating that Dsmurf downregulates Dpp signaling in cells distant from niche cells, and Dsmurf repression of Dpp signaling does not require Bam function (Casanueva and Ferguson, 2004). Indeed, Dsmurf recently has been shown in conjunction with Fused (Fu), a serine/threonine kinase, to downregulate the activated form of Thick veins (Tkv) (Xia *et al.*, 2010). It remains to be determined how niche Piwi downregulates Dsmurf function.

Germline Piwi is responsible for GSC division since germ cell proliferation in *bam* mutants requires germline Piwi (Szakmary *et al.*, 2005). Germline Piwi and Bam are expressed in reciprocal patterns in GSCs and CBs where Piwi is preferentially expressed in GSCs and late mitotic cysts, downregulated in CBs and early mitotic cysts, while Bam expression is high in CBs and early mitotic cysts, less in GSCs and late mitotic cysts (Szakmary *et al.*, 2005). *bam* mutant GSC-like cells all express Piwi. Brief ectopic Bam expression specifically downregulated the germline Piwi accumulation; interestingly, Piwi protein is displaced from the nucleus to cytoplasm prior to degradation (Szakmary *et al.*, 2005). The molecular basis of this regulation is currently not known.

F. Bam repression of microRNAs via Mei-P26

MicroRNAs constitute a small noncoding RNAs and regulate gene expression by controlling mRNA stability and translation (Filipowicz *et al.*, 2005). Mutations in *argonaute 1* (*AGO1*) (Neumuller *et al.*, 2008), *dicer-1* (Jin and Xie, 2007), and *loquacious* (*loqs*) (Park *et al.*, 2007) all cause stem cell differentiation, indicating that microRNAs play an important role in the maintenance of GSCs. Neumuller *et al.* (2008) made an important contribution that Bam could downregulate microRNA genesis and function via Mei-P26, a Trim-NHL (tripartite motif and Ncl-1, HT2A, and Lin-41) protein (Reymond *et al.*, 2001).

Mei-P26 mRNA and protein levels are detectable in stem cells through 16-cell cysts (Neumuller *et al.*, 2008; Tastan *et al.*, 2010). *mei-P26* mutations cause tumorous germarium (Page *et al.*, 2000), which are filled with GSC-like single cells containing spectrosome and early cysts containing branched fusomes (Neumuller *et al.*, 2008; Page *et al.*, 2000). Mei-P26 overexpression in ovaries depleted GSCs and germline, with phenotypes similar to Bam overexpression. Interestingly, the overexpression phenotype of Bam and Mei-P26 requires each other, suggesting that Mei-P26 and Bam are interdependent. The molecular nature of the interdependency between Bam and Mei-P26 is currently unknown.

Interestingly, Mei-P26 physically interacts with AGO1 via the NHL domain of Mei-P26 (Neumuller *et al.*, 2008). The direct interaction appears to be important for Mei-P26 function since Mei-P26 lacking the NHL domain fails to ablate GSCs with its overexpression. In *mei-P26* mutants, most microRNAs examined are upregulated. Interestingly, in the *bam* mutants, most microRNAs examined are also upregulated (Neumuller *et al.*, 2008). Thus, Bam and Mei-P26 appear to promote differentiation by downregulating microRNAs genesis. Consistently, Mei-P26 mutant phenotypes are partially rescued by removing one copy of Loquacious, suggesting that Mei-P26 acts by inhibiting the microRNA pathway in cystocytes. In particular, Mei-P26 inhibits Bantam microRNA activity which is active in stem cells but repressed in cystocytes. Bantam microRNA is active in all germ cells in the *mei-P26* mutant germaria, confirming that Mei-P26 inhibits Bantam microRNA activity in dividing cysts (Neumuller *et al.*, 2008). Heterozygotes for *bantam* null mutations have a reduced number of stem cells, suggesting that Bantam microRNAs play a role for stem cell maintenance (Neumuller *et al.*, 2008). In S2 cells, Mei-P26 expression derepressed expression of a reporter harboring a sequence that can be bound by Bantam microRNAs; importantly, this derepression requires the NHL domain of Mei-P26, a domain required for interaction with AGO1, suggesting that Mei-P26 functions through physical interaction with AGO1 to downregulate microRNA activity.

G. Bam repression of Dpp signaling

In *Drosophila*, *decapentaplegic* (Dpp, a Drosophila homolog of human bone morphogenetic proteins 2 and 4 (BMP2/4)) binds both type I serine/threonine kinase transmembrane receptors (Saxophone (Sax) and Tkv) and type II receptor (Punt). Activated type 1 kinases activate *Drosophila*

R-Smads, Mothers against dpp (Mad, a *Drosophila* Smad1/5 homolog) by phosphorylation, which associate with Co-Smad, Medea (Med, a *Drosophila* Smad4 homolog) and transferred into nucleus to act as a transcriptional regulator. Daughters against dpp (Dad, a Smad6/7 homolog) is an I-Smad which acts as a negative regulator of Dpp signaling which is expressed by Dpp signaling, so it is often used for reporting active Dpp signaling (Raftery and Sutherland, 1999; Tsuneizumi *et al.*, 1997).

Overexpression of Dpp resulted in "tumorous" germaria filled with GSC-like cells, a phenotype similar to *bam* mutant ovaries (Xie and Spradling, 1998). GSCs lacking Mad or Medea differentiate precociously, suggesting that Dpp signaling is essential for maintenance of stem cells (Xie and Spradling, 1998). Germ cells mutant for both Mad and Bam contained round spectrosomes, with phenotypes similar to *bam* single mutant germ cells, indicating that Mad's role in stem cell maintenance is to antagonize Bam function (Casanueva and Ferguson, 2004). In accord with this hypothesis, Dpp signaling directly represses transcription of Bam through a transcriptional silencer element (SE) located in the Bam promoter (Chen and McKearin, 2003b). *Drosophila* Smads, Mad, and Medea, bind to the Bam SE (Chen and McKearin, 2003a; Song *et al.*, 2004). Thus, lack of Bam protein in the GSCs is mainly due to transcriptional repression by Dpp signaling.

To accumulate Bam in the CBs, Dpp signaling should be weakened in the CBs. Indeed Dpp signaling is precipitately reduced in the CBs as shown by absence of activated form of MAD, pMAD (phosphorylated MAD) in GSCs (Kai and Spradling, 2003). Dad-lacZ expression (a marker of active Dpp signaling) is confined to GSCs and is rapidly reduced in CBs (Casanueva and Ferguson, 2004; Kai and Spradling, 2003). Dpp expression together with another BMP ligand Glass bottom boat (Gbb) is widely expressed in the cap cells and escort cells (Casanueva and Ferguson, 2004; Song *et al.*, 2004; Xie and Spradling, 2000; Zhu and Xie, 2003) that surround CBs so that the rapid reduction of Dpp signaling in the CBs was an interesting question to be addressed.

Overexpression of Tkv(ca), the constitutively active Dpp receptor, in the germarium resulted in tumorous ovarioles filled with GSC-like cells (Casanueva and Ferguson, 2004). When Bam is coexpressed in the Tkv(ac) expressing germaria, differentiated cells containing branched fusome appear. Of note, Dad-lacZ expression is reduced in spectrosome-containing single cells, and eliminated in fusome-containing germ cells (Casanueva and Ferguson, 2004), indicating that Bam downregulated dpp signaling prior to overt differentiation. Nos protein was found in all spectrosome-containing cells, but was absent in two- to eight-cell cysts (Casanueva and Ferguson, 2004), indicating that Bam repression of Dpp signaling comes prior to Bam's repression of Nos expression. Unraveling the molecular mechanism of Bam downregulation of Dpp signaling remains to be addressed.

H. Brat, Fu, and Dsmurf repression of Dpp signaling

Brat belongs to the Trim-NHL family (Arama *et al.*, 2000; Harris *et al.*, 2011). Brat transcript is widely expressed in the germline, yet Brat protein is excluded from GSCs, but high in CBs and dividing cysts (Harris *et al.*, 2011). Brat's downregulation in the GSCs is due to Pum/Nos repression at the translational level (Harris *et al.*, 2011). The *brat* mutant germaria had more GSCs expressing pMad and Dad-lacZ (Harris *et al.*, 2011), suggesting that normal Brat functions are to repress Dpp signaling in the CBs. Ectopic expression of Brat in all germ cells resulted in the complete loss of germline (Harris *et al.*, 2011); this Brat's function requires Pum. Consistently Pum protein is codetectable with Brat protein in CBs, where Nos is absent. In S2 cells, Pum and Brat coexpression repressed accumulation of a reporter bearing the Mad 3′-UTR; this repression does not require Nos function (Harris *et al.*, 2011). Thus, Brat in conjunction with Pum might downregulate Mad at the translational level in the CBs.

Xia *et al.* reported an interesting observation that expression of Tkv(ca) in GSCs blocked stem cell differentiation, resulting in tumorous germarium, while expression of Tkv(ca) in the CBs failed to block differentiation, resulting in normal germline development, indicating that in the CBs, Tkv signaling is actively repressed (Xia *et al.*, 2010). Importantly, repression of Tkv(ca) signaling in CB is relieved in the absence of either Dsmurf or Fu, indicating that Dsmurf or Fu plays a critical role in the repression of Tkv downstream signaling. Overexpression of the ubiquitin-resistant form of Tkv (ca) in the CBs resulted in a tumorous germarium, suggesting that ubiquitination of Tkv is critical to Tkv downregulation in the CBs. In CBs, activated Tkv, but not Tkv is downregulated at protein level (Xia *et al.*, 2010). Fu and Dsmurf physically interact with Tkv to form a trimeric complex. Dsmurf ubiquitinates Tkv in a Fu-dependent manner. Knockdown of Fu in S2 cells increased accumulation of the Tkv protein, suggesting that Fu downregulates Tkv protein accumulation. Since kinase activity of Fu is required for Fu-mediated ubiquitination of Tkv, it remains to identify Fu targets that are needed for Tkv ubiquitinylation. Ectopic GSC-like cells in the *fu* mutant ovaries are active in Dpp signaling, indicating that Fu negatively regulates Dpp signaling. It would be interesting to examine whether Bam function is related to the regulation of Dsmurf and Fu in the CBs.

III. Conclusion

Asymmetric division of ovarian GSCs relies on interplay between diverse maintenance factors, including Pum/Nos, microRNAs, Piwi signaling, fs(1)Yb (King and Lin, 1999; King *et al.*, 2001), Dpp signaling,

Pelota (Xi *et al.*, 2005), Otefin (Jiang *et al.*, 2008), Stonewalling (Maines *et al.*, 2007), ISWI (Xi and Xie, 2005), E-cadherin, eIF4A, Dally (Guo and Wang, 2009), and differentiation-promoting factors, Bam, Bgcn, Mei-P26, Brat, Fu, and Dsmurf (Fig. 19.3). Logic unearthed from molecular genetic analysis is that differentiation-promoting factors are repressed in the GSCs by maintenance factors. For example, in the GSCs, Pum/Nos represses Brat translation: Dpp signaling in conjunction with Dally and Ote represses Bam transcription; Piwi represses Dsmurf function. Conversely, in the CBs, differentiating-promoting factors repress functions of maintenance factors: Brat in conjunction with Pum downregulates Mad translation; Bam downregulates Nos and E-cadherin at the translational level. Bam also represses germline Piwi and microRNA function in conjunction with Mei-P26.

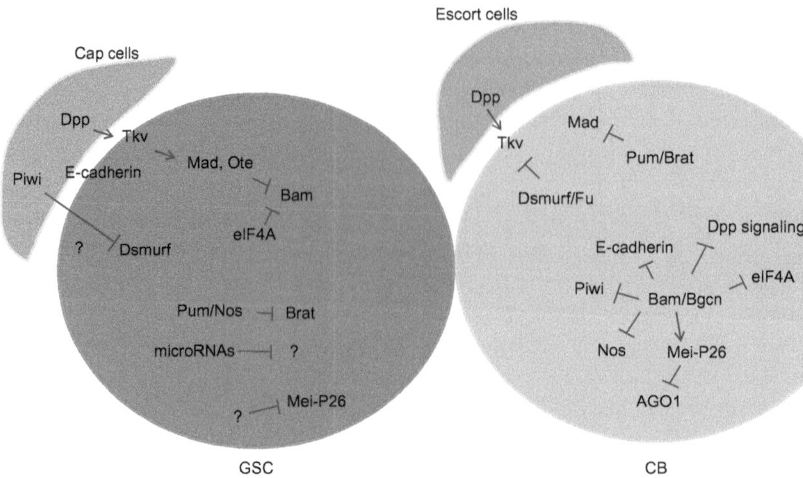

Figure 19.3 Molecular interplay between stem cell maintenance factors and differentiation-promoting factors. In the GSCs, differentiating-promoting factors, Bam, Brat, and Mei-P26 must be repressed. Bam is transcriptionally silenced by Dpp signaling with Ote (Jiang *et al.*, 2008). Recently, eIF4A binds and inhibits residual Bam function in the GSCs (Shen *et al.*, 2009). Pum with its obligatory partner Nos represses Brat translation through 3′-UTR (Harris *et al.*, 2011). AGO1 and microRNAs inhibit unknown genes required for differentiation. Low level of Mei-P26 protein should be blocked by unknown mechanism. Dsmurf is inhibited by niche Piwi by unknown mechanism. In the CBs, stem cell maintenance factors must be repressed by differentiation-promoting factors. Dsmurf/Fu complex downregulates activated form of Tkv (Xia *et al.*, 2010). Brat in conjunction with Pum represses translation of Mad through 3′-UTR. Bam/Bgcn plays a pivotal role in repressing diverse stem cell maintenance factors. Bam/Bgcn represses Nos and E-cadherin at the translational level through 3′-UTR (Li *et al.*, 2009; Shen *et al.*, 2009). Bam/Bgcn downregulates Piwi by unknown mechanism. Mei-P26 in conjuction with Bam/Bgcn binds and inhibits AGO1 to repress microRNA-mediated gene silencing (Neumuller *et al.*, 2008). Exact mechanisms for this regulation are not known. (For color version of this figure, the reader is referred to the Web version of this chapter.)

Bam accumulation in the CBs is a critical event in the transition of GSCs into CBs since Bam is a key differentiation-promoting factor in the CBs. To achieve Bam accumulation in the CBs Dpp signaling should be reduced in the CBs. How is Dpp signaling reduction initiated in the CBs? Recent papers provide two possible ways of reducing Dpp signaling in the CBs. One is an extrinsic way in which Dpp is trapped and concentrated on niche cap cells so that GSCs close to cap cells receive high dpp concentration (Guo and Wang, 2009; Hayashi *et al.*, 2009), while one cell away from the cap cells receives a reduced Dpp concentration. Another is an intrinsic way in which Dpp signaling is actively reduced in the CBs. Dsmurf in conjunction with Fu plays a critical role in this intrinsic way of reduction of Dpp signaling by downregulating activated form of Tkv in the CBs. In this scenario, either Dsmurf or Fu should be inactive in the GSCs and active in the CBs. The molecular nature of this regulation is not known, but niche Piwi might play a critical role in this regulation since Dsmurf function is downregulated in the GSCs by niche Piwi (Chen and McKearin, 2005). Thus, we propose the following scenario (Fig. 19.3). Dsmurf is derepressed from niche Piwi in the CBs since CBs is moved away from localized niche Piwi, even though the nature of this niche Piwi is unknown. Activated Dsmurf in the CBs forms a protein complex with Fu to actively downregulate activated form of Tkv, culminating in Dpp signaling reduction in the CBs. Reduction of Dpp signaling will derepress Bam transcription and thus Bam protein begins to accumulate in the CBs. Accumulated Bam downregulates Nos and E-cadherin through translational repression. Downregulation of Nos will lead to accumulation of Brat since Brat is downregulated by Pum/Nos in the GSCs. Brat in conjunction with Pum downregulates Mad at the translational level, culminating in deeper repression of Dpp signaling.

Critical issues to be addressed include investigating how Bam downregulates diverse regulators. In the case of Nos and E-cadherin, Bam acts through their 3′-UTR to downregulate their translation (Fig. 19.4). Tethered Bam repressed its mRNA translation, raising a possibility that Bam by itself is a translational repressor. Bam forms a complex with Bgcn that is a potentially RNA binding protein, so that possibly Bam is recruited to the 3′-UTR by Bgcn. Since Bam physically and genetically interacts with eIF4A, a component of translation initiation, tethered Bam could interfere eIF4A interacting with other components involved in translation initiation, resulting in downregulation of translation. Bam also could downregulate germline Piwi in analogous manner, but this invites future research.

One important function of Bam is to downregulate the genesis or function of microRNAs that are essential for repressing differentiation-promoting genes in the GSCs. In the *bam* mutant ovaries, a number of microRNAs are upregulated. Bam's differentiation-promoting function requires Mei-P26 and Mei-P26's differentiation-promoting function also requires Bam. So this interdependence suggests that Bam may physically

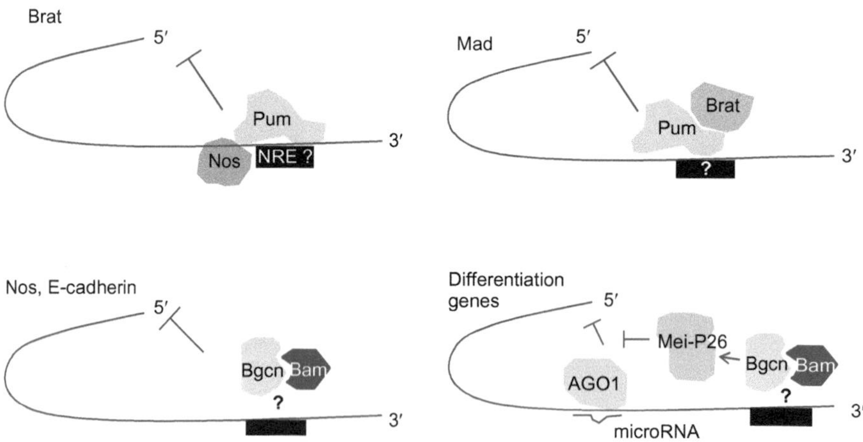

Figure 19.4 Translational regulation between stem cell maintenance and differentiation-promoting factors. In the GSCs, Brat translation is repressed by Pum/Nos through the 3′-UTR (Harris *et al.*, 2011). In the CBs, Mad translation is repressed by Brat in conjunction with Pum through the 3′-UTR by an unknown mechanism (Harris *et al.*, 2011). Translation of Nos and E-cadherin is repressed by Bam and Bgcn through the 3′-UTR (Li *et al.*, 2009; Shen *et al.*, 2009) by an unknown mechanism. Mei-P26 in conjunction with Bam/Bgcn represses AGO1 function (Neumuller *et al.*, 2008). The exact nature of Bam/Bgcn function in these processes requires further research. (For color version of this figure, the reader is referred to the Web version of this chapter.)

interact with Mei-P26. Of note, Mei-P26 physically binds and inhibits AGO1. The exact nature of interdependence between Bam and Mei-P26 and detailed role of Bam in repression of genesis or function of microRNAs should invite further investigation.

Another question to be addressed is how Brat represses Mad translation. Data clearly show that Brat's repression of Mad translation requires Pum, but not Nos. Consistently, Brat is coexpressed in the CBs with Pum, but Nos is absent in the CBs. Given that Brat's NHL domain interacts with the Puf domain (of Pum)-NRE-Nos complex, but not with Puf domain alone nor Nos alone (Sonoda and Wharton, 2001), it remains to be determined that there is physical interaction between full-length Brat and full-length Pum without Nos. In the CBs, a multiprotein complex involving Bam, Bgcn, Pum, and Brat might be formed that could play a role in many aspect of Bam function.

ACKNOWLEDGMENTS

This work was supported by research grants from the Basic Science Research program from the National Research Foundation (NRF) 2010-1398 (to C. K.) and the BK21 Program (to S. P.).

REFERENCES

Arama, E., Dickman, D., Kimchie, Z., Shearn, A., and Lev, Z. (2000). Mutations in the beta-propeller domain of the Drosophila brain tumor (brat) protein induce neoplasm in the larval brain. *Oncogene* **19,** 3706–3716.

Casanueva, M. O., and Ferguson, E. L. (2004). Germline stem cell number in the Drosophila ovary is regulated by redundant mechanisms that control Dpp signaling. *Development* **131,** 1881–1890.

Chen, D., and McKearin, D. (2003a). Dpp signaling silences bam transcription directly to establish asymmetric divisions of germline stem cells. *Curr. Biol.* **13,** 1786–1791.

Chen, D., and McKearin, D. M. (2003b). A discrete transcriptional silencer in the bam gene determines asymmetric division of the Drosophila germline stem cell. *Development* **130,** 1159–1170.

Chen, D., and McKearin, D. (2005). Gene circuitry controlling a stem cell niche. *Curr. Biol.* **15,** 179–184.

Cox, D. N., Chao, A., Baker, J., Chang, L., Qiao, D., *et al.* (1998). A novel class of evolutionarily conserved genes defined by piwi are essential for stem cell self-renewal. *Genes Dev.* **12,** 3715–3727.

Cox, D. N., Chao, A., and Lin, H. (2000). Piwi encodes a nucleoplasmic factor whose activity modulates the number and division rate of germline stem cells. *Development* **127,** 503–514.

de Cuevas, M., Lee, J. K., and Spradling, A. C. (1996). Alpha-spectrin is required for germline cell division and differentiation in the Drosophila ovary. *Development* **122,** 3959–3968.

Filipowicz, W., Jaskiewicz, L., Kolb, F. A., and Pillai, R. S. (2005). Post-transcriptional gene silencing by siRNAs and miRNAs. *Curr. Opin. Struct. Biol.* **15,** 331–341.

Forbes, A., and Lehmann, R. (1998). Nanos and Pumilio have critical roles in the development and function of Drosophila germline stem cells. *Development* **125,** 679–690.

Gingras, A. C., Raught, B., and Sonenberg, N. (1999). eIF4 Initiation factors: Effectors of mRNA recruitment to ribosomes and regulators of translation. *Annu. Rev. Biochem.* **68,** 913–963.

Guo, Z., and Wang, Z. (2009). The glypican Dally is required in the niche for the maintenance of germline stem cells and short-range BMP signaling in the Drosophila ovary. *Development* **136,** 3627–3635.

Harris, R. E., Pargett, M., Sutcliffe, C., Umulis, D., and Ashe, H. L. (2011). Brat promotes stem cell differentiation via control of a bistable switch that restricts BMP signaling. *Dev. Cell* **20,** 72–83.

Hayashi, Y., Kobayashi, S., and Nakato, H. (2009). Drosophila glypicans regulate the germline stem cell niche. *J. Cell Biol.* **187,** 473–480.

Jiang, X., Xia, L., Chen, D., Yang, Y., Huang, H., *et al.* (2008). Otefin, a nuclear membrane protein, determines the fate of germline stem cells in Drosophila via interaction with Smad complexes. *Dev. Cell* **14,** 494–506.

Jin, Z., and Xie, T. (2007). Dcr-1 maintains Drosophila ovarian stem cells. *Curr. Biol.* **17,** 539–544.

Jin, Z., Kirilly, D., Weng, C., Kawase, E., Song, X., *et al.* (2008). Differentiation-defective stem cells outcompete normal stem cells for niche occupancy in the Drosophila ovary. *Cell Stem Cell* **2,** 39–49.

Kai, T., and Spradling, A. (2003). An empty Drosophila stem cell niche reactivates the proliferation of ectopic cells. *Proc. Natl. Acad. Sci. USA* **100,** 4633–4638.

Kim, J. Y., Lee, Y. C., and Kim, C. (2010). Direct inhibition of Pumilo activity by Bam and Bgcn in Drosophila germ line stem cell differentiation. *J. Biol. Chem.* **285,** 4741–4746.

King, F. J., and Lin, H. (1999). Somatic signaling mediated by fs(1)Yb is essential for germline stem cell maintenance during Drosophila oogenesis. *Development* **126,** 1833–1844.

King, F. J., Szakmary, A., Cox, D. N., and Lin, H. (2001). Yb modulates the divisions of both germline and somatic stem cells through piwi- and hh-mediated mechanisms in the Drosophila ovary. *Mol. Cell* **7,** 497–508.

Lavoie, C. A., Ohlstein, B., and McKearin, D. M. (1999). Localization and function of Bam protein require the benign gonial cell neoplasm gene product. *Dev. Biol.* **212,** 405–413.

Lehmann, R., and Nusslein-Volhard, C. (1987). Hunchback, a gene required for segmentation of an anterior and posterior region of the Drosophila embryo. *Dev. Biol.* **119,** 402–417.

Lehmann, R., and Nusslein-Volhard, C. (1991). The maternal gene nanos has a central role in posterior pattern formation of the Drosophila embryo. *Development* **112,** 679–691.

Li, Y., Minor, N. T., Park, J. K., McKearin, D. M., and Maines, J. Z. (2009). Bam and Bgcn antagonize Nanos-dependent germ-line stem cell maintenance. *Proc. Natl. Acad. Sci. USA* **106,** 9304–9309.

Liang, Y. Y., Lin, X., Liang, M., Brunicardi, F. C., ten Dijke, P., *et al.* (2003). dSmurf selectively degrades decapentaplegic-activated MAD, and its overexpression disrupts imaginal disc development. *J. Biol. Chem.* **278,** 26307–26310.

Lin, H. (2002). The stem-cell niche theory: Lessons from flies. *Nat. Rev. Genet.* **3,** 931–940.

Lin, H., and Spradling, A. C. (1997). A novel group of pumilio mutations affects the asymmetric division of germline stem cells in the Drosophila ovary. *Development* **124,** 2463–2476.

Lin, H., and Yin, H. (2008). A novel epigenetic mechanism in Drosophila somatic cells mediated by Piwi and piRNAs. *Cold Spring Harb. Symp. Quant. Biol.* **73,** 273–281.

Lin, H., Yue, L., and Spradling, A. C. (1994). The Drosophila fusome, a germline-specific organelle, contains membrane skeletal proteins and functions in cyst formation. *Development* **120,** 947–956.

Macdonald, P. M. (1992). The Drosophila pumilio gene: An unusually long transcription unit and an unusual protein. *Development* **114,** 221–232.

Maines, J. Z., Park, J. K., Williams, M., and McKearin, D. M. (2007). Stonewalling Drosophila stem cell differentiation by epigenetic controls. *Development* **134,** 1471–1479.

McKearin, D., and Ohlstein, B. (1995). A role for the Drosophila bag-of-marbles protein in the differentiation of cystoblasts from germline stem cells. *Development* **121,** 2937–2947.

McKearin, D. M., and Spradling, A. C. (1990). Bag-of-marbles: A Drosophila gene required to initiate both male and female gametogenesis. *Genes Dev.* **4,** 2242–2251.

Murata, Y., and Wharton, R. P. (1995). Binding of pumilio to maternal hunchback mRNA is required for posterior patterning in Drosophila embryos. *Cell* **80,** 747–756.

Neumuller, R. A., Betschinger, J., Fischer, A., Bushati, N., Poernbacher, I., *et al.* (2008). Mei-P26 regulates microRNAs and cell growth in the Drosophila ovarian stem cell lineage. *Nature* **454,** 241–245.

Ohlstein, B., and McKearin, D. (1997). Ectopic expression of the Drosophila Bam protein eliminates oogenic germline stem cells. *Development* **124,** 3651–3662.

Ohlstein, B., Lavoie, C. A., Vef, O., Gateff, E., and McKearin, D. M. (2000). The Drosophila cystoblast differentiation factor, benign gonial cell neoplasm, is related to DExH-box proteins and interacts genetically with bag-of-marbles. *Genetics* **155,** 1809–1819.

Page, S. L., McKim, K. S., Deneen, B., Van Hook, T. L., and Hawley, R. S. (2000). Genetic studies of mei-P26 reveal a link between the processes that control germ cell proliferation in both sexes and those that control meiotic exchange in Drosophila. *Genetics* **155,** 1757–1772.

Parisi, M., and Lin, H. (1999). The Drosophila pumilio gene encodes two functional protein isoforms that play multiple roles in germline development, gonadogenesis, oogenesis and embryogenesis. *Genetics* **153,** 235–250.

Park, J. K., Liu, X., Strauss, T. J., McKearin, D. M., and Liu, Q. (2007). The miRNA pathway intrinsically controls self-renewal of Drosophila germline stem cells. *Curr. Biol.* **17,** 533–538.

Pek, J. W., Lim, A. K., and Kai, T. (2009). Drosophila maelstrom ensures proper germline stem cell lineage differentiation by repressing microRNA-7. *Dev. Cell* **17,** 417–424.

Podos, S. D., Hanson, K. K., Wang, Y. C., and Ferguson, E. L. (2001). The DSmurf ubiquitin-protein ligase restricts BMP signaling spatially and temporally during Drosophila embryogenesis. *Dev. Cell* **1,** 567–578.

Raftery, L. A., and Sutherland, D. J. (1999). TGF-beta family signal transduction in Drosophila development: From Mad to Smads. *Dev. Biol.* **210,** 251–268.

Reymond, A., Meroni, G., Fantozzi, A., Merla, G., Cairo, S., *et al.* (2001). The tripartite motif family identifies cell compartments. *EMBO J.* **20,** 2140–2151.

Rogers, S., Wells, R., and Rechsteiner, M. (1986). Amino acid sequences common to rapidly degraded proteins: The PEST hypothesis. *Science* **234,** 364–368.

Shen, R., Weng, C., Yu, J., and Xie, T. (2009). eIF4A controls germline stem cell self-renewal by directly inhibiting BAM function in the Drosophila ovary. *Proc. Natl. Acad. Sci. USA* **106,** 11623–11628.

Song, X., Zhu, C. H., Doan, C., and Xie, T. (2002). Germline stem cells anchored by adherens junctions in the Drosophila ovary niches. *Science* **296,** 1855–1857.

Song, X., Wong, M. D., Kawase, E., Xi, R., Ding, B. C., *et al.* (2004). Bmp signals from niche cells directly repress transcription of a differentiation-promoting gene, bag of marbles, in germline stem cells in the Drosophila ovary. *Development* **131,** 1353–1364.

Sonoda, J., and Wharton, R. P. (1999). Recruitment of Nanos to hunchback mRNA by Pumilio. *Genes Dev.* **13,** 2704–2712.

Sonoda, J., and Wharton, R. P. (2001). Drosophila brain tumor is a translational repressor. *Genes Dev.* **15,** 762–773.

Spradling, A., Drummond-Barbosa, D., and Kai, T. (2001). Stem cells find their niche. *Nature* **414,** 98–104.

Struhl, G. (1989). Differing strategies for organizing anterior and posterior body pattern in Drosophila embryos. *Nature* **338,** 741–744.

Szakmary, A., Cox, D. N., Wang, Z., and Lin, H. (2005). Regulatory relationship among piwi, pumilio, and bag-of-marbles in Drosophila germline stem cell self-renewal and differentiation. *Curr. Biol.* **15,** 171–178.

Tastan, O. Y., Maines, J. Z., Li, Y., McKearin, D. M., and Buszczak, M. (2010). Drosophila ataxin 2-binding protein 1 marks an intermediate step in the molecular differentiation of female germline cysts. *Development* **137,** 3167–3176.

Tsuneizumi, K., Nakayama, T., Kamoshida, Y., Kornberg, T. B., Christian, J. L., *et al.* (1997). Daughters against dpp modulates dpp organizing activity in Drosophila wing development. *Nature* **389,** 627–631.

Verrotti, A. C., and Wharton, R. P. (2000). Nanos interacts with cup in the female germline of Drosophila. *Development* **127,** 5225–5232.

Wang, Z., and Lin, H. (2004). Nanos maintains germline stem cell self-renewal by preventing differentiation. *Science* **303,** 2016–2019.

Wang, C., Dickinson, L. K., and Lehmann, R. (1994). Genetics of nanos localization in Drosophila. *Dev. Dyn.* **199,** 103–115.

Wharton, R. P., and Struhl, G. (1991). RNA regulatory elements mediate control of Drosophila body pattern by the posterior morphogen nanos. *Cell* **67,** 955–967.

Wharton, R. P., Sonoda, J., Lee, T., Patterson, M., and Murata, Y. (1998). The Pumilio RNA-binding domain is also a translational regulator. *Mol. Cell* **1,** 863–872.

Xi, R., and Xie, T. (2005). Stem cell self-renewal controlled by chromatin remodeling factors. *Science* **310,** 1487–1489.

Xi, R., Doan, C., Liu, D., and Xie, T. (2005). Pelota controls self-renewal of germline stem cells by repressing a Bam-independent differentiation pathway. *Development* **132,** 5365–5374.
Xia, L., Jia, S., Huang, S., Wang, H., Zhu, Y., *et al.* (2010). The Fused/Smurf complex controls the fate of Drosophila germline stem cells by generating a gradient BMP response. *Cell* **143,** 978–990.
Xie, T., and Spradling, A. C. (1998). decapentaplegic is essential for the maintenance and division of germline stem cells in the Drosophila ovary. *Cell* **94,** 251–260.
Xie, T., and Spradling, A. C. (2000). A niche maintaining germ line stem cells in the Drosophila ovary. *Science* **290,** 328–330.
Xie, T., Kawase, E., Kirilly, D., and Wong, M. D. (2005). Intimate relationships with their neighbors: Tales of stem cells in Drosophila reproductive systems. *Dev. Dyn.* **232,** 775–790.
Yue, L., and Spradling, A. C. (1992). hu-li tai shao, a gene required for ring canal formation during Drosophila oogenesis, encodes a homolog of adducin. *Genes Dev.* **6,** 2443–2454.
Zhu, C. H., and Xie, T. (2003). Clonal expansion of ovarian germline stem cells during niche formation in Drosophila. *Development* **130,** 2579–2588.

CHAPTER TWENTY

The Effects of Mechanical Loading on Mesenchymal Stem Cell Differentiation and Matrix Production

Robin M. Delaine-Smith *and* Gwendolen C. Reilly

Contents

I. Introduction 418
II. Mesenchymal Stem Cells 420
- A. Stem cells 420
- B. Mesenchymal stem cells 421
- C. MSC differentiation *in vitro* 422
- D. Mechanical regulation of MSC function 425
- E. Mechanotransduction 426

III. Mechanical Loading 427
- A. Introduction 427
- B. Tension 428
- C. Compression 429
- D. Fluid flow-induced shear stress 430
- E. Other (direct cell magnetic stimulus, ultrasound, vibration) 432

IV. Experimental Results 433
- A. Tension 433
- B. Compression 441
- C. Fluid flow-induced shear stress 451
- D. Other load types 458
- E. *In vivo* loading 462

V. Conclusions and Future Directions 466

Acknowledgments 469

References 469

The Kroto Research Institute, Department of Materials Science and Engineering, University of Sheffield, Sheffield, United Kingdom

Vitamins and Hormones, Volume 87

ISSN 0083-6729, DOI: 10.1016/B978-0-12-386015-6.00039-1

Abstract

Mesenchymal stem cells or stromal cells (MSCs) have the potential to be used therapeutically in tissue engineering and regenerative medicine to replace or restore the function of damaged tissues. Therefore, considerable effort has been ongoing in the research community to optimize culture conditions for predifferentiation of MSCs. All mesenchymal tissues are subjected to mechanical forces *in vivo* and all fully differentiated mesenchymal lineage cells respond to mechanical stimulation *in vivo* and *in vitro*. Therefore, it is not surprising that MSCs are highly mechanosensitive. We present a summary of current methods of mechanical stimulation of MSCs and an overview of the outcomes of the different mechanical culture techniques tested. Tissue engineers and stem cell researchers should be able to harness this mechanosensitivity to modulate MSC differentiation and matrix production; however, more research needs to be undertaken to understand the complex interactions between the mechanosensitive and biochemically stimulated differentiation pathways.

I. Introduction

Tissue loss or organ failure results in major health problems that can seriously affect the quality and length of a person's life. If the damaged tissue fails to heal spontaneously, then surgical treatments are often needed to restore some function to the affected tissue or organ. The normal practice is to implant graft tissue from a donor site; however, there are complications associated with tissue grafting, dependant on the method. The favored method is to use donor tissue taken from another site in the patient (autologous), but there is a limited supply of available tissue, especially in patients who suffer ongoing disease, and extraction requires further surgical procedures, which can lead to donor site morbidity. Less preferred methods are to take donor tissue from a different individual (allogeneic) or a different species (xenogeneic), but these carry the risk of immunological rejection or the transfer of pathogens (Murugan and Ramakrishna, 2007). An alternative is an artificial replacement, for example, joints often require metal or polymer implants to replace the damaged area, but these do not have the ability to undergo wound healing and remodeling and can also cause mechanical mismatch and undergo wear. These drawbacks are due to the implanted materials being a nonliving tissue which carry a finite lifespan (the average for a total joint replacement is 10–20 years) meaning that future surgery is often needed to revise the implant, especially in younger people.

Tissue engineering (TE) has emerged as a major discipline of regenerative medicine and presents potential solutions to the problems associated with current therapies in the treatment of injuries, diseases, and age-related conditions. The principles of TE involve seeding cells on a suitable scaffolding material, ideally with similar chemical, physical, and mechanical properties to the target tissue. The scaffold and culture environment then promotes the organization, growth, and development of the cells into a fully functioning tissue (Murugan and Ramakrishna, 2007). However, in order for TE strategies to be successful, an appropriate cell type must be sourced that is capable of regenerating new tissue without causing complications. The ideal cell type would be one that is easy to obtain with a minimally invasive biopsy, easily expandable *in vitro*, nonimmunogenic, and possesses the ability to differentiate into a variety of cell types with specialized functions (Stock and Vacanti, 2001).

Mesenchymal stem cells (MSCs) are a promising cell source that match these requirements and have received much attention due to their multipotent differentiation potential, self-renewing ability, and apparent immunosuppressive properties. They can be easily obtained from the stroma of adult bone marrow and cells with similar phenotypic characteristics have been isolated from a range of other tissues such as adipose tissue, tendon, muscle, and skin making them an attractive source of autologous cells. They can also be sourced from fetal tissues such as umbilical cord blood and amniotic fluid for allogenic use. The use of MSCs for regenerative medicine would mean that tissue-specific cells would not be required from the already damaged or diseased tissue. MSCs have been induced *in vitro* to differentiate into cell types of the musculoskeletal system, including bone, cartilage, tendon/ligament, and fat. More controversially, researchers have claimed that MSCs differentiate into other tissue types such as smooth muscle, endothelial, and nervous tissue (for a recent review, see Nombela-Arrieta *et al.*, 2011). There are no strong ethical and political issues associated with MSCs compared to embryonic stem (ES) cells making them an attractive cell source for TE and regenerative medicine therapies.

A key task for tissue engineers is to identify the appropriate culture conditions for development of a tissue engineered construct *in vitro* ready for implantation *in vivo*, that reduces subsequent healing time. External mechanical forces are likely to play a major role in developing these optimum conditions. While there have been advancements in our understanding of the behavior and capabilities of stem cells, we are still unable to precisely control their differentiation, which is a major limiting factor in the development of successful TE therapies. It is well known that biochemical cues, such as cytokines, growth factors, and signaling events (Augello and De Bari, 2010), can control the function of stem cells as well as environmental factors (e.g., surface chemistry and topography) (Dalby *et al.*, 2007; Reilly and Engler, 2010), but it is also becoming clear that mechanical forces can greatly influence stem cell behavior.

Cells and tissues in the human body are subjected to a wide variety of external forces, which influence their growth, development, and maintenance (Vogel, 2006). For example, it is well demonstrated that if bone does not receive adequate loading, as occurs during space flight or prolonged periods of bed rest, then bone mass and density decrease (Janmey and McCulloch, 2007), whereas exercise in the form of high-impact loading increases bone density (Bassey and Ramsdale, 1994). Many researchers have explored these mechanical influences *in vitro* and subsequently, many different cell types have been demonstrated to be highly mechanosensitive (Orr *et al.*, 2006). Recent TE strategies for MSC differentiation have involved attempts to simulate the physiologically relevant mechanical environment by reproducing the forces experienced *in vivo* in an *in vitro* setting. Mechanical stimulation of MSCs *in vitro* has shown that hydrostatic pressure induces chondrogenic differentiation in human MSCs (hMSCs) (Luo and Seedhom, 2007), tensile strain applied to MSC monolayers enhances osteogenesis and inhibits adipogenesis (Sen *et al.*, 2008; Simmons *et al.*, 2003), and fluid shear forces upregulate genes associated with osteogenesis (Arnsdorf *et al.*, 2009; Yourek *et al.*, 2010). However, due to the wide range of types of mechanical stimulus, the large array of possible conditions, and the different chemical stimulants used, the optimum conditions for the differentiation of MSCs along each specific lineage remain elusive. A better understanding of the mechanical control of MSC differentiation is important when designing bioreactor conditions for growing TE constructs. In particular, reproducible dynamic culture conditions, capable of scale-up, will need to be devised in order for such constructs to be engineered in a laboratory for economically viable clinical use (Archer and Williams, 2005).

This review aims to summarize the literature to date on the effect that different forms of mechanical stimulation have on MSCs in terms of lineage preference and matrix expression. First, MSCs will be described, followed by some common lineage-specific differentiation inducers and markers. Then the main types of mechanical stimulus that have been applied to MSC cultures, together with some common apparatus used, will be discussed. The review will conclude with a survey of the literature, categorized into the types of force applied with a summary of key results and findings.

II. Mesenchymal Stem Cells

A. Stem cells

Two of the major types of stem cell suggested to have applications in regenerative medicine are embryonic stem cells (ESC), which are present in the early stages of development from the inner cell mass of the blastocyst-stage embryo, and adult stem cells (ASCs), which reside in adult tissue. ESC are considered pluripotent as they are able to differentiate into all cell types in the body and have an unlimited self-renewal capacity, whereas ASC are termed multipotent as they have a limited self-renewal capacity and can only form a few different tissue types. ESC rapidly proliferate during growth and could be considered an ideal, inexhaustible cell source for tissue replacement, however, the ability for ESC to rapidly form tissue has also lead to teratoma formation (Ben-David and Benvenisty, 2011). There is also ethical opposition to the clinical use of ESC. On the other hand, stem cells in the adult body do not pose the same ethical concerns and being specialized cells should not produce teratomas, although there may be some cancer risk associated with their therapeutic use (Chanda *et al.*, 2010).

B. Mesenchymal stem cells

MSCs are multipotent ASCs located within the stroma of the bone marrow and other tissues and are therefore also referred to as stromal cells. It has been recommended that MSCs grown *in vitro* by plastic adherence selection should be termed "multipotent MSCs," as it is not clear if they possess all the characteristic of true stem cells (Nombela-Arrieta *et al.*, 2011). MSCs are nonhematopoietic progenitors capable of extensive self-renewal with potential immunosuppressive properties. hMSCs have been isolated from numerous tissue types including, adipose tissue (Zuk *et al.*, 2001), blood from the umbilical cord (Erices *et al.*, 2000; Goodwin *et al.*, 2001), peripheral blood (Zvaifler *et al.*, 2000), and amniotic fluid (Soncini *et al.*, 2007). These cells have been shown to differentiate into multiple cell types of the mesoderm, particularly the skeletal tissues, but also may differentiate into cell types of ectodermal and endodermal origin. MSCs were first isolated from bone marrow by Friedenstein *et al.* (1966), who demonstrated their osteogenic potential, but since then populations of cells with similar characteristics have been obtained from other tissues. MSCs are a heterogeneous population and up until now no unique cell-surface markers for MSC identification have been found, although there are some general guidelines for defining what an MSC is. MSCs should adhere to tissue culture plastic and express the specific surface antigen markers CD73, CD90, and CD105; be negative for CD11b, CD19 or CD34, CD45, CD79α, and HLA-DR; and must be able to differentiate into osteoblasts, chondroblasts, and adipocytes *in vitro* (Dominici *et al.*, 2006).

MSCs are most commonly sourced from bone marrow by iliac crest biopsy, and there have been extensive investigations into the potential use of bone marrow MSCs (BMSCs). The potential therapeutic benefits of MSCs extend further than their direct contribution to tissue healing by differentiation into the specific cell type of the damaged tissue, they also participate indirectly in wound healing by secretion of bioactive proteins including growth factors, chemokines, and antiapoptic agents (Rehman *et al.*, 2004). Production of these proteins plays an important role in local cellular dynamics, promotion of angiogenesis, and recruitment of cells including other ASCs, which can then aid in the healing process

(Meirelles *et al.*, 2009). MSCs also act as an immunosuppressant by interacting with B and T lymphocytes, Natural Killer cells, and monocytes (Sioud *et al.*, 2011), which means they may be suitable as a source of allogenic cells (Koc *et al.*, 2002).

While MSCs are commonly sourced from bone marrow, harvesting of bone marrow is an invasive and painful procedure, and the number of MSCs obtained is rather small. This has lead tissue engineers to search other tissue types for a more accessible and abundant source of cells with MSC capabilities. One promising source is adipose tissue, which can be obtained from any donor regardless of body fat content, using an easily tolerated harvesting procedure. These cells, termed by some as adipose-derived mesenchymal stem cells (ADMSCs) or adipose-derived stem cells (ADSCs), are thought to be present in larger numbers than BMSCs, and they also have multi-differentiation potential including osteogenesis, chondrogenesis, and adipogenesis (Seong *et al.*, 2010).

MSCs *in vivo* reside in what is known as the stem cell niche containing many biochemical factors that regulate their behavior. Mechanical forces in the form of compression, tension, and fluid-induced shear are all present in the bone marrow (Gurkan and Akkus, 2008), but very little is known about the nature of these forces and how they affect stem cell mobilization and function. As well as external mechanical stimuli, there are intracellular tensile forces at focal adhesions as a result of cell–extracellular matrix (ECM) interactions and recent studies suggest that both intrinsic and extrinsic forces play a key role in MSC differentiation (Guilak *et al.*, 2009).

C. MSC differentiation *in vitro*

MSC differentiation can be stimulated *in vitro* under controlled conditions and then confirmation of the progression along a specific lineage can be observed using phenotypic makers associated with that tissue type. This generally involves growing the cells in an optimized medium including supplements or growth factors and then observing the expression of differentiation pathway signals and ECM components that form the infrastructure of that tissue, via assays of gene upregulation and matrix secretion. The differentiation of MSCs into osteoblasts, chondrocytes, and adipocytes has been the most extensively studied, and tenocytes have also been investigated to a lesser extent. The following section will briefly describe some of the most commonly used media additives for these lineage-specific differentiation pathways along with typically analyzed differentiation markers.

1. Osteogenesis

Early markers of osteogenic differentiation include a switch in the MSC morphology from a fibroblastic, fusiform to a more cuboidal shape, and mRNA expression of Runt-related transcription factor-2 (RUNX-2), osteonectin matrix protein (ON), the growth factors bone morphogenic protein 2 (BMP-2) and transforming growth factor-beta1 (TGF-β1), the enzyme alkaline phosphatase (ALP), and the abundant matrix protein collagen type I (COLI). Later stage differentiation markers include expression of other matrix proteins such as osteopontin (OPN) and osteocalcin (OC), as well as secretion of the major ECM components, first COLI, and at a later stage bone mineral. The mineral found in bone is carbonated hydroxyapatite a form of calcium phosphate, but most studies identify the presence of a calcium salt by alizarin red staining or a phosphate salt by von Kossa staining without further characterization of the mineral chemistry.

The most commonly used osteogenic medium for MSCs typically contains ascorbic acid (AA) or its phosphate salt, beta-glycerophosphate (βGP), and dexamethasone (DEX), applied for at least 7 days up to several weeks. AA is required for the production of stable collagen, the major matrix protein of bone, and plays a critical role as a cofactor in the hydroxylation of proline residues in the collagen molecule (Kielty *et al.*, 1993). βGP provides inorganic phosphate for synthesis of bone-like mineral by the cells (Chang *et al.*, 2000). DEX is a nonprotein, synthetic glucocorticoid and is a component of multiple differentiation media formulae including osteogenic, chondrogenic, and adipogenic media. In its absence, or the absence of other osteogenic inducers in 2D static culture, there is incomplete osteogenesis (Porter *et al.*, 2003). However, DEX has also been shown to have an inhibitory effect on collagen production and high levels (≥ 100 nM) have been shown to induce adipogenesis (Beresford *et al.*, 1992). Other additives have been used to enhance osteogenic induction in combination with the previously described supplements or alone, including the growth factors, TGF-β1 and bone morphogenetic proteins (BMPs), as well as vitamin D3 (VD3) (Seong *et al.*, 2010; Vater *et al.*, 2011). BMPs belong to the TGF-super family and appear to stimulate osteogenic differentiation *in vitro* and *in vivo*; however, their effect *in vitro* may be species-specific (Diefenderfer *et al.*, 2003). For example, in order to stimulate *in vitro* osteogenesis of hMSCs, BMP-2 requires a specific defined medium in which serum is removed or the extracellular regulated-kinase pathway 1 and 2 (ERK1/2) is inhibited and either insulin or insulin-like growth factor (IGF) is added (Osyczka and Leboy, 2005).

2. Chondrogenesis

MSCs undergoing chondrogenic differentiation can be observed to form large spherical cells and express the chondrogenic transcription factor protein SRY-box 9 (SOX9), which controls the gene expression of collagen type II (COLII), collagen type X (COLX), collagen type XI (COLXI), cartilage oligomeric matrix protein (COMP), and aggrecan, all early markers of chondrogenesis. Later stage matrix proteins can be seen surrounding the cells and consist of a predominantly COLII network with glycosaminoglycans (GAGs) and aggrecan (large aggregating proteoglycan).

Chondrogenic induction of MSCs *in vitro* typically involves a pellet culture or a micromass culture method (Johnstone *et al.*, 1998) with media addition of AA, DEX, TGF-β, and/or BMPs (as with osteogenic culture) but also fibroblast growth factor (FGF) and IGF. AA enhances COLII production, which is the major ECM protein of cartilage (Na *et al.*, 2007). TGF-β has three isoforms, all of which stimulate the expression of aggrecan, COMP, and COLII (Tuli *et al.*, 2003), playing an important role during the early and middle stages of chondrogenic differentiation. BMPs play a key role in triggering the chondrogenic differentiation of progenitor cells during bone morphogenesis (Pogue and Lyons, 2006); however, they do not strongly initiate chondrogenesis of MSCs *in vitro* compared to TGF-β (De Bari *et al.*, 2001). The addition of FGF has been shown to increase the deposition of GAGs and COLII either alone or in combination with other factors (Stewart *et al.*, 2008). IGF has been shown to strongly influence the chondrogenic differentiation of MSCs by the upregulation of specific genes and increases in matrix deposition including aggrecan and COLII (Gelse *et al.*, 2003). Chondrogenic media (CM) usually does not include serum but is supplemented with ITS (insulin, transferrin, selenious acid; Vater *et al.*, 2011).

3. Adipogenesis

Adipogenic differentiation is accompanied by a morphological switch to a large rounded shape. Early differentiation markers include peroxisome proliferator-activated receptor-gamma (PPARγ), which is an adipocyte-specific transcription factor and acts as a key regulator of genes associated with adipogenesis (Takada *et al.*, 2010) including lipoprotein lipase (LPL) and collagen type 6 (COLVI). Late stage differentiation is observed by staining for lipid vesicle accumulation.

Typical adipogenic culture medium may contain combinations of high levels of DEX (compared with osteogenic culture medium), insulin or IGF, 3-isobutyl-1-methyxanthine (IBMX), and indomethacin (Lin *et al.*, 2005). DEX stimulates the production of adipogenic transcription factors such as PPARγ, which act to induce adipogenesis in MSCs. IBMX enhances the effects of DEX on stimulation of PPARγ and has been shown to inhibit the expression of RUNX2 and OPN (Yang *et al.*, 2008), which are important markers of osteogenic differentiation.

4. Tenogenesis

Tenogenic differentiation incurs little morphological change in the cell appearance, which remains fibroblastic and fusiform. Early markers of tenogenesis include expression of the transcription factor scleraxis along with gene upregulation of COLI and COLIII, and the matrix protein tenascin-c.

There are a wide range of growth factors that have been used *in vitro* for inducing tenogenesis in MSCs such as IGF, growth and differentiation factors (GDFs), platelet derived growth factor (PDGF), and FGF; however, optimal media formulations are less well characterized than for the other more common differentiation pathways and at present there is no standard media formulation that has been widely adopted (Yin *et al.*, 2010).

There are disadvantages associated with the use of chemical supplements to induce differentiation. The growth factors used are very expensive and have relatively short half-lives and so repeated administration is required, which makes them a costly method. For all supplements, there is the possibility of adverse side effects if they are applied *in vivo* (van Beuningen *et al.*, 1998). DEX is known to induce osteoporosis (bone resorption) (Scutt *et al.*, 1996) and therefore it is not desirable for any implanted MSCs to be used in a clinical context to contain DEX within the media, or construct, that they are implanted with. Alternative methods for the differentiation of MSCs that do not have the potential side effects of pharmacological additives are desirable, which makes the use of external mechanical forces an attractive option.

D. Mechanical regulation of MSC function

Culturing MSCs on tissue culture plastic or on a scaffolding material under static conditions (no external mechanical forces) in the presence of differentiation media is the simplest and most commonly used method of developing tissue engineered constructs. However, culture under static conditions excludes the important role mechanical forces contribute to the growth, development, and maintenance of tissues under physiological conditions. Cells and tissues in the body are subjected to a wide variety of external mechanical forces including compression, tension, torsion, fluid shear stress, and hydrostatic pressure, each of which can influence their growth, development, and maintenance (Vogel, 2006). As all cell types that MSCs differentiate into are strongly mechanosensitive, it is not surprising that undifferentiated MSCs are also mechanosensitive.

While the mechanical environment *in vivo* of load bearing tissues such as bone, cartilage, and endothelium have been well characterized, little is known about the mechanical environment of the MSC niche *in vivo* (Gurkan and Akkus, 2008). The differentiation of MSCs occurs during fetal development, in children and adolescents during growth and in adults during tissue maintenance, such as bone remodeling. MSCs are also involved in the wound healing response, for instance, in a healing fracture and all of these are environments where extracellular forces are in abundance and are thought to influence the action of the MSCs (Henderson *et al.*, 2007; Lacroix *et al.*, 2002; Morgan *et al.*, 2008).

There are also intracellular tensile forces present in cells caused by the cell pulling action of the myosin motors on the actin cytoskeleton which in turn pulls on the ECM (Pelham and Wang, 1999). This mechanism has been hypothesized to mediate MSC differentiation *in vivo* and has certainly been shown to operate *in vitro* where the morphology of the actin cytoskeleton and cell stiffness vary dependant on the stiffness and roughness of the cellular substrate and in turn regulate the differentiation fate of MSCs (Colley *et al.*, 2009; Engler *et al.*, 2006; Reilly and Engler, 2010; Yourek *et al.*, 2007; Titushkin *et al.*, 2010). Myosin motor activity and actin cytoskeletal reorganization are also modulated by external mechanical forces (Jonas and Duschl, 2010), which may indicate synergistic effects of cell/substrate attachments and extrinsic mechanical forces (Fan *et al.*, 2009; You *et al.*, 2008).

Despite the lack of understanding of the mechanical environment of differentiating MSCs *in vivo*, it has become clear that MSCs respond strongly to mechanical loading *in vitro*. Some of these forces may be considered "supraphysiological"; however, from the TE perspective, the important parameter to elucidate is the loading regime that will ensure precise differentiation and best accelerate matrix production and/or improve matrix quality either *in vitro*, prior to implantation, or *in vivo*, after implantation. A wide variety of mechanical stimuli have been used in order to either induce or enhance cellular differentiation *in vitro* including stretch, compression, fluid shear, ultrasound, vibrations, and magnetic particle stretching, to be described in detail in Section IV.

E. Mechanotransduction

Cells are able to sense and respond to mechanical forces and convert them into biochemical signals using mechanosensitive receptors in a process called mechanotransduction (Liedert *et al.*, 2008; Nomura and Takano-Yamamoto, 2000; Rubin *et al.*, 2006). This phenomenon is believed to occur due to some common mechanisms regardless of cell type. One key theory is that a response results from cytoskeletal deformations caused by integrin pulling (Iqbal and Zaidi, 2005). Integrins are transmembrane proteins that couple the cellular cytoskeleton to the ECM; they cluster at focal adhesion points on the cell surface forming an integrin–ligand bond with the ECM. This integrin-ligand bond resists external forces, which enables the force to be transferred across the plasma membrane and into the cell causing deformation of the cytoskeleton. A second mechanism involves regulation of membrane proteins which may be stretch-activated ion channels, or G-protein coupled receptors, where the plasma membrane is deformed by an external force causing an influx/efflux of ions into/out of the cell (Liedert *et al.*, 2008). Other potential sensors of external forces are the primary cilia and the cell glycocalyx. The glycocalyx is a pericellular GAG-proteoglycan layer, which surrounds the cell membrane, and when fluid moves past the cell, such as during blood flow past endothelial cells or interstitial fluid flow in the osteocyte's lacunar-canalicular system, the glycocalyx creates a drag force resulting in deformation of the cellular membrane and so transducing the applied force (Morris *et al.*, 2010; Reilly *et al.*, 2003; Weinbaum *et al.*, 2007). Primary cilia are immotile microtubule-based organelles, one per cell, that protrude from the cell surface and have been seen to exist in most mammalian tissue types (Malone *et al.*, 2007). They are populated with receptors that participate in numerous signaling events and are believed to act as sensors to fluid flow.

In most cases, it is not well understood how exactly these mechanisms transduce mechanical signals to individual cells, or indeed which cells sense them, and extensive research is ongoing in this area. It is also not clear how universal these mechanisms are and whether there is redundancy between pathways. Only a few studies have investigated mechanotransduction mechanisms in MSCs but investigations that have taken place observe many mechanisms common to that of mature mesenchymal cells (Liedert *et al.*, 2008) such as the cytoskeletal changes described previously. Mechanically mediated upregulation of ERK1/2 is common to osteogenic MSCs (Glossop and Cartmell, 2009; Kim *et al.*, 2007; Simmons *et al.*, 2003) and both precursor (Case *et al.*, 2011) and mature (You *et al.*, 2001) bone cells. Recently, it has been shown that the primary cilia of hMSCs are required for the modulation of osteogenic and adipogenic differentiation pathways in static conditions, opening up the possibility that they may also mediate mechanically activated differentiation pathways in these cells (Tummala *et al.*, 2010).

III. Mechanical Loading

A. Introduction

Cellular responses to mechanical forces can be detected and quantified after just a few seconds, as measured by cellular influxes of calcium (You *et al.*, 2001) or release of nitric oxide (McGarry *et al.*, 2005), after a few days, measured by secretion of hormones or upregulation of genes, or after a few weeks, measured by matrix deposition. The optimum durations, magnitudes, and frequencies of mechanical loading for lineage-specific differentiation of MSCs is not known due to the difficulty of undertaking multiple loading regimens within one set of experiments. The response of MSCs to loading are likely to be age-specific, may be specific to site of origin, and appear to depend on how differentiated the cells are at the time of loading, as well as whether loading is performed in conjunction with biochemical supplements. There are many ways in which researchers have stimulated cells with mechanical forces *in vitro* and while there is a wide range of experimental apparatus, they can generally be categorized into the primary type of stimulation they provide. For each stimulation mode, the stimulus can be applied in 2D (monolayer) or 3D and differences between cells cultured in 2D and 3D have been observed in terms of cellular morphology and migration strategies, matrix adhesion, gene and protein expression, and responses to fluid flow (Pedersen and Swartz, 2005).

A brief account of the major types of mechanical stimulation will be described here separated into four parts; "tension," "compression," "fluid-induced shear stress," and "other," along with some commonly used experimental setups. For more in depth reviews on the modes of mechanical stimulation and the designs of the bioreactors used, see Brown (2000) and El Haj and Cartmell (2010).

B. Tension

When a sample is stretched, it will experience tensile strain and this will result in an increase in length of the sample. Stretching is experienced in the body by many tissue types including bone, tendons and ligaments, skin and vascular tissue, and there are a number of ways in which cells have been stretched *in vitro* but ultimately most methods are based on two primary designs; uniaxial (and multiaxial) grip tension and substrate bending. Longitudinal stretch requires a deformable substrate and can be in the form of static stretch, where the sample is held at a specific strain for a fixed period of time, or dynamic stretching where strain is applied in cycles. Gripped tensile loading is performed mainly on cell-seeded constructs and involves gripping the sample at the ends with the cells seeded in the middle of the sample away from the gripped area (Fig. 20.1B). Substrate bending involves cells seeded

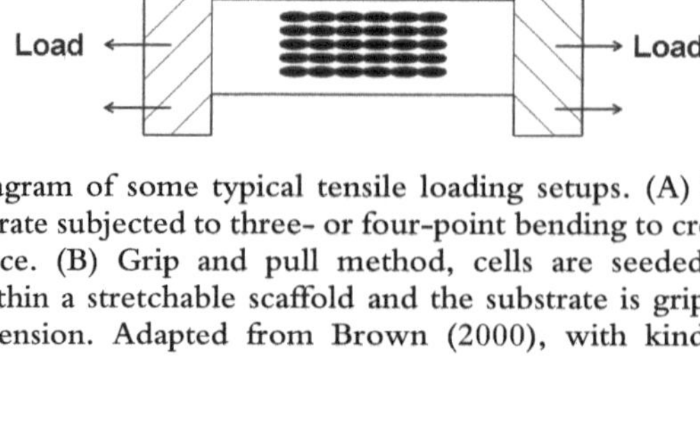

Figure 20.1 Diagram of some typical tensile loading setups. (A) Bending: cells are seeded on a substrate subjected to three- or four-point bending to create tension on the cell-seeded surface. (B) Grip and pull method, cells are seeded on a stretchable membrane or within a stretchable scaffold and the substrate is gripped and pulled to create uniaxial tension. Adapted from Brown (2000), with kind permission from Elsevier.

on a flexible substrate placed over a number of pivot points with an external load or displacement being applied to the ends (Fig. 20.1A). This is normally performed in 2D on a monolayer of cells as the apparatus is often a specially built flat culture dish. An alternative method is to use a cell-seeded deformable circular substrate, which can be displaced using a circular platen or prong for example Gilbert *et al.* (1994). In each of these methods, there will be some compression forces exerted due to substrate deformation in the direction perpendicular to the applied tension (Poisson effect). Also, cyclic tension can subject the cells to major or minor fluid shear stresses as they are moved backward and forward through the media simulating the flow of fluid over the cells (Thompson *et al.*, 2010).

For 3D tensile loading, cells are seeded in a gel or porous scaffold which can also stimulate the flow of nutrients into and out of the scaffold. In 3D culture, the forces that the cells experience will also depend on the scaffold architecture. A gel-based scaffold, for example, will deform almost entirely uniformly and so the force resulting from the applied strain should be relatively homogeneous on each cell throughout the scaffold. This is in contrast to a randomly orientated fibrous scaffold or a foam scaffold, where cells will be attached to fibers or struts that are not in the same plane as the applied strain and so the forces experienced throughout the scaffold will be heterogenous. The applied strain is usually then referred to as a global strain.

C. Compression

When something pushes against a sample, it will experience compressive strain and possibly a reduction in size in that plane. The two main types of compressive loading that have been utilized are hydrostatic pressure and direct contact compression (platen abutment). Hydrostatic pressure involves no direct contact of the load platform with the cells and can utilize a pressurized gas phase on top of the liquid phase (Fig. 20.2A). It has many

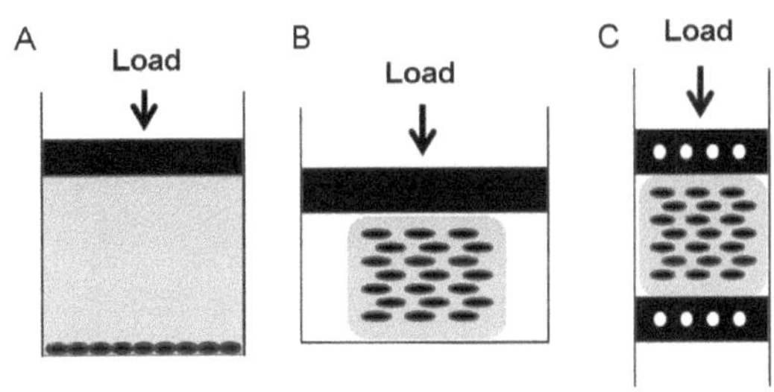

Figure 20.2 Diagram of some typical compressive loading setups. (A). Hydrostatic pressure. (B) Unconfined compression and (C) confined compression. Confined compression requires porous plates to allow media access to the cell-seeded scaffold. Adapted from Brown (2000), with kind permission from Elsevier. (For color version of this figure, the reader is referred to the Web version of this chapter.)

advantages such as using a simple setup, producing a spatially homogenous force, and it avoids any physical impedance of the media to the cells. It is thought that this type of loading may play a significant role in the mechanical environment of bone marrow *in vivo* (Gurkan and Akkus, 2008). Direct contact compression is normally carried out on a cell-seeded scaffold and involves a plate in direct contact with the scaffold providing the compressive force. Direct contact compression is popular because it simulates closely the compression thought to occur in the body when bone or cartilage is loaded. There are two types of setup for this mechanical stimulus and that is unconfined compression, where there is no support around the outer edges of the scaffold (Fig. 20.2B), or confined compression, where the scaffold is tightly housed and not able to deform outward (Fig. 20.2C). Confined compression is more appropriate to physiological conditions but can restrict the transfer of nutrients across the scaffold.

Similarly to tensile loading, there may be side effects from fluid being drawn in and out of the scaffold in the form of fluid-induced shear stresses and nutrient transfer. A study by Tanaka *et al.* (2005) suggests that in the compression of a porous scaffold the fluid-induced shear stress has the dominant role. In addition, compression can also cause hydrostatic forces from the pressure of the media pushing down on the cells. Also similarly to tensile loading, the scaffold architecture will regulate how much of the applied force is received by the cells with a cell-seeded gel transferring most of the full force and a sponge-like scaffold showing a heterogenous distribution where the strains experienced by the individual cells may be tensile rather than compressive while others are not exposed to any strain.

D. Fluid flow-induced shear stress

Fluid flow-induced shear stress (FSS) is created when a fluid passes over the surface of an object. This type of force is thought to be one of the major influencing mechanical stimuli experienced *in vivo* especially in bone where it is thought that shear forces acting on osteocytes, caused by interstitial fluid movement in the lacunar-canalicular system, result in a biochemical response to initiate remodeling (Klein-Nulend *et al.*, 2005). For 2D systems, various experimental apparatus have been used to study the effects of shear stress on monolayers of cells. Commonly used systems include the cone and plate system (Frangos *et al.*, 1988), the parallel plate flow chamber system (Jacobs *et al.*, 1998), rotating disk or radial flow devices (Deligianni *et al.*, 2001), an oscillating orbital shaker system (Hubbe, 1981), and a rocking "see saw" system (Zhou *et al.*, 2010). The first system involves a plate containing cells in which rotation is forced around a cone axis that stands perpendicular to the plate. The fluid shear stresses experienced are spatially homogenous due to the differences in radial position and ultimately the relative velocities of the cells. The parallel plate flow system consists of a relatively long rectangular chamber with an opening at both ends and a pressure gradient created between the two, which causes a uniform laminar flow to develop between the two. These systems are able to apply well-defined parameters allowing for easy modeling of the shear forces experienced. There are many commercially available parallel plate systems, many of which were designed for endothelial cell work, and cell attachment studies, as well as many in house designs (Anderson and Tate, 2007).

More recently, bioreactors for 3D culture have been designed to allow for the cells on scaffolding materials to be subjected to flow-induced shear forces (El Haj and Cartmell, 2010). Some common designs include rotating wall vessels, spinner flasks, and perfusion systems. Perfusion flow bioreactors induce mechanical forces on cells seeded on scaffolds by passing the culture medium through the porous scaffold which passes over the cells exposing them to fluid shear stresses. As well as applying a fluid shear force, nutrient transfer to cells is improved as the media is constantly being forced through the scaffold and therefore cells in the center are able to receive a greater supply of nutrients.

In scaffolds with a complex 3D architecture, it is more difficult to calculate what the cellular level shear forces applied, but these will vary strongly from scaffold to scaffold, for example, Jungreuthmayer *et al.* (2009) found that shear stresses within a collagen/GAG scaffold were 2.8 times higher than in a calcium phosphate scaffold at the same flow rate. How cells are attached to a scaffold will also determine how they experience a fluid flow-induced stress, cells attached in a single layer to scaffold pores may

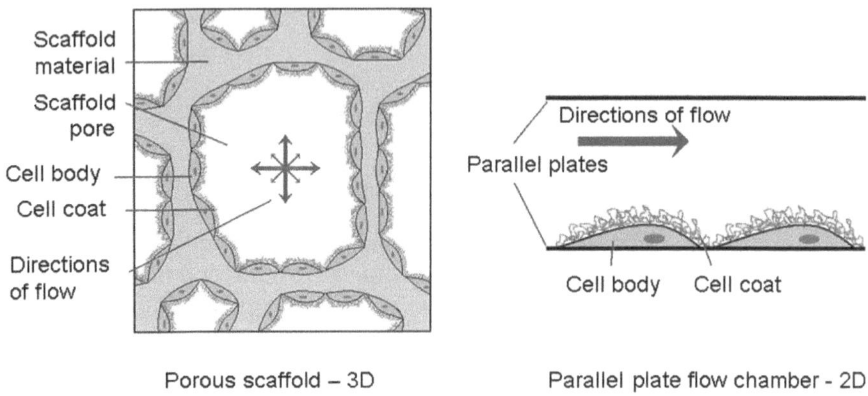

Figure 20.3 Schematic diagrams of fluid flow in a 3D scaffold (left) and a parallel plate flow chamber (right). Fluid flow within a 3D porous scaffold is likely to vary in direction and magnitude in local regions, the flow chamber allows a much more defined flow regimen but does not allow for tissue growth. Adapted from Morris *et al.* (2010), with kind permission from SAGE publications. (For color version of this figure, the reader is referred to the Web version of this chapter.)

experience fluid shear on their apical surface in a similar way to in a 2D parallel plate flow chamber (Fig. 20.3) but cells embedded in a hydrogel or in a 3D mass containing both cells and ECM will experience shear all around the cell body in a more similar way to cells *in vivo*.

In both 2D and 3D fluid flow devices, it is difficult to separate the effects of improved nutrient transfer with a direct mechanical effect of shear stress at the cell membrane. Some studies have attempted to do this for bone cells in 2D by varying the viscosity of the media such that for a given flow rate the shear stress is altered (Bakker *et al.*, 2001; Donahue *et al.*, 2003); these appear to show that a response is shear stress dependent rather than flow rate dependent but does require some nutrient transport in order to occur. Using a similar technique in a 3D, calcium phosphate porous scaffolds Li *et al.* (2009a) showed that mineralized ECM matrix deposition was dependant on both shear stress and flow rate (mass transport) with too high a flow rate being inhibitory to ECM deposition. Therefore, it seems unlikely that biochemical/nutrient movement alone will explain the effects of flow within perfusion systems on MSC matrix production.

E. Other (direct cell magnetic stimulus, ultrasound, vibration)

1. Direct cell membrane magnetic stimulus

Direct cell membrane magnetic stimulus (DCMMS) utilizes magnetic macro- or nanoparticles bound directly to receptors on the cell surface followed by the application of an external magnetic field to directly strain the targeted cell receptor (Hughes *et al.*, 2005). Magnetic nanoparticles are usually made from iron oxides, which are not harmful to cells, and then coated with a biocompatible polymer such as silica or dextran which will readily bind to specific ligands, enabling the particle to bind to receptors on the cell surface. The applied magnetic field pulls on the particles subjecting the ligand–receptor bond to precisely controlled mechanical nanoforces. Two main techniques that utilize magnetic nanoparticles to subject cells to mechanical forces include magnetic drag, which applies translational, stretching forces (Glogauer *et al.*, 1995), and magnetic twisting, which induces rotational forces on the cell membrane (Wang *et al.*, 1993). The advantage of this method is that it allows specific membrane bound molecules to be targeted, such that specific downstream mechanotransductive regulators could be activated once the signaling relationships are better understood.

2. Ultrasound

Ultrasound is a well-established medical imaging technique that has found recent use in the acceleration of fracture healing (Della Rocca, 2009; Watanabe *et al.*, 2010) and in TE applications, one of which is for the mechanical stimulation of cells (Kobayashi *et al.*, 2009;

Korstjens *et al.*, 2008). Low-intensity pulsed ultrasound (LIPUS) is the term commonly used by tissue engineers, which comprises of applying a wave of alternating pressure that translates into a mechanical force by striking the scaffold/ECM or cell membrane, which then is mechanotransduced into a intracellular biochemical signal (Angle *et al.*, 2011). It enables noninvasive loading of cells but many of the commercial systems have limited control of the parameter selections.

3. High frequency, low-magnitude vibration

This technique involves rapidly accelerating the cells or cell-containing construct over very short distances many times per second, the strains applied are very small, to negligible, but the number of cycles is high. There are two main methodologies to apply high-frequency loading. The first is to directly distort the substrate in which the cells are on (Dumas *et al.*, 2010), and the second involves simply moving the entire culture container (Edwards and Reily, 2011). The technique is very simple and noncontact and allows for a large number of samples to be loaded at the same time. Vibration treatment has been shown to have an anabolic effect on bone (Xie *et al.*, 2006) and may inhibit fat production *in vivo* (Rubin *et al.*, 2007); however, it is only recently being used in the mechanical stimulation of MSCs *in vitro*.

IV. Experimental Results

A. Tension

1. Introduction

Tensile loading of MSCs has mainly been used to initiate or enhance osteogenic or tenogenic differentiation with a small number of studies looking at chondrogenic or cardiomyogenic potential (Table 20.1). BMSCs from a number of different species have been stimulated using tension, but there are also a few studies that have sourced MSCs from adipose tissue (Colazzo *et al.*, 2010; Yang *et al.*, 2010). Tensile forces have been applied to 2D cultures mainly by bending flexible substrates and 3D cultures have tended to be stimulated using the grip and pull method, both of which are usually uniaxial and cyclic. Studies that use a 3D, 4-point bending method will subject different regions of the substrate to either tension or compression forces, but they have been included in this section. Common conditions include applying tension in a sinusoidal manner at a frequency of 1 Hz and a global 2–5% strain for osteogenesis, 10% strain for tenogenesis, or higher for cardiomyogenesis, with loading cycles being applied intermittently.

Table 20.1 Tension: Summary table of the literature cited in the text describing the effect of tensile loading on mesenchymal stem cell differentiation and matrix production[a]

MSC source	Type of strain	Parameters	Culture surface	Culture media	Culture period	Lineage	Load effect	Reference
Human BMSCs	Cyclic, uniaxial	1 Hz; 2 or 8% ε; 3*2 h/day (D1–3)	2D—silicone dish	Osteogenic w or w/o DEX	1–7 days	Osteogenesis	↑ ALP activity and COLI mRNA (8% ε ±DEX)	Jagodzinski *et al.* (2004)
Human BMSCs	Cyclic, 4-point bend	3% ε at 5 mm/min; 250 cycles/day	3D—demineralized bone	Osteogenic w 0, 10, or 100 nM DEX	8–16 days	Osteogenesis	↑ ALP activity, OPN mRNA and calcium content (10 nM DEX)	Mauney *et al.* (2004)
Human BMSCs	Cyclic, uniaxial, grip	1 Hz; 10% or 12% ε; 4 h/day	3D—collagen gel	Basal	7 or 14 days	Osteogenesis	↑ BMP-2 gene expression	Sumanasinghe *et al.* (2006)
Human BMSCs	Cyclic, uniaxial, flex plate	0.6 Hz; 7, 10, 14 or 20% ε; Day 3–7 (continuously)	2D—COLI coated BioFlex plates	Basal	7 days	Cardio-myogenesis	↑ Collagen production (14% ε highest). ↑ LOX mRNA	Ku *et al.* (2006)
Rat BMSCs	Cyclic, uniaxial, grip	1 Hz; 5% ε; 4 h/day (D5, 6, and 7)	3D—COL/GAG scaffold	Osteogenic	7 days	Osteogenesis	↓ OPN, BSP and OCN gene expression	Byrne *et al.* (2008)
Human BMSCs	Cyclic, uniaxial, flex plate	1 Hz; 3 or 10% ε; 8 or 48 h	2D—COLI coated surface	Basal	48–96 h	Tenogenesis and Osteogenesis	↑ COLI, III, tenascin-C mRNA (10%). ↑ ALP, OCN (3%)	Chen *et al.* (2008)
Rat BMSCs	Cyclic, uniaxial, grip	1 Hz; 10% ε; 7 days (continuous)	3D—COL/GAG scaffold	Chondrogenic	14 days	Chondrogenesis	Similar GAG synthesis	McMahon *et al.* (2008)
Mouse BMSCs	Cyclic, uniaxial, grip	0.33 Hz; 10% ε; 48 h	3D—COL or GDF-5 coated PCL scaffold	Basal	48 h	Tenogenesis	↑ COLI, III, scleraxis gene expression	Farng *et al.* (2008)
Goat BMSCs	Static uniaxial	6.7 N static load	3D—braided PLGA scaffold	Basal	23 days	Tenogenesis	↑ Collagen production at day 5	van Eijk *et al.* (2008)

Human BMSCs	Cyclic uniaxial, flex plate	1 Hz; 1% ε; 30 min/day	3D—COLI gels	Basal	1–7 days	Tenogenesis	↑ Collagen content. ↑ COLI, III, XII and scleraxis gene expression	Kuo and Tuan (2008)
Rat BMSCs	Cyclic, uniaxial, grip	1 Hz; 10% ε; 3, 6, 12, 24, or 36 h (continuous)	2D—Gelatine coated silicone membrane	Basal	60 h	Tenogenesis	↑ COLI, COLIII and tenascin-C mRNA (36 h highest)	Zhang *et al.* (2008)
Mouse C3H10T1/2 MSCs	Cyclic, biaxial, flex plate	0.16 Hz; 2% ε; 6 h/day (3–5 days)	2D—COLI coated BioFlex plates	Adipogenic or multipotential	5 days	Osteogenesis	↓ Adiponectin and PPARγ mRNA. ↑ RUNX2 and Osterix mRNA (w BMP-2)	Sen *et al.* (2008)
Rabbit BMSCs	Cyclic, uniaxial, grip	1 Hz; 2.4% ε; 100 or 3000 cycles; 8 h/day (12 days)	3D—cross-linked COLI sponge.	Basal	14 days	Tenogenesis	No differences between load conditions	Nirmalanandhan *et al.* (2009)
Human ADMSCs	Cyclic, uniaxial, flex plate	1 Hz; 10% ε; 4 h/day (continuous or 10 s cycle rest)	2D—COLI coated BioFlex plates	Osteogenic	14 days	Osteogenesis	↑ Calcium content	Hanson *et al.* (2009)
Human BMSCs	Cyclic, uniaxial, grip	0.1 Hz; 5% ε; 24 h (continuous)	3D—alginate gel	Basal	24 h	Osteogenesis	↑ RUNX2, BMP-1, ALP	Haudenschild *et al.* (2009)
Human BMSC	Cyclic, uniaxial, flex plate	1 Hz; 3% ε; (1, 3, or 5 days)	2D—COL1, VN, FN or LN coated plates	Basal	1, 3, or 5 days	Osteogenesis	↑ ALP activity, ↑ Calcium deposit	Huang *et al.* (2009)
Rat ADMSCs	Cyclic or continuous, 4-point bend	1 Hz; 2000με; for 6 h (1 day) or 17 min/daily	2D—TCP	Osteogenic	12 days	Osteogenesis	↑ BMP-2 and RUNX2 mRNA (6 h)	Yang *et al.* (2010)
Human ADMSCs, BMSCs	Cyclic, uniaxial, flex plate	0.6 Hz; 14% ε; 3 days (continuous)	2D—collagen coated BioFlex plate	Basal	3 days	Cardio-myogenesis	↑ Collagen production. ↑ matrix cross-linking in ADMSCs	Colazzo *et al.* (2010)

(*Continued*)

Table 20.1 (*Continued*)

MSC source	Type of strain	Parameters	Culture surface	Culture media	Culture period	Lineage	Load effect	Reference
Bovine BMSCs	Cyclic, uniaxial, grip	1 Hz; 10% ε; 1 h load, 3 h rest 6 times/day (24 h–2 weeks)	3D—fibrin hydrogels	Basal or chondrogenic	4, 12, 14, or 21 days	Chondrogenesis	↑ Protein and proteoglycan (24 h). ↑ COL1 (D14 and 21)	Connelly *et al.* (2010)
Human BMSCs	Cyclic, uniaxial	1 Hz; 1% ε (continuous)	2D/3D—PLLA and COLI aligned nanofibers or films	Osteogenic	1–4 weeks	Osteogenesis	↑ ALP activity (fibers week 1)	Ngiam *et al.* (2011)
Rat TDMSCs	Cyclic, uniaxial, flex plate	0.5 Hz; 4 or 8% ε; 4 h	2D—COLI coated plate	Basal	3 days	Osteogenesis	↑ BMP-2 protein and gene expression	Rui *et al.* (2011)

[a] BMSCs (bone marrow derived mesenchymal stem cells); ADMSCs (adipose derived mesenchymal stem cells); TDMSCs (tendon derived mesenchymal stem cells); ε (strain); 2D (two-dimensional); 3D (three-dimensional); COL (collagen); GAG (glycosaminoglycan); GDF-5 (growth and differentiation factor 5); PCL (poly(caprolactone)); PLGA (poly(L-lactide/glycolide) acid); VN (vitronectin); FN (fibronectin); LN (laminin); TCP (tissue culture plastic); PLLA (poly(L-lactic) acid); DEX (dexamethasone); multipotential (consisting of chemical inducing factors for multiple lineages); ALP (alkaline phosphatase); OPN (osteopontin); BMP (bone morphogenic protein); LOX (lysyl oxidase); BSP (bone sialo protein); OCN (osteocalcin); RUNX2 (runt-related transcription factor 2); PPARγ (peroxisome proliferator-activated receptor-gamma).

2. Osteogenesis

When cyclic tensile loading has been applied to MSCs, increased expression of early osteogenic makers has been seen as well as increases in later stage matrix secretion, both in the presence and absence of osteogenic media.

Rat ADMSCs stretched (2% strain) on 2D substrates in osteogenic media showed an increase in RUNX2 and BMP-2 mRNA after one bout of continuous loading (6 h), but these increases were not seen when cells were stretched for 17 min daily (Yang *et al.*, 2010). Human ADMSCs were subjected to a larger stretch magnitude (10% strain) for 4 h daily resulting in an increase in calcium content (Hanson *et al.*, 2009). Continuous cyclic loading was applied to human BMSCs seeded on PLLA/COLI films or fibers over a period of up to 4 weeks in osteogenic media (Ngiam *et al.*, 2011). After 1 week of stretch, ALP activity increased in the cells seeded on the fibrous scaffold but not on the films; however, continuous loading lasting longer than 1 week had no effect on the osteogenic differentiation of cells on either surface. This indicates that prolonged continuous loading may not be suitable for osteogenic differentiation and that the cellular response is partially regulated by the cell-surface adhesion.

The effect of DEX on osteogenic differentiation, in combination with tensile loading, has been observed in a small number of studies. Jagodzinski *et al.* (2004) stretched human BMSCs for 6 h/day, on the first 3 days of culture, at two different strain rates (2% and 8%) with or without DEX treatment. Stretching, with and without DEX, upregulated COLI mRNA and ALP activity, but only at the 8% strain rate, and COLIII mRNA showed an increase with stretching (+DEX). Stretching alone was seen to be as effective as DEX treatment alone and there was a synergistic effect of the combination of stretch and DEX (Fig. 20.4). Mauney *et al.* (2004) investigated the effect of the concentration of DEX (0, 10, or 100 nM) on the osteogenic enhancing potential of loading. Without DEX, loading alone was able to upregulate ALP activity and expression, but it had no effect on OPN and OCN. The addition of 10 nM DEX seemed to cause a synergistic response by further enhancing ALP activity and showing a small, significant increase in OPN gene expression but at higher DEX levels (100 nM) the effect of loading was suppressed.

The effect of tensile loading on adipogenesis was observed by Sen *et al.* (2008), by subjecting mouse C3H10T1/2 MSCs to low frequency (0.16 Hz) tensile loading for 6 h/day (3–5 days), in the presence of adipogenic or multipotential media. In adipogenic media, nonstretched cells expressed PPARγ and adiponectin mRNA but following application of stretch, this expression was lower and continued to decrease as the number of days of stretching were increased. Stretching did not induce osteogenesis in the cells when cultured in adipogenic media; however, stretching followed by the application of BMP-2 caused a significant upregulation of the mRNA expression of RUNX2 and Osterix. This study supports the hypothesis that adipogenesis is inhibited by stretching, whereas osteogenesis is stimulated.

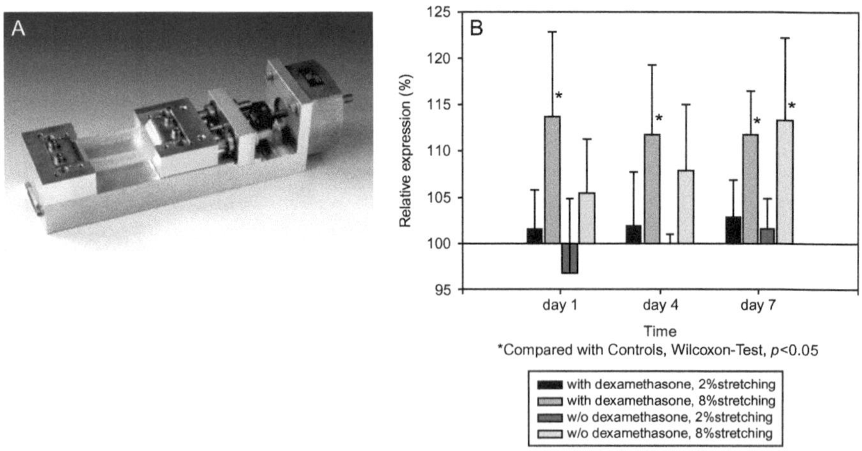

Figure 20.4 Tension upregulates osteogenesis of mesenchymal stem cells. (A) Photograph of a cell stretching devise used to place MSCs under cyclic tension, kindly provided by Michael Jagodzinski, Hannover Medical School (previously unpublished). (B) Cyclic stretching upregulated the transcription factor associated with osteogenic differentiation, Cbfa1(RUNX2) at 8% strain when cultures were supplemented with DEX. Adapted from Jagodzinski *et al.* (2004), kind permission of reproduction from eCMjournal.org. (For color version of this figure, the reader is referred to the Web version of this chapter.)

Human BMSCs seeded on collagen gels in basal media showed an increase in BMP-2 expression after daily loading (Sumanasinghe *et al.*, 2006). Chen *et al.* (2008) also showed that in basal media, human BMSCs seeded on collagen coated surfaces and cyclically stretched over short time periods showed an upregulation of ALP and OCN mRNA. These increases were seen when a strain of 3% was used and not at a higher strain of 10%, which increased tenogenic differentiation markers. Other studies have also shown that ALP activity, as well as mineralized matrix deposition, is higher after tensile loading in the absence of osteogenic media (Huang *et al.*, 2009). In this study, human BMSCs were seeded on substrates with different ECM surface coatings and subjected to continuous short-term stretching at a low (0.1 Hz) frequency. While mechanical loading upregulated osteogenic markers on all surfaces, it was seen that fibronectin and lamanin supported the greatest differentiation potential (Huang *et al.*, 2009), implying that the cell–ECM interactions that determine how the cell senses the load may have an influence on the outcome. Short-term, low-frequency (0.1 Hz), and continuous tensile loading has also been shown to increase RUNX2 and BMP-1 gene expression in human BMSCs seeded in an alginate gel (Haudenschild *et al.*, 2009).

While these previous studies appear to show cyclic tension as an important regulatory tool for osteogenesis of MSCs, Byrne *et al.* (2008) showed that subjecting rat BMSCs seeded in a collagen-GAG scaffold to 3 days of tensile loading in the first week actually had an inhibitory affect on osteogenesis. Gene expression of OPN, OCN, and BSP were all downregulated compared to free-swelling (FSW) controls, and this was also seen with constrained culture, while COLI mRNA was unchanged. The authors of this study suggested that a FSW environment with a free-to-contract scaffold may benefit the osteogenic potential of MSCs.

3. Tenogenesis

It is common-place to use a collagen-based or collagen coated scaffold as a substrate for the cyclic loading of MSCs when targeting tenogenesis. Most studies carried out on the tenogenic potential of tensile loading on MSCs have been in the presence of basal media. Human BMSCs were seeded on COLI coated flexible substrates in the presence of basal media and subjected to 3% and 10% global strains for 8 or 48 h (Chen *et al.*, 2008) and markers for osteogenesis and tenogenesis were analyzed. It was seen that applying 10% strain for 48 h increased COLI, COLIII, and tenascin-C mRNA, whereas 3% strain increased ALP mRNA after 8 h and also OCN gene expression after 48 h of loading and a relaxation period of 48 h. This suggests that different levels of strain alone can influence the differentiation pathway of MSCs, with lower strains influencing osteogenesis over tenogenesis. 10% strain was applied to mouse BMSCs on PCL scaffolds for 48 h and initiated higher expression of the tenogenic genes COLI, COLIII, and scleraxis (Farng *et al.*, 2008). Zhang *et al.* (2008) observed the effect of applying 10% strain to rat BMSCs continuously over different periods of time (3–36 h), and found that after a single bout of stretching for 24 h, COLI, COLIII, and tenascin-C mRNA were all upregulated.

Altman *et al.* (2002) applied a tension strain (10%) combined with a rotational strain (25%) at a low frequency (0.0167 Hz) continuously from day 2 of culture to human and bovine BMSCs in collagen gels, with no media supplements, and showed that loading caused cell and collagen alignment together with upregulation of COLI, COLII, and tenascin–C by day 14 (note in Table 20.1).

Other studies have used much lower strain levels to induce tenogenesis. Human BMSCs seeded in a COLI gel were subjected to a cyclic strain of 1% which upregulated gene expression of COLIII and COLXII and produced more collagen overall (Kuo and Tuan, 2008). Gene expression of scleraxis in static controls reduced over time, but cyclic loading maintained an almost constant level throughout the 7 days of culture. Levels of matrix metalloproteinase-1 (MMP-1) were significantly higher with cyclic loading after 3 days but this decreased at day 7. Rabbit BMSCs were seeded on a cross-linked COLI sponge and subjected to 2.4% cyclic strain for 100 or

3000 cycles with the insertion of rest periods between individual pulses (Nirmalanandhan *et al.*, 2009). However, there were no differences seen in terms of gene expression or construct stiffness between loading regimens.

While most studies look at the effect of cyclic loading, the effect of a continuously applied static load has also been observed (van Eijk *et al.*, 2008). Goat BMSCs were seeded on a PLGA scaffold under a constant static load of 6.7 N applied during seeding, immediately after seeding, or 2 days after seeding. At day 5 scaffolds loaded during seeding had the highest collagen content but by day 21 unloaded scaffolds had the highest collagen levels. This suggests that immediate loading can have a positive effect on matrix production but that a constantly applied load inhibits differentiation.

4. Chondrogenesis

While tensile loading seems to favor osteogenesis and not chondrogenesis, there are a number of studies that have observed that cyclic tension can regulate chondrogenesis in appropriate culture conditions (Connelly *et al.*, 2010; McMahon *et al.*, 2008). Rat BMSCs were seeded on a collagen-GAG scaffold and subjected to 10% cyclic loading at 1 Hz in CM (McMahon *et al.*, 2008). Constraining the scaffolds resulted in a reduction in GAG content compared with FSW controls but cyclic loading returned the GAG content to the normal FSW level. Connelly *et al.* (2010) seeded bovine BMSCs into fibrin hydrogels and subjected them to similar conditions over short (24 h) and longer term (2 weeks) culture in basal or CM. After short-term loading at day 4, both protein and proteoglycan synthesis were upregulated in basal and CM but there was no increase in GAG content and by day 12 only protein synthesis was increased. Longer term cyclic loading upregulated COLI production, which was also seen in constrained samples but COLI was lower in FSW conditions while COLII and aggrecan gene expression were unaffected by loading. These studies indicate that constraining MSC-seeded scaffolds, and therefore inhibiting cellular contraction of the scaffold is detrimental to chondrogenesis but favorable to a fibrous matrix. Connelly *et al.* (2010) suggest that TGF-β can inhibit the effects of loading on chondrogenic differentiation at later time points, which may explain why there was no upregulation of chondrogenic markers.

5. Cardiomyogenesis

MSCs have been shown to differentiate into cardiomyocyte-like phenotypes under a number of different conditions including culture in cardiomyogenic medium containing DEX, insulin, and AA (Shim *et al.*, 2004), treatment with 5-azacytidine (5-aza) (Antonitsis *et al.*, 2007), coculture with cardiomyocytes or in cardiomyocyte conditioned medium (Li *et al.*, 2007). However, the proportion of MSCs that have shown a cardiomyocyte phenotype after these different treatments has been low and so alternative methods such as mechanical loading have been sought with the aim of increasing this response.

Human BMSCs were seeded on COLI coated BioFlex plates (a commercially available cell straining device from Flexcell® International Corporation) and subjected to a range of tensile strains (7–20%) over 4 days (Ku *et al.*, 2006). Collagen production was higher after loading and was highest in the 14% strain group along with an increase in lysyl oxidase (LOX). There was no increase in COLII mRNA or ALP activity suggesting that chondrogenesis and osteogenesis do not occur at these high-strain rates. In another cardiomyogenic study, human BMSCs and ADMSCs were seeded on a collagen coated BioFlex plate and continuously stretched at 14% strain for 3 days (Colazzo *et al.*, 2010). Collagen production was increased in both cell types with stretching but only COLIII mRNA was upregulated in ADMSCs and there was also a higher degree of collagen and elastin cross-linking. COLIV mRNA was also upregulated in both cell types with stretching, but there was no COLII expressed.

6. Conclusions: Tensile loading

Applying tensile loading to MSCs can initiate or enhance osteogenesis, tenogenesis, and also cardiomyogenesis under the appropriate conditions, whereas stretching appears to inhibit adipogenesis and have limited effect on chondrogenesis. The reviewed literature agrees in general that osteogenesis of MSCs tends to occur at strain magnitudes lower than that for tenogenesis and that cardiomyogenesis is stimulated by using larger strains. In the absence of osteogenic media, upregulation of early (ALP activity) and late (mineralized matrix deposition) osteogenic markers have been observed (Huang *et al.*, 2009), while it seems that the application of DEX can have a synergistic or inhibitory effect depending on the concentration used and the marker investigated (Mauney *et al.*, 2004). Tenogenesis is also induced in the absence of any chemical inducers, but like osteogenesis, it appears that static stretching or long term continuous loading has a negative effect on matrix production (Ngiam *et al.*, 2011; van Eijk *et al.*, 2008). Adipogenesis seems to be inhibited by stretching even in the presence of adipogenic media and stretching has been shown to not increase GAG secretion. These findings indicate that intermittent, cyclic stretching of MSCs is beneficial for increasing the production of a fibrous matrix resembling cell types of the osteogenic, tenogenenic, and cardiomyogenic lineages.

B. Compression

1. Introduction

MSCs subjected to compressive loading have mainly been investigated for the potential of compression to promote differentiation toward chondrogenesis (Table 20.2). A variety of BMSCs from different species have been

Table 20.2 Compression: Summary table of the literature cited in the text describing the effect of compression loading on mesenchymal stem cell differentiation and matrix production[a]

MSC source	Type of strain	Parameters	Culture surface	Culture media	Culture period	Lineage	Load effect	Reference
Human BMSCs	Cyclic hydrostatic pressure	5.03 MPa; 1 Hz; 4 h/day at days 1 or 1–7	2D/3D—aggregates	Chondrogenic	14–28 days	Chondrogenesis	↑ Proteoglycan and collagen content (7 days loading; D28)	Angele *et al.* (2003)
Human BMSCs	Cyclic compression, unconfined	4 h/day (7 days)	3D—hayluronan-gelatin composite	Chondrogenic	1, 7, 14, 21 days	Chondrogenesis	↑ COLII and aggrecan gene expression. ↑ proteoglycan and collagen content (D21)	Angele *et al.* (2004)
Human BMSCs	Cyclic hydrostatic pressure	0.1, 1, or 10 MPa; 1 Hz; 4 h/day	2D/3D—pellet culture	Chondrogenic	3, 7, or 14 days	Chondrogenesis	↑ SOX9 and aggrecan (≥0.1 MPa), COLII (10 MPa) mRNA. ↑ GAG and collagen production (10 MPa)	Miyanishi *et al.* (2006a)
Human BMSCs	Cyclic hydrostatic pressure	10 MPa; 1 Hz; 4 h/day	2D/3D—pellet culture	Chondrogenic w or w/o TGF-β3	3, 7 or 14 days	Chondrogenesis	↑ SOX9, aggrecan and COLII w and w/o TGF-β.	Miyanishi *et al.* (2006b)
Human BMSCs	Cyclic hydrostatic pressure (steady or ramped)	7.5 MPa or ramped 0.5 MPa/day; 1 Hz; 4 h/day (14 days)	3D—agarose gels	Basal	14 days	Chondrogenesis	No COLII or aggrecan mRNA. ↑ COLI and SOX9 mRNA (D4)	Finger *et al.* (2007)

Ovine BMSCs	Cyclic hydrostatic pressure	0.1 MPa; 0.25 Hz; 30 min/day for 10 days (at week 4)	3D—filamentous plasma coated polyester	Chondrogenic	38 days	Chondrogenesis	↑ GAG and collagen content (10 days)	Luo and Seedhom (2007)
Bovine BMSCs	Cyclic compression, unconfined	0.33, 1, or 3 Hz; 10% ε; 60/180 min (1–5 days)	3D—agarose construct	Chondrogenic	28 days	Chondrogenesis	↑ GAG deposition (Long term)	Mauck *et al.* (2007)
Bovine BMSCs	Static or cyclic compression, unconfined	1 Hz; 10% ε; for 1, 3 or 20 h (D8 or 16)	3D—agarose gels	Basal w or w/o TGF-β1 and DEX	8 or 16 days	Chondrogenesis	↑ COLII, aggrecan gene expression. ↑ GAG and protein synthesis (Dynamic load + TGF-β and DEX)	Mouw *et al.* (2007)
Goat BMSCs	Cyclic compression, unconfined	1 Hz; 10% ε; 1, 2, 2.5, or 4 h/daily (1–3 weeks)	3D—hydrogel	Chondrogenic w or w/o TGF-β1	1–3 weeks	Chondrogenesis	↑ SOX9 and aggrecan mRNA (w/o TGF-β). ↑COLII and aggrecan mRNA (w TGF-β). ↑ GAG production.	Terraciano *et al.* (2007)
Human BMSCs	Cyclic hydrostatic pressure	1 MPa; 1 Hz; 4 h/day for 10 days	3D—COLI sponge	Osteochondrogenic	10 days	Chondrogenesis, osteogenesis	↑ COLII, aggrecan, SOX9 mRNA; proteoglycan production. ↑ COLI mRNA. No change in RUNX2	Wagner *et al.* (2008)
Human BMSCs	Cyclic compression, unconfined	0.1 Hz; 10% ε; 24 h (continuous)	3D—alginate gel	Basal	24 h	Chondrogenesis	↑ Aggrecan, BMP-6, COLXI mRNA ↓ COLI mRNA	Haudenschild *et al.* (2009)
Equine BMSCs	Cyclic compression, unconfined	0.3 Hz; 2.5% ε; 45 min/6 h (alternate days), or 45 min on/off daily	3D—agarose hydrogels	Chondrogenic w or w/o TGF-β1	21 days	Chondrogenesis	↑ GAG synthesis w/o TGF-β1 (but less than control w TGF-β 1). ↓ GAG w TGF-β1	Kisiday *et al.* (2009)

(*Continued*)

Table 20.2 (*Continued*)

MSC source	Type of strain	Parameters	Culture surface	Culture media	Culture period	Lineage	Load effect	Reference
Rat BMSCs	Cyclic or static hydrostatic pressure	10–36 kPa; 0.25 Hz; 1 h/day (1, 3, or 5 days)	2D—TCP	Osteogenic preload, basal with load	5–12 days	Osteogenesis	↑ ALP activity and RUNX2 mRNA (Static and dynamic)	Liu *et al.* (2009)
Human ADMSCs	Cyclic hydrostatic pressure	0.5 MPa; 0.5 Hz; 7 days (continuous) (week 1)	3D—collagen scaffold	Chondrogenic	4 weeks	Chondrogenesis	↑ SOX9, aggrecan, COLII gene expression. Increasing to week 4	Ogawa *et al.* (2009)
Bovine BMSCs	Cyclic hydraulic pressure	0.5–3 MPa; 1 Hz; 4 h/day	2D/3D—aggregates	Chondrogenic w or w/o BMP-2 and TGF-β1	14 days	Chondrogenesis	↓ SOX9 mRNA all conditions. No change in COLI, II or aggrecan mRNA	Zeiter *et al.* (2009)
Bovine BMSCs	Cyclic compression, unconfined	0.01–1 Hz; 10% ε; 1 or 4 h, 5 days/week (3 weeks)	3D—agarose constructs	Chondrogenic w or w/o TGF-β3	3–6 weeks	Chondrogenesis	↑ COLII and aggrecan gene expression. ↓ GAG synthesis	Huang *et al.* (2010a)
Human BMSCs	Cyclic compression, unconfined	1 Hz; 10–20% ε; 1 h/day for 7 days (D7–14)	3D—fibrin/polyurethane scaffolds	Chondrogenic w or w/o TGF-β1	14 days	Chondrogenesis	↑ GAG content. ↑ COLII and aggrecan expression (highest w no TGF-β1). ↑ TGF-β1 and 3	Li *et al.* (2009b)

Human BMSCs	Cyclic compression unconfined	1 Hz; 5% ε; 2 h/day (D9, 14, and 19)	3D—polyurethane foam scaffold	Osteogenic w or w/o DEX	24 days	Osteogenesis	↑ ALP and COL1 gene expression. ↑ Collagen and calcium content (load and load+DEX)	Sittichokechaiwut *et al.* (2010)
Porcine BMSCs	Cyclic compression, unconfined	1 Hz; 10% ε; 1 h/day, 5 days/week (D0 or 21)	3D—agarose hydrogel	Chondrogenic w TGF-β3 (21 or 42 days)	42 Days	Chondrogenesis	↓ Proteoglycan and collagen content (D0–42). No difference D21–42	Thorpe *et al.* (2010)

[a] TGF-β (transforming growth factor-beta); SOX9 (SRY (sex determining region Y)-box 9). Other abbreviations as per Table 20.1.

used including human (Angele *et al.*, 2004; Li *et al.*, 2009b), bovine (Huang *et al.*, 2010a; Mauck *et al.*, 2007; Mouw *et al.*, 2007), porcine (Thorpe *et al.*, 2010), and equine (Kisiday *et al.*, 2009) using mainly hydrostatic pressure for 2D or pellet cultures, or unconfined platen-abutment compression for 3D cultures. The most common loading conditions used in the studies surveyed have been cyclic compression with a load frequency of 1 Hz and a global strain of 10%, applied intermittently for 1–4 h every day. Nearly, all of the studies have been carried out in the presence of CM and almost half of the studies have also looked at the effects of compressive loading in the presence or absence of typical chondrogenic inducing factors, particularly TGF-β1 or -β3.

2. Chondrogenesis

A number of studies have seeded bovine BMSCs onto 3D agarose gels and subjected them to a common global strain of 10% with a different range of frequencies (0.01–3 Hz), applied either on a one off basis (Mouw *et al.*, 2007), over a short period of 5 days (Mauck *et al.*, 2007), or over an extended period of 3 weeks (Huang *et al.*, 2010a). In each case, changing the frequency and duration of load did not seem to have much effect on differentiation, except for in Mauck *et al.* (2007) who found that 3 Hz for 180 min showed the highest increase in aggrecan promoter activity. It does, however, seem that media conditions and timing of the load are very important. It has been seen that loading initiated before MSCs have undergone chondrogenic differentiation can inhibit chondrogenesis (Mouw *et al.*, 2007; Huang *et al.*, 2010a; Thorpe *et al.*, 2010; Fig. 20.5). Mouw *et al.* precultured BMSCs in basal media or CM (±TGF-β1 and DEX) for 8 or 16 days followed by one bout of loading and noticed increased gene expression of COLII and aggrecan as well as increased protein and GAG synthesis with CM + load (Mouw *et al.*, 2007). However, loading with basal media did not induce any changes in gene expression or protein synthesis, and at day 8, there were no changes in any of the culture conditions. Notably, COLI gene expression was also increased in CM with loading at day 16. Huang *et al.* (2010a) looked at the effects of loading on longer term culture (6 weeks) with the addition of CM ± TGF-β3 under a load period of 3 weeks, initiated either immediately after seeding or after 3 weeks of FSW culture. Loading samples directly after seeding in CM reduced GAG and collagen production relative to FSW + CM controls; however, this level was slightly higher than FSW − CM and load − CM. Interestingly, loading did increase gene expression of COLII and aggrecan relative to FSW with or without CM. MSCs subjected to loading after 3 weeks of CM preculture did not differ in collagen or GAG content but mechanical properties did improve, most likely due to improved collagen distribution throughout the scaffold. Changing the frequency and strain of the load did not have an effect.

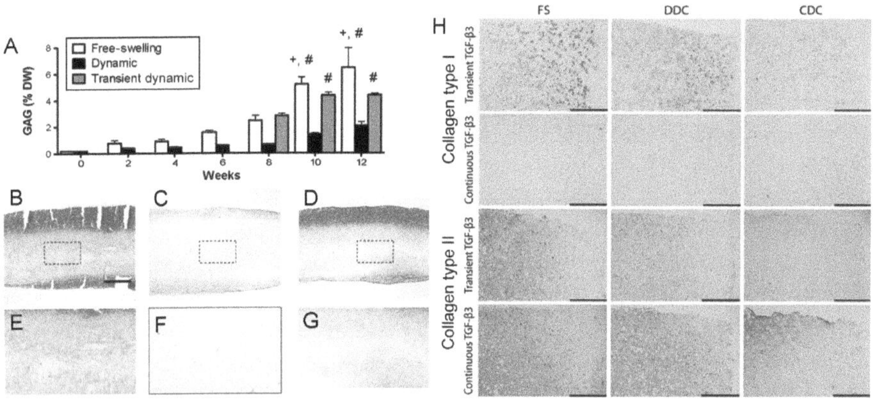

Figure 20.5 Effects of compression loading on chondrogenesis of mesenchymal stem cells. (A–G) Chondrogenesis is inhibited by dynamic culture in electrospun scaffolds, but cell and matrix distribution is improved. Biochemical measurement of GAG content (A) reveals depletion due to dynamic culture conditions, and a partial recovery following removal from dynamic culture. In all three culture conditions, GAG increased significantly with time in culture ($p < 0.05$). Alcian Blue (dark gray in black and white image) stains at 12 weeks confirm biochemical findings, with more intense staining of free-swelling (B) constructs compared to dynamic (C) and transient dynamic (D). Regions indicated in (B)–(D) are enlarged in (E)–(G), respectively. + and # indicate $p < 0.05$ versus transient dynamic and dynamic culture, respectively. Scale bar = 250 m (B–D) and 62.5 m (E–G). Reproduced from Nerurkar *et al.* (2011), with kind permission from Elsevier. (H) Chondrogenesis is inhibited by immediate mechanical loading (CDC) but not by loading in later stages of culture (DDC) in agarose hydrogels and TGF-β3 is required to stimulate matrix production. Immunohistochemical analysis of MSC-seeded constructs at day 42 stained with for the noncartilage specific collagen type I (top) and the cartilage specific collagen type II (bottom). FS, free-swelling. DDC, delayed dynamic compression, initiated on day 21. CDC, continuous dynamic compression, initiated on day 0. Representative images taken of approximately of each construct section. Scale bar: 500 μm. Reproduced from Thorpe *et al.* (2010), with kind permission from Springer Science and Business media. (See Color Insert.)

Thorpe *et al.* (2010) also carried out long term culture (42 days) of MSCs with loading initiated at day 0 or day 21 with or without TGF-β3. They also noticed that initiating loading immediately after seeding (before chondrogenic differentiation is induced) results in the downregulation of chondrogenic markers (proteoglycan, collagen, COLII), but by delaying loading until day 21, there was a similar level of these markers compared to FSW samples. Removal of TGF-β3 resulted in a downregulation of these chondrogenic markers in all conditions. In both these latter two studies, only small amounts of COLI were present compared with COLII (Huang *et al.*, 2010a; Thorpe *et al.*, 2010) indicating a preference toward chondrogenic differentiation. At the latter time points of each study when loading was initiated, the cells were more differentiated, and they may be more responsive to loading at this stage. Together, though, all of these studies conclude that without the presence of TGF-β, loading alone has a minimal affect on chondrogenic differentiation.

Another long term study subjected ovine BMSCs to hydrostatic pressure under relatively gentle conditions (0.1 MPa, 0.25 Hz, 30 min/day) for 10 days after a 4 week preculture in CM (Luo and Seedhom, 2007). An increase in GAG and collagen content was observed after 10 days of loading indicating a positive effect on chondrogenesis. In a study by Kisiday *et al.* (2009), equine BMSCs were subjected to intermittent cyclic load at a lower strain of 2.5% with or without TGF-1. Daily loading without TGF-β caused an increase in proteoglycan synthesis compared to nonloaded without TGF-β controls at day 21; however, loading in the presence of TGF-β showed a decrease in matrix production compared to the TGF-β nonloaded samples. The greatest matrix production was seen in the nonloaded with TGF-β group. This study indicates that TGF-β had an inhibitory effect on compressive loading and that loading on alternate days had a negative effect on chondrogenesis. Cyclic hydraulic pressure (0.5–3 MPa) was applied to bovine BMSCs for 4 h/day in the presence of BMP-2 and TGF-β1 (Zeiter *et al.*, 2009), but after 14 days of load, there was a downregulation of SOX9 mRNA in all conditions and there was no change in COLII or aggrecan mRNA.

In contrast to these previous findings, other studies have shown that early compressive loading can enhance chondrogenic differentiation markers in the presence or absence of TGF-β (Haudenschild *et al.*, 2009; Terraciano *et al.*, 2007). Goat BMSCs seeded in hydrogels were subjected to short (1–2 h) or long (4 h/day for 1–3 weeks) bouts of cyclic compression (Terraciano *et al.*, 2007). In the absence of TGF-β, 1 h of loading upregulated SOX9 and aggrecan mRNA but the difference was not so strong with long term loading in the presence of TGF-β. However, by day 21, compression had upregulated aggrecan and COLII gene expression and GAG content was also higher than without loading. Human BMSCs cultured in basal media and subjected to compressive loading at 0.1 Hz and 10% strain showed an upregulation in gene expression of the chondrogenic markers aggrecan, COLXI, and BMP-6 but also a downregulation in COLI mRNA (Haudenschild *et al.*, 2009).

There have been a number of studies that have shown that early loading of MSCs followed by a prolonged period of unloaded culture can be beneficial to chondrogenesis (Angele *et al.*, 2003, 2004; Ogawa *et al.*, 2009). Mauck *et al.* (2007) found that subjecting MSCs to early loading (5 days) and then placing them in FSW culture until day 28, actually enhanced proteoglycan deposition and GAG content, which was further enhanced by the addition of TGF-β. When human BMSCs cultured in CM were subjected to either cyclic hydrostatic pressure (5.03 MPa) or compression for the first 7 days of culture and then left in static culture for a further 2–3 weeks, increases in both proteoglycan and collagen content were seen (Angele *et al.*, 2003, 2004). Human ADMSCs were subjected to cyclic hydrostatic pressure (0.5 MPa) continuously for the first week of culture and then cultured in static conditions for a further 3 weeks in the presence of CM (Ogawa *et al.*, 2009). Increases in SOX9, aggrecan, and COLII gene expression were all seen after 2 weeks, and these levels continued increasing up to 4 weeks. These studies indicate that an early period of loading followed by a prolonged period of unloading may be beneficial for MSC chondrogenesis, but the mechanism behind this is unclear.

Other studies have observed the effects of hydrostatic pressure on MSCs with varying media and loading conditions (Finger *et al.*, 2007; Miyanishi *et al.*, 2006a,b). In studies by Miyanishi *et al.* (2006a,b), human BMSCs in pellet culture were subjected to different hydrostatic pressures (0.1, 1, or 10 MPa) and media conditions (CM ± TGF-β3), with daily loading over 14 days. They found that in the presence of TGF-β3, the lowest pressure of 0.1 MPa was enough to upregulate gene expression of SOX9 and aggrecan; however, a pressure of 10 MPa was required to increase COLII mRNA. GAG production was increased at 1 MPa and was highest at 10 MPa, whereas collagen production was only upregulated at 10 MPa (Miyanishi *et al.*, 2006a). Loading at 10 MPa in the absence of TGF-β3 increased the expression of the chondrogenic markers COLII, aggrecan, and SOX9 over nonloaded controls (−TGF-β3) but this response was lower than non-loaded (+TGF-β3). A synergistic effect was seen when combining hydrostatic pressure with TGF-β3 with a large increase in these chondrogenic markers (Miyanishi *et al.*, 2006b). Finger *et al.* (2007) employed a steady versus ramped (+0.5 MPa/day) regime for human BMSCs seeded in agarose gels in the absence of TGF-β. They observed no change in COLII or aggrecan mRNA over the 14-day culture period, but increases in gene expression of COLI and SOX9 were observed to peak at day 4 in the steady load group. Steady pressure generally outperformed ramped pressure although some donor variability was noticed.

3. Osteogenesis

A small number of studies have investigated the effect that global compressive loading (Sittichokechaiwut *et al.*, 2010; Wagner *et al.*, 2008), or hydrostatic compression (Liu *et al.*, 2009) of cell-seeded scaffolds may have on osteogenic induction. In our lab, human BMSCs were seeded onto a polyurethane (PU) foam and subjected to short bouts of cyclic compression at 1 Hz, 5% strain, for 2 h/day at days 9, 14, and 19, and cultured to day 24 with and without DEX (Sittichokechaiwut *et al.*, 2010; Fig. 20.6). ALP activity, COLI mRNA, and collagen production were higher in loaded samples − DEX than static controls − DEX. Calcium

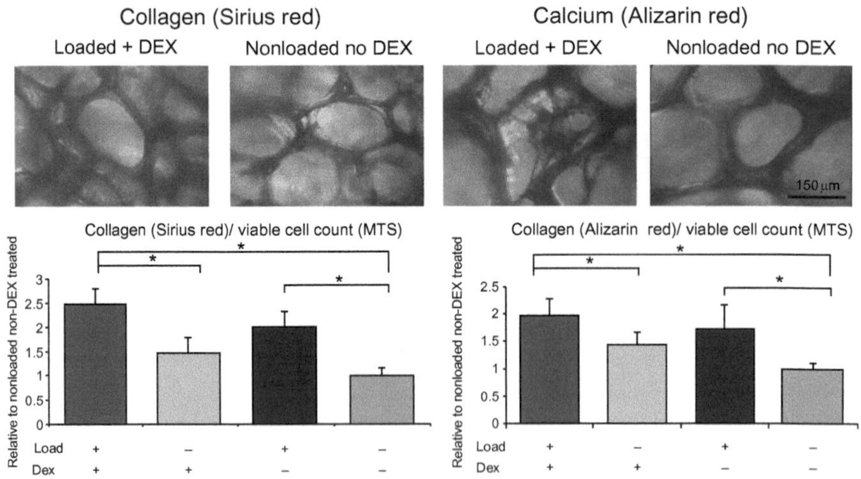

Figure 20.6 Scaffold compression upregulates osteogenesis of mesenchymal stem cells in 3D. hMSCs seeded in polyurethane scaffolds subjected to 2 h of dynamic compression loading (5% strain 1 Hz) every 5 days contained more collagen and calcium by day 24 of culture. Top: light micrographs representative images of Sirius red (collagen) and alizarin red (calcium) staining (black in black and white image). Bottom: quantification by colorimetry normalized to controls. Loading was as effective as DEX in stimulating osteogenic matrix production (*$p < 0.05$ ANOVA with Tukeys pair-wise comparison). Adapted from Sittichokechaiwut *et al.* (2010), kind permission of reproduction from eCMjournal.org. (See Color Insert.)

and collagen content was also higher in loaded + DEX samples than non-loaded + DEX samples, but there were no differences seen in ALP activity, collagen, or calcium content between nonloaded samples with DEX and loaded samples without DEX. This indicates that short bouts of scaffold compression loading can induce or enhance osteogenesis and are as effective as DEX treatment. However, as discussed previously, given the isotropic nature of the scaffold used these results cannot be interpreted as a direct effect of cell compression and may be an effect of scaffold compression inducing different forces on individual cells (compression, tension, and fluid shear) depending on their location within the construct with respect to the applied load.

Wagner *et al.* (2008) seeded human BMSCs onto COLI sponges and subjected them to hydrostatic pressure (1 MPa) in the presence of osteo-CM. After 4 h of loading per day COLI mRNA was upregulated, but there was no increase in RUNX2 gene expression. There was an increase in the gene expression of the chondrogenic markers COLII, aggrecan, SOX9, and also the production of proteoglycan. This indicates that compressive hydrostatic pressure may favor chondrogenesis over osteogenesis. However, when cyclic hydrostatic pressure was applied to rat BMSCs precultured in osteogenic media at lower pressures (10–36 kPa) and less frequently (0.25 Hz, 1 h/day on alternate days), there was an increase in ALP activity and RUNX2 mRNA (Liu *et al.*, 2009).

4. Conclusions: Compression loading

Compressive loading has been shown to have a regulatory effect on the differentiation of MSCs; however, this effect does not appear to influence differentiation in the same manner that tensile loading does. The influence of compression forces on chondrogenesis has been the most commonly investigated, with different studies showing contradictory results, but with the general consensus that timing of the load, along with media conditions are the key parameters influencing the outcome. Subjecting MSCs to continuous compression throughout the culture period actually downregulates key chondrogenic markers compared with FSW controls but by inserting a period of static, FSW culture either at the start of culture (Thorpe *et al.*, 2010) or after an initial loading period (Mauck *et al.*, 2007), this positively influences chondrogenic differentiation. It is widely agreed that TGF-β alone can induce chondrogenesis in MSCs but its affect on differentiation in combination with loading is still not fully understood. Some studies have shown that in the absence of TGF-β, compressive loading alone can induce chondrogenic differentiation (Haudenschild *et al.*, 2009). A small number of studies have also shown that combining TGF-β treatment with compressive loading has an inhibitory affect on chondrogenesis; however, more studies indicate that addition of TGF-β is beneficial and its removal downregulates chondrogenic markers. In a few cases, compressive loading has been shown to upregulate osteogenic markers including the fibrous matrix marker COLI and ALP activity, but osteogenic studies are limited. Overall, compressive loading of MSCs seems to be beneficial for the production of a nonfibrous, cartilage-like matrix, in contrast to tensile loading. However, further studies are required in order to agree on the correct time to initiate this load and to fully determine the regulatory effects it has.

C. Fluid flow-induced shear stress

1. Introduction

A tabulated summary of the reviewed literature for the effects of fluid shear stresses on MSCs can be seen in Table 20.3. The majority of studies have focused on osteogenic differentiation with the main cell source being BMSCs. Osteocytes (terminally differentiated bone cells) are thought to respond to shearing caused by fluid forces in the canaliculi in bone and osteoblasts have been shown to respond to fluid forces *in vitro* (Bonewald and Johnson, 2008; Klein-Nulend *et al.*, 2005; You *et al.*, 2001), so there

Table 20.3 Fluid flow induced shear stress: Summary table of the literature cited in the text describing the effect of fluid flow induced shear forces on mesenchymal stem cell differentiation and matrix production[a]

MSC source	Type of flow	Parameters	Culture surface	Culture media	Culture period	Lineage	Load effect	Reference
Rat BMSCs	Steady perfusion	0.3, 1, or 3 ml/min	3D–titanium fiber mesh scaffold	Osteogenic	4, 8 or 16 days	Osteogenesis	↑ ALP activity. ↑ calcium content with ↑ flow rate.	Bancroft *et al.* (2002)
Rat BMSCs	Steady perfusion	Continuous 0.3 ml/min	3D–SEVA-C and SPCL scaffolds	Osteogenic	3, 7 or 15 days	Osteogenesis	↑ Calcium at day 15 (Static produced hardly any calcium)	Gomes *et al.* (2003)
Rat BMSCs	Steady perfusion	1–3 × 10^{-2} Pa for 4, 8, or 16 days	3D–titanium fiber mesh scaffold	Osteogenic	16 days	Osteogenesis	↑ ALP activity. ↑ calcium content with ↑ shear force	Sikavitsas *et al.* (2003)
Rat BMSCs	Steady, radial	0.36–2.7 × 10^{-1} Pa, 30 min alternate days (D3–13)	2D–fibronectin coated substrates	Osteogenic	21 days	Osteogenesis	↑ OCN protein	Kreke and Goldstein (2004)
Human BMSCs	Oscillatory	2 h at 1 Hz, 1 Pa at days 3 or 5	2D–parallel flow chamber	Osteogenic	4–8 days	Osteogenesis	↑ osteopontin and osteocalcin gene expression. ↓ ALP	Li *et al.* (2004)
Rat BMSCs	Steady perfusion	Continuous flow 1 ml/min	3D–titanium fiber mesh	Osteogenic	4, 8 or 16 Days	Osteogenesis	↑ early ALP, ↑ calcium and osteopontin content	Holtorf *et al.* (2005)
Rat BMSCs	Steady	1.6 × 10^{-1} Pa for 5, 30, or 120 min every 2 days	2D–fibronectin coated substrates	Osteogenic	6 or 20 days	Osteogenesis	↑ ALP activity (D6). ↑ Osteopontin and bone sialoprotein gene expression (D20)	Kreke *et al.* (2005)
Rat BMSCs	Steady perfusion	Continuous 1 ml/min	3D–titanium fiber mesh w or w/o ECM	w or w/o DEX	4, 8 or 16 days	Osteogenesis	↑ Calcium content w and w/o DEX	Datta *et al.* (2006)
Rat BMSCs	Steady perfusion	5 × 10^{-3} Pa for 4, 8, or 16 days	3D–poly(L-lactic acid) nonwoven scaffolds	Osteogenic	16 days	Osteogenesis	↑ Calcium deposition	Sikavitsas *et al.* (2005)

Human BMSCs	Steady perfusion	1–0.1 × 10^{-4} Pa for D1–20	3D–PET matrices	Basal or osteogenic	34 days	Osteogenesis	↑ ALP activity and calcium with higher flow rate	Zhao *et al.* (2007)
Human BMSCs	Steady rotation	16 rpm (6 weeks)	3D–silk scaffolds	Basal, chondrogenic or osteogenic	6 weeks	Osteogenesis	↑ COLI and calcium production (Osteogenic)	Augst *et al.* (2008)
Human BMSCs	Steady perfusion	Continuous flow 0.85–0.01 ml/s, 0.7–10 mPa	3D–decellularized bone matrix	Osteogenic	35 days	Osteogenesis	↑ Bone volume in high-flow group	Grayson *et al.* (2008)
Mouse C3H10T1/2 MSCs	Oscillatory	1 Hz, 1 Pa, 1 h	2D–fibronectin coated glass	Basal	48 h	Osteogenesis, Chondrogenesis, Adipogenesis	↑ RUNX2, SOX9 and PPARγ gene expression	Arnsdorf *et al.* (2009)
Human BMSCs	Steady	1.2 Pa	2D–parallel flow chamber	Osteogenic preflow	30 or 90 min	Osteogenesis	↑ ALP and ↓ COLI gene expression	Grellier *et al.* (2009)
Rat BMSCs	Steady or pulsatile	0.015–0.074 Hz, 0.23–0.43 Pa for 24 h at day 6	2D–fibronectin coated glass	Osteogenic	20 days	Osteogenesis	↑ COLI, OPN, OC, BSP mRNA. ↑ BMP-2,7 and TGF-B1	Sharp *et al.* (2009)
Rat BMSCs	Steady	0.5, 1, 1.5, or 2 Pa for 24 h	2D–parallel flow chamber	Basal media w or w/o 5-azacytidine	7 days	Cardio-myogenesis	↑ Cardiomyocyte gene and protein expression	Huang *et al.* (2010b)
Human hES-MP 002.5	Oscillatory, rocker	0.5 Hz, 1 h/day	2D–gelatine coated TCP	Osteogenic	20 days	Osteogenesis	↑ ALP activity and calcium content	Delaine-Smith *et al.* (2011)
Rat MSCs	Steady perfusion	0.5 ml/min daily (≈0.01 Pa)	3D–PLA foam or nonwoven fiber scaffolds	Osteogenic	4 or 8 days	Osteogenesis	↑ ALP activity (D4 and 8). ↑ Calcium on fiber mesh (D8)	VanGordon *et al.* (2011)

[a] hES-MP 002.5 (human embryonic stem cell-derived mesenchymal progenitors); SEVA-C (Starch and ethylene vinyl alcohol); SPCL (starch and polycaprolactone); ECM (extracellular matrix); PET (poly(ethylene terephthalate)). Other abbreviations as per Table 20.1.

is an natural assumption that fluid shear stress will influence BMSCs. The source of BMSCs most often used has been from rats (Datta *et al.*, 2006; Gomes *et al.*, 2003; Sharp *et al.*, 2009), with some use of hMSCs (Grellier *et al.*, 2009; Li *et al.*, 2004; Zhao *et al.*, 2007), and a small number of MSC cell lines (Arnsdorf *et al.*, 2009; Delaine-Smith *et al.*, 2011). The most commonly used methods for inducing fluid shear have used the parallel-plate flow chamber for 2D culture (Huang *et al.*, 2010b; Kreke *et al.*, 2005) and perfusion bioreactors for 3D cultures (Holtorf *et al.*, 2005; Sikavitsas *et al.*, 2003) with a steady, unidirectional flow being the most popular method. MSCs seeded on 2D substrates are usually subjected to levels of shear stress around (0.1–2 Pa) whereas the shear stresses experienced in 3D constructs are often much lower than in 2D experiments (0.1 mPa–0.03 Pa) (e.g., Cioffi *et al.*, 2006; Maes *et al.*, 2009; and as summarized in McCoy and O'Brien, 2010). The magnitude of the shear forces usually applied in 2D experiments is estimated to be in the physiological range of those experienced by osteocytes in bone (Weinbaum *et al.*, 1994); however, these may not reflect fluid shear forces that are normally experienced by MSCs *in vivo*.

3D perfusion systems are typically used to continuously flow fluid over the seeded constructs for the entire length of the culture period (Grayson *et al.*, 2008; VanGordon *et al.*, 2011), whereas 2D systems tend to be used to subject the cells to shearing intermittently and for relatively shorter amounts of time, for example, from 30 min (Grellier *et al.*, 2009) to 24 h (Sharp *et al.*, 2009). However, recently short bouts of flow applied to 3D scaffolds have also been shown to be effective at stimulating MSC osteogenic differentiation by our group (Matsiko *et al.*, 2009; Fig. 20.7C) and osteoblasts by others (Jaasma and O'Brien, 2008). There may also be differences in response depending on whether the flow is oscillatory (thought to mimic the flow induced in tissues by the repetitive loading of locomotory activity) or unidirectional flow which may be steady or pulsatile (pulsatile flow being the most representative of continuous blood flow). Oscillatory and steady flow upregulate different mechanotransduction pathways in mature bone cells (Jacobs *et al.*, 1998) but have recently been shown to have a similar overall osteogenic outcome in a preosteoblast cell line (Case *et al.*, 2011).

2. Osteogenesis

Overall, fluid shear forces appear to either induce or enhance osteogenesis in MSCs. Human BMSCs seeded in a parallel-plate flow chamber precultured in osteogenic media were subjected to 1.2 Pa shear forces for either 30 or 90 min and an increase in ALP activity was seen with 30 min of stimulation showing the highest levels (Grellier *et al.*, 2009). However, there was no increase in RUNX2 expression and interestingly COLI gene expression was lower after flow. This enhancement of ALP activity was also seen in other 2D systems (Delaine-Smith *et al.*, 2011; Kreke *et al.*, 2005), whereas some studies have noticed an inhibition of ALP activity by

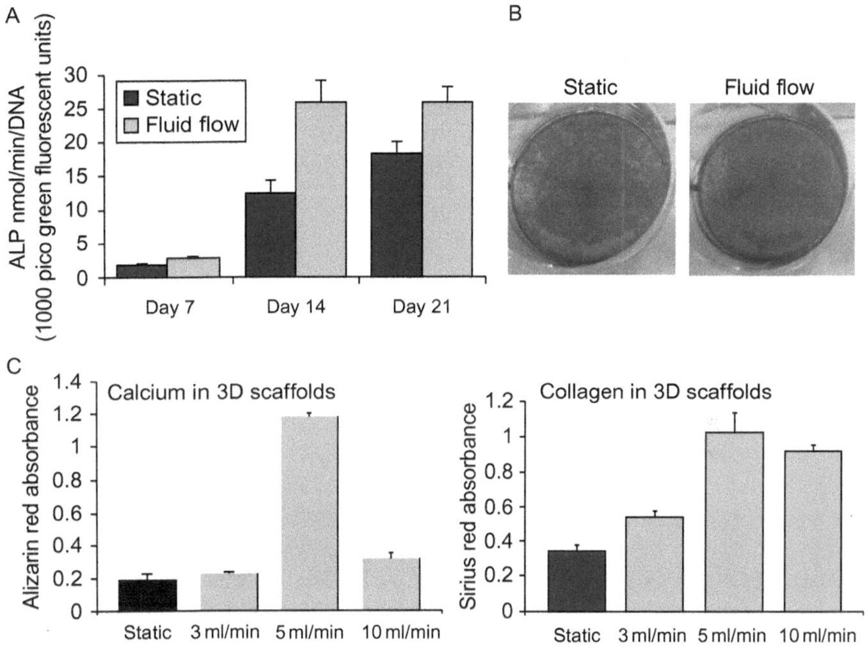

Figure 20.7 Fluid flow upregulates osteogenesis of mesenchymal stem cells in 2D and 3D. (A, B) Mesenchymal precursors derived from embryonic stem cells (hES-MP 002.5, Cellartis, Gothenburg) were subjected to oscillatory fluid flow by placing cell-seeded culture wells on a rocker at 1 Hz for 1 h/day beginning at day 5 of culture. (A) Mean ± SD of alkaline phosphatase (ALP) activity, which was significantly higher at day 7 and day 14 of culture in cells subjected to fluid flow compared to static controls ($n = 6$, $*p < 0.05$). (B) Alizarin red (black in black and white image) staining of culture wells (6-well plate) at day 21 indicated that cells subjected to rocking produced more calcium. (C) Preliminary data indicate that short bouts of oscillatory fluid flow stimulate MSCs matrix production in 3D scaffolds. Immortalized mesenchymal stem cells from mouse bone marrow, BM1 (kindly donated by Ilaria Bellantuono, University of Sheffield), were subjected to oscillatory fluid flow within porous polyurethane scaffolds (described in Sittichockechaiwut *et al.*, 2009) at the peak flow rates shown, at 1 Hz for 1 h on days 4, 7, and 10 of culture. (A) Mean ± SD of Sirius red (collagen) and alizarin red (calcium) absorbance units after staining the scaffolds on day 14 of culture. Previously unpublished data from our laboratory collected by Mohsen Shaeri. (See Color Insert.)

flow (Li *et al.*, 2004). In our laboratory, only 1 h per day of oscillating flow at 1 Hz, beginning on day 5 of culture was sufficient to upregulate ALP, collagen, and calcium production (Fig. 20.7A and B).

Steady perfusion of 3D cell-seeded constructs is almost always reported to stimulate ALP activity, with the greatest effects occurring at the earlier time points of 4–8 days (Holtorf *et al.*, 2005; Sikavitsas *et al.*, 2003; VanGordon *et al.*, 2011) and then often leveling off, although in some studies, significant increases have been seen up to 14–16 days of culture (Bancroft *et al.*, 2002; Gomes *et al.*, 2003). In 3D perfusion systems ALP activity does not seem to be affected by increasing the shear rate; however, calcium production is higher suggesting that increasing the fluid shear force affects later stages of differentiation more than earlier stages or is more important for matrix formation than stimulation of differentiation (Bancroft *et al.*, 2002; Sikavitsas *et al.*, 2003).

3. Other differentiation pathways

In some cases, the fluid shear modulation of other differentiation pathways has been suggested. Bovine BMSCs were seeded on aligned nanofibers of PCL and placed on an orbital shaker at 1.2 Hz for 6 weeks followed by 6 weeks in FSW conditions (Nerurkar *et al.*, 2011; Fig. 20.5A–G). After the initial 6 weeks of dynamic culture, there was a lower GAG content, but this was seen to recover to levels similar to the 12 week FSW if the constructs were not subjected to a further 6 weeks of dynamic culture. This is a similar outcome to many of the previously described compression loading regimes, which some groups showed to be more effective for chondrogenesis only when load was applied at early stages of culture followed by a FSW period. The opposite effect was seen with collagen content which was higher after dynamic culture and lower in FSW conditions. Constructs subjected to just the first 6 weeks of flow did show an increase in scaffold stiffness compared to 12 weeks in FSW conditions. Human BMSCs were seeded on a silk scaffold in a rotating vessel for 3 weeks in the presence of chondrogenic or osteogenic media (Augst *et al.*, 2008). After 3 weeks, the media of each formed osteochondral construct was switched and then constructs were subjected to a further 3 weeks of dynamic culture. COLI was present at both 3 and 6 weeks in all culture groups, whereas calcium matrix was present in only constructs cultured in osteogenic media. In all culture groups, there was very minimal GAG and COLII. Both these studies suggest that this type of fluid shear force has a positive effect on osteogenesis but a negative effect on chondrogenesis; however, in both cases, the shear stress was not calculated.

The murine MSC cell line C3H10T1/2 was seeded onto fibronectin coated glass slides and subjected to oscillatory fluid forces (1 Hz, 1 Pa for 1 h) in absence of any chemical differentiating factors (Arnsdorf *et al.*, 2009). After 48 h of culture, gene expression for osteogenesis (RUNX2), chondrogenesis (SOX9), and adipogenesis (PPARγ) were all seen to significantly increase compared with static controls, suggesting that oscillatory shear forces can regulate multiple differentiation pathways. McBride *et al.* (2008) also showed that multiple genes involved in early mesenchymal differentiation (condensations stage), RUNX2, SOX9, COLI, COLII, could all be regulated by low shear (0.1 MPa) laminar flow in a cell density and flow duration dependant manner, but in their case genes described as "later stage genes" (PPARγ, osterix, and aggrecan) were not affected.

Similarly to tensile loading, fluid shear forces have been shown to modulate cardiomyogenesis. Huang *et al.* (2010b) subjected rat BMSCs to fluid shear forces in a parallel flow chamber at a range of shear stresses, 5, 10, 15, and 20 Pa. At shear forces of 15 Pa cardiomyocyte gene expression was downregulated whereas below <10 Pa it was upregulated, showing a biphasic response. When the authors combined shear stresses with the cardiomyocyte inducing biochemical 5-azacytidine (5-aza), they found a significant increase in cardiomyocyte gene expression compared with each treatment alone suggesting a synergistic interaction.

4. Scaffold pore size

For cells seeded on a porous scaffold, the forces experienced due to flow will be largely affected by the porosity and the mechanical properties of the scaffold. There have been a small number of studies that have looked at the influence of scaffold architecture on the resulting effect of fluid shear forces. Rat BMSCs were seeded on titanium mesh scaffolds with two different pore sizes (20 and 40 μm) and subjected to perfusion fluid forces (1 ml/min) (Holtorf *et al.*, 2005). Under flow, the 40 μm mesh showed the highest levels of calcium at day 8 but this was reversed at day 16 with the 20 μm mesh showing the highest levels. In another study, rat BMSCs were seeded on a PLA foam or fibrous scaffold with differing pore sizes and then subjected to perfusion flow (0.5 ml/min) (VanGordon *et al.*, 2011). In contrast to the previous study, the smaller pored fiber scaffold showed the highest ALP activity and the presence of calcium (not seen in foam scaffold) at days 4 and 8, respectively.

5. Conclusions: Fluid shear force

Fluid flow-induced shear stress is well established as a mechanical force experienced by tissues *in vivo* and has multiple effects on fully differentiated cells *in vitro*. Given the importance of enabling nutrient access and waste removal to occur across a 3D scaffold it is not surprising that continuous flow was one of the first mechanical regimens to be applied to 3D scaffolds containing MSCs. However, much evidence suggests that cells respond specifically to the shear stresses induced by flow and not just the increased mass transport incurred. The work summarized here shows effects on MSC differentiation of both specific flow regimens in well-characterized 2D devices and less defined perfusion regimens in 3D scaffolds. A common finding is that overall matrix distribution and/or production are improved when flow is applied within a tissue engineered scaffold (Bancroft *et al.*, 2002; Sikavitsas *et al.*, 2003). It remains to be seen whether fluid flow alone can induce all the necessary mechanical signals for MSC differentiation and matrix production or whether additional strain stimuli are necessary.

There is evidence from fluid flow regimens as well, as the other mechanical regimens described, to suggest that when cells are constantly subjected to mechanical forces without rest, irrespective of the flow type, this causes quiescence of the cellular response, effectively the cell becomes desensitized to the loading. In osteoblasts including rest or low flow periods between bouts of higher flow rates has shown to upregulate OPN in mature osteoblasts in 2D (Batra *et al.*, 2005) and in 3D scaffold (Plunkett *et al.*, 2010) but not affect other responses to flow such as release of the paracrine factor Prostaglandin E2 (PGE2). In our laboratory we have successfully upregulated matrix production by mature osteoblasts and MSCs with short bouts of flow interspersed with a day or several days of rest in static culture in multiple studies (e.g., Matsiko *et al.*, 2009; Fig. 20.7) but have not directly compared these to continuous flow regimens. Considering the importance of rest periods in animal studies that use mechanical loading to stimulate tissue formation (e.g., Srinivasan *et al.*, 2002), it is likely that rest periods will be important in optimal stimulation regimens for MSCs; however, to our knowledge, no systematic studies have identified the best rest periods to use within any specific MSC-stimulating flow regimen.

D. Other load types

1. Direct cell membrane strain by magnetic force

Stimulation of magnetic nano–micro particles bound to receptors on the cell surface has been used to stimulate MSCs for osteochondrogenic purposes (Kanczler *et al.*, 2009; Kasten *et al.*, 2010). hMSCs seeded onto TCP surfaces coated with or without fibronectin and RGD peptides and cultured in osteogenic or adipogenic media were stimulated at 1 Hz for 15 min by a force of 200 pN/bead, which pulled on β1 integrins (Kasten *et al.*, 2010). Coated surfaces showed an increase in COLI gene expression but, surprisingly, only in adipogenic media and SOX9 was only increased on noncoated surfaces in both media types. Stress generally downregulated ALP activity, and PPARγ expression increased, also, only in osteogenic media on TCP. These results show that integrin pulling can regulate MSC differentiation, and that the outcome is strongly regulated by the surface the cells are adhered to and the media conditions. In another study, human BMSCs were seeded on TCP or in alginate/chitosan scaffolds and subjected to integrin magnetic particle stimulation over a longer culture period of 7–21 days (Kanczler *et al.*, 2009). Loading was induced for 1 h daily on alternate days resulting in an increase in SOX9 and OPN gene expression at day 7 and an increase in COLI and proteoglycan synthesis at day 21. These studies show that direct stimulation of mechanotranducers can enhance differentiation of MSCs favoring osteochondrogenesis.

2. Ultrasound

LIPUS has found popularity in the past couple of years for stimulating 2D MSC cultures toward osteogenic (Angle *et al.*, 2011; Ikeda *et al.*, 2006; Marvel *et al.*, 2010) and chondrogenic differentiation (Ikeda *et al.*, 2006; Lai *et al.*, 2010; Lee *et al.*, 2006). Human BMSCs and ADMSCs were seeded on BioFlex plates, cultured in osteogenic media, and subjected to LIPUS with a power of 30 mW/cm^2 and a range of different pulse repetition frequency (PRF) (1 Hz–1 kHz) for 20 min/day over 1–2 weeks (Marvel *et al.*, 2010). At day 7, there was more calcium present in LIPUS stimulated ADMSCs and this amount increased with increasing PRF with 1 kHz stimulating the highest amount of calcium. This increase in calcium content and increasing PRF trend were also seen in BMSCs but not until day 14 and there were variations between donors. Human BMSCs were seeded on 2D culture plates and subjected to LIPUS in the presence or absence of CM with TGF-β1 (up to 4 weeks), and osteogenic media with BMP-2 (3 days) (Lai *et al.*, 2010). After 3 days, gene expression of RUNX2 and ALP were significantly increased in BMP-2, LIPUS, and LIPUS + BMP-2 groups compared with controls; however, there was no difference between the groups indicating a lack of synergistic response. Gene expression of SOX9, aggrecan, and COLII was increased in LIPUS, TGF-β1, and LIPUS + TGF-β1 groups compared to controls from weeks 2 or 3 onward with LIPUS + TGF-β1 showing the largest increase indicating a synergistic response. Interestingly, LIPUS alone drastically increased RUNX2 gene expression from week 1 to 4 and COLI expression from week 3 to 4. Lee *et al.* (2006) subjected rabbit BMSCs to LIPUS in a 3D environment in the presence of CM (±TGF-β3) and observed increases in SOX9 and aggrecan expression as well as synthesis of proteoglycan and COLII. Interestingly, these increases were noticed even in the absence of TGF-β with LIPUS + TGF-β showing a small synergistic effect. Both these studies suggest that LIPUS alone can induce osteogenic and chondrogenic differentiation in the absence of differentiation media but when combined with TGF-β, chondrogenic differentiation appears to be further enhanced.

A study by Ikeda *et al.* (2006) is in agreement with these findings. LIPUS was applied to the mouse C2C12 MSC cell line cultured in multipotential media resulting in an upregulation of RUNX2 and SOX9 gene expression. In addition to this, a decrease in the adipogenic transcription factor PPARγ was also seen suggesting LIPUS can inhibit adipogenesis. It has also been shown that altering the LIPUS conditions has an effect on matrix production. Increasing the pulse frequency from 1 Hz to 1 kHz showed an increase in calcium deposition in both human BMSCs and ADMSCs (Marvel *et al.*, 2010). Angle *et al.* (2011) investigated the effect of altering the power output on the osteogenic differentiation of rat BMSCs with a range of values lower than those typically used (2, 15, or 30 mW/cm^2). While all LIPUS conditions increased ALP activity and calcium deposition

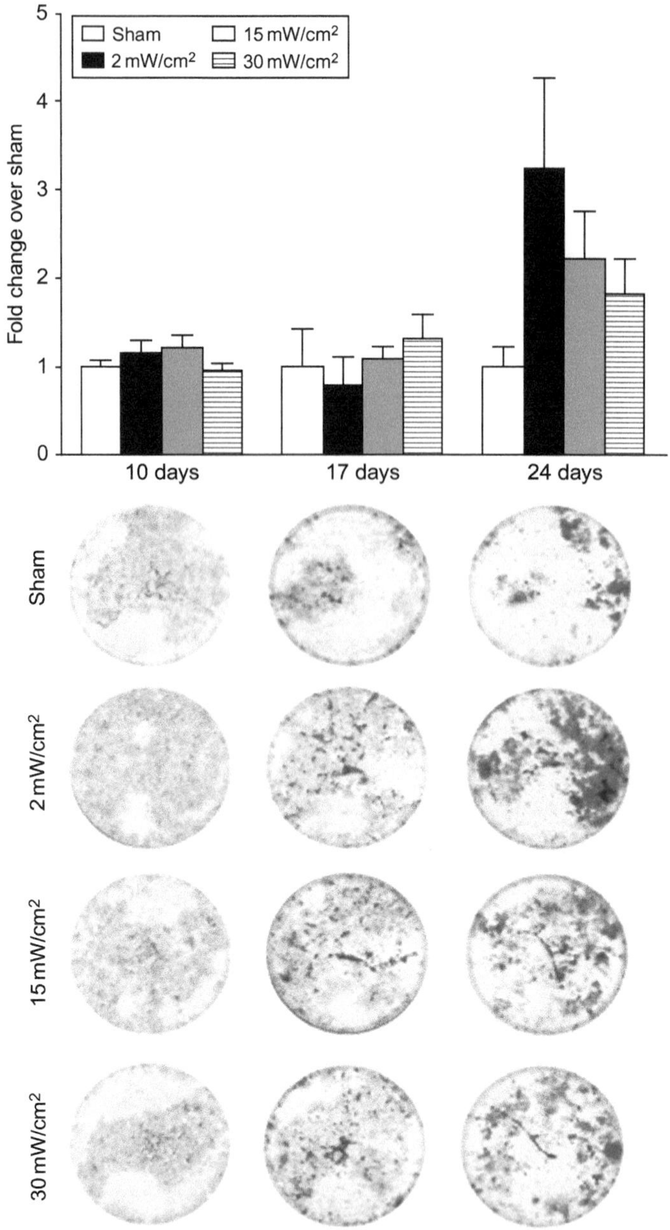

Figure 20.8 Example of the effects of low-intensity pulsed ultrasound (LIPUS) on osteogenesis of mesenchymal stem cells. Relative mineralization by rat BMSCs seen after cultures in 12-well plates were exposed to 0 (sham), 2, 15, and 30 mW/cm^2 LIPUS for daily treatment for 10, 17, and 24 days. Mineralization (calcium) quantified in mM of alizarin red (black in black and white image). Data represented as mean ($N = 4$) $\pm$ SEM. The well pictures are representative images for each treatment group. Reproduced from Angle *et al.* (2011), with kind permission from Elsevier. (See Color Insert.)

over untreated controls, the highest power value of 30 mW/cm^2 showed the largest increase in ALP activity, whereas the lowest value of 2 mW/cm^2 showed the highest increase in calcium deposition (Fig. 20.8). This suggests that the optimal conditions for accelerating early and late stage differentiation are not the same and that the loading regimen needs to be varied depending on what stage of maturity the cell is at.

3. Low magnitude, high-frequency vibration

In a preliminary *in vitro* study conducted in our lab, the effect of vibration treatment on MSCs was observed. The human cell line hES-MP (Cellartis, Gothenburg) was subjected to whole plate vibration (15–60 Hz, 0.2 m/s^2) in osteogenic media (±DEX) at day 6 for 10 or 45 min (Edwards and Reily, 2011; Fig. 20.9). While vibration in the absence of DEX did not significantly enhance osteogenesis, loading with DEX showed an increase in ALP activity after 45 min of stimulation at 60 Hz compared to all other conditions. This suggests that vibration treatment may play a regulatory role in the osteogenesis of MSCs. In another study, LMHF loading was carried out on the mouse C3H10T1/2 MSC cell line in the presence of either adipogenic media or multipotential media with daily application of load (Sen *et al.*, 2011). In the presence of adipogenic media, LMHF treatment initially had no effect on adipogenesis but when rest periods were inserted between daily loading cycles, gene expression of adiponectin and PPARγ was downregulated and this was further enhanced

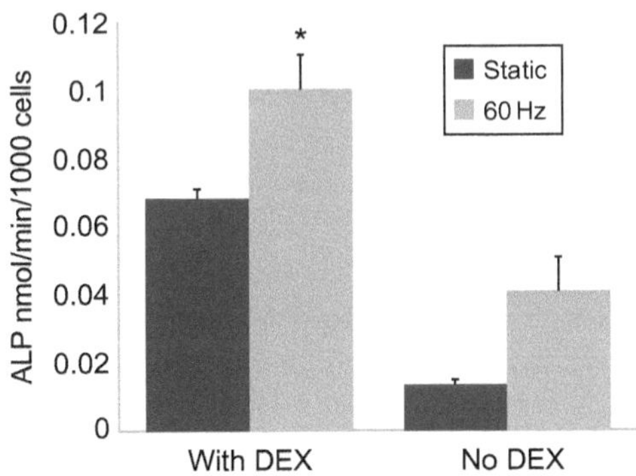

Figure 20.9 Low magnitude, high-frequency (LMHF) vibration loading upregulates osteogenesis of mesenchymal stem cells. Mesenchymal precursors derived from embryonic stem cells (hES-MP 002.5, Cellartis, Gothenburg) were subjected to LMHF vibration by clamping cell-seeded culture wells on a mechanical testing machine (Bose ElectroForce 3200) and subjecting the plate to vertical displacements at 60 Hz for 45 min on day 6 of culture. Mean ± SD of ALP activity, which was significantly higher at day 8 of culture 48 h after stimulation, in cells subjected to vibration compared to static controls ($n = 6$, $^*p < 0.05$) in the DEX-treated cultures. Previously unpublished data from our laboratory provided by Jennifer Edwards.

by increasing the length of the rest period from 1 to 3 h. In the multipotential media, OCN mRNA was upregulated with LMHF treatment.

4. Conclusions: Other loading modes

The three techniques described here have so far had limited use as methods for mechanically stimulating MSCs *in vitro*. This may be because none of these techniques mimic the forces that tissues in the body experience, unlike the major methods of *in vitro* mechanical stimulation including tension, compression, and fluid shear forces. However, the literature to date (Table 20.4) shows that they all appear to have a regulatory affect on MSC differentiation, especially osteogenesis and chondrogenesis. Directly straining the cell membrane using magnetic nanoparticles subjected to an external magnetic field upregulated markers of osteochondrogenesis at the early and later stages, and also the transcription factor of adipogenesis (PPARγ). The application of LIPUS to MSCs showed it could induce osteogenesis and chondrogenesis in the absence of differentiation media and that there was a further enhancement of chondrogenesis when TGF-β was added showing a synergistic effect. The application of LIPUS was also shown to inhibit adipogenic differentiation. The early and late stages of osteogenesis also seemed to be enhanced by different conditions, suggesting that the LIPUS parameters should be tuned depending on the stage of cell development. Our preliminary study of the affects of vibration treatment of MSCs suggests that noninvasive low magnitude, high-frequency loading may be able to enhance osteogenesis and this is supported further by the study conducted by Sen *et al.* (2011), who additionally showed vibration treatment may downregulate adiopogenesis. Future studies utilizing these techniques for the mechanical stimulation of MSCs are required in order to fully understand their capabilities in the regulation of MSC differentiation.

E. *In vivo* loading

1. Effect of *in vivo* loading

A number of studies have been carried out using mouse or rat *in vivo* models for observing the effects of mechanical stimulation on MSCs either by directly stimulating MSC populations within the animal or by inserting an MSC-seeded or nonseeded scaffold (Table 20.5). The two techniques described here include either insertion of a compressive force bioreactor into the animal along with a mounted scaffold (Duty *et al.*, 2007; Roshan-Ghias *et al.*, 2010) or using LMHF vibration treatment (Luu *et al.*, 2009; Ozcivici *et al.*, 2010; Rubin *et al.*, 2007) with the aim of directly stimulating the MSC population.

Duty *et al.* (2007) subcutaneously implanted uniaxial compressive bioreactors into male Fisher rats mounted with BMSCs seeded scaffolds that had been precultured *in vitro* for 8 weeks. They were then subjected to short

Table 20.4 Other: Summary table of the literature cited in the text describing the effects of mechanical loading by direct cell membrane strain using magnetic force, low-intensity pulsed ultrasound, or low magnitude, high-frequency vibration on mesenchymal stem cell differentiation and matrix production[a]

MSC source	Stimulation	Parameters	Culture surface	Culture media	Culture Period	Lineage	Load effect	Reference
Mouse C2C12 MSC	LIPUS	1.5 MHz; 70 mW/cm^2; 20 min	2D—TCP	Multipotential	24 h	Osteogenesis Chondrogenesis	↑ RUNX2, SOX9 mRNA. ↓ PPARγ mRNA. ↑ RUNX2 protein	Ikeda *et al.* (2006)
Rabbit BMSCs	LIPUS	1 MHz; 200 mW/cm^2; 10 min/12 h	3D—alginate gel	Chondrogenic w or w/o TGF-β3	1–2 weeks	Chondrogenesis	↑ Proteoglycan synthesis and COLII, SOX9 and aggrecan mRNA	Lee *et al.* (2006)
Human BMSCs	Integrin and ion channel, magnetic bead+force	1 Hz, 1 h/day (alternate days), 1–100 pN	2D—TCP or 3D—alginate/ chitosan capsules	Osteogenic	7–21 days	Osteochondrogenesis	↑ SOX9 and osteopontin mRNA (D7). ↑ COLI and proteoglycan synthesis (D21)	Kanczler *et al.* (2009)
Human BMSCs	Integrin, magnetic bead+force	1 Hz, 15 min, 2×10^{-10} N	2D—TCP coated w or w/o fibronectin and RGD peptides	Osteogenic or adipogenic	24–48 h	Osteochondrogenesis	↑ COLI on coated surfaces. ↑ SOX9 on TCP	Kasten *et al.* (2010)
Human BMSCs	LIPUS	1 MHz, 200 mW/cm^2, 100 Hz, 20 min/day	2D—culture plates	w or w/o DEX and TGF-B1, BMP-2	3 days–4 weeks	Osteochondrogenesis	↑ SOX9, aggrecan, COLII (Chondro). ↑ RUNX2 and ALP (Osteo)	Lai *et al.* (2010)
Human BMSCs and ADMSCs	LIPUS	1 MHz, 30 mW/cm^2, 1, 100, or 1000 Hz, 20 min/day	2D—BioFlex plates	Osteogenic	7 or 14 days	Osteogenesis	↑ calcium (D7 ADMSCs, D14 BMSCs). 1 kHz largest	Marvel *et al.* (2010)

(*Continued*)

Table 20.4 (*Continued*)

MSC source	Stimulation	Parameters	Culture surface	Culture media	Culture Period	Lineage	Load effect	Reference
Rat BMSCs	LIPUS	1.5 MHz; 2, 15, 30 mW/cm^2; 1 Hz; 20 min/day	2D—TCP	Osteogenic	8–24 days	Osteogenesis	↑ ALP activity (30 mW/cm^2 top). ↑ mineral deposition (2 mW/cm^2 top)	Angle *et al.* (2011)
Human hES-MP 002.5	LMHF vibration	15–60 Hz; 0.2 m/s^2; 10 or 45 min on day 6	2D—Gelatine coated TCP	Osteogenic w or w/o DEX	7 or 8 days	Osteogenesis	↑ ALP activity (w DEX, 60 Hz, 45 min)	Edwards and Reilly (2011)
Mouse C3H10T1/2 MSCs	LMHF vibration	90 Hz; 0.7 g; 20 or 40 min/day; 1 or 3 h rest periods (7 days)	2D—COLI coated silicon membranes	Adipogenic or multipotential	7 days	Osteogenesis	↓ Adiponectin and PPARγ mRNA. ↑ OCN (LMHF + BMP-2)	Sen *et al.* (2011)

[a] LIPUS (low-intensity pulsed ultrasound); LMHF (low magnitude, high frequency); RGD (arginine-glycine-aspartic acid). Other abbreviations as per Tables 20.1–20.3.

Table 20.5 *In vivo*: Summary table of the literature cited in the text describing the effect of mechanical loading *in vivo* on wound healing and mesenchymal stem cell fate[a]

MSC source	*In vivo* model	Type of stimulus	Parameters	Culture surface	Food	Culture period	Lineage	Load effect	Reference
Rat BMSCs	Male Fisher rats, subcutaneous	Cyclic, uniaxial compression	1 Hz; 13.3 N; 30 min (3 times/ week)	Polymeric scaffold	N/A	8 weeks *in vitro*, 6 weeks *in vivo*	Osteogenesis	↑ Mineral content	Duty *et al.* (2007)
Mouse bone marrow transplants	C57BL/6J mice	LMHF vibration	90 Hz; 1.98 m/s^2; 15 min/day; 5 days/week	No scaffold	Normal chow	6–15 weeks	Osteogenesis	↓ Fat production. ↓ Adipocyte MSC commitment	Rubin *et al.* (2007)
N/A	Male C57BL/6J mice	LMHF vibration	0.2 g; 90 Hz; 15 min/day; 5 days/week	No scaffold	High-fat diet	6 weeks	Osteogenesis	↑ Trabecular BVF. ↑ RUNX2 mRNA. ↓ Visceral adipose tissue. ↓ PPARγ mRNA	Luu *et al.* (2009)
N/A	Male C57BL/6J mice	LMHF vibration	0.2 g; 90 Hz; 15 min/day; 5 days/week	No scaffold	N/A	3–6 weeks	Osteogenesis	↑ MSC osteoprogenitor population to marrow adipocytes. ↑ Bone formation	Ozcivici *et al.* (2010)
N/A	Wistar rats–distal femoral epiphyses	Cyclic, uniaxial compression	4 Hz; 10 N; 5 min × 5 (every 2 days)	PLA/β-TCP scaffolds	N/A	13 weeks	Osteogenesis	↓ Bone volume (week 2) ↑ Bone volume and formation (week 13)	Roshan-Ghias *et al.* (2010)

[a] BVF (bone volume fraction). Other abbreviations as per Tables 20.1–20.4.

bouts of compressive loading (30 min, 3 times/week) for 2 weeks and harvested after a further 4 weeks *in vivo* resulting in a marked increased in mineral production compared with nonloaded constructs. Roshan-Ghias *et al.* (2010) implanted poly(L-lactic acid)/β-tricalcium phosphate scaffolds into the distal femoral epiphyses of female Wistar rats and subjected one leg to compressive loading using an external compressive force apparatus, while keeping the other leg unloaded. After short bouts (5 min, 5 times/every 2 days) of compressive loading for 13 weeks, bone volume and formation were seen to increase inside loaded scaffolds.

LMHF applied to whole animals *in vivo* is normally applied for a period of 6 weeks and all three studies have used similar parameters including a vibration frequency of 90 Hz at an acceleration of 2 m/s^2 for 15 min a day, 5 days a week. In all three studies described here similar outcomes were obtained, showing that vibration treatment enhanced bone formation (Ozcivici *et al.*, 2010) and decreased the formation of adipose tissue (Luu *et al.*, 2009; Rubin *et al.*, 2007). It was also seen that a larger number of the MSC population showed a greater commitment to being an osteoprogenitor than an adipogenic progenitor after the animal was subjected to vibration. Luu *et al.* (2009) also showed that, even under a high-fat diet, the BMSC population showed higher gene expression of RUNX2 and a lower expression of PPARγ. These studies indicate that short bouts of daily LMHF vibration treatment applied *in vivo* can increase the osteogenic potential of the MSC population and inhibit adipogenesis.

2. Conclusions: *In vivo* loading

The handful of *in vivo* studies reviewed in the section show that mechanical stimulation of MSC populations in animal models can have a regulatory effect on MSC lineage commitment and matrix production. All of these studies show a similar trend in that MSCs, when mechanically stimulated *in vivo*, are preferentially recruited for osteogenic differentiation and not adipogenic differentiation and that loading can also enhance the formation of bone matrix. While there are numerous *in vivo* animal studies of the effect of mechanical loading on bone or cartilage healing, there are few studies that look at the effect this loading has on MSC recruitment and differentiation, and understanding this may be important for the future success of TE strategies.

V. Conclusions and Future Directions

Our understanding of the capabilities of MSCs and the cues affecting their differentiation and development has advanced greatly over the past few decades. Various factors have been identified that are able to induce or enhance MSC differentiation into a wide variety of tissue-specific cells including chemical reagents, surface chemistry, scaffold composition and topography, and internal and external physical or mechanical forces. This makes MSCs a very promising cell source for future regenerative medicine and TE strategies. However, precisely controlling the timing and outcome of this differentiation process is still a challenge facing tissue engineers. Therefore, there are a number of issues that need to be addressed in order to successfully use MSCs in a clinical context.

A clearer understanding of the heterogeneity of MSC populations is required, along with an improved method for identifying a true MSC in order to enable an understanding of how to better target the chosen stimulate. If cells are being used that are not capable of reaching the desired outcome, then this will not only mislead the researcher but also result in a construct that is not suitable for implantation. It is important that researchers are using the same definitions and are able to unanimously identify MSCs with the ability to form functional tissue (Augello and De Bari, 2010).

While many relevant markers (gene expression and matrix synthesis) for tissue-specific differentiation and their temporal expression have been identified, complete profiling of each tissue-specific differentiation process is required in order to better understand how to fully induce the desired phenotype in MSC. There also needs to be an agreement over the best conditions to achieve successful differentiation, which includes the use of optimal media formulations and additives along with other cues, such as scaffold composition and architecture, and internal and external mechanical forces. These cues must then be initiated at the optimal time, in the correct sequence using the best parameters. Comparisons between unipotent committed cells and their precursors are required, at the early (e.g., preosteoblast) and later (e.g., osteocyte) stages of development to identify how the responses to specific chemical and physical cues alter during the differentiation process.

Methods of regulation of stem cell function have mainly been focused on the use of chemical induction factors but it has become apparent that both intrinsic and external mechanical forces may play a significant role in the induction or enhancement of MSC lineage commitment and subsequent development. Mechanical regulation of MSCs is starting to gain interest with early results indicating mechanical forces can influence the differentiation process and are capable of upregulating key differentiation markers. The understanding of mechanical regulation has lagged behind chemical regulation and there are at present no defined standard mechanical conditions for MSC stimulation. This is in contrast to chemical formulations for cell specific differentiation, which are more defined and have been widely adopted by the TE community, for example, the use of TGF-β for chondrogenesis or DEX for osteogenesis. A reason for this is there are a wide range of external mechanical stimuli available (e.g., tension, compression, fluid shear) each accompanied by an almost unlimited choice of conditions, which has made it difficult to pinpoint the optimal strategies. Future work will need to decipher which type of load is the most appropriate along with the magnitude of this load, the frequency at which it is applied, and at what time point it should be applied in order to design culture systems that are capable of being scaled-up to commercially viable processing techniques. What also must not be forgotten is that a cell-seeded construct *in vitro* initially resembles a healing wound and not a fully differentiated tissue, and so the dynamics of the construct will be very different to that of a fully developed *in vivo* tissue (Willie *et al.*, 2010).

There also needs to be a better characterization of the forces MSCs experience from these loading systems. While most loading systems employ a common force type (e.g., tension or compression), there may also be other forces at work causing significant side effects (e.g., shear forces from fluid flow in compression systems). These other forces along with loading effects on mass transport and biochemical availability may have more of an effect on the MSC differentiation process than the specific regimen the investigator intended to apply, causing a misinterpretation in the reason for the obtained results. Although a tissue engineer may only be concerned with the overall response and not the specific mechanism behind the response, it is important to know what forces are actually inducing the improved tissue formation to help simplify the design of future systems. Mathematical and computer modeling of loading systems can help better characterize the stresses and strains that are being experienced throughout the construct and subsequently provide a better understanding of what the cell is sensing and responding to (Thompson *et al.*, 2010).

Most of the loading systems employed have been on a very small scale or in 2D using non-human MSCs, and while they can provide valuable insights and pilot data into the outcomes of applying various conditions, it is not a straight forward process to convert these directly into 3D human models for clinical applications. Bioreactors will need to be designed on a much larger scale using well-defined parameters and characteristics with precisely predictable outcomes. The main limitation to creation of large-scale implantable constructs from MSCs is the lack of a vascular network, at present, cell survival rate toward the center of tissue engineered constructs are low, even in the smallest cell-seeded scaffolds due to poor nutrient diffusion and gas exchange into the center of the scaffold. It has been suggested that TE constructs should have the rudiments of a vascular system incorporated within them by seeding with endothelial progenitor cells possessing angiogenic differentiation capabilities and using appropriate growth factors to direct these cells toward the center of the scaffold and induce them to form a vascular network (McCoy and O'Brien, 2010). The constructs would then be seeded with MSCs and directed toward the required tissue type using the appropriate cues. Since MSCs also produce angiogenic growth factors themselves they should aid the formation of the vascular network and here mechanical loading may form a dual role since it has also been shown to be a modulator of angiogenic growth factor release by MSCs (Kasper *et al.*, 2007; Kasten *et al.*, 2010).

In conclusion, mechanical loading is a potent modulator of MSC differentiation and harnessing its effects may allow for improved preculture methods of MSCs in implantable constructs or the design of exercise/tissue stimulation regimens that a patient can undergo after MSC implantation. A better understanding of the mechanical environment of MSCs *in vivo* could also lead to noninvasive methods of amplifying the MSC pool or stimulating their migration and differentiation toward a target tissue. Multiple laboratories have designed a wide range of cell loading devices in order to better understand the mechanical responsiveness of MSCs and tissue engineered constructs, which we have attempted to compare and contrast in this review. However, comparison of results between laboratories is very difficult due to the high variety of loading devices, loading regimens, and cell sources available to the investigator. Apparently, contradictory results may be easily explained by the exquisite sensitivity of MSCs to very a specific combination of biochemical factors and physical stimuli at a specific time-point within their life, proliferative and differentiation cycles. The optimum preculture conditions for MSCs will, of course, vary dependant on the clinical procedures they are aimed to treat so it is likely that a range of application-dependant bioreactors will need to be developed over the coming years. In the mean time, it is hoped that some common trigger factors are identified from the range of ongoing work in this field such that reproducible loading regimens with predictable outcomes can begin to be devised.

ACKNOWLEDGMENTS

R. M. D.-S. is sponsored by the Engineering and Physical Sciences Research Council. We thank Mohsen Shaeri for the data shown in Fig. 20.7C, Ilaria Belluantono for the cells used in the study shown in Fig. 20.7C, and Jennifer Edwards for providing the data shown in Fig. 20.8. The Royal Society of London and Bose ElectroForce Systems Group partially sponsored the study shown in Fig. 20.6.

REFERENCES

Altman, G. H., Horan, R. L., Martin, I., Farhadi, J., Stark, P. R. H., Volloch, V., Richmond, J. C., Vunjak-Novakovic, G., and Kaplan, D. L. (2002). Cell differentiation by mechanical stress. *FASEB J.* **16,** 270–272.

Anderson, E. J., and Tate, M. L. K. (2007). Open access to novel dual flow chamber technology for in vitro cell mechanotransduction, toxicity and pharamacokinetic studies. *Biomed. Eng. Online* **6,** 46.

Angele, P., Yoo, J. U., Smith, C., Mansour, J., Jepsen, K. J., Nerlich, M., and Johnstone, B. (2003). Cyclic hydrostatic pressure enhances the chondrogenic phenotype of human mesenchymal progenitor cells differentiated in vitro. *J. Orthop. Res.* **21,** 451–457.

Angele, P., Schumann, D., Angele, M., Kinner, B., Englert, C., Hente, R., Fuchtmeier, B., Nerlich, M., Neumann, C., and Kujat, R. (2004). Cyclic, mechanical compression enhances chondrogenesis of mesenchymal progenitor cells in tissue engineering scaffolds. *Biorheology* **41,** 335–346.

Angle, S. R., Sena, K., Sumner, D. R., and Virdi, A. S. (2011). Osteogenic differentiation of rat bone marrow stromal cells by various intensities of low-intensity pulsed ultrasound. *Ultrasonics* **51,** 281–288.

Antonitsis, P., Ioannidou-Papagiannaki, E., Kaidoglou, A., and Papakonstantinou, C. (2007). In vitro cardiomyogenic differentiation of adult human bone marrow mesenchymal stem cells. The role of 5-azacytidine. *Interact. Cardiovasc. Thorac. Surg.* **6,** 593–597.

Archer, R., and Williams, D. J. (2005). Why tissue engineering needs process engineering. *Nat. Biotechnol.* **23,** 1353–1355.

Arnsdorf, E. J., Tummala, P., Kwon, R. Y., and Jacobs, C. R. (2009). Mechanically induced osteogenic differentiation—The role of RhoA, ROCKII and cytoskeletal dynamics. *J. Cell Sci.* **122,** 546–553.

Augello, A., and De Bari, C. (2010). The regulation of differentiation in mesenchymal stem cells. *Hum. Gene Ther.* **21,** 1226–1238.

Augst, A., Marolt, D., Freed, L. E., Vepari, C., Meinel, L., Farley, M., Fajardo, R., Patel, N., Gray, M., Kaplan, D. L., and Vunjak-Novakovic, G. (2008). Effects of chondrogenic and osteogenic regulatory factors on composite constructs grown using human mesenchymal stem cells, silk scaffolds and bioreactors. *J. R. Soc. Interface* **5,** 929–939.

Bakker, A. D., Soejima, K., Klein-Nulend, J., and Burger, E. H. (2001). The production of nitric oxide and prostaglandin E-2 by primary bone cells is shear stress dependent. *J. Biomech.* **34,** 671–677.

Bancroft, G. N., Sikavitsast, V. I., van den Dolder, J., Sheffield, T. L., Ambrose, C. G., Jansen, J. A., and Mikos, A. G. (2002). Fluid flow increases mineralized matrix deposition in 3D perfusion culture of marrow stromal osteloblasts in a dose-dependent manner. *Proc. Natl. Acad. Sci. USA* **99,** 12600–12605.

Bassey, E. J., and Ramsdale, S. J. (1994). Increase in femoral bone density in young women following high impact exercise. *Osteoporos. Int.* **4,** 72–75.

Batra, N. N., Li, Y. J., Yellowley, C. E., You, L. D., Malone, A. M., Kim, C. H., and Jacobs, C. R. (2005). Effects of short-term recovery periods on fluid-induced signaling in osteoblastic cells. *J. Biomech.* **38**(9), 1909–1917.

Ben-David, U., and Benvenisty, N. (2011). The tumorigenicity of human embryonic and induced pluripotent stem cells. *Nat. Rev. Cancer* **11,** 268–277.

Beresford, J. N., Bennett, J. H., Devlin, C., Leboy, P. S., and Owen, M. E. (1992). Evidence for an inverse relationship between the differentiation of adipocytic and osteogenic cells in rat marrow stromal cell-cultures. *J. Cell Sci.* **102,** 341–351.

Bonewald, L. F., and Johnson, M. L. (2008). Osteocytes, mechanosensing and Wnt signaling. *Bone* **42,** 606–615.

Brown, T. D. (2000). Techniques for mechanical stimulation of cells in vitro: A review. *J. Biomech.* **33,** 3–14.

Byrne, E. M., Farrell, E., McMahon, L. A., Haugh, M. G., O'Brien, F. J., Campbell, V. A., Prendergast, P. J., and O'Connell, B. C. (2008). Gene expression by marrow stromal cells in a porous collagen-glycosaminoglycan scaffold is affected by pore size and mechanical stimulation. *J. Mater. Sci. Mater. Med.* **19,** 3455–3463.

Case, N., Sen, B., Thomas, J. A., Styner, M., Xie, Z., Jacobs, C. R., and Rubin, J. (2011). Steady and oscillatory fluid flows produce a similar osteogenic phenotype. *Calcif. Tissue Int.* **88,** 189–197.

Chanda, D., Kumar, S., and Ponnazhagan, S. (2010). Therapeutic potential of adult bone marrow derived mesenchymal stem cells in diseases of the skeleton. *J. Cell. Biochem.* **111,** 249–257.

Chang, Y. L., Stanford, C. M., and Keller, J. C. (2000). Calcium and phosphate supplementation promotes bone cell mineralization: Implications for hydroxyapatite (HA)-enhanced bone formation. *J. Biomed. Mater. Res.* **52,** 270–278.

Chen, Y. J., Huang, C. H., Lee, I. C., Lee, Y. T., Chen, M. H., and Young, T. H. (2008). Effects of cyclic mechanical stretching on the mRNA expression of tendon/ligament-related and osteoblast-specific genes in human mesenchymal stem cells. *Connect. Tissue Res.* **49,** 7–14.

Cioffi, M., Boschetti, F., Raimondi, M. T., and Dubini, G. (2006). Modeling evaluation of the fluid-dynamic microenvironment in tissue-engineered constructs: A micro-CT based model. *Biotechnol. Bioeng.* **93,** 500–510.

Colazzo, F., Sarathchandra, P., Smolenski, R. T., Chester, A. H., Tseng, Y. T., Czernuszka, J. T., Yacoub, M. H., and Taylor, P. M. (2010). Extracellular matrix production by adipose-derived stem cells: Implications for heart valve tissue engineering. *Biomaterials* **32,** 119–127.

Colley, H. E., Mishra, G., Scutt, A. M., and McArthur, S. L. (2009). Plasma polymer coatings to support mesenchymal stem cell adhesion, growth and differentiation on variable stiffness silicone elastomers. *Plasma Process. Polym.* **6,** 831–839.

Connelly, J. T., Vanderploeg, E. J., Mouw, J. K., Wilson, C. G., and Levenston, M. E. (2010). Tensile loading modulates bone marrow stromal cell differentiation and the development of engineered fibrocartilage constructs. *Tissue Eng. Part A* **16,** 1913–1923.

Dalby, M. J., Gadegaard, N., Tare, R., Andar, A., Riehle, M. O., Herzyk, P., Wilkinson, C. D. W., and Oreffo, R. O. C. (2007). The control of human mesenchymal cell differentiation using nanoscale symmetry and disorder. *Nat. Mater.* **6**(12), 997–1003.

Datta, N., Pham, Q. P., Sharma, U., Sikavitsas, V. I., Jansen, J. A., and Mikos, A. G. (2006). In vitro generated extracellular matrix and fluid shear stress synergistically enhance 3D osteoblastic differentiation. *Proc. Natl. Acad. Sci. USA* **103,** 2488–2493.

De Bari, C., Dell'Accio, F., and Luyten, F. P. (2001). Human periosteum-derived cells maintain phenotypic stability and chondrogenic potential throughout expansion regardless of donor age. *Arthritis Rheum.* **44,** 85–95.

Delaine-Smith, R. M., MacNeil, S., and Reilly, G. C. (2011). Osteogenic media and fluid shear forces can induce osteogenic differentiation in dermal fibroblasts. *Trans. Annu. Meet. Orthop. Res. Soc.* **57,** Long Beach, California.

Deligianni, D. D., Katsala, N. D., Koutsoukos, P. G., and Missirlis, Y. F. (2001). Effect of surface roughness of hydroxyapatite on human bone marrow cell adhesion, proliferation, differentiation and detachment strength. *Biomaterials* **22,** 87–96.

Della Rocca, G. J. (2009). The science of ultrasound therapy for fracture healing. *Indian J. Orthop.* **43,** 121–126.

Diefenderfer, D. L., Osyczka, A. M., Reilly, G. C., and Leboy, P. S. (2003). BMP responsiveness in human mesenchymal stem cells. *Connect. Tissue Res.* **44,** 305–311.

Dominici, M., Le Blanc, K., Mueller, I., Slaper-Cortenbach, I., Marini, F. C., Krause, D. S., Deans, R. J., Keating, A., Prockop, D. J., and Horwitz, E. M. (2006). Minimal criteria for defining multipotent mesenchymal stromal cells. The International Society for Cellular Therapy position statement. *Cytotherapy* **8,** 315–317.

Donahue, T. L. H., Haut, T. R., Yellowley, C. E., Donahue, H. J., and Jacobs, C. R. (2003). Mechanosensitivity of bone cells to oscillating fluid flow induced shear stress may be modulated by chemotransport. *J. Biomech.* **36,** 1363–1371.

Dumas, V., Ducharne, B., Perrier, A., Fournier, C., Guignandon, A., Thomas, M., Peyroche, S., Guyomar, D., Vico, L., and Rattner, A. (2010). Extracellular matrix produced by osteoblasts cultured under low-magnitude, high-frequency stimulation is favorable to osteogenic differentiation of mesenchymal stem cells. *Calcif. Tissue Int.* **87,** 351–364.

Duty, A. O., Oest, M. E., and Guldberg, R. E. (2007). Cyclic mechanical compression increases mineralization of cell-seeded polymer scaffolds in vivo. *J. Biomech. Eng. Trans. ASME* **129,** 531–539.

Edwards, J. H., and Reilly, G. C. (2011). Low magnitude, high frequency vibration modulates mesenchymal progenitor differentiation. *Trans. Annu. Meet. Orthop. Res. Soc.* **57,** Long Beach, California.

El Haj, A. J., and Cartmell, S. H. (2010). Bioreactors for bone tissue engineering. *Proc. Inst. Mech. Eng. H* **224,** 1523–1532.

Engler, A. J., Sen, S., Sweeney, H. L., and Diskher, D. E. (2006). Matrix elasticity directs stem cell lineage specification. *Cell* **126,** 677–689.

Erices, A., Conget, P., and Minguell, J. J. (2000). Mesenchymal progenitor cells in human umbilical cord blood. *Br. J. Haematol.* **109,** 235–242.

Fan, X. F., Zou, R., Zhao, Z. H., Yang, P., Li, Y., and Song, J. L. (2009). Tensile strain induces integrin beta 1 and ILK expression higher and faster in 3D cultured rat skeletal myoblasts than in 2D cultures. *Tissue Cell* **41,** 266–270.

Farng, E., Urdaneta, A. R., Barba, D., Esmende, S., and McAllister, D. R. (2008). The effects of GDF-5 and uniaxial strain on mesenchymal stem cells in 3-D culture. *Clin. Orthop. Relat. Res.* **466,** 1930–1937.

Finger, A. R., Sargent, C. Y., Dulaney, K. O., Bernacki, S. H., and Loboa, E. G. (2007). Differential effects on messenger ribonucleic acid expression by bone marrow-derived human mesenchymal stem cells seeded in agarose constructs due to ramped and steady applications of cyclic hydrostatic pressure. *Tissue Eng.* **13,** 1151–1158.

Frangos, J. A., McIntire, L. V., and Eskin, S. G. (1988). Shear stress induced stimulation of mammalian cell metabolisam. *Biotechnol. Bioeng.* **32,** 1053–1060.

Friedenstein, A. J., Piatetzky-Shapiro, I. I., and Petrakova, K. V. (1966). Osteogenesis in transplants of bone marrow cells. *J. Embryol. Exp. Morphol.* **16,** 381–390.

Gelse, K., von der Mark, K., Aigner, T., Park, J., and Schneider, H. (2003). Articular cartilage repair by gene therapy using growth factor-producing mesenchymal cells. *Arthritis Rheum.* **48,** 430–441.

Gilbert, J. A., Weinhold, P. S., Banes, A. J., Link, G. W., and Jones, G. L. (1994). Strain profiles for circular cell culture plates containing flexible surfaces employed to mechanically deform cells in vitro. *J. Biomech.* **27,** 1169–1177.

Glogauer, M., Ferrier, J., and McCulloch, C. A. G. (1995). Magnetic fileds applied to collagen coated ferric oxide beads induce stretch activated Ca^{2+} flux in fibroblasts. *Am. J. Physiol. Cell Physiol.* **269,** C1093–C1104.

Glossop, J. R., and Cartmell, S. H. (2009). Effect of fluid flow-induced shear stress on human mesenchymal stem cells: Differential gene expression of IL1B and MAP3K8 in MAPK signaling. *Gene Expr. Patterns* **9,** 381–388.

Gomes, M. E., Sikavitsas, V. I., Behravesh, E., Reis, R. L., and Mikos, A. G. (2003). Effect of flow perfusion on the osteogenic differentiation of bone marrow stromal cells cultured on starch-based three-dimensional scaffolds. *J. Biomed. Mater. Res. A* **67A,** 87–95.

Goodwin, H. S., Bicknese, A. R., Chien, S. N., Bogucki, B. D., Oliver, D. A., Quinn, C. O., and Wall, D. A. (2001). Multilineage differentiation activity by cells isolated from umbilical cord blood: Expression of bone, fat, and neural markers. *Biol. Blood Marrow Transplant.* **7,** 581–588.

Grayson, W. L., Bhumiratana, S., Cannizzaro, C., Chao, P. H. G., Lennon, D. P., Caplan, A. I., and Vunjak-Novakovic, G. (2008). Effects of initial seeding density and fluid perfusion rate on formation of tissue-engineered bone. *Tissue Eng. Part A* **14,** 1809–1820.

Grellier, M., Bareille, R., Bourget, C., and Amedee, J. (2009). Responsiveness of human bone marrow stromal cells to shear stress. *J. Tissue Eng. Regen. Med.* **3,** 302–309.

Guilak, F., Cohen, D. M., Estes, B. T., Gimble, J. M., Liedtke, W., and Chen, C. S. (2009). Control of stem cell fate by physical interactions with the extracellular matrix. *Cell Stem Cell* **5,** 17–26.

Gurkan, U. A., and Akkus, O. (2008). The mechanical environment of bone marrow: A review. *Ann. Biomed. Eng.* **36,** 1978–1991.

Hanson, A. D., Marvel, S. W., Bernacki, S. H., Banes, A. J., van Aalst, J., and Loboa, E. G. (2009). Osteogenic effects of rest inserted and continuous cyclic tensile strain on hASC lines with disparate osteodifferentiation capabilities. *Ann. Biomed. Eng.* **37,** 955–965.

Haudenschild, A. K., Hsieh, A. H., Kapila, S., and Lotz, J. C. (2009). Pressure and distortion regulate human mesenchymal stem cell gene expression. *Ann. Biomed. Eng.* **37,** 492–502.

Henderson, J. H., De la Fuente, L., Romero, D., Colnot, C. I., Huang, S., Carter, D. R., and Helms, J. A. (2007). Rapid growth of cartilage rudiments may generate perichondrial structures by mechanical induction. *Biomech. Model. Mechanobiol.* **6,** 127–137.

Holtorf, H. L., Jansen, J. A., and Mikos, A. G. (2005). Flow perfusion culture induces the osteoblastic differentiation of marrow stromal cell-scaffold constructs in the absence of dexamethasone. *J. Biomed. Mater. Res. A* **72A,** 326–334.

Huang, C. H., Chen, M. H., Young, T. H., Jeng, J. H., and Chen, Y. J. (2009). Interactive effects of mechanical stretching and extracellular matrix proteins on initiating osteogenic differentiation of human mesenchymal stem cells. *J. Cell. Biochem.* **108,** 1263–1273.

Huang, A. H., Farrell, M. J., Kim, M., and Mauck, R. L. (2010a). Long term dynamic loading improves the mechanical properties of chondrogenic mesenchymal stem cell laden hydrogels. *Eur. Cell. Mater.* **19,** 72–85.

Huang, Y., Jia, X. L., Bai, K., Gong, X. H., and Fan, Y. B. (2010b). Effect of fluid shear stress on cardiomyogenic differentiation of rat bone marrow mesenchymal stem cells. *Arch. Med. Res.* **41,** 497–505.

Hubbe, M. A. (1981). Adhesion and detachment of biological cells in vitro. *Prog. Surf. Sci.* **11,** 65–138.

Hughes, S., El Haj, A. J., and Dobson, J. (2005). Magnetic micro- and nanoparticle mediated activation of mechanosensitive ion channels. *Med. Eng. Phys.* **27,** 754–762.

Ikeda, K., Takayama, T., Suzuki, N., Shimada, K., Otsuka, K., and Ito, K. (2006). Effects of low-intensity pulsed ultrasound on the differentiation of C2C12 cells. *Life Sci.* **79,** 1936–1943.

Iqbal, J., and Zaidi, M. (2005). Molecular regulation of mechanotransduction. *Biochem. Biophys. Res. Commun.* **328,** 751–755.

Jaasma, M. J., and O'Brien, F. J. (2008). Mechanical stimulation of osteoblasts using steady and dynamic fluid flow. *Tissue Eng. Part A* **14**(7), 1213–1223.

Jacobs, C. R., Yellowley, C. E., Davis, B. R., Zhou, Z., Cimbala, J. M., and Donahue, H. J. (1998). Differential effect of steady versus oscillating flow on bone cells. *J. Biomech.* **31,** 969–976.

Jagodzinski, M., Drescher, M., Zeichen, J., Hankemeier, S., Krettek, C., Bosch, U., and van Griensven, M. (2004). Effects of cyclic longitudinal mechanical strain and dexamethasone on osteogenic differentiation of human bone marrow stromal cells. *Eur. Cell. Mater.* **7,** 35–41.

Janmey, P. A., and McCulloch, C. A. (2007). Cell mechanics: Integrating cell responses to mechanical stimuli. *Annu. Rev. Biomed. Eng.* **9,** 1–34.

Johnstone, B., Hering, T. M., Caplan, A. I., Goldberg, V. M., and Yoo, J. U. (1998). In vitro chondrogenesis of bone marrow-derived mesenchymal progenitor cells. *Exp. Cell Res.* **238,** 265–272.

Jonas, O., and Duschl, C. (2010). Force propagation and force generation in cells. *Cytoskeleton* **67,** 555–563.

Jungreuthmayer, C., Donahue, S. W., Jaasma, M. J., Al-Munajjed, A. A., Zanghellini, J., Kelly, D. J., and O'Brien, F. J. (2009). A comparative study of shear stresses in collagen-glycosaminoglycan and calcium phosphate scaffolds in bone tissue-engineering bioreactors. *Tissue Eng. Part A* **15,** 1141–1149.

Kanczler, J. M., Sura, H. S., Magnay, J., Green, D., Oreffo, R. O. C., Dobson, J. P., and El Haj, A. J. (2009). Controlled differentiation of human bone marrow stromal cells using magnetic nanoparticle technology. *Tissue Eng. Part A* **16,** 3241–3250.

Kasper, G., Dankert, N., Tuischer, J., Hoeft, M., Gaber, T., Glaeser, J. D., Zander, D., Tschirschmann, M., Thompson, M., Matziolis, G., and Duda, G. N. (2007). Mesenchymal stem cells regulate angiogenesis according to their mechanical environment. *Stem Cells* **25,** 903–910.

Kasten, A., Muller, P., Bulnheim, U., Groll, J., Bruellhoff, K., Beck, U., Steinhoff, G., Moller, M., and Rychly, J. (2010). Mechanical integrin stress and magnetic forces induce biological responses in mesenchymal stem cells which depend on environmental factors. *J. Cell. Biochem.* **111,** 1586–1597.

Kielty, C. M., Hopkinson, I., and Grant, M. E. (1993). Collagen: The collagen family: Structure, assembly, and organization in the extracellular matrix. *In* "Connective Tissue and Its Heritable Disorders: Molecular, Genetic, and Medical Aspects," (P. M. Royce and B. Steinmann, Eds.), pp. 103–147. Wiley-Liss, Inc., Chichester.

Kim, S. H., Choi, Y. R., Park, M. S., Shin, J. W., Park, K. D., Kim, S. J., and Lee, J. W. (2007). Erk 1/2 activation in enhanced osteogenesis of human mesenchymal stem cells in poly(lactic-glycolic acid) by cyclic hydrostatic pressure. *J. Biomed. Mater. Res. A* **80A,** 826–836.

Kisiday, J. D., Frisbie, D. D., McIlwraith, C. W., and Grodzinsky, A. J. (2009). Dynamic compression stimulates proteoglycan synthesis by mesenchymal stem cells in the absence of chondrogenic cytokines. *Tissue Eng. Part A* **15,** 2817–2824.

Klein-Nulend, J., Bacabac, R. G., and Mullender, M. G. (2005). Mechanobiology of bone tissue. *Pathol. Biol.* **53,** 576–580.

Kobayashi, Y., Sakai, D., Iwashina, T., Iwabuchi, S., and Mochida, J. (2009). Low ntensity pulsed ultrasound stimulates cell proliferation, proteoglycan synthesis and expression of growth factor related genes in human nucleus pulposus cell line. *Eur. Cells Mater.* **17,** 15–22.

Koc, O. N., Day, J., Nieder, M., Gerson, S. L., Lazarus, H. M., and Krivit, W. (2002). Allogeneic mesenchymal stem cell infusion for treatment of metachromatic leukodystrophy (MLD) and Hurler syndrome (MPS-IH). *Bone Marrow Transplant.* **30,** 215–222.

Korstjens, C. M., van der Rijt, R. H. H., Albers, G. H. R., Semeins, C. M., and Klein-Nulend, J. (2008). Low-intensity pulsed ultrasound affects human articular chondrocytes in vitro. *Med. Biol. Eng. Comput.* **46,** 1263–1270.

Kreke, M. R., and Goldstein, A. S. (2004). Hydrodynamic shear stimulates osteocalcin expression but not proliferation of bone marrow stromal cells. *Tissue Eng.* **10,** 780–788.

Kreke, M. R., Huckle, W. R., and Goldstein, A. S. (2005). Fluid flow stimulates expression of osteopontin and bone sialoprotein by bone marrow stromal cells in a temporally dependent manner. *Bone* **36,** 1047–1055.

Ku, C. H., Johnson, P. H., Batten, P., Sarathchandra, P., Chambers, R. C., Taylor, P. M., Yacoub, M. H., and Chester, A. H. (2006). Collagen synthesis by mesenchymal stem cells and aortic valve interstitial cells in response to mechanical stretch. *Cardiovasc. Res.* **71,** 548–556.

Kuo, C. K., and Tuan, R. S. (2008). Mechanoactive tenogenic differentiation of human mesenchymal stem cells. *Tissue Eng. Part A* **14,** 1615–1627.

Lacroix, D., Prendergast, P. J., Li, G., and Marsh, D. (2002). Biomechanical model to simulate tissue differentiation and bone regeneration: Application to fracture healing. *Med. Biol. Eng. Comput.* **40,** 14–21.

Lai, C. H., Chen, S. C., Chiu, L. H., Yang, C. B., Tsai, Y. H., Zuo, C. S., Chang, W. H. S., and Lai, W. F. (2010). Effects of low intensity pulsed ultrasound, dexamethasone TGF-beta1 and/or BMP-2 on the transcriptional expression of genes in human mesenchymal stem cells: Chondrogenic vs. osteogenic differentiation. *Ultrasound Med. Biol.* **36,** 1022–1033.

Lee, H. J., Choi, B. H., Min, B. H., Son, Y. S., and Park, S. R. (2006). Low-intensity ultrasound stimulation enhances chondrogenic differentiation in alginate culture of mesenchymal stem cells. *Artif. Organs* **30,** 707–715.

Li, Y. J., Batra, N. N., You, L. D., Meier, S. C., Coe, I. A., Yellowley, C. E., and Jacobs, C. R. (2004). Oscillatory fluid flow affects human marrow stromal cell proliferation and differentiation. *J. Orthop. Res.* **22,** 1283–1289.

Li, X. H., Yu, X. Y., Lin, Q. X., Deng, C. Y., Shan, Z. X., Yang, M., and Lin, S. G. (2007). Bone marrow mesenchymal stem cells differentiate into functional cardiac phenotypes by cardiac microenvironment. *J. Mol. Cell. Cardiol.* **42,** 295–303.

Li, D. Q., Tang, T. T., Lu, J. X., and Dai, K. R. (2009a). Effects of flow shear stress and mass transport on the construction of a large-scale tissue-engineered bone in a perfusion bioreactor. *Tissue Eng. Part A* **15,** 2773–2783.

Li, Z., Yao, S. J., Alini, M., and Stoddart, M. J. (2009b). Chondrogenesis of human bone marrow mesenchymal stem cells in fibrin-polyurethane composites is modulated by frequency and amplitude of dynamic compression and shear stress. *Tissue Eng. Part A* **16,** 575–584.

Liedert, A., Claes, L., and Ignatius, A. (2008). Signal transduction pathways involved in mechanotransduction in osteoblastic and mesenchymal stem cells. *In* "Mechanosensitivity in Cells and Tissues," (A. Kamkin and I. Kiseleva, Eds.), pp. 253–265. Springer, Dordrecht.

Lin, T. M., Tsai, J. L., Lin, S. D., Lai, C. S., and Chang, C. C. (2005). Accelerated growth and prolonged lifespan of adipose tissue-derived human mesenchymal stem cells in a medium using reduced calcium and antioxidants. *Stem Cells Dev.* **14,** 92–102.

Liu, J., Zhao, Z. H., Li, J., Zou, L., Shuler, C., Zou, Y. W., Huang, X. J., Li, M. L., and Wang, J. (2009). hydrostatic pressures promote initial osteodifferentiation With ERK1/2 Not p38 MAPK signaling involved. *J. Cell. Biochem.* **107,** 224–232.

Luo, Z. J., and Seedhom, B. B. (2007). Light and low-frequency pulsatile hydrostatic pressure enhances extracellular matrix formation by bone marrow mesenchymal cells: An in-vitro study with special reference to cartilage repair. *Proc. Inst. Mech. Eng. H* **221,** 499–507.

Luu, Y. K., Capilla, E., Rosen, C. J., Gilsanz, V., Pessin, J. E., Judex, S., and Rubin, C. T. (2009). Mechanical stimulation of mesenchymal stem cell proliferation and differentiation promotes osteogenesis while preventing dietary-induced obesity. *J. Bone Miner. Res.* **24,** 50–61.

Maes, F., Ransbeeck, P., van Oosterwyck, H., and Verdonck, P. (2009). Modeling fluid flow through irregular scaffolds for perfusion bioreactors. *Biotechnol. Bioeng.* **103,** 621–630.

Malone, A. M. D., Anderson, C. T., Tummala, P., Kwon, R. Y., Johnston, T. R., Stearns, T., and Jacobs, C. R. (2007). Primary cilia mediate mechanosensing in bone cells by a calcium-independent mechanism. *Proc. Natl. Acad. Sci. USA* **104,** 13325–13330.

Marvel, S., Okrasinski, S., Bernacki, S. H., Loboa, E., and Dayton, P. A. (2010). The development and validation of a LIPUS system with preliminary observations of ultrasonic effects on human adult stem cells. *IEEE Trans. Ultrason. Ferroelectr. Freq. Control* **57,** 1977–1984.

Matsiko, A., Edwards, J., and Reilly, G. C. (2009). Human mesenchymal stem cell responses to steady and oscillatory fluid flow in a porous scaffold. *Regen. Med.* **4,** S159.

Mauck, R. L., Byers, B. A., Yuan, X., and Tuan, R. S. (2007). Regulation of cartilaginous ECM gene transcription by chondrocytes and MSCs in 3D culture in response to dynamic loading. *Biomech. Model. Mechanobiol.* **6,** 113–125.

Mauney, J. R., Sjostorm, S., Blumberg, J., Horan, R., O'Leary, J. P., Vunjak-Novakovic, G., Volloch, V., and Kaplan, D. L. (2004). Mechanical stimulation promotes osteogenic differentiation of human bone marrow stromal cells on 3-D partially demineralized bone scaffolds in vitro. *Calcif. Tissue Int.* **74,** 458–468.

McBride, S. H., Falls, T., and Tate, M. L. K. (2008). Modulation of stem cell shape and fate B: Mechanical modulation of cell shape and gene expression. *Tissue Eng. Part A* **14,** 1573–1580.

McCoy, R. J., and O'Brien, F. J. (2010). Influence of shear stress in perfusion bioreactor cultures for the development of three-dimensional bone tissue constructs: A review. *Tissue Eng. Part B Rev.* **16,** 587–601.

McGarry, J. G., Klein-Nulend, J., and Prendergast, P. J. (2005). The effect of cytoskeletal disruption on pulsatile fluid flow-induced nitric oxide and prostaglandin E-2 release in osteocytes and osteoblasts. *Biochem. Biophys. Res. Commun.* **330,** 341–348.

McMahon, L. A., Reid, A. J., Campbell, V. A., and Prendergast, P. J. (2008). Regulatory effects of mechanical strain on the chondrogenic differentiation of MSCs in a collagen-GAG scaffold: Experimental and computational analysis. *Ann. Biomed. Eng.* **36,** 185–194.

Meirelles, L. D., Fontes, A. M., Covas, D. T., and Caplan, A. I. (2009). Mechanisms involved in the therapeutic properties of mesenchymal stem cells. *Cytokine Growth Factor Rev.* **20,** 419–427.

Miyanishi, K., Trindade, M. C. D., Lindsey, D. P., Beaupre, G. S., Carter, D. R., Goodman, S. B., Schurman, D. J., and Smith, R. L. (2006a). Dose- and time-dependent effects of cyclic hydrostatic pressure on transforming growth factor-beta 3-induced chondrogenesis by adult human mesenchymal stem cells in vitro. *Tissue Eng.* **12,** 2253–2262.

Miyanishi, K., Trindade, M. C. D., Lindsey, D. P., Beaupre, G. S., Carter, D. R., Goodman, S. B., Schurman, D. J., and Smith, R. L. (2006b). Effects of hydrostatic pressure and transforming growth factor-beta 3 on adult human mesenchymal stem cell chondrogenesis in vitro. *Tissue Eng.* **12,** 1419–1428.

Morgan, E. F., Gleason, R. E., Hayward, L. N. M., Leong, P. L., and Palomares, K. T. S. (2008). Mechanotransduction and fracture repair. *J. Bone Joint. Surg. Am.* **90A,** 25–30.

Morris, H. L., Reed, C. I., Haycock, J. W., and Reilly, G. C. (2010). Mechanisms of fluid-flow-induced matrix production in bone tissue engineering. *Proc. Inst. Mech. Eng. H* **224,** 1509–1521.

Mouw, J. K., Connelly, J. T., Wilson, C. G., Michael, K. E., and Levenston, M. E. (2007). Dynamic compression regulates the expression and synthesis of chondrocyte-specific matrix molecules in bone marrow stromal cells. *Stem Cells* **25,** 655–663.

Murugan, R., and Ramakrishna, S. (2007). Design strategies of tissue engineering scaffolds with controlled fiber orientation. *Tissue Eng.* **13,** 1845–1866.

Na, K., Choi, S. J., Kim, S., Sun, B. K., Woo, D. G., Chung, H. M., and Park, K. H. (2007). Enhancement of cell proliferation and differentiation by combination of ascorbate and dexamethasone in thermo-reversible hydrogel constructs embedded with rabbit chondrocytes. *Biotechnol. Lett.* **29,** 1453–1457.

Nerurkar, N. L., Sen, S., Baker, B. M., Elliott, D. M., and Mauck, R. L. (2011). Dynamic culture enhances stem cell infiltration and modulates extracellular matrix production on aligned electrospun nanofibrous scaffolds. *Acta Biomater.* **7,** 485–491.

Ngiam, M., Liao, S., Jie, T. O. J., Sui, X. D., Dong, Y. X., Ramakrishna, S., and Chan, C. K. (2011). Effects of mechanical stimulation in osteogenic differentiation of bone marrow-derived mesenchymal stem cells on aligned nanofibrous scaffolds. *J. Bioact. Compat. Polym.* **26,** 56–70.

Nirmalanandhan, V. S., Juncosa-Melvin, N., Shearn, J. T., Boivin, G. P., Galloway, M. T., Gooch, C., Bradica, G., and Butler, D. L. (2009). Combined effects of scaffold stiffening and mechanical preconditioning cycles on construct biomechanics, gene Expression, and tendon repair biomechanics. *Tissue Eng. Part A* **15,** 2103–2111.

Nombela-Arrieta, C., Ritz, J., and Silberstein, L. E. (2011). The elusive nature and function of mesenchymal stem cells. *Nat. Rev. Mol. Cell Biol.* **12,** 126–131.

Nomura, S., and Takano-Yamamoto, T. (2000). Molecular events caused by mechanical stress in bone. *Matrix Biol.* **19,** 91–96.

Ogawa, R., Mizuno, S., Murphy, G. F., and Orgill, D. P. (2009). The effect of hydrostatic pressure on three-dimensional chondroinduction of human adipose-derived stem cells. *Tissue Eng. Part A* **15,** 2937–2945.

Orr, A. W., Helmke, B. P., Blackman, B. R., and Schwartz, M. A. (2006). Mechanisms of mechanotransduction. *Dev. Cell* **10,** 11–20.

Osyczka, A. M., and Leboy, P. S. (2005). Bone morphogenetic protein regulation of early osteoblast genes in human marrow stromal cells is mediated by extracellular signal-regulated kinase and phosphatidylinositol 3-kinase signaling. *Endocrinology* **146,** 3428–3437.

Ozcivici, E., Luu, Y. K., Rubin, C. T., and Judex, S. (2010). Low-level vibrations retain bone marrow's osteogenic potential and augment recovery of trabecular bone during reambulation. *PLoS One* **5,** e11178.

Pedersen, J. A., and Swartz, M. A. (2005). Mechanobiology in the third dimension. *Ann. Biomed. Eng.* **33,** 1469–1490.

Pelham, R. J., and Wang, Y. L. (1999). High resolution detection of mechanical forces exerted by locomoting fibroblasts on the substrate. *Mol. Biol. Cell* **10,** 935–945.

Plunkett, N. A., Partap, S., and O'Brien, F. J. (2010). Osteoblast response to rest periods during bioreactor culture of collagen-glycosaminoglycan scaffolds. *Tissue Eng. Part A* **16,** 943–951.

Pogue, R., and Lyons, K. (2006). BMP signaling in the cartilage growth plate. *In* "Current Topics in Developmental Biology," (G. P. Schatten, Ed.), Vol. 76, pp. 1–48. Elsevier Inc., San Diego.

Porter, R. M., Huckle, W. R., and Goldstein, A. S. (2003). Effect of dexamethasone withdrawal on osteoblastic differentiation of bone marrow stromal cells. *J. Cell. Biochem.* **90,** 13–22.

Rehman, J., Traktuev, D., Li, J. L., Merfeld-Clauss, S., Temm-Grove, C. J., Bovenkerk, J. E., Pell, C. L., Johnstone, B. H., Considine, R. V., and March, K. L. (2004). Secretion of angiogenic and antiapoptotic factors by human adipose stromal cells. *Circulation* **109,** 1292–1298.

Reilly, G. C., and Engler, A. J. (2010). Intrinsic extracellular matrix properties regulate stem cell differentiation. *J. Biomech.* **43,** 55–62.

Reilly, G., Haut, T., Yellowley, C., Donahue, H., and Jacobs, C. (2003). Fluid flow induced PGE(2) release by bone cells is reduced by glycocalyx degradation whereas calcium signals are not. *Biorheology* **40,** 591–603.

Roshan-Ghias, A., Terrier, A., Bourban, P. E., and Pioletti, D. P. (2010). In vivo cyclic loading as a potent stimulatory signal for bone formation inside tissue engineering scaffolds. *Eur. Cells Mater.* **19,** 41–49.

Rubin, J., Rubin, C., and Jacobs, C. R. (2006). Molecular pathways mediating mechanical signaling in bone. *Gene* **367,** 1–16.

Rubin, C. T., Capilla, E., Luu, Y. K., Busa, B., Crawford, H., Nolan, D. J., Mittal, V., Rosen, C. J., Pessin, J. E., and Judex, S. (2007). Adipogenesis is inhibited by brief, daily exposure to high-frequency, extremely low-magnitude mechanical signals. *Proc. Natl. Acad. Sci. USA* **104,** 17879–17884.

Rui, Y. F., Lui, P. P. Y., Ni, M., Chan, L. S., Lee, Y. W., and Chan, K. M. (2011). Mechanical loading increased BMP-2 expression which promoted osteogenic differentiation of tendon-derived stem cells. *J. Orthop. Res.* **29,** 390–396.

Scutt, A., Bertram, P., and Brautigam, M. (1996). The role of glucocorticoids and prostaglandin E_2 in the recruitment of bone marrow mesenchymal cells to the osteoblastic lineage: Positive and negative effects. *Calcif. Tissue Int.* **59,** 154–162.

Sen, B., Xie, Z. H., Case, N., Ma, M. Y., Rubin, C., and Rubin, J. (2008). Mechanical strain inhibits adipogenesis in mesenchymal stem cells by stimulating a durable beta-catenin signal. *Endocrinology* **149,** 6065–6075.

Sen, B., Xie, Z. H., Case, N., Styner, M., Rubin, C. T., and Rubin, J. (2011). Mechanical signal influence on mesenchymal stem cell fate is enhanced by incorporation of refractory periods into the loading regimen. *J. Biomech.* **44,** 593–599.

Seong, J. M., Kim, B. C., Park, J. H., Kwon, I. K., Mantalaris, A., and Hwang, Y. S. (2010). Stem cells in bone tissue engineering. *Biomed. Mater.* **5,** 062001.

Sharp, L. A., Lee, Y. W., and Goldstein, A. S. (2009). Effect of low-frequency pulsatile flow on expression of osteoblastic genes by bone marrow stromal cells. *Ann. Biomed. Eng.* **37,** 445–453.

Shim, W. S. N., Jiang, S., Wong, P., Tan, J., Chua, Y. L., Tan, Y. S., Sin, Y. K., Lim, C. H., Chua, T., Teh, M., Liu, T. C., and Simr, E. (2004). Ex vivo differentiation of human adult bone marrow stem cells into cardiomyocyte-like cells. *Biochem. Biophys. Res. Commun.* **324,** 481–488.

Sikavitsas, V. I., Bancroft, G. N., Holtorf, H. L., Jansen, J. A., and Mikos, A. G. (2003). Mineralized matrix deposition by marrow stromal osteoblasts in 3D perfusion culture increases with increasing fluid shear forces. *Proc. Natl. Acad. Sci. USA* **100,** 14683–14688.

Sikavitsas, V. I., Bancroft, G. N., Lemoine, J. J., Liebschner, M. A. K., Dauner, M., and Mikos, A. G. (2005). Flow perfusion enhances the calcified matrix deposition of marrow stromal cells in biodegradable nonwoven fiber mesh scaffolds. *Ann. Biomed. Eng.* **33,** 63–70.

Simmons, C. A., Matlis, S., Thornton, A. J., Chen, S. Q., Wang, C. Y., and Mooney, D. J. (2003). Cyclic strain enhances matrix mineralization by adult human mesenchymal stem cells via the extracellular signal-regulated kinase (ERK1/2) signaling pathway. *J. Biomech.* **36,** 1087–1096.

Sioud, M., Mobergslien, A., Boudabous, A., and Floisand, Y. (2011). Mesenchymal stem cell-mediated T cell suppression occurs through secreted galectins. *Int. J. Oncol.* **38,** 385–390.

Sittichockechaiwut, A., Scutt, A., Ryan, A., Bonewald, L., and Reilly, G. (2009). Use of rapidly mineralising osteoblasts and short periods of mechanical loading to accelerate matrix maturation in 3D scaffolds. *Bone* **44,** 822–829.

Sittichokechaiwut, A., Edwards, J., Scutt, A., and Reilly, G. (2010). Short bouts of mechanical loading are as effective as dexamethasone at inducing matrix production by human bone marrow mesenchymal stem cells. *Eur. Cells Mater.* **20,** 45–57.

Soncini, M., Vertua, E., Gibelli, L., Zorzi, F., Denegri, M., Albertini, A., Wengler, G. S., and Parolini, O. (2007). Isolation and characterization of mesenchymal cells from human fetal membranes. *J. Tissue Eng. Regen. Med.* **1,** 296–305.

Srinivasan, S., Weimer, D. A., Agans, S. C., Bain, S. D., and Gross, T. S. (2002). Low-magnitude mechanical loading becomes osteogenic when rest is inserted between each load cycle. *J. Bone Miner. Res.* **17,** 1613–1620.

Stewart, A. A., Byron, C. R., Pondenis, H. C., and Stewart, M. C. (2008). Effect of dexamethasone supplementation on chondrogenesis of equine mesenchymal stem cells. *Am. J. Vet. Res.* **69,** 1013–1021.

Stock, U. A., and Vacanti, J. P. (2001). Tissue engineering: Current state and prospects. *Annu. Rev. Med.* **52,** 443–451.

Sumanasinghe, R. D., Bernacki, S. H., and Loboa, E. G. (2006). Osteogenic differentiation of human mesenchymal stem cells in collagen matrices: Effect of uniaxial cyclic tensile strain on bone morphogenetic protein (BMP-2) mRNA expression. *Tissue Eng.* **12,** 3459–3465.

Takada, I., Kouzmenko, A. P., and Kato, S. (2010). PPAR-gamma signaling crosstalk in mesenchymal stem cells. *PPAR Res.* Article No.: 341671.

Tanaka, S. M., Sun, H. B., Roeder, R. K., Burr, D. B., Turner, C. H., and Yokota, H. (2005). Osteoblast responses one hour after load-induced fluid flow in a three-dimensional porous matrix. *Calcif. Tissue Int.* **76,** 261–271.

Terraciano, V., Hwang, N., Moroni, L., Park, H. B., Zhang, Z., Mizrahi, J., Seliktar, D., and Elisseeff, J. (2007). Differential response of adult and embryonic mesenchymal progenitor cells to mechanical compression in hydrogels. *Stem Cells* **25,** 2730–2738.

Thompson, M. S., Epari, D. R., Bieler, F., and Duda, G. N. (2010). In vitro models for bone mechanobiology: Applications in bone regeneration and tissue engineering. *Proc. Inst. Mech. Eng. H* **224,** 1533–1541.

Thorpe, S. D., Buckley, C. T., Vinardell, T., O'Brien, F. J., Campbell, V. A., and Kelly, D. J. (2010). The response of bone marrow-derived mesenchymal stem cells to dynamic compression following TGF-beta 3 induced chondrogenic differentiation. *Ann. Biomed. Eng.* **38,** 2896–2909.

Titushkin, I., Sun, S., Shin, J., and Cho, M. (2010). Physicochemical control of adult stem cell differentiation: Shedding light on potential molecular mechanisms. *J. Biomed. Biotechnol.* **2010,** Article No.: 743476.

Tuli, R., Tuli, S., Nandi, S., Huang, X. X., Manner, P. A., Hozack, W. J., Danielson, K. G., Hall, D. J., and Tuan, R. S. (2003). Transforming growth factor-beta-mediated chondrogenesis of human mesenchymal progenitor cells involves N-cadherin and mitogen-activated protein kinase and Wnt signaling cross-talk. *J. Biol. Chem.* **278,** 41227–41236.

Tummala, P., Arnsdorf, E. J., and Jacobs, C. R. (2010). The role of primary cilia in mesenchymal stem cell differentiation: A pivotal switch in guiding lineage commitment. *Cell. Mol. Bioeng.* **3,** 207–212.

van Beuningen, H. M., Glansbeek, H. L., van der Kraan, P. M., and van den Berg, W. B. (1998). Differential effects of local application of BMP-2 or TGF-beta 1 on both articular cartilage composition and osteophyte formation. *Osteoarthritis Cartilage* **6,** 306–317.

van Eijk, F., Saris, D. B. F., Creemers, L. B., Riesle, J., Willems, W. J., van Blitterswijk, C. A., Verbout, A. J., and Dhert, W. J. A. (2008). The effect of timing of mechanical stimulation on proliferation and differentiation of goat bone marrow stem cells cultured on braided PLGA scaffolds. *Tissue Eng. Part A* **14,** 1425–1433.

VanGordon, S. B., Voronov, R. S., Blue, T. B., Shambaugh, R. L., Papavassiliou, D. V., and Sikavitsas, V. I. (2011). Effects of scaffold architecture on preosteoblastic cultures under continuous fluid shear. *Ind. Eng. Chem. Res.* **50,** 620–629.

Vater, C., Kasten, P., and Stiehler, M. (2011). Culture media for the differentiation of mesenchymal stromal cells. *Acta Biomater.* **7,** 463–477.

Vogel, V. (2006). Mechanotransduction involving multimodular proteins: Converting force into biochemical signals. *Annu. Rev. Biophys. Biomol. Struct.* **35,** 459–488.

Wagner, D. R., Lindsey, D. P., Li, K. W., Tummala, P., Chandran, S. E., Smith, R. L., Longaker, M. T., Carter, D. R., and Beaupre, G. S. (2008). Hydrostatic pressure enhances chondrogenic differentiation of human bone marrow stromal cells in osteochondrogenic medium. *Ann. Biomed. Eng.* **36,** 813–820.

Wang, N., Butler, J. P., and Ingber, D. E. (1993). Mechanotransduction across the cell surface and through the cytoskeleton. *Science* **260,** 1124–1127.

Watanabe, Y., Matsushita, T., Bhandari, M., Zdero, R., and Schemitsch, E. H. (2010). Ultrasound for fracture healing: Current evidence. *J. Orthop. Trauma* **24,** S56–S61.

Weinbaum, S., Cowin, S. C., and Zeng, Y. (1994). A model for the excitation of osteocytes by mechanical loading-induced bone fluid shear stresses. *J. Biomech.* **27,** 339–360.

Weinbaum, S., Tarbell, J. M., and Damiano, E. R. (2007). The structure and function of the endothelial glycocalyx layer. *Annu. Rev. Biomed. Eng.* **9,** 121–167.

Willie, B. M., Petersen, A., Schmidt-Bleek, K., Cipitria, A., Mehta, M., Strube, P., Lienau, J., Wildemann, B., Fratzl, P., and Duda, G. (2010). Designing biomimetic scaffolds for bone regeneration: Why aim for a copy of mature tissue properties if nature uses a different approach? *Soft Matter* **6,** 4976–4987.

Xie, L. Q., Jacobson, J. M., Choi, E. S., Busa, B., Donahue, L. R., Miller, L. M., Rubin, C. T., and Judex, S. (2006). Low-level mechanical vibrations can influence bone resorption and bone formation in the growing skeleton. *Bone* **39,** 1059–1066.

Yang, D. C., Tsay, H. J., Lin, S. Y., Chiou, S. H., Li, M. J., Chang, T. J., and Hung, S. C. (2008). cAMP/PKA regulates osteogenesis, adipogenesis and ratio of RANKL/OPG mRNA expression in mesenchymal stem cells by suppressing leptin. *PLoS One* **3,** e1540.

Yang, X. M., Gong, P., Lin, Y. F., Zhang, L. R., Li, X. Y., Yuan, Q. A., Tan, Z., Wang, Y. Y., Man, Y., and Tang, H. (2010). Cyclic tensile stretch modulates osteogenic differentiation of adipose-derived stem cells via the BMP-2 pathway. *Arch. Med. Sci.* **6,** 152–159.

Yin, Z., Chen, X., Chen, J. L., and Ouyang, H. W. (2010). Stem cells for tendon tissue engineering and regeneration. *Expert Opin. Biol. Ther.* **10,** 689–700.

You, J., Reilly, G., Zhen, X., Yeollowley, C., Chen, Q., Donahue, H., and Jacobs, C. (2001). Osteopontin gene regulation by oscillatory fluid flow via intracellular calcium mobilization and activation of mitogen-activated protein kinase in MC3T3-E1 osteoblasts. *J. Biol. Chem.* **276,** 13365–13371.

You, L., Temiyasathit, S., Coyer, S. R., Garcia, A. J., and Jacobs, C. R. (2008). Bone cells grown on micropatterned surfaces are more sensitive to fluid shear stress. *Cell. Mol. Bioeng.* **1,** 182–188.

Yourek, G., Hussain, M. A., and Mao, J. J. (2007). Cytoskeletal changes of mesenchymal stem cells during differentiation. *Asaio J.* **53,** 219–228.

Yourek, G., McCormick, S., Mao, J., and Reilly, G. (2010). Shear stress induces osteogenic differentiation of human mesenchymal stem cells. *Regen. Med.* **5,** 713–724.

Zeiter, S., Lezuo, P., and Ito, K. (2009). Effect of TGF beta(1), BMP-2 and hydraulic pressure on chondrogenic differentiation of bovine bone marrow mesenchymal stromal cells. *Biorheology* **46,** 45–55.

Zhang, L., Tran, N., Chen, H. Q., Kahn, C. J. F., Marchal, S., Groubatch, F., and Wang, X. (2008). Time-related changes in expression of collagen types I and III and of tenascin-C in rat bone mesenchymal stem cells under co-culture with ligament fibroblasts or uniaxial stretching. *Cell Tissue Res.* **332,** 101–109.

Zhao, F., Chella, R., and Ma, T. (2007). Effects of shear stress on 3-D human mesenchymal stem cell construct development in a perfusion bioreactor system: Experiments and hydrodynamic modeling. *Biotechnol. Bioeng.* **96,** 584–595.

Zhou, X. Z., Liu, D. W., You, L. D., and Wang, L. Y. (2010). Quantifying fluid shear stress in a rocking culture dish. *J. Biomech.* **43,** 1598–1602.

Zuk, P. A., Zhu, M., Mizuno, H., Huang, J., Futrell, J. W., Katz, A. J., Benhaim, P., Lorenz, H. P., and Hedrick, M. H. (2001). Multilineage cells from human adipose tissue: Implications for cell-based therapies. *Tissue Eng.* **7,** 211–228.

Zvaifler, N. J., Marinova-Mutafchieva, L., Adams, G., Edwards, C. J., Moss, J., Burger, J. A., and Maini, R. N. (2000). Mesenchymal precursor cells in the blood of normal individuals. *Arthritis Res.* **2,** 477–488.

Index

A

Adipose derived mesenchymal stem cells (ADMSCs)
  BMSC, 441
  LIPUS, 458–459
ADMSCs. *See* Adipose derived mesenchymal stem cells
Adult cardiac-derived stem cells
  bone marrow stem cell therapy, 112
  c-kit-positive
    $GATA_4$, 120
    IGF-1 and Akt pathway, 120
    LA-PCs *vs.* other stem cells, 115–116
    left atrium-derived, 115
    long-term cultures, 113–114
    microarray and pathway analysis, 116–118
    myogenesis and adipogenesis, 118–119
    noggin and cardiac myocyte development, 119–120
    transcription factors, 116
    Wnt and TGF-β, 118
  isolation, 112–113
  myocardial infarction, 112
  rodent and murine stem cells, 121
  spheroid, 112–113
  TGF-β functions, 121
Adult stem cells (ASC), 292–293
Alkaline phosphatase (ALP)
  activity, 133–134, 437
  COLI, 451
  OCN mRNA, 438
ALP. *See* Alkaline phosphatase
Alzheimer's disease (AD), 146
ARNT. *See* Aryl hydrocarbon receptor nuclear translocator
Aryl hydrocarbon receptor nuclear translocator (ARNT), 368

B

Bag-of-marbles (Bam) and benign gonial cell neoplasm (Bgcn)
  brain tumor, fused and Dsmurf repression, 409
  Brat repression, 412
  cystoblasts, 411
  decapentaplegic signaling
    cystoblasts, 408
    description, 407–408
    thick veins (Tkv), 408
    "timorous", 408
  E-cadherin translation, 405
  expression, cysts, 401, 402–403
  extrinsic and intrinsic factors, 401–402
  germline Piwi expression
    Dpp signaling, 405–406
    Dsmurf, 406
    germline, 406
    piRNAs, 405
  GSC, 400
  microRNAs, Mei-P26, 406–407
  molecular interplay, 409–410
  Nos translation
    Pum, 403
    role, 403–404
    30'-UTR, 404
  protein expression, 403
  Pum function, 404–405
  somatic niche cells, 400–401
  translational regulation, 411, 412
$Bcl-X_L$ effects
  BAX protein, 187
  cell death
    brain development, 189
    embryogenesis, 189
    immature cells, 189–190
    oligodendrocytes, 190
    pathways, 188–189
    PD model, 190
    progenitor cells, 189
  cell proliferation
    calcium homeostasis, 192
    cell cycle, 191
    cytosolic calcium, 192
    intermediate progenitors (IPs), 191–192
    neurogenic effect, 190–191
    p53–p21 pathway, 193
    p53 role, 192
    synapse formation, 191
  hematopoietic system, 187
  Mcl-1 protein, 186
  neuronal parameters, 186–187
  progenitor population, 187
  proneuronal gene Ngn2, 193
  rat striatum, 188
  TH-positive generated neurons, 188
BMPs. *See* Bone morphogenetic proteins
Bone marrow MSCs
  adipose tissue, 128, 129

Bone marrow MSCs (*cont.*)
cell types, 128–129
differentiation potency
chondrogenic, osteogenic and adipogenic lineages, 129–131
mesenchymal lineages, 131
signaling networks, 131
differentiation, TGF-β1
adipogenic, 135
chondrogenic, 133–134
embryonic heart development, 135–136
molecular mechanism, 136
osteogenic, 134
SMCs and cardiomyocytes, 135
and TGF-β signaling, 131–133
direct plating and heterogeneous nature, 128–129
plastic-adherent cells, 128
Bone morphogenetic proteins (BMPs)
adenoviral expression, 47
adipogenesis, 49
chondrogenesis, 48
components, 299–301
description, 46
mediating MSC self-renewal, 52–53
MSC, commitment, 299–301
self-renewal process, 42
signaling (*see* Signaling pathways, pancreatic β-cells)
survival and proliferation, MSC, 50
treatment, 301–302
Wnt pathway, 45–46
Bulk-cultured cardiac stem cells (CSC-BCs)
description, 113
$GATA_4$, 120
gene expression, 113, 114
vs. LA-PCs, 115

C

Calcineurin/NFAT signaling (*see* Signaling pathways, pancreatic β-cells)
Chondrogenesis, MSC
BMP2, 48
chondrocyte maturation, 48
TGF-β, differentiation induction, 48
c-kit-positive cardiac cells
$GATA_4$, 120
IGF-1 and Akt pathway
enhancement, cardiac myocytes, 120
role, 120
LA-PCs
cytokine, 115–116
vs. mesoangioblasts, 115
vs. peripheral blood mononuclear cells, 115–116
left atrium-derived, 115
long-term cultures
CSC-BCs, gene expression, 113, 114
description, 113
RT-PCR, 113
microarray and pathway analysis
LA-PCs culture, 116–118
muscle contraction signaling, 116–118
Wnt and TGF-β, 116–118
myogenesis and adipogenesis, TGF-β effect
LA-PCs, 119
role, 118–119
Smad7 activation, 118
noggin and cardiac myocyte development, 119–120
transcription factors, 116
Wnt and TGF-β, 118
CME. *See* Crushed muscle extract
Crushed muscle extract (CME), 256–257
CSC-BCs. *See* Bulk-cultured cardiac stem cells

D

DAn generation. *See* Dopaminergic neurons generation
DCMMS. *See* Direct cell membrane magnetic stimulus
D-glactosamine model, 100
Direct cell membrane magnetic stimulus (DCMMS), 432
Dopaminergic neurons (DAn) generation
basal ganglia, 176–177
cell replacement, PD
fetal mesencephalic neurons/cells, 178–179
stem cells, 179–185
hNSCs differentiation, A9 DA phenotype
epigenetic cues, 185–186
genetic manipulations, 186–197
potency loss, 177
source, standards, 176–177
tissue transplantation, 176–177

E

ECM. *See* Extracellular matrix
Embryonic stem (ES) cell regulation
BMP pathway
differentiation, 349–350
embryos, 349
ERK, 348–349
growth differentiation factor, 347–348
helix-loop-helix proteins, 348
Id factors, 348
LIF, 347
role, 347
Nodal pathway
Brachyury expression, 352
differentiation, 353
embryo, 352–353
Id genes, 351–352
Lefty factors, 352

Nodal–Smad2 signaling, 351
Smad2 signaling, 350–351
Embryonic stem (ES) cells, pancreatic β-cells generation
endocrine progenitors, 80
final stages, 86–87
genetic manipulation, 80
insulin therapy, 80
islet transplantation, 80
signaling pathways
BMP, 84–86
Calcineurin/NFAT, 83
Epac, 83–84
Hedgehog, 82–83
manipulations, 86–87
PI3K, 81–82
type 1 diabetes mellitus, 79–80
Epac signaling. *See* Signaling pathways, pancreatic β-cells
ERK. *See* Extracellular receptor kinase
Extracellular matrix (ECM)
FGF receptors, 258–259
fibronectin, 253
glycosaminoglycans (GAGs), 253
growth control and differentiation, 255
IGF-1 and-2
expression, 259
myoblast proliferation and differentiation, 259
in vitro cultured myoblasts, 260
MMPs, 254, 255
skeletal muscle repair, 252–253
slow-twitch soleus muscle, 249
TGF-beta, 253, 259–260
Extracellular receptor kinase (ERK), 348–349, 354

F

FGF. *See* Fibroblast growth factor
Fibroblast growth factor (FGF)
receptors, 258–259
syndecan-1 and glypican, 255
Fluid-induced shear stress (FSS)
2D and 3D, 432
description, 430–431
differentiation pathways, 456–457
3D perfusion systems, 454
3D scaffold, 431–432
effect, 451–454
osteogenesis
ALP activity, 455–456
2D and 3D, 454–455
scaffold pore size, 457
Folbp1. *See* Folbp1-folic acid binding protein
Folbp1-folic acid binding protein (Folbp1)
cell surface proteins, 152
folate receptors, 152
Folic acid (FA), maternal intake
AD, 146
asthma, 148
autism, 146
cardiac diseases, 147–148
cerebral ischemia, 147
depression, 147
mouse models, NTD, 148–149
neural crest development, 150–162
NTDs, 145–146
PD, 147
FSS. *See* Fluid-induced shear stress

G

GCPs. *See* Granular cell precursors
Gene regulatory networks (GRNs)
maternal folate intake, 150
NC development, 149–150
potential folate responsive, 160–162
Germline proliferation, *Caenorhabditis elegans*
counter-intuitive assay
cell development, 66–67
DTC-intrinsic migration program, 66
screening strategy, 67–68
description, 62
development
cell nuclei, 63
larval stages, 63
somatic gonad rearrangement, 63–64
sperm production, 64
transition zone, 63–64
diet/metabolic signaling, 75
germline cell cycle arrest, IIR role, 73–74
IIR pathway, identification
DAF-2 activity, proliferative zone, 68, 69
description, 68, 69
IIR role, 74
IIR signaling, 68–69
insulins
characteristics, 71
ligand-encoding genes, 71
larval germline cell division cycle
DAF-2 activity, proliferative zone, 69–70
daf-2/IIR mutants, 69–70
parameters, 69–70
ligand activity, 76
Notch-independent soma germline signaling mechanisms, 64–65
Notch signaling pathway, 64–65
nutrition impacts, 75
target tissues, IIR signaling
daf-16 activity, 72
FOXO transcription factor *daf-16*, 71–72
nuclear localization, 72–73
sheath cell-ablation phenotype, 72, 73
Germline stem cells (GSCs), 400
Glucocorticoid hedgehog agonists, neurogenesis
cell growth, 210–211

Glucocorticoid hedgehog agonists, neurogenesis (*cont.*)
mood disorders, 208
potential effects
β-arrestin-GFP reporter, 209–210
dexamethasone, 210
GCP proliferation, 208, 209
hedgehog signaling pathway, 208
SAR
β-arrestin2-GFP translocation assay, 211, 212, 213
hydroxyl group, 211–212
working model, 211–212, 214
seven transmembrane receptors, 208
steroids, 210–211
Granular cell precursors (GCPs)
cell growth, 210–211
proliferation
cerebellum, 209
dexamethasone, 210
stem cell population, 209
treatment, 209–210
GRNs. *See* Gene regulatory networks

H

hADSCs. *See* Human adipose tissue derived MSCs
Hedgehog signaling. *See* Signaling pathways, pancreatic β-cells
Hematopoietic stem cells (HSCs), 242, 245–246
Hepatocyte growth factor (HGF), satellite cells
activation, ERK1 and ERK2, 256
cell cycle, 257–258
c-fos and c-jun genes, 258
CME, 256–257
cyclin D1 and CDK inhibitor, 257–258
exogenous administration, 257
muscular disorder, 256–257
proliferation, 255–256
regeneration, skeletal muscles, 255
hESCs. *See* Human embryonic stem cells
hiPS cells. *See* Human induced pluripotent stem cells
HLA-G. *See* Human leukocyte antigen G
HLSCs. *See* Human liver stem cells
HSCs. *See* Hematopoietic stem cells
Human adipose tissue derived MSCs (hADSCs), 298–299
Human embryonic stem cells (hESCs)
challenges, 180
clinical application, 179
coculture, 179
feeder-based protocols, 180
Human induced pluripotent stem (hiPS) cells
challenges, 181
DNA methylation, 181
parkinsonian rats, 181
transcription factors, 180
tumor formation, 181
Human leukocyte antigen G (HLA-G) expression, progesterone
agents, 226–227
CD4+ T lymphocytes, 226
cell cultures, 227–228
cytotrophoblast cells, 226
endothelial cells, 227
immune interactions, 228–229
immunomodulatory factors, 226
MSCs cultures, 228
PIBF, 230
steroid effects, 228
Human liver stem cells (HLSCs), 299
Human neural stem cells (hNSCs)
cell differentiation, 182–185
cell division-related genes, 181–182
cultures, precursors, 181–182
differentiation, A9 DA phenotype
Bcl-X_L effects, 186–194
bHLH transcription factors, 195
cell therapy strategies, 194
cytokines, 185
embryonic development, 185
hVM1 cells, 186
Lmx1a, 195, 196–197
notch signaling, 196
Nurr1 and Foxa2, 196
orthodenticle homeobox 2, 195
Pitx3, 195
tissue samples, 185
transcription factors, 194–195
VM neurosphere cultures, 197
Wnts, 195
hVM1 cell line, 182–185
Hypertension
Ca(L), 332
description, 331–332
vascular system, 332, 333
Hypoxia-inducible factors (HIFα) proteins
angiogenesis therapies, 374
ARNT, 368
and cancer
genetic instability, 369
proliferation and metastasis, 369–370
cardiac ischemic diseases, 373
cell proliferation and differentiation pathways, 374–375
description, 368
neurodegenerative diseases, 372–373
stem
cancer, 371–372
Oct4, 370–371
Sox2 and Klf4, 370–371

I

IIR. *See* Insulin/IGF-like receptor
Induced pluripotent stem (iPS), 358
Inner cell mass (ICM), mammalian embryos
  blastocyst stage, 4
  cell lineage segregation mechanisms
    implantation rodent, 14–15
    Nanog, Cdx2 and *Oct4* expression, 14
    reciprocal inhibition, 14
  epiblast (EPI) and primitive endoderm (PE) formation
    blastocysts, 12
    egg cylinder stage, 12
    FGF signaling, 12–13
    GATA4 and GATA6, 11–12
    mosaic "composition", 12–13
    Nanog expression, 11–12
    positional signals, 11
    Sox7 expression, 13
  and trophectoderm (TE)
    blastomeres division, 4–5
    Cdx2 mutant cells, 9
    8-cell stage, 4–5
    chromatin remodeling protein BRG1, 6–8
    FGF/ERK, 10–11
    KLF5, 9–10
    lineage, 5–6
    perturbation, MAPK signaling, 11
    "polarity" factors, 4–5
    SOX2, 9
    TEAD4 and YAP1, 8–9
    transcriptional networks, 5–6, 7
    transcription factors, 5–6
Insulin/IGF-like receptor (IIR)
  germline cell cycle arrest, 73–74
  larval germline proliferation, role, 74
  mutants, analysis, 69–70
  pathway, germline proliferation, 68
  signaling, *Caenorhabditis elegans*
    description, 68–69
    target tissues, 71–73
iPS. *See* Induced pluripotent stem

K

Krüppel-like transcription factor (Klf5)
  core pluripotency network, 391–392
  description, 382
  expression, 382
  function, ESCs
    knockdown (KD), 385–386
    LIF absence, 386
    proliferation rate, 386–387
  gene expression network, ESC pluripotent
    Nanog, 385
    Oct3/4, Sox2, and Nanog, 384
    Sox2, 384–385
    STAT3 suppression, 384
    Wnt proteins, 384
  reprogramming, 383
  role, embryonic development
    blastocyst outgrowth, 387–389
    expression, 387, 388
    Klf2 and Klf4, 390
    phenotype, null blastocysts, 387–389
  targets, 390–391

L

LA-PCs. *See* Left atrium-derived pluripotent cells
Left atrium-derived pluripotent cells (LA-PCs)
  cytokine, 115–116
  vs. mesoangioblasts, 115
  microarray and pathway analysis, 116–118
  vs. peripheral blood mononuclear cells, 115–116
  TGF-β1, 119
Leukemia inhibitory factor (LIF)
  BMP4, 348–349
  presence and absence, 347–348
LIF. *See* Leukemia inhibitory factor
LIPUS. *See* Low-intensity pulsed ultrasound
Low-intensity pulsed ultrasound (LIPUS), 459, 460
L-type calcium channel [Ca(L)], 332

M

Mad-homology 2 (MH2), 345
Mammalian embryos and embryo-derived stem cells
  BMP/SMAD/Id signaling, 22
  cell fate determination, 26–27
  FGF2 and TGFb/ActivinA/Nodal signaling
    culture medium, 25–26
    differentiation markers, 23–25
    neural induction, 23–25
    neuroectodermal differentiation, 25–26
    self-renewal, 23–25
    SMAD2/3 transcription factors, 23–25
  human origin, 2–3
  ICM
    blastocyst stage, 4
    cell lineage segregation mechanisms, 14–15
    EPI and PE formation, 11–13
    and TE, specification, 4–11
  "immortal" stable cell lines, 15–16
  LIF/Jak/STAT3 signaling
    description, 19–21
    receptor (LIFR), 19–21
    self-renewal and differentiation, 19–21
  LIF/Ras/MEK/ERK signaling, 21
  lineage choice, 15
  PI3K/Akt and Src signaling, 21–22
  pluripotency, 2
  pluripotent stem cell lines
    embryonic stem (ES) cells, 16–17

Mammalian embryos and embryo-derived stem cells (*cont.*)
epiblast stem (EpiS) cells, 17
pre-and post-implantation, 15
totipotency and pluripotency, 3
transcriptional regulators
extrinsic and intrinsic factors, 18
gene expression, 18–19
ground and transient intermediate state, 18
OCT4 transcription factor, 18
SSEA1 and PECAM1, 18–19
WNT/β-catenin signaling
activation, 22–23
description, 22–23
"3i conditions", 23
self-renewal, 20, 22–23
teratoma formation, 23
MAPK pathway. *See* Mitogen-activated protein kinase pathway
Matrix metalloproteinases (MMPs), ECM remodeling
growth factors, 255
MMP-9 and-2 activity, 254
TIMPs and enzymes, 254
Mesenchymal stem cells (MSCs)
adipogenic differentiation
miR-31 and BMP-2, 302–304
Wnt signaling, 302–304
bone marrow, 422
CD271, 222–223
cells and tissue, 419–420
chondrogenic differentiation
in vitro model, 304–305
microarray profile, 304–305
compression
chondrogenesis, 446–449
3D cultures, 441–446
effect, 441–446
osteogenesis, 449–451
DCMMS, 458
definition, 222
description, 419
differentiation *in vitro*
adipogenesis, 424
chondrogenesis, 423–424
description, 422
osteogenesis, 422–423
tenogenesis, 424–425
features, isolated cells, 223
FSS, 451–458
function, mechanical regulation, 425–426
heterogeneity, 223
homogeneous population, 223
human endometrium
basal layer, 225
CD146 and PDGF-Rb, 224–225
changes, menstrual cycle, 224
chimerism, 224
clonogenicity, 224–225
lymphocytes and NK cells, 224
niche, 224
reproductive hormones, 225
in vivo loading
effect, 462–465
LMHF, 465
low magnitude, high-frequency vibration, 461
mechanical loading
compression, 429–430
DCMMS, 432
description, 427–428
FSS, 430–432
high frequency, low-magnitude vibration, 433
tension, 428–429
ultrasound, 432–433
mechanotransduction, 426–427
osteogenesis
adipogenesis, 437
ADMSC, 437
BMSCs, 441
dexamethasone, 437, 438
tenogenesis, 441
osteogenic differentiation
BMPs, 299–301
hADSC, 302
miR-206 expression, 301–302
plasticity
HNF4A, 306
miR-181, 305
neuronal-like cells, 306
role, miRNAs, 300, 305
progesterone
HLA-G expression, 226–229
PIBF expression, 229–230
SSEA-4 cells, 223
Stro-1 antibody, 222
TE, 418–419
tension
cardiomyogenesis, 440–441
chondrogenesis, 440
description, 433
osteogenesis, 437–439
tenogenesis, 439–440
tensile loading, 433, 434
tissue culture, 421
types, 420–421
ultrasound
BMSCs and ADMSCs, 459
LIPUS, 458–459
MicroRNA (miRNA) and vascular smooth muscle cells (VSMC)
application, 334–335
biogenesis and mechanism
description, 324–325
mRNA arrays, 325–326
differentiation

Dicer, 327
ES cells, 326
miR-143/miR-145, 326–327
diseases
aneurysm, 332–334
atherosclerosis, 331
diabetic vascular complications, 334
hypertension, 331–332
histone modification, 323–324
neointima hyperplasia
description, 330
microarray analysis, 330
vascular injury models, 330–331
phenotypic switch
KLF4 and KLF5, 329
miR-143/145, 328
PDGF, 327–328
PDGF and TGFβ pathways, 328–329
proliferation, 327
TGF-β and BMP, 328
target pairs, 324, 325
MicroRNAs (miRNAs)
ASCs and MSC, 292–293
biogenesis, 293, 294
cell-to-cell communication
MVs, 307–308
tissue injury, repair, 306–307
description, 292
ESCs, 292
expression profiles
hADSC, 298–299
HLSC, 299
mammals, 297
miRISC, 295
MSC differentiation
adipogenesis, 302–304
chondrogenic, 304–305
osteogenic, 299–302
plasticity, 305–306
role, 299, 300
P-bodies, 295
profile, MSCs, 298
role, stem cell biology, 296–297
Microvesicles (MVs)
biological effects
genetic information, 310
tissue repair, 310–311
description, 307–308
miRNA transfer
ESCs, 308
gene ontology analysis, 308–310
human bone marrow, 308–310
miRISCs. *See* miRNA-induced silencing complexes
miRNA. *See* MicroRNA
miRNA-induced silencing complexes (miRISCs), 295
Mitogen-activated protein kinase (MAPK) pathway
components, 10–11
MEK1 inhibitors, 21
signaling perturbation, 11
MSCs. *See* Mesenchymal stem cells
MSCs, molecular mediator
adipogenesis
adipogenic differentiation, 48–49
cell lines, 49
BMP, 46
canonical Wnt and BMP signaling, 46–47
chondrogenesis
BMP2, 48
chondrocyte maturation, 48
TGF-β, differentiation induction, 48
definition, murine, 40–41
description, 40
differentiation, 41
ex vivo expansion, HSCs, 53
immunomodulatory properties
anti-inflammatory role, 43
autoimmune encephalomyelitis model, 44
clinical trials, 43
proinflammatory cytokines, 43–44
roles, 44
mediating MSC self-renewal
description, 52
engraftment, 52–53
model, sFRP2 role, 52
mode of action, 52–53
MI therapy
challenges, myocardial repair, 44
myocardial repair and ventricular dysfunction, 44
osteogenesis
adenoviral expression, 47
canonical Wnt signaling, 47
self-renewal process, 42
sFRP
cytoprotection and proliferation, 51–52
description, 51
netrin (NTR) domain, 51
survival and proliferation
BMP, 49–50
enhancement, wound, 50–51
Wnt, 49–50
therapy
clinical approaches, 42
intracoronary administration, 43
Wnt pathway
BMP and Wnt signaling, activation, 45
inhibition, 46
noncanonical and canonical, 46
Muscle stem cells activation
ECM, 252–255, 258–260
growth and repair
Dicer-null satellite cells, 244

Muscle stem cells activation (*cont.*)
  microRNAs molecules, 244
  MRFs, 244
  MSCs and HSCs, 245–246
  MyoD-positive cells, 242–243
  myofibers, 240–241
  myogenic precursor cells, 240, 241
  myostatin, 242–243
  Pax7 transcription factor, 242–243
  pericytes, 245–246
  "replacement" stem cells, 246
  SDF-1 receptor, 242
  side population (SP) cells and PICs, 244–245
  Sprouty1, 242–243
  transcription factors, 242
 HGF, 255–258
 regeneration, skeletal muscles
  fast-twitch and slow-twitch, 249
  fibrotic tissue, 248–249
  macrophages, 248
  myolysis and reconstruction, 246–247
  neutrophils, 247–248
  protein degradation, 247–248
  recruitment, inflammatory cells, 247–248
  soleus-and edl-derived myoblasts, 249
 satellite cell niche
  aging muscle, 252
  asymmetric stem cell division, 250
  Delta-like protein and Numb, 250–251
  described, 250
  m-cadherin, 250
  myofiber damage, 252
  syndecan-3, 251–252
  Wnt proteins, 251
MVs. *See* Microvesicles

N

NCSCs. *See* Neural crest stem cells
Neural crest development, NCSCs and FA role
 Dicer and miRNA
  cell death, 153
  KDM6B expression and chromatin, 153
 EMT, 149
 Fgfr4
  expression, 159
  Pax3 downstream, 158–159, 160
 Folbp1
  cell surface proteins, 152
  folate receptors, 152
 GRN genes, 149–150
 Hcy levels, NCC formation, 155
 Hes1 and Neurog2
  H3K9Ac, H3K18Ac and H3K27me2, 157
  stem cell proliferation and neurogenesis, 158, 159
  WT embryos proliferate and neurospheres, 157, 158
 Lrp6-Wnt coreceptor
  β-catenin levels, 151–152
  mutations, 151–152
 miRNAs, 156–157
 neuroepithelial cells, transformation, 154–155
 Notch signaling
  cell proliferation, 160
  cranial development, 154
  potential folate responsive GRN, 160–162
 p53
  craniofacial defects, 153
  diabetes, 153–154
  expression, 153–154
 Pax3, 155
 Tead2
  Pifithrin-α, 150–151
  transcription factors, 150–151
 transcription factors, 149–150, 151
 Wnt signaling
  β-Catenin, 155–156
  FA supplementation, NTD, 156
Neural crest stem cells (NCSCs)
 life history, 144
 and maternal FA intake
  folate levels, 144–145
  homocysteine (Hcy) levels, 145
  mouse models, NTD, 148–149
  neural crest development, 149–162
  role, human health, 145–148
Neural tube defects (NTDs)
 autism, 146
 diabetes, 153–154
 folate nonresponsive genetic mouse models, 162
 humans, 145–146
 Lrp6 mutations, 151–152
 mutant mice
  FA nonresponsive, 148–149
  folate responsive, 148
Notch-independent soma germline signaling mechanisms
 distal Sh1 sheath, 66
 L3 and L4 stage, structure, 65
 mutant analysis, 65
NTDs. *See* Neural tube defects

O

Osteogenesis, MSC
 adenoviral expression, 47
 canonical Wnt signaling, 47

P

Parkinson's disease (PD)
 cell replacement, human DAn
  donor tissue and cell composition of the grafts, 178–179
  hESCs, 179–180

hiPS cells, 180–181
hNSCs, 181–185
host tissue/environment, 178–179
limitations, 179
transplantation, stem cells, 179
safe and functional transplantation, 176–177, 178
P-bodies. *See* Processing bodies
PD. *See* Parkinson's disease
PDGF. *See* Platelet-derived growth factor
PIBF. *See* Progesterone-induced blocking factor
Platelet-derived growth factor (PDGF), 328–329
Pluripotent stem cell lines
embryonic stem (ES) cells
chimeras and teratomas, 16–17
isolation, research groups, 16
molecular characteristics, 16
population-doubling time, 16
epiblast stem (EpiS) cells
characteristics, 17
description, 17
Processing bodies (P-bodies), 295
Progesterone
extra-reproductive tissues
epithelial cells, 219–220
experimental data, 220
mammary gland, 219–220
neuroprotective and promyelinating effects, 220
NK cells, 220–221
interactions, 218
MSCs, 222–223, 224–225, 226–230
receptors
genomic and nongenomic effects, 221–222
knockout approach, mice, 221
protein isoforms, 221
reproduction
decidualization, 219
endometrium transformation, 219
granulosa cells, 218–219
inhibitory effect, 218–219
target cell populations, 218
Progesterone-induced blocking factor (PIBF) expression
cell culture, 230
described, 229
HLA-G, 230
immunomodulatory effect, 229
monoclonal antibody, 230

R

Reserve liver progenitor cells, activation and regulation
adult rodent, structure, 97, 98
cell populations, 97
hierarchical responses, liver disease
cellular responses, 102
description, 102, 103
IL6 role, 102
mature hepatocytes, liver regeneration
description, 94–95
gene expression, phases, 95
mitogenic growth factors, 96
periportal location, 94–95
surgical partial hepatectomy (PH), 94–95
TGFβ and activin, 96–97
TNFα, 95–96
transcription factors, activation, 95–96
oval cells and SHPCs, 97–98
and proliferation, oval cells
cellular events, 99
compartment, 99
cytokine-mediated priming stimulus, 100
experimental models, 98–99
D-glactosamine model, 100
pathophysiological circumstances, 98–99
Solt-Farber model, 100
and proliferation, SHPCs
cellular responses and time course, 101
molecular mechanisms, 102
phenotype, 101
proliferative stimulus, 101
retrorsine, 100–101
stem cell lineage system, 94
Reverse transcriptase polymerase chain reaction (RT-PCR)
long-term culture, 113, 114
measurement, TGF-R1 expression, 116–118
RT-PCR. *See* Reverse transcriptase polymerase chain reaction

S

SAR. *See* Structure–activity relationships
Secreted Frizzled-related protein 2 (sFRP2)
cytoprotection and proliferation, 51–52
description, 51
mediating MSC self-renewal, 52–53
NTR domain, 51
sFRP2. *See* Secreted Frizzled-related protein 2
SHPCs. *See* Small hepatocyte-like progenitor cells
Signaling pathways, pancreatic β-cells
BMP
description, 84–85
endocrine differentiation, 85
Noggin, inhibitor, 85
Pdx1 and *Hnf6* expression, 86
Pdx1 and insulin expression, 85
Calcineurin/NFAT
description, 83
NFATc proteins, 83
Epac
cAMP, 83–84
description, 83–84
dosage, exendin-4, 84

Signaling pathways, pancreatic β-cells (*cont.*)
regulation, insulin secretion, 84
Hedgehog
activation level, 83
description, 82
downregulation, Sonic (Shh), 82–83
inhibition, 82–83
PI3Ks
activators, 81
description, 81
maturation, role, 81
regulation, 81–82
Small hepatocyte-like progenitor cells (SHPCs)
cell populations, 102
description, 97
proliferation
cellular responses and time course, 101
molecular mechanisms, 102
phenotype, 101
proliferative stimulus, 101
retrorsine, 100–101
SMCs. *See* Smooth muscle cells
Smooth muscle cells (SMCs)
development, TGF-β1 signaling, 135
transcription factors, 136
Solt-Farber model, 100
SSEA-4 cells. *See* Stage-specific embryonic antigen-4 cells
Stage-specific embryonic antigen-4 (SSEA-4) cells, 223
Structure–activity relationships (SAR)
β-arrestin2-GFP translocation assay, 211, 212, 213
hydroxyl group, 211–212
working model, 211–212, 214

T

TE. *See* Tissue engineering
TGF-β1. *See* Transforming growth factor-β1
Thymosin-β4 (Tβ4)
chemotactic activity
description, 285
F_1-F_0 ATP synthase, 285, 286
classification
description, 278–279
structures, 278–279
extracellular
Ac-SDKP, 282
sulfoxide, 281–282
VEGF, 281
gene and mRNA expression, 279
intracellular, 281
roles, heart regeneration, 282–283
roles, skin tissue regeneration, 282
skeletal muscle regeneration
injury, 283
myoblasts, 283–284
therapeutic applications, muscular dystrophy, 284–285
in species, 280
structure and posttranslational process, 280
Tissue engineering (TE), 418–419
Transforming growth factor-β1 (TGF-β1)
chondrogenic differentiation
cartilage development, 133
pellet culture system, 133–134
type II collagen hydrogel, 133–134
embryonic heart development, 135–136
inhibitor, adipogenesis, 135
molecular mechanism
MAPK activation, 136
signaling pathway, 136
transcriptional activity, Runx2 and TAZ, 136
osteogenic differentiation
ALP activity, 134
bone regeneration, 134
SMCs and cardiomyocytes, 135
TGF-β signaling
biological processes, 132
downstream target genes, 132–133
mammalian isoforms, 131–132
transmembrane receptor, 132–133
Transforming growth factor-beta (TGF-beta)
epiblast stem cells, 356–357
ES, 342
ES cell regulation, 347–350
human ES cells
FGF signaling, 356
LIF and BMP, 354–356
Nanog expression, 356
iPS cells, 357–358
ligands and transmembrane receptors
BMP and Nodal, 343, 344
type I and II receptors, 343
pluripotency pathways, 354, 355
signaling modulators, 346–347
smad proteins
DNA, 345–346
members, 346
MH1 and MH2 domains, 345

V

Vascular endothelial growth factor (VEGF), 281
Vascular smooth muscle cells (VSMC). *See* MicroRNA (miRNA) and vascular smooth muscle cells (VSMC)
VEGF. *See* Vascular endothelial growth factor

W

Wnt pathway, MSCs
BMP and Wnt signaling, activation, 45
inhibition, 46
noncanonical and canonical, 46

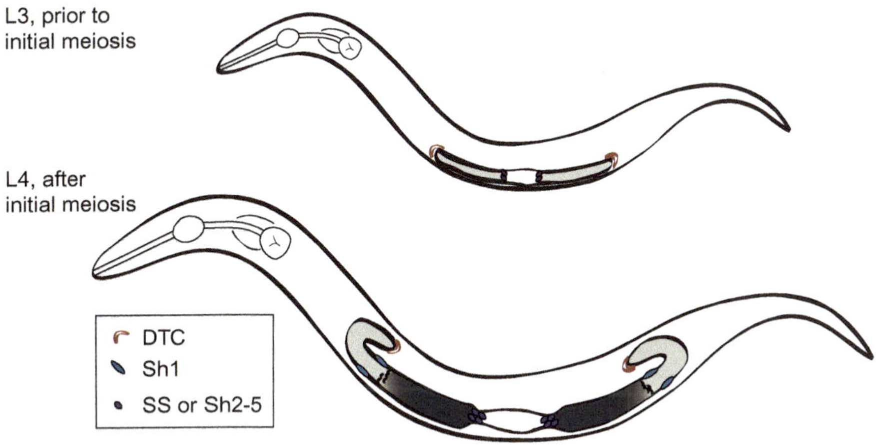

E. Jane Albert Hubbard, Figure 3.1 Cartoon depictions of L3 and early L4 stage *C. elegans* hermaphrodites. Bilobed gonad is indicated with proliferative germ cells in light gray, meiotic cells dark gray to black. Crescent-shaped nuclei characteristic of the transition zone are depicted. White central portion indicates position of future spermathecae and uterine cells. Distal tip cells are shown in red and distal and proximal sheath cells in lighter and darker blue, respectively.

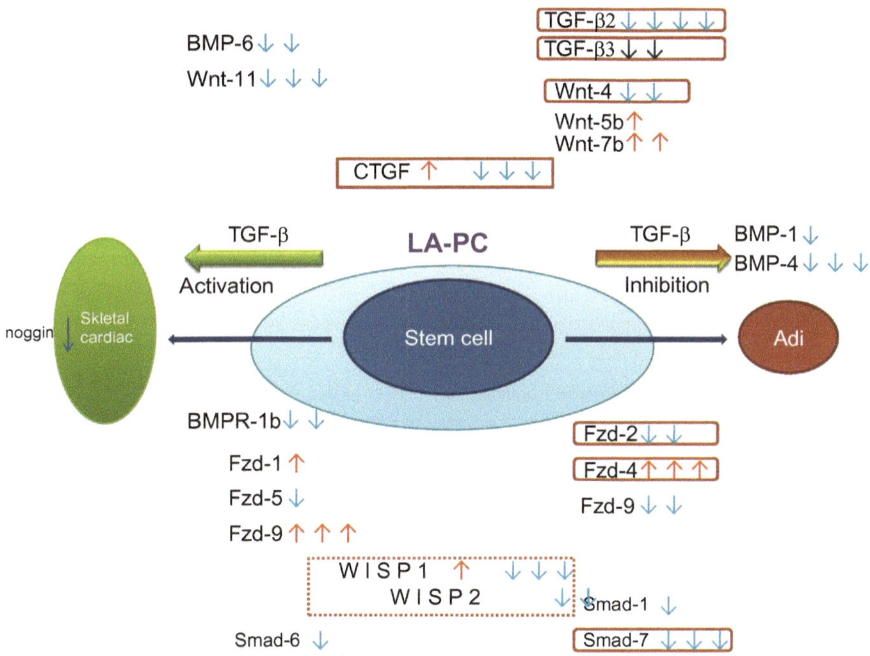

Nanako Kawaguchi, Figure 6.1 Transforming growth factor (TGF)-β and Wnt signal transduction in left atrium-derived pluripotent cell (LA-PC) differentiation. Genes marked with a red line were differentially expressed, as determined using both Agilent and Torey 3D-GENE methods. Blue arrows indicate expression levels lower than the levels in the control undifferentiated cells (↓ twofold, ↓↓ fourfold, and ↓↓↓ eightfold lower). Red arrows indicate expression levels higher than the levels in the control (↑ twofold, ↑↑ fourfold, and ↑↑↑ eightfold higher).)

Longmei Zhao and Basil M. Hantash, Figure 7.2 Differentiation potential of MSCs. Human bone marrow MSCs were seeded on tissue culture plate and induced to differentiate into adipocytes by culturing the cells in adipogenic media (Cell Applications, San Diego, CA) (B), or regular α-MEM media (A, control) for 14 days and then stained with Oil red O. For osteogenic differentiation, human adipose-derived MSCs were seeded on tissue culture plates and cultured in osteogenic media (Cell Applications) (D), or regular DMEM media (C, control) for 21 days and then stained with alizarin red. Cellular nuclei were count-stained with hematoxylin (C and D). To induce chondrogenic differentiation, human adipose-derived MSCs were cultured as pellets in chondrogenic media (Cell Applications) (E). After 21 days, differentiation into the chondrogenic lineage was visualized.

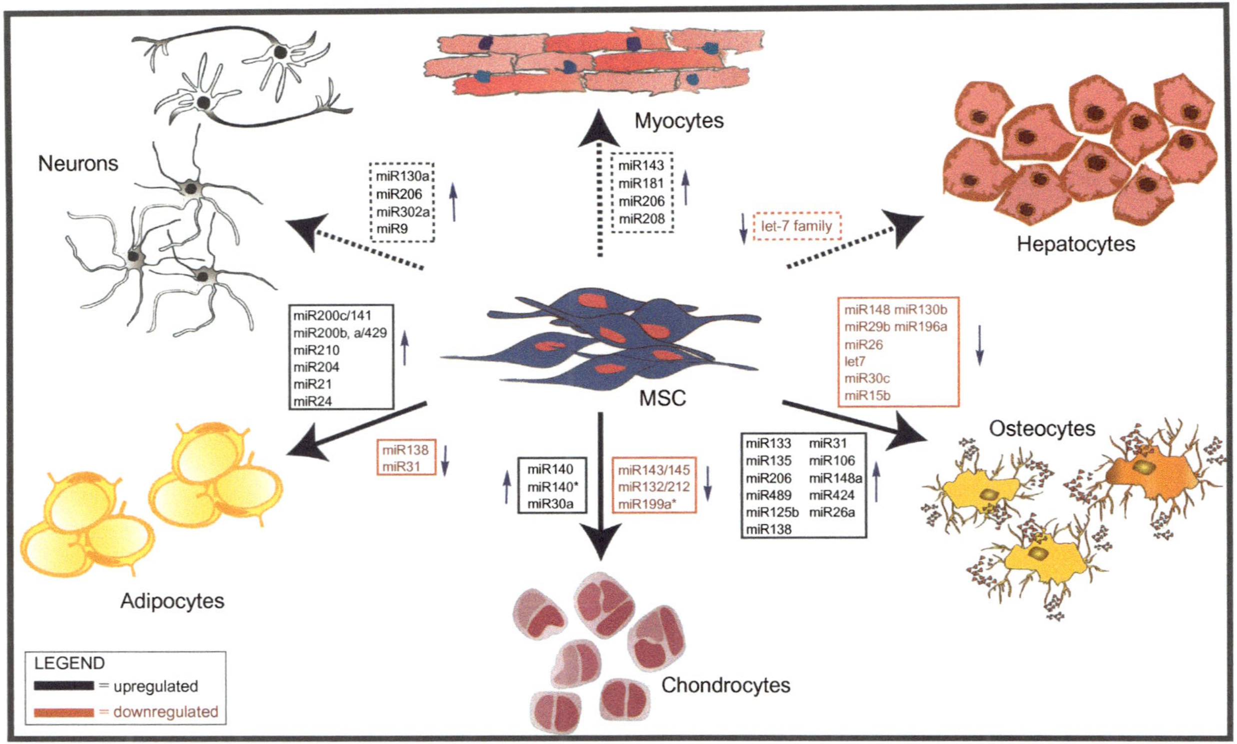

Federica Collino *et al.*, Figure 14.2 Role of miRNAs that regulate differentiation of MSCs in mesodermal and nonmesodermal tissues. Red-labeled miRNAs are downregulated (↓), whereas black-labeled miRNAs are upregulated (↑) during the process of differentiation.

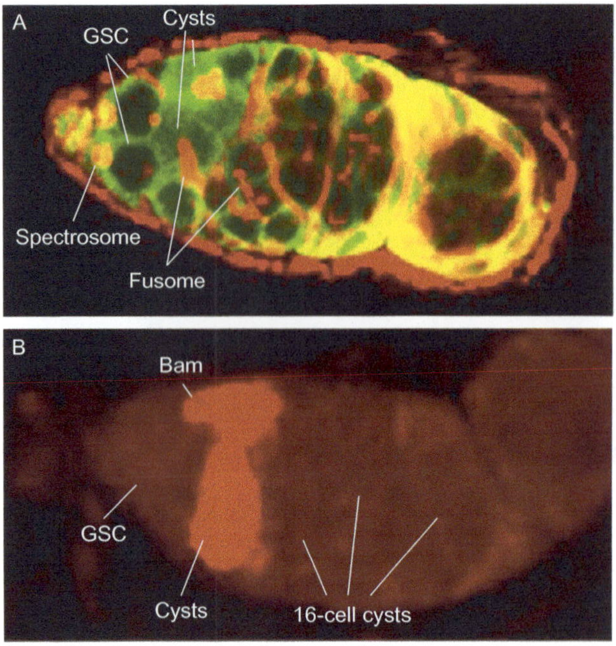

Sreejith Perinthottathil and Changsoo Kim, Figure 19.2 Bam expression in dividing cysts. (A) A confocal image of a germarium carrying a GFP reporter bearing 3′-UTR of Bam (Pek *et al.*, 2009) that was double stained with anti-GFP (green) and 1B1 (red) (Lin *et al.*, 1994). 1B1 stains spectrosomes and fusomes. GFP (green) marks GSCs and dividing cysts. (B) Wild-type germarium stained with Bam-C antibody (Ohlstein and McKearin, 1997). Bam is not detected in the GSCs and late 16-cell cysts, but highly expressed in the dividing cysts (red).

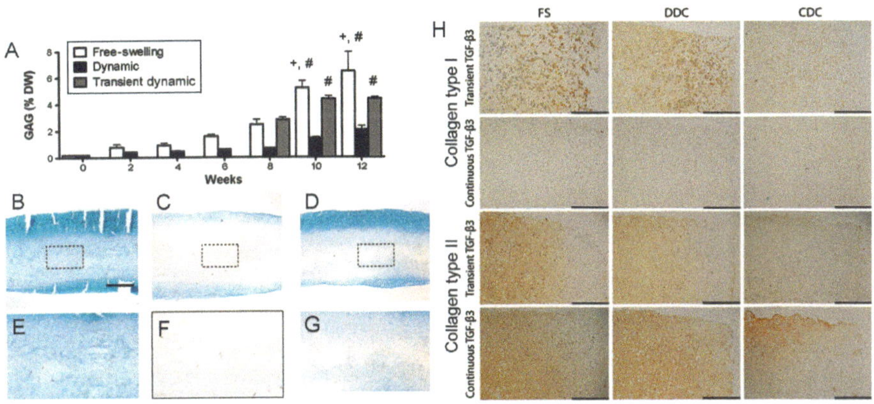

Robin M. Delaine-Smith and Gwendolen C. Reilly, Figure 20.5 Effects of compression loading on chondrogenesis of mesenchymal stem cells. (A–G) Chondrogenesis is inhibited by dynamic culture in electrospun scaffolds, but cell and matrix distribution is improved. Biochemical measurement of GAG content (A) reveals depletion due to dynamic culture conditions, and a partial recovery following removal from dynamic culture. In all three culture conditions, GAG increased significantly with time in culture ($p < 0.05$). Alcian Blue (dark gray in black and white image) stains at 12 weeks confirm biochemical findings, with more intense staining of free-swelling (B) constructs compared to dynamic (C) and transient dynamic (D). Regions indicated in (B)–(D) are enlarged in (E)–(G), respectively. + and # indicate $p < 0.05$ versus transient dynamic and dynamic culture, respectively. Scale bar = 250 m (B–D) and 62.5 m (E–G). Reproduced from Nerurkar *et al.* (2011), with kind permission from Elsevier. (H) Chondrogenesis is inhibited by immediate mechanical loading (CDC) but not by loading in later stages of culture (DDC) in agarose hydrogels and TGF-β3 is required to stimulate matrix production. Immunohistochemical analysis of MSC-seeded constructs at day 42 stained with for the noncartilage specific collagen type I (top) and the cartilage specific collagen type II (bottom). FS, free-swelling. DDC, delayed dynamic compression, initiated on day 21. CDC, continuous dynamic compression, initiated on day 0. Representative images taken of approximately of each construct section. Scale bar: 500 μm. Reproduced from Thorpe *et al.* (2010), with kind permission from Springer Science and Business media.

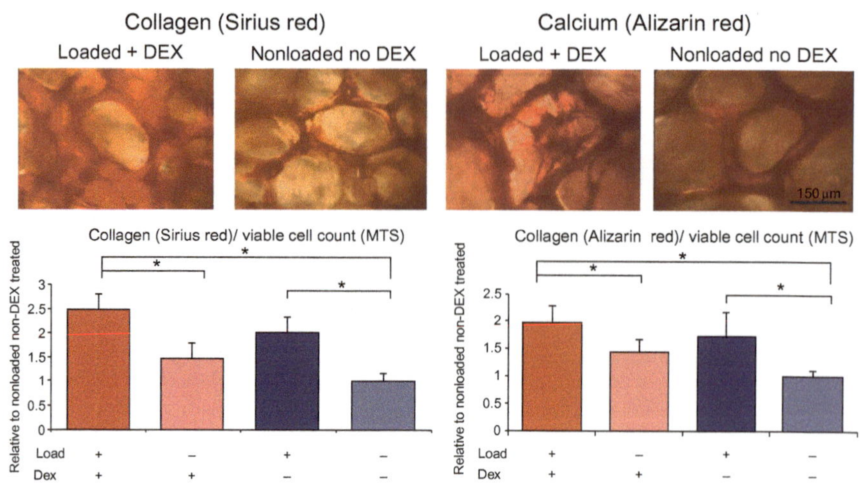

Robin M. Delaine-Smith and Gwendolen C. Reilly, Figure 20.6 Scaffold compression upregulates osteogenesis of mesenchymal stem cells in 3D. hMSCs seeded in polyurethane scaffolds subjected to 2 h of dynamic compression loading (5% strain 1 Hz) every 5 days contained more collagen and calcium by day 24 of culture. Top: light micrographs representative images of Sirius red (collagen) and alizarin red (calcium) staining (black in black and white image). Bottom: quantification by colorimetry normalized to controls. Loading was as effective as DEX in stimulating osteogenic matrix production (*$p < 0.05$ ANOVA with Tukeys pair-wise comparison). Adapted from Sittichokechaiwut *et al.* (2010), kind permission of reproduction from eCMjournal.org.

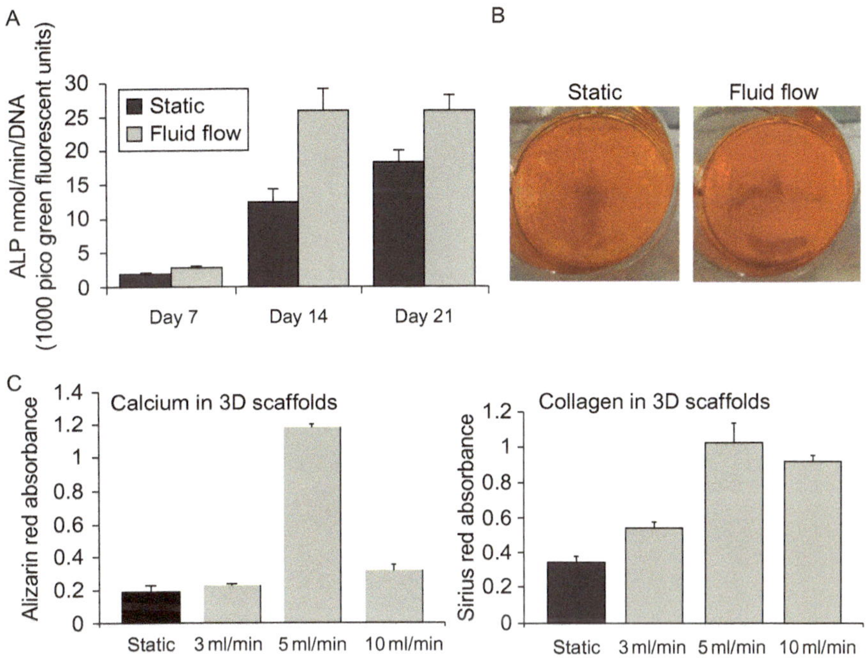

Robin M. Delaine-Smith and Gwendolen C. Reilly, Figure 20.7 Fluid flow upregulates osteogenesis of mesenchymal stem cells in 2D and 3D. (A, B) Mesenchymal precursors derived from embryonic stem cells (hES-MP 002.5, Cellartis, Gothenburg) were subjected to oscillatory fluid flow by placing cell-seeded culture wells on a rocker at 1 Hz for 1 h/day beginning at day 5 of culture. (A) Mean ± SD of alkaline phosphatase (ALP) activity, which was significantly higher at day 7 and day 14 of culture in cells subjected to fluid flow compared to static controls ($n = 6$, $^{\star}p < 0.05$). (B) Alizarin red (black in black and white image) staining of culture wells (6-well plate) at day 21 indicated that cells subjected to rocking produced more calcium. (C) Preliminary data indicate that short bouts of oscillatory fluid flow stimulate MSCs matrix production in 3D scaffolds. Immortalized mesenchymal stem cells from mouse bone marrow, BM1 (kindly donated by Ilaria Bellantuono, University of Sheffield), were subjected to oscillatory fluid flow within porous polyurethane scaffolds (described in Sittichockechaiwut *et al.*, 2009) at the peak flow rates shown, at 1 Hz for 1 h on days 4, 7, and 10 of culture. (A) Mean ± SD of Sirius red (collagen) and alizarin red (calcium) absorbance units after staining the scaffolds on day 14 of culture. Previously unpublished data from our laboratory collected by Mohsen Shaeri.

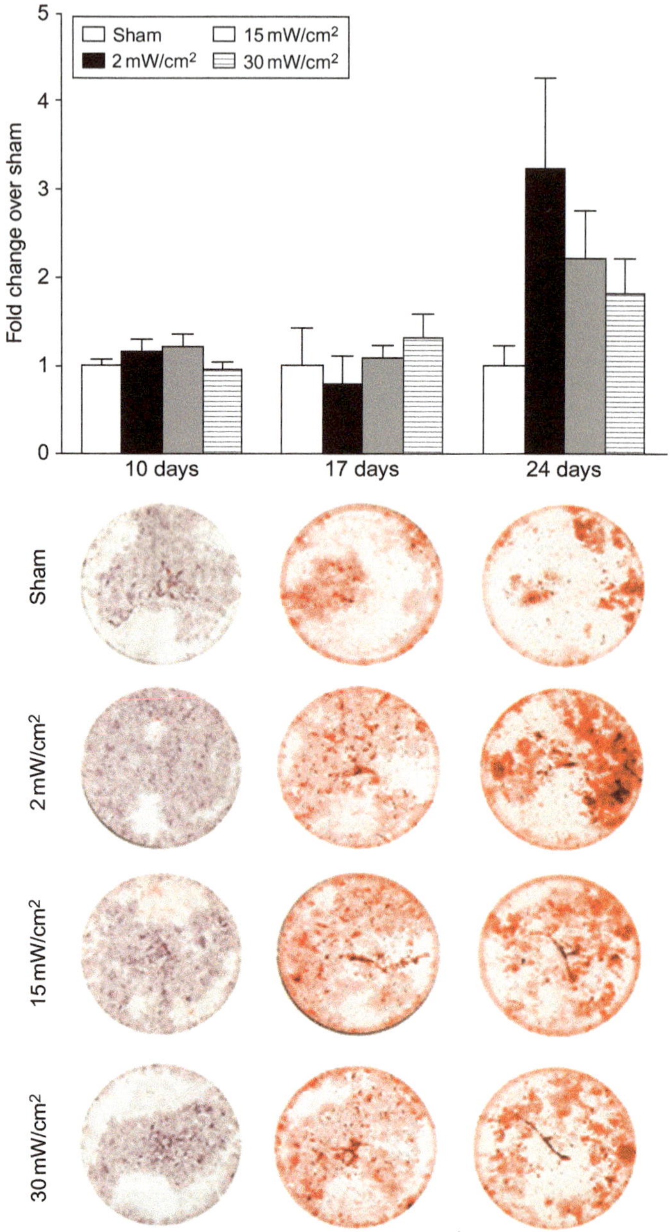

Robin M. Delaine-Smith and Gwendolen C. Reilly, Figure 20.8 Example of the effects of low-intensity pulsed ultrasound (LIPUS) on osteogenesis of mesenchymal stem cells. Relative mineralization by rat BMSCs seen after cultures in 12-well plates were exposed to 0 (sham), 2, 15, and 30 mW/cm^2 LIPUS for daily treatment for 10, 17, and 24 days. Mineralization (calcium) quantified in mM of alizarin red (black in black and white image). Data represented as mean ($N = 4$) ± SEM. The well pictures are representative images for each treatment group. Reproduced from Angle *et al.* (2011), with kind permission from Elsevier.

Kolibri 360 GmbH
Ferdinand-Jühlke-Straße 7
99095 Erfurt, Deutschland
produktsicherheit@kolibri360.de